# Archibald Liversidge, FRS

## Imperial Science under the Southern Cross

# Archibald Liversidge, FRS

## Imperial Science under the Southern Cross

Roy MacLeod

THE ROYAL SOCIETY OF NEW SOUTH WALES       SYDNEY UNIVERSITY PRESS

Published by
The Royal Society of New South Wales
and Sydney University Press

National Library of Australia Cataloguing-in-Publication entry
Author:              MacLeod, Roy M
Title:               Archibald Liversidge, FRS: Imperial Science Under the Southern Cross
ISBN:                9781920898809 (pbk)
Edition:             1st ed.
Notes:               Includes index and bibliography
Subjects:            Liversidge, Archibald, 1846-1927
                     Scientists–Australia–Biography
                     Scientists–England–Biography
                     Science–Australia–History
                     Science–Study and teaching–Australia–History
                     Science–Study and teaching–England–History
                     Science–Methodology

Dewey Number:   500.81

Design:                  Colin Rowan
Editorial Co-ordinator:  Kimberley Webber
Picture Research:        Penelope Grist; additional research: Edmund McMahon
Production:              Bruce Welch

Printed in Australia

Cover:

Archibald Liversidge, an oil on canvas by the Hon. John Collier, painted in London in 1909, following a subscription
among friends and members of the Royal Society of New South Wales. Collier, a son-in-law of Thomas Henry Huxley,
painted a number of contemporary men of science, including Huxley himself. This portrait, for which he charged
£157.10s., was exhibited at the Royal Society of Portrait Painters in London in November 1909. Today, it hangs in the
Great Hall of the University of Sydney. A photograph of the painting is held in Christ's College, Cambridge.

Liversidge is shown in characteristic pose, pointing to a geological map of New South Wales, and lecturing on the three
principal processes by which gold is reduced from its ores. He is holding a group of gold crystals that resemble the shape
of Great Britain, which were found in New South Wales, and which are now in the National Museum of Scotland. The
crystals are embossed on the cover of Liversidge's *The Minerals of New South Wales* (London: Trübner & Co., 3rd edition,
1888). To Liversidge's left stands a flask containing a solution of 'Purple of Cassius,' its distinctive colour owing to the
formation of colloidal gold – then and now a convenient qualitative test for the presence of gold.

Inside front and back covers:

*Leptospermum Liversidgei*: a tea tree dedicated in 1905 to Liversidge and named in his honour by R.T. Baker and Henry
G. Smith.

*Aurum irrepertum et sic melius situm.*

Horace, *Odes*, III, iii, 49

For Alex, the Navigator

# Table of Contents

# Preface

It has been over thirty years since Geoffrey Serle, in his *From Deserts the Prophets Come* – that fine study of the 'creative spirit' in Australia – spoke of the 'lamentable lack' of research in the history of Australian science; and nearly forty years since Ann Moyal reminded Australian historians of their neglect of science in the teaching of the nation's history. Until then, Australian science seemed, in Patrick White's phrase, a landscape without figures. There were familiar names, to be sure, but their work was forgotten or misunderstood. With notable exceptions, leading figures were relegated – respectfully, to be sure, with that unspoken deference an active people accords certain classes of prophets – to the periphery of that larger canvas of political and economic events that most Australians recognise. It is instructive that even the most recent edition of the *Shorter Oxford History of Australia* scarcely mentions the sciences on which the nation's wealth was built, and neglects even to list 'science' in its index.

Slowly, however, we are beginning to recognise the significance of those men and women who, in their time, loomed so large as to form part of the landscape itself, and who merit more of the public recognition that Australians accord their heroes of exploration and discovery. Especially since the Bicentennial in 1988, biography has lent its talents to history, and has recaptured the lives of Ludwig Leichhardt and W.B. Clarke, Baldwin Spencer and David Orme Masson, J.T. Wilson and T.W. Edgeworth David, David Rivett and Mark Oliphant. These have helped illuminate continuities between Australia's cosmopolitan present and its colonial past, and have opened the way for assessments of Australian science before Federation and since.

In this company, Archibald Liversidge held for a time a notable place. Yet, his name is known to few outside his inner circle. A short, retiring scholar, with round spectacles, a slight stammer, and a city suit – a style more obedient to custom than climate – he is not at first sight the stuff of which Australian legends are made. No mention of him will be found in G.V. Portus' *Fifty Famous Australians*, or in Clive Turnbull's *Australian Lives*. He is not among the *One Hundred Famous Australian Lives* found on suburban library shelves, nor among Robert Macklin's *100 Great Australians*, which at least includes his student, Douglas Mawson. He is not even in Rigby's *1000 Famous Australians*, which draws the scientific line at Macfarlane Burnet and Howard Florey (and excludes Mawson). His absence from a bicentennial book devoted, simply, to *The Greats* – which includes no scientist at all except Mawson – is predictable.

Like many 'transplanted Britons' who avoided the limelight, Liversidge has enjoyed at best a fugitive celebrity. Visitors to Mount Victoria, in the Blue Mountains, will pass by a Liversidge Ridge, but will find no other memorial save an ancient view of the Megalong Valley, of which he was fond. Tourists in Sydney's Botanic Gardens may admire the lemon-scented tea tree outside the Herbarium Library, without noticing it is a *Liversidgei*. Visitors to Canberra will find that Liversidge Street takes them to and beyond University House, or to the new visiting academic flats which, without local explanation, bear his name. Talented students at Sydney University, Christ's College, Cambridge, and Imperial College London win Liversidge Scholarships, but few hear much about their benefactor.

Yet, as with much in science, reputation lies less in celebrity than in the careful application of method. And in this, Liversidge's influence is everywhere. For many, his name is inextricably linked to the establishment of the Australasian Association for the Advancement of Science (or ANZAAS, as it became known). But far wider and deeper ran his influence. To him we owe the idea of an Australian Academy of Science, a national journal of science, and a national metric system. To him, we owe the first systematic recording of minerals in New South Wales, a model for all subsequent mineral registers in this country. In an age of improvement, he lent energy to efficiency in public science. As a student of T.H. Huxley, John Tyndall and Michael Foster, Liversidge transplanted the scientific culture of London to Sydney, and projected a progressive adaptation of British influence, combining colonial nationalism with imperial loyalty, and promoting a vision for Australia that was international in scope and global in influence.

Small wonder that, in 1976, Michael Hoare, in recalling the 'heroic age' of Australian science, should point to Liversidge as 'unquestionably the most influential scientist in Australia between 1870–1890'. 'To him,' Hoare rightly observed, 'much of the original advanced thinking and planning in the nation's science can be traced.' His story can scarcely be told by two columns in the *Australian Dictionary of Biography*, or even by two pages in the *Oxford Dictionary of National Biography*. From his pen flowed over a hundred reports, monographs, catalogues and papers. He helped turn colonial Australians from a tradition of descriptive 'inventory science', sustained by 'gentleman collectors', to a world of analysis, sustained by professionals. His laboratory work spurred the economic exploitation of mineralogical and geological discoveries, and helped launch the discipline of geochemistry. His analyses of meteorites and trace elements contributed to scientific speculations that have found renewed interest today.

Above and beyond his service to science, came Liversidge's work for the colony of New South Wales and the University of Sydney. For over a quarter of a century, he led a small circle of those who taught students what they knew of chemistry and mineralogy. He was the architect of the University's Faculty of Science, and was for nearly two decades its first Dean. Beginning with a small room, with next to no equipment, he built a department that entered the twentieth century with a sizeable staff, laboratories, lecture-rooms, and over two hundred students. Tirelessly, he established a school in mining, metallurgy and applied chemistry that stood comparison with the best in Britain and North America. His students swelled the ranks of 'science militant' in the schools, professions, and public services of New South Wales. Through his influence on appointments, he crystallised traditions that were to characterise the chemical sciences in Australia for at least a generation.

For over thirty-five years, there was little in Sydney's scientific life that his hand failed to touch, or having touched, failed to quicken. From his personal papers, we see his influence on Sydney's Technical College, and its Museum of Applied Arts and Science (today, the 'Powerhouse' in Darling Harbour), and on the Royal Society of New South Wales. Before him, as Edgeworth David wrote, scientific men in Australasia were as *quot homines tot sententiae*; at his departure, they spoke with a corporate voice.

For twenty years after his retirement, and through the Great War, Liversidge continued to serve the interests of Australia as Sydney's 'man in London'. At his death in 1927, tributes flowed in over twenty British and Australian journals and newspapers. Eighty years later, chemists still deliver the Liversidge Lectures he endowed at the University of Sydney, the Royal Society of New South Wales, and the Royal Chemical Society in London.

In 1909, the Hon. John Collier, noted for his celebrated portrait of T.H. Huxley, was commissioned to paint the Londoner whom Huxley had taught. His portrait was brought to Sydney, and today, visitors to the Great Hall of the University will see Liversidge as he wished to be remembered – in academic dress, lecturing, inevitably, on the location and extraction of gold. Remarkably, he has never attracted a biographer. These pages seek to remedy that neglect.

Roy M. MacLeod
University of Sydney

# Acknowledgements

Biography is an intrusive medium, and in England and Australia, America and Europe I have imposed on friends and colleagues almost too numerous to mention. In return, I can but offer the thanks of a traveller who has followed his subject across two continents and over many years. This book began life as a seminar paper at the Australian National University, when I was on research leave from London; and I am grateful to Professors Ken Inglis and Barry Smith, and the late Professor Oliver MacDonagh, for embarking me on a long voyage. The adventure has continued, with interruptions, for many years, and has greatly benefited from the increasing professional interest now accorded the history of imperial and colonial science. With colleagues in many countries, I have shared the experience of arriving where we started, and of seeing a familiar place for the first time.

Among those who wish to be remembered chiefly by their published work, Liversidge holds pride of place. Few personal letters survive in his papers, nor are there many references to family or social life, or indeed to his many interests outside science, the university, and public affairs. For guiding me to the Liversidge Papers at the University of Sydney, I am grateful to Mr Ken Smith, formerly University Archivist, Dr Peter Chippindale, and Dr Ursula Bygott, of the University History Project. But for a wider view of Liversidge's life, one must look elsewhere – to relatives, colleges and clubs, and to his contemporaries. A bachelor don, Liversidge had many but few close friends. Yet, he was a family man by extension, and was well remembered by a dozen nephews and nieces. To Mr William Liversidge of Abingdon, I am grateful for information about his great-uncle, and for access to family records. Other family details were kindly provided by a great-niece, Mrs Priscilla Macartney, of Charminster, Dorset, and by a great-nephew, Mr Terence Liversidge, of Elizabeth Grove, South Australia. Mr S.A. Liversidge of Billericay, Essex and Ms Christine Liversidge of London helped my early searches, as did Mr A.J. Hughes, a later owner of Liversidge's London home, 'Fieldhead' (now called 'Hampton Spring', Coombe Warren, Kingston-upon-Thames), who is also, by happy coincidence, an Australian chemist.

For their help in tracing Liversidge's life in London, I am grateful to Ms Maureen Castens, Deputy Head of Library Services, City of London Polytechnic; Mr John Pryce Jones, Chiswick District Library; Mr C.A. Cornich, Local History Officer and Mrs Margaret Vaughan-Lewis, Hon. Archivist, Borough of Kingston-upon-Thames; Mr. B Curle, Local Studies Librarian, Royal Borough of Kensington and Chelsea; and the staff of the Guildhall Library, the Hounslow Record Office, and the Hackney Public Record Office.

For help in tracing Liversidge's career, I am grateful to Mr Norman Robinson, former Librarian of the Royal Society of London; Mr John Kennedy, Librarian of the Royal Society of Chemistry; Mr A.E. Kilby of the Society of Chemical Industry; and the library staff of the Linnean Society, the Physical Society, the Geological Society of London and the Library and Archives of the Senate House, University of London. For Liversidge's special relationship with South Kensington, I am indebted to the former Archivist of Imperial College, Mrs Jeanne Pingree and Ms Anne Barrett; Dr E.A. Jobbins, Curator of Minerals and Gemstones, and Ms Sylvia Bracknell, Library of the Institute of Geological Sciences, in the Geological Museum; and to Dr A.R. Wooley and Miss D.L. Morgan of the Mineralogical Department of the British Museum (Natural History). For news of Liversidge's ethnological collections and their fate, I am grateful to Ms Dorota Starzecka, Assistant Keeper in the Museum of Mankind. On many historical matters, Professor W.H. Brock, then of the University of Leicester, offered helpful suggestions.

As the book progressed, Liversidge coincidentally became a subject of interest to the Australian Joint Copying Project, and I am grateful to Dr Graeme T. Powell, former Project Officer, for supplying me with references. For unearthing records of Liversidge's membership and residence, thanks go to the librarians of the Commonwealth Institute and the Royal Commonwealth Club; Mrs J.M. Chandler, of the United Oxford and Cambridge University Club; Captain Denys Wyatt, OBE, RN, of the Athenaeum; Mr R.N. Linsley of the Carlton Club; Mr Peter Aldersley of the Savile Club; and Ms Jane Cullis of John Walker and Sons Ltd., later occupants of the premises of the Royal Societies' Club at 63, St James's Street, London, SW1.

In Cambridge, I am indebted to Professor Hans Kornberg, FRS, former Master of Christ's College; and to Dr P Sykes, Vice-Master, and Dr E.A. Carson, College Archivist, who answered many questions about undergraduate life. Mrs Michele Courtney and Mrs M. Katvars retrieved Liversidge's books from the darkest recesses of Christ's Old Library. Timely advice on Cambridge chemistry came from Mrs C.M. Cook, of the University Chemical Laboratories; Dr D.A. Chinner, of Trinity College; and the librarians of the Cambridge Philosophical Society and the Departments of Earth Sciences, Zoology, Physiology and Anthropology. For assistance with Cambridge University Archives, I am indebted to Dr. Elizabeth Leedham-Green.

Elsewhere in Britain, I received generous assistance from Mr P.B. Hetherington, Secretary of the Association of Commonwealth Universities; Mr C.A.H. James, Secretary of the Royal Commission for 1851; Dr Irene O'Brien, of the Library, University of Glasgow; and Mr Peter McNiven, Archivist of the John Rylands Library at the University of Manchester. For their courtesy and skill in preparing a cast of Liversidge's gold nugget from the original in their possession, I am grateful to Mr Robert J. Reekie and Dr Harry Macpherson of the Department of Geology, National Museum of Scotland, Edinburgh.

Among the many organisations in Sydney with which Liversidge was associated, I must thank several. The late Dr J.R. Willis, former Director of the Museum of Applied Arts and Sciences (the 'Powerhouse'), allowed me to consult his unpublished history of the Museum, and Mrs Vanessa Mack helped me find the Museum's records. In the early stages, I also received generous help from the late Dr Harley Wood, former Director of the Sydney Observatory; from staff of the Australian Museum, particularly Dr R.O. Chalmers, formerly

Curator of Rocks and Minerals, and Ms Gwendoline Baker and Ms Carol Cantrell, Librarians; from Ms Joy Thompson of the National Herbarium, Royal Botanic Gardens; and from Mr I.H. Cartwright and Colonel N.H. Marshall, past Secretaries of the Union Club.

In Melbourne, I thank the librarians of the Royal Society of Victoria; Mr P.W. Woodhouse of the Royal Australian Chemical Institute; and Mrs Cecily Close, of the University of Melbourne Archives. Mrs Susan Woodburn of the Registrar's Department, University of Adelaide, gave me space and sources on her university's history; while the late Ms Joan Radford, of the University of Melbourne, gave me in her work a model of what a departmental history should aim to be. Among many others, Mrs Beatrice Cosh of Killara, daughter of Professor Charles Fawsitt; Mrs Clarice Morris of Mount Victoria, granddaughter of Edward Hufton; and Mrs Nancy Douglass, of the Blue Mountains Historical Society, helped with Liversidge's personal life in Australia. For their assistance, I am indebted to the staff of the Basser Library of the Australian Academy of Science in Canberra, and to the Mitchell Library in Sydney. For assistance in New Zealand, I am grateful to Mr Gordon Maitland, Auckland Institute and Museum; to Mrs Beverly Booth, Hocken Library, University of Otago, Dunedin; Mrs Caroline Etherington, Canterbury Museum, Christchurch; and Dr Michael Hoare, formerly of the Alexander Turnbull Library, Wellington. For details of Liversidge's connections with France, I am grateful to M.J. Fabries, of the Muséum National d'Histoire Naturelle and M. Pierre Berthan, Archivist of the Institut de France, Académie des Sciences, Paris; and for his American links, I am grateful to Dr H.L. Hornbaker, of the Mineral Resources Section, Colorado Geological Survey; the State Archives of Colorado, Denver; and Dr Marc Rothenberg, formerly of the Smithsonian Institution, Washington, DC.

In reprising long-neglected fields in the history of science, a prospective biographer can confront an interpretative desert, lacking reliable points of reference. Tracking is difficult, and prospecting, even more so. For technical advice on some of the many issues that Liversidge made his own, I am grateful to Mr M.B. Duggan, Curator of Minerals, Bureau of Mineral Resources, Geology and Geophysics, Canberra; Associate Professors Ken Williams and the late Tom Vallance and Ian Threadgold, of the Department of Geology and Geophysics, University of Sydney; and Dr Alex Jenkins, formerly of the Department of Mining Engineering at Sydney. The late Professor R.J.W. Le Fèvre, who pioneered the history of colonial chemistry, received me at his home; while Professor Ian Rae, then of the School of Chemistry, Monash University, was always ready with advice. Professor Robert Cahn, FRS, then of the Cavendish Laboratory, and Professor I.J. Polmear of Monash University were generous in their advice on metallurgy; while for their assessments of Liversidge's theories of ore genesis, I am grateful to Dr Neil Phillips of the University of the Witwatersrand, Johannesburg; Professor Allan Wilson of the Department of Geology and Mineralogy, University of Queensland; and especially to Dr Charles Butt of CSIRO, who is currently writing in this field himself.

For the history of the Australian and New Zealand Association for the Advancement of Science (ANZAAS), I am indebted to the late Mr James Davenport, formerly General Secretary, and the late Dr Rupert Best, of the Chemistry Department, University of Adelaide. On architectural points, I am indebted to Mr Robert Bland, formerly the University of Sydney's Architect, and Dr James Kerr, consultant to the National Trust of Australia.

In the Easter Term, 1986, I was a Visiting Scholar at Pembroke College, Cambridge, and wish to express my thanks to the Master and Fellows, especially Professor Jay Winter. A.D. Buckingham, FRS, then Professor of Theoretical Chemistry at Cambridge, and formerly a Liversidge Scholar, greatly encouraged my research. In the cordial environment of Clare Hall, Cambridge, I drafted the first parts of the book, which were finished during a Regents' Fellowship at the Smithsonian Institution, Washington, DC. For their support, I am grateful to Dr Arthur Molella and Mr Jim Roan, of the Smithsonian's Library, for many rare and curious references, and to Dr Daniel Appleman of the Natural History Museum, for his comments on Anglo-Australian and American approaches to mineralogy.

In Canberra the book began, and so in Canberra it was finally completed, thanks to the kindness of the late Professor Paul Bourke, Director of the Research School of Social Sciences, and Dr Anthea Hyslop, of the Australian National University, who read the full text. The book also benefited from the generous comments of the late Associate Professor Ken Cable, AM; Dr Peter Chippindale and Dr Ursula Bygott of the University History Project; Dr Jim Eckert; and my scholarly comrade-in-arms, Emeritus Associate Professor David Branagan.

Material arising from the book has been delivered in seminars and lectures at several institutions, including the Australian National University, the University of Sydney, the University of New South Wales, the Royal Australian Historical Society, Griffith University, University College London, the Royal Society of Chemistry, the National Institute for Science, Technology and Development (New Delhi), the University of Hawaii, the University of Pennsylvania, the University of Texas, and the Royal Institution of Great Britain. I am grateful for comments received on all these occasions. Parts of the book were foreshadowed in *Science across the European Empires, 1800–1950,* edited by Benedikt Stuchtey in 2005, material from which is used by permission of Oxford University Press. Parts of chapters 7, 8, 9 and 10 have appeared in *Ambix, Australian Cultural History, Minerva, History and Technology,* and *Earth Sciences History,* and in Graeme Davison and Kimberley Webber (eds.), *Yesterday's Tomorrows: The Powerhouse Museum and its Precursors, 1880–2005* and in the *Oxford Dictionary of National Biography.* I am grateful to the editors and publishers of these papers for permission to use this material. Grateful acknowledgement also goes to many sources for the use of illustrations, which are cited at the end of the book. The cover portrait of Liversidge has been reproduced by courtesy of the University of Sydney Art Collection.

In 1984 and 1985, I convened a series of three, one-day public workshops on the history of 'Scientific Sydney' under the auspices of the Royal Australian Historical Society and the Royal Society of New South Wales. Sessions were devoted to 'Artisans and Managers: the History of Technical Education', 'Culture and Learning in the Colonial Metropolis', and 'The University and the Community'. These events brought Liversidge to a wider audience, and I am grateful to the two societies for their cooperation. Papers from the second workshop appeared in the *Journal and Proceedings of the Royal Society of New South Wales,* while a volume arising from the third – *University and Community in Nineteenth Century Sydney: Professor John Smith (1821–1885)* – was published in 1988 by the University of Sydney History Project. Liversidge's life also found its way into an edited volume on *The Commonwealth of Science*, prepared for the centennial of ANZAAS, and published by Oxford University Press; and into the new *Oxford Dictionary of National Biography*, while many of its themes featured prominently in the Honours course I taught at Sydney University on 'Science, Technology and the "Imperial Idea"'.

As with most scholarly works, this book has not escaped the need for subsidy, and in its early stages, was generously supported by the Australian Research Grant Scheme. For grants-in-aid of publication, I owe thanks to the Chancellor's Committee of the University of Sydney, the Dean of the Faculty of Science, the Council of the Royal Australian Chemical Institute, the Trustees of the Sydney Mechanics' School of Arts, the Earth Sciences History Group of the Geological Society of Australia, Inc., and the Royal Society of New South Wales. None have been more fortunate than I in the student assistants this support has enabled me to employ. Working part-time, Ms (now Dr) Shirley Saunders and Mrs Marilyn Orr were followed by Mr Kurt Oppliger, Ms Christa Ludlow, and Ms (now Dr) Melanie Oppenheimer. Ms Ruth Bennett and Dr Lea Beness kept control of burgeoning records, and Ms (now Dr) Leanne Piggott helped edit the first draft. For the book's design, I am grateful to Mr Colin Rowan, and for production to Mr Bruce Welch. The index has been prepared by Ms Mary Coe. At the close, it fell to Ms Penelope Grist and Mr Edmund McMahon to help with formating. Mr Ron Brashear and the staff of the Chemical Heritage Foundation gave much valued encouragement and support in the final stages, as did Mr John Hardie, President of the Royal Society of New South Wales. My warmest thanks are reserved for Dr Kimberley Webber, who has come to know all too well the life of at least one bachelor scientist.

In recalling a figure of imperial vision, such as Liversidge, it is tempting to make bold claims; but for one, at least, no apology is required. In its passage across the heavens, the Southern Cross is visible to many lands, cultures, and countries. It is not the special property of New South Wales, whatever the confident assertion on the State's flag. However, in certain respects, its symbolism conveys a shared meaning, with a message that expresses precisely the duality of Liversidge's loyalties and lifework – a life in science canvassed against a southern sky, but quartered by the Union Jack.

# Abbreviations

| | |
|---|---|
| **AAAS** | Australasian Association for the Advancement of Science |
| **AM** | Australian Museum (Sydney) |
| **ANZAAS** | Australian and New Zealand Association for the Advancement of Science |
| *ATCJ* | *Australian Town and Country Journal* |
| **BAAS** | British Association for the Advancement of Science |
| **BM (Nat. Hist.)** | British Museum of Natural History |
| **BSG** | British Science Guild |
| **CSIR** | Council for Scientific and Industrial Research |
| **CSIRO** | Commonwealth Scientific and Industrial Research Organisation |
| **JRAHS** | *Journal of the Royal Australian Historical Society* |
| **JPRSNSW** | *Journal and Proceedings of the Royal Society of New South Wales* |
| **LA (C) NSW** | Legislative Assembly (or Council), New South Wales |
| **MAAS** | Museum of Applied Arts and Sciences (Sydney) |
| **ML** | Mitchell Library, State Library of New South Wales (Sydney) |
| **NMNZ** | National Museum of New Zealand (Wellington) |
| **RACI** | Royal Australian Chemical Institute |
| **RSL** | Royal Society of London |
| **RSM** | Royal School of Mines |
| **RSNSW** | Royal Society of New South Wales |
| *SMH* | *Sydney Morning Herald* |
| **TISM** | Technological, Industrial and Sanitary Museum, later the Museum of Applied Arts and Sciences, now the Powerhouse Museum (Sydney) |
| **TRSNSW** | *Transactions of the Royal Society of New South Wales* |
| **UCL** | University College London |
| **VPLANSW** | *Votes and Proceedings of the Legislative Assembly of New South Wales* |
| **VPLCNSW** | *Votes and Proceedings of the Legislative Council of New South Wales* |

# Introduction

Bust of Archibald Liversidge.

'Not only the University of Sydney and the state of New South Wales, but the whole of the Commonwealth will be the poorer by the departure for England of Professor Archibald Liversidge, MA, FRS.' Thus Sydney's *Evening News* on Boxing Day, 1907, solemnly mourned the loss of a man who had inspired the culture of science and higher education. His arrival from England in 1872 marked, as the *Illustrated Sydney News* put it, 'a new era in the history of the University'. His departure in 1908 'removed a link' with a past of colonial visions and imperial sympathies. For thirty-five years, Liversidge had given voice, as one crying in the wilderness, to Australia's need for science, and for a science appropriate to the needs of Empire and Commonwealth. His life illuminates the ways in which British science continued – well beyond early exploration, discovery and settlement – to be a medium of colonial exchange and imperial purpose.

Liversidge's life also occupies a significant place in the historiography of science and empire – a domain which in recent years has proceeded from an uncertain infancy to robust maturity.[1] Just as the concept of 'empire' has yielded to an expanding conception of the world as a 'laboratory' for European observation and theory, so the role played by science in consolidating global power has inspired Australians – as it earlier inspired Americans and Canadians[2] – to grasp its significance as an instrument of imperial enterprise.[3] But while many historians have now mapped the coastlines of scientific exploration from the voyages of Cook,[4] fewer have studied the developments in science that followed the first few decades of European settlement.[5] The task now is to recover the special character of imperial science – with its concomitant ideology of 'scientific imperialism' – and its application to what becomes a distinctive form of colonial science, revealing specific features of life and culture at the 'periphery'.[6] In so doing, we begin to see more clearly the ways in which the culture of science accompanied the 'moving metropolis' of imperial settlement, and acquired – not always without difficulty – a vital if unsung role in Australia's self-styled 'march to nationhood'.[7]

As a historiographical enterprise, this endeavour is not new, nor are its general tendencies unfamiliar. Since the fifteenth century, beginning with what Alfred Crosby has called the 'Columbian exchange', Europe has embraced, and often controlled, the means whereby natural knowledge of the world is conceptualised, codified, and commercialised.[8] Imagination inspired discovery on the titlepage of Francis Bacon's *Instauratio Magna,* where, sailing between the Pillars of Hercules towards the imperium of the unknown, Europeans embraced the expanding boundaries of knowledge and the possibilities of power that it implies.[9] With the voyages of discovery, came glimpses of a world *outre-mer*, sometimes presenting giant 'anomalies', and puzzles for the philosophers of Europe. Nowhere was this more relevant than in Australia, with its unfamiliar flora and fauna. From what William Goetzman has called the 'Second Great Age of Discovery'[10] arose the need to order and define nature, creating descriptive systems and categories, revising ideas of territorial space and historical time, and so transforming the disciplines of astronomy and anthropology, geography and geology, botany and zoology, mineralogy and chemistry.

The expansion of Europe altered both the consciousness of Europeans, and their world system. But as ideas sail from the 'centre' to the 'periphery', eventually building reciprocities of mutual influence, our attention is drawn from the process of metropolitan expansion in itself, towards the 'peripheral' expression of that process, in domains that operate at first within imperial rules of engagement, but later become representative of colonial interests. The 'reconstruction' of this domain remains a neglected priority in the history of Western science, and in colonial historiography. Yet, such a reconstruction is vital to an understanding of colonial governance, colonial institutions, and the rise of colonial nationalism.[11] In Australia, the reconstruction of colonial science is also key to an understanding of the factors that – especially through higher education and research – helped shape an emerging cultural identity. In beginning this process of reconstruction, the life and times of Archibald Liversidge is of remarkable value, offering as it does a significant illustration of the ways in which science assisted Australia in general – and the city and University of Sydney, in particular – in becoming a 'moving metropolis' on an imperial frontier.

***

I have used the term 'reconstruction' in a twofold sense: first, as the act of reassembling dates, personalities and institutions into new interpretative patterns; and second, as a means to show how certain aspects of Anglo-Celtic culture, transmitted to Australia, acquired a particular form of 'dependent independence'.[12] In part, this exercise requires us to lower barriers that have long divided historians of science and technology from historians of Australian culture, politics and economics.

The role accorded science in colonial Australia finds its justification in the motivations of imperial and colonial economic interest. It is debatable whether 'imperial economics' formed a necessary stage in capitalism, as Hobson and Lenin argued.[13] But if we may take as read the 'frontier capitalist' nature of the Australian economy, the ethos of which corresponds to a phase in the history of 'settler capitalism',[14] and to what Warwick Armstrong, looking to New Zealand, has called 'dominion capitalism',[15] then we find colonial science cast within an economy that is based on the land, its productive capacities based on the agricultural, mineral and extractive industries, its trade driven by primary producers whose fortunes, when not ravaged by drought, disease, ignorance and misadventure, are regulated by markets and investment houses in Asia, Europe and North America. In this context, the role of colonial science in Australia came to be shaped by the interests of a mercantile bourgeoisie which, as early as the 1860s, had won representative government, rejected convictism, encouraged education and culture, and joined with the 'squattocracy' in asserting political leadership.

Within these broad boundaries, the practice of colonial science came to occupy a recognised but ill-defined status – respected, but not precisely admired; sometimes controversial, but mainly routine; cultivated, but not widely encouraged.[16] We know something of its dimensions.[17] But we know much less about its contours, its ambitions. How far could what was known of austral nature actually serve practical economic interests? Among men of science – transplanted Britons for the most part, but not always – how far could one speak of a 'colonial mentality'?[18] In the United

States, Daniel Boorstin has argued, geographical distance from Britain brought about a 'therapy of distance', with attitudes conducive to invention in the professions, manufacturing and the arts of government.[19] But did Australia's much greater distance from British and European colleagues, libraries and laboratories create, or reinforce, a 'carrier conservatism' among its custodians of science? How indeed did the Australian experience compare with American settlers' experience of the frontier, the 'Wild West', which afforded, we are told, 'an open scene, permitting almost total freedom, encouraging eccentricity', creating a 'habit of freedom from habit', a 'tradition of independence from tradition'?[20]

A generation ago, historians debated whether the concept of a frontier is applicable to the Australian experience.[21] Today, historians of science are acquiring a sense of an extremely complex Australian frontier, one conspicuously populated by unusual flora and fauna, but which also raises environmental questions that elude easy answers. How did men (and latterly, women) of science deal with the environment they found? How did they bring its special features to notice, in a professional and commercial world governed by metropolitan systems of recognition and reward? How did the fact of distance shape their assumptions? Did separation bring not emancipation, as in America, but rather a 'psychology of abdication', an attitude encouraging adaptation, but discouraging innovation – a state of mind defensive about theoretical generalisation, one which actively preferred to leave theoretical initiatives to older, wiser and wealthier institutions of 'Home'. Did the 'tyrannies' of isolation, loneliness and introspection make it easier, if not indeed necessary, for Australia's scientists to become the 'hewers of wood and drawers of water' for the learned nations of Europe?[22]

Such questions suggest the need for a fresh look at the changing relationship between imperial and colonial science. Traditionally, the colonial viewpoint has been left to the margins.[23] Yet, there is abundant evidence that the 'imperial idea', to borrow A.P. Thornton's phrase,[24] inspired not only ambitious Europeans, but also ambitious colonials.[25] And as such imperial interests emerge, so the institutions of science – from museums to learned societies, universities to international exhibitions – become key categories of imperial activity in themselves with research sites and networks, moving with the metropolis, and extending their scope across the globe.[26]

There is no reason, of course, to deny the dominance of the metropolis. Lucile Brockway and Richard Drayton have shown how Kew Gardens dominated British colonial botany,[27] and a similar case can be made for the Astronomer Royal and Greenwich Observatory in colonial astronomy and for the Royal School of Mines in colonial geology.[28] For a time, the arguments of expedience justified a division of labour, governed by the assumption that scientific Marys and Marthas both served a common goal. Thus, Australia was early and easily conceived as a large outdoor laboratory, garden and quarry for London – as, had La Pérouse been quicker and survived – it would have been for Paris. The vast continent seemed an inexhaustible mine of information, and the right to natural knowledge of its flora, fauna and minerals – whatever the views of the Indigenous population – was taken for granted by men of science, no less than by colonial governments. To this extent, we find Australian artifacts furnishing the glasshouses and libraries of Kew; its skies, the star charts and tide tables of the Admiralty; and its mineral specimens, the museum display cases of Jermyn Street and South Kensington.

However, the presence of power invokes questions of colonial agency. The intellectual monopolies of London needed men at the periphery willing to sustain them. But how long could Australians, or Anglo-Australians, cede these privileges to a distant governance? Indeed, even leaving aside Aboriginal claims, when did colonial Australians begin to assert intellectual property rights over knowledge of Australian flora, fauna, water, wind and land? Canada, it is said, used the imperial idea as a nationalist counterweight to offset pressures from the United States.[29] How far was Australian scientific dependency on Britain self-inflicted?

Such questions now quicken attention. Regrettably, they have long been neglected. It was not always so. Eighty years ago, it was customary to begin lectures in Australian history at Sydney University, in the manner of Professor George Arnold Wood, with accounts of voyages of discovery, which were largely scientific in content and intention.[30] Wood built upon a nineteenth-century tradition that celebrated a connection between the fortunes of science and nation building. On the centennial of European settlement in 1888, J. Steel Robertson echoed a similar sentiment, when he claimed that 'Natural science stood godmother to Australia'.[31] In 1909, J.H. Maiden even credited the nation's paternity to Sir Joseph Banks – the 'Czar of the Philosophers' and the 'Father of Australia'.[32] And Professor Ernest Scott of Melbourne, welcoming ANZAAS to his city in 1939, announced that the advancement of science in the country conformed 'to the general historical development of Australia'.[33]

These assumptions conjure powerful associations. However, they were interrupted by postwar tendencies in Australian scholarship that, while admitting to the importance of science and technology in modern society, largely neglected the contribution of science to colonial culture.[34] The pages of *Historical Studies*, for example, reflected a growing interest in ideas and issues that were deemed vital to understanding Australia's changing place in the world. Through the 1980s, the history that most Australians met at university emphasised political, economic and social themes, underlined by Manning Clark's influential titles – *The Age of the Bourgeoisie, 1861–83*; *Radicals and Rationalists, 1883–90*; *The Earth Abideth Forever 1851–88* and *The People Make Laws 1888–1915*.[35] In the meantime, the history of science marched to a different drummer, established its own disciplinary boundaries, and concentrated on questions in European science, while gravitating away from mainstream social and intellectual history.

Whilst these pressures towards specialisation made it difficult to establish a consensus, it remained clear – as the Americans Hartley Grattan and George Nadel argued – that something was missing from a country's history that neglected its science.[36] From the 1970s onwards, this view gradually gained ground, thanks in large part to a few historians, including Ann Moyal and Michael Hoare, and a few scientists, notably David Branagan and Thomas Vallance.[37] Both sides cultivated an extension of 'practitioners' or 'scientists' history, notable for its emphasis upon colonial biography and colonial achievement. In celebrating pioneering 'firsts', this tradition gave us new epic sagas of explorers and discoverers, their lives clothed in Homeric language of portents and sieges, wandering heroes and sirenic adventures. Like all literatures whose purpose is to retrieve from neglect, and to commemorate more than to criticise, this tendency stressed filial piety, and the veneration of what Marc Bloc called the 'idol of origins'. At the same time, however, this literature left a valuable legacy, in that it helped place science securely in the context of social and political history, its values enriched by narratives of enlightenment, endurance, individualism and practicality.

Early attempts at writing the history of science in Australia cherished these values, often using the organic language of family and nation. Thus we have Alexander Macleay heralded as the 'Father of Australian entomology', and W.B. Clarke given a similar place in Australian geology.[38] Other historians depicted Australian science as a sequence of stages, or epochs, evoking images of progressive evolution, loosely resembling Herbert Spencer's 'march of the mind'.[39] Ernest Scott described the history of Australian science as occurring in four sequential, linear and 'progressive' periods – with voyages of exploration, the act of collection, the creation of learned societies, and the establishment of universities.[40] Once a 'plateau' is reached, then a biological (or perhaps, Durkheimian) metaphor arrives to explain how, by intellectual 'osmosis' and institutional acculturation, science contributes to a stable equilibrium.

Contemporary sources were easily found to support the notion. Thus, Richard Twopenny's architectural history of Australia – from slab-hut to weatherboard house to brick and stone – seemed to rest comfortably alongside a trajectory from convict origins to a colonial aristocracy-in-the-making.[41] It was easy to infer a national history unfolding from a state of childhood, under British care, to ones of adolescence, maturity, and eventually, independence. A similar metaphor was applied to Australian science, which was said to have 'come of age' in 1914, when the British Association for the Advancement of Science visited the country for the first time, and bestowed its blessing upon its colonial offspring.[42] Others might prefer to situate the 'age of consent' in 1926, with the establishment of the Council for Scientific and Industrial Research (also, in certain respects, an imperial gesture);[43] some might prefer a later date; and some might decline to see the process as complete even now.[44] But whatever 'rite of passage' is chosen, the evolutionary model invited Australians to associate themselves, and even to identify, with the emergence of science as a feature of nationhood. The history of science became a matter of appealing simplicity, little more challenging than a history of institutional gradualism.

\*\*\*

This view, implicit in the literature for many years, helped render the history of Australian science innocuous, and perhaps for that reason, unattractive. In the 1970s, however, these certainties were disturbed by George Basalla's ambitious attempt to gird the earth with a single, embracing hypothesis to describe the 'spread of Western science'.[45] Taking examples from several cultures, Basalla used the image of a 'core' (or set of cores) and a circumference, or a hub and a rim – metaphors already well established in anthropology and the sociology of science.[46] The result was to fix in the minds of historians the image of science as a 'projected' enlightenment, with ideas derivative from supposedly advanced cultures in Europe and North America. In Australia, this had the force of recasting earlier, descriptive and softly organic models of development within a powerful, virtually deterministic framework of historical explanation, in which the history of Australian science was simply a case *sub specie aeternitatis*.

In Popperian terms, the 'Basalla model' has stood the test of a good theory; by encouraging attempts at falsification, it has thrown new light on old problems.[47] As an explanatory device, it has perhaps worked better in non-Western cultures than for 'settler' science. However, even in such cases, its weaknesses have become apparent. In suggesting that science has an implicitly linear thrust, the model avoided political and economic contingencies, and the frequently fragile institutions of science at the 'periphery'. Its inherent 'Whiggishness', with its linear and deferential 'centre-periphery' relationship, failed to account for differing opinions held by 'scientific colonists', or their impact on debates at 'Home'.[48] It said nothing about science as a cultural agency, producing sometimes not accommodation, but dissent; not harmony, but competition. As such, it neglected the tensions implicit in colonialism, and the motives of colonial nationalism. It failed to explain the relative roles attributed at different times to science as 'material practice' and cultural symbol, and neglected the functions served by both. Above all, it failed to recognise that on the 'periphery' (for our use of the term is inevitable), new scientific information was useful both to support and subvert European theoretical positions, whether in matters of human biology, the age and structure of the earth, or the movement of the heavens.

Today, we are less dogmatic. We view the discourse of imperial and colonial science as matters of adaptation, assimilation and reciprocal influence. We also see that ways in which scientific information was collected, argued, theorised and recognised, can tell us much about the assumptions that underlie imperial history. As Lewis Pyenson has observed, waves attenuate as they travel from their source and interact with local disturbances.[49] What appear truths at the centre may be seen quite differently at the rim. Ultimately, the 'metaphors of transmission' are not explanations in themselves, but merely signifiers, leaving us to explain by other means how natural knowledge, gained at the rim, came to confer prestige, codify influence and legitimise control. We now examine, rather than take for granted, propositions holding that knowledge is always manipulated by the 'mother country'; that colonial science is inherently derivative; and that a colonial posting involved such isolation as inevitably made international science virtually impossible to achieve. We are also compelled to enquire about the respective, contemporary experiences of other 'settler colonies', as set against those of 'alien rule', and as against those of empires other than the British.[50]

To view the world of science as signifying the irradiation of barbarism by civilisation, is to reject an explanation that is so general as to be meaningless. However, we must weigh carefully what we put in its place. Science and its culture are often implicit in social practice, and for that reason, are understated. Overall, the contingencies of politics continued – in the colonies as at 'Home' – to determine what was encouraged and undertaken in the name of science. We see this in the history of the early Australian universities, no less than in other fields of activity.

However, we also find that the practical interests of government and private enterprise are increasingly guided by men of science who understand the tensions between utility and knowledge for its own sake.

In Australia, this process took its cue from a unifying premise: that knowledge, like survival, required struggle and sacrifice. In the words of the pioneering chemical industrialist, Sir Russell Grimwade:

> Carefree Australians today are apt to forget that at the time of first entry their land produced no orthodox food and its soils had never been cultivated and that the abundance of foods produced within its boundaries today all have their origin overseas ... When these very first needs were fulfilled, even in a rudimentary manner, the obligation to posterity became revealed. We are fortunate that from our very beginnings there have been far-seeing and intellectual giants amongst us who, almost fanatically, have sought to gain a full knowledge of their surroundings.[51]

His was not an unfamiliar message. A decade earlier, the geologist Ernest Andrews recalled for his readers that:

> the history of geological progress in Australia is bound up with a wealth of heroic deeds; with herculean struggles with Nature when in unbending and savage mood; it is a record of triumphs won in the teeth, as it were, of grim and armed antagonisms ... even the plain unvarnished recital of their epic struggles is redolent of romance ... which imparts life, pathos, grace, dignity and grandeur to Australian history.[52]

Thus, we have the cockleshell heroes, Flinders and Bass; the indomitable Eyre of the Bight, and the gentle Sturt, descending the Murrumbidgee and Murray. Australian history would be inconceivable without Kennedy, plagued by insects and Aborigines; the impetuous Burke; the amiable, high-souled Wills, dying before Mt Hopeless; and the tragic John Gilbert, whose memorial in Horace's posthumous eloquence proclaims to visitors of St James' Church in Sydney precisely that symbolic nationalism that colonial science chose to honour – *Dulce et decorum est pro scientia mori*.

Such an emphasis on the 'struggle' of discovery, of survival and conquest – with all its imperialistic overtones – is not accidental in Australian literature. It is surely part of the history of Europeans in their encounter with an alien, and at first unyielding land.[53] However, some would argue that this view is also epistemologically central to the Western scientific heritage. And if to be 'masters and possessors of Nature', as Descartes taught, is to reflect the scientific ethos, perhaps it also asserts an underlying *leitmotif* in the history of 'frontier societies'. For in these, struggle at the hands of Nature acquires an almost mimetic significance. If Australian political history can be construed as a process of maturation, a biological unfolding, its institutions in a continuous process of creation, so Australian science is well represented as a continuous act of self-exploration and self-discovery. In this 'coming of age', there is an implicit transition, by degrees, from a state of comparative error and ignorance, to one of enlightenment and understanding. In the 1890s, Sydney's *Daily Telegraph* used the metaphor of homesteading: once the ground was cleared and crops were sown, then Australians could reflect on ways of 'furnishing the national home'.[54] The individual was to accompany, and shape, this transition. Confronted by a harsh, ungiving environment, the natural philosopher turns in upon himself, becoming aware of the world around him, and acquiring an enlarged sympathy with Nature. The model of Patrick White's *Voss,* if taking liberties with the historical Leichhardt, nonetheless reflects Bunsen's trilogy of God, Man and Humanity.

In this history of Australia, there are deep implications for science, of which the most obvious was to be among the least understood. Within the protective carapace of their metropolitan assumptions, the 'transplanted Britons' who reached into the depths of the continent learned that unless and until all that is known about Australia – and the skies and seas and islands that surround it – is known by Australians themselves, and lies firmly in the hands of their own museums, archives, libraries and galleries, Australia can never be truly independent. Natural knowledge of the continent became self-knowledge for a nation and science, a prerequisite of sovereignty.

***

In the history of Australian science, there are certain features that can reasonably be regarded as 'colonial'. The most obvious is a continuing philosophical Eurocentrism, an attitude of mind and method that dates from the earliest years of naval settlement, and persists for generations. Such conditions were (and are) not exclusive to Australia, and indeed, feature in most settler colonial contexts.[55] In Australia, however, distances were great, discovery was hard work, and interpretation, difficult. Difficulties in interpretation followed in part from the fact that, in European eyes, Australia was a puzzle. As Darwin put it, 'An unbeliever in anything beyond his reason might exclaim, "Surely two distinct Creators must have been at work".'[56]

But what rules of Creation did such exceptions prove? The strange features of Australia that had taken the early navigators by surprise – the land where 'all things are queer and opposite', as P.P. King put it – were slowly understood as 'natural' in their own right. The platypus was called *Ornithorhynchus paradoxus*, although there is nothing inherently paradoxical about an animal so highly and logically evolved.[57] Such Eurocentrism was embedded in the prevailing tradition of natural theology, and in the beliefs of colonial naturalists, who, like George Bennett, were deferential to their British mentors, sometimes to a degree that T.H. Huxley and J.D. Hooker could not countenance, and did not expect. But they reflected the attitudes of a generation educated in Europe, which asked questions that Europeans asked everywhere they travelled. Colonial explorers drew upon precedents and died searching for non-existent inland lakes. Comparisons with European 'maps of learning' stressed the 'deviance' of Australian landforms, until new maps, drawing comparisons with the Americas and Africa, made some sense of austral singularities.

If Australia were to arbitrate, let alone uncouple, the Great Chain of Being, it would take men of independent vision, as well as secular science. By the late 1850s, beyond the era of transportation and the early gold rushes, such men were beginning to arrive. By the 1870s, they included men who resisted the image of colonial science as a 'branch office', managed from London – Britons who were neither solely British nor wholly Australian, but deeply both; whose loyalties may now seem ambiguous, but were then, simply, bivalent. For a century and more, the metropolis refused to relinquish its grip upon Australian attachments, which in any case continued to welcome them. From the 1820s and 1830s, colonial institutions demonstrated the capacity of Australia to serve, like Ireland and India, as a social laboratory for the mother country.[58] But by the end of the century, both colonial and imperial governments welcomed the creation of 'scientific services' to increase and channel wealth from toil – including the cultivation, extraction and sale of minerals, cotton, sugar, wheat and wool. Within the Empire, governments from Madras to Melbourne followed similar paths, creating implicit networks for scientists living on the rim of the imperial wheel.

Imperial science fostered features that were not precisely derivative, but rather constituent of what might be called a 'colonial ethos'. Of these, at least three are readily accessible. First, there was a state of mind that celebrated the practical and venerated the empirical.[59] Australia was by no means unique among Anglo-Celtic societies in giving primacy to 'fact', thereby attracting both criticism and praise for unselfconscious 'practicality' and scepticism of theory. Surely 'Aussie improvisation' must rank next to 'Yankee ingenuity' as one of the most tired phrases in our history. Many colonial Americans and Australians had theoretical leanings.[60] Yet, there is evidence of a tendency that actively preferred taxonomic classification to conceptual argument, a feature familiar to what Robert Bruce has called the 'theory-dodging tendency' of Australia's American cousins, for whom 'only the pressure of myriad discrete facts could warrant a hypothesis – just as some English economists believed that only innumerable individual decisions could properly govern an economy'.[61] In Australia, arguments from scientific principles were coupled with moral dictates from Baconian method, and proved resistant to the theocracy of theory on 'practical grounds'. Well into the twentieth century, unswayed by the German research ethos, and unmoved by its adoption in America, the traditions of practicality retained a powerful hold on Australian universities, as well as on the marketplace more generally.[62] The lobbies of professionalism in chemistry, geology, architecture, agriculture, engineering and medicine were to keep them in office for many years to come.[63]

However, there were two sides to this coin. With a bias towards the empirical, we find a preference for expertise applicable to problems encountered in the 'real world'. It is in this sense that Denholm speaks of Australia's 'descriptive draughtsmen, architects, and analytical naturalists'.[64] This reflected the emergence of the 'practical idealist' of which Liversidge was a notable example. Such idealism embraced a wide range of skills and experience. Until the late nineteenth century, for example, it was possible for men of science to be generalists.[65] Faraday's generation could embrace most of physics and chemistry; Tyndall's generation, with greater difficulty, did likewise. By the 1890s, however, specialisation triumphed in the universities, as in professional life more generally. Still, Australian scientists, like their Canadian *frères*, worked to a broader gauge. The idea of 'problem-solving' by 'scientific generalists' became a colonial specialty – a function of small numbers and large tasks, as well as a by-product of new mineral discoveries, varieties of wheat and breeds of cattle. The tendency persisted long after specialisation and professionalisation gave shape to discipline-formation in Britain and the United States.[66]

A second characteristic of colonial science was its commitment to the twinned 'values' of science as both material and moral agency. In a sense long and well understood in Britain, science received government patronage because of the material and strategic benefits it promised. But in colonial life, science had a special place. Its language drew upon the same eighteenth-century vocabulary of reason, stability, and order (but not, as it happens, the language of liberty, equality and fraternity). In a society where a criminal conviction was at best only a rough indicator of public disapproval, and where religion had a tenuous hold, science was a providential means by which to rid the colony of 'stigma and slur'. Indeed, upon arriving as Governor of New South Wales in 1821, Sir Thomas Brisbane prophesied that the worth of Australia would rise in public estimation as man unfolded and proclaimed its natural beauty.[67] As the *Sydney Gazette* wishfully put it:

> The sons of Australasia's clime
> Shall soon redeem their country's shame,
> No more the penal 'land of crime'
> But nurse of science, truth and fame.[68]

Michael Roe has rightly observed that in an improving sense, 'science *was* culture'; in fostering enlightenment, belief in science strengthened belief in progress.[69] This was a vital and cherished necessity. From the harsh penal settlements of Van Diemen's Land in the 1820s to New South Wales in the 1850s, mechanics' institutes and learned societies taught this text. A corollary followed after the discovery of gold, when it was less commonly, but no less truly observed that the real enemies of enlightenment were no longer the penal system, nor governments nor religion, but the crass materialism that 'progress' was bringing in its wake. These circumstances required a new moral guide. As the Rev. John Woolley, the first Principal of Sydney University wrote in 1861, 'We have learned that no accidental impulse can precipitate an infant colony into a nation. Boundless pastures, bottomless depths of alluvial soil, inexhaustible mines are not the sole conditions of greatness; material, without moral, resources are less than of avail.'[70] In Australia, perhaps more than anywhere else, as Alan Moorehead has put it, humanity had a chance to make a fresh start.[71] Science was there to provide not only the means to progress, but also the moral metric by which to measure its contradictions.

For all these reasons, from the beginning of the University of Sydney – whose founders exhorted its professors 'to enlighten the mind, to refine the understanding, and to elevate their fellow men'[72] – the teaching of science was to feature in the curriculum. In 1850, the 'fellow men' who were to benefit were among a population which, in Sydney, numbered only 54,000, in a colony of 189,000, one-third of whom were ex-convicts. To these new Australians, science was to impart what John Williams of the London Missionary Society called a 'practical divinity', an idealism that would guide a 'new gentry' of 'natural aristocrats', bearing the badge not of birth and privilege but of education; men equally at home in the world of politics and literature, whose fortunes would be devoted to rural development, civic improvement and social betterment. Such men, conversant with the practices of science, would attain heights above the conceits and pitfalls of philistinism, and become the country's principal benefactors.

This vision – which Archibald Liversidge embodied and endorsed – reflected a commitment that linked university and museum, library and learned society, medicine and the law, schools and, less frequently, the churches, in an unwritten contract of progressive self-interest, invoked not only by 'cultivators' and 'amateurs' in science, as Nathan Reingold once defined them,[73] but also by commercial and civic interests similar to those in contemporary British provincial cities of comparable size. The 'bunyip aristocracy' of New South Wales was to give borrowed forms a local content, and secure the twin goals of utility and culture. Indeed, the learned societies and university would be hard pressed to distinguish between the two, just as the wider public could not easily distinguish between the membership of either.

Given its size, Sydney's men of science were to encourage a degree of practical idealism such as that seen in colonial Boston or Philadelphia a century before. From the 1850s, with the devolution of representative government from London, progressive opinion in Sydney looked to the establishment of an 'Australasian Metropolis' of intelligence and civility, through the importation of what Donald Horne has called the 'memorised culture' of Britain.[74] In practice, moral and material improvement rarely followed from the same investment; and editors and educators unfailingly met philistinism high and low.[75] Nevertheless, what we see now as practical idealism flourished in the foundation of the Australian Society for the Encouragement of Arts, Science, Commerce and Agriculture[76] – flattering by imitation London's eighteenth-century (Royal) Society of Arts. The *Sydney Magazine of Science and Art* testified to a growing confidence that 'A man with some scientific knowledge in the bush is a benefactor to his neighbours for miles around'.[77] In the 1850s, the editor of the newly founded *New South Wales Magazine* clasped together the hands of literature and science, as between them having the means to 'Advance Australia'.[78]

A similar spirit impelled higher education.[79] Whether in the museums – the 'cultural cathedrals' of Australia – or in the learned societies; whether in sandstone Gothic or rented rooms, science was taught in the vernacular.[80] Science was meant to have a part in liberal education. In this respect, if not in wealth and tradition, the Australian universities were in advance of England. For a moment, they had almost within their grasp the means of matching and outpacing developments in London, Manchester and Cambridge, and even Harvard and MIT.[81] By the 1870s, greater wealth – fuelled by gold, good harvests and the land boom – bred a sense of optimism in New South Wales; a 'springtime, adolescent period', as Brian Fitzpatrick wrote, which seemed to presage 'the ending of a [colonial] childhood'.[82] In 1871, Anthony Trollope saw Australians trying to 'get to heaven all at once'.[83] Charles Dilke thought

Australia's prospects nearly limitless, and J.A. Froude supposed that in proportion the six colonies already had more educated men than Britain[84] – most, he might have added, looking for suitable employment. As knowledge and skills were 're-radiated' to Britain, so emerged a new sense of cultural identity that saw Australia as part of an imperial circle, and not as the end of the line.

From the same sentiment that championed mechanics' institutes, universities, schools, public libraries and art galleries, emerged a fourth characteristic of colonial science – the importance assigned to 'public science'.[85] By this is meant not the self-serving 'public relations' now routine in seeking public funds, but rather the implicit understanding between governments and science that, where relevant knowledge exists, it must be placed at the disposal of the state. Given these assumptions, the colonial government co-opted science both as instrument and ideology. Whereas before the 1850s, London appointed government botanists and mineralogists, surveyors and astronomers; representative governments thereafter continued, albeit unevenly, the custom themselves. Professors were irresistibly drawn into civic activities. Classicists set texts for teachers and school leaving examinations. Mathematicians set up departments of vital statistics, advised insurance committees and chaired public enquiries. Chemists and medical men reviewed sanitary policy, catalogued toxic plants and proposed improvements in the water supply. Physicists advised on the electrification of railways and the storing of explosives. What was thus already customary (if not always observed) in municipal England became commonplace in Australia; and from such beginnings grew a tradition in which private interests expected, and often had, to rely upon the public purse.

As Australia passed from boom to bust, from the depression of the 1890s to Federation in 1900, as science became part of the state and federal bureaucracy, and disciplines assumed corporate identities, so the business of 'practical science' remained a university responsibility. Science became a plank of Australian progressivism, mimicked in the catchphrase of national and imperial efficiency, in which the teachings of science and 'scientific method' had a special role.[86] In turn, colonial science placed its confidence not in politics or party, but in a business-like ethos that linked the Technological Museum with research in essential oils, departments of agriculture with wheat farmers, secondary education with domestic hygiene, and mining districts with the Technical College. Whether in the 'search for order', or in extending the tropical limits of white habitation,[87] the conviction that progress was best achieved by government support would remain central to Australian science for more than half its history.

\*\*\*

Seen in this light, it is clearly ill-advised to fasten upon simple diffusionist models to describe the development of imperial and colonial science. Rather, it is more useful to constitute a new set of categories, linking imperial and colonial interests. Eurocentrism, practicality, moral agency and public service – these were among the distinguishing features of colonial science, certainly in New South Wales and

Victoria, and to some extent in Queensland and South Australia, and helped frame the organisation of science in Australia at large, both then and later.

In 1837, Ralph Waldo Emerson urged 'the American scholar' of his generation to look for inspiration in the indigenous materials and local character of democratic American life.[88] Fifty years later, a similar message was taken up in Australia, heralded by the *Bulletin* of Jules François Archibald, and the voices of colonial nationalism. Just as literature and the decorative arts and crafts celebrated Australian creativity, so science celebrated the applications of natural knowledge. And just as science and technology united the country by rail and telegraph, so the curiosity and ambition of meteorologists, mineralogists, anthropologists and explorers – some inspired by 'manifest destinies' in the Antarctic and the tropics – gave New South Welshmen, Queenslanders, South Australians, Victorians and Tasmanians a sense of common cause. Where earlier generations of surveyors and explorers had opened the country by boat and on horseback, now scientists, their power symbolised by the telegraph and steamship, united it.[89] Beginning in the 1870s, colonial cooperation in science paralleled the labour movement and the professions, and foreshadowed federalising developments in other aspects of national life – and did so in a similar way – by asserting a commonality of interest achievable only by cooperative action. Australia developed a sense of 'federated science' that united Australia and New Zealand long after talk of political union had faded away.

Australia's industry might remain derivative, trailing Europe and America and dependent upon them; yet, a spirit of eager adaptation proclaimed a deep faith in new technology and its benefits. International exhibitions were making Sydney and Melbourne world capitals. The consciousness of progress, blended with colonial pride, made for a nation in all but name.[90] 'Our colonies are rapidly coming abreast of us', *The Saturday Review* of London announced in 1888. 'There is a freshness and a breadth about the work of Australian science which, alas! we rarely find now in that of the old country.'[91] However, just as energy and promise reached a pedestal, the vision vanished. The successes of science found few ready markets. Given its slowly gathering potential, science had only begun to work for Australia. There were few jobs for scientists. New horizons were opening in anthropology and astronomy, anatomy and meteorology, but up and coming scholars would wait years for recognition.[92] The 'take-off' prophesied in the press, failed to materialise.

Instead, science became hostage to economic crisis, and research languished in the depression.[93] Federation brought encouragement to agriculture and the chemical industry, but little to the universities. In the meantime, the country was caught up in evangelical imperialism and confounded by drought and class strife. The high promise of science, its energies harnessed to the expansion of Australia's domain from the tropics to the pole, was frozen in suspended animation. The coming of war in 1914 rallied Australian scientists to the Lion and Kangaroo, but put an end to the idealism. Slowly and by degrees, through another depression and across another war, Australian science steadily advanced, but struggled to find its way, cultivating a progressive demeanour, but deprived of many young men who would otherwise have been part of its future.

\*\*\*

In this history of hope and struggle, ambition, disappointment and survival, the life of Archibald Liversidge holds a special place. As much as anyone who lived in his 'heroic' generation of Anglo-Australian science, Liversidge was an exemplary figure.[94] Englishman by education, colonial by adoption, he was a paradigmatic 'transplanted Briton', in W.K. Hancock's phrase, who travelled in the service of Empire. An account of his life falls conveniently into three periods, corresponding to his three 'careers' – England before Australia; Australia between 1872 and 1908; and England again, from retirement to his death two decades later. Liversidge, as we shall see, was a son of the commercial middle classes, who made his way up the slopes of academic life by industry and perseverance. His journey led him through a national examination system that was new to his day, assisted by grants aimed by the British government precisely at men of his class and station. In London, and subsequently in Cambridge, he had the good luck to study with some of the most influential 'coming men' of his day. Arriving in Sydney, he was 'transported' into an ostensibly classless world bounded by privilege, and lived among men who saw the world through British eyes. Committed to an ideal of public service, he served Australia, and Australia served him. As we shall see, in the end, it was fulfilling, but not quite enough. With his eventual return 'Home' at the age of sixty, he took up other aspects of public life in imperial chemistry. When war came, he did what he could for his country, for the Empire, and for his discipline. His home in London became an unofficial chancellery for Australian science, and he, its informal ambassador. His life was a fitting advertisement of those who had trained him: *scientia imperii, decus et tutamen*.[95]

Given this history, Liversidge's biography must pay due attention to his upbringing in England, his life after retirement, and the four year-long intervals he spent on leave. It is, however, chiefly to his Australian career between 1872 and 1908 that this book is directed: for it was in Sydney that he left his mark, endorsing principles of 'practical idealism', and sketching the landscape of what may be called 'Liversidgean science'. His collected papers, willed first to his college in Cambridge, and now at the University of Sydney, while detailing much of his professional life, betray little of the private man, or of his personal philosophy. For this reason, it has been necessary to reconstruct his background from many sources, and to follow his contributions to many different domains through a series of sequential essays.

Mirroring the social and scientific movements of his day, Liversidge gave shape to a 'third generation' of men of science in Australia, following the early European explorers, most of whom returned to Europe, and the gentlemen collectors, whose zeal created the first learned societies and museums in the Australasian colonies. Liversidge contributed to these traditions, but added a more innovative dimension, bringing to Sydney's university and public institutions the hallmarks of the 'scientific movement' in Victorian England. Not for the first time, but in a way that made an indelible mark, Liversidge codified a way of thinking that linked academic science with civic benefit.

Liversidge left Sydney in 1908, partly in the hope of resuming his research. He never returned to Australia. Then, it was unclear whether Australians would ever throw off the 'colonial cap and gown'.[96] But that Australians would contribute significantly to science and scholarship was already abundantly clear. During his lifetime, and for decades beyond, Australia's future would be strengthened by the bonds of tradition that tied its people to their British roots. In that knowledge lay his destiny, and through it, flowed his legacy to imperial science. Three generations since, that legacy remains part of Australia's scientific heritage.

# CHAPTER 1

# Science in the City: Family and Youth in Victorian London

View of Newington Turnpike on Newington Causeway, with the Elephant and Castle Inn,
G. Yates (watercolour), 1825.

In the Harleian Manuscripts at the British Museum, there appear records of one Edmund Leversedge who, in the fifteenth century and the reign of Henry IV, became Lord of the Manor of Frome in Somerset. His ancient seat was Vallis House, from the Old French *la falaize,* signifying a cliff, bank, or hill; and his descendants' coat of arms is still to be seen at the Victoria and Albert Museum. In our time, there is a township called Liversedge near Dewsbury, not far from Leeds. The name may derive from Leofric's boundary ('sedge') or hill, after Leofric, Earl of Mercia, less well known as Lady Godiva's husband, who died in 1057. If Liversidge is not at first a name to conjure with, it is one frequently misspelled. Authorities differ as to which misspellings are correct. There are branches of the family, dating from the sixteenth century near Sandbach, and from the seventeenth century near Wheelock in Cheshire, which spell their name Leversage; and others, in the North Riding of Yorkshire, which prefer Leversidge. But the significant connection we seek is that of a Liversidge family which honoured – with a Victorian stained-glass window in Harringay Parish Church – its father, John Liversidge, of Bexley in Kent.[1]

## 1. The Liversidges of London

John Liversidge's parents, George and Elizabeth, had left Bethnal Green in East London in 1809, and moved to the parish of St George the Martyr, Southwark, just north of the Elephant and Castle, and south of the Thames. Off Borough High Street stands The Tabard, one of the few surviving coaching inns in the district, and the point from which Chaucer's pilgrims departed for Canterbury and immortality. Centuries later, at the corner of Great Dover Street and Borough High Street, amidst a busy swirl of tumbledown cottages and little shops, offset by the prosperous filigree ironwork of Georgian townhouses, grew a transient population, bred to progress and poverty. Not far away was Little Dorrit's church, made famous by Dickens, who as a child lived in Lant Street nearby, while his father reposed in the debtors' prison at the Marshalsea. Work for a wheelwright was always easy to find there, and especially after 1820, along the Great Dover Street turnpike, built to carry the bustling traffic of Southwark Bridge and London Bridge towards the principal roads leading south and southeast from London.

In 1830, open four-wheeled coaches first ran on the new Liverpool to Manchester Railway. The steam engine would not quickly replace the horse-drawn carriage. But for a time, both highways and railways would determine the working rhythms of the carriage trade and, that same year, George Liversidge set up his wheelwright's yard, as did many of his trade, in association with a pub, along the turnpike that linked the growing population of London with the hop fields and fruitgardens of Kent. Trade was booming, and George grew sufficiently prosperous to style himself a 'coach builder'. By 1832, he expanded beyond his original works into two new yards, one on George's Road, and another, across the Elephant and Castle, on the new Kent Road. His family had grown to three sons: William (1809–54), John (1810–87) and George (1813–51); and two daughters, Elizabeth and Mary. William and George remained unmarried and lived near their father in George's Road; but the middle son, John, had other ambitions.

About 1831 or 1832, George acquired as a neighbour in Great Dover Street, a modest Sussex farmer, John Jarratt, who had opened a van and wagon office in the Borough to serve the forest and farming trades of his coastal county. He befriended the Liversidge sons, particularly John, inviting them to his home in Buxted, just south of the Ashdown Forest, about six miles north-east of Uckfield, on the old road between East Grinstead and Eastbourne. Either in the Borough or in Buxted (probably the latter) John, in late 1832 or early 1833, met Jarratt's daughter, Caroline Sophia, a handsome girl then approaching twenty. Jarratt, in middle age, had just lost his wife, and was left with five children; it was perhaps with special satisfaction that he consented to the marriage of Caroline and John by licence, in Buxted, on 13 May 1834.

Only two years later, Caroline's father died. Her links with the countryside died with him; for the young Liversidges became an urban family, riding the crest of a wave of prosperity overtaking the cities of Europe, and bringing money into the pockets of men who put coaches to horses. As long as George lived, he stayed in the Borough, next to the teeming but profitable traffic of Southwark. John lived briefly with his bride in Buxted, but within months he, too, had found a rented house near his father in St Andrews' Road, Newington. There in November 1835, was born their first son, John George, named after his father and grandfather, and baptised, suitably, at St George the Martyr's, Southwark.

As publicans and coachbuilders moved in synchrony, so wheelwrights followed fashions in travel and trade. As opportunities arose, the young family of John and Caroline Liversidge followed suit. Between July and December 1837, they crossed the river to Turnham Green in Middlesex, to be near the great arterials then beginning to link London with the West of England. In September 1837, their first daughter, Caroline Sophia was born. In 1839, John opened a works next to the popular White Bear in Great Smithfield, near the city end of the Holborn Viaduct. But the family stayed in Brentford, where their second daughter, Elizabeth, was born, and in 1840, their second son, William. As tenancies were difficult, by March 1845, John moved back to Turnham Green, where he established a prosperous business on Chiswick High Street. Into this semi-rural city life four other children followed, at almost yearly intervals – Fanny in December 1841, Jarratt in September 1843, James (who died in infancy) in September 1845, and in November 1846, Archibald. Three years later, the family moved not far away to Stamford Brook, in Hammersmith, where baby Jane completed the family. In the way of the times, Caroline had her first child at twenty-one, and her ninth by the age of thirty-five.

In this large and close family of second generation Londoners, 'Archie', as the family called him, was thus the youngest son of a second son, and eighth of nine children. Unlikely to inherit the family business, he was in some ways the favourite. As there was no school nearby, John and Caroline apprenticed their sons to the trade, and kept their daughters at home. Private education was reserved for Archibald. The 1850s were harsh for the family. In 1851, Archibald's uncle George died, followed in 1854, by uncle William, in 1855 by his elder sister, Elizabeth Harriet, only sixteen, and in 1858, by his grandfather George. The parents and their seven surviving children made their way as best they could.

View of Elephant and Castle Inn, Newington Butts, Southwark showing busy traffic and street life, James Pollard (aquatint), 1826.

In or around 1858, when Archie was about twelve, the Liversidges moved again, this time to Hackney, in the east of London, lately pasture and countryside, where much new building was taking place. It was an important move, bringing the family near Kingsland Road, the principal highway leading north to Cambridge, and near the schools of the City of London. A substantial semidetached terrace house was found on Buckingham Street, and there the family settled, together with Archibald's grandmother, Caroline Jarratt. Between 1859 and 1866, his father and brothers William and Jarratt established a 'superior coach building business' in Lavender Grove, a ten-minute walk from home and several steps up from the wheelwright's trade.

The move to Hackney brought prosperity to the family, but marriage then began to divide it. John George married in 1864, and moved to the well-to-do western outskirts of London, near Kew Bridge, where the middle classes were settling along the new District Line railway. In Brentford, he established a printing, dyeing and bleaching works and eventually became a sidesman and churchwarden, a Poor Law Guardian and a member of the Burial Board. He had little thereafter to do directly with his father's family. Only at his death, in 1908, would they all come together again; by that time, his firm, united in 'Balfern and Liversidge dyers, bleachers and laundrymen, club and hotel contractors' of Kew Bridge,[2] had launched the family into a future that would ultimately form part of the Sketchley organisation. Of his thirteen children, ten survived infancy. All five sons were sent to the famous Central Foundation Boys' School, founded in the City in 1866 for the sons of artisans.[3] The children kept together in spirit and in name. In 1873, one year after he left for Australia, Archie's brother christened his third son Archibald, and later, his fourth son Sydney.

Just as John George, the eldest of John's sons, preferred to leave the family business, so, too, did William, who went into the poultry trade. There he catered to the expanding appetites of the City and its connections overseas. William married in 1868, and between 1869 and 1879 had six children; within twenty years, he owned the business of Howard and Sons in the Leadenhall Market, where he became Master of the Poulterers Company, a long-serving Common Councillor of the City of London, and purveyor of fine fowl to the Cunard Line. William ultimately retired to leafy Surrey.[4] Only George's third son, Arthur Jarratt, warmed to the carriage trade, and it was to him that the entire business increasingly fell. Arthur Jarratt had six children, including two sons, Jarratt and Percy Balfern. Both brothers declined to be coach-makers, but during the First World War formed a company with an American, Thomas Allen, for the commercial production of dissolved acetylene. 'Allen and Liversidge' prospered.[5]

In 1865, Archibald's mother Caroline died, aged fifty-one, and was buried in the family plot in Brompton Cemetery. With William and Arthur Jarratt married and gone, only Archie and the girls – Caroline Sophia, Fanny and Jane – remained with their father in Buckingham Street. With John's prosperity came expansion, however, and in 1868, his works, now run jointly with Arthur Jarratt, moved to new premises in Tabernacle Street, just above elegant Finsbury Square, and near the heart of the City. Some years later, by happy coincidence, Tabernacle Street became the site of the new Finsbury Technical College. By 1869, John's circumstances were comfortable enough for him to move Archibald and his sisters to a fine new leasehold house, at 119 King Edwards Road, Hackney, not far from the sylvan pleasure grounds of Victoria Park. From here, Liversidge would begin his long journey to classes in Jermyn Street; here John would live until 1881, when old age, and retirement, led him to part company with London.

By a partnership agreement dated 6 August 1881, Jarratt effectively succeeded his father in the coach-building business, then worth some £10,000; and within the next twenty years, 'Liversidge and Sons Ltd., coach, cart and carriage builders' occupied a succession of premises – in Grove Street (1872) and Lauriston Road (1883), both in South Hackney; and in Old Street in the City (1885). The firm's trajectory culminated in the acquisition of six and a half acres at 561 Old Kent Road (1902), where it remained for twenty years. There, Arthur Jarratt adapted to modern times. By the 1920s, the company diversified into automobile body work, and built delivery vans for Harrods and Selfridges. The heyday of Liversidge and Son came to an end only in 1925, when it was taken over by rival Glover Webb, makers of dustcarts, removal vans and brewers' drays, and later coach-builders to Her Majesty the Queen. However, the new firm kept the Liversidge name; and although badly blitzed in 1940, when it was making military vehicles, 'Glover, Webb and Liversidge' continued to operate at the same premises in Old Kent Road until a Japanese company bought the site in 1975.[6]

The Liversidge family home in Hackney, built about 1858.

But this is to look ahead. After the death of their mother, Archibald's eldest sister, Caroline Sophia, in traditional fashion, looked after the family: a fact that did not advance her marital prospects. Despite a twenty-five-year engagement, she cared for John until his death in 1887. Then, at the age of forty-nine, she at last married John Edward Balfern, but her happiness was short lived, as in 1893, he, too, died.[7] Widowed, she lived with William and his family for a time,[8] then returned to her home at Buxted Lodge, where she lived until her death in 1928.[9] It was the ever-dutiful Caroline Sophia, who, in 1927, would eventually raise a headstone over the grave of her distinguished younger brother Archibald, on the main drive of Putney Cemetery, just five miles from his last home in Kingston-upon-Thames.

The picture of the Liversidge family that survives is not dissimilar to the one we find of the vigorous, upwardly mobile Forsytes, through whom Galsworthy immortalised a generation. But while that saga revealed a declining way of life, the Liversidges were rising, born into their time, and moving adaptively from a political economy built along lines of class and craftsmanship into one shaped by steam and refrigeration, by the heavy metal industries and the internal combustion engine. Theirs was not a Trollopian world of property and politics, nor was their history a slow progress through the traditional professions. Instead, within just two generations, they rose to success in both commerce and scholarship. In doing so, they remained committed churchmen; doing well (but not ostentatiously) as they did good. The women in the family feature rarely as pioneers, but often as dutiful partners. They and their men reflect the opportunities and limitations of Victorian possessive individualism, as aspiring artisans moved into the reaches of wealth and stability that Edwardian England offered the 'middle class'. That tradition, that process, rested heavily upon individual self-help, aided by the knowledge that the world was wide open to young English men and women with marketable skills and a sense of adventure.

## 2. Evening Classes for Working Men

The year that the Great Exhibition opened in Hyde Park, 1851, saw several events that were to shape young Archie's life. Lecture courses of the Government School of Mines began in the Museum of Practical Geology on Jermyn Street, and the first undergraduates to sit the Natural Sciences Tripos took their examinations at Cambridge. In Australia, the Royal Assent was given to an act of the colonial legislature that created the new University of Sydney; and payable gold was discovered near Bathurst in NSW. Aged four, Liversidge was taken to the Crystal Palace in South Kensington; but he could have had little appreciation of the influence that all these events would have upon his life.

Just as the Great Exhibition celebrated Britain's industrial and commercial prosperity, so the new Royal School of Mines opened pathways to young men, hitherto excluded from the groves of higher education. While ancient Cambridge admitted new subjects, infant Sydney University set to marry science with practical knowledge, for the betterment of a colony on the threshold of responsible government, less than a decade removed from penal transportation. The Empire needed young men like Liversidge.

We know little of Archibald's life in the 1850s. But by the early 1860s, we see him take steps that led far beyond the City of London. With his elder sons in trade or apprenticeships, and the family firm in good condition, John could accept a different career for young Archie. When he was perhaps fifteen or sixteen, Archie and his brother William entered into the wonders of microscopy, sharing their generation's fascination with slides of insect and rock specimens and metal objects, and reading avidly such popular journals as the *Universal Instructor,* and Hardwicke's famous *Science Gossip*. Both boys became knowledgeable members of the informal network sustained by the microscopical and natural history societies, which by the mid-1860s traversed London from Croydon to Highbury, and from New Cross to Tower Hill. In 1877, five years after Archie's departure for Australia, William would become a charter member of the Borough of Hackney Microscopical and Natural History Society.[10] The formal literature beloved of such societies, exemplified in the 'Notes to Collectors' sections of natural history magazines, extolled the virtues of 'habitual observation' and individual resourcefulness. In the words of Dr Mordecai Cooke, the famous entomologist, these skills, coupled with the methods of comparison and the practice of friendly criticism, would safely tend 'to the diffusion of knowledge, the correction of error, and the advancement of science'.[11]

Lantern Slide of the Great Hall of the Royal Polytechnic Institution,
painted in the 1860s by W.R. Hill.

Natural history was valued for reasons both moral and practical. In Baconian
tradition, the object was to lift one from the tedium of business life, so as to gaze more
intelligently upon the wonders of the Creator. From evening microscopical meetings and
visits to the library of the Workingmen's Institute in Bishopsgate, Liversidge acquired a keen
interest in rocks and living things, and in the pleasures of collecting, exhibiting and exchanging
them, that were by no means limited to children of his age. In terms of sheer interest, these
pursuits easily outmatched coach-building. But natural history was not a paying prospect for
a coach-builder's son. As Archie approached seventeen, he and his parents looked for further
education, near the City, to complement mornings spent at the works. This they found in
London's evening classes, which during the previous twenty years had become a familiar
aspect of commercial and working-class culture in the cities of England.

From the second quarter of the nineteenth century, London saw many attempts
to capture science for display, instruction and rational entertainment. Balancing the
famous Royal Institution in the West End, founded by Count Rumford in 1799 – where
Davy and Faraday dazzled polite audiences, and laid the foundations of chemistry and
physics – was the London Institution, founded in Finsbury Circus in 1819 to bring
to the people of the City of London 'the advantages which science and commerce

# The Mirror

### OF
### LITERATURE, AMUSEMENT, AND INSTRUCTION.

No. 910.]     SATURDAY, SEPTEMBER 1, 1838.     [Price 2d

**THE POLYTECHNIC INSTITUTION, REGENT STREET.**

The Façade of the Royal Polytechnic Institution,
from the *Mirror of Literature, Amusement and Instruction*, 1 September 1838.

derive from each other'.[12] Science, commerce, utility – the Trinity of civilising, secular religion – unified the Institution's afternoon and evening lectures on topics ranging from steam power and domestic chemistry to electricity and forensic medicine. From the 1840s, the London Institution boasted, as its Professor of Experimental Philosophy, W.R. Grove, inventor of the 'Grove cell', together with a galvanic series of speakers, including Fellows of the Royal Society, sanitary reformers, naturalists and inventors. It served a broad clientele, from City clerks and accountants, to bored ladies and off-duty soldiers, who came to be informed, or more simply, amused, by the demonstrations of optics and hydrostatics by W.T. Brande and George Birkbeck; or who came to see the models of planetary action explained by Edmund Clarke. Others would hear scientific sermons by Dr Peter Mark Roget, Secretary of the Royal Society, of Thesaurus fame and knowledgeable about the 'evidences of design' afforded by Nature. Some audiences sought little more than casual knowledge and a dry roof on a wet afternoon. But others obtained, at a shilling a time, the best science education in the metropolis. With systematic instruction still in its infancy, these 'institutional lectures' were the principal means by which young men of the calibre of Michael Faraday, a bookbinder's son, and Roderick Murchison, a veteran of the Peninsular War, were first introduced to science. By the 1850s, lecturers and 'Institutions' both held a secure place in the affection of Londoners.

But the times were changing, and with the advent of secondary schools, open competition in the civil service and the credentialist pressures increasingly demanded by professional England, their educational influence was being overtaken. In their place came institutions which could offer 'certified' knowledge normally made examinable. Liversidge came upon the scene just as the old order was giving way, but before the new one had quite arrived. His formal education resembled both the experience of Faraday a generation before, and the expectations of a generation still to come. Those expectations were circumscribed by the facts of metropolitan life. With its vast population, its burgeoning problems of poverty and prostitution, housing and sanitation, water supply and transportation, neither London nor any other great city had yet found a way of providing education to those millions of young men and women between sixteen and twenty-four, who '... are earning their living during the day, but whose general culture and technical training ... has in some way to be provided for'.[13]

Despite the efforts of utilitarian and Broughamite reformers in the 1820s and 1830s, mechanics' institutions had failed to attract significant numbers of young working men as students, and all failed to earn subscriptions sufficient to meet their costs. With one exception – the Birkbeck Institution, established originally at Chancery Lane in the City, and later to become Birkbeck College – all these had faded out of existence by 1865. Of alternatives, there were only two: the Working Men's College, established by F.D. Maurice in 1854, and the City of London College. As late as 1880, Sidney Webb calculated that not more than two per cent of young men in London between the ages of sixteen and twenty-five, and an 'infinitesimal handful' of young women, were enrolled in any educational institution whatsoever.[14] They had little choice. Even for an enterprising young man, it required energy and determination to gain a polytechnic education mainly through evening lectures.

Between 1864 and 1867, Liversidge, aged between seventeen and nineteen, attended classes at the City of London College and lectures at the Royal Polytechnic Institution in Regent Street. These two establishments embodied rival, but coalescing motives. The 'Royal' was descended from the Polytechnic Institution, founded in 1838 by Sir George Cayley, inventor, pioneer of aviation, and popular educator. The Institution's Royal Charter, granted the following year, endorsed Cayley's passion: 'to teach engineering and science and to draw public attention to new discoveries in these fields' at a place where 'the Public, at little expense, may acquire a practical knowledge of the various arts and branches of science connected with Manufactures, Mining Operations and

Rural Economy'.[15] In its first decades, the 'Poly' trained some teachers preparing to give secondary science classes, but it was better known for its spectacular displays and lectures for inquisitive minds on the 'science of common things', from photography and railway engineering, to electricity and magnetism.[16] By 1845, thousands annually flocked to its Great Exhibition Hall, set in an elegant arcade off Nash's triumphal terrace where, in 'department store' fashion, they were shown 'by the most simple and interesting method of illustration, those sound and important principles upon which science is based'. The Hall's most famous attraction was a great iron diving bell, which for a shilling took adventurous onlookers into a tank of water twelve feet deep. But it also specialised in the new science of photography – Fox Talbot's process for producing pictures on iodized paper was on show from 1841 – and in its patented displays of those new 'dissolving views' that heralded the arrival of moving pictures to Regent Street.

Twenty years on, however, the 'Poly' began what would prove to be an irreversible decline. Although it offered twice-weekly evening lectures – at sixpence each for the industrial classes – and styled itself an 'Educational Department for the Study of Art, Science and Literature', its popularity nonetheless waned, and its income fell. Finally, in 1881, after a disastrous fire, the 'Poly' briefly closed but was miraculously rescued the following year. Quintin Hogg bought its premises, and there unfurled under the banner of the Regent Street Polytechnic, the modern movement in tertiary education.[17] Later, it was amalgamated within the Polytechnic of Central London, and is now part of the University of Westminster.[18] In Liversidge's day, it served a valuable purpose, and possibly introduced him to photography. But such pleasant diversions were not examinable paths to a future career; and for that, the serious student had to look elsewhere.

The City of London College, now part of the City of London Polytechnic, is today the largest college of business studies in Britain.[19] A century ago, J.H. Levy, the economist, took classes there. In its precincts, Sidney Webb began his famous studies of poverty and municipal government, and there he also lectured after the First World War. But this secular bastion built on the Roman walls of London owed its foundations to the much earlier efforts of a group of Anglican clergy, motivated in the 1830s by the Rt. Rev. C.J. Blomfield, Bishop of London, to 'improve the intellectual and moral condition of the industrial classes'. Among those answering his call was the Rev. Charles Mackenzie, who founded in 1849 an association dedicated to providing 'Metropolitan Evening Classes for Young Men'.[20] Beginning with a small library and coffee room in Crosby Hall, Bishopsgate, in 1851, there emerged by the late 1850s, under the patronage of Prince Albert, a program to which over 750 young men a year subscribed. Between 1846 and 1860, the 'classes', as they were familiarly known, catered to the needs of some 6000 registered students;[21] and in 1862, they were reconstituted into the City of London College (motto: 'wisdom better than rubies'). Success led in the early 1860s to the purchase of larger premises from the Bricklayers' Company at Sussex Hall in Leadenhall Street,[22] where for the next twenty years,[23] the College opened its doors every evening to 'young men who employ their leisure well, and who, while they are among the working classes, are also the most intelligent members of the community'.[24] From Liversidge's perspective, it was ideally located in the heart of the City, a half-hour horse-tram ride from Buckingham Street, and only a short walk from the Liversidge works in Tabernacle Street.

A Scientific *Conversazione* at the Apothecaries' Hall,
*The Illustrated London News*, 28 April 1855.

Unlike the utilitarian Birkbeck College, where workingmen at first managed their own instruction, the City of London College set out to offer an improving Anglican curriculum – 'to entertain and benefit the young men of the day'. In some ways, its purposes were not unlike those of the many working men's clubs – begun in 1862, and numbering around 300 by 1867 – that set out to 'rescue' young men from 'intemperance, ignorance, improvidence and religious indifference'. In this noble project, the mechanics' institutions had notably failed,[25] and likewise some of the clubs suffered from the 'too positive pressure of the parson's hand', or from the distracting attentions of political agitators, especially during the franchise reform debates of the late 1860s. For advancement, serious pupils had to look elsewhere. The City of London College was a convenient place to begin.

Christian spirit, intellectual vigour and moral purpose; these were the messages radiating from Sussex Hall to the hundreds of young men who by the early 1860s flocked annually to its doors. The College's hero was unmistakably the self-made man: as the College journal put it in 1859, 'genius without industry has rarely won a lasting reputation; perseverance, with but one talent, has rarely failed'.[26] The College did well by its students. Typically, they took only individual courses; but if one were sufficiently dedicated to spend a total of nine terms, and win a First Class in each of three subjects, and pass an elementary examination in arithmetic and English besides, he would qualify for the Associateship of the College. For those wishing to attend the new Government School of Mines, there were scholarships endowed by Joseph Whitworth, and Royal Exhibitions and Scholarships offered by the government.

Lectures were given over three terms, from October to June. In the beginning, these were free, but by the 1860s, pupils were charged a shilling a lecture; their choice, if expensive, was still freely made, between mathematics, chemistry, English literature, language, art and business, and such undefined subjects as were considered 'suitable for persons intending to emigrate'. Subjects of religion and political controversy were excluded. By the 1860s, the College specialised in languages and commerce;[27] with a small established staff of fourteen, including a chemist (Dr A.C. Maybury), and a number of visiting lecturers in geology, botany and applied mechanics. Some of these held appointments elsewhere in London, but for financial and professional reasons, taught for the College on a peripatetic basis. Sir Roderick Murchison, Director of the Geological Survey, was one; E. Ray Lankester, later KCB and FRS, Thomas Eltoft, FCS, and Sir John Lubbock, the noted banker and naturalist, were others. The College's principal lecturers and examiners in geology were H.G. Seeley, FRS, of King's College, London, and Robert Etheridge Sr, Curator of Palaeontology at the Geological Survey of Great Britain. Whether addressing the mysteries of dyes or Dryden, lectures were directed, in the words of the Rev. T.H. Bullock, MA,

> not so much with a view to thoroughly instructing [the] audience, as to induce the young men ... to investigate and think for themselves. They were, in fact, intended to be rather hints for study, than study itself ...[28]

That study would have its own rewards.

## 3. The Scions of South Kensington

'It is the fashion among writing men of the present day', the City of London College announced in 1859, to 'characterise the age we live in by some comprehensive title. The age of progress, the age of invention, the steam age are cases in point; and we may expect to hear shortly of the "age of examinations".'[29] Competitive, impartial, written exams offered a window on the future for those 'deprived of alternative careers'.[30] In 1852, in the wake of the Great Exhibition, Lord Aberdeen's government, acting on a proposal by Prince Albert advanced in the Queen's Speech, established a new ministerial portfolio, the Department of Science and Art, whose task was to supply 'scientific and artistic instruction to the industrial classes of this country in a more systematic manner than has hitherto been possible'.[31]

In June 1859, under its energetic secretary John Donnelly, the Department embarked upon a program of 'payments by results' in science classes. From 1862, following the recommendations of the Newcastle Commission, the Education Department introduced a similar system throughout the country's primary schools. But as early as 1852, coinciding with the extension of honours examinations at Oxford and Cambridge, the College had already established a Board of Examiners; three years later, in the wake of the Northcote-Trevelyan reforms in the civil service, the College made it possible for its pupils to sit the examinations of the Society of Arts. Indeed, of all eighty-two institutions in association with that robust organisation, the College soon presented the largest number of candidates: of the 800 students enrolled in 1862–63, fifty-five took examinations, and twenty-five received First Class certificates.[32] Prizes of ten guineas were awarded for the best performers, who could look to careers in the clerical ranks of H M Customs, the Civil Service, or the City. Perhaps some pupils took seriously the topic of the College's English prize – 'Colonisation, illustrated from the Colonies of Great Britain'[33] – and looked to prospects overseas. At home, many emerging professions aspired to the high fashions of credentialism, and by the 1880s, College students were preparing for the examinations of several different boards.

For the academically-minded, there was the University of London matriculation; for the technically-minded, the examinations of the Pharmaceutical Society, or (after 1879) of the City and Guilds of London. But the most popular examinations – if such is not a contradiction in terms – were those of the Science and Art Department, conducted each May from the Department's famous address in South Kensington. What Baker Street became to great detectives, South Kensington was to the examiner. By virtue of its rapid expansion of science classes, the Department created an unprecedented demand for science teachers. For an academic young person (customarily, male), the 'smart money' lay in the classroom, well before the hero of *Love and Mr Lewisham* ventured north of the Thames.

Liversidge entered Sussex Hall in 1864. What he had lacked in formal schooling, he soon made up for in self-instruction. The experience was pivotal. With only four others from his year, Liversidge persisted until winning the Associateship of the College; three others went on to read for degrees at King's College, while a fourth, H.T. Pollard, became a distinguished geographer and librarian.[34] The College gave them and others (including women, by the 1880s) a library and freedom to read at their own pace. By favouring enterprising families, the College engendered a special *esprit de corps* among the artisan aristocracy. Liversidge's father became a 'donor' of the College in 1867, the year his son took the Associateship, and gave ten guineas to its building fund in 1882. Sons William and Jarratt took individual courses (but not the full Associateship), and from 1868 were regular subscribers, each giving ten guineas a year as late as 1884. These were considerable sums, but the Liversidges were appreciative. If Liversidge had acquired a technical education by the age of nineteen, he remained in the minority. As late as 1881, over a million students annually registered at the nine London polytechnics which offered evening classes, but while some 39,786 of these took 'domestic science', only 383 sat for examinations in technology or the natural sciences.[35]

For this persisting imbalance, reasons were not hard to find. Civilisation, as W.T. Brande told audiences in Albemarle Street and Finsbury Circus, grew from the union of chemistry and commerce. But at the City of London College, this marriage of civilisation was not without friction. The College prospectus gave little attention to contemporary debates about the value of science either as mental training or as a basis for understanding the natural world. Liversidge arrived too early to benefit from the improvements in secondary science teaching, inspired by such men as James Wilson of Rugby. Learning at the College was by rote and text. Reading lists were few and brief: by the 1880s, chemistry required students to find Fownes' *Chemistry*, originally given as lectures at the Middlesex Hospital Medical School; Henry Roscoe's recent *Chemistry*, and Eltoft's concise *Qualitative Analysis*, 'specially arranged for students'.

As no lists for the 1860s survive, we are left to guess at Liversidge's wider reading.[36] Edward Frankland, then of the Royal Institution and later of the Royal College of Chemistry, examining for the Science and Art Department, noted the candidates' sparing use of new chemical notation, and their poor understanding of the new doctrines of atomicity and chemical equivalence. A.W. Williamson, of University College London, who examined for the Society of Arts, similarly found that 'many of the papers leave a good deal to be desired in the way of accuracy'. From the answers, it was clear that 'teachers of chemistry are apt to include in their course of instruction a wider range of facts than their students can thoroughly master'.[37] Worse, there was no opportunity to demonstrate manipulative ability, or analytical skill through laboratory work: the nature of 'examination science' was such that the practical analysis of salts, acids and bases had to be confined to a written essay.

If Liversidge and the dozen or so others reading chemistry came away with a narrow vision of science, it was not much different from that received by their upper-class contemporaries at any of the nine great public schools just then visited by the Clarendon Commission. In giving evidence to the Commissioners in 1862, Michael Faraday, speaking *ex cathedra experimentiae*, described the continuing tragedy of school masters representing science as a gymnastic activity, with facts carted in tumbrels. As long as science lessons were conceived as memory work, and not as the exercise of what Dr George Moberley, Headmaster of Winchester, called 'the highest faculties', the schools of England could well doubt whether the teachings of science could ever form a basis of knowledge 'synthetical enough for general education'.[38]

If students at the College lacked the education thought fit for the public schools, at least they shared some of the advantages of personal tuition. In science, small numbers made individual attention unavoidable. In 1865 and 1866, we may speculate that Liversidge was tutored by Alexander Morrison Thomson, a brilliant young scientist who had graduated BA (1862) and BSc (1864) from the University of London; who had received one of London University's first doctorates of science in 1866, and who was in the 1865–66 session taking courses at the Royal School of Mines in Jermyn Street. Thomson had begun work as a civil servant, but was not in full-time employment, and possibly took pupils to make ends meet.[39] In 1866, on the recommendation of the Director of the RSM, he was appointed to a new position at Sydney University, where his fortunes would, within only five years, shape the destiny of young Archie.

Whilst Thomson's connection remains speculative, we are on surer ground with another graduate of King's College London, the Rev. Burford Gibsone, MA (1828–1896), who was also a Wrangler of Trinity College, Cambridge, and in 1854 one of the first to take honours in Cambridge's new Natural Sciences Tripos.[40] Gibsone took holy orders in 1856, but chose to follow a path in teaching that took him from Guernsey to Birmingham, and eventually back to London. Between 1862 and 1866, he was a mathematics master at the Mercers School in the City, where he supplemented his income by preparing thirty or so external pupils a year for the exams of the Science and Art Department. Gibsone had studied the 'elements of chemistry' in evening classes at King's, and produced an impressive record of attendance at the College's practical chemistry classes between 1863 and 1868, later publishing his experiments on the measurement of hydrogen sulfide gas.[41] While engaged as a tutor, he took up Archie the coach-builder's son, encouraging him in mineralogy and inorganic chemistry, and introducing him to the meetings of the Chemical Society of London.

What intellectual influence Gibsone exerted, we do not know; but overall it was positive, as by 1866 Liversidge was leaning towards a chemical career, possibly combined with school teaching. Faraday had preached in 1862 that the poor standard of teaching in English schools was neither inevitable nor irremediable. 'You want men who can teach [science]', he said, 'and that class has to be created.'[42] Liversidge was among those he nearly persuaded. Before 1865, the City of London College rarely produced science students, but in Gibsone's class at Christmas 1866, there were twenty-seven: nineteen reading inorganic chemistry and eight reading mineralogy.[43] With good coaching, London's interconnecting network of examiners ensured that once a young man was 'noticed', his prospects could advance rapidly. In this, at least, science had learned much from commerce. By the close of 1866, as he began to prepare for the May examinations of the Science and Art Department the following year, Liversidge's experience of science teaching had been bookish, brutish and short. Yet, from these limited beginnings his fortunes would take a momentous turn.

# CHAPTER 2

# Education for Empire:
# The Royal School of Mines
# and the
# Royal College of Chemistry

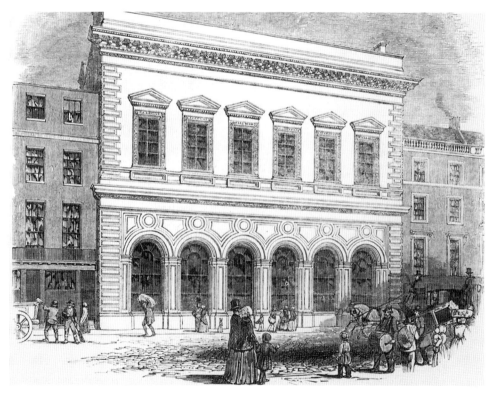

The Museum of Practical Geology in Piccadilly, the first home of the Royal School of Mines,
*The Illustrated London News*, 1848.

If any single event may be said to mark the beginning of Liversidge's scientific life, it must surely be the day in early February 1867 when he first entered the Royal School of Mines, located in the Museum of Practical Geology, on Jermyn Street. Just off Piccadilly, surrounded by the rush and traffic of central London, the RSM was the lineal descendant of the 'Metropolitan School of Science applied to Mining and the Arts', created by administrative union of the former Government School of Mines and the Royal College of Chemistry. The School had its origins in 1835, with the beginning of the Geological Survey of Great Britain, which was the creation largely of its first Director-General, Sir Henry Thomas De la Beche.[1] For two decades, this unusual research school – in part academic, in part governmental – staffed surveys and either directed or inspired the mapping of vast regions throughout the British Empire.[2]

In 1837, the Survey established the Museum of Practical Geology, in which from 1841 members of its staff were given irregular instruction in the 'sciences of mining', including analytical chemistry, metallurgy and mineralogy. It was not, however, until 1851, in the wake of the Great Exhibition, that Prince Albert opened the new Government School of Mines and Science Applied to the Arts, in the Museum in Piccadilly – where, it was hoped, staff and students could 'direct the researches of science and ... apply their results to the development of the immense mineral riches granted by the bounty of Providence to our isles and their numerous colonial dependencies'.[3] Over the next century, the graduates of the School built upon the foundations laid by De la Beche, forming one of the most extended, influential and cohesive networks in British science.

## 1. A School of Science in the Metropolis

Historians find it ironic that, at a time when Britain was producing nearly half the mining wealth of Europe, and her expanding Empire promised far more, there was no school in Britain to equal the level of instruction available on the Continent, notably at the Bergakademie in Freiberg and the École des Mines in Paris.[4] An even greater irony persisted through the School's early years, when the lectures of Edward Forbes, Robert Hunt and Lyon Playfair attracted no more than ten matriculants and twenty or so 'occasionals', who were chiefly an assortment of soldiers (from the British Army in India), naval officers destined for voyages of exploration, incoming Geological Survey staff and pupils from the government's Schools of Design. Certainly, attempts were made to place 'survey science' before the public. In 1851, for example, when news reached London of Hargraves' discovery of gold in NSW, the School put on a series of evening *Lectures on Gold, for the Instruction of Emigrants about to Proceed to Australia*.[5] But despite gathering evidence of fresh discoveries throughout the world, and substantial growth in the Museum's work,[6] mining education remained a fugitive concept, necessary for a few, but unattractive to most. In 1852, in hopes of encouraging a wider audience, the School broadened the scope of its lectures to include 'the more important applications of the sciences to the arts and useful purposes of life'.

Given that Londoners did come to hear lectures in natural history and geology; and given what the School's earliest historian describes the doubtful 'quality of those engaged in the English mining and metallurgical industries', it was surely 'impractical' to neglect the demand for 'advanced instruction in general science' that its staff could supply.[7] Even so, the School failed to attract large numbers. In concept and practice, its curriculum differed fundamentally from the liberal education offered by University College and King's College London, in that it gave 'a practical direction to the course of study, so as to enable the student to enter with advantage upon the actual practice of mining, or of the Arts which he may be called upon to conduct'.[8] But ease of access and the lure of wealth seemed insufficient to induce many young people, at a time when few livings were open to science in any case, and fewer still in the applications of science to industry; and when polite society gave its prizes to those who held university degrees.

In March 1853, following a Treasury enquiry into the reasons for the School's apparent failure, Lord Aberdeen's government decided to unite the Survey, the Museum and the School under the jurisdiction of the newly-created Department of Science and Art. Later the same year, Lyon Playfair, who had held a lectureship in chemistry at the School before becoming the Department's first Secretary, invited the Royal College of Chemistry (RCC) to join the condominium. The RCC was privately founded in 1845, in the wake of a visit by the distinguished German agricultural chemist, Justus von Liebig, to England in 1842. Liebig was courted by Sir Robert Peel and other statesmen, and feted in the great houses of the land. Playfair had studied with Liebig at Giessen; and on returning to England became his devoted disciple, translating his celebrated text on *Chemistry in its Relation to Agriculture and Physiology*, and preparing the way for his visit. The 'immediate effect' of that 'triumphal tour', as Playfair later recalled, was 'to make chemistry a popular science'.[9] Joined with the interests of the 'chemical constituency' of England,[10] its wider effect was to 'induce colleges to open laboratories' in ways as yet unfamiliar in Britain.[11]

By its charter, the RCC was to be one such institution, offering public lectures and laboratory training to pupils of all ages at low fees, as well as chemical analyses for commercial clients. On Liebig's recommendation, its directorship was given to August von Hofmann, then at the University of Bonn, where Prince Albert had studied. Coming to England in 1845 – again, through the influence of the Prince Consort – in eight years Hofmann trained some sixty pupils, many of whom were to make their mark in academic and industrial chemistry.[12] But the College's high hopes were always clouded by financial uncertainty, so in July 1853, Hofmann accepted Playfair's invitation, and in October, the Royal College of Chemistry, with its fifty students, became the Chemical Department of a new Metropolitan School of Science.[13]

Given the publicity surrounding practical chemistry and government interest in the Jermyn Street Museum, the School's prospects should have brightened. For the next decade, however, matriculants averaged fewer than twelve, with fifty-four occasional students a year. Not surprisingly, an institution with such a complex lineage did not escape disagreement about aims and objectives. However, stronger leadership came with the soldierly Sir Roderick Murchison, who succeeded De la Beche in 1855. Murchison was as ambitious for the School as he was for imperial science, and re-dedicated it to the practical sciences bearing on mining and agriculture.[14] In this, he had the support of his professors of geology, mining and metallurgy – Andrew Ramsay, Warington Smyth and John Percy.

Interior of the Museum of Practical Geology, Piccadilly, 1851.

In 1857, for these and other reasons, the 'general division' of the School, which included the biological sciences, was closed, and the curriculum was restricted to mining and metallurgy, with natural history available only to students of geology. To enhance the School's status, Murchison asked the University of London to award degrees to its successful students. However, the University would not alter its matriculation requirements, which required passes in Greek and Latin, an almost insurmountable obstacle for precisely those pupils of the working middle classes whom the School hoped to attract. Accordingly, the School decided to give a 'certificate of competency' to those who attended nine courses and passed the examinations. This qualification became, from 1862, the 'Associateship' of the Royal School of Mines.[15]

For the next decade, true to Murchison's word, the School focused on the sciences and engineering relevant to mining.[16] Its aim was to become to Britain 'what the Bergakademie of Freiburg and the École des Mines of Paris are to Germany and France'.[17] But this noble vision failed to win a wide following, and between 1853 and 1866, only fifty-two men took the 'Associateship'. In government circles, the School was seen as a disappointment. Murchison protested that more numbers understated the significance of those 'Associates' who had gone to the Geological Surveys of Britain, India, Ireland, and the colonies, and who had taken up key academic and commercial positions overseas. By the turn of the century, over a third of the Associates worked overseas, mostly in the Empire, or in the 'informal empire' of South America.[18] As the editor of the RSM's magazine put it, 'we all know that those ... who leave our School to fight the battle of life, seldom stay in England to wage that warfare'.[19] As scientific emissaries for Britain, they formed an imperial 'elite' comparable with the well-travelled officers of the Royal Engineers.[20] Possibly for the same reason, they were less well known at home, than honoured abroad.

Despite their unusual status, they were well educated. T.H. Huxley – who replaced Edward Forbes on the staff in 1854 – had few reservations about this, and although he predictably opposed (and would one day reverse) the separation of biology from mining, he vigorously defended the School's reputation. By 1871, over 1000 occasional students and 870 science teachers benefited from its courses.[21] And as for the Associates, his praise was unqualified. 'Those', Huxley said, with eloquent simplicity, '... are instructed men. Most of them have found and are filling important positions, and act as centres for the diffusion of science throughout the country and the colonies.'[22] They were an advance guard, leading the people of England 'everywhere, though slowly, to admit the importance of professional training' in the sciences.[23]

What Huxley called 'professional instruction' was exactly what Liversidge wanted, and what at least some of his contemporaries – lacking a taste for Greek and Latin and the privileges of class – could find nowhere else in England. In these close quarters, Liversidge and his friends formed not only an upwardly mobile *corps d'élite*, but also a philosophical fraternity, which was to have a formidable effect on science education. In 1867, some thirty students enrolled in the second and third years, including Henry Armstrong, later Professor of Chemistry at Finsbury College, and a pioneering figure in science education;[24] and Alexander Pedler, who served as a Demonstrator in Chemistry at the School for three years before going to a chair at Calcutta in 1871.[25] Both eventually received honours (and Pedler, a knighthood) for their services to imperial education. Fifty years on, Armstrong, Pedler and Liversidge would come together again, around a wartime committee table rather than a laboratory bench.

## 2. Scientific Circles of the Sixties

In the spring term of 1867, Liversidge began his formal instruction at the RSM. That year and the next marked a turning point in British science. By August, Michael Faraday and William Whewell had died, and Charles Babbage and Roderick Murchison were terminally ill. The 'young Turks' of the 1850s – Huxley, Frankland and Tyndall – had reached their prime. A new generation was needed to take up their banner. At the time, nearly all eyes focused on the Second Reform Bill, then making its way through Parliament, and ultimately pointing towards an expanded democracy. But many were also concerned about what later historians have called the 'decline of the English industrial spirit'.[26]

In May 1867, alarmed by what seemed a disappointing performance in the manufacturing categories at the Paris Exhibition, at which he was a principal juror,[27] Lyon Playfair wrote to the Schools Enquiry (the Endowed Schools) Commission, then sitting under the chairmanship of Lord Taunton, saying a dire fate awaited industrial Britain if no greater provision were made for scientific and technical education.[28] The Commission, which had by its charter no remit for technical education, solicited evidence from sixteen other jurors, MPs, men of science (including Frankland, Tyndall and Smyth) and manufacturers, all of whom broadly echoed Playfair's warnings; and saw to it that the correspondence was published by the Society of Arts in June, and reprinted as a parliamentary paper in July.[29]

In August, just one week after J.S. Mill addressed the University of St Andrews on the values of a liberal education, appeared the celebrated *Essays on a Liberal Education,* edited by Frederick W. Farrar, which restated the case for science. Its message was underlined by the BAAS, meeting in Dundee later that summer, which had appointed a committee the previous year to 'consider the best method of extending Scientific Education in Schools'. The committee – which included F.W. Farrar, J.M. Wilson, Huxley and Tyndall – addressed government on the 'urgent necessity for introducing science teaching into our schools, if we are not willing to sink into a condition of inferiority as regards both intellectual culture and skill in art, when compared with foreign nations.'[30]

Supporters leapt to their aid. In November 1867, Bernhard Samuelson, the barn-storming MP for Banbury, recently returned from a survey of manufacturing industry on the Continent, persuaded Lord Montague, Lord President of the Council, to enquire more deeply into Britain's capacity for educational reform.[31] His appeals led the following March to the appointment of a Select Committee, under his chairmanship, which searched for evidence of technical instruction (or its absence) in British schools and universities.[32] In proselytising Parliament, Samuelson cited the model of the Metropolitan School of Science. 'Give a thoroughly scientific training in Jermyn Street to a small number of young men, chosen, if you like, in part from amongst your most promising "science teachers", in order to qualify them as professors of science.'[33]

In December, the Taunton Commission endorsed Samuelson's case, and prompted the government to begin a scheme of Local Scholarships and Exhibitions 'for the encouragement of science instruction and for the support of students of the industrial classes'.[34] For those in need, even the fees of the Royal School of Mines were waived. 'Associates' would become 'merchants of light' for the empire.

Young Liversidge was precisely the product Samuelson hoped the RSM would produce. He began at Jermyn Street in February 1867 with Andrew Ramsay's lectures in geology, and won a mark of 83 and a credit for the course.[35] He also sat the May examinations of the Science and Art Department, for which he had been preparing since December.[36] Some 4920 pupils across the country sat different sections of these exams,[37] but the twenty-seven candidates from the City of London College that year chose only the questions set in mineralogy and inorganic chemistry.[38] In these, they were asked to list the chief ores of lead, zinc, tin and other metals; to explain the reduction of metals from their ores and alloys; and to identify the symbols used to signify chemical compounds.

Liversidge scored brilliantly. Thanks to Gibsone's tutoring (for which 'results' Gibsone received an extra £10 from the Education Department), he took the Gold Medal in mineralogy, and a Bronze Medal (for seventh place among 900) in inorganic chemistry. His marks gained him one of the six Royal Exhibitions, created by the Government in 1862 to assist men of 'the artisan classes who may show an aptitude for scientific instruction'.[39] Three were given for the Royal School of Mines, and three for the Royal College of Science in Dublin; each paid £50 per year – the equivalent of a workingman's wage (or that of a Demonstrator in the College's laboratory) – for three years.[40] As the winner of a Gold Medal, Liversidge was also entitled to free tuition, worth perhaps £90 over three years. By living at home, and saving rent, Liversidge could easily see the course to completion.

Liversidge was one of thirteen who entered the RSM in October 1867. His year was one on which the stars would fall. Seven took the Associateship, of whom three became successful in commercial mining and government, and four in academic life. Three won distinction in imperial service. Of particular note were Francis Bayly, later an Assayer at the Royal Mint, and Henry Renwick, a mining consultant; William Gowland, chemical and metallurgical adviser to the Imperial Mint of Japan, and Thomas James, Professor of Chemistry at Woolwich.

Next to Liversidge sat two younger men who had won the other Royal Exhibitions that year: William Garnett, aged sixteen, and William Johnson Sollas, aged seventeen – both described in the register simply as 'students' – and A.W. Bickerton, aged twenty-five, a 'cabinet maker of Birmingham'. All four would win distinction: two became Fellows of the Royal Society, and three would be professors by the 1870s. Garnett became James Clerk Maxwell's Demonstrator at the new Cavendish Laboratory, before turning to administration as Principal of the College of Newcastle-upon-Tyne (later part of the University of Durham, now the University of Newcastle). Bickerton would go to the chair of chemistry at Canterbury University College in Christchurch, New Zealand, where he became noted (or notorious) for heretical views on the origins of the solar system.[41] Sollas (whose uncle, William Johnson, was the Secretary of the Royal College of Chemistry) progressed through chairs of geology at Bristol, Dublin, and Oxford, and only narrowly missed (as we shall see) the chair of geology and physical geography at Sydney.[42] Twenty years on, Liversidge and Sollas shared an interest in testing Darwin's theory of coral atoll formation in the Pacific.[43]

*Day & Haghe Lith.rs to the Queen .*

THE LABORATORIES
of the ROYAL COLLEGE of CHEMISTRY,
the first stone of which was laid by
HIS ROYAL HIGHNESS PRINCE ALBERT,
in the presence of the Council & Members,
June 16th 1846.

James Lockyer, Architect

The Royal College of Chemistry in 1846.

# Liversidge's teachers at the Royal School of Mines

Andrew Ramsay
(1814–1891)

Robert Hunt
(1807–1887)

T.H. Huxley
(1825–1895)

Edward Frankland
(1825–1899)

Lyon Playfair
(1818–1898)

Warington Wilkinson Smyth
(1817–1890)

John Tyndall
(1820–1893)

If architecture shapes behaviour, Liversidge and his companions could be forgiven for having a dipolar view of the RSM. Following the reorganisation of 1853, chemical practicals were given at the cramped Royal College of Chemistry in Oxford Street, but all lectures were given at the Museum of Practical Geology.[44] Despite their links, the two institutions seemed 'sufficiently separate to leave an impression of co-operation rather than of fusion'.[45] Both buildings had fine neoclassical frontages, but interiors scarcely adequate to the needs of their students, let alone the desires of their staff – 'totally unfit', in Huxley's verdict, 'for the purposes of a school'[46] – and within only two years, additional accommodation was being sought in London.

Nevertheless, the two institutions, made one, shared the cream of English science. 'What a splendid education we had', Liversidge recalled fifty years later, as he and Sollas stood before a warming winter fire at the Royal Society in Burlington House. As Sollas observed, 'He [Liversidge] was usually very reticent and not a bit enthusiastic, so I was delighted at this outburst, which just expressed my own feeling'.[47] A generation later and wiser, some of the School's students recalled certain shortcomings. Henry Armstrong, in particular, came away with a different approach to science education. For the moment, however, Oxford Street and Jermyn Street had to suffice for young Englishmen who needed to earn a living.

There had never been any question of the staff's dedication. When, in 1865, Edward Frankland succeeded von Hofmann, his salary was only £200, at a time when Owens College, Manchester, paid its chemistry professor £350; and when even brewers paid their chemists £700 or more.[48] Others – notably Andrew Ramsay – made sacrifices to remain in the service of School and Survey. There was little dispute about their standing in the scientific community. By the time Huxley, Tyndall, Percy, Frankland and Hunt arrived at the RSM, all were Fellows of the Royal Society. Indeed, of the nine earliest professors, only three – Smyth, Willis and Edgar – were not. Of the eight professors in Liversidge's day, three were also members of that famous 'invisible college' known as the X Club, founded just three years earlier, which for over twenty years helped shape liberal policies towards science and science education.[49]

By the late 1860s, the staff had navigated successfully the turbulence of the preceding decade. Although not meeting everyone's wishes – mathematics, for example, were omitted[50] – the curriculum by 1867 incorporated a version of the 'Huxleyan programme', whereby students were grounded in 'primary scientific principles' upon which practical applications depended, and were taught to deploy the superiority of science over 'merely empirical and rule-of-thumb knowledge'.[51] Regular lectures were supplemented by courses of evening lectures, given routinely for workingmen and others, at which audiences of 700 were known.[52] Lectures were open to all for £3 per course, plus laboratory fees. To receive the Associate certificate, after three years, cost an additional £30.

The first year concentrated on inorganic and organic chemistry and physics; the second, mineralogy and geology (including natural history). According to Liversidge's notebooks, students consulted the Museum's collection of fossils, minerals and rocks, took geological excursions, and in the third year, specialised in one of three divisions: mining, metallurgy or geology. In mining, Liversidge learned the 'art', if not the engineering, of mine construction and operation; in metallurgy, he was expected to 'acquire facility and exactness' in knowledge of ores 'so that he may not be at a loss in an unsettled country'; in geology, he took natural history and palaeontology; in all three, he learned how, 'more or less, to improvise'.

## 3. Laboratory Life in London

If not all Liversidge's lecturers were inspiring, all were serious. In October 1867, Liversidge walked to Oxford Street to begin the first-year course of twenty lectures in mechanical drawing,[53] alongside a set given by Edward Frankland, best known today for the discovery of valency and the chemical bond, and for beginning the discipline of organometallic chemistry. [54] Frankland was among the most active public scientists in England. A friend of Huxley, Frankland began his career as the first Professor of Chemistry at Owens College in Manchester, before moving in 1857 to London where, a noted 'pluralist', he held a variety of appointments – at one point, no fewer than three at once. In 1863, he became Professor of Chemistry at the Royal Institution, where he did research on lactic acids while lecturing to large popular audiences, continuing the tradition of spectacular demonstrations for which the RI is justly famous. It is highly likely that Liversidge attended these lectures as a private scholar – offered from 1853 to 1867, they formed a broad base of subjects across the entire spectrum of applied chemistry, probably the best introduction that could be had in England.

In 1865, Frankland was invited to the Royal College of Chemistry, first as a temporary replacement for August von Hofmann, then building new laboratories in Bonn and Berlin, and in September 1868, became his permanent successor. With Lyon Playfair at Edinburgh, and Henry Roscoe, his successor at Owens, Frankland was among the leading chemists of England. All three played key roles in transforming chemistry from a craft-based handmaiden of medicine, agriculture and the textile industry, into an academic discipline and a corporate profession. All had received their research training in Germany, and all saw chemistry as an agency of reform. To Frankland, research was needed, as Colin Russell recalls, to 'retrieve the national character'.[55] However, it was lecturing, not research, that paid the bills – a cautionary lesson for an impressionable young student from Finsbury.

As Liversidge discovered, his teachers at the RSM held strong views about the relationship between abstract knowledge, or 'higher scientific culture', and its practical applications. Implicit in this distinction were conflicting ideas about the education deemed suitable for the different social classes.[56] By the 1870s, the curriculum had become an educational battleground, with deep divisions about the nature of knowledge and the best way to impart it.[57] From this, eventually arose the practical conclusion that a common core of factual information should be fundamental to all learning, and available across social divisions.[58] But the question remained whether that information was best imparted by lectures, demonstrations, object lessons, textbooks, or practice. Frankland's educational philosophy, nurtured during a year that he and Tyndall spent at Queenwood College in Hampshire, enjoyed an enormous influence upon science teaching throughout England and Wales.

These views reflected his experience at Owens College, where his attempts to teach 'chemical technology' had enjoyed mixed success. At the RCC, he was able to parlay his knowledge of principles to the solution of problems. As a result, his students were given no cause to distinguish between 'science education' and 'technology education'; and distinctions between 'pure' and 'applied' were consigned to the textbooks. The RCC's curriculum bypassed abstractions and created its own vernacular. By doing much work for government, Frankland brought this 'language' into everyday use. And insofar as the RCC promoted, as J.F. Donnelly has put it, an 'industrial representation', it produced portable skills needed by both.[59]

Within a decade, specialisation increasingly overtook the discipline. But at the beginning of a period that saw the formation of the Society of Public Analysts (1874), the Institute of Chemistry (1877), the Society of Chemical Industry (1881), and the Society of Dyers and Colourists (1884), the routes taken by academic and 'works' chemistry' were signposted more by differences in emphasis than by disciplinary boundaries, and more by career choices than formal qualifications. The reputation of the College was high; whatever its limitations, its curriculum offered the best training in applied chemistry available in England.

It was in this expectation that Liversidge – in a group of forty, along with many 'occasionals' – entered the Oxford Street laboratory in the autumn of 1867, a year before Frankland became permanent, and when he was still juggling lecture commitments at the RI. Frankland lectured on Mondays, Wednesdays and Fridays at 10 a.m., and practicals were given in three sessions, five afternoons a week, and between 10 a.m. and 2 p.m. on Saturdays, during all three terms. His lectures ranged from the properties of gases to spectrum analysis, to organic radicals and alcohols and organometallic compounds.[60] They may have lacked the élan of von Hofmann, and some, including Armstrong, found them 'altogether lacking in humour and sparkle'.[61] But to Liversidge they were 'the best and most thorough to be found anywhere'.[62] Frankland's class notes, carefully published the year after Armstrong heard them, and a year before Liversidge arrived, were designed to 'relieve' pupils from the distraction of writing down formulae and reactions, in the already antiquated analytical notation.[63]

But it was in the laboratory that his teacher's virtues really shone. Chemistry was the only subject at the RSM where laboratory work was required – a benefit, Armstrong later recalled, 'we owed to the great Liebig's foresight'.[64] Frankland's philosophy, summarised for the Devonshire Commission in 1871, was simply to 'induce every student to study his science experimentally, and by bringing himself personally into contact with the phenomena, so that he may afterwards become himself a worker in the science, and perhaps an investigator'.[65]

Pupils were assigned benches and exercises designed to illustrate the 'general properties of simple substances' and the 'analytical properties' of 'more complex bodies'. From November to March, pupils were marched through routine acid and base reactions and methods in qualitative and quantitative analysis – learning the craft of 'distinguishing, separating and estimating'. The syllabus – 'deducing fundamental laws of chemistry from experimental facts' – embraced flame tests, aided by the blowpipe, to determine composition, and wet tests to determine solubility. The Bunsen burner was an important innovation, as was the spectroscope – used to detect the presence of a given element by its particular absorption lines in light passed through a mineral specimen, and dispersed by a prism. Spectrum analysis was a wonder of the age, and was just coming into use, and appealed to the imagination.[66] But nothing identified the RCC-man more distinctly, than his ability to manipulate the analytical balance, with its two pans, beam and set of references – 'sensitive, accurate, and tedious to use'.[67] Fresenius' *Quantatitive Chemical Analysis* was the text, presenting in great detail the rules that would mystify present day electronic instruments.

The bustling laboratory – built for forty, but serving over one hundred – was heavily congested. If spirits were willing, the air was weak; and the conditions would not meet today's health and safety standards. In the event, the advanced men occupied the benches nearest the open windows – an arrangement which, it was said, served as 'both an aid to the recruits and an incentive to their progress'.[68] Whether 'serious' students, 'freelances', or 'occasionals', Liversidge's generation met two guiding lights in the laboratory – Herbert McLeod – Hofmann's former assistant and Frankland's 'right-hand man'[69] – and Wilhelm Valentin – Hofmann's senior assistant, and Frankland's 'particularly exact' Principal Demonstrator. Neither had yet produced the texts that would make them famous, but both were on their way.[70] Good 'methods' were their message, underlined by such standard texts as Hofmann's edition of Friedrich Wöhler's *Handbook of Inorganic Analysis*.[71] Quantitative techniques were of greatest importance.[72] The key instrument was the 'balance', and praise went to those who could use it well. There was, however, no balance room, so weighing was done in the library, 'where a floor vibrating through the constant passage of people but an effective stop to accurate readings'.[73] It was a salutary lesson Liversidge took to heart.

For all his generation, the task was the same – 'work for work's sake', as Armstrong recalled – where at least no externally imposed examinations 'prevented him from learning'.[74] Indeed, RSM students at the College were 'not sent to be taught, but come to learn'.[75] Laboratory classes emphasised individual learning, at an individual's pace. Self-help and sobriety set the tone. Tyndall found the chief virtue of the place its 'spirit of earnestness'.[76] There were some, such as Armstrong, who chose not to work for the Associateship.[77] But everyone came away prepared. George Stokes, Lucasian Professor at Cambridge, and Huxley's counterpart as Physical Secretary of the Royal Society from 1854–85, taught at the RSM with Huxley from 1854 until 1860. His Jermyn Street students, he recalled, had a great advantage over university men, 'as we had just sufficient assistance and not too much'[78] – a good training for postings where there would be no assistance at all.

With all the emphasis on hard work, it was hard to find fun. Bickerton, who earned his living by part-time coaching, disagreed with Armstrong's dusty view of Frankland, and learned from his teacher how 'to make classes as attractive as music halls, exactly as a circus'. He, for one, studied every chemical reaction that would 'go off'.[79] But Liversidge, as Sollas recalled, was probably more typical – a diffident lad not far removed from a world of commerce, wooden wheels and wagons, who 'looked on science as a serious matter, and could not stand heckling' from what Bickerton called the 'witty artisan students'. To no one's surprise, Liversidge 'retired in disgust',[80] revealing a disposition for self-discipline, verging on the severe, which remained his throughout life. Frankland's examination questions reflected an intense desire to get the 'right answers'.[81] Ironically, it was Bickerton who topped chemistry with 97 marks; Liversidge came second, with 90; and Sollas, third, with 88.

For want of space, Frankland opened his laboratory to some 'half dozen of the better students',[82] and Liversidge was among them. There he met many visitors, such as Hermann Kopp of Heidelberg, who stimulated his interest in the history of chemistry.[83] Students were involved in public work – analyses of sulphur from gas lamps at the South Kensington Museum, reports on adulterated food for HM Customs, and for the Registrar General on inks used in recording births, deaths and marriages.[84] Less lucky lads might be given explosives. On one occasion, a cloud of nitrogen dioxide, evolving from decomposing guncotton, sent the class fleeing into the busy human (and animal) traffic of Oxford Street. Above all, Frankland 'set an example'. As Armstrong later recalled, 'Under [him] I was handling things all the time: really getting to know them; an intuitive understanding of the method of discovery was creeping in'.[85]

In 1866, a year after he took charge of the Royal College of Chemistry, Frankland was appointed a member of the Rivers Commission,[86] which sat under the chairmanship (until his death in 1871) of Sir William Denison, RE, not long returned from Sydney, where he had been Governor of New South Wales. The Commission seemed to sit forever. When Liversidge arrived, Frankland was preoccupied with its proceedings, and with a parallel Royal Commission on Water Supply, to which he gave evidence in February 1868.[87] For over seven years, Frankland attempted to 'place this branch of chemical analysis upon a more satisfactory basis'[88] – not least by exposing the ambiguous definitions of 'impurity' used by the water companies, and municipal failures to treat contaminated water.

While much work was done in the laboratory set up by the Rivers Commission in Westminster, some came home to Oxford Street – Bickerton remembered how residents in neighbouring Hanover Square complained about the smell.[89] Frankland set his keenest students, including Liversidge, to study samples of stream and spring water from the Thames valley. Careful microscopy assisted his growing belief in the 'zymotic', or germ theory of disease, and his eventual rejection of the 'oxidation' theory of purification. Two years before Liversidge began, Frankland took up the young Henry Armstrong, then only eighteen, and assigned him the task of devising tests for detecting organic impurities. Within two years, the work of Frankland and Armstrong led to methods of detection that began to rescue London from enteric disease, and contrived to render water 'a safe drink the world over'.[90] In 1897, not before time, Frankland was knighted.

From Frankland, McLeod and Valentin, Liversidge 'acquired a good knowledge of qualitative and quantitative analysis, both of mineral and organic bodies'.[91] He put all this to good effect, and in January 1868 published his first article – a small item in the 'Notes and Queries' section of *Chemical News,* on a simple analytical test for the presence of precipitated magnesium,[92] which he proudly showed to his father. In April, both William and Jarratt married, and moved to Hackney, but 'Archie' continued to live at home with his father and sisters, well cared for by a family that now knew middle-class comfort. Blessed with a father who evidently favoured his youngest son, he was left ample room to do science.

Beginning in his second term, from March 1868, Liversidge took Frankland's lectures in organic chemistry, and devoted alternate mornings, when there were no lectures, to his own work, using a bench not needed by the demonstrators. His enthusiasm for science was infectious; at least once, and probably more often, he took his sister, his brother William, and his prospective sister-in-law, Eliza Wilmott, to a 'microscopical *soirée*' at the Cannon Street Hotel, near Liverpool Street. That summer, so great was his dedication, and so good his technique, that Frankland appointed him to an unexpected vacancy at the Royal School of Naval Architecture. There he began his first job, as an Acting Instructor in Chemistry, for twenty-six students.[93] They were poorly prepared, and had little time; the prospects of learning chemistry were remote. Yet, helped by the resourceful Valentin,[94] Liversidge put what he had learned to advantage.

More good fortune befell him in October 1868, when, on the basis of his examinations in June, he was invited to work in Frankland's private laboratory. This was a distinction at once rare and commonplace – rare, in that it was offered to few; commonplace, in that those thus hand-picked became famous. Armstrong left to study in Germany,[95] and was followed by W.J. Sollas,[96] and then by Liversidge, to whom Frankland gave the task of analysing the sulphur content of domestic coal gas.[97] Frankland encouraged his *protégés* to publish, and in April 1870, Liversidge published his second note in *Chemical News*. Possibly, his teacher's conscience tinged his careful prose, when he cautiously concluded that 'the gas was rather too rich in sulphur to be an unmitigated good to the consumer'.[98] Such language would hardly shock or surprise the average Londoner, or shake the coal gas industry; but it was revealing, just the same. What it signified was a man careful in small things, who could be trusted in larger ones.

In the spring term of 1868, Liversidge and his class walked between organic chemistry lectures in Oxford Street and John Tyndall's lectures on physics in Jermyn Street.[99] Tyndall's performances were hugely popular – 'always a great treat', Sollas recalled – particularly for the legions of schoolteachers who came. 'The teachers of the world ought to be its best men', Tyndall told students in 1868, 'by the fullness and freshness of their own lives and utterances they must awaken life in others'.[100] And so he did. In Armstrong's phrase, Tyndall was a 'born actor', who performed to the 'great profit and delight' of his pupils. His demonstrations of the wave theories of light and sound were made to 'live before our eyes'.

That year, as before, Tyndall delivered the substance of his essays on *Heat as a Mode of Motion* and his treatise on *Sound* (a copy of which Liversidge kept for the rest of his life).[101] That year, Tyndall also entertained students with demonstrations of his work on the scattering of beams of light by particles, and on the decomposition of vapours by arc lamps. He was at a turning point in his career. On the threshold of publishing his first papers on chemical reactions produced by light,[102] he would soon begin the experiments that would rout Henry Charlton Bastian, Hughes Bennett and other believers in spontaneous generation, and strengthen the case for Pasteur's germ theory of disease.[103]

The sweep of Tyndall's syllabus – from crystallography to biology – left his rivals gasping and his students amazed.[104] In the physics examination, Liversidge scored only 80 marks and a credit, coming behind Garnett, Bickerton and Sollas. But, as his notes reveal, this did not lessen his admiration for the iconoclast of Albemarle Street, Tyndall, an outspoken critic of what natural theologians liked to call 'miracles and special providences',[105] who went to the British Association at Norwich that August, claiming the sufficiency of material explanations, and arguing that science thrived best where the imagination was left unfettered. Liversidge took careful notes.[106] The year 1868 was Tyndall's last at Jermyn Street; he left in June, reluctantly giving up his salary, but explaining to Murchison that 'during my time lecturing, I can do nothing considerable in the way of research'.[107] No one failed to regret his leaving, or failed to respect the reasons for it. It was some compensation that Liversidge soon met Tyndall's successor, Frederick Guthrie, destined also to become an FRS; within a decade, his son would follow Liversidge to Sydney. The 'family' of British science was a small one.

## 4. A Turn to Minerals and Mining

During the summer of 1868, Liversidge's new-found laboratory life gave him a special opportunity to explore the relationships between chemistry and mineralogy. In this, he came under the wing of the meticulous Sir Warington Smyth, Professor of Mineralogy, whose forty lectures Liversidge began in October. Smyth, the son of an admiral, had spent four years travelling in Asia and the Middle East before returning to the Geological Survey in 1844. He was neither a theorist, nor a researcher, and played to a simple *coda*: 'mining is a practical art depending upon the application of different sciences'. However, he also held that 'the methods of applying these sciences are so peculiar that their proper use can only be learnt by actually assisting in mining operations';[108] and he taught students to disregard theory in favour of experience.[109] The fullest knowledge, he said, 'can only be acquired by practice in the field': the task was to combine the 'gradual acquisitions of practical men' with 'regular method and system'. His course stressed techniques that were 'serviceable to the miner, the geologist, the scientific traveller and the general observer when in the field'.[110] In the centre of London, his precept was easy to accept, if difficult to follow. Nonetheless, Liversidge found it attractive, particularly because what he said about mineral species led to the elegant science of crystallography.

A symbol of the age of measurement, crystallography sought to understand the mathematical rules underlying mineral structures.[111] As recently as 1837, ten years before Liversidge was born, the basis of modern mineral classification was laid by James D. Dana, in his celebrated *System of Mineralogy*; and this was the work of 'reference' that RSM students consulted. With the passage of years, as more minerals were discovered, so Dana's taxonomies were steadily revised.[112] In the middle of the nineteenth century, France and Germany dominated the field, but earlier, the discipline owed much to England. In 1809, W.H. Wollaston invented the reflecting goniometer, a 'simple, cheap and portable little instrument' that, in John Herschel's words, gave the subject 'all the characteristics of one of the exact sciences'.[113] With increasing interest in crystal morphology came a proliferation of mineralogical 'cabinets', such as Liversidge began himself. W.H. Miller, Professor of Mineralogy at Cambridge, took the subject seriously, and developed a unifying concept of the crystal as a geometrical entity, within a 'taxonomy' of structures. He was not alone, but despite competing systems, by the

1870s minerals were being systematically classified by their optical characteristics into thirty-two structural classes and six or possibly seven systems.[114]

This was a good beginning, but as late as 1875, Nevil Story-Maskelyne, who fostered the subject at Oxford from 1851,[115] deplored the 'extreme paucity of really scientific mineralogists and crystallographers' in the United Kingdom – men sufficiently competent 'to undertake the most elementary crystallographic calculations'.[116] And those that did naturally concentrated on optical and physical properties. Mineral chemistry, dating from the work of Mitscherlich in the years just after Waterloo, was greatly neglected,[117] and neither Oxford nor Cambridge taught the subject in a 'practical manner.'[118] Even at the RSM, Andrew Ramsay lectured without reference to chemistry, or to the microscope, at all;[119] indeed, Ramsay told the Devonshire Commission that he wished he had a chemist to do analyses for him.[120]

What could be done? For centuries, chemists had known that minerals are crystal structures, and that each has a distinct chemical composition. When Liversidge first listened to Sir Warington Smyth and Andrew Ramsay, few in England were attempting to classify minerals by chemical structure. In 1867, Clement Le Neve Foster, later Inspector of Mines and Smyth's successor, gave Liversidge a copy of notes taken from the lectures Smyth delivered in the late 1850s. In these, the chemical composition of minerals was merely guessed at.[121] Once, after a lecture in mineralogy, Sollas recalled a conversation with Garnett and Liversidge on the arrangement of the carbon atoms in diamond and graphite. 'You might test that by a comparison of the atomic volumes!', Garnett suggested; and so began an interest they shared for the next fifty years.[122] But the theoretical initative remained in Germany, where, at the turn of the century, the study of crystal-chemical structures was taken up by P.H. von Groth. Modern crystallography then began, from the first experimental X-ray analyses of sphalerite crystals (zinc sulphide) by Max von Laue in 1912, followed by W.L. Bragg and the invention of the crystal spectrometer in 1913.

The mineral chemist's task was essentially practical – the identification of substances by chemical means. In this, he joined hands with geology. In the second edition of his *System of Mineralogy*, popular during the 1840s, James Dana introduced students to the relationship between chemistry and geology, and described the role of the mineralogist as a 'go-between'. In his third edition, in 1850, he introduced a mineralogy based upon a system of chemical classification.[123] However, as late as 1862, and the second edition of J. Beete Jukes' *Student's Manual of Geology*, it seemed necessary to comment that the 'tendency of late years has been to neglect ... the bearing of mineralogical knowledge on Geology. There are many subjects on which we have still to ask the chemist and mineralogist to enlighten us'.[124] And before the word must come the object: the empirical examination of rock specimens (today called petrography), as the first step in explaining their origin (petrogenesis). To study either chemistry or geology was a respectable choice; to study both was demanding. Liversidge seemed destined to take the harder path.

Andrew Ramsay succeeded Murchison as Director-General of the Geological Survey in 1872 and first taught Liversidge much of what he knew (and would later teach) about geology. From February to April 1869, Liversidge heard Ramsay lecture on morphological and topographical classes, topics normally intended for officers of the Indian Army and other 'occasional students' for whom geological information was a practical necessity.[125] By some accounts, Ramsay's flair for the anecdotal enlivened his dusty delivery, with stories from his time with the Survey in the North of England.[126] But others remembered him as didactic; his style, better suited for taking dictation (notes of which Liversidge religiously preserved) than provoking debate (except, notably, on the theory of river beds and, more particularly, glacial lakes on which he held pronounced views).

If to progressive education Ramsay made no copious difference, much the same could be said of contemporary geologists elsewhere in Britain. Outside the memorable lectures of William Buckland and John Phillips at Oxford, of Adam Sedgwick and (after 1869) T.G. Bonney at Cambridge, and (until his sudden death in 1854) of Edward Forbes, it remained a neglected subject well into Liversidge's adolescence.[127] In later life, the venerable Sir Joseph Prestwich recalled that the only instruction in geology and mineralogy available to him as an undergraduate at University College London in the 1830s, took the form of three lectures by Dr Edward Turner, delivered at the end of forty lectures in chemistry. A chair of geology was established at UCL in 1841, but not filled by a geologist until Ramsay was pressed briefly into service in 1848. In 1855, John Morris was appointed to a combined chair of geology and mineralogy, but for years struggled to find students.[128] Even in 1877, when Bonney was elected to the chair at University College London, his salary was too small, and his students too few, even to justify his moving house to London.[129]

Lacking laboratory exercises and offering excursions which, apparently, few took, the RSM's geology course failed to convey what Armstrong called an 'impression of reality'. Ramsay's examination – on which Liversidge came a poor seventh among twenty – typically asked for 'book' answers to questions on which the last word had not been said. Thus, 'What do you consider to be the cause of the ejection of lavas and ashes from volcanoes?'; 'Explain the theory of slaty cleavage'; 'What is the origin of the coal measures?'; or, more intriguing, 'How would you distinguish between silurian rocks and carboniferous limestone by their fossils?'[130] And, in a field to which Liversidge would turn, 'How do coral reefs prove that islands in the Pacific Ocean are gradually sinking?'[131] Each question had a subtext. Presumably, Ramsay worked from a 'hidden curriculum'; but if so, nothing comes down to us. In the meantime, learning from Ramsay came by rote.

Aside from Huxley's lectures, Liversidge found that reading was preferable to listening, and he could choose from a list of texts that began (and nearly ended) with De la Beche's *Geological Observer*, Lyell's *Elements of Geology*, Phillips' *Manual of Geology* (1855) and J. Beete Jukes' *Student's Manual of Geology* (1857), already standard references. Not until the early 1870s did the famous classbooks of Macmillan and Longmans begin to appear; not until 1882 did Archibald Geikie rewrite Jukes, and not until 1885, did Robert Etheridge Sr and H.G. Seeley revise Phillips.[132] A.H. Green's *Physical Geology* (1876) had not yet appeared; nor had Prestwich's two-volume *Geology* which, with Geikie's *Classbook of Geology*, became standard reading for the next generation.[133]

For the beginner, Jukes' manual, although ancient, was stimulating. 'There is no natural science', Jukes explained to his students in Dublin – in language more eloquent than that used at the RSM – 'to which the geologist has not to appeal for information upon some point or other'. Ambitiously, he proclaimed geology 'not so much one science, as the application of all the physical sciences to the examination and description of the structure of the earth, the investigation of the agencies concerned in the production of that structure, and the history of that action'. He imparted a sense of mission. It was comforting to be assured that its mastery 'involves neither profound study, nor requires any great power of mind above the average of human intellect. It is, indeed, what every well-educated man ought to possess.'[134]

Jukes' message Liversidge took to heart. At the same time, he became expert in techniques. First, it was the balance, then came the microscope and the spectroscope – both taught in a rudimentary way. In 1858, H.C. Sorby of Birmingham pioneered the study of rock slices using the field microscope.[135] In his textbook of 1862, Jukes predicted its widespread adoption, and Bonney used microscope slides in his geology classes at Cambridge from 1869. Liversidge's youthful interest in microscopy was easily captured by its applications.

So, too, was he captivated by the new science of spectroscopy which, in the hands of a skilled chemist, could identify substances by their chemical composition, and even lead to the identification of new elements. In 1860, with a spectroscope, the German chemists Robert Bunsen and Gustav Kirchhoff had discovered a new element, caesium, in the waters of the Dürkheim mineral springs, and the following year discovered a new alkali metal, rubidium, in mica lepidolite.[136] In 1868, Norman Lockyer's solar observations with the spectroscope revealed the existence of an unknown element, which he called helium. In March 1869, Dimitri Mendeleev introduced the periodic table,[137] and for chemists everywhere, the great 'detective' hunt was on. If chemistry allied to astronomy could show the composition of the stars, surely chemistry allied with mineralogy could find new elements in the earth. In 1871, the chemist William Crookes agreed, suggesting that rare elements might 'turn up expectedly' when new minerals were found.[138]

All this was tantalising. That chemistry might help rewrite the history of evolution was a tempting prospect.[139] Theories of mineral formation were commonly used to patch up conflicting explanations, or to refute opposing claims.[140] If the subject could be rescued, Hunt believed, it would be 'by the individual effort of a single mind – who [sic] can free itself from the trammels of the schools'.[141] With this possibility dancing before him, Liversidge sailed through his middle year at the RSM (1869).

In April 1869, he joined a spirited correspondence in Hardwicke's popular *Science Gossip,* on the curious nature of dendritic spots, which began a lifelong interest in organo-metallic substances.[142] Between April and June 1869, he bravely took issue with a leading analytical chemist, Charles Tomlinson, FRS, on the theory of solutions, a field of great interest to many prominent chemists, including Marcellin Berthelot and Dimitri Mendeleev.[143] Liversidge attempted to identify experimentally the 'nuclei' that he believed brought about supersaturation. Tomlinson preferred to think of these agents as chemical impurities, but Liversidge argued for the presence of microscopic organisms, or 'germs', using the biological metaphor that Tyndall proposed. In the event, a satisfactory answer eluded them all. 'All we can do', Liversidge advised his readers, is 'apply questions and extort [sic] an answer from nature'.[144] He reported his results in a series of notes in *Chemical News* between August and December 1870;[145] within two years, parts were read into the *Proceedings of the Royal Society.*[146] His career had begun.

In the autumn of 1869, Liversidge began his third and last year at the RSM. For the most part, this was spent at Jermyn Street. For reasons having as much to do with curriculum as choice, he chose the Mining and Metallurgical Division. And for the next term, he took sixty lectures on mining by Sir Warington Smyth (which he dutifully copied verbatim and kept forever),[147] on applied mechanics (thirty-six by Thomas Goodeve), and on metallurgy (fifty by John Percy). Smyth was a known quantity; Goodeve impressed him not at all. But in Percy's laboratory, a new world of practice opened before him. Every weekday afternoon between October 1869 and March 1870, Liversidge learned procedures for analysing the ores of iron, copper, tin, zinc and lead, and gold and silver.

Percy, a tall, spare man and a popular lecturer, intrigued his pupils. Well known for his views on Home Rule for Ireland he also performed – as adeptly as Frankland but with less fanfare – as an expert witness on enquiries ranging from the ventilation of Parliament to the role of metals in the defence of the realm. In future, he would be better known as the author of a mammoth (and never completed) project on *Metallurgy* – some 3500 pages, as his biographer dutifully recalls, of 'terse and exact discussion of metallurgical processes and of minute and scientific discussion of the chemical problems they involve'.[148] According to Chambers' official *éloge*, Percy found 'Metallurgy practised as an empirical art, and by his teaching and researches ... secured for it a scientific basis'.[149] For years, Liversidge and he remained in contact, and Liversidge even kept a photograph of Percy in his office. Percy taught him that public science and politics must mix, and that science had a commitment to society. From Percy also came the 'collector's habit', and with it, a lifelong addiction to the collection of objects and specimens.[150]

While a role model in many ways, Percy's examination questions, like those of Smyth, conveyed matter to be memorised, rather than understood. It must have been a relief to listen, between October and February, to the eighty-two lectures on natural history given by T.H. Huxley in the Geology Division. These were not compulsory for metallurgy students, but Liversidge was captivated by them. For Liversidge and Sollas, Huxley towered above the rest of the staff. Years later, Sollas recaptured the moment, in the lecture room of the Museum, when 'a few earnest students were seated round the green baize table immediately below the lecturer's desk ... [and], at the stroke of the clock, Huxley entered and began to lecture on the "Science of Living Things"'. His diagrams, his language, spoke to 'facts', passed on personal knowledge and organised into a natural system, fully expecting his listeners to have accepted the 'fact' of evolution. His delivery was 'deliberate', there was 'nothing *a priori*', and conveyed with such a 'singleness of aim' that every student (except possibly Liversidge, whose notebook is lost) took careful notes.

Armstrong, an uncompromising critic, remembered Huxley's lectures as 'formal and descriptive, without living illustration'.[151] But Bickerton, always the showman, found Huxley's style 'striking' and personable, and his coloured blackboard drawings 'wonderful'. 'I well remember,' he recalled, 'the feeling of vandalism when he rubbed them out. It actually hurt me, and I felt they ought to have been done on prepared paper and kept.'[152] Sollas sided with Liversidge and Bickerton, and thought

Huxley arranged his subject matter so logically that 'our interest was captured and maintained, even through the most complicated dissertations ... there was such an exactitude in his choice of words that it produced ... a sense of ... something unattainable ... It gave a mark to aim at.'[153] There was no laboratory, but there was always conversation after class. In this way, Sollas recalled, Huxley taught us his subject, but 'a vast deal more besides'.[154]

After three years, Liversidge and his classmates were no strangers to evolutionary theory. James Secord has noted the prominence of Geological Survey (and School of Mines) men among the earliest and most enduring of Darwin's advocates. Indeed, he argues, 'almost all of the specifically geological support in Britain' for Darwin came from within the Survey.[155] In May 1870, Liversidge sat in on the special lectures given by Robert Etheridge Sr, the Survey and Museum's Curator of Palaeontology. Like Huxley, Etheridge impressed upon Liversidge the significance of the fossil record for the resolution of the great geological debates.[156] From Ramsay and Huxley (in person), and Geikie and Jukes (by text), Liversidge imbibed an understated Darwinism. 'What arguments in favour of, and against, the doctrine of the progressive modification of animal forms in time, are deducible from the order and occurrence of Echinodermata and Crocodilia, in the fossil state?' 'Through what stages does the vertebrate cranium pass in the course of its development?' Such were among the questions he was asked in the examination on Natural History. But where evolutionary arguments touched dangerous ground, Liversidge had no need to tread.

## 5. Careers at the Crossroads

In 1870, as Liversidge neared the end of his course, he faced a crossroads. The future for a professional man of science was none too bright. Overall, Liversidge presented well. He emerged from the RSM groomed in Huxley's preference for 'facts'; in Huxley's phrase, he was 'agnostic' in matters of theory, and so far as we know, in matters of religion as well. Excepting the Tomlinson episode, he showed a disinclination to engage in controversy, and was genuinely liked by his fellow students. It was the serious Liversidge, not the prankster Bickerton, who was chosen to represent the RSM students at the inaugural meeting of the London Union Society in 1869. There he met Henry Newell Martin, representing the medical students of UCL, and there he found a way of 'improving' student attitudes that he might later commend to Sydney.[157] However, 'popularity' seemed a pointless pastime; it was enough that, among his fellow students, he worked hard and did well.

Liversidge had serious academic competition. The best metallurgist of his generation was probably George Snelus (RSM, 1864–67) of Manchester, who, like Liversidge, had won a Royal Exhibition to Jermyn Street. With Percy Gilchrist, another RSM associate a year ahead of him, Snelus transformed the Bessemer process, and with it the iron and steel industry. Other nearcontemporaries were also highly successful. T. Jeffery Parker – the son of Huxley's friend, W.K. Parker – also in Gilchrist's year, went to a chair at Otago, and there co-authored (with W.A. Haswell, later Liversidge's colleague at Sydney) an internationally famous textbook on biology.[158] Parker did better

# ROYAL SCHOOL OF MINES.

### Sessions 1867-68 to 1869-70.

#### *Mining and Metallurgical* DIVISIONS.

*It is hereby certified that*

## *Archibald Liversidge*

*has passed the examinations which entitle him to be an*

## ASSOCIATE OF THE ROYAL SCHOOL OF MINES.

*Approved*

*Rod: I Marchison* Director

*de George Ripon*
*Lord President of the Council.*

London, 14ᵗʰ of July, 1870.

*Entered Trenham Reeks*
*Registrar*

Liversidge's Associateship Certificate from the Royal School of Mines,
14 July 1870.

in Huxley's course; both Gowland and Snelus got better marks in mining, metallurgy, and applied mechanics; and Garnett swept the board in physics. Of course, marks alone do not maketh the man. But no one did better than Liversidge in mineralogy, and few – his contemporaries agreed – shared his talents in the laboratory.

Fortunately, with his father's firm prospering, the choice of a career was not immediately pressing. There was time for the twenty-four-year-old to look around. In early 1870, Liversidge, with his friends Sollas and Garnett, did just that. None wished to work in the mining industry; and for different reasons all were spared a future in trade. Prospects of government work looked good. Indeed, Alexander Williamson of University College London enviously told the Devonshire Commission, 'It is said, that if you go to the Government School you will get the Government appointments; and if not, not.'[159] Perhaps some envy was justified.[160] But for Liversidge, an academic career would have priority.

This ambition did not seem unreasonable. Returning to England in 1870, Henry Armstrong had gone to the chair of chemistry at the London Institution in Finsbury at the age of only twenty-three.[161] A half-century later, Bickerton recalled that there were 'plenty of posts to be had'; that the 'eminence of the staff caused offers of appointments absolutely to pour in'.[162] He was placed in New Zealand by Ramsay, after declining offers of chairs in Japan and Canada, and a job as a mineral sorter in the Cordilleras at £1000 a year.[163] But opportunities of this kind were rare. Leaving aside the expanding international market for mining engineers and chemists, the 'Wages and Wants of Science Workers', as one author put it,[164] were by no means equally matched; and the poverty of appointments that confronted Huxley's generation in the 1850s, and contributed to their sense of persecution in the 1860s, had not yet disappeared.[165]

It was William Garnett who, early in 1870, proposed to Sollas and Liversidge the idea of going for an open scholarship in natural science at Cambridge, as were being advertised by some of the colleges. There were precedents. Thomas W. Danby, an ARSM and Forbes Medallist in 1860, had gone up to Queens', Cambridge, in 1862, and then migrated to Downing, where he led the First Class in the Natural Sciences Tripos of 1864 and was elected to a Fellowship.[166] There he gave college lectures, supporting George Liveing, Professor of Chemistry, who had himself studied at the Royal College of Chemistry in the early 1850s. As there were others. Among them, was Richard Daintree, one of Danby's contemporaries at the RSM, who had matriculated at Christ's in 1851, before succumbing to ill-health and 'gold rush' fever in Australia. Daintree had worked with the Victorian Geological Survey both before and after a year spent at the RSM in 1856, and was well known in both geological and chemical circles. He shared many interests with Liversidge, and fate would soon draw them together.

On 8 April 1870, well before taking his final examinations at the RSM, Liversidge took the train to Cambridge to sit the scholarship examinations at Christ's. About the same time, Garnett and Sollas visited St John's. Cambridge marked a turning point for many like them. Graduates from Manchester, London and the Scottish universities were now coming regularly to do second undergraduate degrees at Cambridge, and their coming was inspiring reform. Lyon Playfair had for years lamented that Oxford and Cambridge had not yet risen to the demands of a 'changed civilisation', transformed by 'the rapid advance of science and its numerous applications to industrial life … and the liberalisation of political institutions'. Unless they did, they would survive only as 'venerable monuments of a past age'.[167] But Cambridge was also a place that celebrated 'social education', in Armstrong's phrase, that could transmute metallurgists into gentlemen. Fortunately, Playfair's *protégés* were on their way to a place that, even before the arrival of the Devonshire Commission, was beginning to see a 'revolution of the dons', and a coming transformation in British science.[168]

# CHAPTER 3

# College Life and Chemistry

King's Parade in Cambridge, 1887.

Cambridge, in the autumn of 1870, was familiar and changing at the same time. To the casual visitor, it was as it always had been, a busy East Anglian market town of small shops and noisy streets, ancient dwellings bordered by farmers' fields, green meadows bounded by graceful poplars. But walking from the railway to the city centre, change was clearly underway. Muddles of shops were being pulled down near King's and Magdalene, and congested alleys were giving way to more open thoroughfares. In 1873 the town's Board of Improvement Commissioners described the Cam as 'a huge cesspool',[1] and the *Cambridge Review* complained that the stench was so bad 'it seems a wonder that more boating men are not laid up with typhoid fever'.[2] But newly swept streets were stretching from Hills Road to Great St Mary's, and plans were being laid for improved drainage, lighting and sewerage in the terraces lying beyond Lensfield Road and Midsummer Common. Town and gown, not always at peace, coexisted. The cityfolk, subsisting on rural industries, agriculture and college employment, steadily increased. As ever, poverty coexisted with privilege; yet, if the 'age of improvement' had come late, it had not passed Cambridge by.

Safe amidst this bustle, serene behind their sheltering walls, stood the seventeen colleges of Cambridge University, the earliest dating from the thirteenth century; the largest, among the wealthiest and most powerful institutions in the land. Trinity and St John's had, between them, half of all the university's 1600 undergraduates, and dominated the intellectual life of the nation. For centuries, they and the others, serving the interests of Crown and clerisy, had enjoyed their cobblestone courts and gracious gardens, exquisite libraries and elegant lodgings, customs and privileges long ordained by statute and usage. But beginning in the 1820s, they had been stirred by threats of parliamentary enquiry and liberal interests seeking reform. From the 1840s, these rumours became real, and the 'mediaeval islands', as Gibbon called them, were asked to account for their monopolies, and for the pastoral indolence that could sometimes masquerade as education. Following a Royal Commission in 1853, Cambridge embarked upon changes in ways that, thirty years later, would be described as little less than marvellous.[3] Against critics who recalled Coleridge's image of 'quiet ugliness' –

> where deep in mud Cam rolls his slumbrous stream
> And bog and desolation reign supreme.[4]

– arose a new generation of scholars and men of science, seeking the rejuvenation of academic life.

J.B. Priestley later spoke of the 'almost idyllic Cambridge of the seventies and eighties' – the Cambridge of Sir Henry Maine, Henry Fawcett, Leslie Stephen, Lord Acton, J.R. Seeley, Henry Sedgwick, W.H. Bateson, Montagu Butler and Michael Foster.[5] Despite fluctuating rents and railway shares,[6] the colleges were active in building or rebuilding, with styles embracing less the extravagances of Oxford's Ruskin or Butterfield, than the ornamental utility of Waterhouse.[7]

A similar spirit of change spread through the work of admissions, teaching and examining. By the 1870s, a new Cambridge was emerging; beneath the gown of tradition was beating the heart of reform.[8] Since 1856, matriculating undergraduates no longer had to assent to the Thirty-Nine Articles of the Church of England.[9] In the summer of 1871, Gladstone's government abolished the 'tests' completely, except for heads of houses and candidates for divinity degrees and college fellowships, and university chairs could be held by non-conformists and non-Christians. In 1869, the University accepted its first non-collegiate undergraduates, and the same year opened lectures to women.[10] Within three years, a new University Extension movement took learning to the four corners of the kingdom. For centuries, change had been glacially slow, evolutionary not revolutionary, embodying proposals long in the making and outcomes hesitantly agreed upon. Now, Joseph Mayor, the Latinist and Fellow of St John's, found 'everywhere ... movement, growth, expansion ... improvement externally in the widespread influence of the University ... and ... internally in the range of studies, the methods of teaching, the position of the teacher and the encouragement offered to students'.[11]

Into this vibrant environment, on the cusp of sweeping change, arrived Archibald Liversidge, and his friends W.J. Sollas and William Garnett – a bold step, with far reaching consequences for them all.

## 1. Reading Science in 'Reformed' Cambridge

Since the 1830s, the advocates of reform, allied with reforming interests in parliament and the professions, had begun to prepare the way for the 'modern University'.[12] The natural sciences provided the motor and the motive of change, and a method that other disciplines looked to, with a mixture of admiration and concern. From the days of the 'Cambridge network' – William Whewell, John Herschel and George Airy[13] – came internal pressures to expand the syllabus, improve teaching and respond to the needs of the country. In 1851, prompted by the findings of a Royal Commission, [14] the University conceded the need for several reforms. The Mathematical and Classical Triposes, which dominated undergraduate life for generations,[15] were joined by new triposes in the moral and natural sciences,[16] which were assigned the task of fashioning a new liberal education for the modern mind.[17] In 1849, there were but two paths to an honours degree; by 1882 there were nine, and those who lived through these years were astonished by their achievement.[18]

Liversidge arrived on the cusp of this new age. In 1867, a House of Commons Select Committee deplored what seemed slow progress in the teaching of science at the University, despite the fact that the Natural Sciences Tripos had been in existence for fifteen years, and an honours degree in its own right for seven.[19] In October 1870, he would have found Cambridge still nearer the past than part of the future. This his generation was set to remedy.

In 1870, there were thirty-four chairs at Cambridge; of which only eleven fell outside the humanities. There were four in mathematics, one in anatomy, and two in medicine. All lectures in the natural sciences were given by just five: C.C. Babington in botany, Adam Sedgwick in geology, W.H. Miller in mineralogy, George Liveing in chemistry and Alfred Newton in zoology. Of these, two were recent appointments: chemistry in 1861[20] and zoology in 1866. In 1870, as in earlier years, professors gave between forty and fifty lectures annually to small audiences who were aiming principally for medicine.[21] Those who were reading Natural Sciences to become medical practitioners had no physiological laboratory, and outside chemistry, no experimental instruction.[22] Everyone suffered from a lack of accommodation. If the twenty-three 'literary professors' of Cambridge had to be fitted into the Old Schools and borrowed rooms, the 'scientific professors' were even less well housed.

In the decade before Liversidge, new buildings were proposed for the site of the Old Botanic Garden, bounded by Downing Street, Corn Exchange Street and Free School Lane. This property had been bought by the University in 1760 for use as a herbarium, but in 1786 the Senate had built a small brick lecture room for the Professor of Botany and the Jacksonian Professor of Natural and Experimental Philosophy. For many years, this room, capable of holding ninety uncomfortably, was also used by the Professor of Chemistry and Mineralogy, after that chair was established in 1818. Until 1840, geology lectures were given in a room connected to the Old Library, a short distance away at the north end of the Arts School, a space which Adam Sedgwick described as 'small, damp and ill-lighted', and 'utterly unfit for a residence or a lecture room'. There was no students' laboratory in any subject. In 1828, William Whewell, then Professor of Mineralogy, protested against this state of affairs, but it was not until 1832 that the Senate found funds to add two small structures on the central site. These buildings were to house 'Cambridge science' for more than thirty years.

In 1854, the University again commissioned new plans – for three lecture rooms to hold 400 students, museums, and a practical laboratory – from Anthony Salvin, the distinguished architect; but of the estimated cost of £23,000, the University could afford only £5,000. The need for a comprehensive 'science site' remained. In 1861 the issue was revived, and after prolonged discussion, involving three further syndicates, a compromise (eventually costing £30,000) was approved.[23] In 1863, construction began, and was completed in two years. Along three sides of a square, facing south, this 'new museum' contained a museum of human anatomy and pathology, a botanical museum, a museum of chemical preparations, a mineralogical museum, and a museum of comparative anatomy – albeit in rooms which, when finished, were already too small for their purpose.

In 1866, to oversee all such future developments, the Senate established a Museums and Lecture Rooms Syndicate, which soon became the 'headquarters' of the University's Registrar, John Willis Clark, who doubled as Superintendent of the Museum of Zoology. Willis Clark was to play a central role in university developments for many years. In October, when Liversidge arrived, the Senate expressed an interest in having a laboratory for the physical sciences, and a syndicate was set up to consider a proposal from William Cavendish, seventh Duke of Devonshire, Chancellor of the University, and a Trinity man, to endow a new building for this purpose.[24] From the inauguration of the Cavendish Laboratory in 1874, and the appointment of James Clerk Maxwell as its first professor, the 'city of science' acquired a modern face.

In 1870, however, the vision was far from complete. To be sure, the geology museum, holding a gift of 6000 specimens from J.L. Woodward, had received notable donations in the last two decades. But Woodward's careful instructions for collecting and describing rocks and fossils were compromised by neglect.[25] As for chemistry, the laboratory occupied two floors, 'well-lighted with sand bath and dressers, water and gas', with a separate professor's laboratory, open from 10 a.m. to 6 p.m., 'for the use of any members of the University who may desire to acquire a knowledge of chemical manipulation, or to pursue original researches'.[26] But chemistry practicals for undergraduates took place in a single room in the Old Anatomy School, where there was space for forty students, when over seventy applied. In teaching the new subjects, the colleges were compelled to cooperate, and by the late 1860s, Caius and Trinity Hall, and later Trinity, St John's and Sidney Sussex began their own laboratory classes.

It was this partial, segmented and scattered reality that made up Cambridge science, as seen by the Samuelson Committee and the Taunton Commission. In August 1869, as Liversidge entered his last year at the Royal School of Mines, the Council of the BAAS, meeting at Exeter,[27] appointed a committee to enquire whether there was in the British Isles 'sufficient provision for the vigorous prosecution of Physical Research' and, if not, 'what measures should be taken to secure it'.[28] Their report echoed the views of Liversidge's teachers. With the full weight of scientific opinion behind them, the BAAS invited them, with their allies from Scotland (Sir William Thomson and Henry Jenkin) and Manchester (Balfour Stewart) to give fresh evidence. They rewrote the committee's recommendations and asked government to act.

In 1870, as Liversidge made plans for Cambridge, Gladstone's first administration set up a Royal Commission to 'make enquiry with regard to Scientific Instruction and the Advancement of Science', under the chairmanship of the Duke of Devonshire, with the young astronomer and civil servant Norman Lockyer as Secretary.[29] The famous Devonshire Commission began its sittings in June, just as Liversidge sat his examinations; and continued for five years, gathering evidence from 14,500 witnesses. Its eight massive reports, appearing at intervals between 1871 and 1875, constituted what R.V. Jones aptly called the 'Domesday Book' of British science.[30] Its recommendations laid the basis of Britain's civilian science policy for the next century. It foreshadowed the consolidation of scientific institutions in South Kensington, the extension of science in secondary schools, the provision of national fellowships and scholarships, the institution of a DSc degree at Oxford and Cambridge, the establishment of a national science museum, a national physical laboratory, a national scientific advisory council, and a Ministry of Education and Science.[31]

Of these developments, the young Liversidge was no more than an observer. But in one sense, he was a participant. Just as he left London, the Commissioners turned their eyes to the ancient universities. Their Third Report – on Oxford and Cambridge – did not appear until August 1873.[32] But its message was common knowledge: college fellowships in science were few, and restricted by the rule of celibacy; lecturing was limited to 'hand to mouth teaching, and presentation for exams'; museums were underfunded, and laboratories had little space and poor equipment. There was no constituency of science undergraduates and the Natural Sciences Tripos was taken mostly by those destined for medicine.[33] Liversidge may well have wondered how he was to benefit from its eclectic tutorials after the rigorous training he had received in London.

Christ's College, Cambridge, about 1870.

## 2. College Life and Christ's

Of the eight colleges offering scholarships in Liversidge's year, only Christ's gave candidates a choice of subjects, holding that a 'good clear knowledge of one or two subjects ... [is] more esteemed than a general knowledge of several'.[34] This gave men of unusual backgrounds more than an even chance. On 30 April 1870, *The Lancet*, always a careful observer of university events, announced that 'Mr A. Liverside [*sic*]' had been awarded a scholarship; on 9 May, the college admitted him to read for the Natural Sciences Tripos.[35] Thus, before his final exams in London, his Cambridge career had begun.

Christ's, the college of Milton and Cudworth, Paley and Darwin, was in 1870 the eighth largest among the seventeen colleges, with eighty-two resident members.[36] Liversidge, with twenty-one other freshmen, arrived in October, at the close of a memorably beautiful

Second Court, Christ's College, Cambridge, about 1870.

summer, marred only by the rumblings of war between Prussia and France. With him came W.J. Sollas and William Garnett (who chose to read for the Mathematical Tripos), both bound for St John's. All three had praises from London ringing in their ears.[37] Liversidge was required to take no entrance examination, and had only to supply a certificate of good conduct and character from a 'graduate of the University or a beneficial clergyman', before he could enter a realm 'distinguished', in the words of the *Student's Guide*, 'by the cleverest youths in the country'. To take a 'good degree' required ambition and 'abilities above the average level or a course of steady industry pursued through many years'.[38] Liversidge and his friends easily possessed the former, and had few illusions about the latter.

If Liversidge's reasons for choosing Christ's remain conjectural, Christ's reasons for wanting him are clear enough. Since the late 1860s, several of the colleges had been making plans to include a science side. Downing built a small laboratory in 1867, as did Sidney Sussex in 1871, Trinity and Caius in 1873, Girton in 1877, and Newnham in

1879. But there were no teaching fellowships. According to James Stuart, later Professor of Mechanism and Applied Science, the problem was circular: there was no reason to appoint teachers until there was a demand, and there would be no demand as long as there were no jobs. In 1873, of the 350 fellowships in Cambridge, only thirteen were in mathematics and science, and of these, six were in mathematics.[39] In 1867, Downing, by no means the richest, broke ranks and appointed its first teaching fellow in natural science in advance of student demand – an unprecedented event, as Winstanley records[40] – and the following year, the far wealthier Trinity confirmed its good judgment by doing the same.

Other colleges then followed suit. King's, Gonville and Caius, Clare and Peterhouse began to offer scholarships. In 1869, St John's announced it would make natural science a regular part of college teaching, and Trinity and St John's made a combined appointment.[41] By 1870, St John's had two science fellows, Sidney Sussex had one, and Downing appointed another in medicine, making six in all. Writing in *Nature*

Slaughter House Lane (now Corn Exchange Street), near the site of the Chemical Laboratory, photographed in 1870.

on 3 March 1870, an article which Liversidge probably saw, T.G. Bonney of St John's described the various 'college inducements' on offer. Clare advertised one scholarship and Caius and Sidney Sussex, two of £50 each. Peterhouse, St John's, Trinity and Downing gave one each, at £40 and £50. By 1870, between two and fifteen men a year were thus encouraged to read natural science.[42] Among the colleges soon to join was Christ's.

The year 1870 marked nearly four decades since Darwin had been up at Christ's – 'the most joyful in my happy life', as he later recalled.[43] Since then, especially since the 1850s when J.R. Seeley knew it, the college had improved its academic reputation and decided to direct to the sciences some of an unexpected windfall from the discovery of minerals on its estates. In 1854, the college voted to give £500 to the new museums proposed for the Old Botanic Garden site, if a sufficient sum were subscribed from

An undergraduate's room in Christ's College, in the 1860s.

other sources. The following year, when this did not materialise, the college renewed its offer. Sixteen years later, the intention still survived, and in 1870, when a syndicate was appointed to obtain college contributions for university purposes – notably including new appointments – Christ's was again among the willing.[44] To open scholarships and exhibitions to outside competition was consistent with this policy, and in 1869 the college agreed to designate between one and four scholarships – of £30 to £70 each – to the natural sciences.

Giving evidence to the Devonshire Commission in November 1870, George Liveing conceded that 'every man who has made tolerable proficiency in science would be almost sure of obtaining a scholarship ... as the competition is so limited'.[45] The scholarship papers sat in April 1870 covered familiar ground, with predictable questions in botany, anatomy, geology, physiology and chemistry. There was little choice, and answers were meant to be short. Chemistry dealt with the properties of hydrogen, the various reactions of metals and water, and the methods of detecting and separating metals in solution, subjects familiar to anyone who had been through Frankland's laboratory. Questions in mineralogy and geology required little more than recitations of Ramsay's lectures or Smyth's lecture notes. If the candidate knew how to account for hot springs like those of Bath, Bristol and Buxton; could explain the movement of glaciers; could say how chalk cliffs were formed; and could articulate the law of symmetry applicable to the rhombohedral system, then he was home and dry. Looking to the future, one question Liversidge met was significant: 'Describe,' he was told, 'the usual mode of occurrence, otherwise than in drifts, of gold, lead and iron.'[46] No record of his answer survives; but the question would return to haunt him many times in the years to come.

Liversidge and Martin were among the more conspicuous members of Christ's new aristocracy of talent.[47] Liversidge matriculated as a Pensioner (subsequently altered to Exhibitioner) in what would be the last tutorial side kept by the Rev. William Gunson, a strong-willed North Country man, a well-known Cambridge character, and one of the college's fourteen dons. Gunson had come up in 1843, and took his degree as fourth Classic in 1847, on the strength of which he won a fellowship in 1849. A tutor from 1851 to 1870, he remained a Fellow of Christ's until 1881.[48] He had – as a hard-pressed undergraduate recalled – one principal object, which was to get two men from the college each year into the first class of the Classical Tripos.[49] He was assisted on the mathematical side by the young John Fletcher (later Lord) Moulton, a brilliant scholar, who had taken his BA in 1868 and was immediately elected a Fellow. Moulton, as Lloyd George's Director of Explosives Supply, would play a central part in the scientific war of 1914–18; on the committees of wartime Whitehall, he and Liversidge would meet again.

Their timing was perfect. Meanwhile, in his new world of grace and gardens, far removed from the soot and mud of Finsbury and Piccadilly, Liversidge was assigned rooms in staircase E (Second Court), not far from those that Darwin had occupied. As he soon found, Cambridge was expensive.[50] Christ's annual fees were £110, of which £50 were defrayed by his Exhibition, but to this were added the matriculation fee (£5) and the caution money deposit (£15). College meals cost £2 a week, and lodgings, £5 a term, while private tutoring, which most undergraduates arranged, cost another £8 to £10 a term.[51] Some lectures, and all laboratory work, incurred additional charges. A prudent Pensioner (or Exhibitioner) was assured that he could keep his bills down to £80 a year.[52] But at a time when the total cost of a science degree at Owens College in Manchester was no more than £120 a year, in fees and expenses, Archie's father had to find £200 a year to keep his youngest son at Cambridge.

In hall, Liversidge met men of predicable social backgrounds. Many were sons of clergymen, but by the 1870s fewer were sons of the landed gentry, and more were sons of business. Liversidge's neighbours were destined for commerce and the liberal professions.[53] Martin lived upstairs from Liversidge. In October 1871, both moved to less expensive lodgings in the Old Court (now First Court), M Staircase. There, for nine months of his second academic year, in rooms said to have been Milton's, Liversidge looked out over a round lawn on one side and on the traffic of St Andrew's Street on the other. Plagued by a childhood stammer that never completely left him, Liversidge attended, but did not join that nursery of statesmanship, the Cambridge Union.[54] There is little evidence that Liversidge enjoyed, in Leslie Stephen's phrase, 'the advantages to be derived from the neglect of his teachers'.[55] But he and Martin did join the college boat club, where both, being short of stature, became coxes of the college's Eights. Liversidge's boat was bumped early in the preliminary races; Martin's was more successful, but not enough to redeem Christ's from eventual defeat at Trinity's hands in the Lent races of 1871.[56] Liversidge's bills for damaged college furniture suggest at least one successful Bump Supper – whether in celebration of triumph or (more likely) mourning defeat. He gave no sign of what John Peile, chronicler and later Master of Christ's, described as a 'general apathy', a 'diminishing patriotism', towards college life, evident among undergraduates of the day.[57]

The Geological Museum, Cambridge.

## 3. The 'Naturals' and their Nurture

In October 1870, the Natural Sciences Tripos had just begun its twentieth year. In more than a formal sense, it had yet to reach maturity. That month, 616 undergraduates matriculated at Cambridge, but only twenty entered to read the natural sciences.[58] Before 1860, the 'Naturals,' as Aldis Wright called them, had first to take honours in mathematics before reading for honours in science; in the first nine years of the Tripos, only forty-three had done so. Since 1861, men could read natural sciences without first taking mathematical honours, but only ninety-five had done so.[59] Eventually the Tripos would overtake classics and mathematics in undergraduate numbers, but it did not rival them in social acceptability.[60] Still, T.G. Bonney boldly claimed that 'the coldness and even dislike with which the study of Natural Sciences was once regarded here, is rapidly passing away, that the number of earnest students in the various branches is annually increasing, and that the University is fully alive to the wants of the age'.[61] His was both hope and expectation. Just how could Cambridge satisfy the needs of Liversidge's generation?

C.W. Merrifield, FRS, Principal of the Royal School of Naval Architecture at South Kensington, painted for the Devonshire Commission an interesting picture of a typical young man taken 'from the lower middle class with the ordinary education that he brings'. After three or four years' instruction, 'he will come out having gone through the course, but deficient in general knowledge, even of the subjects which he has learned'.[62] The London student, while being 'able to find iron or copper or nitric acid or sulphuric acid in a particular mixture, [may] have very little actual knowledge of chemistry'.[63] This was probably the situation in which Liversidge found himself. From Frankland and Smyth, he had gained technical skills, and from Huxley and Tyndall, a scientific creed. What he lacked was a comprehensive understanding of nature and the opportunity to approach natural phenomena from an variety of perspectives.

For decades, Cambridge had nourished the belief that the classics and mathematics, including the Newtonian science of applied mathematics, would together sustain a concept of liberal education based upon the encouragement of reason, memory and generalisation. The Natural Sciences Tripos embraced a program that was appealingly comprehensive. In 1870, the curriculum included chemistry, mineralogy (with crystallography), geology and palaeontology, botany, comparative anatomy, physiology and zoology – encyclopedic even before Clerk Maxwell added physics to the list in 1874.[64]

Like all triposes, the NST came to stress competitive display of knowledge, in what amounted to an annual academic Derby.[65] Results were measured by aggregate scores; and as all subjects were required, its coverage was inevitably superficial. The Tripos rewarded the student who could conjoin 'the cultivation of facts by observation and experiment', and by the 'classification of facts thus obtained … trace the operation of several laws by deduction and reference again to the phenomena of nature'.[66] By this means, the Tripos would produce 'liberally educated' chemists.[67] The difficult task of drawing together this 'circle of the sciences' was made easier by cooperation among the five professors, and by the belief that a synthesis would flow from the subject matter itself: the universal laws of chemistry would appeal to the physicist and the botanist; the geologist required a knowledge of physics, and so on.[68]

However, the undergraduate faced several obstacles to climb to honours from the lowlands of the first year. First, there were college lectures in classics and mathematics – every day between nine and eleven – and term examinations.[69] Begun in the late eighteenth century to ensure a standard of literacy, by the 1870s these college classes had become more a form of discipline than a source of knowledge. For decades, they were the object of intense criticism: the 'grumbling undergraduate complains, the angry tutor storms, the plaintive poet pours forth a piteous lament'; but still they remained.[70] According to John Venn, compiler of *Alumni Cantabrigienses,* the practice of repeating the same lectures annually, to men put side by side – whether from the best public schools, destined to be wranglers and take Holy Orders, or men from business, who 'regarded Greek letters and Euclid's diagrams as woeful mysteries' – was among the worst features of the Cambridge system. Not surprisingly, he recalled, 'it was quite natural to say, as one came out of the room, "Now I can begin my work"'.[71]

Under Gunson, Christ's College lectures prepared undergraduates for the Previous Examination, or 'Little-go', which the University required before an undergraduate could attempt honours. This examination, only slightly altered since Darwin's day, was elementary enough for public school boys, but far more threatening to artisan lads from London. It was small consolation to be told, in the words of W.H.H. Hudson, the rebarbative mathematics tutor of St John's: 'that the standard of the Preliminary Exam is deplorably low, and that a deplorable number fail to attain that standard, are facts that have not been contested'.[72] There were two parts which could be taken separately. The first required a translation of passages from a gospel in the original Greek (St Luke, in 1871); from one of the classical Latin authors (Terence that year), and from one in Greek (usually Herodotus), together with a paper on Latin and Greek grammar, principally using set texts. The second part covered Paley's *Evidences of Christianity*,[73] books of Euclid, arithmetic and algebra. Following the 'Previous', for honours students there was also an examination in 'Additional Subjects', including mechanics, trigonometry, French and German. Both in timing and in content, the 'Previous' was a tiresome hurdle, traditionally taken in the fourth or fifth term, at the end of the second year, which meant that the honours student had little else to show for his first year's work. By 1870, there was a tendency to push both parts back into the first term of residence. In Liversidge's first Lent Term, the Senate was asked to reform the system, and by 1879, undergraduates were allowed to take the 'Previous' in their first term if they wished.[74] But for the moment, the obstacle remained.

All being well, under the regulations then in force, the new Bunting Scholar could sit the Tripos in December 1873 and receive his BA degree the following year. Liversidge's first five terms (that is, October 1870 to May 1872) were to be spent (or misspent) in preparing for the exams. Gunson resigned his tutorship in 1870, and in all likelihood Liversidge had recourse thereafter to a 'crammer' for the purpose of getting up his mathematics, Latin and Greek, none of which had figured in his early education. Darwin had agonised over the 'Little-go', but having deferred at his tutor's insistence, finally passed.[75] Liversidge had an even more difficult time. Taking Paley's premises on trust, Darwin had been 'charmed and convinced by the long line of argumentation.'[76] Liversidge was neither 'charmed nor convinced' and he failed at his first attempt. This was not uncommon; of the 408 who sat the exam on that occasion, only 232 passed. He sat again the following Michaelmas Term (December 1871), but failed. Not until Lent Term, March 1872, did he pass in the 'Ordinary Subjects' (in the 'First Class'); although even then, so far as records reveal, he failed to pass his 'Additional Subjects'. By contrast, both Martin and Sollas passed the Previous in March 1871; Sollas passed his Additional Subjects in December 1871, and Martin, in March 1872.[77]

At the end of his first 'long vac', Liversidge and Martin joined the Cambridge University Rifle Volunteers, whose two qualifications for membership were an oath of allegiance to the Queen, and an annual subscription of one guinea. There is no record of rifle drill for Private Liversidge and Colour Sergeant Martin, of 'D' Company (Christ's), nor did they pass the Napoleonic height standards set for Trinity's college company.[78] Martin outshone his London classmate to some extent, becoming a Captain of Volunteers, and President of the Union, while gathering materials for the book on practical biology that he produced for Foster and Huxley in 1875.[79] Apart from Martin, Liversidge's closest friends were foreigners, including Syed Mahmood, son of a supreme court judge in Benares, and later a conservative opponent of the nationalist movement in India. Through them,

Liversidge was introduced to the world overseas and to the excitements of travel. His nephews had entered the Royal Navy and he was to give them a close second.

For Liversidge, university life was a serious business and he soon recognised the distance between 'honours' and 'pass' (or 'poll') men – in the popular phrase, between those who 'read' and those who 'rowed'. In 1867, about 200 men matriculated to read for honours (over 150 to read mathematics); but another 300 entered as 'poll men', not all of whom would take degrees.[80] For the latter, student life had not changed greatly from the caricatures of 'Mr Golightly' and Verdant Green. In the caustic phrase of T.D. Acland, Oxford provided for the 'social education' of 400 men a year, but the 'mental education' of only 100.[81] Cambridge was comparable. With a traditional curriculum, an unambitious undergraduate could live a sunny existence, darkened only by the thought of 'ploughing' the final exams. But those Exhibitioners of Christ's, men with some experience of the world, were in a class apart. For them, distinctions in marks overlooked distinctions of class.

The year 1871 began bravely. Liversidge visited the Mineralogical Museum, which used Gustav Rose's familiar classificatory system;[82] and in March 1871 saw there a recent gift of minerals from San Francisco, including specimens of Californian gold. Second, there were public lectures. In May, Norman (later Sir Norman) Lockyer, then the young editor of *Nature*, Macmillan's newly launched journal of science, came from his work as Secretary to the Devonshire Commission to deliver the Rede Lecture on 'Recent Solar Discoveries'. The same month, Liversidge heard W.B. Carpenter, Registrar of University College London, speak on the first reports from the *Challenger*, not long embarked on its epochal circumnavigation of the globe.[83] There were debates at the Cambridge Union, and walks along the meadows, and the fraternity of college life and conversation. Beginning in October 1870, and continuing through the Lent Term of 1871, every Monday, Wednesday and Friday, Liversidge heard Alfred Newton in zoology, Charles Babington in botany, Adam Sedgwick in geology, and George Liveing in chemistry. On Tuesday, Thursday and Saturday, it was W.H. Miller in mineralogy.[84] The natural history classes left much to be desired. Newton of Magdalene, best known as an ornithologist, was famous for the annual repetition of his lectures – verbatim – no doubt for ease of undergraduate note-taking. Gerald Geison has nicely observed that, as an expert on the extinct dodo, Newton was among the better-preserved representatives of 'a certain species of Cambridge don'.[85] His appointment to the new chair of zoology and comparative anatomy in 1866 was widely attributed as owing more to his friends on the Senate than to his scientific reputation.[86] Babington, a pious, eclectic don of St John's – amateur antiquarian, archaeologist and missionary – had been a student of descriptive and systematic botany all his long life. Popularly known as 'Beetles Babington', he gave lectures that were disappointingly descriptive, 'wanting both in spirit and in substance'.[87] Under him, botany, as a required subject, had sunk, in the phrase of William Thiselton-Dyer, a long-time friend of Liversidge, into 'a sort of back door for the poll degree'.[88]

Both Newton and Babington were bastions of the scientific establishment. Newton was forty-one, Babington was sixty-two; neither had yet learned of the revolutions in botany and physiology that were taking place in Germany. They shared the Salvin buildings with the venerable Sedgwick of Trinity, then eighty-five. Sedgwick was a Cambridge institution. When elected Woodwardian Professor in 1818, he was virtually ignorant of geology, but he dedicated himself to three goals: to build a geological collection, to establish a museum and to bring students into the field.[89] After fifty years, he succeeded in the first, even if he failed in the second, and very nearly forgot the third.

However, despite their intellectual conservatism, and the conceptual senility of much they taught, Newton, Babington and Sedgwick were steadfast allies of the 'new sciences', and stood united in their defence. Their service to the University was equalled if not surpassed by W.H. Miller, whose classic *Treatise on Crystallography* had appeared over thirty years earlier.[90] Miller, architect of the 'Miller Indices,' was arguably the best crystallographer in England.[91] Liversidge was one of the seven or eight who attended his lectures in 1870, sharing his fascination with crystal structures, and an open mind.[92] Then aged sixty-nine, Miller had reached the end of a long career (although he stayed in his post until his death ten years later).[93] Story-Maskeleyne, his Oxford contemporary, called him one of the 'patriarchal minds' of Cambridge, and praised his mechanical ingenuity, his technical contributions to government commissions on weights and measures, and his committee work for the university.[94] As a younger man, he had travelled widely in Europe, met Alexander von Humboldt, and kept in regular contact with continental work. When the Prussians besieged Paris in the summer of 1871 Miller stayed in France, collecting minerals, and attending an international conference. There was no doubt that he, too, became an influential role model for young Liversidge. From his lectures and museum sessions, Liversidge gained a keen interest in mineral collecting. Miller's models fascinated him with a passion that he took to the study of chemistry.

## 4. Chemistry for Gentlemen

One great advantage of Cambridge, as Venn recalled, was that the Long Vacations were entirely unhindered by examinations. For the 'Naturals', this meant time free for experimental work. Liversidge arrived in Cambridge with a scientific name, his two articles on supersaturation having just appeared in the *Chemical News*.[95] In November and December the same journal carried a lengthy essay by him reviewing the work on chemical supersaturation and detailing 120 experiments with thin films and 90 experiments with crystals. Liversidge signed the last of these articles from Christ's in December 1870, and then left the field to others. These early experiments failed to make any major impact but they had an immediate effect on his prospects.

One of only twenty 'Naturals', Liversidge caught the eye of his chemistry professor, George Downing Liveing.[96] The 1850s and 1860s saw the arguments of natural theology groan under the weight of biblical and scientific criticism, but by the 1860s, Liveing had begun to make a name for himself in university reform. Legends survive of his hot temper and his fierce red hair, yet also of his courtesy and his shrewd skill at teaching.[97] Liveing was not a man of the cloister. Before coming to university, he had known Italy in the time of Garibaldi, and America during the Mexican War. Among the pioneering dons to marry (in 1881, so forfeiting his college fellowship), he supported the cause of women's education.[98] Repeatedly, he urged his contemporaries to break away from the tradition, as he put it, of 'too much collegiate, too little public

George Downing Liveing (1827–1924).

spirit'.[99] He was equally zealous in reaching out beyond the traditional catchments of the University: 'We want a different class of students here', he told the Samuelson Committee in July 1868, 'it is not likely or desirable, that the actual artisan class should assemble here in numbers, but the master manufacturers might well be educated here.' He could well have had Liversidge's father in mind when he reflected:

> At present the traditions and habits of the place are such that few who are engaged in trade think of sending here those sons whom they intend for like pursuits. I want to see these traditions broken though, and many of that class coming here.[100]

Cambridge, in Liveing's view, offered students the chance to see science 'philosophically treated' – 'not in technicalities, but in principles'. His words were almost written for Liversidge.

When Liveing arrived, Cambridge offered virtually no chemical instruction. The subject's chequered history dated from 1682, in Newton's day, with the arrival of G.F. Vigani from Verona, who settled in Cambridge as a private tutor in chemistry and pharmacy.[101] For over a century, the subject had declined, under a series of undistinguished successors.[102] It was with Liveing that a new era dawned. Before the 1850s, St John's – Liveing's college – was 'more likely to be found in the rear-guard than in the van of reform'.[103] Liveing threw himself into the front lines, becoming a magistrate, an organiser of local examinations, and a member of countless university boards and syndicates. Revered and sometimes feared, as a 'maker of a department, builder of a laboratory, acute, incisive, practical',[104] W.H. Heitland of St John's thought him a 'liberal and individualist of the old school', a man who found it hard to read the signs of his times.[105] But to others, he was 'a man whom all men trusted, one to be depended on for good works, good faith, and common sense'.[106]

All these qualities he applied to his chemistry. In 1852, at his own expense, he fitted up a small laboratory for medical students in Slaughterhouse Lane (now Corn Exchange Street) on the eastern edge of the old Botanic Garden. From these beginnings the 'chemistry school' steadily grew.[107] In 1853, with the support of W.H. Bateson (St John's Senior Bursar) the college built a laboratory in New Court – open daily during term from ten to four, charging fees of two guineas per quarter – and retained Liveing as its superintendent.[108] This tiny building – two rooms with benches, two smaller private rooms, and a preparation room – cost £511, and survived until 1914.[109]

In 1861, Liveing was appointed the University Professor of Chemistry, but continued to teach at St John's, and in his rooms on the east side of the 'New Museums' site. In the Salvin scheme, he was offered a building 'capable of standing violent explosives and as uninflammable as possible, containing a series of vaults', but this would have put the teaching of chemistry largely into cellars, as at Oxford, and this he resisted, moving into the rooms vacated by the Jacksonian Professor, and modifying one as a students' laboratory. In 1872, he persuaded the university to build an upper storey, with accommodation for fifty students. There he bided his time until the Museums and Lecture Rooms Syndicate agreed to his terms. After a further thirteen-year-long struggle with cold draughts and an irregular water supply,[110] he won approval for a new laboratory, a massive stone structure designed by J.J. Stevenson, costing £34,000 and erected between December 1885 and the autumn of 1888 at the corner of Free School Lane and Pembroke Street.[111]

In May 1889, the Professor of Chemistry and the Jacksonian Professor at last came into possession of the whole building, which completed the south-western corner of the New Museums site. In briefing his architect, Liveing had made a special tour with Coutts Trotter to the new laboratories at Strasbourg, Munich, Leipzig and Aix-la-Chapelle; for Cambridge, the result was one of the best in the kingdom.[112] With its coming, the importance of college laboratories began slowly to fade.[113] Ironically, its long-delayed appearance neatly coincided with the new chemistry laboratory built at the much younger University of Sydney, designed from plans prepared by Liveing's student, Liversidge, and with the experience of his Cambridge teacher well in mind.

Although the author of several papers, and elected FRS in 1879, Liveing made little stir in the scientific world. A man described as the 'first teacher of experimental science in the University',[114] he disliked self-advertisement and resented the 'sulky or captious opposition' that denied him the money and materials to make chemistry progress at Cambridge.[115] For many years he taught alone. A single professor, at a salary of £300, he was accountable for all lecturing and most practicals in both inorganic and organic chemistry, as well as for teaching on 'the properties of substances, ponderable and imponderable' – that is, heat, light and electricity – for almost three decades. As he knew little organic chemistry, his pupils learned little before the appointment of a former student of von Hofmann as his laboratory assistant in 1876.[116] This proved too late for Liversidge, whose education reflected these gaps in knowledge. Moreover, it was not until Liveing brought a collaborator, James (later Sir James) Dewar, to Cambridge as Jacksonian Professor in 1875 – partly to take over the teaching of elementary physics – that at last he gave evidence of an interest in research. With Dewar, he published prolifically after 1877. But he never established a 'research school'. Liversidge was among the many who felt his influence, but among the few who almost copied his example.

Liveing's intellectual legacy was a mixed blessing. He shared contemporary criticism of the examination structure and its emphasis on 'book learning'.[117] His lectures, illustrated by experiments, were carefully rehearsed. His laboratory discipline was memorable. An undergraduate in 1871 recounted the epic tragedy of an assistant (whose name has been lost to history), who was caught one day cleaning a platinum crucible by rubbing it with a wet cork and sand. The hapless man was made to march the length of the lecture room 100 times, each time repeating a promise never to clean a platinum crucible 'with sand or other gritty substances without the express permission of the Professor of Chemistry'.[118] If the story is not apocryphal, it was unnecessary advice for any veteran of Frankland's Royal College of Chemistry.

In the Lent and Easter terms of 1871, Liveing's lectures continued on three mornings a week, with practicals on three afternoons. And as all lectures had to reach the lowest common denominator, reading became a dietary supplement. To a list beginning with Roscoe's familiar *Lessons,* Liversidge added Charles Bloxam's useful *Chemistry, Laboratory Teaching, and Metals*, together with W.A. Miller's older *Elements* and his new textbook on inorganic chemistry.[119] Liversidge had seen in London the English edition of Hofmann's lectures on *Modern Chemistry*,[120] then out of print. Thanks to Liveing and the library of St John's, he had access to books that had never been part of his examination syllabus, including William Buckland's exquisite Bridgewater Treatise, *Geology and Mineralogy*,[121] Adolphe Wurtz' *Leçons de Philosophie Chimique*,[122] Marcellin Berthelot's *Chimie Organique*,[123] and, from America, Josiah Cooke's *First Principles of Chemical Philosophy*.[124]

At the beginning of 1871, after years of discussion, the University agreed to increase Liveing's salary from £300 to £500, so that he could in turn appoint a demonstrator at £150.[125] The number of undergraduates reading chemistry, whether for natural science or medicine, exceeded fifty that year, and their poor supervision was an embarrassment.[126] Accordingly, Liveing appointed Thomas W. Danby, the man earlier elected to be the first science Fellow at Downing, and the first college science lecturer at Trinity.[127] Danby was also a former student of the RSM, some six years senior to Liversidge, who had come up to read the natural sciences after taking his Associateship in 1860.[128] Danby occupied himself in college teaching, principally in chemistry and geology for two years, but left his appointments at Downing and Trinity in 1869. As one of the few eligible 'Naturals', he first accepted Liveing's offer but, in April 1871, concluding that he had little future in Cambridge (given Liveing's experience, a reasonable expectation), he changed his mind and joined the national Education Inspectorate.[129] His sudden departure left Liveing short of an assistant for the Easter term and Long Vacation. Liversidge, who had just completed his first year as an undergraduate, was the right man in the right place – just as in 1867, when Frankland needed a 'locum' for his naval architects – and became Liveing's Acting Demonstrator.[130] As H.N. Martin was already Foster's Demonstrator at Trinity, Liversidge thus became the third Demonstrator in the history of Cambridge science, and arguably the first in Cambridge chemistry.

The experience of being in constant attendance, mornings and afternoons, for the twenty or thirty students who came during that warm summer to use burners and fuming chemicals in the small, ill-ventilated workrooms, can only be imagined; but during those middle four months of 1871, Liversidge not only showed his ability to supervise chemical practicals, but also his capacity to manage a laboratory. At the end of the summer, Liveing made a permanent appointment of a slightly senior man, now forgotten to science – J.W. Hicks, a Fellow of Sidney Sussex, who had taken a First in the NST in 1870 and a degree in theology in 1871.[131] Liversidge watched, as Liveing's laboratory was renovated, furnaces rebuilt, pavements relaid, a new galvanometer installed, and a new electric lamp set in place; all useful experience for one who might one day fit out a laboratory of his own. He later shared the experience of his teacher who, after twenty years' struggle, still worked in surroundings palpably incomplete, in an institution that laboured under the double disadvantage 'of being poor, and of being thought rich'.[132]

Liversidge's experience with Liveing polished his laboratory skills. In the summer of 1871, continuing his enquiries on dendritic spots, Liversidge believed their origins were inorganic, but could not perform a convincing quantitative determination. Another question emerged in a short paper he gave to the Chemical Society of London in November 1871, on a new method for determining traces of fluorine in minerals. The paper by 'Mr Liversidge of Cambridge' was noted in the press; and Frankland, in the chair, welcomed it, as a step towards the solution of a difficult problem. But praise was dimmed by the criticism that Liversidge had not developed comparative data for his analyses so as to 'standardise' his results.[133] Whether it was due to shortcomings in technique (which he had learned at Frankland's bench), or to the limitations of his instruments, was not clear. Possibly, it was too early to publish. Liversidge was in a hurry. Technical procedures were debatable, and in the bullish market of the day, where so many new questions were ready for the asking, mistakes could be pardoned among even competent researchers.

## 5. Experimental Biology and the Foster Circle

In the spring of 1872, Liversidge girded himself for his second attempt at the 'Little-go'. It must have been with relief that, with the ordeal before him, Martin introduced him to the talented group of undergraduates who, since early 1871, had begun to form around Michael Foster in the rooms at Trinity.[134] Foster's unusual appointment in 1870 to a praelectorship in physiology – a college, as distinct from a university office – followed agitation dating from 1857 when the Statutory Commissioners appointed under the University Act of 1856 had proposed that Trinity (as among the richest colleges) appoint at least six praelectors, to give lectures in the subjects of the new triposes and in other subjects as required by the university. For over a decade, nothing was done. In 1866, a second proposal was put to the vote, and lost. In 1869, however, Coutts Trotter, a leading Trinity reformer, succeeded to Danby's vacant college lectureship, and renewed the campaign.[135]

Trotter had given up a curacy in 1865 to study physics and chemistry in Germany. With John Willis Clark, also of Trinity, he proved one of the most effective 'co-conspirators' in the revolution that was overtaking university studies. Trotter deplored much contemporary practice. 'Exam mania', in particular, attracted his wrath, and he was pleased in 1868 by Trinity's agreement to award a fellowship, every three years, based on research, rather than on Tripos results. 'I doubt', he argued before the Devonshire Commission in February 1871, 'whether by mere examination ... you can find out who is likely to do really good work'. While the Tripos might 'test a man's reading and clearness of head', in science 'it is not easy to test his original power, or to find out whether he is likely to be an original teacher'.[136] This move led to what may have been the first 'research fellowship' in England. Trotter's intention, and that of Trinity, was to find an original teacher.

However, there remained the need to teach undergraduates. In November 1869, acting under its revised statutes, Trinity approved a teaching fellowship in the natural sciences. Adam Sedgwick proposed Sir William Thomson of Glasgow, who declined on the grounds of his wife's illness. Whether the college then actively sought a biologist, or simply appointed the best man they could find, remains debated.[137] But it seems reasonable to suppose that, were a new college appointment to be made, it should be made in a field of natural science where, given current developments in medical education, the teaching demand was likely to increase. As early as 1866, members of college – led, some say, by W.G. Clark, philologist and Vice-Master of Trinity – had taken advice from T.H. Huxley.[138] In 1870, advice was also taken from W.B. Carpenter, the Registrar of London University, who recommended Michael Foster, then an instructor in practical physiology at University College.[139] Huxley – Foster's mentor and friend – warmly concurred. Aged thirty-four and 'relatively untried, if not unknown', Foster had left medicine for physiology only three years earlier. But in that short time he had emerged as the rising star in experimental biology, succeeding Huxley as Fullerian Professor at the Royal Institution, and being promoted to professor at University College.[140] In Cambridge, Coutts Trotter had known him for some time, and welcomed his appointment by Trinity in May 1870 to its first praelectorship in natural science.[141]

Foster arrived – coincidentally with Liversidge in October 1870 – and his lectures began in November 1870. Although these were technically for Trinity men, undergraduates from other colleges were permitted to attend and did.[142] Fresh from a visit to the physiological laboratories in Germany, they marked a turning point in British science. It also marked the beginning of a lifelong friendship between Foster, Liversidge and imperial science.[143]

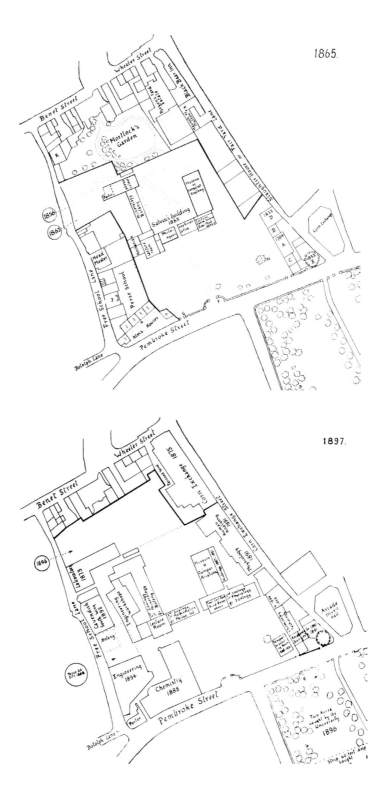

Site and buildings of the 'Museums of Natural Science', Cambridge, in 1865 and 1897.

Foster may have met Liversidge in London in 1870, during the brief lecturing 'locum' that he served for Huxley at the RSM. But they could also have met during the early months of 1871, through their mutual friendship with Martin. Like Sollas and Martin, Liversidge planned to take the Tripos in December 1873. His prospects thereafter depended on his exam success, and Foster was likely to be one of his examiners. For this reason, at least, he took time from 'cramming' his Paley and Euclid to acquaint himself with Foster's research methods, in the one-room laboratory given Foster in the basement of the Philosophical Library (now the Scientific Periodicals) Library.[144] As events were to prove, it was a wise decision.

For a young man who had taken delight in the microscopic world, who had learned analysis from the best experts in London, and who found fascination in the relations between inorganic and organic life, Foster added weight to Liversidge's growing list of role models. Writing in 1867, following England's poor showing at the Paris Exhibition, Foster had protested against the neglect of experimental science, a situation which would continue 'so long as physics and chemistry play hide and seek with dancing and gymnastics ... so long as science is offered to boys on half-holidays as an obstacle to cricket and football'. He insisted on giving undergraduates laboratory experience and training in independent research.[145] Liversidge was drawn to the Trinity Praelector, who was inventing what George (later Sir George) Humphry, his companion at Trinity and Professor of Anatomy, called the 'very highest branch of physical science ... dealing with the highest and hardest of physical problems ... the complexity of the animal machine'.[146] For a student already familiar with minerals and chemicals, what could be more exciting than an encounter of minds in the biological world?

It was once fashionable to think that Foster's original contributions to science were 'few and not important'.[147] But history has since qualified Gaskell's left-handed tribute, that he was 'a discoverer of men rather than of facts, and worked for rather than at physiology'.[148] Today, we view Foster as one (with Paget and Humphry) of the 'Great Triumvirate' of Cambridge medical history, and among the 'great schoolmen' of his age.[149] Where Foster triumphed was in the art of research and the discovery of men like Liversidge.[150] Hale Bellot recalled his 'instinctive perception of ability and a power to inspire others' which 'enabled him to gather round himself a band of capable and devoted disciples',[151] in whom he inspired 'an almost filial affection'.[152]

Liversidge was the second to join Foster's charmed circle, and the first to leave. Matriculating the following year (1873) were F.M. (Frank) Balfour, A.G. Dew-Smith, W.H. Gaskell, A. Sheridan Lea, and John Newport Langley – a galaxy of stars who ultimately transformed British experimental medicine, biological chemistry and precision instrumentation. With them, Foster began the Physiological Society in 1876.[153] In 1872, soon after Liversidge joined, Foster was elected a Fellow of the Royal Society; and in 1883, long after Liversidge left, he was elected Cambridge's first Professor of Physiology. He made sure his *protégés* advanced with him. In 1877, Foster nominated Liversidge for an honorary MA degree, and supported his election to the Royal Society of London in 1882. J.N. Langley became a demonstrator in 1874, took the Tripos the same year, and ultimately succeeded Foster as professor in 1903.[154] Through Foster's influence, F.M. Balfour and P.H. Carpenter were elected to Trinity's new fellowships in the natural sciences. Langley,

Gaskell, and Lea had lectureships by 1883, and Balfour was elected to Cambridge's first personal chair of comparative morphology in 1882, a month before his untimely death.

Into his 'Cambridge group', Foster infused the spirit of research. 'It was the air we breathed', Langley recalled, 'he was of the widest catholicity, as to the kind of research to follow'.[155] Foster's charter was uncompromisingly Baconian, rejecting 'flights into the regions of imagination'. It was 'only by reasoning, by induction from ascertained facts that physiology is to be studied and advanced'. Detailed methods allied to problems would lead research from the 'investigation of the fine processes of the animal organism', as George Humphry put it, 'to the consideration of the forces by which they are brought about'.[156]

Foster's group was famous for its work on nerve and muscle physiology, and on the heart and circulatory system, but he was also interested in physiological chemistry and chemical theories of nervous action. This is where Liversidge's expertise came in. As the *Journal of Anatomy and Physiology* put it, 'the chemical side of physiology has been and is, grievously neglected'. The old proverb, 'what is everybody's business, is nobody's business', seemed to apply with particular force.[157] With the exception of William Rutherford at King's College London, and Alexander Crum Brown at Edinburgh, physiological chemistry had been actively taken up in France and Germany, but virtually neglected in Britain outside the medical schools. 'It is obvious', wrote Crum Brown in 1868, 'that there must exist a relation between the chemical constitution and the physiological action of a substance, but as yet scarcely any attempts have been made to discover what the relation is'.[158] The famous *Handbook* of Klein, Burdon-Sanderson, Foster and T. Lauder Brunton had yet to appear.[159]

The agenda was endless, with unanswered questions ranging from the composition of tissue, blood and serum and the phenomena of gastric and pancreatic digestion, to the components of bile and urine, and the properties of albumens, carbohydrates and fats. Humphry and Foster agreed that the prizes would go to those who could apply physical concepts to the solution of biological problems. At the time, it was from chemistry that 'probably more information as to the nature of the organic processes is to be expected than in any other way'.[160] What was known was limited to the medicinal (or worse) consequences of using certain compounds, such as mercury and arsenic, and the physiological effects associated with the salts of strychnine, codeine, morphine and nicotine. Some had sought to establish connections between chemical and physiological activity through studies of inorganic salts and poisonous substances.[161] Overall, there was need for better techniques of purification, and better ways of elucidating chemical composition (and eventually, chemical structure). Theoretically, the subject was open to debate. A century later, James Watson and Francis Crick, working around the corner in Free School Lane, might well have felt the same.

The problem appealed directly to Liversidge, as the 'chemical member' of the team.[162] At Foster's suggestion, Liversidge (at last passing the 'Little-go', but still not his 'Additional Subjects') began work in March 1872, testing the procedure for preparing and purifying kryptophanic acid. The procedure had been proposed in March 1870 by Dr John Thudichum, the chemist attached to the Medical Officer of the Privy Council (after 1871, the Local Government Board), so as to identify a 'transparent, amorphous, gummy solid mass, almost or entirely colourless' that appeared after filtration of known acids from human urine. Thudichum had prepared a number of toxicological and pathological procedures.[163] Liversidge's analyses were completed in April, and published in Humphry's *Journal of Anatomy and Physiology* in May.[164] He analysed the substance, rendered in calcium salt and purified by copper and lead acetate, and in so doing revealed discrepancies in Thudichum's results, which suggested that Thudichum had not isolated a discrete acid after all.[165]

Foster was pleased with Liversidge's careful work, and moved him on to the analysis of what were called 'amylolytic ferments' (the enzymes that convert starch to maltose in the pancreas). The role of these 'ferments' in metabolic processes was not well understood. The pancreas secretes both insulin and pancreatic juice that contains digestive enzymes. In France, Claude Bernard had begun to analyse digestive enzymes, but in many cases their composition and mode of action remained obscure. Foster had broached the subject in 1867;[166] he now asked Liversidge to take it up again, to isolate the digestive juices, and to determine their chemical properties. Again, he worked in counterpoint to Thudichum. In June 1872, *The Lancet* reviewed Thudichum's new *Manual of Chemical Physiology*, which purported to define the 'state of the art'.[167] But the same month, Liversidge analysed samples of pancreatic juice, and developed a new technique to extract what he believed to be one of its three major enzymes. His results were complete that summer, written up for Humphry's journal, and summarised by *The Lancet*.[168] He was clearly a man to watch.

These early months of 1872 were a time of pure excitement. Walking the short distance from his rooms in Christ's to the Philosophical Library, Liversidge travelled imperceptibly from college life, embedded in a custodial view of learning, to a 'Faculty of Science', looking for new ways of seeing nature. Then and later, Liversidge demonstrated a capacity to bestride the two worlds and to see value in both.

## 6. ORGANISING SCIENCE: PRECEPT AND PRACTICE

The changing spirit towards science and research, typified by Foster, reached downwards from the dons, and into the mainstream of undergraduate life. Mutual encouragement was generally lacking, but in the natural sciences, an inclination to band together became all the stronger among men making common cause against the 'pass' men on one side, and wranglers and classicists on the other. In the early 1860s, a few undergraduates reading for the NST established the Cambridge University Natural Science Society (CUNSS),[169] to serve students in the life sciences, just as the prestigious Cambridge Philosophical Society served the physical sciences. At its peak, CUNSS enrolled fifty members, but after a few years, in the absence of college inducements, it had languished. When Liversidge arrived, the time was ripe for revival; and he was to be its organising genius.

In January 1872, about the same time that he was recruited to Foster's circle, Liversidge persuaded six friends to conspire in establishing (or, strictly, in re-establishing) the Cambridge University Natural Science Club (or CUNSC), whose objects were to 'promote natural science by means of friendly intercourse and mutual instruction'.[170] The club continues today. Alex Hall, FRS, a contemporary of Liversidge, and later a distinguished physiologist, recalled its sessions as the most 'enjoyable feature of his university career'; in days dominated by exams, it was, 'one of the most important educational influences at work in the Cambridge Natural Science School'. Bearing more than a faint resemblance to the natural history gatherings that Liversidge knew in London, the Club met on Saturday evenings during term, in members' rooms by turns, to hear short papers on subjects from their own research.[171] To guard against the 'degradation of meetings into social gatherings, which is so often the destruction of University Clubs', as the rules earnestly warned, only tea and coffee were served. *Nature* gave the young men an encouraging press, and soon they were nine.[172]

By Lent Term 1872, weekly CUNSC meetings had become a regular part of the informal curriculum for aspiring 'Naturals'. By May 1872, their number had increased to twelve; including that brilliant biologist of Liversidge's generation, F.M. Balfour (FRS, Fellow of Trinity, and Professor of Comparative Morphology, all before his sudden death in 1882), R.D. Roberts (later DSc of London, and a pioneer in the University Extension movement), and J.J. Harris Teale (later knighted, elected FRS, Fellow of St John's, and Director of the Geological Survey). When the Club celebrated its tenth anniversary in 1882, its early members already numbered ten college Fellows and seven professors.[173] By January 1920 and its 1000th meeting, of the 330 men who had belonged to CUNSC to date, fifty-five (seventeen per cent) were Fellows of the Royal Society.[174]

Liversidge became club Secretary. His organising talents were recognised, a foretaste of things to come. His circle were all in their twenties, and willing to give papers which, if specialised, caught the flavour of experimental research. In April, J.C. Saunders, working at the Botany School, spoke on 'Conscious Movements in Plants', and C.T. Whitmell, an early physical chemist, who took the top First in 1872, spoke on 'Isothermals and Adiabatics'. At the fifth meeting in May, C.J.F. Yule (who afterwards took a distinction in physiology) described the 'Anatomy of Pyrosoma', and Newell Martin spoke 'On the Modes of Reproduction in Animals and Plants'. On 18 May, it was Liversidge's turn to play host, and he entertained the twelve at Christ's with a paper on supersaturated saline solutions, borrowed from his research in London. Summarising three years' work, he set out what was known about the mechanism of precipitating supersaturated solutions and the nature of crystallisation. W.H. Miller, his teacher, thought well of it, and presented it that June to the Royal Society of London. It was duly published in the Royal Society's *Proceedings,* a signal honour for a young man still an undergraduate.

Set forth upon this sea of scholarly enterprise, it must have come as a blow to the members of the Natural Sciences Club, meeting in Carpenter's rooms in Trinity on 1 June, 1872, to learn that their organising Secretary was suddenly resigning, and leaving Cambridge within the month.[175] Unexpectedly, after keeping only six terms, and before he could face the Tripos, Liversidge had been offered the Readership in Geology and Mineralogy, associated with a Demonstratorship of Chemistry, at the distant University of Sydney. He was expected in Australia by November. The news was remarkable. None of his generation had won preferment so early, or so quickly. At their monthly meeting on 20 July 1872, the Club immediately, and by acclamation, elected Liversidge their first Honorary Member.

Ironically, this was to be, for some years, Liversidge's highest accolade. He was leaving too soon to make his mark in the group that was winning fame for science and the university. To his abiding regret, Liversidge's project in physiological chemistry would remain incomplete. His departure left the group without an experimental chemist until the arrival of Arthur Sheridan Lea, who came up to Trinity after Liversidge left, and later continued his work on pancreatic secretions.[176] By 1876, there were three biological texts emerging from Cambridge: the *Handbook of Physiology* edited by Edward Klein and Michael Foster, and Martin's *Elementary Biology*, and Foster's *Course of Elementary Practical Physiology*. In the *Handbook,* the section on digestion and secretion fell to T. Lauder Brunton, Lecturer on Materia Medica at St Bartholomew's Hospital.[177] Liversidge had lost his place on this research front.

During the next several years, the Cambridge group moved towards the precise measurement of nerve and muscle stimuli, and to the studies of the cardiovascular system for which it became celebrated. Of that first generation, Dew-Smith, who had as much difficulty passing his 'Little-go' as Liversidge, remained to carry on chemical work; but his polymathic temperament declined the narrow path of chemistry, and he turned to photography, the arts, and the Cambridge Scientific Instruments Company.[178] No one quite replaced Liversidge, though many succeeded him, and the waters soon covered his memory. His second paper, published in the *Journal of Anatomy and Physiology in* November 1873, was reprinted that month with other pioneering essays in the first volume of Foster's *Studies from the Physiological Laboratory.* There it remains today, nestled between papers by Foster and Martin, among the forgotten cornerstones of British experimental physiology.[179]

Liversidge was the second to join Foster's charmed circle, and the first to leave. Matriculating the following year were F.M. Balfour, A.G. Dew-Smith, W.H. Gaskell, A. Sheridan Lea and John Newport Langley. All were to lead in the development of experimental medicine, biological chemistry and precision instrumentation.[180] Of the twenty who sat the Tripos in 1873, six won Firsts, and four became professors. Among them, Sollas, Garnett, Martin and Balfour remained companions on the research frontiers to which Foster had introduced them. Liversidge's eighteen months in Cambridge had gone quickly. College routines, laboratories, museums and lectures, were all easily remembered. But perhaps the strongest impression left (to paraphrase Matthew Arnold) was in what he remembered after he had forgotten what he learned. To the public science of Frankland, the creed of Tyndall and Huxley, and the practicality of Smyth and Ramsay, were added the role models of Miller, Babington, Sedgwick, Liveing and Foster. He had seen the problems of creating a department and a research school in an environment dominated by tradition, where resources could only be won – grindingly slowly – by tact, patience and diplomacy.

Liveing's methods dominated the provinces of scholarship, but Foster's set a cosmopolitan example, shaping fortunes of men throughout the English-speaking world. Above all, Foster had style. As Huxley's successor in the Biological Secretaryship of the Royal Society for twenty-two years, he stood centre right in the developing drama of imperial science. As General Secretary of the BAAS from 1872 to 1876, and President in 1899, he helped move the association along lines of reform.[181] His hand was at work in the establishment of the Meteorological Office and the National Physical Laboratory; and his services were rewarded by a KCB. Through his influence (as we shall see), the International Association of Academies and the International Catalogue of Scientific Papers had their origins.[182] As chairman or member of four Royal Commissions, and as Member of Parliament (1900–1906), he was the model 'statesman of science'.

Liveing's hallmarks were punctuality and purity. But Foster had what Liveing lacked: vision and political acumen. Liversidge was the beneficiary of both. His career in Australia became a successful blend of their talents, adapted to the circumstances of colonial life.

# Chapter 4

# 'Scientific Sydney' in the 1870s

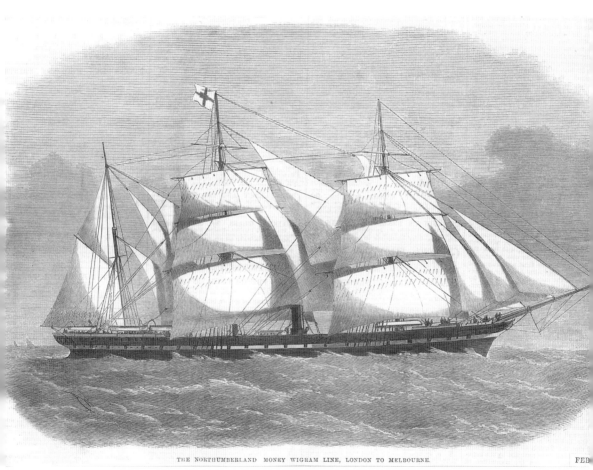

THE NORTHUMBERLAND MONEY WIGRAM LINE, LONDON TO MELBOURNE.

The *Northumberland*, sailing from London to Melbourne, *The Illustrated London News*, 3 February 1872.

The sight of Sydney University was stunning on the warm spring September day when, after heavy showers had cleared, a hansom cab brought Liversidge the four miles from the harbour quayside to The Glebe, as it was then called, and up the dusty driveway that led from Newtown Road. Formally established by Act of Parliament in 1850, the University had moved in 1857 from its modest beginnings at Sydney College, in the city facing Hyde Park, to 128 acres of bleak pasturage called Grose Farm.[1] Fifteen years later, raised above the horizon in splendid isolation, on a slight ridge above a rolling paddock, stood Edmund Blacket's magnificent, 400-foot long Gothic-revival building, its Great Tower rising in the austral bush like the barbican of a crusader fort. 'We must build not for today, or for to-morrow,' Francis Merewether had said, 'but for Futurity'. So 'Futurity's Folly', it was called.

On the north side of the main building, and sharing its commanding view of Sydney, stood the majestic Great Hall, inspired by London's Westminster Hall, its massive doors guarded by sentinels of stone. Soft light filtered through stained-glass windows, illuminating historical figures in British literature, the arts, law, the church and science.[2] On the northern wall was a stained-glass montage of the kings and queens of England since the Conquest, centring on the young Victoria, a window that she and Albert had personally admired. Suspended from great cedar arches, wooden angels bore symbols of the medieval trivium and quadrivium. Corbels, decorated with the arms of the English and Scottish universities, gave fealty to higher learning, and vast windows at the western and eastern ends portrayed the founders of the colleges of Oxford and Cambridge. The arms of the University, granted in 1857, united the book of Oxford, the lion of Cambridge, and the Southern Cross. Sydney's University was, in the words of Sir Charles Nicholson, its first Chancellor, to be a 'reflex in all its higher attributes and associations, of that land from which we spring'.[3] Was this not a fine place, in a land otherwise unfamiliar, for a man to serve both science and empire?

## 1. The Gargoyles of Grose Farm

When the University opened its doors in October 1852, there were three professors, selected by a committee in London, and devoted, respectively, to classics, mathematics, and to chemistry and 'experimental philosophy'.[4] In the first two subjects, they were chosen to 'bring with them the prestige of high Academical distinction at one of the Universities of Oxford or Cambridge'; for the third, they wished a scholar familiar with courses given in Edinburgh or London. From Oxford, came John Woolley, who also became the first Principal; from Cambridge, came an American-born Senior Wrangler of St John's, Morris Birkbeck Pell;[5] while from Scotland, came John Smith, a medical doctor and lecturer in chemistry at Marischal College, Aberdeen. If these three were, in Manning Clark's enigmatic phrase, 'exiles from the old world',[6] at least two were also in the bloom of youth: Woolley was by far the senior, at thirty-six; Smith was twenty-nine, and Pell, only twenty-five. They came into an academic ambience uncommonly free of religious tests and compulsory observances, and from 1857 in a splendid building, to teach a perhaps twenty-five students drawn from a colony four times the size of Britain, whose population in 1851 scarcely exceeded that of Bristol.[7]

From 1849, when W.C. Wentworth first proposed the idea of a university to the colony's Legislative Council, a mixture of motives surrounded the concept and its justification. On the one hand, there were good arguments for training an educated class, providing, in Wentworth's phrase, an opportunity for 'the child of every man, of every class, to become great and useful in the destinies of his country'. The British-born elite, with fresh memories of the 1848 revolutions in Europe, saw the importance of instilling traditional values. At the very least, the sons of landowners, squatters and merchants should represent the interests of stability in a colony on the threshold of representative government. Schools were needed for the professions of law and medicine, and for school teachers, to cap the non-sectarian system of 'national' (that is, state) secondary education begun in 1849. All these arguments were sound – but in their interplay were rival, frequently conflicting aspirations.

It was all very well to have a university of 'comprehensive design and character, limited to no sect, and confined to no class'.[8] But in a place which had seen the last convicts transported to NSW arrive only nine years earlier, its future was not to be taken for granted.[9] Was it to be an austral version of Oxford or Cambridge, a finishing school for wealthy colonists 'emulous of literary honour and the rewards of scholarship'?[10] Was it to be dominated, as at Oxford and Cambridge, by colleges, run by men in holy orders? Or was it to be a place of secular learning and self-discipline, a place to turn colonial boys into civic leaders?

In another way, was it to be a school for what Daniel Deniehy, the Irish-Australian 'darling of the Sydney proletariat', and the enemy of Wentworth, called the 'bunyip aristocracy' – its main purpose, to educate a gentry suitable for an hereditary Upper House?[11] Or was it to be principally a place of useful learning, like that favoured by liberals such as Henry Parkes, resembling Owens College in Manchester, dedicated to the interests of the commercial and professional bourgeoisie, and open to the 'people'?

Such questions were in large part political, and were ultimately answered politically. The University's 'architects' had before them several ruling models – the Scottish, urban, non-residential system; the University of London, with its professorial lectures coordinated by central examinations; and Oxford and Cambridge, with their strong, independent colleges. The University's founders were not unaware of the doctrinal debates that coursed through Oxbridge in the 1830s and 1840s; the foundation of the Queen's University of Ireland in 1844, with colleges at Belfast, Cork and Galway;[12] and the emergence of colleges in London and the provinces. At the same time, they were conscious of doing something different – a 'first university' for Australia.

*The University of Sydney Act* of 1850, which coincidentally received royal assent within weeks of the announcement of Royal Commissions to examine Oxford and Cambridge, took the form of a constitutional compromise. The University would combine the virtues of London and the Queen's Colleges to have both a teaching and examining function. Its management was to be entrusted to a Senate of sixteen Fellows, appointed by the Governor on the advice of the Legislative Council, at least twelve of whom were to be laymen (including, to Wentworth's dismay, members of the clergy). Drawn from the ranks of property, parliament, the bench, and the cloth, Senate was to be self-perpetuating: future vacancies were to be filled by the election of 'fit and

The Main Building of Sydney University, 1872.

proper persons' by the remaining members. Its rule was law.[13] Dragon's teeth were sown by the provision that the professors, chaired by their Principal, had no power beyond the classroom and the examination. The University's chief executive officer was the Chancellor (a Vice-Chancellor was his deputy), who reported to Senate. To meet recurrent expenses, the colonial government fixed an annual grant (called the 'endowment') of £5000. No provision was made for its increase. Any additional funds were to subject to special votes, introduced to and passed by parliament.

Like London, the University was both to examine and teach. The curriculum – in which Nicholson, an Edinburgh man, took a special interest – was to conform to the conventional model of an Arts degree, based on classics and mathematics, but one to which science was a compulsory addition. From 1856, candidates for the BA studied chemistry and experimental physics in their first and second year, and logic in the third. Professional training was left for the future. For the first twenty years, there was to be only one Faculty, that of Arts – awarding degrees not only in Arts but also, remarkably, in Medicine and Law. Until 1883, there was but one Dean.

These early arrangements were easily hostages to fortune. In the wake of the gold rushes, the 'aristocracy' that appeared to govern the colony was overtaken by a Legislative Assembly dominated by shopkeepers, publicans and traders, many of whom held quite different expectations of a University whose professors were paid from the public purse. Much of what David Macmillan calls a 'conservative, traditional and exclusive' vision of the University was to survive down the years.[14] In a place where philosophical principles clashed with political preferences, debate inevitably turned on the role of the colleges, the content of the curriculum, the standard of examinations, and the place of what were called 'modern', or 'useful studies', including science, medicine, and languages other than Latin and Greek.[15]

When the first professors arrived in 1852, the respective roles of the teaching university and the residential colleges were unresolved, but within the next two years, a compromise was reached, and as codified by the *University Amending Act* of 1854, lectures were made compulsory for all students, with supporting tutorials offered by the colleges, aided by government funds, and conducted under their own Acts of Parliament.[16] The first curriculum set out to give a comprehensive, if not Arnoldian vision of the best that had been thought and said, with a view to reproducing the 'memorised culture' of the classical tradition, such as was seen from Oxford and Cambridge. Manning Clark has described the founders' wishes as confined by a vision of education that limited undergraduates 'to the civilization of Europe, never thinking to encourage them to be pathfinders of a new civilization in their adopted country'.[17] In fact, for boys destined to become Australia's elite, Sydney's BA represented a significantly 'modern' development. In requiring a matriculation examination that demanded Greek and Latin, arithmetic, algebra and Euclid, the University set a standard that was as rigorous as the best colleges of Oxford and Cambridge. In these beginnings, Sydney came to reflect not only an imported culture, but also one of principle and promise.[18]

Almost immediately, however, the vision splendid became fragmented, as the three professors wrestled, both among themselves and with Senate, with the future of classical, professional and modern studies. In his report to the Legislative Council in 1849, Wentworth recommended the establishment of chairs not only in classics and mathematics, but also in natural history, civil engineering, anatomy, physiology and medicine; and that new chairs in political economy, modern history and modern languages should be created when the size of the University and its revenues permitted.[19] There is a legend that, after three chairs were accepted, a fourth chair, in natural history, was proposed; and that T.H. Huxley, who knew the colony from the long visit of the HMS *Rattlesnake* in 1847 had made enquiries about it. When the chair did not come about, England kept what Australia might have gained.[20] The natural history sciences were postponed, and the construction of the Blacket building, at a cost of over £80,000, in pre-gold rush revenues, fuelled fears that there would be no money left for 'modern' studies or for the appointments they required.

Professional appointments – in medicine, law, or theology – were not a part of the founders' project, and when the first professors arrived, there was doubt whether they should be made at all. John Woolley, a friend of Dean Stanley, and quite familiar with contemporary reforms at Oxford,[21] advocated the 'idea of a University' that distinguished between a 'school of *liberal* and *general* knowledge' and 'a collection of special schools, devoted to the learned professions'.[22] He desired Sydney to concentrate on the first. John Smith, on the other hand, saw the need for professional studies. In the struggle between 'ancient' and 'modern', the natural sciences inevitably bore the burden of proof, and were blamed for everything that could be attributed to progressive, practical, or useful knowledge. At the Queen's University in Ireland, the curriculum had science and professional studies by 1847.[23] But Sydney was not to be in advance of them. At Oxford and Cambridge, indeed outside the 'Godless institution of Gower Street', as London University (later University College London) was known, English higher education viewed the natural sciences as ancillary, to be studied only after sensibility had been served by inoculations of classics and mathematics. Cambridge, in creating the Natural Sciences Tripos in 1850, limited entry at first to men who had already read honours in mathematics.[24]

If there were doctrinal differences among the professors, there was common cause on two issues – the centrality of teaching and representation on Senate. To ensure a 'central standard', they argued, 'which it is the prominent object of the Royal Commissioners to restore at home', the professors were to teach and not merely examine. Senate agreed, and the role of the colleges was limited to tutorial support of professorial classes.[25] As for Senate representation, Britain offered conflicting precedents. Edinburgh University, ruled by city magistrates with the Lord Provost as Rector, was almost a caricature of lay interference in academic affairs. Glasgow University, on the other hand, was in the hands of its professors, who fostered a reputation for 'self-patronage'. London University was in many respects a government department; its professors were civil servants, and its Senate was nominated by the Privy Council. At Owens, the professors were not at first admitted to university government. In Ireland, professors began asking for representation in 1857. The question was clearly one of not merely British, but imperial concern.

In 1859, in response to widespread criticism – pointing, not least to the University's failure to attract more than a handful of students – the Legislative Assembly set up a Select Committee to enquire into its administration. This was an opportunity to set many things right. But its proceedings evolved, as Ken Cable once put it, into a 'congerie of frustrations'.[26] Nicholson, Woolley and Smith wished to reform the curriculum, but all in different ways.[27] Nicholson took issue with a neglect of new disciplines, and the 'mercenary view which deprecates as unsuitable to the Colony any outlay on objects of a not strictly utilitarian nature'.[28] Woolley argued that the University could do no better than offer a 'training in mathematics and grammar as given through the classical languages', but agreed that the curriculum should be expanded. Smith wanted more space for science and beyond: 'I think these are the two leading defects', he said, 'the want of natural history, and the want of moral philosophy. I would also like to see, by and by, a Professor of English Literature.'[29]

Caught in a crossfire between liberals who wanted change and conservatives who feared the cost, the Select Committee recommended that the University introduce no new faculties, but that it widen the Arts curriculum, to improve 'its capacity for adapting its teaching to the spirit of the age'. In the event, there was no immediate change either in the curriculum or the range of studies. Only on the point of representation did the Committee concede to admit 'no fewer than three, nor more than six' professors as *ex-officio* members of Senate; and recommended giving University graduates a voice in Senate elections. Both recommendations were secured by the *University Amendment Act* of 1861,[30] three years before Belfast, Cork and Galway won similar concessions. Henceforward, professors were to have at least a vote on academic developments, but they were not yet to have a determining say.

John Smith
(1821–1885)

Charles Badham
(1813–1884)

Morris Birbeck Pell
(1827–1879)

The Professors of Sydney University in 1872.

Early accounts of the University have tended to favour the impression of Manichaean battles between the 'whiggish' interests of a 'powerless professoriate' and the 'obstructionism' of an oppressive Senate, dominated by politicians and laymen. In reality, the professoriate was caught in minefields of intrigue, sometimes having little to do with the classroom. And as long as Senate was dominated by William Macarthur, who feared developments 'democratic in their tendencies', conservatives kept a close hold on the curriculum.[31] The course of studies remained narrow. There were optional lectures in logic and jurisprudence, and even in French and German, but classes not compulsory were poorly attended. The result was not a great University, but something closer to a public school. Woolley's attempts at reform exhausted him; he went to England on leave in 1865 in a state of depression. When, on his return voyage, his ship went down in a storm in the Bay of Biscay, death put a tragic end to a promising career.[32]

In 1865, the Senate's annual report still made unimpressive reading. In its first fourteen years, the University awarded only sixty-five degrees, and only thirty-two in the preceding five years.[33] New enrolments averaged only twelve a year. In its defence, a few graduates had already made a mark. William Windeyer (BA 1856), a 'very impulsive, original and independent man', was destined to become the University's first MA and its first barrister; for a time, a liberal reformer; and eventually Attorney-General in two of Henry Parkes' administrations. He was to figure prominently in education, and in 1876, became the University's first Member of the Legislative Assembly, a member of the Senate from 1866, and eventually its Chancellor.

Other graduates included Edmund Barton (BA 1868), later a Fellow of Senate, the University's third (and last) Member of the Legislative Assembly, and Australia's first Prime Minister, and Henry Chamberlain Russell (BA 1859), one of the first Australian-born graduates, who studied physics with John Smith, and whom Smith sent to work at the Sydney Observatory. There he would become the first Australian-born Fellow of the Royal Society of London. But this is to look ahead. The reality was that the University had so far produced a very small number of graduates. If, in 1865, the University was widely regarded, in David Macmillan's phrase, as 'a failure, or, at best, an expensive novelty',[34] the best case for its future was yet to be heard.

Any university serving a city of fewer than 100,000 people, even one drawing upon the country districts of NSW and southern Queensland, which insisted upon matriculation in Greek and Latin, but drew upon schools that taught few boys either, was unlikely to attract many students. However, enrollments between 1852 and 1862 point to the rising importance of merchants, traders and artisans' sons, whose fathers were replacing pastoralists and landowners in the political arena, and who might be thought to favour 'useful knowledge'.[35] Such parents might send their sons to a university that offered 'practical' subjects and professional degrees.

Certainly, this strategy seemed to be the secret of Victoria's success. In the early 1860s, Melbourne was larger and richer that Sydney, with a better system of secondary education, and in regular communication with the educated families and schools of Tasmania and South Australia. In establishing its University in 1853, the Victorian government copied Sydney's legislation, but doubled Sydney's endowment, thereby creating more chairs and a broader curriculum. In addition, Melbourne offered professional courses in law (1857), engineering (1861), and medicine (1862).[36] In NSW, politicians who demanded easier access for students, and greater 'utility' in their teaching, were usually reluctant to meet the cost of either.

In 1866, the press, outraged at the University's ratio of expenditure to return, obliged Senate to propose the appointment of Readerships in a range of 'modern subjects', including French, political economy, geology and mineralogy; and to widen access to matriculation by allowing optional papers in French and German. To encourage competition, the University instituted a system of colony-wide Junior and Senior Examinations, modelled on Oxford and Cambridge precedents,[37] with chemistry and geology included among the optional subjects.[38] Sydney's professors became their examiners, and so a bridge was established between secondary and higher education.[39]

The Junior and Senior Public examinations were part of the solution, at least until the turn of century. But they were slow to work. By 1871, there were still only seventy undergraduates in all three years, and the proposed Readerships had been abandoned. The University, as a later archivist put it, was scarcely 'surviving'.[40]

This was the legacy that Liversidge was to inherit. Constrained by their curriculum, yet liberated from large classes, Sydney's professors had time for many things. Their salaries and allowances were generous.[41] Neither Pell nor Smith did much research, but both were active in the colony's commercial and educational life. Pell became an actuarial consultant for the Australian Mutual Provident (AMP) Society, and chairman of several public boards and commissions. Smith became a Director of AMP, and augmented his daily lectures to undergraduates with afternoon lectures to the general public.[42] From 1853, he was a member of the Board of National Education. He became a Member of the Legislative Council, and helped secure passage of Parkes' residential, denominational colleges *Public Schools Act* of 1866, the first step towards a system that was to become 'free, compulsory and secular'. For ten years from 1870, he was President of the Council of Education, set up by Parkes in 1867 to supervise the colony's growing system of national (that is, state) schools. He followed in the footsteps of Nicholson, who adapted to NSW the principles established in Ireland, and contributed vastly to the extension of literacy in the colony.[43] In 1878 his educational work was recognised with an imperial CMG.[44]

Both Smith and Pell made early contributions to public science, a tradition by now well established in Britain where, by the late 1860s, London, Edinburgh, Manchester, Birmingham, Liverpool and Bristol were accustomed to hearing tributes paid to science as the source of civic improvement.[45] 'It is only when commerce and science are united', as one contemporary editor put it, 'that the one is dignified and the other [made] useful' in the ultimate victory of mankind.[46] The railways symbolised the union, as did schemes for housing, public safety and sanitation. By the time Liversidge arrived in Sydney, Pell had done important work in vital statistics, and Smith had served on two enquiries into Sydney's water supply.[47] But there was still much to do.

Since the 1820s, Australia's governors had regularly endorsed science as a means of advancing colonial development, and the 1830s had seen many attempts to raise Sydney's cultural expectations, sustained by the Macleay family, the Cunninghams, and the gentlemen who established the Sydney Mechanics' School of Arts in 1833. The establishment of a botanical garden and a museum – the Australian Museum – in the 1820s spoke to a belief in the improving virtues of natural knowledge.[48]

However, the course of colonial science never ran smooth. As early as 1821, the colony's first scientific society, the short-lived Philosophical Society of Australasia, was encouraged by the Governor, Sir Thomas Brisbane, but attracted no more than a dozen members.[49] When Smith arrived in 1852, there were two principal scientific societies – the Australasian Agricultural and Horticultural Society and the Australian Philosophical Society.[50] The latter, established in January 1850, was inspired by that ubiquitous pair, Henry Douglass and Charles Nicholson, together with Edward Deas Thomson, the Colonial Secretary, and others active in the University movement. Its goal was to encourage the development of manufactures and new industries of wool, wine, silk, 'perhaps sugar and cotton, certainly the olive'. To exploit the resources of the colony appealed to 'practical men' of the stamp of T.S. Mort, the wool and beef merchant.[51] For eight monthly meetings the Society continued, under the patronage of the Governor, Sir Charles Fitzroy; but soon, attendance withered, and the Society fell into what its historian has called a 'state of suspended animation'.[52]

Smith thought 'gold fever' had sapped its vitality.[53] Whatever the cause, it was not until 1855 that twenty-two of its members resolved to make a fresh start. In 1856, with the encouragement of the new Governor, Sir William Denison, recently arrived from Van Dieman's Land, the gentlemen reconstituted themselves as the Philosophical Society of NSW, and began monthly meetings, eight months a year, with 'original papers on subjects of science, art, literature and philosophy'.[54] Denison, a Royal Engineer and Fellow of the Royal Society of London, took a keen interest in its proceedings, and from 1855 to 1861, first as ruling Governor, then as the Queen's representative, did much to restore its sense of self-confidence. 'I have got my Philosophical Society to work at last', he wrote Sir Roderick Murchison in 1856, and set a good example by giving the first paper (on 'railways').[55] But even Denison – a 'habitual man of action', as John Ward has described him, 'who owed little to powerful connections, and who found a challenge in every opportunity of exercising his authority',[56] felt, as he put it, 'constantly obliged to keep a check on one's wishes and aspirations, and to accommodate one's place to the jog-trot of society'.[57] His policy was 'to generate, first an appetite for writing, and then a taste for observation, in order to have something to write about'.[58]

The gold rushes of the 1850s underlined his point. That any state must know the extent of its natural resources was not at issue. But Australia had a special history of starts and stops. In 1803, fully three decades before the Geological Survey of Great Britain was established, the Secretary of State for War and the Colonies appointed A.W.H. Humphrey to be 'H.M. Mineralogist', and sent him from England to survey the mineral resources of NSW. His adventures came to grief in 1812 when Governor Macquarie decided that he had found nothing 'worthy of notice' and sacked him. Humphrey's dismissal cast a long shadow. 'A different man who has real Scientific Knowledge,' Macquarie agreed, 'might be very Useful and Make Very Important Discoveries in Various parts of this Widely extended Colony.'[59] But it was not until 1850 that Governor Fitzroy appointed Samuel Stutchbury as the first Geological Surveyor of NSW, at a salary of £600.[60]

Stutchbury surveyed for only five years, and for a time, there was no official geological expert in NSW.[61] Given the circumstances, it was fortunate that Sydney could boast the presence of Rev. William Branwhite Clarke. Clarke had read geology with Adam Sedgwick at Cambridge, and had become a Fellow of the Geological Society of London in 1826, at the age of only twenty-eight. Coming to Australia in 1839, as headmaster of The King's School at Parramatta and incumbent of the parish of Castle Hill and Dural, he moved to the parish of St Thomas' in North Sydney in 1846. From this vantage point, the 'remarkable Reverend Clarke' explored the vast countryside on foot and horseback, and became the colony's authority on the stratigraphy of NSW.[62] A founding member of the Australian Philosophical Society, he was then perhaps the colony's leading man of science.[63] A friend of John Smith, he would also prove a guiding light for Liversidge.

The Main Building and Great Hall, with cows grazing on what is now Victoria Park, about 1872.

Clarke helped establish colonial geology at a critical time. In 1851, Sir Henry Young, Lieutenant Governor of South Australia, obtained permission from the Colonial Secretary, Lord Grey, to survey that colony, 'and make known its mineral resources to the colonists'. He wanted a scientific man of zeal and ambition to 'promote the practical applications of geology to mining purposes', but also one of 'such already high and established professional reputation, and disinterestedness, as would impart to his Report the stamp of conclusive authority'.[64] Sir Henry De la Beche, Director of the Geological Survey, recommended Stutchbury, already *en route* to NSW for a similar purpose.

However, such men as Stutchbury were hard to find. Indeed, unless Stutchbury could do the job, warned De la Beche, 'it would be excessively difficult to procure the services of a person qualified in the manner required, who should be willing to proceed to Australia'. There were, he added, 'very few persons so qualified as respects both mining and geological knowledge, and they are usually in receipt of good salaries'.[65] When the Government School of Mines, (later RSM) opened its doors in 1851, De la Beche hoped its graduates would meet the demand.[66] A salary of £550 was the going rate. When the government of the newly-separated colony of Victoria appointed a Mineral Surveyor in 1852, they paid £500.[67]

In 1856, with the gold rushes underway, Murchison urged the Colonial Office to survey NSW, and to take advice from J. Beete Jukes, who had visited NSW whilst serving as naturalist on HMS *Fly*, and who had prepared the first geological map of the Australian continent. [68] Every colony, Murchison argued, required a geological surveyor to advise government and private enterprise how best 'to open out advantageously the mineral resources of distant regions, many of which are as yet entirely "unknown lands" as respects their internal productions'.[69]

There was no suggestion that in supporting geology the imperial government would compromise the principles of free trade, or burden the public purse with costs that commercial interests should bear. However, it was clear to the far-sighted that both interests could be served by the universities. In Melbourne, geology was taught from the University's inception in 1853 by Frederick McCoy, Professor of Natural Science.[70] But at Sydney, outside brief mention in Smith's chemistry lectures, geology had no place in the first curriculum. If mineral wealth had not inspired the foundation of the University of Sydney, mineral revenues would help pay its upkeep; and in 1852 and 1853, there were rumours of an appointment in geology. Although these came to nothing, the demand for payable minerals made an overwhelming case.

This proposition was well understood by Edward Deas Thomson, the Colonial Secretary, who in 1865 would become Chancellor of the University. In 1854, Deas Thomson, who had earlier encouraged the Philosophical Society, and overseen the grant of Grose Farm to the University, gave £1000 for scholarships in the natural sciences.[71] The first was to be given in physical science, but the next would be in geology and mineralogy. A chair in the field was also needed. 'Although we may startle some', nudged the *Sydney Magazine of Science and Art* in 1859, 'when we suggest that the classical halls of our University *should* be invaded by intending diggers', the colony had a clear interest in acquiring a 'school of mines', together with a museum of mineralogical specimens for the miners, who were otherwise 'so plainly demonstrating how much such information is wanted'.[72]

Inevitably, the University's critics looked to John Smith. According to his terms of appointment, Smith was to teach chemistry and physics as given at the universities of Edinburgh and London, as well as the 'experimental philosophy and mathematical physics that fell to the Plumian and Jacksonian Professors at Cambridge'. Even by the standards of the day, this was an impossible demand. As he was not especially interested in geology, little use was made of the small mineralogical collection that had come into existence in the Blacket building. Moreover, he found experimental work uncongenial – except in photography, in which he became expert.[73] Otherwise, 'philosophical equipment' brought from England at great expense lay neglected. Well after the University migrated from College Street to Glebe in 1857, Sydney Grammar reported that two of its four downstairs rooms were 'cluttered' with chemical apparatus that Smith had left behind.[74] Judge Backhouse, a graduate of the University (1872), and a later Vice-Chancellor, about whom Liversidge would have much to say, recalled learning from Smith 'enough of the elements of chemistry and physics to induce me to a further study later on; but his lectures were as dry as were his experiments uncertain. As the judge said of counsel who had not been conspicuously brilliant, "He may have done his best, but I hope not!"'[75]

As early as 1862, Smith thought of returning to Britain. His serial letters, written on study leave, and published in the *Herald,* reveal a man longing for 'Home'.[76] He coped with isolation by shifting his priorities away from the University. His conduct in public controversies – on the place of religion in secondary education, the pollution of Sydney's water supply, and the exclusion of 'dogmatic theology' from University affairs – won many friends. By contrast, at the University, where he was for a time both Dean of Arts and Dean of Medicine, he made enduring enemies. Suffering for years what he called the 'burden of moral slavery',[77] he was a man with a grievance. For the first seven years, he had encompassed the entire 'circle of the sciences' for a salary of £675, plus a house allowance of £120. This was ultimately rectified, but the experience left a sour taste.[78] His early terms contrasted vividly with the £925 (plus free accommodation) given Woolley and his successor, and the £825 (plus housing allowance of £150) given to Pell, not to mention the large salaries paid to government scientists like Stutchbury and Selwyn.[79]

Many years later, Smith's successor as Dean of Medicine remembered him as a 'zealous, but *cautious*, promoter of public education, and the public good'. Others knew him as 'conservative to the point of short sightedness'.[80] Geoffrey Serle has called him 'undemonstrative', and 'not brilliant and successful socially'. Certainly, he could be difficult. He opposed, for example, the move from College Street to Grose Farm, arguing, like a good Aberdonian, that a university *should* be in the centre of a city. Famously, as Dean of Medicine, he opposed the establishment of a medical school, on the argument – which laymen would appreciate – that untrained doctors did as little damage as trained ones. Such sentiments were unlikely to earn him sympathetic treatment in the annals of medicine.

As his first biographer put it, Smith was 'fortunate' to be appointed just before his subject began to develop rapidly.[81] But he also saw his shortcomings, notably in geology and chemistry. Both were important; neither could he supply. By 1865, he had seen the significance (as who could not) of the minerals rush; and saw the advantage – both to the University and himself – in having a chair in geology and mineralogy, subjects which otherwise would fall to him.

After Woolley left for England, never to return, Smith pressed Senate to accept a modification in the Arts curriculum, so as to enable third year students to choose two among four optional courses. One of these four was described as 'Chemistry and Experimental Physics, and such branches of Natural Science as may at any time be taught in the University'.[82] By this means, Smith hoped to attract students into advanced work, especially in chemistry, which would strengthen his case for a new appointment. Senate agreed, and the following year, prompted by Deas Thomson, the Vice-Chancellor, tried to outflank critics by establishing new readerships in a range of 'modern subjects' – two in political economy and English, and a third in geology and mineralogy. It was unlikely that the government would countenance two new appointments, but one that combined geology, mineralogy and chemistry was just possible. For such a person, the colonial government would inevitably look to the RSM.

Twelve thousand miles away, and unaware of these events, nineteen-year-old Archibald Liversidge had just embarked on his first year at Jermyn Street. In December 1865, Deas Thomson invited Murchison, Director of the RSM, to nominate an 'assistant in chemistry and lecturer in geology', at a salary of £450. A month later, Murchison replied, recommending a former pupil, twenty-five year old Alexander Morrison Thomson, as 'better entitled to hold the appointment in question than many other men I am acquainted with whose services could now be obtained'.[83]

With experience of commercial and academic chemistry, in both Scotland and England, Thomson was an admirable choice. Andrew Ramsay, who taught him at the RSM, considered him 'just the well-informed, noble and enthusiastic youth who is likely to fulfil eventually all the conditions required'. Born in 1841, Thomson, a bookseller's son, was educated in London and Aberdeen before taking work in 1859 with consulting chemists in Islington. In 1860, he received a matriculation prize in chemistry at King's College London, where for two years he attended night classes, studying chemistry, mathematics, Latin, Greek, and French. He received the BA (1862) and BSc (1864) of London University, and worked briefly for Inland Revenue and Customs, before returning to science and a year at the RSM. In the early 1860s, he worked as a private tutor, and if Deas Thomson is correct, actually may have tutored Liversidge.[84] In 1866, he took one of the first DSc degrees awarded in chemistry by London University – marking him, by any measure, as a capable man.

Thus, in December 1866 – seventeen years after a chair of natural history was first proposed Sydney acquired its first Reader in Geology and Mineralogy, and fortuitously, a Demonstrator in Practical Chemistry as well. Thomson sailed to Sydney with his wife, children and a collection of instruments and mineral and fossil specimens, and began teaching at the beginning of Lent Term in March 1867. He was an instant success. His notes, preserved in the Fisher Library, reflect a thorough understanding of Ramsay's courses, coupled with a keen curiosity in his new environment. Sponsored by John Smith, he joined the Royal Society of NSW, where he met W.B. Clarke, and in 1868, rode through the colony, investigating mineral deposits that Clarke recommended. The following year, he became a Trustee of the Australian Museum; published a *Guide to Mineral Explorers;* geographically mapped the district around Goulburn; and explored the Wellington Caves for fossil remains with Gerard Krefft, the Australian Museum's dynamic curator.[85]

In November 1869, in recognition of the growing importance of mining, Thomson was elevated to a personal chair, making him the first professor of geology in Australia. Tragically, his success was short-lived. In early 1871, chronic rheumatism and chest pains, made worse by exposure to the elements on a field trip to the Wellington Caves, forced him to bed in his small Newtown terrace.[86] There, on 16 November 1871, at the age of thirty, he died of complications from pneumonia. His death stunned the small community of science, and left his family bereft. No one grieved more than Clarke, for whom 'poor Thomson's' friendship had been a comfort.[87] Deas Thomson knew that the impulse he had given to geology and chemistry, and to the University's image, must be sustained. On the last day of November, scarcely a fortnight after Thomson's death, Senate requested Smith – and Sir Charles Nicholson – to find a successor.[88]

In effect, it fell to Smith to take the practical steps. In May 1871, Senate had given him permission to take his second round of overseas leave in 1872. Replacement teaching was arranged, and in December 1871, he sailed from Sydney with instructions to bring back Thomson's successor.[89] For reasons that remain obscure, Senate did not advertise its vacant Readership in Geology and Mineralogy in the British press. Nor was there, apparently, a London Committee, such as had decided the destinies of Pell, Smith, Woolley and many others. As in Thomson's case, a successor had to be found by personal contact. Smith and W.B. Clarke wrote to Richard Daintree, a well-known geologist, who had recently been appointed Special Commissioner for Queensland to the London International Exhibition of 1871–72, and who for this reason was resident in London. Daintree, whom we have met before, was a man at home in both hemispheres, and likely to know eligible candidates.[90]

Briefly serving with Alfred Selwyn and the Victorian Survey,[91] he had returned to London in 1856–57 to study geology and chemistry at the RSM, and brought this knowledge back to Victoria. His appointment in 1871 as Queensland's Agent-General in London gave him leisure, and the facilities of Jermyn Street gave him a chance to analyse specimens that he was sent to display at the Exhibition. Content in London, Daintree had no personal interest in the Sydney post. But he had known Thomson,[92] and few in London better understood the situation that would face Thomson's successor.

Daintree offered to make enquiries. As De la Beche had prophesied, the job was difficult to fill. 'There is a dearth of good chemists and mining managers for the management of mining properties', Daintree wrote W.B. Clarke. 'Three times the number of applications are made to [John] Percy [at the RSM] than he can find men for'. 'Nevertheless', he added optimistically, 'I expect … to hear from Judd and another man who have passed through the full course at the School of Mines, and are good chemists into the bargain'. Either, he added, 'would do well, but the pay is small for really good men and is only bread and cheese'.[93] John Wesley Judd, a talented *protégé* of Murchison, and a student at the RSM in 1863–64, had recently accepted an offer from his friend, Matthew Arnold, to join HM Schools Inspectorate. Within a year, he returned to geology, became an authority on the tertiary volcanoes of the Hebrides, and in 1876 succeeded Andrew Ramsay as Professor of Geology at the RSM.[94] Judd was not interested in Sydney. The 'other man', we may presume, was Archibald Liversidge.

It is not clear how contact between Daintree and Liversidge came about. Daintree knew Sir Warington Smyth and Percy, and presumably Frankland, all of whom would have remembered Liversidge from his student days. Perhaps Daintree had heard from dons at Christ's about the ambitious young chemist, then still at Cambridge, who was making such a good impression in Liveing's laboratory. It is possible that Daintree and Liversidge met in London; certainly, they shared interests in photography and microscopy. And it can be no coincidence that, a year later, on 11 March 1873, Daintree nominated Liversidge, 'from personal knowledge', to be a Fellow of the Geological Society of London, with the support of Liversidge's teacher Robert Etheridge Sr, and F.W. Rudler, the Geological Museum's Assistant Curator (later Curator and Registrar of the RSM).

In the absence of conclusive evidence, the following reconstruction of events must remain conjectural. From his *Wayfaring Notes,* we know that Smith visited London in the spring of 1872,[95] and attended the International Exhibition, where a meeting with Daintree was unavoidable. According to Smith's later wife, who may not have been disinterested, a 'large proportion of his time was spent in searching for a successor to Thomson'.[96] Whatever the case, by early April, Sir Charles Nicholson had received £150 to pay the passage of the successful candidate, and £50 to purchase geological specimens for his use; and Nicholson promised to devote his 'best exertions' to the task of finding a man.[97]

By early May, an approach to Liversidge had been made, and within a month, the matter was settled. On 5 June, Senate received a letter from Smith, announcing that he had selected 'a Mr A. Leversage [*sic*] of Christ's College, Cambridge'. Coincidentally, Smith also met the woman who became his wife, and in hopes of a European honeymoon, told Senate he had decided to remain in England until August, beyond the end of his appointed leave. When the Senate unromantically stopped his salary, Smith reluctantly agreed to return by the fastest mail. In July, he and the new Mrs Smith took passage for Sydney. With them in saloon class travelled their new colleague, Archibald Liversidge.

## 2. THE COLONIAL CHOICE

'I must make up my mind to [emigrating]', the young surgeon-naturalist T.H. Huxley wrote from London to his *fiancée*, Henrietta Heathorn, in Sydney, in August 1852, 'if nothing turns up'.

> However, I look upon such a life as would await me in Australia with great misgiving. A life spent in a routine employment, with no excitement and no occupation for the higher powers of the intellect, with its great aspirations stifled and all the great problems of existence set hopelessly in the background, offers to me a prospect that would be utterly intolerable but for your love ...[98]

As all the world knows, Huxley did not pursue the project, and remained in England. Huxley was in good company. During the *Beagle's* visit in 1836, Charles Darwin, deploring the evils of convictism, said that 'nothing but sharp necessity' would induce him to emigrate to Australia;[99] while Dickens' *Household Words* created for the English public a picture of isolation and shame. Many who had emigrated in the 1840s had stayed to suffer; some, if they could, returned.[100] Popular prejudice favoured Charles Lamb's caricature – that Australians must be good, because they had been picked by the best judges – and literary men posted occasional signals: 'Nine times out of ten', Henry Kingsley cautioned likely emigrants, 'a man does not improve himself by emigration, and the proof of the pudding is in the eating: all men come back if they can'. Many agreed that so far in the colonies, 'There is no *life* for an educated man'; and perhaps, at worst, there was only a 'choice between rowdyism and Lotus eating'.[101]

By the early 1870s, with the advancing tide of prosperity, Australia was no longer the slough of despond of Darwin and Dickens, but a destination for Anthony Trollope, who visited in 1871, and celebrated the beauties of the place and the energies of its people. For Liversidge, emigration was not a choice made for love or health, but for professional advancement. With elder brothers earning high wages, and the family firm in good hands, he could safely choose an academic life, and in the twenty years since Huxley pushed aside the thought of leaving England, Australia's image had improved, especially for a chemist, geologically inclined.

In any case, what were the alternatives facing him? By 1872, Huxley had a secure job in London, but few of his generation had been as lucky. Students were multiplying, but academic posts were not.[102] As Liversidge could not yet know, expansion in higher education would not begin until the 1880s. Meanwhile, Liversidge's Cambridge prospects were problematic. His progress had been delayed by Prelims, and at best he could not sit the Tripos until December 1873. Had he received a First and stayed a bachelor, he had a respectable chance of a college fellowship, if he were prepared to wait. But there was no guarantee of success. Of the ninety-five men who had taken the Natural Sciences Tripos since 1851, only fourteen had been elected to college fellowships at Cambridge (and none at Oxford), and they had been selected for distinction in fields other than science.[103] So deep-seated was the preference for Wranglers who could teach the Mathematical Tripos, that most 'Naturals' left Cambridge as soon as they could. The picture would change, as the NST overtook the Mathematical Tripos in the 1890s; but in 1870, such developments lay well in the future.

Liversidge, who in his youth had been on the cusp of so many changes in London, found himself on the brink of change in Cambridge as well. It was precisely his generation – 'Foster's men' – who pushed the curve. His classmate Henry Newell Martin took the highest First in the Tripos their year (December 1873), and was elected a Fellow of Christ's in October 1874. Two years later, he became the first Professor of Biology at the newly-established Johns Hopkins University in Baltimore. Another classmate, Frank Balfour, just behind Martin in the Tripos, overtook him in research. Helped, like Martin, by Huxley, Balfour won a Prize Fellowship in Natural Science at Trinity in 1874, and within half a decade had an international reputation in comparative morphology. Elected a Fellow of the Royal Society in 1881, Balfour was destined, in Darwin's words, to 'some day be the chief of the English Biologists'.[104] Cambridge gave him its first personal chair in animal morphology in 1882.[105]

These were the 'superstars' of Liversidge's year. For others, the going was less easy. Some were to know years of professional uncertainty. Of Liversidge's two closest friends at the RSM who went up with him in 1870, W.J. Sollas, who also worshipped Huxley, was bracketed third among the Firsts after Martin and Balfour, but lectured for six years in the University Extension Scheme before getting a job as Curator at the Bristol Museum in 1878. Only later would he achieve academic respectability as Professor of Geology and Zoology at University College Bristol in 1880, and at Dublin in 1883.[106] A second friend, William Garnett, passed out as Fifth Wrangler in the Mathematical Tripos of 1873, and was elected a Fellow of St John's in 1874. He became James Clerk Maxwell's first Demonstrator in Experimental Physics at the Cavendish for the next six years, but then married and in consequence vacated his Fellowship. Not until 1882 did he become Professor of Physics and Applied Science at University College Nottingham. Much later, he would become the Principal of Nottingham – but this outcome was not obvious in 1871.

Surveying Sydney's advantages, Liversidge might also have weighed other factors. With the support of Foster and Huxley, he could go far, but his chances of international recognition were limited. The brilliant Balfour would be elected to the Royal Society in 1878, only five years after taking his degree; but Martin had to wait twelve years, until 1885; and Sollas, sixteen. Garnett was never elected. Although he could not know it, a period of colonial science might improve his chances. Indeed, his turn did come early, in 1882. By taking up Sydney's offer, he advanced his chances of professional success, and his prospects of becoming an FRS, by six to eight years.

In any case, the alternative pathways awaiting a chemist were few. School teaching and the Education Inspectorate held few attractions for him. Trade did not appeal, nor did industry. Becoming a mining manager was not to his taste. True, most of his contemporaries at the RSM thought otherwise. By the end of the century, over two-fifths of the Associates were working overseas, and mostly for mining companies.

However, it was also true that nearly a tenth of the RSM's Associates were going to Australia and New Zealand, and a colonial chair at least promised an interesting life. 'There is a field of mind to be cultivated [in Australia]', he read, 'as well as wide tracts of land'.[107] *Christ's College Magazine* praised the generation of young colonial professors, 'who have more power in their hands, perhaps, than they are aware of' – and more money, too. Sydney offered scope to 'make a name', without great hardship.[108] The starting salary of a Reader was £450 per annum, not a fortune, but almost as much as the £500 that his teacher, George Liveing, earned after twenty years as a professor at Cambridge.[109]

Years later, Liversidge's City of London Polytechnic's student magazine, *The Londonium,* declared Australia a stage 'wide enough to work out the problems of life'. 'There', Londoners were told, men could 'do big things', if only, perhaps, because 'they had not the time to think them'.[110] Whether because of  or in spite of  such recommendations, the case for emigration was persuasive. In any event, as Daintree's experience showed, an Australian sojourn need not be permanent. England would still be 'Home'. It was important merely to be sure that, if one went abroad, anything one did, should be done visibly. 'Out of sight, out of mind', as the Governor of NSW wrote to Sir Roderick Murchison, 'is a proverb of general application, and unless [London] is reminded of one's existence, they are apt to bury one without asking questions'.[111] For a younger son of the City, with no family to bind him, a taste for travel and adventure, and a better than even chance of making his fortune, Australia seemed a reasonable choice.

Having decided to go, what did Liversidge take with him? Certainly, from London and Cambridge, the ingredients of a 'memorised culture' of practical science. In London, he learned analytical procedures, inventory methods, a technical vocation and a sense of civic duty. The 'balance' was his attribute. In Cambridge, he acquired a deepened respect for culture, an attitude of mind, laboratory technique and social skills. He also saw the disadvantages that universities could place in the way of artisans' sons who lacked Latin and Greek. If Huxley gave him passion, Foster gave him determination. Both gave him a professional, practical philosophy of nature, based upon the careful analysis and assembly of facts, caring little for abstract argument. Leslie Stephen caricatured the 'Cambridge mind', as one which took 'the strongest objection to look far beyond our noses. We take what lies next to us, and don't trouble our heads about its remoter bearings'.[112] The same moral order was implicit in academic life: 'We introduce', Stephen wrote, 'reforms when we want them, and not before'. This strictly 'practical view' afforded good grounds for experiment, and left theology to the theologians. It was the basis of a 'practical idealism' that would shape Liversidge's vision of himself and his science.

Whether Liversidge's training best suited him for Empire, or whether the needs of Empire best suited the training he received, is a difficult question. Certainly, Australia's expanding economy favoured the applications of applied science. If Liversidge had no prior knowledge of 'imperial science', the RSM had prepared him for imperial service. His

Sydney Harbour photographed in 1872, the year Liversidge arrived.

experience of technical education and public science, polished by collegiate Cambridge were a preparation for life in a place where technical education hardly existed, where a 'laboratory' was a euphemism, where leisured learning was open to few, and where academic science was always under siege and never beyond debate.

Liversidge brought to Sydney many qualities – a reputation for getting on with the job, good health (not, as his predecessor had discovered, to be taken for granted), an outgoing disposition, and freedom from family obligations. There was scope for a man who had risen from the ranks of artisan London, who had sat at the feet of England's ablest lecturers, who was abreast of the latest science and who had some familiarity with the gentlemanly arts. On the other hand, Sydney in the 1870s was well suited to a man of reasonable if not exceptional attainments; of wide, if not profound reading; of liberal, if not radical views; of confident, if not dogmatic opinions on race, religion and politics. Above all, Liversidge offered an open mind. In an age of vaguely worded texts, where 'descriptive' geology was burdened with inconsistent

terminology, where scientific writers too often soared from selective evidence to epic generalisation, Liversidge instinctively had recourse to the specific, the empirical, the test tube and the balance.[113] Orderliness, energy and a Carlylean capacity for taking pains; these, in their way, constituted genius in the colonies.

With a suddenness that surprised his family and friends, on 1 June 1872, Liversidge settled his College bills. On 21 July, he bade farewell to his brother's family in Hackney. The following day, in the company of Smith and his new wife, Minnie, Liversidge boarded the 2200-ton steamship *Northumberland* in the Thames and saw off Plymouth Sound three days later. The blacksmith's son from Aberdeen and the carriage-maker's son from London could have had much in common. They certainly had time to get to know one another. Sailing via Brindisi, Suez and Point de Galle, they made Melbourne in fifty-two days and twenty hours, said to be the quickest passage on record.[114] Transferring to the *Bangalore*, five days later, on 23 September, they arrived safely in Sydney. The Smiths bade Liversidge both farewell and welcome. He was just three weeks short of his twenty-sixth birthday.

# 3. Science in the City

The forest of masts that greeted newcomers entering Port Jackson in the southern spring of 1872 fully confirmed the fine advertisements of London shipping agents. Certainly, the Circular Quay onto which the passengers of the *Bangalore* disembarked had changed greatly from the days, safely beyond memory, when men and women struggled against the inhospitable prospects of Australia's 'Fatal Shores'.

Indeed, the Sydney of the 1870s was far different from the city that John Smith had met for the first time twenty years earlier. Since the discovery of gold, the population of NSW, even discounting the separation of Victoria, had doubled; her revenues had doubled twice; and her trade, increased sevenfold.[115] Wide streets led from the quay's woolstores and trading depots, and wooden pavements partitioned stately banks, clubs and churches. Carriages plied a handsom trade across a bustling Pyrmont Bridge, while heavy wagons lumbered out Parramatta Road, destined for the Blue Mountains and the rich pastoral lands beyond. Familiar names hid unfamiliar meanings. The generous expanse of Hyde Park, planned by Macquarie to beautify the city, offered pale eucalypts in lieu of the oaks and elms of its London namesake. Instead of the yellows and whites of Regent's Park, were purple bougainvilleas and the occasional haunting, evanescent mauve of the flowering jacaranda, imported from the Cape.

After a long sea voyage, the newcomer met a city in motion. Reaching beyond the crowded quay and teeming Rocks, stretched a population of 138,000, a city-state two-thirds the size of Melbourne, boasting few of great wealth and many in misery, but most in employment, and all in such conditions of odour and insanitation, animal and human, as would have been familiar to any resident of Liverpool or Glasgow, Belfast or London. Reaching beyond the confines of first settlement, the city had long since 'pushed to the bush', making suburbs of Newtown, Darlinghurst, Burwood, Paddington and Surry Hills, where another 63,000 people lived – not all surrounded by gardens of night-scented jasmine, but many in small sandstock or brick houses, with iron railings that came cheaply as ballast, built beneath shadeless gumtrees on sandy scrubland where, except in the heaviest rains, little grass grew. Life expectancy, except for children prone to summer diarrhoea, was improving. This was, after all, the 'Workingman's Paradise'.[116]

This much would have been familiar to the English reader of Australian advertisements and no shocks greeted the newcoming professor. The University's splendid building was a welcome relief. No picture in the *Illustrated London News* could quite convey its beauty, or its horse-cab ride distance from everything else. There was space in the Blacket building to teach, and for the Principal, to live.[117] Alexander Thomson and his family rented a terrace in nearby Newtown, in sight of one of the colony's two railway lines, which took trade after a fashion towards the frontier at Albury, and brought country men and their families to the city, its shows and its University.

With Smith's help, Liversidge found rented rooms, first in Dynever Terrace on College Street, near the Sydney (or, more correctly, the Australian) Museum,[118] and then in one of the many boarding houses that jostled next to shops on Bridge Street. Neither was far from the Botanic Gardens or the Domain, and both were a pleasant walk from the University. Early in 1873, Liversidge joined the Union Club, which then occupied

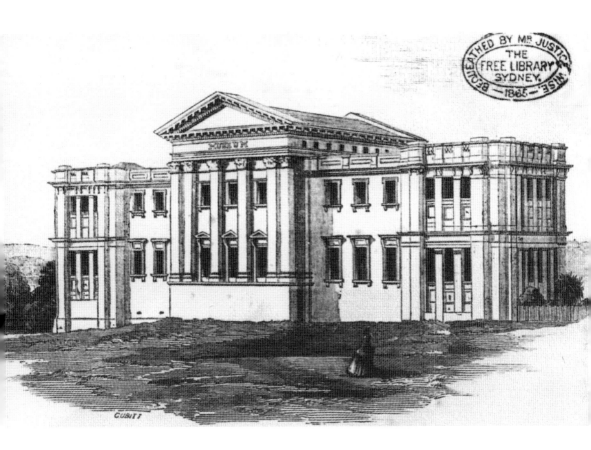

The Australian Museum, about 1870.

the 'Campbell residence' in Bligh Street. Into these comfortable lodgings, he moved by the middle of the year. There he was to live, not far from the Smith's family in Macquarie Street, until 1889.

The Union Club was not the preserve of the 'bunyip aristocracy', nor of the 'Establishment' as defined in England by the Church, the law and medicine. That privilege went to the neighbouring Australian Club, the first club in Sydney and the second in Australia (after Melbourne), founded in 1838 and meeting (until 1891) in the Pulteney Hotel in Bent Street. Among its members were those who formed the inner circle of 'scientific Sydney', including William J. (later Sir William) Macleay, the 'resident Sir Joseph Banks of Australian science'; Dr James Cox, successful physician, keen shell collector, and Crown Trustee of the Australian Museum; and Alexander Scott, an 'entomologist and entrepreneur' as he is remembered, and another Trustee of the Museum.[119] The Club's president, since 1857, was the University's Chancellor, Sir Edward Deas Thomson.

The slightly less imposing Union Club, established in 1857, by – among others – dissident members of the Australian Club, prided itself on its desire to 'accommodate the ideals of both Tories and Whigs'.[120] Some members of the Australian, including William Manning, Charles Nicholson, Thomas Holt, and George and William Macleay, belonged to both. But the circle Liversidge joined included a more progressive, and possibly a more entertaining group, including Alfred (later Sir Alfred) Stephen, then Chief Justice; the flamboyant Robert Hoddle White, bank manager, station owner and politician; and William (later Sir William) Windeyer. He and his friends introduced Liversidge to the issues of the day – protectionism *versus* free trade, colonial nationalism *versus* federalism – and shared views on the scandals that daily rocked Sydney society.

As W.J. Gardner has wisely observed, a new professor in the colonies required a 'stable social base; he could hardly fly in the face of established attitudes and values of the circle from which his students were drawn'.[122] The colonial census of 1851 listed slightly over 2700, or about five per cent of the population, as 'professionals', a definition of doubtful precision; many of whom, in any case, were 'visitors' to the colony. By 1871, the percentage had not much changed. Arguably, 'high culture' had to appeal to a class numbering about 6000, of which over four-fifths were craftsmen and mechanics, and a further fifteen per cent were counted as squatters, bankers and merchants.[123] Russel Ward suggested that 'for the most part members of the colonial upper class came from the middle or lower middle class in England'.[124] To this extent, Liversidge was in his element, although to meet his students' fathers, the commercial hotels and public houses might have been a better place to start. At the Union Club, alongside pastoralists and squatters from outback stations sat self-made 'merchant princes' of metropolitan business and trade – a bachelor existence in every way different from home in Hackney or college life in Cambridge.

By 1872, the Union Club and Sydney society generally were familiar with the usages of men of science – the 'creators and servants of civilisation', as the *Sydney Morning Herald* put it.[125] Science, as both agency and ideology, was said to be the instrument by which culture was introduced into a society 'from which', as George Nadel put it, 'the convict taint had yet to be erased'.[126] The founders of the University may have debated the point. But if such claims had relevance, they embodied a sense of obligation, reinforced by deference. Since the First Fleet, cultural leadership reflected British priorities, and although costs once borne by imperial revenues were now charged to a colonial treasury, the 'agenda' was still defined in London. Just as the University looked to 'London committees' for its professors, so government appointments were determined by the metropolis – in botany, by Sir William and his son, J.D. (later Sir Joseph) Hooker of Kew; in geology, Murchison and Ramsay at the RSM; in astronomy, Sir George Airy of Greenwich; and in natural history, Professor Richard Owen, of the Hunterian Museum of the Royal College of Surgeons and later of the British Museum of Natural History.

By the early 1870s, this hegemony had begun to change. But for most of the colonial period, Englishmen held the patronage of Australian (and other colonial) appointments as theirs by right. When, for example, in 1855 Earl Grey, at the Colonial Office, appointed the young Charles Moore to go out as Superintendent of Sydney's Botanic Garden, on the recommendation of Sir William Hooker's rival, Dr John Lindley, Hooker could not conceal his annoyance. 'I can scarcely congratulate you', Hooker wrote Moore, 'in as much as the appointment has been in my hands for the last three years, and you appear to have stepped in and taken it away from me'.[127]

ON CLUB SYDNEY. MARCH 1871

The Union Club in Bligh Street, 1871.

Liversidge was prepared to follow an imperial agenda, but also to set his own priorities. The history of science in Australia had been a matter of fits and starts, of bold designs and mixed outcomes.[128] As Liversidge found, success owed everything to luck, enterprise and a good press.[129] The *Sydney Times,* which once helped form the Debating and Literary Society, printed lectures on scientific themes, and until its demise in 1842, the *Sydney Gazette* heralded science as a liberating force. During the 1850s, its place was taken by the *Empire,* famously edited by Henry Parkes, whose writers missed few opportunities to emphasise the value of practical knowledge. The *Sydney Mail* published a regular section on science, while the *Australian Town and Country Journal* featured columns called 'Science Notes', 'The Cultivator' and 'Science-Invention'. The fortnightly *Illustrated Sydney News* devoted pages to what it called 'Science Gossip' and 'Scientific Jottings'. Even the *Sydney Morning Herald,* the colony's newspaper of 'record' – its pages otherwise devoted to British royalty, European wars, trade figures and parish-pump politics – had for a time a Wednesday column on 'Science and the Arts' and rarely missed reporting major scientific events.

Fuelled with such enthusiasm, the Australian public welcomed, in Michael Cannon's fine phrase, 'science coming as a friend'.[130] Steam engines, railways, photography, gas-lighting, hot-air ballooning, telegraphy and the novelties of electricity helped shape expectations. Not only in the city, but also in goldfield and agricultural towns beyond the mountains, through scientific and literary societies and field clubs and the libraries of mechanics' institutes, science was advertised as a route to both material and moral improvement. The University was not behind in making a similar point. Scientific practice, John Woolley told a School of Arts meeting at Maitland in 1857, furnished 'a most powerful, though indirect, motive to good conduct, by providing ready employment and an honest field of ambition to the poorest and most laborious'.[131] More important, the 'great Truths of Science which are the Laws of the Creator', as the *Empire* put it, formed 'the solid basis of true religion'.[132] Such sentiments resonated with the Rev. W.B. Clarke, who rode through the country reading Homer and the gospels alongside the Book of Nature.[133]

For anyone new to science, or to the colony, the Rev. Clarke set an unimpeachable standard of diligence, honesty and self-discipline. By the 1870s, he was only one of a distinguished company of gentlemen collectors, migrating from Britain, who flourished in Australia. For some, their passion was natural history, while others collected what Asa Briggs has called the 'wonders of common things'.[134] In the wake of London's Great Exhibition of 1851 came a series of colonial, intercolonial and international exhibitions which, in appealing to a public seeking entertainment, instruction and profit, captured and diffused the spirit of scientific enterprise. Collection, classification, arrangement and display were the methods by which nature could be better understood and made profitable.

The role of the 'exhibition movement' in presenting a positive view of these methods cannot be overstated. 'Only', Graeme Davison has observed, 'if we appreciate the capacity of objects to stir the curiosity and imagination', can we understand their appeal.[135] In 1855, with a fanfare of trumpets and a military parade passing before Governor Denison – standing, significantly, outside the Australian Museum – Sydney farewelled a shipment of animal and mineral products from NSW to the International Exhibition in Paris. Behind this demonstration lay belief in an 'enlightened combination of Practice with Science'. Only £132 worth of Australian products went to the Great Exhibition in 1851; but in 1855, NSW alone sent to Paris produce worth £6,00 – assisted by a government grant of £6000.[136] This marked the beginning of a rush that was open-ended.

In Melbourne and Sydney, 'Exhibition mania' gave a particular focus to colonial patriotism.[137] Melbourne held exhibitions in 1854 and 1861, followed by its first Intercolonial Exhibition in 1866–67 – events taken as evidence of civic maturity and (or so it was unworthily suggested) superiority to the mother colony. Not to be outdone, Sydney matched these upstart ambitions by holding its own 'Metropolitan Intercolonial Exhibition' in 1869, and by planning a spectacular show for 1870, to coincide with the centennial of Cook's discovery of eastern Australia. Then, it was the Agricultural Society of NSW that took the initiative, and in Prince Alfred Park, not far from the University, arose a temple of wood and iron, the largest structure of its kind in the colonies. This 'crystal palace' was to NSW, the *Herald* proclaimed, 'what the Great Exhibition of 1851 was to England', and 'the best evidence that many are on the right road to solid fortune'.[138] Half the population of the colony came to see the exhibits of livestock, farm produce, horticulture and agricultural machinery. At least some went on to admire the astronomical and surveying instruments and pharmaceutical products, alongside the elegantly decorated gallery of fine art.[139]

Such events gave easy support to the argument that science had an important place in popular esteem. However, as Liversidge discovered, between the rhetoric and reality fell a long shadow. Behind the fine façades of the Australian Museum, the Sydney Mechanics' School of Arts and the Free Public Library dwelt visible proof of stop-and-start building and boom-and-bust management, caused by irregular government funding and unpredictable donors. Institutions were run by boards of unpaid trustees, often representing vested interests, and maintained by overworked staff. Since 1861, the Australian Museum – founded in 1827 – at last could pay a curator, but he could not keep up with its mounting collections. The Sydney Observatory, the first of its kind in Australia, opened by Governor Denison as early as 1858, managed magnetic and tidal observations and kept daily rainfall records, but struggled for want of staff and equipment, and depended on amateurs for routine observations.[140]

Elsewhere, the situation was much the same. The colony's Botanic Gardens, begun with such promise, had fallen into decay. 'There was not a single plant labelled in that Garden', wrote the young Charles Moore, when he sailed from London in 1855 to bring system into its administration; 'there was not a single effort at any arrangement'; indeed, 'it was no credit to the Colony'.[141] By the time Liversidge arrived, Moore had done much to rescue the gardens, but suffered from a lack of economic botanists. In time, Liversidge saw these as the beginnings of a new discipline of botanical chemistry, but this was not on the horizon in 1872.

First would come the question of mines and minerals. In the wake of the gold rushes, Sydney became a focus of chemical metallurgy. The Sydney Mint, founded in 1853, was, like its later Melbourne counterpart, a branch of the Royal Mint in London. Its mission was to assay and process bullion for shipment to England, and to coin and issue sovereigns and half-sovereigns for imperial use.[142] These procedures required careful skills, and the Mint, located between Hyde Park Barracks and the Sydney Hospital, was at first sensibly run by Royal Engineers, who were fortunate in their choice of chemical staff. There the young W. Stanley Jevons, fresh from University College London, worked as an assayer from 1854 to 1859, filling his leisure with walks through the city, and excursions into journalism, photography and social statistics, before returning to England and international fame as a political economist.[143] As with Liversidge, Jevons' icon was his chemical balance and he left behind a model and a metaphor that inspired utility theory.[144]

Jevons came to Sydney with no intention of staying. However, he was a close friend of Henry Roscoe, the Owens College chemist who later featured prominently in Liversidge's life, and left his traces. Among Jevons' colleagues at the Mint were Francis B. Miller, formerly of King's College London, and Robert Hunt, FGS, who had been sent from the RSM to be a 'Practical Chemist and Clerk'.[145] Hunt went on to run the melting and refining department and in 1877, after a short deviation to Melbourne, returned as the Mint's Deputy Master (Director).[146] Hunt was also to play a key role in Liversidge's future – from the Board of Trustees of the Australian Museum, to the management of the Technological Museum.

The Sydney branch of the Royal Mint in 1870.

In 1859, when Jevons returned to England, Hunt was joined by Adolph (later known as Carl) Leibius, a German-born, Heidelberg-educated chemist, who had studied at the Royal College of Chemistry under Augustus von Hofmann. Appointed as Senior Assayer by Sir John F.W. Herschel, then Master of the London Mint, Leibius was a man of wide culture, who welcomed newcomers with invitations to golf and lawn bowling.[147] In 1860, the Mint wanted to build a machine for crushing auriferous quartz, so he rode for six weeks with John Smith to the gold diggings in the Snowy River, to see what techniques miners were using. For the NSW display at the 1862 International Exhibition, he assayed £5000 worth of gold specimens. In 1868, he and Hunt collaborated with Francis Miller in developing methods of purifying Australian gold for coinage.[148] Over the years, he introduced several new instruments to the colony's scientific repertoire. With Smith's encouragement, he joined the Philosophical Society of NSW, where he hoped to find men of a similar persuasion.

The Philosophical Society was lit by an intermittent flame, which passed unsteadily from hand to hand. In 1859, the society counted 186 members, and had £300 in the bank.[149] But despite the Governor's endorsement, the support of Nicholson, Deas

Archibald Liversidge

Thomson, Douglass, and the cooperation of Smith, Woolley and Pell, the *Lazzaroni* of Sydney languished. To monthly meetings, ten times a year, as John Smith put it, came few who had new things to report. Worse, those who ventured papers had difficulty in publishing them. From 1857, the Society's proceedings were noticed in the *Sydney Morning Herald* and the *Empire,* and briefly in James Waugh's *Sydney Magazine of Science and Art,* but these served only a local, ephemeral audience. Between 1862 and 1865, the Council managed to produce only one volume of scientific papers, and it was not until 1866 that, thanks to careful economies, it was able to launch its *Transactions.* Worse, the Society lacked a visible presence. Its members met, like a social club, in borrowed rooms, sometimes at the Australian Library, but without a permanent place, or a library, to call its own.[150]

Viewed demographically, the scarcity of what John Smith called 'learned leisure' is easy to understand.[151] In the two decades before Liversidge arrived, Sydney's population had risen sharply – from 53,900 in 1851 to 95,700 in 1861 – and then nearly doubled again by 1871.[152] In 1877, the population of NSW exceeded 550,000. But what might be called the cultural leadership drew upon a thin veneer of educated men, from medicine, law and the Church; resident or retired officers of the army and navy; a handful of amateur naturalists; and wealthy gentlemen who either held high positions, or who aspired to culture, or both.[153] It would take more than a dedicated Governor and a 'booster' press to persuade members of the materialistic majority, its eyes fixed firmly on 'getting on', that the fellowship of science held something of value.

As much was admitted, by Society members who knew that the subjects they discussed 'are of that abstruse and abstract character that few have had time or opportunity to study; and that there are no general or useful results to be derived from it'. In 1865, the Society pondered whether the its 'languishing condition' might be traced to its title, which assigns to it 'an exclusiveness by which many are deterred'.[154] Following the precedent of Victoria in 1859, the Society successfully petitioned the Governor to sanction the 'Royal' prefix,[155] and from 1866, the re-christened Royal Society of NSW welcomed contributions on 'Science, Art, Literature and Philosophy, and especially on such subjects as tended to develop the resources of Australia, and to illustrate its Natural History and Productions'.[156] However, royal titles brought no fresh blood, and the year Liversidge arrived, its numbers had fallen below 100. The Rev. W.B. Clarke biblically lamented the fate of a city whose people were 'generally given to the frivolities of ephemeral excitement, or whose mental occupation is only exercised by sensational novels or railway literature'.[157]

What some saw as bleak despair, Liversidge saw as fresh opportunity. There was an emerging consensus about what needed to be done. And the city's circuit of science, tracing a line on foot from the Observatory, to the Library, to the Museum and the University, made Sydney a bowl of possibilities, especially if one had the ability to be in three places at once. To make the most of the day, required special talent. Fortunately, Liversidge, the founder-secretary of the Cambridge University Natural Science Club, had come fully prepared. 'If, in a few words', T.H.S. Escott observed in 1887, 'the contrast between the England of the later and the earlier part of the Queen's reign were to be summed up, it might be expressed by the single word, organisation'.[158] The principle of 'organisation' was to be his *vis anima.* Give Liversidge a lever and he would move the world.

The Sydney Mechanics' School of Arts, about 1870.

First, he would need help. On arrival, Liversidge was greeted warmly by W.B. Clarke, fellow Cambridge scholar, then seventy-four and in failing health, but still capable of delivering a stirring address. With Clarke as his guide, he was guaranteed access to Sydney society. And when he met Henry Russell – then aged thirty-six, a student of Smith, a graduate of the University, and the first Australian-born Government Astronomer – he knew he was off to a good start.[159] The colonial boy was to inspire the 'transplanted Briton'. Within a month, Russell had proposed Liversidge to the Royal Society, and by December 1872, Russell had him reading his first Australian paper: a chemical analysis of the remarkable Deniliquin meteorite, a fragment of which Russell had obtained the previous year.[160]

As the Michaelmas term of 1872 drew to a close, the Royal Society of NSW and other bodies would open their arms, and welcome his energy. But as he walked from the Union Club to Glebe in the sunshine of that first spring, his knew his first responsibilities lay with the University. There, the problems were not to be solved by additional lectures and demonstrations. There, his colleague-professors were chained to a curriculum that neither stimulated the keen nor encouraged the dullards, and locked in an embrace that inauspiciously suggested a conspiracy against science.

## 4. 'No inaccessible shrine, no sullen fortress'

On 20 March 1872, William Charles Wentworth, first among the University's founding fathers, died in England. His body was returned to Sydney and buried in Vaucluse. When Liversidge arrived that September, he found the University – Wentworth's noblest creation – approaching crisis. As the *Herald* put it, 'We can scarcely escape from the conviction that, so far, the University has lamentably failed to realise the general benefits that were expected from it'.[161] To a public unable to see beyond a façade of Greek and gowns, the University was only four miles from the city, but a world away from life.

When Trollope visited Sydney in 1873, he wrote approvingly of a University, which knew nothing of 'gaudy days, high tables, and Latin graces'. To a world-weary chronicler of the English upper classes, it was fresh and appealing – 'not its social charms, perhaps, but its sheer bravado'. It had not yet, he wrote, 'had time for success'. But certainly 'there is no institution in the colonies which excites and deserves the sympathy of an English traveller more completely than does the Sydney University'.[162]

In fairness, the University deserved better than sympathy, but during its first twenty years, it had added few laurels to its crown. Whatever it had done to raise the tone of colonial conversation, its curriculum was far from Huxley's fêted 'ideal'.[163] Sydney in 1850, followed by Melbourne in 1853, were both founded during the 'deep twilight', as Gardner puts it, of unreformed Oxbridge.[164] But since Woolley's death, while both Oxford and Cambridge had begun major reforms, Sydney had stood still. It was instructive that a visiting Edinburgh naturalist commended Sydney to the New Zealanders of Otago, then contemplating a university of their own, as 'less of a model for imitation, than as a beacon of warning, against dangers liable to be incurred'. No medical school, few students, 'a want of confidence in its stability' – and above all, shocking 'in a country which owes so much of its prosperity to its gold, its coal, its sandstone' – no professor of Geology or Mineralogy'.[165]

We have already seen reasons for the University's poor showing. For years, the magnificence of its building contrasted with the realities of its situation. Above all, student numbers remained vanishingly small. The *Electoral Act* of 1858 provided that as soon as there were 100 graduates, the University would return its own member of parliament; but this did not happen until 1876.[166] Only seventy students enrolled in 1871, and only sixteen new students matriculated that year.

In exasperation, the University pointed to the absence of secondary schools and pupils who could meet its matriculation requirements. By the 1870s, little had changed since 1854, when Woolley had written, 'The anxious father finds, indeed, a university but in vain he looks for a high school'.[167] Although by 1872 there were 524 private schools in the colony, few prepared pupils for university matriculation.[168] Fewer still prepared boys for the Junior and Senior Examinations in chemistry and geology, which had been introduced, although without laboratory practice, in 1867. The Governor, Sir Hercules Robinson, attributed the shortfall to the diversions of colonial life: wealthy colonists who could afford higher education for their sons simply did not value whatever benefits might be derived from a liberal education.[169]

A conservative curriculum was confounded by the system of public endowment. In 1872, the University's income from government was only £6,200, as contrasted with the colonial elementary education vote of £157,500.[170] The cost to a student was lower than the cost of attending Oxford or Cambridge. On the other hand, the government outlay of £125 per undergraduate was high by Irish, Scottish and provincial English standards. Such arithmetic could not escape criticism. And whilst the revenues of NSW amounted to more than £2.2 million in 1871, there was also a public debt of £10 million,[171] and few looked upon the University's monumental edifice without also thinking of its cost.

Meanwhile, the three professors had other reasons to be unhappy. Since the compromises of 1861, relations with Senate had normalised, but 'lay control' remained complete. Since 1862, the gentlemanly reformer Edward Deas Thomson, first as Vice (Deputy) Chancellor and later as Chancellor, enjoyed little success in improving the situation. Moreover, there were irritating signs of disunity among the staff, which surfaced before Liversidge arrived, and which were to become part of his life for the next twenty years.

In 1867, the University appointed the distinguished classicist Charles Badham to replace Woolley as Principal and Professor of Classics and Logic. Unlike Pell and Smith, who came from working-class backgrounds, 'Badham of Wadham' was the son of a Scottish medical professor and a literary mother, educated in Switzerland and at Eton before going up to Oxford in 1830. Although he took a mediocre third in classics, a source of lifelong embarassment, he made up for it by taking holy orders, a Cambridge DD (1852), and an honorary doctorate from Leyden, before devoting himself to classical studies in Italy. After returning to England, he served for fifteen years as a classics master in Lincolnshire and Birmingham, building up a reputation for classical textual criticism, notably, with editions of Plato and Euripides. Preferment was denied him, seemingly because of what were regarded as heterodox views on religion, so he turned to Sydney, secure in the gift of 'magnificent' testimonials from Cardinal Newman, T.H. Huxley (with whom he served as an examiner in London), and George Grote, the philosopher.[172] His departure from England was accompanied by anguished laments for a prophet who was not without honour save in his own country; and by a snide slander from the *Athenaeum*, who mourned the loss of 'our foremost scholar ... for the purpose of presiding over a Colonial University'.[173]

Arriving in Sydney in 1867, Badham was – to Trollope, who met him – the 'prince of professors and greatest of Grecians'. To his students, a 'reincarnation of Demosthenes'.[174] To his reputation as the best classical critic in England, he added a powerful personality and a formidable voice, a quick temper and a readiness to take offence. He genuinely believed in the University as a 'monument of extraordinary wisdom and public spirit' and did much to cultivate the wider community.[175] He also easily dominated the intellectual life of the University. He could be an impressive ally. But he became an ideological enemy, both for Smith and Liversidge. He held strongly the view, based on his years of teaching, that 'the spirit of industry and … adventure cannot create civilisation; neither can they avert its inevitable decay if the conditions of its existence are not carefully maintained'.[176] The debate between him and his scientists was not simply a familiar rendering of the 'two cultures', of science vs. the humanities. Rather, their views as individuals reflected complex levels of difference in British education – as between Oxbridge and London, certainly, but

FINAL INSTRUCTIONS: THE RACE FOR THE UNIVERSITY STAKES.

*Trainer Badham :* CUT HIM DOWN FROM THE START, MY BOY; YOU KNOW YOU CAN'T STAY. REMEMBER YOU CARRY OFF MONEY.

'Final Instructions',
*Sydney Punch,* 8 September 1876.

also as between Oxford and Cambridge; between Scotland and England; between Eton and evening schools; and between middle-class scholars and working-class professionals. The next twenty years would see these elements constantly interwoven in Senate debates, curricular changes and future appointments. Their outcome reflects the uniqueness of colonial higher education in its particular Australian form.

Charles Badham was passionately committed to elevating the University in public esteem. The institution was to play a key role in colonial identity. If he did his duty, 'fathers of families, instead of looking upon this colony as a mere place of temporary resort during the accumulation of their wealth, will be more disposed to look upon it as the home of their final adoption'. At the same time, however, he feared that the institution 'which alone could give colonial civil society its dignity and permanence', was being eroded from within, even when not being assailed from without. His task, as he saw it, was to 'educate the taste and therefore the candour, forbearance, gentleness and compassion' of its students, who in turn would 'arrest the encroachments of a barbarism that is visibly coming among us'.[177]

The integrity of education depended first and foremost on an appreciation of the classics. Science had its place in the curriculum, but it was ancillary to language, literature and logic. 'Robinson Crusoe and a few Catherine-wheels will answer all the purposes of a chemical lecture', he told students at Commemoration in 1868.[178] For Badham, the University was no place, as one student put it, for 'Professors of Artificial Manures'.[179]

To caricature Badham's position, or to liken his battles with Smith and Liversidge to those of their contemporaries Matthew Arnold and T.H. Huxley, would do injustice to both. But their conflict reveals similar contours. To set a high standard to which the wise and just could repair was consistent with the wishes of the University's founders, and for Badham the principal objective; but to do this at the cost of modern subjects, was to keep the University out of reach of the majority – if only that majority could be inspired to study classics. Rejecting charges that the University was 'aristocratic' and resisted reform, Badham's retort was: 'What can we concede without injuring education?'[180]

For Liversidge, who had probably had less sympathy with Greek the more he swotted for his Cambridge Prelims, the 'idea of the University' was a concept in flux. Huxley had proclaimed a new charter for higher education and the provincial cities of England were joining the chorus. Surely Sydney should follow suit. Smith's new 'find' was, in F.M. Cornford's immortal phrase, a 'young man in a hurry'. But he was also, at twenty-six, by far the least experienced of the staff. Badham was twice his age, and treated Liversidge with courtesy, but not warmth, and perhaps warily as well. Like many early colonial academics, Liversidge had been appointed on the basis of promise rather than performance. He required grooming. He also needed stamina. Although Badham – the eldest at fifty-nine in 1872 – flourished in Sydney's climate, his younger colleagues fared less well. Pell, only forty-five, who had come to Australia for his health, was away on sick leave in 1872, laid low by illness that would oblige him to retire in 1876.[181] He pleaded with Senate for a reduction in his lecturing duties.[182] From at least 1871, Smith, then aged fifty, who had ostensibly come to Sydney for his health, complained of 'increased action of the heart'. Without protest from students, he ultimately folded his courses in half, and gave chemistry and physics in alternate years.

The professors taught, and sometimes lived, in close proximity. Woolley, and Badham after him, lived with their families in the section of the Blacket building between the north vestibule and the Tower, while Pell's family of five lived near the centre of the city, and Smith and his wife, in Macquarie Street. Their lecture rooms were spacious, but not well adapted to their purpose. To science, as to classics and mathematics, the University had, in the *Herald's* phrase, merely 'resigned one of its halls'.[183] Smith occupied the southern third of the building, in rooms never designed for science. One (now known as the History Room) was used for lectures, and a second (now Latin I) for experiments; a third room (now Latin II) was used for a minerals collection – much space, but unsuitably arranged. Although students at first had no required laboratory practicals, Thomson gave demonstrations in chemistry, and on days when southerlies blew, fumes of sulphur mingled with the furies of Aeschylus. Cambridge was Elysium by comparison.

Such was the place that greeted Liversidge at the beginning of Michaelmas Term 1872. Three matters soon dominated his working life: the improvement of science teaching, the purchase of apparatus and the construction of a proper laboratory. The

The Chemistry Laboratory in the Main Building.

first could be achieved quickly; the second would come with time; but the third would take almost an eternity. The texts Liversidge found on Smith's syllabus were familiar from Cambridge: Chambers' *Practical Chemistry*, Bloxam's *Laboratory Teaching*, *Practical Chemistry* by Harcourt and Madan, and Fresenius' *Qualitative Analysis.* Some new works had appeared within the last few years and these he added to the list. Next, he began to overhaul the laboratory notes, which he sold to his students for a fee of two guineas.[184] On 4 December, Senate, extending a 'honeymoon' to its new Reader, gave him £50 to order scientific instruments from London, and shortly afterwards, voted him another £40 to purchase Thomson's personal books and apparatus for his use.[185] Over the summer of 1872–73, Liversidge developed an ambitious design for a new laboratory and lecture theatre, to be created by an extension of the southern end of the Blackett building, at a cost of £6000 – a modest sum, in comparison with the cost of the original building, even if equivalent to the University's entire annual grant.[186]

But this was just the beginning. Since the gold rushes of the 1850s, 'careless and ignorant miners' had, by want of good management, failed to find valuable lodes and ruined the prospects of others. To improve mining practice, the Legislative Assembly had debated the establishment of a School of Mines, but nothing had been decided. In 1869, the Sydney Mechanics' School of Arts began informal classes in geology and mineralogy, but these were discontinued for want of students, a fact that encouraged those who opposed them. In 1872, however, mining education again hit the headlines.

Anthony Trollope recalled that on his first visit to Sydney in December 1871, no one talked of mining; but that in July 1872, no one talked of anything else. Gold production had been declining since the early 1860s, but with reports of copper discoveries in the Bathurst district and around Orange, and tin in New England, miners and speculators were clamouring for information.[187]

In June 1872, Edward Dowling, a member of the School of Arts committee that sponsored the classes, recalled seeing mistakes and avoidable accidents among gold miners and mine owners, and wrote to the *Herald,* stressing the 'recent impetus given to the mining enterprise' and the desirability of teaching mining engineering. Whether the University would accept the challenge was the question. Dowling was not above taking a pot shot: 'Would not', he asked in June 1872, 'a knowledge of the elements of mineralogy be of as much use to our Colonial youth as a smattering of a classical language?'[188]

Given his experience of mining education, Liversidge found it impossible to remain silent. The problem seemed clear and the remedy, obvious. On 6 December 1872, not yet two months in the colony, and a fortnight before giving a maiden address to the Royal Society of NSW, Liversidge sent the first of three long letters to the *Herald*. This first outlined the curriculum of the RSM, making a case for having something similar in New South Wales.[189] The second, published during the summer in February 1873, went beyond London, to describe mining schools at Bristol and Truro, and the larger Dublin School of Mines (soon to become the Royal School of Science for Ireland). Private interests had established a School of Mines at Columbia University in New York in 1864; but state sponsorship of mining education, and the sciences upon which it was based, had a long pedigree in Europe. He described the famous Bergakademie at Freiburg, the Royal Mining Academy at Berlin, the École des Mines in Paris, and the remarkable Sheffield Science School at Yale, with its endowment of $100,000 and staff of twenty-three.

'No thoughtful person who has the interests of the colony at heart', Liversidge observed, 'will for a moment deny the need of some such institution in a country abounding with unlimited mineral wealth'. Such a school was no mere self-indulgence. 'Heaven only knows', he declaimed, 'where the self-styled assayers and mining engineers' available in NSW 'obtained their so-called science':

> the ignorance of many is profound, and it might even be termed, exhaustive, so absolutely ignorant are they of that knowledge for which they arrogate to themselves the position of authorities.[190]

Malfeasance and corruption had undermined mining revenues. False advertisements misled luckless investors and undermined overseas confidence. To Liversidge, the 'reckless speculation of unprincipled adventurers, who neither know nor care whether their alluring and specious prospectuses are founded in fact or not',[191] was avoidable.

In his third and final letter, on 27 February 1873, Liversidge drew a longer bow, and outlined his solution to the larger problem: the creation of a 'science school' at Sydney University. The term 'school' was in common usage in America. Harvard and Yale had for decades had 'science schools', and the land-grant colleges of the mid-West had recently followed suit. Princeton was beginning a School of Science in 1873.[192] 'It must be almost universally admitted', Liversidge concluded, 'that a Science School of some sort is absolutely necessary for the well-being and proper progress of the colony'.[193]

To bring this about, the University needed a set of laboratories – 'under one roof and under one direction' – for instruction, for research and for analysis; courses for training in theoretical, practical and analytical chemistry, and provision for research on vegetable products and minerals, which could best be studied on the spot, and not sent back to laboratories at 'Home'.[194] 'What is really known respecting our iron ores?' Liversidge demanded, 'their nature, quality and extent?' And what provision was made for the study of flora and fauna, 'not only from a purely scientific point of view, but also on economic grounds'? A mining school would pay immediate dividends. In Victoria, Ballarat had already established a School of Mines in 1871, followed by Bendigo in 1872. Should not NSW follow suit?

And what of the other sciences, from which no immediate return could be expected? Here, justification by faith was required: because the sciences formed part of 'universal education' and found a place in all progressive schemes for higher education. Taking his cue from University College London, Liversidge's plan for a science school, on a 'by no means extended scale', included the appointment of professors in 'mechanical and biological science'. Necessarily, the science school should be established 'not only in close association with the University' (like the Sheffield School at Yale, or the Lawrence School at Harvard), but should 'form part and parcel of it'. However, he personally preferred not a direct copy of the RSM of the 1860s, but a place 'rather of the character of the Royal College of Science for Ireland, and of the continental institutions'.

As to the curriculum? A new university in a new country, Liversidge said, 'has not the excuse which an old university has: it is not, or should not be trammelled in any way by obsolete customs'.[195] Sydney's Arts degree, he added, 'has not yet any great prestige, although it is devoutly to be hoped that it one day may have'.[196] In his proposed School, all Arts students would take the same lectures and practical classes. In addition, there would be a new faculty – a Faculty of Science – with degrees to match. This would add depth to breadth, and make the University the central focus of science education in the colony.

To those who shared Badham's views, these were dangerous ideas. What would the new science degree comprise? First, it would be comprehensive. In physics, Liversidge proposed that it should embrace the 'non-Newtonian' subjects – the subjects of heat, light, sound, magnetism and electricity – that Cambridge had introduced in the Natural Sciences Tripos twenty years earlier. That very year, James Clerk Maxwell had begun to lecture in these subjects at the just opened Cavendish Laboratory. Next, to Sydney's existing courses in chemistry, mineralogy and geology, Liversidge wanted to add the whole spectrum of the Tripos, including zoology and botany, mechanics, metallurgy and natural history. There should be a professor in each new subject, for if his 'whole time and energy is occupied in merely teaching, he cannot hope nor be expected to advance science by original research'. Rejecting the caricature of the gentlemanly amateur, Liversidge also rejected the idea that 'science is one and the same, or nearly so'. A professor, he said, should be a 'man who excels in one subject and should devote his heart and soul to that one and not fritter his time away by passing from one to another'. In just this way, the German universities were making rapid advances.

All this was, he wrote, for the public good: 'We want men to whom we can apply for scientific information and advice' – not only when anything extraordinary occurs, as France had found herself in need of Pasteur during the recent silkworm epidemic – but for 'the everyday operations' of the 'agriculturist, the miner, the engineer [and] the manufacturer'. In iron works and metallurgical manufacturing, at least:

> ... it is highly probable that we should have by this time successful iron smelting works in the colony, had the operations been checked by some competent chemist, who could have ascertained if any, and what, elements were present.[197]

It was transparent to Liversidge that an extension of the sciences would benefit the University's reputation and popularity. Cambridge and London had learned this. And even Melbourne, he observed, had been successful in attracting students, 'more from the greater completeness and comprehensiveness of its systems of lectures, than from any other reason'. But he would also maintain tradition. His program, with its emphasis upon discipline, observation and utility, would build upon a 'well-grounded foundation' in mathematics and classics.

For a young man at the beginning of his career, scarcely in the country two months, who had not yet actually lectured in the subjects he was engaged to teach, to voice such searching proposals in the press and before the public was, to say the least, ambitious. Liversidge's ideas were sound. But he was little aware of the difficulties they would raise. His 'school of science' would take a decade to realise. His mining school would take a further decade. Yet, what he proposed was not new. Universities in America were already moving in this direction; in France, new laboratories were being built, partly in reaction to the country's defeat at the hands of a Prussia ostensibly forged on the anvil of science. Liversidge could also unashamedly draw upon contemporary British experience, and point to the first three massive Reports of the Royal Commission on Scientific Instruction and the Advancement of Science – the Devonshire Commission – which he had read before leaving England. The Commission's first report, devoted to the establishments at Jermyn Street and South Kensington, had laid down a similar scheme.[198] Liversidge had merely the wish to do in Sydney what Britain could not do quickly itself.

In some ways, Liversidge's 'manifesto' of 1873 moved the discussion of scientific research and education onto a higher plane. His suggestions were relevant even to England. In London, the RSM was caught in the tendrils of its transfer to South Kensington, where a 'College of Science' was being created, to give courses in biology, chemistry and physics. The Museum of Practical Geology, the Geological Survey and the Mineralogical Collections of the Mining Record Office, were to remain in Jermyn Street. A considerable distance would thus separate 'theory' from 'practice'. Like Huxley, Liversidge wanted a school of science that contained the two. Eventually, he would succeed. Huxley got his Royal College of Science, but London was not to achieve a combined mining and science school until the establishment of Imperial College in 1907.

In hindsight, such foreknowledge might have dampened his enthusiasm. Perhaps his letters could only have been written by a young man, one as yet unfamiliar with the slow progress of scientific development in NSW. But there was no *a priori* reason why Sydney, in adopting British practice, had to duplicate British delays. A favourable wind caught his sails, and the Sydney press endorsed his ideas. The *Herald* applauded his plan for a school of science, as it announced a report of the Science and Art Department on the promotion of science in English schools. The University must be stirred, it said: 'A liberal hand is wanted – faith and courage are wanted'. And money, too. '[A]lthough they can analyse and describe air [professors of physics] cannot live on it'. To meet the Badhamites, in their almost certain opposition, the *Herald* added reassuringly: 'There is no desire to degrade the forces of the University to professional and trade purposes, but to make it subserve the purposes for which it was intended.'[199]

As Liversidge discovered, the support of Sydney's press was helpful, but not decisive. In March 1873, Senate first deferred, then rejected Liversidge's application for a laboratory fund of £6000. Instead, its minutes counselled, 'the professors should consider and report on how present accommodation could be altered to cope with the increased demand for laboratory accommodation'.[200] There was a good deal of space in the Blacket building; perhaps it could be adapted. By the time of Commemoration Day, 5 April 1873 – marking the beginning of the academic year – it was clear he had gone too far, too fast. The customary Address by the Chancellor, Deas Thomson, courteously welcomed Liversidge's arrival, and noted his qualifications and suitability.[201] But in the Principal's Address, which immediately followed, Badham was less polite.

'I cannot conceal from myself', he thundered, the dangerous tendency 'which I see manifesting itself in more than one quarter – I mean the tendency to turn the University straightaway to account; to adopt it as a means of helping ... mining operations, and to turn that which was intended by William C. Wentworth and all his compeers into a mere machinery for bringing out of the bowels of the earth the riches they contain, and turning them into marketable cash. That, I may say, is not the mission of the University'. What Badham feared most was not the idea of lectures on metallurgy, but the implication that the University 'instead of pestering us with Latin, Greek, Mathematics and a heap of things we cannot understand, has at last condescended to make itself useful in a way we can appreciate'. The cardinal sin was 'utility'. 'I bear no grudge against any science', he told a bemused audience; 'I admire the chemist and the metallurgist as well as the mathematician ... for I believe that all knowledge is intended for the improvement of man'. But scientific knowledge was one thing; the mere 'conveyance of so many facts' was another.[202]

Translated, this meant that Badham would be likely to oppose a school of mines; and on Senate, he resisted any proposal for new subjects or in the hours given to science. A report on the need for new apparatus was duly considered, but only £250 could be spared, and only £100 from current funds.[203] Liversidge consoled himself with an 'immediate expenditure' of £15 on specimen trays and turned back to the text. This was no phoney war, but the beginning of a long campaign. Building on his *Herald* articles, and noting recent developments at Oxford and Cambridge, and the new colleges of Manchester, Leeds and Birmingham, Liversidge drafted a memorandum to Senate in which, instead of pressing immediately for a school (or faculty) of science, he proposed to increase the range of science taught in Arts, and to review the requirements for matriculation. Thanks, predictably, to Badham, his proposals met a lukewarm response. Smith was supportive, to a degree, but Pell was invisible, and Badham stood opposed, on the grounds that any greater number of subjects under the heading of physics and chemistry would inevitably bias the timetable against classics.[204] It was an argument of self-interest which, as Henry Parkes' reforming *Empire* rumbled, 'publicly put both Greek and practical science to shame'. [205]

There for some months the case rested. But – possibly with the encouragement of Deas Thomson – Liversidge refused to let the matter drop. In effect, a compromise was reached, turning partly on the twin themes of access and specialisation. Badham wanted to increase access to students and to introduce advanced language courses. Earlier, he had accepted that abolishing compulsory Greek at matriculation would tend 'not to diminish Greek scholarship, but to increase it'.[206] So, in 1874, with his encouragement, Senate agreed to alter the by-laws to exempt, on application, matriculants from the requirement of Greek, whilst retaining Latin.[207] At the same time, Senate, however, agreed to give all forty-eight students then enrolled a wider choice of subjects. Those who had done chemistry and physics during their first two years would be exempted from further science, if they preferred to do classics or mathematics. But those who displayed proficiency in any one of the three subjects, including chemistry and physics, could devote themselves in their third year to that one subject alone. By this stratagem, Liversidge foresaw the possibility of making more time for science and recruiting students, without resorting to 'conscription'.

The compromise was workable and Senate was pleased. The following summer passed uneventfully, and on 10 February 1874, on a motion by Deas Thomson seconded by Badham – Liversidge was promoted to a chair, and given the title of Professor of Geology and Mineralogy.[208] Five days earlier, Liversidge was elected in Thomson's place as a Trustee of the Australian Museum.[209] He had followed in Thomson's footsteps, although moving two years faster. The following month, in his Commemoration Address, Deas Thomson praised the 'zeal and efficiency with which he has performed his duties', and noted Senate approval for spending £500 to alter two rooms in the main building for a lecture room and laboratories, and 'in providing necessary means of classifying and exhibiting valuable collection of mineral specimens held by the university'. As a gesture to Badham (who followed him on the dias), Deas Thomson added:

> Although the primary and most important objects of a University should be to afford instruction in classics and the higher branches of learning, yet I am not insensible to the advantage of affording to any of the students who may desire to avail themselves of it, an opportunity of cultivating a knowledge of geology mineralogy and the other natural sciences, especially in a new country where there is a wide and unexplored field for discovery and research.[210]

Faint praise, perhaps but welcome enough. It was a red letter day when his promotion was celebrated by an invitation to Sunday dinner at 'Barham', Deas Thomson's home in Woolloomooloo. Liversidge sat next to the Badhams and their two daughters,[211] who joined the company in toasting William J. Macleay, Deas Thomson's son-in-law, who had marked that day's ceremonies by announcing the gift of his collections and a legacy of £6000 to endow a museum for the study of natural history.[212]

In March, Liversidge had good reason to be satisfied. Less than twenty months in the colony, he was already a professor, on the Council of the 'Royal', and a Trustee of the Museum. He was becoming known as a reformer, but one who spoke softly against Greek. As a Mining Bill made its way through parliament, with a clause recommending the establishment of a school of mines, he appeared even to have won over the Colonial Secretary. In some sense, the youngest professor had 'arrived'.

But his victory was short-lived. Just two days after the Chancellor's luncheon, Senate, on a motion by Badham, decided to ease the pressure on classics, by relieving students from compulsory examinations in third year geology and mineralogy.[213] Liversidge' optimism suddenly evaporated. His subject would remain available, indeed compulsory, at an elementary level, but the new rules would not encourage many to go on to advanced work. The *Empire* was outraged – and recommended the overthrow of classics as a 'uniform system of culture' as a means of promoting the 'great end' for which a University exists.[214] But this did not mean that a 'culture of science' would be its immediate successor. And precisely what 'great end' the University should serve remained debatable. In any revision of the curriculum, Badham's voice was crucial. Against his vote, a school, a faculty, and a degree in science – indeed, the whole Liversidgean program – could not proceed.

In his early brushes with Badham, Liversidge learned that reform – reduced to votes in Senate – would require a larger and wider base of influence. Colonial science was a puzzle of many pieces, which had to be assembled patiently. A decade was not too long to wait for a laboratory; fifteen years were not too great for a school of mines. In any case, as British experience showed, he had many companions in adversity. Most British universities found 'useful' and modern subjects difficult to accommodate. Traditionalists were not wrong to view science as a curricular rival, however well put the arguments of mental and moral training. On the contrary, it was well understood that the sciences challenged the order of things, threatening the hegemony of classical learning. New disciplines also made demands upon time, space and resources, and science, properly taught, meant not only books and blackboards, but also apparatus and specimens, laboratories and museums, and all the expense of maintaining them. Britain's experience was too recent to show Australia how such problems could be solved, or how their solution would respond to pressures from industry and the professions. Clearly, science would make demands that, in the long run, could become irresistible. In the meantime, Sydney would have to find solutions for itself.

# CHAPTER 5

# The Perils of Public Science

**A GLASS OF (SYDNEY) WATER.**

*Sydney Punch's* view of the 1874 water crisis.

During his first six years in Sydney, Liversidge initially followed a path charted by others. From Thomson, Smith and Clarke, he inherited established practices, some worth continuing, others whose time had passed. He deployed precisely those qualities of energy and ambition that might be commonplace in commerce, but in academic life constitute genius. His teaching commanded his first attention, but this left him ample leisure. That spacious freedom, which could so easily have reduced him to indolence or despair, he turned to advantage.

The learned society of the colony yearned for leadership. With this went an expectation that Liversidge, like his teachers at the RSM, would use science in the public interest. Sydney, no less than Manchester and Liverpool, required 'public scientists' in ways that called for unbiased expertise, a role for which London had admirably prepared him. As Smith and Pells had found, such work was more demanding than flattering. But service to the community and to the government (which, no professor needed reminding, paid most of his salary) was an essential part of the colonial contract. Like Thomson, he had the task of 'being useful', helping those who were opening up the country, with the added task of making it better understood. Like Smith, he was the colony's sole source of expertise on a wide range of subjects, about which he knew little, and indeed about which little was known.

It was in his threefold aspect – as teacher, mineralogist and expert – that Liversidge first came to terms with life in Sydney. Between 1872 and 1878, stepping through minefields of controversy, he established a reputation for industry, objectivity and reliability. The overall effect was to create and sustain an image of science as a professionally neutral activity, serving the interests of fact, at its best when left untrammelled by politics. The culture of science was open to all, serving all equally, within the broad lines of settler capitalism. Liversidge's task was made easier by the absence of rivals. Thanks to his position, he was accepted into society by right of occupation, not conquest. Thanks to Clarke and Thomson, he arrived to find a mantle lying at his feet; he had only to wear it with aplomb. By the end of the 1870s, he was ready for reform. But this not before he had become thoroughly acquainted with the perils of public science.

## 1. Water Supply and Sugar Cane

For the first century of European settlement, a poor and unreliable supply of drinking water rendered life in Sydney a seasonal hazard. In the heat of February 1874, Professor Liversidge, lunching at the Union Club, was dismayed to find his table water distinctly brown. He had already complained, both to the club, and to the press.[1] The Club Secretary wrote to Edward Bell, the City Engineer: 'The mere sight of this sample should satisfy you that the complaints of the quality of water supplied us are not groundless';[2] and followed his complaint with a letter to the *Sydney Morning Herald*.[3] Liversidge took a sample to his laboratory as well. The saga of Sydney's water supply was to enter a new phase.

As early as 1820, the Tank Stream, which had first supplied the colony in Sydney Cove, was badly polluted the year round. A tunnel, called Busby's Bore, was built between 1827 and 1837 to bring water to the city from the Lachlan Swamps (now part of Centennial Park), but this proved inadequate as, over the next thirty-five years, the city rapidly grew – from 54,000 people in 1851, to 95,000 in 1861, and to 138,000 by 1871. This put immense pressure on metropolitan housing and created problems of sewerage and supply familiar to the industrial cities of Britain.[4]

In the 1850s, Sydney's areas of greatest population density were bounded by George Street and Darling Harbour, but by the 1870s, new developments extended the 'sanitary dilemma' south to Broadway, west to The Glebe and Newtown, and south-east to Surry Hills. Dust and disease dominated the city, and Sydney's mortality statistics mocked the colony's reputation as a 'workingman's paradise'.[5] Even as the suburban population doubled each decade between 1861 and 1891, the city was left with large pockets of badly ventilated, congested terraces, huddled along narrow, undrained lanes, with little or no running water or sewerage. Poor housing, pollution from manufacturing wastes, and bacterial diseases, particularly summer diarrhoea, and infant mortality from enteric disease, all made the 'water question' an issue of concern, but one which the colony approached slowly, lethargically and late.[6]

In 1852, the year that Professor John Smith arrived in Sydney, the City Council chose to draw upon water from the area known as the Botany Swamps. However, the difficulties of expense and engineering were not overcome until 1859, and even then, the continuity and purity of the water supplied were not above suspicion. All these proceedings, familiar to contemporaries overseas, were further exacerbated in Sydney by the colonial tradition of weak local government. Until the City Council was created in 1843, the colonial government had abstained from taking responsibility for water supply; and even then, it would not give the City Council powers to raise funds for sewerage improvements. When it did, the council was reluctant to pursue its responsibilities, especially where these conflicted with the interests of property; and involved expense.[7]

Mortality reports from medical practitioners, made worse by mounting fears of cholera and typhoid, came to a head in 1853, when the Colonial Secretary replaced the City Council with three City Commissioners. Four years later, the City Council was reappointed; but in the meantime, the Commissioners had sought a report from Smith on the city's water and on the sources of its contamination. Smith advocated the closure of abattoirs near Botany, and advised against the use of lead pipes and cisterns. He also conducted a few experiments and reported his fears on lead poisoning to the Philosophical Society in 1856.[8] Stirred by his example, the Society and its successor, the Royal Society, began a tradition of 'water debates' that lasted over twenty years.

The problem of pollution was universal. With the despoliation of rivers and waterways, chemists and sanitary authorities had shown an increasing interest in the pathology of organic decomposition, and in the etiology and specificity of disease.[9] In Sydney, there were few to speak knowledgeably. Before the germ theory of disease became widely accepted, and before modern science distinguished between biological and chemical purity, there was much room for debate.[10] For example, in 1862, speaking to a Select Committee of the Legislative Assembly, Smith asserted the common belief that running Botany water over sand would cleanse it of organic impurities. Representatives of more recent sanitary opinion, persuaded of the connection between bacteria and waterborne diseases, were less sanguine, and Smith became embroiled in controversy.[11] But because he was a medical doctor, the University's Professor of Chemistry, and a leading public figure, he continued as the colony's principal source of technical advice.

It was not until 1869 that the government appointed a medical officer with responsibility for public health. In the meantime, and during 1867 and 1869, it was Smith who was asked to chair a Royal Commission set up to investigate the Botany-Lachlan system.[12] In 1870, with Smith in the chair, the Royal Society of NSW heard papers on the subject at no fewer than eight meetings.[13] It was literally the question of the hour; and it was no coincidence that, at the same moment, Edward Frankland (with the help of his pupils, such as Liversidge) was conducting his analyses of London's water for a Royal Commission in England.

In line with enlightened English opinion, the report of the colonial Royal Commission recommended the abandonment of the Botany scheme, and foreshadowed a solution to the question of supply by recommending what would later become known as the Upper Nepean Scheme.[14] This conceded Sydney's expanding needs, but factional politics delayed a solution to the pollution problem. Not until 1874 did Henry Parkes introduce a Bill to create a metropolitan authority. His Bill failed at the end of the session, but the following year, following revelations of deaths attributable to waterborne diseases, compounded by the threat of a smallpox epidemic, an angry public forced the government to act.[15] Parkes' Bill of 1874 was passed by the Robertson government in April 1875, and in February 1876 the Sydney City and Suburban Sewerage and Health Board came into being. The Act of 1875 coincided almost exactly with the passage of the great Public Health Act in Britain, which consolidated a decade of legislation, and created a framework for public health in local government. As the municipalities of Britain proceeded with their reforms, so Sydney struggled not to be left behind.[16]

Under the chairmanship of Professor Morris Pell, also enlisted as a 'public servant', the Board set about making extensive enquiries.[17] These led to a series of twelve reports on problems of inadequate housing, poor drainage, industrial pollution, night soil disposal and sewerage engineering. In the meantime, Smith's reports on the purity of the Lachlan-Botany water supply had failed to persuade the government, and other reports reached the Board of polluted water coming from Busby's Bore and the Crown Street Reservoir. It was surely time to submit the 'water question' to the colony's leading chemist.

Accordingly, in May 1875, over a year after he had first tasted unsavory drinking water in the Union Club, Liversidge was invited by Pell to analyse samples of Sydney water drawn from a number of locations, to determine the extent of their contamination. He acted quickly, applying techniques he had learned in London. By July, he had sent in his first report, and a bill for £130.[18] His samples showed gross faecal contamination; he recommended the immediate installation of cisterns, and a running supply of water to avoid sewerage backflow to the mains. The report was political dynamite and the press made a field day of it.[19]

Writing to Pell, Liversidge agreed with Smith that 'certain of the aerated waters sold in Sydney contain dangerous quantities of lead', but he differed from Smith about the overall safety of the supply.[20] Liversidge was better read in contemporary chemical literature.[21] There were, and remain, differences of opinion concerning acceptable definitions of bacterial impurity. But Smith's opinion that testable pollution arose principally from decaying vegetation was undermined by Liversidge's analysis.

Smith and Liversidge clashed before the Committee: the elder man, recognising the inadequacies of the existing system with which he had long contended, seeking a publicly acceptable, long-term solution; the younger, alarmed at the immediate danger, seeking instant redress. Their differences caught the public eye and the press made much of them.[22] Liversidge, however, did not; and good relations between the two men, momentarily strained, were not deeply damaged. Pell, adjudicating, looked for a solution in a fresh source from the Upper Nepean scheme, and for a system of ocean outfall sewers, supplied by copious drains. This he achieved in part through the Water Pollution Prevention Bill, introduced into parliament following Liversidge's report. The Bill passed through all its stages in one day, and became law (together with the *Nuisances Prevention Act*) in August 1875.[23]

Thanks to his analytical acumen, Liversidge became momentarily the 'Frankland' of Sydney; and enhanced by his work for the Department of Mines and the mining companies, his reputation spread around the colony. In August 1875, while still up to his neck in water questions, he briefly took leave from the University to respond to a call from Queensland. There, in the sugar cane fields, a desperate blight or 'rust', first recognised in Mauritius and Réunion in the late 1840s, and considered to be a fungus – had appeared north of Brisbane, and was devastating the Mackay and Mary River districts.[24]

Facing losses amounting to seventy per cent of the colony's crop, the Queensland Government established a Diseases in Animals and Plants Commission, and ordered Robert Muir, a planter, and Karl Staiger, Queensland's Government Chemist, to conduct a study of the afflicted areas. Muir and Staiger reported on fourteen of fifteen farms in the district, but one, the Jindah plantation, owned by Edward Ramsay and his brother, was passed over. Annoyed, the Ramsays called in Liversidge as a private consultant. His report, appearing the following May as six weekly installments in the *Moreton Bay Courier* and the *Queenslander*, was a model of investigative chemistry.[25] Muir and Staiger had failed to agree; Muir blamed the weather, and Staiger, the fungus. The particular fungus, in fact, was identified in 1877 by Dr Joseph Bancroft.[26] But Liversidge suspended judgment, and rejected the concept of a 'single cause'. Instead, he examined all the factors he could identify – soil types, seasonal incidence and the susceptibility of different plant varieties – and drew a bead on doubtful plantation methods. In the end, he blamed the growers who had failed to plant without careful attention to such details.[27]

Liversidge's observations were widely reported in the West Indies, where British planters were similarly afflicted and in New Caledonia, where French colonists shared Queensland's concerns.[28] The Acclimatisation Society of Mauritius, where sugar was a basic export crop, read Liversidge's paper carefully and promptly elected him an Honorary Member.[29] In retrospect, he was technically mistaken in asserting 'there does not exist any true specific disease of the sugar cane'.[30] Indeed, there are several diseases, but the identification of individual rusts was (and is) by no means easy. Extensive failures of the widely-grown Bourbon cane variety were commonplace in the tropical world.[31] Liversidge's tests pointed not to a single chemical cause, but to a multiplicity of causes. In 1880, his opinions were widely endorsed by authorities who conceded that 'inferior cultivation' and the use of 'unsuitable soil' had weakened the cane, lowering its resistance to the fungus.[32] His pioneering survey, the most exhaustive of its kind to date, helped open a new era in Australian agricultural chemistry.[33] In Queensland, where no preventative could be found, the variety was eventually abandoned.

Returning to Sydney in November, Liversidge was asked by Pell to conduct another analysis of the city's water, and especially of the 'peaty and other vegetable matter' found in the Lachlan Swamps and the Paddington and Crown Street reservoirs. His results showed a significant presence of 'albumenoid ammonia', a measure of nitrogen products and an indication of contamination that, if not precisely dangerous, was certainly distasteful. The clinical language of his report could hardly conceal his disgust at finding water 'of a pale brown tint', full of 'moving organisms', few of which were removable by filtration. Precise measurements were unobtainable, but Liversidge believed that a substantial proportion of nitrogenous matter over 0.1 ppm must have its origin in 'other than natural sources'.[34]

The Board accepted Liversidge's findings – although not, at first, his bill. Liversidge apologised to the mathematical Pell that he was 'entirely ignorant' of what his fee should be, but had put a figure which 'is not greater than could be rendered to a properly qualified and competent chemist in England'. 'I am at all times', he added, 'willing and ready to give my services gratuitously where necessary, but I cannot allow anyone to appraise them at what would be less than a fair and adequate sum'.[35] He left the matter to Pell; and in due course received a cheque for a hundred guineas.

This was not to be the end of it. In 1876, the government brought in a British expert to give a second opinion. He confirmed Liversidge's results, and supported the recommendations of the Board. But the Board took no action. In early October 1876, Liversidge took water samples from Port Jackson after sailors from the visiting HMS *Sapphire* fell ill with typhoid.[36] The experience persuaded him that sporadic sampling could not replace systematic monitoring.[37] But this lay beyond the powers of the Board; and not until 1879 did the Parkes government fulfill its plans of 1874 to bring water politics under the jurisdiction of a new Metropolitan Water Board, given responsibility for both supply and sewerage.[38] In the meantime, Liversidge and Smith had done what they could. The problems of water supply and sewerage could not be solved by chemistry. Sydney simply did not have sufficient water of any quality, let alone pure water; and another seventy years would pass before the construction of the Warragamba Dam completed the present network of reservoirs.[39] As for sewerage, the problem remained one of engineering, expense and priorities.

Despite their contretemps in 1875–76, Liversidge's relations with Smith remained civil, even cordial and by 1880, their differences seemed forgotten. Then, both were asked (along with Charles Watt, Smith's assistant) by Henry Parkes to advise on the credibility of claims advanced by a commercial salesman for a 'deodorizing agent', whose alleged purpose was to fix the 'volatile salts' in night soil, and preserve the dehydrated substance for agricultural fertilizer. Their report dismissed the claims. But Liversidge minuted on his copy, 'Prof. Smith and Dr. Watt took no part in the analyses'.[40] Liversidge had demonstrated that he, not Smith, wore the chemical crown in Sydney; and for all the remaining years of Smith's life, Liversidge would find ways of reminding him. Smith, for his part, was not averse to quitting a scene that had proved increasingly unrewarding, both financially and personally.

Liversidge spent the last months of 1876 completing his 'overhauling', as Le Fèvre put it, of the University's chemical course and extending Smith's reading list.[41] Students were soon required to read Thorpe's *Quantitative Analysis,* Fownes' more modern *Manual of Chemistry*, and Dana's *Manual of Mineralogy.* As Smith withdrew from the laboratory, it was the Professor of Mineralogy and Demonstrator in Chemistry, not the Professor of Chemistry and Experimental Physics, who set the pace in chemical studies. Within the year, Liversidge introduced experiments on common gases and acids, the use of reagents and exercises in qualitative chemistry. Within two years, he again updated his reading lists and extended his practical subjects. He spoke for the new ways of doing science, and for its public practice. It was a pity he had (in 1876) only fifty-eight first-year students to teach.

## 2. Of Men and Mining

Liversidge's training well suited him for Empire. Since the 1850s, geology, mineralogy and applied chemistry had overtaken botany and natural history as the principal sciences of empire, not least because of the discovery of gold and the expanding needs of industrial society. Into the bush, Liversidge took a set of tools for collection, classification and analysis. Into the classroom, he took his London lecture notes, and those that Thomson had left behind. In his laboratory, was the 'balance', a symbol of his scientific method, and a simile of his approach to life.

Minerals are the building blocks of all rocks, which are defined in terms of the types and proportions of minerals they contain. Mineralogy and crystallography are the disciplines that describe their physical and chemical properties and structure. Neither in Europe nor in Britain was the period between 1860 and 1870 particularly productive in the theoretical development of mineralogy.[42] Mineralogical concepts taught to chemists had changed little since the days of Mitscherlich and Berzelius.[43] In 1875, Oxford's Professor of Mineralogy, Nevil Story-Maskelyne, bewailed the 'extreme paucity of really scientific mineralogists and crystallographers, men sufficiently competent to undertake the most elementary crystallographic calculations'.[44] Their task was made no easier by the 'muddle' – as Robert Hunt, Keeper of the Mining Records at the RSM put it – in which mineralogy was found, caught between conflicting geological opinions. On both sides, there was an air of impatience. 'I find mineralogical papers to be unintelligble to geologists, who can talk of nothing but ice sheets, pluvial periods and other myths', wrote Samuel Haughton, Professor of Geology at Trinity College, Dublin, in 1876.[45]

Rescue from this impasse, Hunt believed, could come only 'by the individual effort of a single mind – who [sic] can free itself from the trammels of the schools'.[46] However, making mineralogy into a discipline of inorganic chemistry faced severe obstacles. Minerals display great diversity and complexity in combination. When confronted by natural impurities, analysts find it difficult to obtain accurate, reproducible results. In the nineteenth century, these limitations were real. However, if Europe had no perfect methods, Australia at least offered the prospect of new discoveries.

This was possibly better understood on the Continent than in Britain. In the early days of settlement, when mineralogy was neglected by the British, it fell to Europeans – including the Pole, Jan Lhotsky, and the Swiss, J.R. Gygax, who worked briefly in the Australian Museum – to put Australian minerals on the 'map'. Lhotsky studied with Werner in the Bergakademie of Freiburg, and is said to have 'brought to Australia a thorough grounding in the Wernerian school of thought'.[47] His essay of 1833 on the 'Mineralogy of Australia' was the first of its kind published in NSW.[48] Another European, Johann Menge, who studied at Heidelberg, came to Adelaide as a geologist to the South Australian Company before the colony itself was born. One of the earliest members of Adelaide's German community, Menge did much to develop metal mining there.[49] But he was inspired by European authors. Vallance has observed that no contemporary British mineralogical work equalled that of Des Cloizeaux on quartz, J.F.C. Klein on *atacamite*, and Albrecht Schrauf on *azurite* and *brochantite*. Even those Australian minerals deposited in the British Museum relied upon the Austrian, Viktor von Lang, to bring them to 'the attention of the mineralogical world'.[50]

At their foundation, both Sydney and Melbourne universities had mineralogical museums, but gave little attention to mineralogical research. John Smith and Frederick McCoy brought geology into the curriculum (and it became in NSW, from 1867, an optional matriculation subject), but neither gave classes in mineralogy. The establishment of the Sydney Mint in the 1850s introduced the technology of mineral analysis,[51] but instruction in mineralogy came to the University only with Alexander Thomson's appointment in 1869. In Melbourne, McCoy built an impressive collection of minerals in the Natural History Museum; so from 1853, when the Government of Victoria established a Geological Survey, Melbourne became the centre of mineralogy in Australia. This was speeded by the appointment to the Survey of George Ulrich, a graduate of the Bergakademie in Clausthal, who arrived in Victoria in 1853. To his scientific training, Ulrich added practical experience of the goldfields; and his observations led to the description of minerals found alongside gold in the quartz reefs.

In 1862, Alfred Selwyn, Director of the Victorian Survey, enlisted his staff's first chemists, C.S. Wood and his successor, J.C. Newbery (a graduate of Harvard), both former students at the RSM.[52] In 1868, the same year in which the Survey was dissolved by a short-sighted administration,[53] they recorded the discovery of the first mineral species in Australia definitively new to science, an auric compound that Ulrich named *maldonite*.[54] With the demise of the Survey, Ulrich and Newbery were transferred in 1870 to Melbourne's new Industrial and Technological Museum, where they became curators at the Natural History Museum; and although Ulrich soon left for Otago, what they had done in Victoria commanded respect throughout Australia.

It has been said that Sydney's 'style' reflected more the chemistry of London, and less the crystallography and physical mineralogy of the mining schools of Europe.[55] However, this can be overstated. Neither Thomson nor Liversidge had been trained in chemical mineralogy, which was a European specialty, but rather in the classical optical and physical methods used in London and Cambridge. What they brought to the study of rocks and formations was a chemist's eye, and the desire to make the 'vestiges of creation' subjects of laboratory analysis. Although they can scarcely have been aware of it, they were on the leading edge of an international movement in mineralogical classification, begun in 1850 by James Dwight Dana's *Descriptive Mineralogy*, and consolidated by its fifth edition in 1868. This distilled a 'comprehensive view of the characters of minerals as species in the inorganic kingdom of nature', giving emphasis first to chemical, then to crystallographic and physical characteristics.[56] Thereafter, newly found, possibly newly discovered minerals, ordered by chemical composition, might bring an understanding of remote processes at work in the earth itself.

To meet this challenge, the mineral analyst needed energy, passion and patience. Liversidge's training was based on guided reading in the Book of Nature; but once in the field, he had to build quickly on the experience of others. Within weeks of arriving, he carefully assimilated the writings of Thomson and Clarke. Fortuitously, Thomson had compiled a magnificent *Guide to Mineral Explorers*, fifty-five pages of close description of seventy-five of the 600 minerals then known in Australia, with details of their crystalline forms and geological and chemical characteristics. Ostensibly for the 'intelligent amateur, the explorer, and many other persons, striving to develop the Mineral Resources of New South Wales', this was also a practical textbook on a subject virtually unknown to students in England.[57] With this, and Thomson's lecture notes, he could begin work immediately.

From W.B. Clarke, Liversidge received an even more generous legacy. Clarke had worked closely with Thomson and grieved for his loss. But he welcomed Thomson's successor with an open hand, and gave Liversidge all the paternal advice one could expect from the 'father of Australian geology'. Moreover, Clarke's Presidential Address to the Royal Society of NSW in 1869 set out a plan that occupied Liversidge for the next twenty years. On many theoretical questions, from meteorite showers to the 'plurality of worlds', Clarke was refreshingly open-minded, more so than Liversidge's teachers in London and Cambridge – approaching nature with Sedgwickian devotion, and frequently endorsing field reports with a sermon text or a reference to the Almighty.[58] Overall, Clarke's writings gave Liversidge an idea of the geological questions that were important to ask, and the reasons for asking them.

The localisation of gold, in the discovery of which he had played a notable part, was one of Clarke's central interests. Liversidge read up the story of gold with the zeal of a digger. Either in London, or soon after reaching Sydney, he consumed accounts of the 'gold rushes' in California and Australia, authored by a miscellany of writers, including Patrick Stirling, John Calvert, Edward Hargraves, Paul de Strzelecki and Simpson Davidson.[59] Clarke gave him a copy of his *Researches on the Southern Gold Fields*,[60] and introduced him to the works of the Victorian Geological Survey, including the *Notes on the Physical Geology and Mineralogy of Victoria,* prepared by Selwyn and Ulrich for the Intercolonial Exhibition in Melbourne in 1866. Selwyn suggested that gold was to be found not only in quartz reefs, where all miners looked; but also in combination with other minerals, notably iron pyrites, in localities less obvious to the untrained eye, but which a chemist could help find. To work out the history of ore deposition was to fathom the prospects of the colony. In 1869, Robert Brough Smyth, Secretary of Mines in Victoria and later Director of the Victorian Survey, was 'compelled to confess that nothing certain is known' about the mineral features of Australia.[61] In helping to clear away what Smyth called a 'fog of doubts and uncertainties',[62] Liversidge was to make his name.

Liversidge was brought up with conceptual models of geological theory that had changed little in fifty years. Broadly, there were two popular theories of ore formation, both deriving from late-eighteenth-century debates between 'Neptunists', led by Abraham Werner of Freiburg, and 'Plutonists', who took their cue from James Hutton in Edinburgh. Neptunists argued that all rocks, including mineral deposits, had been precipitated from solution in a universal primeval ocean. Mineral veins were formed by precipitation from water, which carried minerals in solution, seeping through the ocean floor. Plutonists arued that rocks and ores resulted from volcanic action and from the deposition of molten magmas.

**TWO SCHOOLS OF MINES.**

MELBOURNE            AND            BALLARAT.

The 'Two Schools of Mines': practical and theoretical,
*Melbourne Punch,* 2 June 1870.

Historically, these explanations were influenced by work on quite different local rock formations – the metamorphic rocks of Freiburg, and the largely igneous environment of the mid-Lothian district. As such, they proved difficult to apply universally, and many geologists took up modified positions between the two schools. In 1847, Elie de Beaumont observed that mineralised veins were often related to intrusive igneous rocks, but conceded that different types of ore deposits may have been formed by different processes. Certain ores could have been deposited as precipitates during sedimentation.[63] The possibility that both sedimentation and volcanism were involved in ore genesis moved the debate away from speculative geognosy towards evidence-based arguments, drawing upon actual localities.[64]

As the expanding world searched for minerals, so fresh generalisations emerged. One of the more interesting was the theory of 'lateral secretion', derived ultimately from Neptunism, and advanced by the American chemist and geologist, Thomas Sterry Hunt. To explain low concentrations of widely dispersed metals, Hunt suggested that as the ancient earth had cooled, and familiar elements and minerals had formed, these (after certain catastrophic changes), were taken up in solution and brought to the earth's surface, 'after the manner of modern springs', there to be deposited in crystalline layers. Vein deposits, he supposed, were produced by 'leaching' minerals from their surrounding rocks. It was a provocative thesis and held many clues to a theory of inorganic evolution, but outside a small group of chemists, including Liversidge's friends Norman (later Sir Norman) Lockyer and William (later Sir William) Crookes, it attracted few followers in England.[65]

Indeed, for most of the nineteenth century, central theory remained unsettled. Some came to accept versions of the 'lateral secretion' hypothesis, arguing that, as fractures harden, disseminated particles migrate towards regions of lower pressure, producing veins in which ores are concentrated. But others preferred the view that mineralisation is associated primarily with igneous and hydrothermal activity. This seemed demonstrably the case with nickel ores, for example, as well as with tin and copper. Possibly, consensus favoured a 'modified Plutonist' position – an igneous-hydrothermal theory, according to which ores are deposited by precipitation from circulating aqueous solutions of magmatic fluids that orginated in molten granite-like bodies lying deep within the earth. By this interpretation, gold might have separated as granitic melts cooled and crystalized. For any argument, however, there remained counter-arguments. Gold is, after all, where you find it. But there was reason to believe that the principles of mineralisation, once fully understood, would lead to the prediction of gold deposits.

To Australia, resplendent in its unique flora and fauna, European geologists were prepared to grant an analogous uniqueness to the mineral kingdom.[66] Evidence was confusing and sometimes contradictory. Most Victorian gold reefs bore marks of igneous origin. But Alfred Selwyn found fossils of the Silurian period in the quartz-rich strata where gold was found, which argued against a universal igneous explanation.[67] Clarke's *Southern Goldfields of 1860*,[68] a book otherwise free of theoretical speculation, referred to the 'problem of gold in granite', as if volcanism were inadequate to account for its presence. A decade later, he observed that 'there is nothing whatever to justify the belief that dry heat or direct igneous forces have, as some persons surmised, been the chief or solitary agent in the production of gold-bearing reefs'. Nevertheless, he admitted, the question remained open.[69]

On the formation of gold nuggets, there was even more room for debate. In 1860, Simpson Davidson suggested an explanation that involved the deposition of gold by igneous processes, but agreed that it was still the 'formation of metallic grains *in situ* that requires elucidation'.[70] Many accepted a 'commonsense erosion' theory, holding that nuggets were produced as a consequence of disintegrating strata. In 1864, however, Selwyn proposed an alternative 'chemical' theory, according to which nuggets, including particles of alluvial gold, were formed by the deposition of metallic gold, by a process analogous to electroplating, from thermal saline waters carried through the earth. From this theory, two questions arose, both potentially reducible to observation: first, was gold actually contained in sub-surface driftwaters; second, could gold in fact be deposited in a metallic state from such waters?

The first question seemed answered by the discovery of gold associated with pyrites in fossilised plants of recent geological date. An answer to the second seemed to follow from the accidental discovery by Richard Daintree, then Selwyn's assistant, that a speck of gold, lying in a solution of auric chloride, would increase in size when a small piece of cork was dropped into the solution.[71] Daintree concluded that the presence of organic matter caused gold deposition. In 1866, his colleague, Charles Wilkinson confirmed that gold could deposit from solution in pyrites, galena (lead ore), wolfram (tungsten ore) and other minerals when organic matter was present.[72]

Liversidge arrived to find these questions unresolved; nor was it clear whether organic matter deposited gold only from saline solutions, or also from salts, including silicates. Grappling with such imprecision and speculation was not to his taste. In 1866, Ulrich suggested the 'most conclusive test' would be to cut nuggets in half, and analyse them to see if they contained gold nuclei, of a lower standard derived from the reefs than the gold allegedly formed by deposition from what he termed 'meteoric', or thermal waters.[73]

'A vast and very promising region', J.H. Gladstone told the Chemistry Section of the British Association meeting at Brighton in 1872 (just before Liversidge sailed for Sydney), 'is the origin and mode of formation of different minerals ... but in order to investigate it properly, the geologist and the chemist must travel hand in hand'.[74] Over a decade later, the President of the same Section foresaw that 'the elucidation of the chemical processes whereby minerals have been formed' remained, with the physical properties of matter, 'one of the two most important directions for future fundamental research in chemistry'.[75] For Liversidge, this was already received wisdom.

The presence of gold in eastern Australia had been established by Strzelecki and Clarke as early as the 1830s, and from a comparison of geological formations in Russia and elsewhere, Murchison in 1844 had foreseen the likelihood of profitable veins, seven years before Hargraves struck it rich. But as alluvial deposits became depleted, the prediction of gold-bearing sites became increasingly important.[76] Beyond this, to determine how much gold is concentrated in the earth's crust, to specify the geological environments where minerals are concentrated, could also have vast economic implications. And to determine *how* ores are concentrated, the source of mineralising elements, the choice of site for deposition and the transport mechanisms – such discoveries could profoundly affect science.[77] As Brough Smyth put it: a thorough understanding of gold-quartz veins and their origins 'would throw more light on the structure of the crust of the globe than all the information afforded by the palaeontologist, important and valuable as it is'.[78]

In an era that celebrated payment by results, miners put a practical premium on scientific advice, and Liversidge's talents were focused not on speculations, but on specimens. His work was descriptive and might never lead to major discoveries. On the one hand, it was pleasing to know that thirteen new mineral species were found in Australia between 1869 and 1900; and that of the four in NSW, one was credited to him.[79] But the questions that confronted him in 1872 had much deeper implications. Two in particular troubled him: how to explain the growth of nuggets in placer deposits, and how to explain the presence of gold in organic substances. The first dominated his 'gold portfolio' in the 1880s; the second preoccupied him for the rest of his life.

In 1876, on returning from an expedition to the Mary River, Liversidge set out on his yellow brick road. In his first 'gold' paper, delivered to the Royal Society of NSW in September that year,[80] he announced he was 'performing an experiment' to ascertain whether the gold present in a certain specimen of mispickel, existed in the crystallised state, or was merely disseminated in amorphous particles. The question, he said, was of 'great interest and importance in the chemical geology of mineral veins and deposits'.[81] As Carl Leibius, the chemist at the Sydney Mint suggested, 'crystallisation at a low temperature would explain many things' about the origin of native gold.[82] As this was Liversidge's first excursion into the subject, he reserved his position; he was 'not at present prepared to put forth any very definite and final theory'. Instead, he simply reminded his audience that 'the greater portion of the gold found *in situ* in NSW occurs in quartz veins running through the older and metamorphic rocks. It is also said to occur under similar circumstances in true igneous rocks'.[83] Not for the last time, he carefully placed an each-way bet.

From this time onwards, minerals testing became his special *métier*. In March 1875, Robertson's new Minister of Mines, John Lucas, a reformer who shared many modern views on sanitation and education, asked him to undertake certain tasks for his new department.[84] First, he was appointed to a 'minerals committee' of three, with W.B. Clarke and Charles Wilkinson, to prepare a volume on the *Mines and Mineral Statistics of New South Wales* for the Intercolonial Exhibition in Sydney to be held that year, and for the American centennial exhibition that was to be held in Philadelphia in 1876. Liversidge's earlier observations on the diamond fields, and on the iron ore and coal deposits at Wallerawang, were included in the department's report.[85] Second, he was asked to analyse some 120 ore specimens, including coal from the northern districts, iron from Manly and Eskbank, limestone from Wallerawang, and copper from Wiseman's Creek, near Bathurst, to determine their composition and likely economic value.

Styling himself 'Professor of Mineralogy', Liversidge prepared two careful reports. In the first, he demonstrated that coal from NSW had a lower sulphur content than British coal, which suggested a profitable future for the Australian industry.[86] He later added certain 'calorific' measurements on fourteen different specimens of coal, from which he devised a system of ranking for potential fuel use.[87] His second report, completed in February 1876, surveyed over 100 specimens of quartz-tailings, and concluded that vast quantities of precious metals, of a value to the colony 'almost incalculably great', were being lost by inefficient miners. The necessity for 'more thorough and scientific methods, similar to those employed elsewhere', as he put it, 'is in certain cases most glaringly apparent'.[88] He also examined tin tailings, and predicted that, as prices rose, the use of scientific methods could reclaim profitable quantities.[89]

His small laboratory at the University soon became a hive of industry. From specimens sent him by the Rev. George Brown, he compiled perhaps the first chemical accounts of the minerals of New Britain and New Ireland.[90] Intensive efforts went into an encyclopaedic summary of what was known of NSW minerals. Building upon his presentation to the Royal Society in December 1874, he prepared an expanded version of his paper, which the department printed as a book in 1876, and which became for a generation the standard reference on the subject. A second edition appeared again under the imprint of the Department of Mines in 1882, and a third, at Liversidge's expense, in London during the centennial year of 1888[91] (see Chapter 10).

Given his public work, the case for a school of mines was never far from his thoughts, and Lucas shared his belief that minerals were important to Australia's economy. But Liversidge was obliged to recognise the truism that, beneath a general rhetorical approval of science, lay a deep-seated indifference to its public support. In Sydney that September, T.S. Mort celebrated this paradox, in the ceremonies surrounding the opening of – not, it may be said, a new school – a new slaughterhouse, to prepare for the overseas market meat products preserved by the miracle of 'scientific refrigeration':

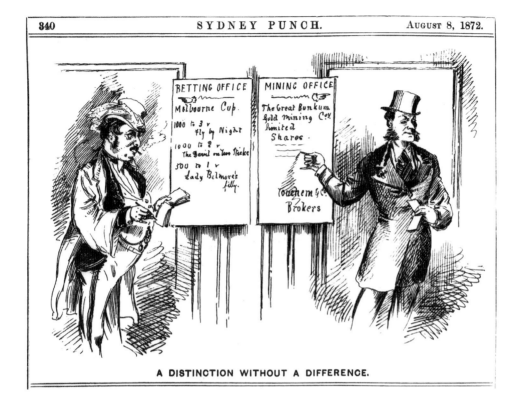

A DISTINCTION WITHOUT A DIFFERENCE.

'A Distinction without a Difference': The School of Mines debate in the 1870s,
*Sydney Punch*, 8 August 1872.

> Science has drawn aside the veil and the plan stands revealed. Faraday's magic wand
> gave the keynote and invention had done the rest. Climate, seasons, plenty, scarcity,
> distance will all shake hands and out of the comingling will come enough for all:
> for the 'earth is the Lord's and the fullness thereof', and it is certainly within the
> compass of man to ensure that all the people shall be partakers of that fullness.[92]

Repeatedly, generous appeals to science masked praise for practical invention. In many
respects, science had yet to reveal its full potential.

## 3. Science Pure, Applied and Political

As Liversidge looked from the gateway of the University's Great Tower to the city below,
there was no question where his responsibilities lay. As Badham insisted, he was a public
servant as well as a professor. Like Thomson, he had the task of 'being useful', helping
those who were opening up the country, with the added task of making the country
better understood. Apart from Smith, he was the colony's sole source of expertise on a
wide range of subjects about which he knew little, and indeed, about which little was
known. Like his predecessors, he was bound to play a key role in creating a colonial

scientific culture. In his own special way, however, he would create new pathways between academic life and the practice of 'civic science'.[93]

By the close of his first year in Sydney, Liversidge's weekdays, apart from lectures in the morning and practicals in the afternoon, were filled agreeably at the club and in the city. Weekends were spent in travelling out of town, to visit the Blue Mountains and the caves, and to view the landforms that Clarke and Thomson had described. Analysing and commenting on mineral discoveries, meeting travellers and prospectors, sending 'Home' what he called 'out of the way' specimens of Australian life and minerals – all this was part of the colonial calling.[94] There were also reputations to be won. Within a year, he had his first brush with fame.

In 1845, a shower of stony meteorites, the first ever observed by Europeans in Australia, occurred near the small town of Deniliquin. In 1871, Henry Russell, the Government Astronomer, obtained a fragment for his Observatory, and gave it to Liversidge for analysis. In December 1872, this specimen became the basis of his first contribution to the Royal Society of NSW. Liversidge had little to go on but the classificatory system suggested by Nevil Story-Maskelyne (Professor of Mineralogy at Oxford). Not until G.T. Prior of the Natural History Museum in London introduced the first systematic description of meteorites ('Prior's Rules') in 1920 were mineralogical and chemical criteria widely applied. Nevertheless, summarising the three known classes of meteorite substances – metallic iron, non-metallic and mixed iron-rich stone – Liversidge applied his chemical tests. He drew few conclusions, other than that water was absent when the meteorite was formed. But he was aware of the significance of the fragment.[95] Meteorites are thought to be fragments of planetary bodies, resembling 'genesis rocks', formed at the beginning of the solar system. Liversidge had no means of doing more than a detailed description.[96] But by 1902, he would write five more papers on the minerals found in meteorites and meteoric dusts, and would begin to speculate that meteorites may once have conveyed organic life from the heavens to earth.

If riddles of the universe remained for the future, there were terrestrial problems that demanded his attention. In June 1873, he took leave from the University, and rode twelve days through mud and rain to visit the first important tin lode in the colony, worked at the Elsmore Tin Mine (near Inverell, in New England).[97] Clarke had discovered tin along the Murrumbidgee in 1849, and others had since made claims, but commercial mining had begun only in 1871. Liversidge was asked by the company to examine the deposit and advise upon the techniques being used. He did so, and after a thorough analysis, recommended that, in the interests of 'judicious management', the owners should send 'some trustworthy person to Cornwall to acquire the necessary information, and to bring back details of the latest improvements introduced there'. Significantly, Liversidge signed himself not as 'Reader in Geology', but as 'Associate of the RSM'.[98] He was fully prepared to wear practical, as well as academic dress.

While in New England, Liversidge stopped off at the Gwydir Diamond Company. The lucky Hargraves had found the first diamond in Australia near Guyong, NSW, in 1851, and other deposits were found at Two Mile Flat near Mudgee in 1867. Their exploitation had proceeded slowly, but Thomson and Clark had recommended closer attention to the district, and in 1873, gold miners made fresh finds at Hill End and at Bingera, some five miles south west of the township. Liversidge prepared analyses of the specimens he was sent; but his report on the difficulty of extracting the diamonds threw cold water on the prospect of quick riches, and stifled the rumours that had produced a short-lived speculative 'bubble' in Thomson's time.[99]

Liversidge's early reports were balanced and, in their way, decisive; he began to acquire a reputation for fairness and objectivity. The work was time-consuming, taking him out of his way on questions that were far too plentiful. However, Liversidge could not avoid the problem of supply and demand, not least because the public expected him to find a solution. In the early 1870s, as alluvial gold production fell, mining interests looked for other minerals. 'If iron can be found in her hills', Trollope had prophesied in 1873, Australia 'will probably become as populous and as rich as the United States'.[100] Enlightened miners railed against the government for failing to provide metallurgical works to process ores as they were discovered, and for failing to find means of rescuing ores lost for want of economic methods of extraction. In the absence of a school of mines, blame was laid at the door of the University. 'What are our learned professors about, that they can stand by and see the country robbed of its due?' asked a critic in the *Australian Town and Country Journal*. 'Where is Professor Liversidge? Surely Dr Percy of the School of Mines, London, taught him something about the extraction of the precious metals?' 'I fear', the same critic added, 'his learning partakes more of the laboratory than the workshop'. With practice and theory combined, 'we have economic and perfect results; with either alone, the converse would be the case'.[101]

Such criticism of the University was unfair, and of Liversidge, doubly so, given his program of 1873. But it struck a chord with the political establishment, and in July 1874, after many false starts, and well behind Victoria, the ministry of Henry Parkes established a Department of Mines and a Geological Survey. The minister was a political appointment, but the Directorship of the Survey was given to Charles S. Wilkinson, an Associate of the RSM a few years older than Liversidge, who had been Daintree's assistant on the Victorian Survey's expedition to the Bellarine District between 1861 and 1864. The Victorians had their own staff of chemists; but until the NSW Survey could hire its own, Liversidge offered his services. The Department accepted. For a year or so, this was to increase his burden by an order of magnitude, even if it gave him work he enjoyed.

Fortunately, for most of 1874, Liversidge was not required to stir outside Sydney for his minerals, as specimens were brought or sent to him from mining sites throughout NSW and southern Queensland. His reports on newly discovered coal, iron, limestone and copper ore deposits soon found their way to England, where chemists were quick to see their importance. In July, the Chemical Society in London discussed his analyses of minerals, some of which were to prove economic, that had been sent to him from New Caledonia.[102] In October, he was given leave to visit the west coast of Tasmania, where he took specimens of tin from Mt Bischoff and Mt Ramsay (named by Charles Gould after Liversidge's teacher of geology at the RSM). The lode had been predicted by Clarke in 1849, and discovered in 1853, but had only recently attracted commercial interest. Liversidge reported on the likely extent of the ore bodies, and brought specimens back to Sydney. In what became a common pattern, he sent his results first to the Royal Society of NSW, then to the *Mining Journal* in London.[103]

'Inside the Mining Department – where a good blast or two is badly wanted to "wake them up",'
*Sydney Punch*, 11 May 1871.

Archibald Liversidge

As his collection of minerals grew, so grew the need for a museum to house and display it. Thanks to the work of Alexander Thomson, Liversidge found in the Blacket Building a small assortment of Australian and foreign minerals. But this was not enough, and to augment the teaching collection, he left no stone unturned. By the end of 1874, he had persuaded the University to part with £50, later raised to £100, to buy specimens he could not obtain by donation.[104] At the same time, he turned his attention to the city. The public of Sydney had an interest, and possibly a need, greater than the Arts students whom he lectured at the University.

From the 1840s, the Australian Museum had made intermittent mineral exchanges with the British Museum, and expanded its collections by gifts from benefactors, notably Alexander Macleay and Edward Deas Thomson. In 1851, with the discovery of gold and the appointment of Samuel Stutchbury as Colonial Geologist, the museum voted to form a collection in economic geology. However, matters proceeded slowly until, between 1859 and 1862, the government granted £1200 for the purchase of silver and gold specimens, and a further £200 to be spent by W.B. Clarke on overseas specimens for study and comparison.[105] Slowly a larger collection took shape. But a 'minerals boom', producing a flood of specimens without proper storage or curatorial attention, could quickly undermine the museum's ability to serve either science or commerce.

In September and October 1874, claiming that Sydney had nowhere 'which at all approaches completeness' to illustrate the geology and mineralogy of NSW, Liversidge persuaded, first, the Department of Public Works, and then the newly-formed Department of Mines, to accumulate any 'minerals and fossils which may be met with' during excavations for foundations, wells and railway tunnels. By sheer flattery, he persuaded John Sutherland, Minister of Public Works, and Harrie Wood, Under-Secretary of Mines, that if they were 'at all times ready to aid the cause of science', they would cause records to be kept of the 'depth, thickness, position, etc, of such rocks', and send samples to both the University and the Australian Museum. In return, Liversidge wrote a set of instructions for prospective 'diggers', printed specimen labels for their use, and volunteered his services in analysing any specimens they found.[106]

This aspect of Liversidge's public service reached a peak in March 1876 when, to general applause, the Department of Mines opened a Mining and Geological Museum in Bridge Street.[107] At a stroke, this step was intended to give Sydney an economic minerals museum on a par with the Museum of the Mining Bureau of San Francisco, the Geological Museum of Montreal (later, of Ottawa), and the Johannesburg Chamber of Mines. Much smaller than the Geological Museum in Calcutta, it lacked the facilities of the Museum of Practical Geology in London. But the prospect of putting a great many specimens on public view was encouraging to Liversidge at a time when the colony's 'scientific' collections, like his own infant subject, were cradled in a corner of the Blacket Building. There, in a room called the Mineralogical Museum (now Latin I), Liversidge began to catalogue specimens sent him for analysis, which he afterwards kept. 'The amount of exact information published', he told the Royal Society in December 1874, 'is surprisingly small', as he presented a list of NSW minerals he had studied, most of which had never been analysed. It was path-breaking work.[108]

If minerals first took Liversidge to the Australian Museum, many other considerations kept him there. In 1874, he began a thirty-year association, first by succeeding to the vacancy on the Board of Trustees caused by Thomson's death. Like much else in the colony, he soon discovered, the museum owed its existence to a small group of promoters, whose influence in its affairs was manifest, overwhelming and exclusive. From 1827, and for its first quarter-century, the museum was run informally by a 'committee of superintendence', dominated by the Macleay family, and assisted by such gentlemen naturalists as Dr George Bennett (who served as its part-time curator and secretary between 1835 and 1841), and W.B. Clarke (who succeeded him between 1841 and 1843). In 1853, the first colonial Legislative Assembly voted the museum an annual budget of £1000, in return for which its administration came under a Board of Trustees. Eleven were 'official', representing the public service and the Crown, and twelve were 'elective', chosen by the others for their 'scientific attainments' and appointed without limit of age or tenure.

Adopting practices similar to those of the British Museum, the Australian Museum's day-to-day running was vested in a Curator and Secretary.[109] Although it had a public mandate, it tended to serve the interests of the few, largely self-perpetuating Trustees who regularly attended meetings. These included Clarke, Bennett and Dr James Cox, a pioneering physician, conchologist, sometime Smith's assistant and member of the Philosophical Society.[110] The inner circle also included the extended Macleay family, including William John (later Sir William) Macleay, Sir William Macarthur and Captain Arthur Onslow, RN, Macarthur's nephew by marriage. In some ways, these Scottish scions considered the museum as their private fiefdom, and found little difficulty in using museum specimens to fill gaps in their private collections, and *vice versa*. Indeed, Macleay found no conflict of interest in employing the Curator's assistant, George Masters, to collect specimens for him privately;[111] while Bennett, the Curator, and the colony's leading zoologist, found nothing wrong in sending to other museums and collectors various specimens he received and did not want. The museum kept no record of exchanges, and indeed the Trustees had no 'exchange committee', until 1862. Even then, such matters were left chiefly to Bennett, as few expressed concern.

That such amiable but unbusinesslike practices were potentially – and politically – explosive, became evident not long after the appointment to the Curatorship of Gerard Krefft, an Austrian educated in Germany, who emigrated to Victoria at the height of the gold rushes in 1852. Moving to Sydney as an assistant in the museum in 1860, he was made Curator in 1861, somewhat against the wishes of the Trustees.[112] Whatever their reservations, his first ten years saw him build an enviable reputation for the museum, while he became the best vertebrate zoologist in Australia.[113] In 1870, he recognised the Queensland lungfish, commonly known as the 'Burnett Salmon', to be a 'living fossil of great importance, and named it *Neoceratodus forsteri,* in honour of William Forster, at whose table in the Union Club he first saw it being eaten. It was a 'transitional' mesozoic amphibian, with lungs that fed on vegetable matter, yet looked like a fish. Louis Agassiz at Harvard called it 'the greatest discovery in Icthyology' since *Lepidosiren* was found, as it suggested a transitional species between fish and amphibian.[114] Such discoveries brought Krefft praise from Darwin, whose colonial advocate he soon became.[115] These conquests did not endear him, however, to those ontological *protégés* who came from the camp of Darwinism's great adversary, Professor Richard Owen.

Beginning in the 1830s, Richard Owen had built much of his formidable reputation upon the description of the 'dead and living' fossil fauna of Australia, and upon their testimony to the wisdom of God in the Creation.[116] In 1838, Owen identified the fossil, thought previously to be an elk, elephant or hippopotamus, as an extinct marsupial, which he called the *Diprotodon*, and placed it within the 'law of succession of types', an idealist, developmental hypothesis which fell short of evolution, but which offered openings to other interpretations. Bennett revered Owen, agreed with his explanations, and in return for his opinion, was frequently content to send London unique specimens of Australian fossils, keeping only plaster casts for Sydney.[117]

With Krefft's coming, this relationship was to change. Krefft took issue with Owen over the interpretation of an extinct marsupial, the *Thylacoleo carnifex*. On the basis of its dentition, Owen, following Cuvier, had classified the animal as a lion-like carnivore; but Krefft thought it must have been a vegetarian, occupying an evolutionary position somewhere between the wombat and the koala. The question was technical, but its interpretation was highly political. Owen, as the metropolitan expert, held the upper hand; Krefft, in disputing that expertise, threatened the primacy of London as well as the taxonomy of Europe.

As taxonomic debates revealed, colonial science turned upon a kaleidoscope of philosophical preferences.[118] For a generation, most in NSW who seriously contemplated natural history, like their English counterparts, favoured theories of catastrophism, or 'special creation', which sought to explain discontinuities in the fossil record by divine intervention. However, there was room for debate where the evidence warranted. In 1831, Sir Thomas Mitchell discovered fossil bones in the Wellington Caves, which he sent for examination to Edinburgh, thence to the Hunterian in London, to see whether they demonstrated in Australia the same diluvial geology that Georges Cuvier and William Buckland had ascribed to Europe. In 1846, long after Charles Lyell had introduced uniformitarianism into English geology, the evangelical John Dunmore Lang preached a text in which Australia's fossil record might be explained in terms of biblical belief. It seemed more than possible that Australia could supply evidence of separate and continuous lines of faunal development; and that living species could be linked with extinct ancestral forms.

In this context, fossils were not merely natural artifacts, but also forms of cultural capital, vital to reputations in Australia and overseas. When displayed internationally, as at the Paris Exhibition of 1867, their interpretation was a matter of global interest. In the relationship between metropolis and periphery, the Owen-Krefft affair highlighted the problem of establishing an independent Australian perspective on questions about which Australians might be said to have the 'right of first sight'. Eventually, European theory would be supplemented, even corrected, by such evidence. But for the moment, colonial deference won the day, especially when such issues touched domestic, as well as imperial, interests.

If the Australian Museum's Trustees found Krefft's manner, style and Darwinian sympathies offensive, they were outraged when they found Krefft currying favour with

DIPROTODON !!

Professor Krefft—the Great Wizard of the North:—

'*Diprotodon*: Professor Krefft, the Great Wizard of the North',
*Sydney Punch,* 26 June 1874.

their opponent, Henry (later Sir Henry) Parkes, whose liberal views on education and reform were well known. Macleay and Parkes had crossed swords politically in 1868.[119] When Parkes formed a government in May 1872, he used Krefft's allegations as a stalking horse against Macleay and his kin in the Legislative Council.

To accuse Trustees of abuse of privilege was bad enough, but far worse when the charges were laid by a foreigner. The audacious Krefft compounded his felonies in 1873, when he appealed to Parkes over the heads of his employers, asking for the government's encouragement to write a 'thorough history of our animals'. Such an account, he argued, could 'only be written *in this country*, and in this colony':

> There is no Museum in the world which has the Australian collections we have and not one Professor can command such a series of skulls and skeletons. We had to supply Professor Owen with his most important skeletons ... and regret to say that Professor Owen goes on teaching what is not correct.[120]

Krefft was right. Unhappily, such colonial nationalism, aggravated by an ill-concealed disrespect for Owen and for the Trustees, degenerated into what Ronald Strahan has aptly called 'guerilla warfare' between the Curator and his governors. In late 1873, simmering contempt erupted into overt fury. The *casus belli,* now long forgotten, turned on the allegation that Krefft had mishandled a theft that had occurred in the Museum. Krefft had probably been incautious, but poor judgement left him open to attack. Worse, it occasioned proceedings that, over the next two years, rent asunder the scientific world of Sydney, doing great injury to the museum and worse to the tradition of fair play.

In February 1874, responding to criticisms that had reached the daily press, Parkes set up a Select Committee, under the chairmanship of his electoral agent, barrister and fellow journalist, Walter Cooper, MLA for East Macquarie, to enquire into the management of the 'Sydney Museum'.[121] Colonial science could not escape party politics, or so it seemed. Following protests from the conservative opposition, the proposal was widened to include Macleay and Onslow, as members of the Legislative Council. The committee sat eighteen times between March and May, and interviewed twenty witnesses, including Bennett, Cox and Krefft himself. Macleay, although a member of the committee, was also called to give evidence.

The documentation was formidable, but inconclusive. In his draft report, Cooper devoted space to the state of the building and to methods of administration. There was still no proper catalogue of the collections and the museum's displays were poor. This could be blamed in part on the Curator. But in management, Cooper vindicated Krefft, and reserved his vitriol for the Trustees, who, he said, had acted with 'perfect irresponsibility'.[122] 'They are unpaid', he noted, correctly; 'they contribute nothing to the expense of the museum', which was technically true; and 'they have no interest whatever in the institution beyond that which an unselfish public spirit and a devotion to science may engender' – which was a mixture of truth and spite. The Trustees had been, he continued, 'inattentive to their duties [and] divided into contending factions'; this had 'diminished their Curator's authority, and thus destroyed discipline among [his] subordinates'.[123] Cooper pointed to the latent potential for abuse where institutions were managed by unelected, self-appointed guardians, unaccountable to parliament, who nonetheless exercised power over the spending of public money. In that condition, as Cooper observed,

> where the apathy of most places power in the hands of two or three active persons, whose zeal has perhaps some personal element in it, the system is fraught with danger.[124]

As the inquiry demonstrated, public and private space were not well defined. It was well known that museum attendants, although they worked for Macleay, also prepared a stuffed wallaby for Henry Parkes and a possum for the Chief Justice, Sir Alfred Stephen. Macleay defended himself against charges of abuse, saying that the balance of any exchange between him and the museum had always been to the museum's favour.[125] He may well have been right. Krefft, by his own admission, also gave specimens to other museums and to the *Challenger* expedition, without accounting for them. Professional men like Cox, Bennett and Smith also worked the system to their advantage. Inherent conflicts of interest were not lessened by factional intrigue. Politics took precedence over principle.

Overall, the committee proceedings were a dismal affair. The allegations against Krefft ranged from open drunkenness and dealing in indecent photographs, to maliciously smashing and losing the jawbone of a fossil *Diprotodon*.[126] James Cox, who was later remembered by a student for his 'great kindness of heart, his hatred of anything mean, and his extreme care in avoiding any possible hurt to anyone's feelings',[127] was out of place in mentioning Krefft's weakness for alcohol. On the other hand, George Bennett who, as a doctor, was then treating Krefft's wife for fever and his son for whooping cough, bravely defended him against Cox's charges of drunkenness, which he put down to depression. Bennett further stated that Krefft had broken up not the fossil *Diprotodon* jawbone, but the plaster cast made of it, which was to be given to the museum, so that Bennett could send the original to Owen. The alleged 'vandalism' was nothing more than Krefft's way of putting his nemesis on an equal footing.[128] As for his abrasive tongue, it was indicative of nothing more 'than the usual excitement which is perceptible in all Germans and foreigners on occasion'.[129] Krefft had been incautious, but 'it would be difficult', Bennett wrote Owen, 'to find a Curator to work like Krefft; he has made our Museum the admiration of scientific visitors'.[130]

The Cooper Committee made three recommendations, two concerning the premises and one, the management. The latter called for the abolition of the Board of Trustees, and for placing the Curator under a minister answerable to parliament, advised by a Board of Visitors. The model was that of the Royal Observatory in Greenwich. Coincidentally, just two months earlier, the fourth report of the Devonshire Commission had recommended an identical plan for the Natural History Collections of the British Museum.[131] Well may they have cited imperial precedent. But nothing would save Krefft.

In parliament, Cooper's draft was sanitised and most of the criticisms of the Trustees were qualified or deleted. But their fury was not appeased. At their monthly meeting in June, angered by Krefft's accusations under oath, the Trustees, on a motion by Onslow, appointed a subcommittee of four to enquire into the charges arraigned against him. An unusually full meeting of the Board summoned the Postmaster-General from his work, called the police, and closed the museum to the public. The press had a field day, with *Sydney Punch* caricaturing Krefft as the 'magician' who had made the fossil disappear. To their 'gang of four', the Trustees called Liversidge, the youngest, the most junior and the least experienced of their number.

The record suggests that Liversidge was both surprised and discomfited to find himself summoned to a jury of such doubtful peers. His fellow 'judges' were Dr H.G. Alleyne, a noted icthyologist of Port Jackson and President of the Medical Board (thus an 'official' Trustee);[132] Edward Hill, a gentleman collector and close friend of Macleay; and Christopher Rolleston, the Auditor-General, who had recently served Parkes on the Board of Asylums for the Infirm and Destitute. One official, one conservative, one liberal, and Liversidge – presumably to add 'balance'.

Liversidge had been elected to the Trustees on Bennett's recommendation at the end of 1873, and had taken his seat in February. He had therefore been on the Board scarcely five months. However, he had been an interested observer of museum politics for some time. With no facilities for display at the University, and with no Department of Mines yet in sight, Liversidge had cultivated the mineral collections in College Street.

He met Krefft at the Royal Society during 1873, when the Curator was completing a catalogue of the Museum's mineral holdings;[133] and as late as October, before he was elected a Trustee, Liversidge was cordially corresponding about the 'tin and diamond drift specimens' obtained during his ride to New England, which he was 'putting up' for the museum. But mineralogy was a Cinderella in the museum, and despite Clarke's influence, Krefft had devoted little space to mineral specimens, even in the College Street extension completed in 1867, preferring to concentrate on fossils of mammals and rare fish.[134] Moreover, as he wrote to Parkes in March 1874, he refused to be reduced to a 'caretaker of specimens' for scientific men.[135] It followed that, in practice, Liversidge cultivated Krefft's assistant, George Masters. But Masters also served Macleay. And when, in April 1874, Masters fell out with Krefft, and resigned to work full-time for Macleay, Krefft felt betrayed. His injured pride made fresh hostages to fortune.

In an undated memorandum sent to Parkes, Krefft warned Cooper to be 'weary [sic] of McL[eay].' 'He is *very* deep, pretends to be *my* friend, and lets me know it by deputy'. But, he added, 'I know him better, and now is the time to expose him and O[nslow]'. Krefft thought that 'McL[eay] is afraid of Masters for some reason or other'; and appeared 'not to want to hurt M[aster's] feelings, and shelters him wherever he can'. Nothing could reassure Krefft of Macleay's fair-mindedness. 'It is not a bit of use telling me McL[eay] is my friend. He hates me, and I return the compliment, but we have kept civil.'[136]

What threat, if any, Liversidge posed to anyone is not clear. His opinion as a Trustee would matter; if he were on Krefft's side, he must be treated as a possible enemy. Macleay, Cox and W.J. Stephens (classicist, naturalist and later Liversidge's colleague)[137] monitored the affair, and counted on his vote.[138] In fact, Liversidge was not present at the Trustees' meeting on 5 March 1874, when Onslow produced the notorious indecent photographs; nor had he animus against Krefft on either political or scientific grounds. His presence on the committee of enquiry was perhaps the one redeeming feature of proceedings that were otherwise, in Strahan's words, a 'pathetic farce'. Over a dozen people were interviewed, but evidence was not taken on oath, and inconsistencies were not examined. Macleay at first opposed publishing the evidence at all, but he was advised to do so by Parkes' Attorney-General, who kept open the question as long as he could.[139] Quite properly, Krefft refused to participate, but silence was taken as an admission of guilt, and absence sealed his fate.

When the subcommittee met, it held its sessions in camera. Krefft's allies were few. Clarke, the one Trustee whose influence would have counted on his behalf, was ill and absent from most of the relevant meetings. Liversidge's questions show a keen desire to set matters straight, persistently seeking objective information and showing a deep concern with the accuracy of Krefft's record-keeping. He was thrown by the discovery that Krefft had grossly exaggerated museum attendances.[140] He was also worried about the security of specimens, especially the sacrosanct fossils.

In August, the subcommittee reported. On ten of the twelve charges, they found the evidence 'extremely meagre', and of guilt, little more than 'irregular conduct'. In other circumstances, this might easily have been overlooked. Yet, Macleay and others insisted that false statements and insubordination constituted a mortal sin. Krefft was asked to show cause against dismissal. When he refused, the Trustees instructed Edward Hill to evict him and his family from their museum apartment; and on a sunny September day, they were put into the dust of College Street. Macleay sighed with relief and turned back to his bird collection.[141] Bennett was the only Trustee to vote against dismissal.[142]

In the museum's annual report, appearing six months later over the signature of Alexander Scott, as chairman of Trustees, Krefft was unequivocally condemned. The case against management was whitewashed, Cooper's report was rejected and any thought of doing away with the Trustees vanished. It was not clear how far Macleay's influence affected the result; it probably sufficed that Scott, who drew up the report, was a close friend of his.[143] In protest against their colleagues – most of whom, as Bennett wrote, 'only look after their own private collections and are a great impediment to the advance of the Museum'[144] – both Bennett and Clarke resigned from the Board. The museum lost face. But their resignations merely strengthened the hand of Krefft's opponents.

At home and overseas, the event made headlines.[145] When news of the dismissal reached London, *Nature* devoted a lengthy feature supporting Krefft: 'That institution seems to be in the hands of a few collectors of the old school', wrote Norman Lockyer, 'who treat it as a plaything of their own, rather than a public institution, supported by public funds. They have a curator ... of whose very high scientific position in the mother country they cannot be fully aware, or they would be more liberal to him, and give him more opportunities for the employment of his abilities'.[146]

In November, the NSW parliament was dissolved, and Macleay, having represented the Murrumbidgee district for fifteen years, decided to stand down, in order, as he put it, to give his 'attention henceforth entirely to Natural History and the improvement of my Museum'.[147] In the election that December, 'Parkes' pets', as Macleay called them, lost office, and Krefft's hopes of obtaining restitution fell with them. Although he won a civil action for £250 against Hill for unlawful dispossession, he lost his job.[148] 'Crushed out of existence', as he put it, the following months were filled with appeals to anyone who would listen.[149] Parkes distanced himself – the museum issue had lost its political momentum – and even William Forster fell away.[150] With no one left to take his part, Krefft languished in Sydney for the next five years, dying in poverty in 1881. He was succeeded as Curator by Edward Ramsay, a thirty-two-year old ornithologist, and the first Australian-born naturalist to head the museum. He was also a *protégé* of Macleay.

It was probably fair to summarise, with Bennett, that Krefft was 'not fitted by temper' for the position he held and perhaps he was 'wanting in method' as well.[151] But the whole business left a bitter taste. Liversidge could have won few marks, as both sides might have seen him to be conspiring with the enemy. But other matters were on his agenda, not least the inauguration of Macleay's new Linnean Society of NSW, which had recruited Alleyne, Forster and Ramsay, and which threatened to undermine the membership of the Royal Society. In any case, museum reforms might come about in other ways. In October, Henry Russell was elected to the Board of Trustees, and Ramsay, although conservative, might be led in more progressive directions. The world could not stop for Krefft. 'My numerous friends and scientific correspondents will be properly staggered', Krefft had written Parkes at the height of the affair.[152] But he had underestimated his competition. Cultural politics required a delicate touch and a calculated commitment to the middle course.

For Liversidge, the Krefft experience was instructive. A student of Huxley, a young man with a name to make, who had just been made a professor by a conservative Senate, remained impartial. His disinclination to take sides did not go unremarked. At worst, he had let down a friend and a man deserving of respect; at best, he had simply played the game. There were lessons for a newcomer in Krefft's tragedy. Perhaps it was as well to repair to the laboratory, avoid theory and seek the consolation of minerals.

After the turbulence of these months, Liversidge ended the year on an uncontroversial note. In October 1874, he travelled beyond the Blue Mountains to investigate newly-found iron and coal deposits at Wallerawang, near Lithgow.[153] The importance of finding iron ore of industrial quality, was, in the prospectus of the Irondale Iron and Coal Company, 'evident to every reflecting mind', and Liversidge's analyses were relied upon to attract investors. Another 'first' came with the local presentation of the New Caledonian mineral he had earlier named *Noumeïte*, and which Clarke and he had distinguished from *Garnierite*.[154] But the most glamorous event of the season came when he looked from the earth to the sun, putting the unhappy Krefft affair behind him, and rode out with Russell to observe the long awaited transit of Venus.

## 4. The 'Transiting' of Venus

To determine the distance from the earth to the sun – the 'fundamental astronomical unit', in technical terms – had been for nearly 200 years a key ambition of Western astronomers. In 1716, Edmond Halley had shown that observations of Venus as it transits the face of the sun might be used to determine that distance. If observers from widely scattered locations could measure precise intervals as Venus crossed the sun's face, this information could be used to calculate the angle subtended at the centre of the sun by the earth's radius. This is the solar parallax, which is measured in seconds of arc. Once this is known, the distance between the earth and the sun can be determined by geometry. And once this distance is known, Kepler's third law can be applied to work out all the planetary distances in the solar system, and so produce a scale of the universe.[155]

To observe the transit of 1769 was a principal article in the instructions given to Captain James Cook, whose first voyage to the Pacific took him to the 'southern hemisphere platform' chosen by the Admiralty, and conveniently furnished by the island of Tahiti.[156] The international observations taken that year produced over 250 scientific papers, but the results of the effort remained inconclusive. For this reason, the next transit pair – in 1874 and 1882 – were to be viewed with even greater care, and from stations set up at strategic positions all over the world.[157] Overseas expeditions were mounted by Russia, Germany, France and Britain; while observations within their own frontiers were to be made by the Netherlands, Denmark and Italy. Colonial astronomers were pleased to learn that the transit should be visible from NSW.

For Henry Russell, the event was 'one of the potentially most valuable in astronomical history'.[158] It was also a triumph for 'practical' men. Russell's predecessor, the Rev. William Scott, had begun the colony's first systematic rainfall recordings in 1859, and from its opening in 1858, had used Sydney's new observatory to pioneer meteorological forecasting.[159] From his arrival at the observatory in 1862, Russell took an interest in these

'The Transit of Venus as seen by our special astronomers',
*Melbourne Punch, 1874 Almanac.*

methods, and in 1870, when he became Director, organised an elaborate program of research. On the one hand, he set out a scheme for the determination of star clusters and the observation of double stars; on the other, he championed the science of weather forecasting. In 1871, Russell and Robert Ellery, the Government Astronomer of Victoria, conducted an expedition to Cape Sidmouth in North Queensland to observe a total eclipse of the sun.[160] They were frustrated by cloud, and gained few scientific results, but they gained valuable experience of organization. With the 1874 transit, Russell had an opportunity to bring this experience, and the colony of New South Wales, to the attention of the scientific world.

Russell began planning for this in 1871 when, at his prompting, the Royal Society of NSW sent a deputation to John Robertson, Colonial Secretary, requesting government funds. 'It was obviously,' Russell argued, 'for the honour of the Colony, as well as for the advancement of science, that the observations and photographs ... should be as complete as possible.'[161] Robertson agreed, and gave £1000 to the project. 'Such observations,' said the *Town and Country Journal*, could not be expected to secure a 'transit of gold' from the planet to the Treasury; however, such 'practical proof of the interest of our Legislature in the case of "pure science" was welcomed as giving an impulse to researches that will directly or indirectly increase the material wealth of the colony' and so attract support for science in general. It was, supremely, 'a practical recognition of the duty of the state to aid in the prosecution of scientific research'.[162]

As European makers could not deliver the necessary apparatus in time, Russell approached 'mechanics in the colony, for the most part unused to such delicate work.'[163] They repaid his efforts. And when, in mid-1874, Sir George Airy, the Astronomer Royal, asked Russell to organise sightings from NSW, he enlisted thirteen men to form, in the *Herald's* happy phrase, the 'New South Wales Corps of Astronomers'.[164] Four groups of observers were chosen, one at the Observatory, one at Goulburn, a third at Woodford in the Blue Mountains, and the last at Eden on the south coast. Each position was carefully assessed; Eden was chosen, for example, because a point in southeastern NSW was thought to be the best position to observe the egress of the planet in its journey across the sun.

Liversidge was keen to join this company of science and spirit. Lawrence Hargrave, later the aeronautical pioneer, was assigned to the Woodford detachment, which Russell favoured for its clear and steady atmosphere. William Scott went to Eden. Liversidge was assigned to the 'naval' group at Goulburn, consisting of Captain Francis Hixon, RN, the forty-one-year old President of the Marine Board,[165] Captain Arthur Onslow, RN, MLC, whom he knew from the Australian Museum, and Mr Angelo Tornaghi, a photographer and instrument maker (and mayor of Hunter's Hill). Their expedition was equipped with heavy wagons carrying food, supplies, and telescopes and plates for 200 photographs. Their success depended on taking precisely-timed measurements and photographs of the tangential contact, beginning about noon on the appointed day. Everything was carefully rehearsed. Russell was optimistic.

'Never before in the world's history did morning dawn on so many waiting astronomers as it did on the 9th of December 1874', Russell told the Royal Society of NSW in January 1875. His observers shared 'an overpowering sense of responsibility which every true worshipper of science must feel, when he knows that the answer to half a century's questionings are depending upon him; not lessened because he feels he is the observed of all observers, determined to do his best in the noble cause of science'.[166] When the time came, only three of the thirteen observers, including Russell, managed to take exact timings. The others, including Liversidge, failed.

Exactly why some succeeded, and others did not, remains unclear. The Sydney station enjoyed the best weather and made the best observations.[167] At Eden, cloud obscured the view.[168] The Goulburn observers thought themselves even less lucky. According to Russell, their weather was fine, clouds were numerous, 'but not sufficient to interfere with observation'. According to Liversidge, however, it was 104 degrees in the shade, the wind produced a 'tremulous outline',[169] and the only position he could assume:

> ... was a reclining one, with my feet to the East and my head to the West. I may mention that it was an uncomfortable and unsteady one, and prevented me from taking such full notes at the time as I had wished.[170]

As a result, Liversidge made premature estimations of the 'contact period', with fatal effects for the record. He did, however, produce detailed notes on the 'halo' of Venus, indicating the presence of an atmosphere, which fuelled popular speculation about life on the planet. Liversidge failed in the immediate task of producing a conclusive measurement, while at the same time fully appreciating the significance of what he was doing. This frustration he would know again, when he turned from the planets of the heavens to the minerals of the earth.

Sydney Observatory in 1871.

Modern historians concede that the atmospheres of Venus and the Earth had 'conspired to render the observations imprecise'. The result observers achieved was a mean solar parallax of 8.8455 seconds of arc, signifying a distance of 92,400,000 miles, within a range of 92,570,000 to 93,000,000 miles. The result was not conclusive; the observations were not 'all the scientific men wished them to be, for then we should have known the solar parallax within a mile'.[171] In the judgement of a contemporary historian, 'the grand campaign had come to nothing'.[172] But it held great promise for the future.[173]

In February 1875, Russell sailed to England with copies of the Australian reports, photographs, plates and drawings, which he delivered to Airy at Greenwich in July. When the NSW materials were exhibited at the Royal Astronomical Society in London, Russell proudly reported, 'they were believed to be the best and most complete'.[174] Considering the extent of the British expeditions to Egypt, India, New Zealand, and the Pacific, this was surely, in Russell's words, 'no small honour for Australia'.[175]

In November 1875, when Russell returned, Liversidge and the other veterans of Venus held a dinner in Sydney to celebrate the anniversary of their adventure.[176] For years, the affair would be remembered with a mixture of pride and disappointment. As time went by, Russell was more inclined to attribute their relative failure to the instruments they used, rather than to the competence of his observers. Even the English photographs, he later recalled, had produced distortions, 'impossible to measure ... with anything like the required accuracy'.[177]

Cameo portraits of the Transit of Venus observers.

For Liversidge, the important point was to see the work appear in print. 'They ought to make you an FRS for your work on this transit', he wrote Russell; 'many a man has got it for far less. Did they not give it to Ellery [of Melbourne]', he added, 'for ordering and importing a big telescope merely?'[178] The book of observations from NSW had, moreover, to be well produced. 'I mention the matter', he counselled his friend, 'because I know that nowadays there is a tendency to honour work which is of any value by sending it forth with a good appearance'. Russell knew the same. Indeed, he told the Royal Society of NSW, every observer held 'a faint hope that his name and his work will appear ages hence in the records of science'.[179] Recognition followed. In 1886, Russell was to become the first Sydney University graduate elected to the Royal Society of London. Ironically, however, not until 1892 did the NSW Government Printer finally publish his report and photographs, including a drawing of the planet's procession, the only known example of Liversidge's artistic talents.[180] The frontispiece of the final report reproduced sepia portraits of the happy band of thirteen brothers who had seen Venus, but who had missed the sun.

CHAPTER 6

# Colonial Science:
# At Home and Away

The seal of the Royal Society of New South Wales.

The same month that Liversidge arrived in Sydney, the international telegraph joined Sydney to the world. Along its wires the electric message came. 'The intellectual light which before shone at a distance' – proclaimed the *Herald* – 'dazzling rather than illuminating ... [now] penetrates every class of population with some portion of its heat and radiance'.[1] In consequence, the *Australian Town and Country Journal* told its readers 'Science has a title to the public aid of this and every other civilized community':

> A very large proportion of the comforts, enjoyments and defences of our daily life are plainly traceable to science; and not merely to what is sometimes called distinctively 'practical science' ... but to 'pure science' or the pursuit of scientific knowledge for the mere love of truth.[2]

The applications of science were not yet to shower their bounty upon the land. Yet, any visitor to Sydney who cared, within a decade, to look up to the allegorical statues of 'Science', 'Land' and 'Capital' set one above the other on the corner of the Chief Secretary's Office on Phillip Street, could take away the simple faith that this trinity embraced.

Liversidge had left England an untried student; six years later, he was a respected professor. In that short time, he had assumed, as Charles Badham said of himself, a 'somewhat hazardous prominence'.[3] Like Thomson, he was bound to play a key role in fostering scientific culture.[4] To do so, however, required diplomacy, tact and perseverance. At the University, reforms were easy to propose, but difficult to achieve, not least because of its dependence on government, and because of the special characteristics of governance in New South Wales. During Liversidge's first five years in Sydney, the government changed three times. In consequence, as Allan Martin has put it, 'politics ... drifted in chaos', and governments 'did little more than business essential to the conduct of administration'.[5] Proposals approved by one ministry could be rejected or deferred by the next; and proposals passed over in one year had to begin all over again the next. Parliamentary inertia was made worse by political fratricide and by the lack of a civil service to provide continuity.[6]

The opening of the year 1875 saw Liversidge extend his power base. The water supply debates, the mining question, the transit expedition and the Krefft affair – all had been instructive. He saw the need to reach out to a wider audience. For years, Smith had given public lectures to popular audiences, such as his memorable evening lectures on 'Electricity for Young Ladies'.[7] In February 1875, Liversidge similarly launched a course of popular lectures on practical chemistry, open by fee to all who wished to attend. 'This is a step', the *Sydney Mail* smiled, 'in the direction of popularising the institution'. A pity only that the classes could not be given in the evening, as well.[8] To such lengths the University was not yet prepared to go, but Liversidge was willing to try.

Moreover, he was willing to work with everyone. In such a small community good relations were essential and, without losing the support of liberals, he had avoided giving offence to conservatives. With the anti-Darwinist William Macleay, for example, he formed an alliance of expedience. He was among those invited in January to join the inaugural Council of the Linnean Society of NSW, founded by Commander T. Stackhouse,

RN, but with Macleay's financial support and blessing.[9] He also met Macleay through Edward Deas Thomson, his father-in-law, who in March 1875 invited Liversidge to a Commemoration Day luncheon for the professors and their families at his home, 'Barham'.[10] There, he mixed with Macleay's powerful friends, Dr James Cox and Dr Alfred (later Sir) Roberts, both Trustees of the Australian Museum and also members of the Royal Society. There he also met Edward Ramsay, Krefft's successor as Curator of the Museum, and W.J. Stephens, one of Macleay's closest friends,[11] from whom he learned about the difficulties Smith had faced in promoting scientific education in the colony. He became good friends with Ramsay, and when he became a Trustee of the Museum himself, began to acquire a collector's taste for Aboriginal and Pacific island artifacts.

Liversidge was never a close friend of Sydney society in the seventies. He was notably not among the 200 guests, including the novelist Anthony Trollope, who were invited to Circular Quay in May 1875 to celebrate the launching of Macleay and Onslow's *Chevert* expedition to New Guinea.[12] But he was destined to need the support of the great and the good on Senate, whose interest in reform could never be counted on. For much of the next twenty years, he watched progress punctuated by failed attempts to revise arrangements fixed by the University's founders long before. It was as well that he took, when he could, opportunities to work in the city. From Smith's experience, he knew that the battle for science could never be won entirely on the paddocks of Grose Farm.

## 1. The Renaissance of the Royal Society

From his earliest experience of science as a youth, Liversidge knew the role that learned societies played in encouraging a public interest in science. To win both town and gown, it was necessary to go outside the lecture room and laboratory. And if there was one institution that anchored his enthusiasm, it was the Royal Society of NSW.

When Russell brought Liversidge onto the Council of the 'Royal' in July 1873, the Society was on the verge of collapse. In the twelve years since the departure of Governor Denison in 1861, its fortunes had evaporated. Members stopped coming, or failed to pay their subscriptions; one after another, debentures were sold to cover expenses, and by 1866, 'it became a question whether the Society could be carried on'.[13] W.B. Clarke and Smith had done much to keep meetings awake, but there were few who joined in their scholarly enterprise. In 1866, they attempted to revive the Society's image, by promoting it as not as an amateur, frock-coated, mildly eccentric club of middle-aged gentlemen, who liked to talk about rocks and stars, but as an 'improvement' society for middle-class men of general culture. Their plan worked, to a degree, and for a time the Society grew, reaching 170 in 1876.[14] But it still lacked a program, and had no premises to call its own.

Historians recall the Society's furtive meetings in rented rooms belonging to the Chamber of Commerce, near the Sydney Exchange. With no space to keep their books, even those the Society received as donations had to be given to the Australian Museum. Less than ten years from its re-foundation, some thought it should be wound up – to die in dignity, rather than in failure.[15] On 27 May 1876, the *Sydney Mail* was brutally frank: 'The experience we have had of the doings of the Royal Society have [sic] not tended to its popularity. Its proceedings are at variance with everything lively, nor is their tendency to elevate clearly perceptible'. What could be made of a Society in which it was 'not uncommon' said one disenchanted scribe, for members to 'paddle to their assembly room through dripping rain',

> a fit prelude to so humdrum an entertainment as listening to philosophical disquisitions on the coruscations of Jupiter, the weight of the dew drop or the exemplary behaviour of the man in the moon.[16]

By 1876, Clarke and Smith, Vice-Presidents for a decade, were exhausted volcanoes. Their efforts had not stimulated the public to take an active interest in the Society's proceedings; nor had their two Honorary Secretaries, the Rev. William Scott, Director of the Sydney Observatory, and Charles Moore, Director of the Botanic Gardens. Worse, the Society's transactions were in arrears: in eleven years and eighty-three meetings, it had published 107 papers that were scarcely visible outside Sydney.

Such was the situation that greeted Liversidge. Yet, by the time he came on board, he was already plotting with Russell to change the order of things. With a membership that included many of Sydney's professional and business leaders, the Society lacked not influence, but energy. Russell and Liversidge moved a pliant Council to embark upon reform. Just as the City societies of London kept their clients by appealing to civic interests, so the 'Royal' would do the same. The first step was to combine annual meetings with *conversazioni*, just as in London, and as in the metropolis, open by invitation to the public.

To the first of these events, Liversidge brought his experience of 'evening science'. Hastily digging out his lecture notes on the famous dendritic spots, he realized that he would be regularly called upon to speak, and began to make the Society's program in a sense his own. Anything new to science and Sydney he scooped into the Society's agenda. Thus, in April 1875 the Society borrowed the officers of the celebrated HMS *Challenger*, arriving for a two-month visit during its historic round-the-world cruise. 'I despair of being able to convey to the reader my own impression of the beauty of Sydney Harbour', Captain Spry, RN wrote,[17] and the *Challenger*'s officers, led by Captain Nares, RN, and its scientific staff – including Professor Wyville Thomson of Edinburgh and H.N. Moseley of Oxford – were welcomed to the Society's meetings. Liversidge also arranged for them to be shown the University, given honorary membership of the Union Club,[18] and entertained with picnics in the Blue Mountains.[19] The experience was well remembered by a young German naturalist in the party, who made it known in Europe that Liversidge was '*ein sehr Artiger Mann*'. It was just a pity that '*das grosse Universitätsgebande ... auf ... sind nur 48 studenten*'.[20]

The glow of prestige from the *Challenger* became palpable in May 1874, when Liversidge booked rooms in the Masonic Hall in York Street for the Society's second *conversazione*. It was a sell-out. The Governor came, as did the President of the Legislative Council, the Speaker of the Legislative Assembly, and a number of army and navy officers, to see tastefully displayed (on trestle tables borrowed from the University) a selection of scientific objects and instruments, ranging from the spectroscope that Smith used in his lecture demonstrations, to the Duplex telegraph apparatus used by Edmund Cracknell (the colony's Superintendent of Telegraphs), and Russell's newly acquired self-regulating thermometer. The *Challenger* officers, in turn, showed instruments and objects from

Archibald Liversidge
(1846–1927)

John Smith
(1821–1885)

H.C. Russell
(1836–1907)

Carl Adolphe Leibius
(1833–1893)

Reverend W.B. Clarke
(1798–1878)

The inner circle of the Royal Society of New South Wales in 1873.

their deep sea harvests, including a specimen of the remarkable lungfish that Krefft had identified in 1870. 'Considering how many things in Australia are first notified in Europe,' W.B. Clarke remarked, 'it was fitting that the creature was first exhibited in Sydney.'[21] The Council burst with colonial pride.

In July, Liversidge's efforts were rewarded by the addition to the Society of twenty-two new members, including two later premiers, P.A. (later Sir Patrick) Jennings and Alexander (later Sir Alexander) Stuart.[22] The situation was improving, but progress was slow. Some monthly meetings went by without papers at all; and in September 1874, Liversidge found himself convening the meeting alone. In October, he recruited Charles Wilkinson from the Geological Survey, and Edward Combes, the entrepreneur, whom he knew slightly from the Union Club; and in November, gave two papers – one on the iron and coal deposits at Wallerawang, and another on specimens of *Garnierite* from the newly discovered nickel mines of New Caledonia, which Clarke had given him to examine. But these were matters of special interest and not wide appeal. The Society's prospects for 1875, when Russell sailed with his transit reports to England, remained uncertain.

It was time to make even more sweeping changes. In April 1875, Liversidge took advantage of movements in the Council arising from Russell's absence to shift the two Honorary Secretaries sideways, and to replace them with himself and another short, kindly man, Carl Leibius, Senior Assayer of the Mint and a veteran member of the Philosophical Society since 1859. Like Krefft, Leibius was a German-born 'outsider' to Sydney's Anglo-Celtic society; unlike Krefft, however, his salary was reliably paid from London. Like Liversidge, Leibius belonged to the Union Club and shared a similar background in analytical chemistry and metallurgy. Their mutual interest in gold paralleled their passion for hard work. Leibius, known (like Krefft) for his 'enthusiasm, thoroughness and directness', but also (unlike Krefft) for his ability to get on with his people, was an invaluable ally, well known and respected in the Society. His value to Liversidge was enhanced by a singular lack of personal ambition.[23] He immediately became Liversidge's offsider, and for the next eleven years shared the Society's administration. With his help, Liversidge assumed the editorship of the Society's *Transactions*, delegating routine matters to him, and creating precedents in an office that he would make his own for the next thirteen years.

The pair of Honorary Secretaries soon made their first significant move – literally. By agreement with the newly founded Academy of Arts and its moving spirit, E.L. Montefiore, Liversidge exchanged the Society's cramped quarters at the Chamber of Commerce for an upstairs room and shared use of the Hall in the larger building leased by the Academy in Elizabeth Street, just five minutes' walk from the Union Club and the Public Library. Liversidge bought gas fittings and furniture for the new 'boardroom', and in May 1875, convened the Society's first meeting there. 'The change', observed the *Herald*'s correspondent, 'from the somewhat sombre apartment at the Exchange to the lofty, well-ventilated and illuminated hall, busy with pictures, was a very pleasant one, and must have been appreciated by the members'.[24]

The same month, Liversidge read the Anniversary Address prepared by the ailing W.B. Clarke, who was unable to preside. Until now, said Clarke, in the language of the Old Testament, they were 'like dwellers in the desert, living in tents, without a spot of earth to call our own'. Now they had crossed the Jordan. But, he stressed, the Society had yet to 'make the figure which is suitable to our assumed dignity'; its members risked becoming merely 'annual subscribers for the purpose of an evening's entertainment'. Above all, Sydney had become 'too insulated', having no regular contact with the learned societies of the other colonies. Liversidge took the sermon to heart. The Society's influence was not yet felt in high places. Sydney had fallen behind Melbourne, which in 1858, not long after the separation of Victoria from NSW, had established a Science Board to advise the colonial administration.[25] Sydney had no such machinery of government.

As he heard the words of the venerable Rev. Clarke, Liversidge preached the same text – faith must be accompanied by good works. New premises had given the Society a 'fresh start',[26] but this was just a beginning. Henry Parkes had been lukewarm about giving the Society a government grant; but in June 1875, Liversidge and Russell persuaded Parkes' intermittent successor, John Robertson, to begin indirect sponsorship, along the lines accorded Victoria's Royal Society, by printing 400 copies of the Society's *Transactions* through the Government Printer at public expense. This helped, and to put the Society on a more businesslike footing, Liversidge asked Professor Smith, then President, to announce that meetings would be held whether there were formal papers or not, so that members would keep a regular commitment to the first Wednesday of every month. Liversidge himself agreed to come to the Hall every Wednesday afternoon at 4 p.m., where, from the fragments left by his predecessors, he began assembling a library.

Next, to put the Society on a wider map, Liversidge created a new category of Honorary Members, to which he invited scientists from the other Australian colonies and New Zealand. The Council briefly paused when distinguished Englishmen were also suggested – 'as not sure whether such a step might not be considered somewhat presumptuous' – but Liversidge pressed on, and the principle was conceded.[27] The Society would be recognised by those whom it recognised: by honour given, so honour would be received. Finally, Liversidge convened a special meeting at which he presented new regulations that gave the Honorary Secretaries complete power to run the activities of the Society.[28] The revolution was nearly complete. Liversidge had become an enlightened autocrat.

In October 1875, when Russell returned from overseas, he found the Society much changed. A letter from the Royal Colonial Institute in London, offering reciprocal facilities, was seen as a sign from above, that the Society could now see itself as a *sodalitas inter pares*. When Julius von Haast and James Hector of New Zealand accepted invitatons to be Honorary Members, even Smith was impressed.[29] The following month, Russell's report on Greenwich Observatory's favourable reception of the NSW transit observations was so well received that, said Smith excitedly, it 'knocked all reaction into a cocked hat'.[30] Russell gave substance to the truism that travel is the passport to success. Nothing better demonstrated 'the advantage of a person going about and seeing with his own eyes and listening to discoveries at first hand', especially if, as in this case, the Government could be persuaded to pay.

From this time forward, Smith and Clarke let their Honorary Secretaries take the reins. 'Russell moved and Liversidge seconded', or vice versa, became an increasingly familiar entry in the Council's minute books. In November, they selected twenty-four journals for subscription and categorically approved £30 from the Society's budget to buy them. Leibius, normally a cautious man, was astonished that 'the more the Society had spent, the greater had become its vitality and the larger the increase of members'.[31] What they could not buy through 'cultural Keynesianism', they obtained by gift or barter; subscribing to the practice of exchange inaugurated by Joseph Henry at the Smithsonian,[32] they began to acquire series of periodicals from sister societies in Canada and the United States.

Early the following year, even while the Society discussed proposals for the city's water supply, Liversidge pressed ahead with co-opting support outside the membership. In his inaugural Presidential Address to the Linnean Society in February 1876, William Macleay charged the 'Royal' with failing to meet Sydney's expectations. Many of its papers, he alleged, had not been of a 'scientific character', and others possessed 'no interest except of the most local kind'. In particular, he pointed out, the publication of its *Proceedings* 'had not been conducted with the celerity and regularity to be expected from a society not deficient in point of means'. That alone, Macleay argued, made the Society 'useless as a record for zoological, botanical or geological discovery'.[33]

Liversidge knew that the Linnean Society had been founded partly in protest against the 'Royal'. He found nothing objectionable in Macleay's precepts of observation and classification – a 'good as well as a useful way', as Macleay had put it, of laying foundations upon which a 'superstructure of high science may be reared'.[34] Nor did he object to the naturalists' earnest prose, evidenced in the Linnean's *Proceedings* from 1877. He personally disliked the anti-Darwinian views of Macleay and his circle. But he feared any division that might threaten the small and vulnerable community. The very existence of the Linnean was a reminder of the fissiparous fate that befell the Royal Society of London in the early nineteenth century.

In fact, fears of fission were unnecessary. Of the 124 founding members of the Linnean Society, only eighteen were already members of the Royal Society, and fifteen who joined the Linnean in 1875, also joined the Royal in either 1875 or 1876.[35] But he was correct to see that dual loyalties – easy enough for the well-off medical practitioners who composed the majority of the pluralists – were difficult to sustain among the rest.

Liversidge served himself for a time on the Council of the Linnean, but focused on the central task of restoring the senior society. Partly for this reason, he cultivated society membership in Britain, Europe and elsewhere. To some he already belonged, including the Chemical Society of London, which he joined soon after he knew he was to leave England in 1872; and the Geological Society of London, to which Daintree had nominated him in 1873. In January 1875, he was elected a Corresponding Member of the Royal Society of Tasmania,[36] and in 1876, a member of the Senckenberg Institute of Frankfurt, the Physical Society of London, the British Association for the Advancement of Science and the Mineralogical Society.[37] He had little prospect of attending their meetings, but by belonging to them, he kept abreast of their activities, and brought their news to his Council.

Most important, however, Liversidge set a good example. During the eight monthly meetings of 1876, he delivered three papers, which contained his first analyses of moss gold and tin, and his observations on siliceous deposits from the Richmond district.[38] In

September, he reported on a mineral that Charles Wilkinson had found, which he thought to be *Laumontite*, hitherto known only in Europe.[39] To the Society's library, he donated reports from the Physiological Society in England and catalogues of minerals in the British Museum. He displayed novel specimens of magnesium oxide found on the Pacific Ocean floor by the *Challenger*, and given him by Wyville Thomson. The result was to create around him an atmosphere of enquiry, a kind of 'senior common room' for colonial science.

In April 1876, Liversidge launched yet another offensive. For the Society's largest *conversazione* to date, held on 3 May 1876, he printed 500 catalogues, booked the Masonic Hall, and arranged for a new portrait of Clarke, commissioned by the Society, to be hung prominently at the front door. Turkey twill covers hid the imperfections of battered tables borrowed from the University, and sofas and chairs were provided for the ladies. Some 300 visitors came and viewed over eighty-five pieces of apparatus, photographs, books and *Challenger* artifacts. Liversidge outdid himself with spectroscopic illuminations, and showing specimens of aluminium, gemstones and Aboriginal weapons.

With ecstatic reviews ringing in their ears, Russell and Liversidge then turned to other substantial reforms, scarcely troubling members in the process. On 17 May, Clarke delivered what proved to be his valedictory Anniversary Address, in which he broadly endorsed the changes they had proposed. Clarke had set the Society on its course of reconstruction in July 1867; he had inspired Liversidge and Russell to action in May 1875; now it was Clarke who carried the tablets from the mountain. Four great reforms were needed. First, the Society needed specialist committees, from physics to sanitary science, to encourage research by 'mutual assistance'. Second, the Society must reorganise its finances, and rid itself of non-paying members. Third, it must seek a charter of incorporation, so that it could own property. Finally, it must approach the Government for an annual grant, and funds to buy or build permanent premises.

Shortly before the May address, it became known that Clarke had been elected to the Fellowship of the Royal Society of London, the highest honour that British science could bestow.[40] The news spurred Liversidge to act quickly on Clarke's program, possibly as much in his honour as in his ambition. Within five days, Liversidge, Leibius and Russell had met and agreed to ask the Government for £2500 towards a building, and an annual endowment of £300, equivalent to that received by the Royal Society of Victoria. Four days later, they organised a deputation of eight – including Liversidge, Russell and Moore, with Clarke and their 'parliamentary members', Francis Lord, a major landowner, and J.S. Farnell, an ex-gold miner who began the Mines Department, and who would soon (in 1877) become the first Australian-born premier of NSW– and kept an appointment with Joseph Docker, a wealthy grazier, and John Robertson's Minister of Public Instruction.

Docker had been a member of the Philosophical Society in 1858; but had since played no ascertainable part in the Society's life. Now he was asked to shape its future. When he became minister, the *Miners' Advocate* described him as 'the weak point in the ministry's armour … given to old fogeyism and a desire to retard rather than advance good legislation'.[41] But perhaps a degree of 'fogeyism' was required. In return for Government help, the Society agreed to 'go public'. In a draft prepared by Liversidge, and signed by Clarke, the Council promised that it would not only create 'working sections' for research, but also give popular lectures, establish and make available a scientific library, and in other respects become 'a central institution in NSW for the collection of scientific work from all parts of the world'. As private resources were 'totally inadequate to carry out their extended scheme of usefulness', he added, only government aid could make 'their past labours and present capabilities of more use to the public'. To a ministry already facing charges of willful indifference towards the needs of the University and the Australian Museum, this was a chance to gain quick publicity. Docker greeted the proposal courteously, and to the Society's amazement promised an annual endowment of £200.[42]

It seemed as if the Council had for once knocked upon an open door; what it sought, it had been given. It rejoiced, little knowing that it would take another year – and another government – to see the money actually materialise. Without waiting for the windfall, however, the Honorary Secretaries set about restructuring the Society's program. Chairmen were appointed for the nine (later, eight) research sections, and each assigned a given day of the week for meetings and a monthly timetable. The idea appealed to more than thirty members who had never spoken at the Society 's meetings, and who now became involved. Membership – 176 in 1875 – leapt to 308 in 1876.[43]

Among members who took up the 'sectional' interests, the keenest were the medical doctors of Section I (Sanitary and Social Science and Statistics). Perhaps they were a little too keen. In October, for example, at the instance of Dr Alfred Roberts, the Section sent a deputation to the Colonial Secretary with a petition, signed by thirty-eight, urging the passage of the *Public Health Act* of 1875. Acting in the Society's name, Roberts would have had the Society draft the Bill. Robertson went so far as to offer him the services of a parliamentary draftsman.[44] Fully aware of the crisis besetting Sydney's water supply and sanitation, no one on the Council doubted the need for an act to bring Sydney into line with Britain. Russell was prepared to agree. However, Liversidge decided that the Society should avoid direct intervention with parliament, and limit itself to offering 'general suggestions to the Government' – following the model not of the British Association, nor of the Social Science Association in England, but rather that of the Royal Society of London. There was reason to suppose that by giving more informal, less partisan advice, the Society would win greater respect and wield more influence. After several meetings, Liversidge's view was endorsed by Clarke and Scott, and ultimately prevailed.[45] With it, came the prospect of receiving state support while remaining free from political entanglements.

This independence Liversidge fully exploited. In June, members were told that the Colonial Secretary had promised an annual grant.[46] The same month, Liversidge and Leibius promulgated new by-laws, consolidating powers they had exercised informally since 1875 to set up provincial branches (one at Newcastle was considered), and to introduce new administrative practices.[47] At the Council meeting in June, Liversidge announced (on Russell's authority) that the library he had assembled 'would not be open to the members generally until the Society has rooms of its own'.[48] What would have been seen as high-handedness in a Vice-President was tolerated in the Honorary Secretary and Treasurer. If members fell in arrears with subscriptions, their *Transactions* were cut off.[49] To ensure that defaulting members would not merely read the Society's proceedings reprinted in the daily newspaper, Liversidge cancelled long-standing agreements to send material to the *Herald* and the *Empire*.[50] Hereafter, the newspapers had to prepare their own features, which, obligingly, they did.

In other ways, Liversidge parlayed his position to the Society's advantage. When, for example, the Sydney Botanical Society offered to become a Section, it was quickly assimilated, together with its property.[51] When the Agricultural Society proposed to produce a cooperative section on public health, Liversidge was quick to agree.[52] Meanwhile, he and Russell kept up a steady stream of scientific papers – Liversidge in September on moss gold,[53] Russell on rainfall prediction – setting standards of productivity few members could match.

By far the most dramatic innovation came when Liversidge, as editor of the Society's proceedings, morphed its *Transactions* into a *Journal*. The older title, Russell and he found 'rather pretentious';[54] with the new, came an image of a dynamic, continuing organ of research, which at the same time kept sight of tradition. In changing the name, Liversidge also more than doubled the price – from five shillings to ten shillings and sixpence per annum – and set about making the Society's *Journal* work for its members.[55] The last *Transactions* (volume IX for 1875, appearing in 1876), was the first to print the Society's business and its papers, the first to be indexed, and the first to print abstracts of its Sections.[56] The vitality of the Society would be evident at a glance. Who could protest?

Inevitably, some thought Liversidge's pace too fast. The Society's Assistant Secretary of nineteen years standing, W.H. Latlett, found it onerous to attend the Society's rooms between 4 to 6 p.m. every Wednesday and three evenings a week from 7 to 10 p.m., so that members could call into the library after dinner. He turned down an offered rise to £100, and resigned, leaving Liversidge to deal with all the details of membership.[57] But this he did and more, and by the close of this breathless year, Liversidge was virtually running the Society. In October, Liversidge again took up with John Robertson the promised government grant, left in abeyance since June; and organised a second deputation for December. He received reassurances, but six months is a long time in politics, and when Robertson's Government fell in March 1877, the promise fell with him. Frustration replaced excitement, as attempts to meet Henry Parkes, Robertson's successor, failed. Parkes had little time for the Society; and even absented himself from a promised meeting on 13 April. Worse, as Liversidge discovered, when caught on the political turntable, neither Parkes nor Robertson were always capable of implementing their promises. By the Society's Anniversary meeting in May 1877, the future of state support, apparently so neatly sewn up in 1876, had again become problematic.

Liversidge refused to admit defeat. In May 1877, it was Russell's turn to give the Anniversary Address and a progress report. All agreed the last years had seen a transformation. Thanks to Liversidge's changes in the by-laws, a new Council of Management had been created, and chaired by James (later Sir James) Fairfax, part-owner of the *Sydney Morning Herald*, who took steps to put the Society's finances onto a business-like footing. The Sections had become a 'most important agency' for original research; twenty-five new periodicals strengthened the library; and the journal for 1876, although delayed at the printers, was on its way to the scientific capitals of the world. In two short years, the Council had acquired international visibility; and so 'become one of the most effectual agencies for making this colony factually known abroad'.[58]

In driving this transformation, Liversidge had taken 'the lion's share'; as Leibius was fond of saying: 'we never got a move on till Liversidge came'.[59] Such was his galvanising effect that later generations would speak of him – still only twenty-nine years old – as having 'practically refounded' the Society, 'organised its activities on proper lines, and made it the power for good it is today'.[60] Joseph Maiden ranked him with Denison and Clarke as one of the three most influential men in the Society's history, and in 1918, accorded him the title of 'our greatest living benefactor'. With his coming began what A.P. Elkin, in his centenary address to the Society, called simply the 'Liversidge Phase' in the Society's history. The very phrase evokes energy. To speak of Liversidge's 'scientific prestige, administrative skill and wide vision' was possibly premature.[61] But he had shown a remarkable capacity for translating visions into reality. Long after Smith and Russell, Rolleston and Leibius had departed, his enterprise remained a model few cared to question. It was in this that he became a symbol of the Society and of colonial science.

For the moment, however, there remained the realities of scientific life in the colonies. 'Building science', as Liversidge had discovered, began with getting a building first. Without premises, 'its members', Liversidge observed, 'remained merely tenants where they met'.[62] The Society's annual *conversazione* in September 1877 welcomed 500 ladies and gentlemen to the Masonic Hall on York Street, to experience a variety of entertainments, played to the accompaniment of the Band of the NSW Artillery. But neither they nor the Society's membership wielded votes enough to sway Henry Parkes. W.G. Murray proposed that the Government should construct a new building to accommodate all of 'cultural Sydney', including the Free Public Library, the Academy of Art and the Royal Society along lines similar to Burlington House in London.[63] The proposal would be revisited many times in the future. For the moment, the Society preferred to be under its own roof. But Parkes supported neither option, and countered the Society's request for £2500 towards a building with an offer of only £500, and even that on condition that the Society find the first £1000 itself.

In September 1877, the Liversidge-Russell partnership was briefly distracted by a bizarre attempt on Russell's life.[64] A bomb was sent to him at the observatory under mysterious circumstances, possibly having to do with staff claims for better pay.[65] Happily unhurt, Russell returned to Liversidge's side in October, to circulate subscription lists among their 300 members, and to print an appeal in the *Herald*. Meanwhile, they made a fresh application for the grant they had been promised eighteen months earlier.[66] From the record, it is difficult to know whether Parkes was persuaded by the Council's promises, by invidious comparisons with Victoria, by the sheer tenacity of Russell and Liversidge, or by a combination of all three. Whatever the case, within another month, the Council reported a modicum of success. The building appeal had failed, but Parkes had agreed to honour his predecessor's promise of £200, which would bring the Society's income to almost £800 per annum.

At the last monthly meeting of the year, in December 1877, Christopher Rolleston, veteran of many such campaigns, explained the details of the agreement. By the following May, the Government had accepted Liversidge's proposals of 1876. The Society's activities were no longer to be dominated by gentlemen-amateurs. Instead, according to Rolleston, one of the most gentlemanly among them, the Council would aim at nothing less than the 'spreading of a taste for scientific enquiry' by an outward show of public activity.[67] With its library growing by leaps and exchanges, 'I think,' Russell announced in delight, 'we have arrived at a most important period of our history, and much of our future progress will depend upon the course we now adopt.'[68]

If Liversidge enjoyed success in the city, he was destined to wait years for the university to realise his ambitions. Reform at Sydney University resembled a gentlemanly version of trench warfare, with tangled and opposing lines of principle swaying back and forth, producing great cost for little gain, and giving no one side final victory. Indeed, it is difficult to identify 'sides', as there were so few involved, their backgrounds, so dissimilar, and their personalities, so strong.

At Sydney, as elsewhere in British higher education, there were the customary frictions between the sciences and the humanities; but at Sydney these were accompanied by differences such as were especially visible in the colonies, as between representatives of traditional Oxbridge and London, between Scotland and England, and between Eton and the evening school. The key issues, as elsewhere, were access, breadth and depth. Access was vital and its achievement united all the professors. As Charles Badham told press and public in his Commemoration Address in 1876, the University must make it clear that it was:

> no inaccessible shrine for the glorification of a few, and no sullen fortress, in which a certain privileged band was to batten on the public revenues; but that we were a beneficent guild – a corporation of thoughtful and patriotic men, yearning for opportunities of usefulness, exhibiting every inducement that we could devise to allure the youth of this colony, sending forth our invitations to him that was near and to him that was far off.[69]

How best to achieve this, was another matter. While high schools remained few, student numbers remained small. And while matriculation required Latin and Greek, few could be expected to succeed. Even fewer could be expected to offer science. The King's School in Parramatta, it is said, taught science for fifty years 'before it was found possible or necessary to build a laboratory'.[70] For those that did, the curriculum was the issue that, with space, repeatedly divided the professors. For twenty years before Liversidge arrived, and continuing well past his time, the central question was, whether an unalterable, irreducible core of subjects should comprise the first or final stages of a liberal education; or whether that core should be altered or enlarged. In parallel, there were running battles over the need to encourage specialisation. It was popularly said, as the ADB recalls, that Liversidge won a place for science only 'after Homeric struggles' with the Faculty of Arts. In fact, the sciences had a good innings in the Board games that consumed the University. But if we are to follow Homer, Liversidge could not take Troy without a siege, and until his death in 1884, Charles Badham, the Principal, was his Hector, and classics held the field.

The first of several rolling volleys of curricular reform in Liversidge's time were fired at the beginning of 1875, when Deas Thomson, entering his fourth term as Chancellor, invited all four professors to express their views on the future of the University's courses and examinations. The request was prompted, as earlier, by criticisms of the University's failure to attract larger numbers of students, and by its apparent lack of 'utility'. In response, Liversidge wrote three pages; each of the other three limited himself to one. His was the enthusiasm of youth, pitted against the experience of age. Badham, as Principal, led the field with proposals to reduce the time allotted to chemistry and physics. Students, he said, were taking too long over Smith's lectures, and now there were Liversidge's classes as well. Having eliminated the science examination requirement in the third year, Badham now willed compulsory science away altogether.

'The scholarship of this University,' he declared, 'will never be what it ought to be, as long as the Curriculum requires that students of the First and Second year should take up the three subjects, of Classics, Mathematics and Physical Sciences.' He suggested cutting science by half: what survived would be left to others, more familiar with 'a subject which is foreign to me', he admitted, 'in more senses than one'. Fewer lectures would hardly hurt, Badham continued: 'a knowledge of the merest elements of these sciences will, in my judgement, be quite enough to satisfy public opinion and it is only in deference to public opinion that such subjects form any part of University training. They are ornaments of the memory which may be acquired at any time of life.'[71]

Smith declined to rise to Badham's red flag, refusing to refight a battle that he had already won. Instead, he sketched an ambitious division of duties among the professors, creating new and separate chairs for mathematics, for natural philosophy and for chemistry. He also spoke of requiring chemistry at matriculation, a romantic notion, given the rare traces of chemistry visible in the colony's few secondary schools.[72]

Liversidge was less romantic. First, he insisted that all undergraduates must study chemistry and physics in the first year, and that geology and mineralogy lectures should be moved from third to second year, where they could become the basis of advanced work. Second, he proposed that, in the third year, there should be a choice of specialisation in any one of three 'Honours Schools', so defined, for classics, mathematics and natural science. These suggestions, he indicated, were along the lines of those recommended by the Devonshire Commission and already accepted in London. To staff the three Schools, Liversidge recommended lecturers in a range of subjects, including law, history, geography, English literature and biology ('the want of well-grounded instruction in [which] … ,' he observed, 'is a great drawback to the Geological student'). To round out these renovations, he added detailed suggestions for timetabling and for the addition of 'modern science books' to the library.[73]

By the 1870s, the University had acquired considerable wealth by private donations, but this could not be used for new appointments, nor new faculties. If Liversidge desired to implement the Devonshire recommendations, new income would be required. For this, he needed the help of his closest colleague. But by the late 1870s, his closest colleague had become his least certain friend. It is unclear to what extent his interventions in the 'water question', or in the curriculum debates, or at the Royal Society had unsettled his relations with John Smith. But by the end of 1876, if there were no animosity, there was little warmth. Smith's energies were distracted by the Council of National Education, and by growing disenchantment with Senate, which continued to back Badham. Despite heart-rending appeals from his wife, the lodgings in the Blacket Building that Pell vacated when he retired went to the large family of the classicist, and not to the smaller family of the chemist.[74] The fact that Smith believed he was paid less than Badham was also an irritation and interpreted as a slight against science.[75] Personal tensions mounted when, in 1876, the Board of Studies, led by Badham, asked Smith to 'give two lectures a day instead of one at present'. He declined, and offered to retire, on the same terms as Pell.[76] Senate was not amused. Gradually, Smith began to withdraw, and by the early 1880s, when he went on overseas leave for a third time, he had given way to a lingering desperation, made worse by chronic illness.

Even as Smith declined, so Liversidge rose. His prospects improved in May 1876, when two vacancies on Senate were filled by William Macleay and Henry Russell. Macleay was a generous patron, if not always a reliable ally; but Russell was both, and in April 1877, following a recommendation by Deas Thomson and Russell, Senate appointed Liversidge the Hovell Lecturer in Physical Geography and Geology. This added only slightly to his teaching but usefully to his salary, which was now an agreeable £800.[77] Russell, already a friend at the 'Royal', became an important friend in the Senate chamber as well. As a Sydney graduate, he spoke with authority on issues where Liversidge could not. It was Russell who helped Liversidge get Senate approval of £150 for his geological laboratory in 1874, and of a further £250 in 1875.[78] It was also Russell who ensured that he was repaid £244 for laboratory supplies he had ordered – apparently without permission.[79] The partnership that reformed Elizabeth Street worked as well at Camperdown.

In 1876, Deas Thomson asked the Parkes Government to increase the University's endowment from £5000 to £9000, to enable it to make new appointments of a 'useful' kind. This, the first big push for funds since 1859, would raise Sydney to the level of its sister university in Melbourne. It would permit junior appointments in law, history and English literature, as well as a medical school and a Faculty of Science 'as in the London University', such as Liversidge had recommended in 1873. With new chairs in biology and anatomy and physiology, junior appointments in chemistry and mechanics, and lecturers in 'subjects necessary for the medical profession',[80] the University could 'enter upon a new and much-extended course of instruction of great practical value to the youth of this Colony intending to engage in any [of] the technical or scientific professions'.[81] Parkes hinted that he would support the proposal, but fell from power on 16 August 1876. Undeterred, Deas Thomson sent the same proposal to the new Premier, John Robertson. The idea was politically attractive, and was received sympathetically by Joseph Docker, Robertson's Minister of Justice and Public Instruction, who happened to be a friend. In August, however, Docker announced that he could not sanction an increase before the end of 1876; and that was too late for the next academic year. For Liversidge, it was time to take a holiday, and reflect.

## 3. The Trans-Tasman Connection

After four eventful years, Liversidge had learned how to play the rudiments of harmony in New South Wales. Even so, there were places where science was better established, and where experienced practitioners could give him master classes. He looked across the Tasman. For some time, William Macleay had bewailed the difficulty of getting information from brother scientists in New Zealand.[82] Liversidge took the cue. During the Christmas and summer vacation of 1876–77, while the Royal Society rested from its revolution, and the University from its struggles, he took his first overseas trip since emigrating to Australia, and the first of three journeys he made to New Zealand. Hitherto, his experience of the Pacific had been vicarious. Sometimes, he received mineral specimens to analyse from travellers and missionaries – in October 1876, for example, the Rev. George Brown, the famous Methodist missionary, sent him a group of native figurine objects found in New Ireland, carved in what appeared to be chalk. He read up the literature and found that James Dana, in his *Corals and Coral Islands*, had speculated on the nature of limestone among the atolls. He had gone further, examined the chemical composition of the specimen, located it in the cretaceous period, and offered a link between the geologies of Queensland, New Guinea and New Ireland. In passing, he had commented on the possibility of commercial use.[83]

Such pleasant armchair analysis had its rewards, but it did not replace personal discovery; and in any case, Liversidge wanted to see things for himself. He had formed a high opinion of scientists in New Zealand – many of whom, as in Australia, had been appointed through the ubiquitous influence of Murchison and Sir Joseph Hooker at Kew – the 'spiritual overlords' of colonial science[84] – and thanks to his exchange of periodicals, Liversidge knew much of their work. Within the first thirty-five years of settlement, New Zealanders had established a Geological Survey, a Colonial Museum, a Colonial Laboratory and a Colonial Observatory, together with museums in the four largest towns.[85] There were good reasons to admire James Hector – since 1865, Director of the Geological Survey and Museum in Wellington, and the 'czar of New Zealand science' – and Julius von Haast, discoverer of the Moa and director of the Christchurch Museum.[86] Both were top of Liversidge's list, when it came to offering honorary memberships of the Royal Society.

Moreover, university science in New Zealand seemed as well or better provisioned than in Australia. The colony's oldest university at Otago, founded as recently as 1869, had established a chair in geology, and chemistry flourished at the University of Canterbury (established in 1873) in the person of Alexander Bickerton, Liversidge's classmate at the Royal College of Chemistry – a world away and a decade earlier.[87] And in the politics of 'federalism', which Liversidge watched with growing interest, New Zealand had much to teach NSW. There was nothing in Australia to match the federal New Zealand University, with its constituent colleges, and the New Zealand Institute, a federation of the scientific societies of the different provinces, established by Hector in 1867.

Liversidge sailed to Auckland via Fiji on the RMS *City of Sydney*. Sitting among saloon passengers escaping the summer's heat, he read Darwin's account of his Pacific voyage in 1836 and conversed with the Rev. John Webster, the Methodist missionary. His collecting instinct turned to the ethnography of Kadavu in Fiji, where the ship made a coaling stop, and where a visiting naval surgeon gave him samples of mineral waters from the hot springs at Savu Savu on Vanua Levu. These had legendary healing powers and later made for an evening's entertainment at the Royal Society of NSW.[88] The artifacts Liversidge collected on Fiji became the basis of his 'exhibition collection' and some proved to be the subject of the last scientific paper he would ever write.[89]

On board ship between Fiji and Auckland, Liversidge made the acquaintance of Constance Gordon Cumming, the artist and travel writer, who immortalised their journey in a series of notes and water colours now held by the Museum of the Anthropological Department in Cambridge.[90] They seem to have made an agreeable party. Arriving in Auckland, they coached to the hot springs at Ohinemutu, on the shores of Lake Rotorua, where Liversidge collected samples of siliceous deposits and cinnabar; and to the celebrated Pink and White Terraces near Lake Tarawera, whose origins he later explained to the

PINK TERRACE. 257

The famous Pink Terraces at Lake Rotomahana, near Rotorua in New Zealand, January 1877.

Royal Society.[91] There he reflected on the role of hydrothermal action in ore genesis. The discovery that iron pyrites at one of the hot springs contained traces of gold raised intriguing questions, which went for the moment to the back of his mind.

Liversidge then went on alone, first to Wellington, where he met Hector, and then to the South Island and to Dunedin, where he met Captain Frederick W. Hutton, Professor of Geology and Natural History at the University of Otago, and a man of similar interests.[92] As a favour to Hutton, Liversidge analysed some rock samples held by the Otago Museum; and two resulting papers were read on his behalf to the Otago Institute, and published in the New Zealand Institute's *Proceedings*.[93] Then, sailing north to Christchurch, he shared chemical notes with Bickerton and met von Haast.[94] From the famous trio – Hector, Hutton and von Haast – Liversidge learned much. He rejoiced at the prospect of receiving New Zealand minerals – 'as you are aware', he told von Haast, 'it is quite a private matter of my own, there being no funds whatever in the University for making a collection'.[95] Looking to send gifts in return, Liversidge had less luck in finding a *Ceratodus* for von Haast, than Japanese curtains for his wife.

'Tua Pua, a celebrated wood carver' and two unidentified women, New Zealand,
photographed by Liversidge in 1877.

This trip to New Zealand left an indelible impression. It was the first time since leaving England that Liversidge had been entertained by professional men outside the circle of Sydney. He was eternally grateful.[96] By taking the initiative, Liversidge opened doors. When Hector first came to Sydney, in 1877, he was greeted with specimens of Australian gem sands, fossils and moss gold. A similar welcome awaited von Haast.[97]

Liversidge returned to Sydney on the *Wakatipu*, through heavy seas, arriving in time for the first week of Lent Term, 1877. That year, although he delivered only one paper to the Royal Society – on the occurrence of chalk in New Britain,[98] he completed another twenty papers for the Department of Mines, including a review of methods for extracting gold from sludge.[99] Returning to the University, he found his plans had made little headway. A fresh approach was necessary. Extracting gold from sludge could well be easier than extracting funds from government.

## 4. Return Engagements

The opening of 1877 saw fresh battle between University and Parliament. Deas Thomson, re-elected Chancellor for a fifth consecutive term, was determined to win the increased endowment for which he had fought in 1876. After a bitter political campaign, Robertson's government fell to Parkes in March 1877, but Parkes fell again in August 1877. To keep the politicians on side was a herculean task. Liversidge lobbied Macleay over whisky at Elizabeth Bay House in June, and over dinner with Badham, Roberts and Stephen at 'Barham', and at the Macleays again in August and October.[100] But the government was too busy to listen. When set against the reforms in secondary education, which saw a dozen bills tabled before Parkes' successful Act of 1880 – or of public health, which failed to win a consolidating measure until 1885 – the University's financial plight was a mild matter.

In June 1877, Deas Thomson renewed his attack, insisting that means, not ends, were at issue; and that the University was willing to open its doors to new subjects, if means were forthcoming. Months passed before Senate sent another deputation, which this time included William Macleay and Arthur Renwick, together with Smith (in this case, speaking for the Legislative Council) and Badham, as Principal. This time the lobby went to John (later Sir John) Lackey, a convict's son, pastoralist and the Minister of Public Instruction in Robertson's fourth administration. And this time, Deas Thomson asked for £10,000.[101] Lackey, well known to Senate, a Vice-President of the Agricultural Society, and a member of the Union Club, heard his appeals – but again, to no avail, as in December, his government lost before it could act. The University could only bide its time against the day when a sympathetic ministry remained in power long enough to see an increase through parliament. Not until 1881 did this happen, when at last the Government managed to find £5000, bringing the endowment for 1882 to £10,000, and paving the way for the first major expansion in the history of the University.[102]

But all this was yet to come. Outside the University, the year 1877 went Liversidge's way. He was pleased when the venerable Baron von Mueller of Melbourne sponsored him for the Fellowship of the Linnean Society in London, and when W.B. Clarke and Arthur Todd Holroyd (another Cambridge man) backed him for election to the Royal Geographical Society.[103] As if to claim all knowledge as his province, Liversidge also joined the Royal Historical Society, and began adding journals of history to those of science in Elizabeth Street.

In April, he cultivated Dr Joseph Hooker, Director of Kew, sending him samples of diseased sugar cane he had collected in Queensland, and by offering his services 'in any other matter in the colonies'.[104] In May, he welcomed as Pell's replacement, Theodore Gurney, a twenty-eight-year old bachelor, Fellow of St John's, Cambridge, and Third Wrangler in the Mathematical Tripos of 1873 – the year Liversidge should have sat the Tripos himself. Gurney was appointed on the recommendation of Sir Charles Nicholson and Sir George Stokes, after what Deas Thomson called 'a most careful enquiry, and an immense amount of trouble', and Liversidge recruited him immediately into the Royal Society. Although Gurney eventually deserted bachelor life, he remained loyal to the Society, and gave early promise of raising the University's reputation in mathematics.[105]

By the end of 1877, Liversidge could see some signs of progress. His mineralogical knowledge had benefited from his work for the Australian Museum and the Department of Mines which, he told Hector, had much more money than he to spend.[106] The Royal Society was coming along well – its membership had grown, its meetings were better housed; and thanks to an agreement with the Government Printer, it had a well-produced journal, which now appeared regularly. The Society, Russell marvelled, now resembled in Australia nothing less than 'the position held by the Smithsonian Institution in America'.[107] In the twilight of the year, Liversidge's shadow lengthened. Now it was time, as he put it, to 'look about me generally'.[108]

In October, while awaiting news from Macquarie Street, Senate approved Liversidge's request for a year's sabbatical. Officially, his plan was to attend the International Congress of Geology, to be held in Paris between 29 August and 4 September 1878. Unofficially, he wished to see England again. His request was uncontroversial. During the previous twenty-five years, Smith had taken overseas leave twice, both on half-pay. The first was to 'inspect the principal laboratories of Europe' and obtain new equipment.[109] Liversidge would do the same, but with a difference. This was the first time a Sydney professor was given overseas leave specifically to attend a conference; and it was the first time that a professor was given leave on full salary, and therefore not obliged to pay for his replacement.[110] As it turned out, it was also the first time that, such leave having been granted, it would be extended without penalty – in this case to a total of eighteen months, from December 1877 to May 1879.

The need to remain visible, a consideration when Liversidge left England, was underlined by his colonial experience. Liversidge now moved easily along the axis that linked the city with Camperdown. He had become a celebrity in a modest way, when lists of his publications, together with his mineralogical and ethnological specimens appeared at the Melbourne and Philadelphia Exhibitions of 1875 and 1876. In 1877, he began sending reports to *Nature* in London.[111] He had kept his network in good repair.

As for the leave, there were no complications. Senate found £150 to bring his friend Captain Hutton from Dunedin to teach the geology course in the Lent Term of 1878. Two men from Wilkinson's Mines Department, Mr Muir and Mr Dixon, covered his practical classes. And to help him make the most of his trip, Senate promised £1000 to cover purchases of geological specimens and apparatus.[112] As a roving commissioner, he would do more for his adopted home than many enquiries would later produce. And when the time came, he would return with a comprehensive program for scientific and technical education in NSW.

## 5. England Regained, Europe Discovered

The 1870s are rightly remembered as a climacteric in the history of geology. Giants of the heroic age of the 1840s and 1850s – Elie de Beaumont, Murchison, Lyell, Sedgwick and Agassiz – were passing from the scene, and in their place came a generation for whom the controversies of the 1840s were already a matter of history. Modern geologists deferred theory, pending the discovery of new evidence. And they were not disappointed. New data were emerging in what Mott Greene has called the 'third great age of discovery', as explorers ranged from the poles to the last unexplored islands of the Pacific. In the process, evidence began to outrun theory.[113] Around the world, geologists faced the problem of organising the information they had generated. Definitions of 'systems' and 'series' were vital, as essential as mapping techniques. This need for clarity and precision in description was felt nowhere more than among the geological surveys that were filling in the blanks of a shrinking world.

Global surveys could often be best approached internationally; and beginning in the 1870s, the nations of Europe and North America moved steadily towards practical forms of scientific cooperation. Postal, telegraph and patent services, weights and measures, eventually an agreed prime meridian (1884), were brought into harmony in ways that predated – perhaps presaged – the international courts of justice of the next generation. Early steps towards geological cooperation came as a by-product of 'fringe' meetings at Philadelphia's Centennial Exhibition in 1876, when American geologists proposed international agreements to govern geological mapping, nomenclature and classification. They also proposed that a meeting to work this out should take place at the next international exhibition, which was scheduled for Paris in 1878 – the first of a series that would continue for the rest of the century.[114]

Between May and October 1878, the French government held the first Universal Exhibition in Europe since the Franco-Prussian War of 1870–71. The Paris Exhibition served political as well as scientific ends, demonstrating the resilience and recovery of France from military defeat, and confirming French leadership in culture and design. As was customary, certainly since the Great Exhibition of 1851, other countries were invited to appoint commissions to send exhibits. As in 1867, when Paris held its last Exhibition, Britain invited the participation of all her colonies. Unlike 1867, however, this invitation met with an enormous response, and the cost blew out to over £80,000.[115]

Nonetheless, the idea of 'selling' the Empire was high on Britain's agenda. In 1867, the Australian colonies had been principally represented by Victoria and by Victorian gold.[116] In 1878, the premier colony of NSW set out to redress the balance. Planning began in November 1877, when the Robertson government agreed to contribute £5000 towards costs, and appointed Sir James Martin to be the President of the NSW Commission. The Executive Commissioner was to be Edward Combes, MLA for Orange, Liversidge's friend, and currently Robertson's Secretary of Public Works. The choice of Combes, a 'Martinian' opposed to Parkes, was politic and logical. Born in England, Combes had been educated at

The *Exposition Universelle* in Paris, 1878.

the École des Mines and the Conservatoire des Arts et Métiers in Paris, and after working in the Victorian goldfields, had returned to Paris to study art. In 1855, he had served as a commissioner to the Paris International Exhibition, and as an engineering aide to the Emperor, Louis Napoleon.[117] Returning to Australia, he was an obvious choice to lead the NSW delegation to the Intercolonial Exhibition in Melbourne in 1875. He was an equally obvious choice for 1878.

In late 1877, Martin and Combes, assisted by Jules Joubert – the polymathic secretary of the Royal Agricultural Society[118] – began by requesting NSW producers to send objects in five categories – Art, Education, Furniture, Textiles, and Raw and Manufactured Products. The first, third and fourth categories produced little; and under the second, came only a few photographs; but under the fifth, came a considerable quantity of fossils, rocks and minerals – including many specimens from Wilkinson and the Department of Mines.

The choice of an impresario to administer a portfolio dominated by minerals could have seemed odd, certainly to Liversidge and W.B. Clarke, both of whom had served

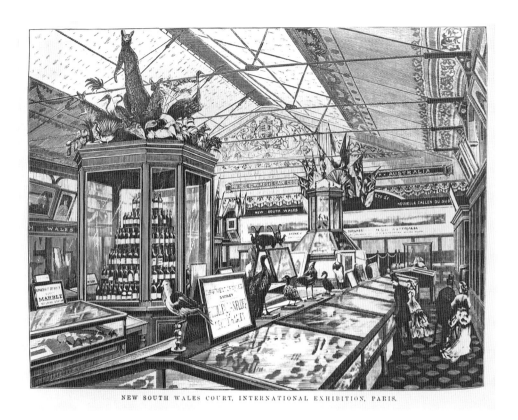

NEW SOUTH WALES COURT, INTERNATIONAL EXHIBITION, PARIS.

The NSW Court at the *Exposition Universelle:* raw materials jostle with stuffed animals,
*Illustrated Sydney News*, 31 August 1878.

on the 'minerals committee' for NSW's Philadelphian Commissioners in 1876. For Clarke
to show NSW to Paris would honour the clergyman's pioneering contribution. 'You would
be the "lion" amongst the geologists, as the companion of Murchison and Sedgwick in
the early days of geological work', Liversidge wrote his mentor and friend.[119] But this
was not to be. Clarke was deemed too old to make the trip and, as a trained engineer,
Combes felt competent to deal with mining. At a stroke, he therefore eliminated the
'mineral committee', discarding both Liversidge and Clarke.

Liversidge could scarcely conceal his disappointment. The decision seemed all the
more remarkable when, in December, as news of Liversidge's departure for the geologists'
congress became known, Combes had the gall to comment on the 'exceptionally favourable
opportunity' this presented 'for distributing information respecting the geology and
mineralogy of New South Wales'. Liversidge's experience in preparing for the Philadelphia
Exhibition – and what the Colonial Secretary called his 'inspiring diligence ... zeal and
sound judgement' – were suddenly in demand. These qualities were surely needed,
and Liversidge was belatedly asked to join the NSW Commission as a 'representative
commissioner'.[120] At short notice, Liversidge dropped everything to help Clarke prepare
the last revision of his classic *Sedimentary Formations of New South Wales*, and to complete
a catalogue of the minerals and objects that were destined for Paris.[121]

The voyage home was planned with customary care. Sailing north along the Australian coast, Liversidge passed through the Torres Strait to Singapore, then to Calcutta and across India to Bombay, visiting friends from Jermyn Street days *en route*.[122] But the NSW collections were delayed, and he was forced to delay his departure. Thoughts of India were postponed for a decade. It was not until Christmas Eve, 1877, that Liversidge sailed on the *Avoca* to Melbourne, trans-shipping to the RMS *Siam* for the long run via Point de Galle and Suez.

In the event, the NSW exhibits were sent separately, on the *Hankow*. But with Liversidge went a cheque for £1280 from the Australian Museum, with instructions to buy a microscope, bottles, mineral specimens and drawings of models.[123] From the Museum, he also took several specimens (including a stuffed kangaroo, birds and casts of fowl) to exhibit and to exchange. From the Royal Society, he took fifty copies of the *Journal*, together with £20 for the purchase of stationery and £50 for books.[124] From the University, he had £1000 to spend on equipment. Finally, he took the blessings and publications of his trans-Tasman colleagues, Hector and von Haast. Unable to be present themselves, they asked Liversidge to exhibit for New Zealand a case of chemical products, books and pictures.[125]

The voyage passed safely and in pleasant company. Among several dignitaries in First Class, Liversidge met Edward Knox of Sydney, chairman of the Colonial Sugar Refining Company, a Fellow of Senate and a member of both the Royal Society and the Union Club.[126] Knox, one of Sydney's wealthy industrialists, knew John Smith well, and in 1873, had been among the first to seek his help in using the polariscope to measure the sugar content of solutions.[127] For the next four weeks, saloon conversation turned to business, science and the future of the colony. Liversidge may have mentioned his teaching of chemists.[128] Knox needed over twenty for his technical staff by 1886, but until the turn of the century, all his senior men would be recruited overseas.[129]

Liversidge enjoyed the company of the financially successful, and may have been pleased to know that the ship's cargo held £355,635 in gold and gold sovereigns from the Sydney Mint, destined for the Bank of England. In January 1878, the passengers disembarked at Brindisi, and Liversidge stepped for the first time on the Continent. For the next four weeks, he travelled by rail. His Baedeker guided him to the principal museums, universities and bookshops of Naples, Rome, Florence, Bologna, Turin, Vienna and Salzburg, and then Munich, Leipzig, Dresden, Freiburg (with its celebrated School of Mines), Bonn, Cologne, Brussels, and finally Paris, where he arrived on 12 March. He took rooms in a small hotel at 1, Rue Desaix, near the Champ de Mars, and around the corner from what is now the Australian Embassy. There, above a pleasant *tabac*, he found the headquarters of the 'colonial mission', an easy walk to the Pont d'Iéna and the Trocadéro Palace, where the Geological Congress was to be held.

Before the Congress, however, came the Exhibition, and for this, the NSW Commission was woefully unprepared. Between 15 and 19 March, Liversidge learned the art of crisis management. The NSW display was to have been Joubert's responsibility; and indeed the secretary and his family left Sydney in good time, taking the *Avoca* on 22 January, and the connecting *Assam* from Melbourne to London. But owing to ferry delays in the Channel, they did not reach Paris until 17 March, and missed the first meeting of the Commission,

'The Mineral and Metals Trophy of New South Wales at the Paris International Exhibition',
*Illustrated Sydney New*s, 7 September 1878.

which by this time faced a disaster of major proportions. Of the exhibits sent on the *Hankow*, only a few had arrived – some tin and copper ingots, and a few cases of wool and wine.

After a worrying fortnight, the rest of the objects and minerals turned up, trans-shipped from England on the *Stadt Amsterdam*; but when they reached Paris, there was no place to put them. The pressure on accommodation for British and imperial entries, especially from Canada, was so great that NSW was assigned a double-booked space in the grand vestibule of the huge pavilion already allocated to India. It fell to Joubert to handle the diplomatic negotiations. Almost all the work of arrangement was left to Liversidge and his co-commissioner, Jacob Montefiore. Combes was nowhere to be seen. When the chief finally did arrive, he was too late to help.[130]

Although little more than a 'conscript commissioner', Liversidge was called upon to save the day. When the mineral labels were lost on the railways, he redid them from scratch. When, owing to Combes' absence, finances fell into disarray, it was Liversidge who set matters right, crossing to London to enlist help from William Forster, NSW's Agent-General.[131] He borrowed a day to visit his brothers, and to register himself as a Fellow of the Institute of Chemistry.[132] But when on 28 April he returned to Paris, he found Joubert battling with French printers to finish the NSW catalogue. On 1 May 1878, the Paris Universal Exhibition opened to great fanfare. But as diplomats lunched in the Grand Hôtel du Louvre, and as the first visitors entered the pavilion, the mounting of objects was still incomplete.

Even so, Liversidge was pleased to be there. Crowding out front page news of Kaffir Wars in South Africa, skirmishes on the North-west Frontier, and war between Turkey and Russia, the marvellous white stone Trocadéro Palace and the main building – erected in the Parc du Champ de Mars, across the Seine – were spectacular. The weather and the homilies of May Day lent a holiday atmosphere. Journalists surrounded the palace, and the major newspapers produced fully illustrated issues.[133]

For a month, the NSW court was only partly occupied, and in June, Liversidge went again to England to meet the *Assam*, with its long-awaited additional cargo of mineralogical maps, 300 copies of Clarke's book, and a description of NSW, prepared by Charles Robinson and prudently translated into French.[134] Eventually, the cases were finished and the exhibits were mounted. When the jurors made their reports, Canada and the colonies of the Caribbean and the Cape made good showings, and several of the smaller colonies were complimented on being represented for the first time. But the greatest praise was reserved for Australia. When the Prince of Wales was called upon to summarise the achievements of the Exhibition, he drew particular attention to the exhibits of coal and wool, for which NSW received a 'Grand Prix'; and to 'her scientific display of mineralogical and natural history collections, by Professor Liversidge'. Even Combe's report commented generously on the mineral exhibits, where 'a scientific and intelligible arrangement became a matter requiring patient care, and the work could not have been done, except by a mineralogist'.[135] This was pleasant enough. It was perhaps less pleasing for Liversidge to hear the jurors also praise Combes – despite his eleventh-hour arrival – for his 'continuous care' of the colony's exhibits.

At the end of August, prizes were handed out and Liversidge received a modest bronze medal for his services as convenor and juror in chemistry and mineralogy. It galled him to learn that the Colonial Office had recommended Combes, as Executive Commissioner, for a CMG. The French government created both Combes and Joubert Officers of the Légion d'Honneur.[136] Forster knew the facts, and as early as 15 May, cabled the Secretary of the Commission in Sydney, with a copy to the University, requesting the Colonial Secretary to acknowledge the government's appreciation of Liversidge's special services. The 'show', Forster knew, had consumed at least two months of his life. But it was too late. The moment had passed.

Not for the last time, public honours would elude him. But perhaps now, at least, Liversidge could get on with his work. The International Congress of Geology, one of eight professional meetings held during the Exhibition, opened at the Trocadéro in the heat of an unusually hot August, a 'fringe' attraction while the Exhibition was still in full swing. The geologists worked through a long agenda, the main purpose of which was to establish uniform rules for geological nomenclature and map colours.

For Liversidge, the benefits of the Congress transcended technical matters. For the first time, he had the intoxicating pleasure of meeting geologists from across the world. In this community, he was the source of news from Australia, introducing colleagues to NSW, its mines and its museum. Delegates made a point of noting his books, maps and specimens, the principal exchange commodities in the political economy of science. Their sheer novelty, coupled with an interest in Australian ores, was undoubtedly a contributing factor to the Committee's decision to elect the young Liversidge a Vice-President of the Congress. The geologists would not meet again until 1881, but Liversidge remained their Australian representative for the rest of the century.

That September, Liversidge was still in Paris to greet *la rentrée des classes*. From his hotel, he walked through the city of Degas, Renoir and Monet, buying works by Balzac and Hugo on the Left Bank, and strolling along Haussmann's magnificent boulevards on the Right. He learned of the patriotism of French *savants*, determined that another *debâcle* like Sedan should never occur – a movement that led in 1872 to the establishment of the Association Française pour l'Avancement de la Science, with its motto: '*par la Science, pour la Patrie*'.[137] He visited the fine mechanical museum models, poorly displayed in the ancient, ill-lit corridors of the Conservatoire National des Arts et Métiers, and in the sepulchral museum of the École des Mines. He met the great chemist Jean-Baptiste Dumas, dined with mineralogists at the Sorbonne, and was elected to the newly-established Société Minéralogique de France. He saw the mineral specimens of the Muséum d'Histoire Naturelle and arranged for exchanges with Sydney. It was all quite pleasant.

However, just as he was enjoying Paris, Sydney was finding new ways for him to be useful. In January 1878, Edward Deas Thomson, aged seventy-eight, Colonial Secretary for nineteen years, and the University's Chancellor, finally retired. His death the following year marked the end of an era in colonial affairs. He was succeeded as Chancellor by Sir William Montague Manning, KCMG, aged sixty-seven, a graduate of University College London, sometime Crown nominee in the Legislative Council, Solicitor-General in the colony's first Legislative Assembly, a Supreme Court judge, and a man who had helped run the University as a member of Senate since 1861.

Manning, who was appointed against the wishes of Badham and Pell,[138] was in some respects Sydney's first modern Chancellor. As such, he faced the twin tasks of defending the idea of a University, whilst dealing with the fact that it was largely inaccessible to the public. As an item of NSW colonial revenue, then exceeding £5 million, the government grant of £6000 (1878) to the University was trivial. In pounds per graduate, the NSW government seemingly ranked among the world's more generous patrons. But this argument had limited appeal to the practical people of NSW. Even worse, the Chancellor had to deal with the fact that, outside the legal profession, Sydney graduates had so far made little impact upon the world they were meant to serve. The four professors, despite their large salaries, might not see themselves as well paid. But before their prospects could improve, the Chancellor had to show how the University could better serve the public.

Manning sensibly began his tenure with another attempt to make the University accessible. The number of students, he said was 'greater than at any former time, but far below what might have been expected for the wealth and intelligence of the country'.[139] Mindful of Deas Thomson's setbacks in the early 1870s, Manning made some headway with John Robertson's government, but in November 1877, Robertson lost to J.S. Farnell, the son of a brewer, a drover and ex-goldminer, who led a 'third party' government between the rival factions. Farnell was not overly sympathetic to the University, nor was his Minister of Public Instruction, Joseph Leary.[140] However, whilst Deas Thomson appealed to noble sentiment, Manning appealed to political self-interest. In language that might find favour in Canberra today, he sought 'ere long' to see the University's ranks 'swelled, not only by a more general appreciation of the benefits of "higher education" … but also but also by an extension of our teaching-power in directions which may attract new classes having different views as to their pursuits in life'. But to 'take upon itself additional labour and responsibility', so as to reach 'the best intellectual capacity spread through all ranks of the people, it is plain that opportunities must be afforded for a greater variety of instruction, embracing professional and even the higher branches of technical education'. 'I doubt the sufficiency of £9000 for all these purposes', Manning added; the University actually needed £10,000.[141] If Farnell was not forthcoming, Parkes took the point. Perhaps, he reasoned, numbers could be boosted by courses aimed at the world of work. Manning took the hint, completing his project with an appeal for scholarships and the admission of women.[142]

On 31 July 1878, just a month after his first Commemoration Address, Manning wrote to Liversidge, asking him to review the systems of technical education operating in Britain and Europe, and of evening classes available in skilled occupations.[143] A week later, on 5 August, Manning sent Farnell his proposal to increase the endowment to £10,000 – to be followed by funds for new buildings. Liversidge alone needed £3000 to bring his laboratory 'up to the modern standard'.[144] Manning's conclusion followed directly upon his 'manifesto' of 1873:

> Having regard to the great strides that have been made of late years, which are daily being made in the various departments of science, and to their increasing value to society, and also to the greater estimation in which science is now held as a part of higher education, the Senate earnestly desires to be placed in a position to establish courses of instruction in its most prominent and useful branches.

'But', Manning continued, 'it is pointless to do so without further endowment.' 'In the direction of applied science and technical education,' he added, 'the Senate is not less anxious to make the University more serviceable to the country.' 'Its members,' he conceded, 'do not abate anything of their estimates of high classical attainments, both for their own sakes and as a foundation for various intellectual superstructures;' but they were persuaded:

> that the University of such an industrial community as ours should adapt itself more closely to the various views of its students concerning the occupations which shall constitute the future industry of their lives.[145]

Parkes agreed and, later that month, asked William Windeyer – judge, scholar and Member for the University in the Legislative Assembly – to use his influence on Senate in 'bringing about a thorough reform in that University'.[146]

Liversidge was not in Sydney to see his proposals debated, but Russell ensured that they were, and on 25 September, Senate approved and sent to parliament a draft 'Faculty Bill' (the University Increased Endowment Bill). In outline, this followed Liversidge's proposals of 1873 and Senate's recommendations of 1877, both providing for the establishment of a 'School of Science'. As before, however, Badham opposed any measure which, he said, would reduce the University to 'mere teaching for commercial gain'. Senate stopped in its tracks. Given Badham within and badinage without, Manning was undermined, and the Bill's doom was sealed.[147]

Manning had lost a battle, but perhaps not the war, and deliberately recruited Liversidge to his cause. In his instructions, he asked to be briefed on the whole subject of technical education, and not from the British perspective alone. He wanted to know best practice, particularly in the training of engineers. Engineering was the key that might open the gates to increasing numbers. The University needed a new approach to technical subjects. Elsewhere, reformers had won backing for a new technical college, and had the prospect of a government grant. Any such development had obvious implications for the University. And if he could not beat the 'technical push', perhaps he could join it. When they drafted the 'Faculty' Bill, Windeyer moved and Russell seconded a request that Senate consider affiliation 'with any literary or scientific institution, or any School of Applied Science'.

The ends of this ecumenicalism became clear at a special session of Senate on 13 November 1878, when a letter was read from the Engineering Association of NSW, a body formed in 1870, and meeting that week in conjunction with the Trades and Labour Council. The engineers asked the University's cooperation in promoting technical education. In return, the University expressed 'a warm interest in the question' and a desire to 'promote it by every means in its power'. There might be new money to be had. Between 1875 and 1877, John Lucas, Secretary for Mines under Robertson, had tried to persuade his colleagues to spend an unexpected surplus on new schools of mines and design in NSW.[148] These might go to the University. Senate went no further than to say that it had commissioned Liversidge to obtain information, without which it could not 'enter more fully into the question with the cooperation of the engineering and other Associations directly interested'.[149] But clearly some such development was on the cards.

The engineers' approach to the University was made with the nominal approval of Norman Selfe, President of the Engineering Association, and a staunch advocate of practical, rather than academic technical training.[150] But the gesture was not repeated. By January 1879, Manning's opportunity had passed, and the Association turned towards the Sydney Mechanics' School of Arts. In 1873, the School of Arts had established a Working Men's College, with which Selfe was connected.[151] Together, these two bodies began discussions on the 'best means of improving technical scientific instruction … and the expediency of applying to the Government for the sum of £1000 for payment of lecturers and teachers'.[152] The timing was important. In December, Farnell's government fell, and the factions of Parkes and Robertson fused to form a coalition which turned out to be the strongest administration in NSW since 1856. This gave Manning and Liversidge another chance to bring technical and university education into harmony, before rival interests tore them apart.

Thousands of miles away, Liversidge was making the most of his stay in Europe. Leaving Paris at the end of the Congress, he took rooms in London at Byron House, Savile Row, just off Piccadilly, a convenient base from which to visit former teachers and colleagues in Jermyn Street, and to attend the autumn 'season' of scientific society meetings in Burlington House. In September, he attended the British Association's Congress at Plymouth, and spent some weeks as a 'merchant' of materials, buying cases and chemicals for Hector and Smith. For the Australian Museum, he bought 1700 'carefully selected and valuable specimens, some of them almost unrivalled',[153] and spent up to the Senate's limit of £1000 on laboratory apparatus.

Aside from scholarly shopping, Liversidge also set up a number of exchanges, trading with museums for mineral specimens in return for stuffed kangaroos and fossil specimens left over from Paris. Some NSW minerals he gave to the Museum of Natural History, but received disappointingly few in return.[154] A better response came from Cambridge, where he sent a *Ceratodus* to the Zoological Museum, via a mutual friend, Arthur Dew-Smith, the gentleman-scholar and instrument maker. He promised to look out other specimens – including a platypus 'in spirit' – to join the one that Frank Maitland Balfour, Trinity's morphologist (and Liversidge's contemporary) had already acquired. Negotiations with Cambridge were always to Liversidge's liking. The Chairman of my Trustees, wrote Liversidge (thinking of W.B. Clarke) 'was a Jesus [College] man, and would pursue any such prospect [of exchange] with great interest'.[155]

In October, Liversidge joined the Savile Club, one of the newer and more fashionable clubs in London, founded in 1868 as a 'young man's Athenaeum', and frequented by Alexander Macmillan, T.H. Huxley, Norman Lockyer and other writers, publishers, artists and scientists. The Savile was recommended not 'as an environment of silence behind newspapers, but as a place where conversation, wit and argument were considered necessities of existence'.[156] Robert Louis Stevenson was a visitor, as were many medical friends. At the club, Dew-Smith invited Liversidge to the monthly meetings of the Physiological Society in nearby Jermyn Street. Beginning in 1876 under the leadership of Michael Foster in reaction against anti-vivisection sentiment, the Society had in four short years become a leading force in experimental biology.[157] At its meetings, Liversidge met David Ferrier, the psychologist and philosopher; and such leading physiologists as T. Lauder Brunton of Cambridge, J.R. Burdon Sanderson and E. Ray Lankester of Oxford, and Francis Darwin and George Romanes. Later he dined with William (later Sir William) Thiselton-Dyer, his contemporary at the RSM, and Sir Joseph Hooker's son-in-law, now heir apparent to the Directorship of the Royal Botanic Gardens, Kew.

From Dew-Smith came news of Newell Martin, their brilliant classmate at Christ's who, after much prodding by Huxley, had gone to the foundation chair of biology at Johns Hopkins in 1876. After months of homesickness, Martin had at last begun to enjoy his American life, and had recently married the widow of a Confederate officer, a popular Baltimore hostess. Martin had once playfully denounced their mutual friends Edward Schäfer, William Gaskell and Thiselton-Dyer for deserting what he called the 'noble army of bachelors'. Now, he wrote, 'I shall have to hurry up and try to do likewise or I shall find myself in the unenviable position of sole representative of a once flourishing body.'[158] He need not have worried: Liversidge, aged thirty-one, showed no sign of deserting the ranks of single men.

That October, Liversidge began to write his report for Manning. But he was not to be left in peace. Quite unexpectedly came a request from the Agricultural Society of NSW. The Agricultural Society had enjoyed great success at Sydney's Metropolitan Colonial Exhibition in 1870 and looked enviously at the press surrounding the Paris Exhibition. Spurred on by rumours that Melbourne proposed to stage a similar event, the Society had decided out to sponsor another, grander and this time, 'international' exhibition in Sydney. In November 1878, Farnell's government gave the Society an amber light.

Exhibitions were costly. However, they seemingly did not require much justification. The Great Exhibition of 1851 had produced a cornucopia of objects that then graced the South Kensington Museum, and the Intercolonial Exhibition in Melbourne in 1866 had left objects to the Victorian Industrial and Technological Museum in 1870. There was every evidence that a pragmatic government, already leaning towards the support of technical education, would contribute to an event that would bring Sydney an array of benefits such as South Kensington had brought to London.

Some credit for this line of reasoning can be given to Sir Alfred Roberts, Trustee of the Australian Museum since 1858 and Liversidge's friend since 1874. Roberts had been the driving force behind the Royal Prince Alfred Hospital and other worthy causes, but he was not always easy to deal with. He was Parkes's personal physician; but Parkes called him 'a fussy officious dilettante in all matters of sanitary reform, who spoils his own efforts to be useful by his desire to be *the* authority on all occasions'.[159] Be that as it may, he spoke with authority. In August 1878, Roberts convened a special meeting of the Trustees of the Australian Museum to consider the proposition that:

> a Technological and Industrial Museum, with classes for Instruction, would afford much valuable and practical information to a large class of the community; that it may be advantageously associated with the institution; and that the necessary accommodation might be provided in the building about to be erected for the exhibition of works of art.[160]

Liversidge was a Trustee, and so certain to help. On 15 August, the Colonial Secretary's office wrote London, asking the Agent-General, William Forster to form a 'collecting committee' of three, including Forster and Liversidge's *nemesis*, Edward Combes who, having arrived late in Europe, was currently enjoying Paris to the full.

These letters were slow to reach London, but by 9 October 1878, both Liversidge and Forster had received Manning's letter of 31 July, followed on 2 November by the Colonial Secretary's letter of 15 August. These were followed two days later by a telegram from Farnell's office, explaining that £500 had been placed on the estimates for 1879 to cover Liversidge's expenses incurred in collecting information 'relative to the working of English and foreign technology museums and colleges, with a view to forming similar institutions in Sydney'.[161]

The two letters caught Liversidge by surprise. 'It was my intention,' he explained to Forster, 'to leave England about the middle of November so as to spend six weeks or so in India on my way out to Sydney.' However, he added, 'under the circumstances, I am ready to forego that engagement in the hope that my time may be more profitably spent for the colony by endeavouring to assist in this attempt to advance the cause of scientific education in Sydney'.[162] The following day he wrote Manning in similar terms:

> [U]nder the circumstances I am ready, if necessary, to forego, although with considerable regret, this engagement [in India], in the hope that the time thus saved may be sufficient for the purpose, and that no interruption to the proper discharge of my duties at the University may occur at the beginning of the year.[163]

Although Senate had in August granted him two months' additional leave, the delayed mails did not bring him this news until December.[164] In the meantime, he might have acted on his own initiative, but he was reluctant to proceed until the promised money was in hand. Thus he wrote to Forster:

> I am afraid that unless the matter can be proceeded with very shortly, I shall have to return to Sydney without taking part in it. This I should regret very much as the subject is one in which I take a very great interest.[165]

Inevitably, the government's response was delayed by fresh parliamentary elections, and by yet another turn in the continuing game of musical ministries. In December 1878, Farnell was replaced by Parkes. But given time, Parkes was likely to approve the project.[166] Possibly realising this, Liversidge agreed with Forster that 'the necessary preliminary arrangements ... should be proceeded with inasmuch as they involve a very small expenditure'. A week later, he wrote to the Trustees of the Australian Museum, accepted their commission, and reached again for his Baedeker.[167]

Liversidge's work for the Geological Congress and the Paris Exhibition had finished. But his work for the University and the government had just begun – at short notice and at his own expense, at least until funds arrived. Under the circumstances, it was fortunate that, even before receiving the official letters, he had already:

for my own private information and satisfaction, visited many of the principal Museums, Universities, and other institutions in the chief towns of Europe ... as well as those of the United Kingdom. I had also previously taken advantage of the many favourable opportunities to obtain information afforded by the Paris Exhibition of 1878.[168]

Through November and December 1878, he collected information, living in London as long as possible and waiting for his funds.[169] He visited and 'carefully examined' six museums in London, and the museums of Edinburgh, Glasgow, Dublin, Newcastle, Liverpool, Oxford, Leeds, Salisbury, Exeter and Cambridge. He also visited Owens College, Manchester; Masons College, Birmingham; the Yorkshire College, Leeds; and the College of Physical Science at Newcastle.

In January, Liversidge summarised for Hector the outcome of his eighteen months' leave. All the jobs he had been asked to do had left him with 'but little time at my own disposal'.[170] Worse, he had now to leave England without completing 'the most important part of the work entrusted to the [Technical Education] Committee – the purchase of the models and specimens, and the preparation of plans for the new museum'.[171] In parting, he gave Forster a list of things to buy and send to Sydney, as and when the promised grant reached London.[172] He also arranged with Hooker and Thistleton-Dyer for eight cases of specimens to be sent to Sydney from the Museum of Economic Botany and the India Museum. These – eventually, the first specimens to be received by the Technological Museum – followed him from England in December.[173]

The English winter of 1879 proved hard, and when illness and exhaustion caught up with him, Liversidge postponed his departure until 26 March. But without extended leave, he could not delay much longer, and on 4 April 1879, he finally left Dover for the Continent – to Paris, where he again visited the Conservatoire National des Arts et Métiers; then to Turin and the Royal Italian Industrial Museum; and finally, to Brindisi, whence he sailed on 13 April 1879. At Suez, he joined the familiar *Assam*, and at Aden, the *Avoca*, for the six-week voyage to Sydney. The trip was long but uneventful, and gave him time to finish his report and reflect on what he had done.

Liversidge arrived in Sydney on 26 May, too late to teach chemistry to the twenty first-year students who had finished Lent Term without him. Smith had covered for his Demonstrator, it seems, and was grateful for his return. So, too, was the Royal Society, which three weeks later gave him a formal dinner. Smith, in the chair, buried any resentment that might have lingered from the 'water debates' of three years before. Indeed, he congratulated Liversidge on the completion of his 'persistent labours, which', he trusted, 'would receive some fitting recognition'.

In the event, no such recognition came. The resourceful Combes had stolen the glory. And while imperial honours went to him and the Victorian commissioners, and while even New Zealanders who had not attended were thanked, there was nothing for Liversidge beyond a fulsome paragraph in the NSW commissioners' report. 'Hereafter', they said, 'the Colony will reap benefit from [the government's] expenditure, by attention being drawn to its mineral resources in the classrooms of the Universities and Technical Colleges of Europe and America ...'.[174] Scant praise, one thought. As in all good theatre, the production staff remained invisible. Liversidge had, he said, 'simply done his duty'.[175] His reward lay in the notes he had taken, the instruments and artifacts he had acquired, in the places he had seen, and in the contacts he had made. Much of this surfaced in his report to Manning and the government. Beyond this, the future was quite unclear. Like the prophet, he might not live to see the promised land. But at least he held the tablets in his hands.

# CHAPTER 7

# The Politics of Practical Men

The Australian Museum, about 1887.

After an absence of sixteen months, Liversidge sailed into Port Jackson on 26 May 1879 with three compelling ambitions. First, he was determined to promote a closer association between men of science in Australasia and the rest of the world. Memories of months of fruitful conversation in London and Paris were a compelling reminder of his isolation in Sydney. Second, he was determined to establish at the University a 'School of Science' to equal any in Europe. Third, he was determined to extend the scope of technical education, linking teaching and practice such as he had known in England. In all three domains, Liversidge applied the same draughtsman-like adherence to accuracy and dispassionate analysis that he had demonstrated in the Krefft affair, in his mineral surveys, in the 'transit' observations, and in his reports on water pollution and sugar cane disease. Now, however, he was to expand his brief even wider, exploring the limits of colonial politics, and becoming a 'practical man' in all but name.

## 1. Museums, Mining and Technical Men: The Liversidge Report

'It is probably unnecessary for me to urge anything as to the necessity which exists for the establishment of a Technological and Industrial Museum in Sydney.'[1] With these confident words, Liversidge began what would become his famous report to the Minister for Justice and Public Instruction, and to the Chancellor of Sydney University. His confidence was not misplaced. Liversidge was born to the transforming power of science and industry. His education was a direct consequence of the enthusiasm that flowed through the Great Exhibition of 1851 and the proposed 'Albertopolis' for South Kensington. Nonetheless, few months passed without popular Cassandras lamenting the country's alleged neglect of science. In 1867, the year that Liversidge began at the RSM, chastening reports from the Paris Exhibition and the Taunton Commission were used as evidence of Britain's failure to make national provision for technical education, innovation and design.

The causes of Britain's 'industrial malaise' were exaggerated, oversimplified and misunderstood.[2] But reformers saw cause and effect in the absence of a system of science and technical education. Before 1870, formal education for the industrial working class rarely reached beyond the mechanics' institutes and evening classes that Liversidge attended. For the middle classes, science seldom featured in secondary, let alone primary education. By the 1860s, thanks to the efforts of the British Association, but just a little too late for Liversidge, there was consensus that science should occupy a place in every secondary school.[3] The science to be taught might differ, but it would reflect relatively clear categories of method and use. Huxley preferred the image of a vineyard and its fruit. Technical education, however, was another matter. At a time when most (like Liversidge's brothers) learned their trade by apprenticeship, there were no universally accepted views as to what constituted a suitable education for trade, industry or manufacturing. This problem was notorious, but by no means unique, in the teaching of chemistry.[4] As long as 'scientific and technical education' had no unique definition, debates about content and methods would lead to semantic quicksands from which educational policies were impossible to retrieve.

Nonetheless, by the 1880s, certain beliefs, when confirmed by the Royal Commission on Technical Instruction in 1884, acquired growing credibility among progressives of Liversidge's generation.[5] These held, first, that different skills required different ways of learning. Second, that no discipline possessed an inherent superiority. Insofar as mental training was involved, a pupil could benefit as much from physics as from philosophy, from crystallography as from classics.[6] Third, preparation for industry required a knowledge of principles. Workers must learn science as part of general education, before 'trade schools' taught them special skills. Finally, 'science as content' could not be separated from 'science as method'.[7] For Karl Pearson, scientific method was as essential to the understanding of art, philosophy, ethics and music, as to the understanding of nature.[8] Science as 'method' was the special message of the laboratory, and also the museum, with its emphasis on 'learning by hand and eye'.[9] Underlying this belief were the principles of political economy, and the presumption that knowledge was power and profit.

These were the assumptions upon which Liversidge based his report of 1880 – the first systematic survey of museums and technical education undertaken for an Australian government, which became a bold blueprint for policy in New South Wales for nearly a century. Within its pages lay a vision that, once realised, would bring South Kensington to Sydney. Liversidge's report was submitted on 20 April to Francis Sutor, Minister of Justice and Public Instruction in Parkes-Robertson's coalition. It had taken over a year to write and another three months to publish. His task, as he saw it, was not to argue a brief, but simply to tell the government – and Sir William Manning – what others were doing. The evidence spoke for itself.

The report was set out in two sections. The first, on museums, described ten institutions that Liversidge had visited – eight in England, the Conservatoire Nationale des Arts et Métiers in Paris and the Industrial Museum in Turin. The second, on Scientific and Technical Instruction, canvassed by mail 107 courses and institutions, including many places he had not seen – forty-two in Britain, sixty on the Continent, and one – the Massachusetts Institute of Technology – in the United States. He began with a description. A technological museum, he wrote, should begin by assembling typical materials of economic value, whether animal, vegetable or mineral. Then, he set out a framework of categories that should feature in the museum, ranging from foods to minerals, sanitary appliances, mining and agricultural machinery and photography, even a reference library of trade catalogues – a complete 'university' for working men. Everything in such a museum should be properly labelled, illustrated by lectures and accompanied by loan collections for circulation. If all categories could not be accumulated at once, they should be gathered systematically, 'naturally', and 'not prematurely, ascending to possibilities which may in due course reasonably be expected to become accomplished facts'.

Next, a technological museum should have a chemical laboratory, so that 'new and other raw materials, as well as waste products likely to be useful … may be examined and made the subject of experiment'.[10] Given his experience of the Mines Department, it is unsurprising to read his recommendation that 'the capabilities and commercial value of new or but little known products, whether of natural or artificial origin, should be fully ascertained by the researches of trained experts, rather than imperfectly from the crude experiments of incompetent and inexperienced persons'. It was clear that 'the losses

The crowded interior of the Technological Museum, about 1886.

sustained by unsuccessful manufacturers fall upon the community at large, and are not confined solely to the individuals at fault'.[11] The state owed better science to the people.

On the question of governance, views differed.[12] The first and oldest public museum in England was the Ashmolean in Oxford, but this was managed by its founder, before being given to the University. Within the government domain, one could look to the Royal Observatory at Greenwich (and its counterpart in Scotland), whose management was vested in an expert director (the Astronomer Royal), and reviewed by a lay Board of Visitors, reporting to a department of state (in this case, the Admiralty), which was answerable to parliament. The Museum of Applied Geology in Jermyn Street and the South Kensington (later the Victoria and Albert) Museum had professional directors, appointed by the Science and Art Department, who reported in this fashion. This model was common in the colonies, and corresponded to the system followed by Government Astronomers and Botanists.

A second model was followed by the British Museum which, while having its origins in private collections, was ordered by an Act of Parliament. Although the BM received public money, it was governed by a board of forty-eight unpaid Trustees, twenty-three appointed by the government of the day, nine representing interests in the Museum's collections, and sixteen elected (if not self-elected) by the others. The three so-called 'Principal Trustees', who made staff appointments, including the Director, were the Archbishop of Canterbury, the Lord Chancellor, and the Speaker of the House of Commons.

The Laboratory and Ladies' Reading Room, Sydney Mechanics' School of Arts, about 1880.

The merits of these models could be, and were, debated. The trusteeship model – combining lay, professional and governmental interests – served well for quasi-public institutions, including hospitals and municipal charities.[13] It reflected Treasury cheese-paring to be sure, but also a willingness among ladies and gentlemen of independent means to serve the community.[14] It was also widely copied in the colonies; and in Sydney, the Australian Museum was a prime example: run by a committee of public spirited individuals from its foundation in 1827, and by trustees – elective and *ex officio* crown nominees – from 1854.

For Sydney's technological museum, Liversidge favoured the latter model – a committee of unpaid trustees, some nominated, some elected, with powers to appoint a paid 'Curator' or 'Director'. The Free Public Library (1869), the (National) Art Gallery of NSW (1876), and other cultural, literary and charitable institutions were similarly governed. The Sydney Mechanics' School of Arts, established in 1833, was run by a Committee of Management (effectively, Trustees).[15] In such cases, voluntarism could (whether by rule or exception) be aided by public grants.

However principled, this model could fall victim to personalities, especially in small communities, where public aid to private institutions was virtually a condition of their existence. As accountability could not be left in a state of ambiguity, such as to require legal opinion at every turn, public opinion turned to favour greater public control over what were in effect public institutions.[16]

Liversidge's experience of the 'Krefft affair' disposed him to put questions of governance permanently beyond dispute. He was a product of the state-administered, professionally-run Jermyn Street model. Yet, he saw virtue in what the Devonshire Commission recommended in 1874 as befitting national museums generally – namely, a 'director appointed by and responsible to the government', but with a 'Board of Advice' to provide a buffer between professional and departmental interests.[17] As for a name – why not the 'Science and Art' Museum, or perhaps the 'Albert Museum', in memory of the 'originator of all English Museums of Science and Art'? Knowing what he did of the parentage of South Kensington, Liversidge recommended that colonial governments take charge of the proposed Sydney International Exhibition, and of any subsequent Melbourne exhibition, to secure for posterity the collections they might charitably acquire.[18]

The Museum's management thus settled, how best to organise its collections? On this question, Liversidge was silent. The reason was clear. 'Cathedrals' of science they were called, but Liversidge knew better than to express an *ex cathedra* opinion.[19] There were at least three traditions in the history of European museums. The first, descended from Renaissance *cabinets de curiosité*, celebrated objects, natural or man-made, as much for their singularity and appeal. The second brought a secular view of knowledge presented for a reason and conveying a message. The organisation of artifacts in the Jardin des Plantes (later the Muséum d' Histoire Naturelle) in Paris,[20] and emulated by museums and zoological and botanical gardens throughout Europe, told the story of nature. Fossils and minerals joined stuffed animals and plants in narrating the history of life on earth.

A third tradition, born of the eighteenth century but consolidated in the nineteenth, reflected a similar message in man-made artifacts. In France, collections of objects and models derived from the progressive spirit of the *Encylopédie* and the revolutionaries who created the Conservatoire National des Arts et Métiers in the abbey gardens of St Denis.[21] By the 1880s, industrial museums were by-products or consequences of the international exhibitions that moved across the world. What stories would they tell? On this subject, there was much to be said. Different messages could be conveyed in different ways. Mineral specimens, for example, could be set out according to geological significance, chemical composition, or economic utility. Inevitably, presentation alters meaning. On the story to tell, Liversidge took a neutral line. The first task of Sydney's new museum was to label and classify – to inventory knowledge. Interpretation would come later.[22]

What this policy lacked in sophistication, it made up for in common sense. When his report moved from 'Museums' to 'Scientific Education', it was clear why a practical strategy was necessary. Liversidge distinguished between the museum interests of the 'professional scientific man' and those of the 'skilled artisan'. The first presupposed a 'good general or liberal education', followed by instruction 'of a general scientific character, to serve as the necessary groundwork ... upon which his special professional education can be built'. The second required a general education, followed by

> elementary instruction in certain branches of science according to
> his future occupation, sufficient to give him an intelligent interest
> in the principles of his trade ...

For the professional aiming at 'higher technical education', Liversidge favoured the technical schools of Europe, notably those of Zurich and Munich, but for these, Sydney was not ready. 'Even on a reduced scale,' he wrote, 'it would be utterly hopeless to entertain the idea at the present time.' The colony's small population and lack of 'feeder' schools would defeat the idea. Indeed, he ventured, 'if a special institution for high technical education were to be started at the present time ... it would not prove a success for some years'. To 'avoid the discouragement to which this might naturally give rise', he suggested that, as an alternative strategy, greater use might be made of the University,[23] especially in engineering, mining, agriculture and forestry, surveying and architecture – subjects which had it 'in their power to do much towards developing the resources of a new country such as this' – and which 'could and should be taught ... [at the University] without any fear of losing caste'.[24]

Whether these words were written with Manning in mind is not known, but they made a salient point – namely, that giving new tasks to the University would appeal to the government. In any case, Liversidge saw no conflict between the aspirations of working men and the future of the University. Refreshing the memories of those who had forgotten his letter of 1873, Liversidge set out a three-year syllabus for a School of Mining and Geology, combining subjects that the RSM had taught him, with the mathematics, natural history and foreign languages that it had not.

When it came time to offer his recommendations, Liversidge drew upon the Devonshire Commission, and particularly its second, third, fifth, sixth and eighth reports. It made sense to introduce a school-leaving examination like that proposed for England; to require students to study both literary and classical subjects; to expect research as well as teaching from professors; and to institute a pension scheme for their security. Fellowships should be awarded for original research, and higher degrees, on the basis of research, as in France and Germany, rather than by examination.[25] All these views reflected a consensus emerging in Britain.

With personal experience of evening classes that most academics lacked, Liversidge recommended that the voluntary efforts of the Sydney Mechanics' School of Arts and the Working Men's College, both of which received public money, should be supplemented by 'a regular system of assistance and supervision', amounting to a Science and Art Department, such as that in England, which gave grants for rooms, laboratories, and apparatus, paid teachers, and awarded scholarships to needy students. As for subject matter, 'plans for evening classes must not be too ambitious', they should be limited to eight of the 'cooler months of the year', and kept at a pace that suited men and women at the end of a working day.

Liversidge reserved his final comments for trade schools that, in Europe, had generally replaced apprenticeship schemes. Australia, like Britain, preferred to train apprentices, but opinion was moving towards formal schooling. Liversidge cited a lecture four months earlier at the Society of Arts in London, in which T.H. Huxley called for a system of education in drawing, design, and geometry such as was offered in Paris, Le Havre and Douai.[26] Similarly, Liversidge recommended training in a variety of practical skills, from wool weaving to masonry and stone carving. Above all, he argued for a flexible approach, suited to the resources of the colony and responsive to its needs.[27]

The Intercolonial Exhibition in Prince Alfred Park, 1870.

Liversidge's report of 1880 confidently expected rational outcomes. But even as it appeared, his recommendations fell prey to politics. On the one hand, the Devonshire Commission, on which Liversidge based so much of his text, turned out not to be what politicians wanted. On the other hand, there was tension between what Richard Yeo has characterized as the 'popular' and the 'profound', and no clear definition as to which subjects should be part of which.[28] Nor was it clear what relationship the subjects to be taught would bear to the economy of the privileged few, or of the representative many. Finally, if Liversidge's views appealed to educators, they faced a parliament reluctant to pay for them.

Undermining his efforts ran a deep vein of cultural pessimism, whose influence was difficult to gauge. The same year, the Rev. Julian Tenison-Woods, explorer and naturalist, warned the Linnean Society that, despite the 'scientific tendencies of the age', the prospects of the colony, never good, were unlikely to improve. His audience was discouraged (or confirmed in its prejudice) to be told that:

The Garden Palace viewed from Farm Cove, about 1879.

the circumstances of young colonies are so peculiar and exceptional
that it would not be fair to compare our literature with that of any old
established country ... our habits are not those of a studious people.
Men of real learning have no place among us, and are consequently
rarely to be found.[29]

With such a description of the past, Liversidge did not disagree. Yet, his eyes were set on the
future. His faith resided in a comprehensive strategy that would bridge differences of class
and politics. It was a brave plan and elements would be debated repeatedly during the next
ten years. Yet, for the next several months, Liversidge's zeal, and even his students, had to
wait. In the southern autumn of 1880, Sydney's first International Exhibition was rising in
the Domain, and all the world, it seemed, was coming to see the 'Paris of the Pacific'.

## 2. Sydney on Show: The International Exhibition

'Of all events in recent history,' John Allwood has reminded us, 'only wars have had more dramatic influence than Expositions upon the experience of civilization.'[30] International exhibitions were statements of the continuity of trade and the display of culture, international pride and public endeavour.[31] Like industrial museums, with which they were often associated, they recited in vaulting structures of iron and glass the lessons of progress – the respected message of learning by doing, and the expectation that prosperity flowed naturally from invention.

To manufacturers, they were events whose success was measured in prizes rather than profits; to governments, they were invitations to invest. To Marxists, they stood for the 'fetishism of commodities'; to Benthamites, the moral virtues of discipline and competition. To Henry Adams, they represented a 'new religion', substituting the icons of faith by the dynamos of technology.[32] To historians, they are part of the Victorian search for order, projecting a common frame of reference for both civilised and civilisers. They spoke both to imperial solidarity and for countries without colonies, 'visions of empire'.[33] To governments everywhere, they were assertions of national pride, made diplomatically safe by use of the language of universal brotherhood, promoting – in Tennyson's words – the 'Parliament of Man, the Federation of the World'.

In Australia, exhibitions were all this and more. Nowhere was their influence stronger, with colonial interests seeking international recognition and colonial governments seeking public approval.[34] For Liversidge's generation, they were also roundhouses of reform, containing both the actual engines of transformation as well as mythopoeic symbols of progress. Each presented a 'university of the world'.[35] Those who 'enrolled' would encounter, in the words of an American enthusiast, not merely displays of peoples and products, 'but the teachings of science and experience as regards their value, importance and use'.[36]

By 1878, the idea of holding an international exhibition in Sydney had been gestating for some time. Thanks to the Agricultural Society of NSW, the Intercolonial Exhibition of 1870 – commemorating the 'discovery' of Australia by James Cook – proved that Sydney could stage major events. In 1872, it was decided to encourage an intercolonial exhibition every three years, and in 1875 and 1876, the Society opened its doors to exhibits from Canada and the United States. A fully international exhibition was the next step.

In July 1877, amidst rumours that Melbourne planned to stage a major international exhibition in 1879, the Agricultural Society sent a delegation to Henry Parkes seeking government support.[37] Parkes greeted the proposal with caution. An exhibition would be large and uncontrollable and, since the Great Exhibition of 1851, almost every national event of the kind had made a loss. So the government procrastinated and no decision was made for months, during which time Parkes lost office, was returned and then lost again. Ultimately, in February 1878, the proposal went to the ministry led by James Farnell (1877–78), a member of the Linnean Society, and formerly Parkes' first Minister of Mines, whom Liversidge had known since 1874. Farnell, the first Australian-born Premier of NSW, was destined to be in office only until December, but he favoured the idea, and sanctioned the proposal – provided that the Agricultural Society sought no public money.

The question of cost was not easily answered. Some recommended a public appeal. Sir Hercules Robinson, Governor and conveniently also the President of the Agricultural Society, viewed the matter as one that belonged to government. Apparently without consulting his ministers – nor they, the Legislative Assembly – Robinson committed the government to a matching, pound-for pound contribution.[38]

Farnell was put in an embarrassing position. Having approved the project in principle, his government was now committed to it in practice. In December, a Royal Commission was created to oversee the project, and a vote of £50,000 was placed on the estimates for 1879. The chairman of the commissioners was Patrick (later Sir Patrick) Jennings, pastoralist, and later Colonial Secretary.[39] Jennings cut a fashionable figure in colonial science and culture, and was a vice-president of the Agricultural Society of NSW, for which he helped obtain a permanent site in Moore Park. The Colonial Architect, James Barnet, was given the task of designing a building that would bring Sydney fame.

The interior of the Garden Palace, 1879.

Australians were 'never so derivative', Graeme Davison has argued, 'as when they attempted to express their sense of cultural identity',[40] and Barnet's drawings were duly inspired by the Crystal Palace of London. Initially, he provided for halls built of glass and iron, as in London, but these the colony's industry could not supply, so he revised his plans to use timber, corrugated iron, and imitation stone. From January 1879, the 'Garden Palace', as Robinson called it, grew from paddock to palace – eventually covering three and a half hectares. John Young, the builder, employed as many as a thousand men, working by day and night with the aid of arc lights, Sydney's first public use of electricity.[41] The result, nine months later, was a modern wonder.[42] Its interior was richly painted – ceiling in sky blue, woodwork in buff, and decorations in red and gold. Charles Badham supplied a motto: *Orta recens quam pura nites*.[43] The massive building dominated the skyline. Nearby, two halls of industry turned the Domain into a cultural *souk*, with Turkish bazaars, a Maori House, and an Austro-Hungarian beer-tasting hall. For those with more refined tastes, a fine arts building was erected just inside the Domain's gate.

The style of the main building was cruciform, wedded to an Italianate design that registered the union of master and man, industry and progress. Its long central section, with vaulted ceilings, two storeys tall, ran north and south parallel to Macquarie Street. At the junction of the transepts and the 'nave', soared a mighty tower, 210 feet high, topped by a magnificent dome, which bore the inscription, 'The earth is the Lord's and the fullness thereof'. Under the dome was a huge fountain, 150 feet in diameter, from which jets of water cascaded to a basement tearoom below. In the middle, stood a bronze statue of Queen Victoria, cast by Marshall Wood. 'Under the dome' and next to Victoria, was for a time Sydney's most popular *rendezvous*. The final cost exceeded £200,000 – more than the University – making it the most expensive public building in Sydney to date.

The opening ceremonies took place in September 1879. Sir William Manning, the President of the Agricultural Society, and the Governors of NSW, Victoria, Tasmania and South Australia heard Henry Kendal recite a poem, pardonably entitled, 'Australia'. A choir of 700 voices sang the 'Hallelujah Chorus', to the accompaniment of an organ, built for the occasion and sent from London. Among the invited guests, stood Liversidge. Although admission was expensive – 2/6d. at first, and 1/- after September[44] – the gatekeepers recorded over 1,107,000 visitors, a remarkable result for a city of only 250,000, in a colony of just over 620,000.[45] On Boxing Day 1879, the Garden Palace posted a record attendance of 27,000.[46] Country crowds used Sydney's first steam tram, drawn by horses until its motor arrived from the United States. To lift public spirits even higher, the first American steam passenger 'elevator' to reach Australia was installed in the northern tower.[47]

For seven months, the Exhibition was open six days a week. Twenty countries were represented, including Britain, Germany, Austria, Japan and India; together with all six Australian colonies and New Zealand. Some 9000 exhibits were listed in a catalogue of over 1000 pages. To judge the exhibits, the Executive Commissioners were advised by sixteen committees. John Smith, one of the commissioners, chaired the Education and Science Committee; and on two of his subcommittees – wines and ethnology – Liversidge's name appears. On the first, Liversidge was accompanied by Dr Alfred (later Sir) Roberts and W.J. Stephens – fellow Trustees of the Australian Museum, and members of the Royal Society of NSW.[48] They were joined by James Hector from New Zealand and a visiting Professor Reuleaux from Germany.

Objects were displayed along national lines, but exhibits were judged according to the classification scheme devised for London in 1851 and confirmed in Philadelphia in 1876.[49] In no other way could one compare the furniture and machinery, the ostrich feathers from Lewinsohn's of Berlin, the crystal from Webb and Co. in London, and the statuary from Messrs. de Christophe in Paris. Switzerland sent cuckoo clocks; the Straits Settlements, ethnography. Germany sent machinery and Britain, steam engines.[50]

From the Australian colonies, there were few manufactured goods, but formidable quantities of raw materials – timber, wool 'ingeniously formed into massive shapes', photographs and emu eggs, pearl shells and sugar cane.[51] From the Mines Department, came pyramids of minerals, struggling to the ceiling. Wilkinson's Mining and Geological

Museum filled 15,000 sq. ft and 160 glass cases with 50,000 specimens of minerals and coals, including many from W.B. Clarke's collection, together with core samples produced by diamond drills, and the largest collection of palaeozoic fossils ever displayed in the colony.

Together with 300 specimens of indigenous woods, collected by Charles Moore of the Botanic Gardens, came the Australian Museum's collection of technological models and drawings that Liversidge bought in Europe in 1878. To these were added the Museum's collection of over 2000 Pacific and Australasian Indigenous artifacts, set out in the northern gallery of the eastern transept. Their arrangement was assigned to the Ethnological Committee of the Museum, including Liversidge, Robert Hunt and Alfred Roberts.[52] The art section was entrusted to Edward Combes, whom the *Bulletin* ambiguously described as 'the best amateur painter in Sydney'.[53]

The Exhibition closed on 20 April 1880 – the same day Liversidge's report was published. For eight months, the Garden Palace had been a source of pleasure and pride. Pride was difficult to measure. But even if receipts came to only £45,000, no one questioned its success. The greatest compliments were paid not in flattery, but in imitation. Melbourne opened its own International Exhibition five months after Sydney's closed, and others followed in Brisbane (1883), Adelaide (1887), Launceston (1891), Hobart (1894), and even Kalgoorlie (1898). In 1888, Melbourne would steal a march on NSW in the celebration of the Centenary of European settlement, erecting an exhibition building that still stands, whose opulence only Victoria's croesal economy could afford.

The Garden Palace had its day. What was to be its future? Liversidge was among those who thought that it should house a museum, along the lines of South Kensington. With this in mind, the gallery of art and the ethnology and technology collections were kept *in situ*; the Department of Lands stored its records in the tower; and the Linnean Society, its library. In time, Liversidge hoped, the Garden Palace might become Sydney's Science Museum, and perhaps its Natural History Museum as well.

## 3. From Domain to Powerhouse: The Technological Museum

During one of the many functions held in celebration of Australia's bicentennial year in 1988, Neville Wran, Premier of NSW, opened the second stage of the stunning new 'Powerhouse', otherwise and officially known as the Museum of Applied Arts and Sciences. A century after Liversidge's report, his vision was realised. A few hundred yards south from the new building, along Harris Street and in the former industrial and woolstore area known as Ultimo, stands its predecessor, the first Sydney Museum building, completed in 1893, and used by the Museum until the early 1990s. Next to it stands Sydney's Technical College. Both institutions, and the spirit that animated them, flowed from Liversidge's report of 1880. To them, Liversidge devoted much time during the next twenty years. Between them, lies a tale of battles and blunders that rivals even the history of the University.

The journey from Macquarie Street to Ultimo took far longer than Liversidge foresaw. As it was, the idea of a technical museum in NSW was old when Liversidge was young. Like the Sydney Observatory and the Philosophical Society of Australasia, proposals for technical education in the colony dated from at least the 1850s. In 1855, following the Great Exhibition, Sir William Denison suggested that an economic museum be built in association with the Sydney Mechanics' School of Arts.[54] But his opponents considered that the 'Colony was not then ripe for such an institution',[55] and nothing came of it. Indeed, but for Liversidge's exertions, the idea might have again fallen victim to colonial cupidity.

On 5 June 1879, not long returned from Europe, Liversidge set to work, seeking a permanent home for a technical museum. The same memorable week in which he announced reforms in the Royal Society, and relaunched his campaign for a faculty of science, he sent a draft of his report to the Trustees of the Australian Museum. The previous year, the Trustees had asked government for £500 to help set up a museum. When the money was finally sanctioned in September 1879 – the same month, the Garden Palace opened – the Trustees appointed three members: Liversidge, Alfred Roberts and Sir Alfred Stephen (who had just completed several months as Acting Governor) to a 'Technological and Industrial Museum Committee'. There were already suggestions, incubating in Liversidge's report, that, when the Exhibition closed, its building might become its principal site.

On 18 September, Liversidge delivered Roberts and Stephen a 'Progress Report', requesting a vote of £1000 to develop plans in the coming year. In December, while the Trustees dithered, Robert Hunt joined the Committee.[56] With Stephen distracted by business, the Trustees asked Hunt and Liversidge to begin selecting specimens. Just as Leibius had become Liversidge's *aide-de-camp* at the Royal Society, so Hunt became his offsider in the museum.[57]

In January 1880, at the height of the summer success, Roberts sent Liversidge's proposal to the Parkes-Robertson coalition, asking for the establishment of an 'Industrial, Technological and Sanitary Museum', formally a branch of the Australian Museum, to be housed in the Exhibition building. Roberts wrote in his double capacity as chairman of the Ethnological Committee of the Museum and the Ethnological Sub-committee of the Exhibition, on both of which he and Liversidge served. While the Government considered the request, the 'Committee of Three' negotiated with foreign exhibitors to leave behind specimens and objects that were uneconomic to ship home, and which 'would be a means of imparting practical information to the industrial classes upon technical and industrial matters'.[58] Not all objects met Liversidge's criteria, as the Netherlands and Ceylon found when their offer of 'courts' *en bloc* was rejected. Liversidge wanted a collection, arranged by 'natural' rather than 'national' categories.

In February 1880, two months before Liversidge's report was published and the Garden Palace closed, his vision was circulated to the Trustees. The museum was to 'occupy a similar position and fulfill the same purpose' for the colony that no fewer than five museums – South Kensington, Bethnal Green, the Museum of Practical Geology in Jermyn Street, the Patent Office Museum and the Parkes Museum of Hygiene – served for London. To this end, Liversidge planned to obtain 'typical collections' of 'all materials of economic value belonging to the animal, vegetable and mineral kingdoms, from the raw material through the various stages of manufacture, to the final product or finished article ready for use'. His experience of building by barter was to guide collecting in fifteen categories of specimens – an 'education in themselves'. 'Nothing', he said, 'should be admitted unless adapted for educational purposes'. Specimens were not to be sought solely for 'their supposed rarity, or ... other equally doubtful reasons'.[59] This was to be a practical museum, not a *cabinet de curiosité*; a museum for teachers, not gentlemen of leisure. The museum was to be in Sydney, but would lend to regional centres, and would organise courses and lectures. Administered by its own Director and 'Board of Advice', it would need 'a general, less restrictive name' than the one he had proposed.

These were the principles; the rest was politics. In March, still without a budget but lacking nothing in resolve, Roberts asked the Australian Museum's Trustees to meet with Parkes, to ask for an annual vote of £1000, space in the Domain, and a board of management. They met Parkes on 15 April; five days later, Liversidge presented his report. His recommendations dovetailed neatly into Parkes' emerging educational agenda, and in May, Parkes agreed to set aside a section of the Garden Palace for what would be officially known as the 'Technological, Industrial and Sanitary Museum of New South Wales' (hereafter, TISM).[60] There were conditions. The promised £1000 was confirmed, a third of which was paid to the Australian Museum, and a vote was created. But public accountability was required, so in the absence of a Department of Public Instruction (which Parkes established later that year), the TISM was placed under a Committee of Management, technically a subcommittee of the Trustees of the Australian Museum, and made answerable to them. The Committee was to comprise Liversidge, Roberts and Hunt.

Such a troika would work until a better solution was found.[61] But what might have been seen as a brake on 'Liversidge and Co.' took off almost as soon as it started. Although legally a creature of the Trustees – with an annual report to parliament appearing as an appendix to that of the Australian Museum – the new Committee of Management took on a life of its own, pushing its barrow carefully between opposing interests. In December 1880, when Liversidge went to the Melbourne Exhibition to collect objects for NSW,[62] Roberts organized another deputation to the Premier, which this time stressed the Museum's role in education. The University was represented by Sir William Manning and John Smith, and John MacIntosh represented the School of Arts.[63] Parkes said he was in 'cordial sympathy' with their objectives, 'and with the cause of technical education generally', so the Committee requested £5000 for the Museum for 1881. Only £2000 was granted, half what they wanted, but twice what they had received the year before.[64]

Not least for this reason, the year 1881 opened for Liversidge in a burst of optimism. Despite the demands of his report, the Exhibition, the Museum and the University, he had just enjoyed one of the most productive periods of his scientific life. Between September and December 1880, he published fourteen articles and presentations, more than any he would ever again publish in any one year. These ranged from a study of *Moa* eggshell for the New Zealand Institute,[65] to four analytical papers for the Royal Society in September,[66] another in October, two in November,[67] and six in December[68] – all of which he ensured would reach the eyes of British readers. He ended the year by sending two more papers to Britain – on Queensland soils, to the Chemical Society in London; and on *stilbite*, to the Royal Society of Edinburgh.[69]

Travelling for Christmas to Melbourne, Liversidge began spending his new museum money, ordering from Europe an array of objects likely to interest Sydney. His appeal to commerce met a gratifying response. 'Donations, or rather promises continue

The Museum in the Domain, photographed in 1886.

to come in and I dread the arrival of each English mail', moaned Charles Buckland, the Australian Museum's Secretary,[70] reluctantly adding industrial arts to ethnography. The Earl of Dudley gave a showcase, and Royal Worcester, a table service; and machinery came from companies in Germany and the United States. To manage these collections, Liversidge secured the appointment in October 1881 of Joseph Henry Maiden, a newly emigrated, twenty-three-year-old Londoner, as his first part-time Curator, at £350 a year.[71]

Only ten years younger than Liversidge, Maiden was to become one of his closest and most enduring friends. In social and educational background, they had much in common. Educated at the City of London School and the Birkbeck Institution, Maiden was talent-spotted by a visiting teacher, Frederick Barff, a graduate of Christ's College, Cambridge, clergyman, lecturer, chemical author, and sometime Professor of Chemistry at the Royal Academy of Arts.[72] Maiden turned a budding chemical interest to profit, working as a part-time demonstrator at evening classes for working men in the City, on the

Joseph Maiden (on the right) with a prized acquisition, the first train to run in NSW.

basis of which he was offered a scholarship to Cambridge (coincidently, Christ's College) by the Fishmonger's Company. This he declined on grounds of ill-health, but in 1879, he matriculated at the University of London, where he met pioneers of technical education whose views were close, if not identical, to those of Liversidge. While at University, Maiden was offered a job in the laboratory of the Royal Arsenal at Woolwich, but again fell ill, and in search of a healing climate, sailed to Australia. He arrived in Sydney with no money, but with a recommendation to a bishop and a letter of introduction to Liversidge.[73]

Maiden arrived in Sydney – described as 'a block of civic England, detached from the parent mass, and planted on a virgin soil, beneath a brighter sky, and in far more genial atmosphere'[74] – in late January 1881, to find virtually all of Liversidge's proposals already in place. In September 1880, the government gave the new Technological Museum 30,000 square feet in the southwestern part of the Garden Palace formerly occupied by the 'foreign courts'. Into this vast 'depot' went the specimens that Liversidge had bought in Europe, together with a magnificent collection of fossils from the Darling Downs. The Government also gave Liversidge £600 to purchase from August Krantz, the leading minerals dealer in Bonn, a set of 8000 fossil types from European localities. 'I have spared neither cost nor labour to prepare and furnish you an exquisite collection,' Krantz wrote.[75]

Maiden began his new life in Sydney as Liversidge's laboratory assistant, preparing classes in the 'interim' building Liversidge had been given to move chemistry out of the Blacket Building; but nine months later, Liversidge arranged for him to be appointed to the Museum to 'classify and arrange the collected specimens and generally to perform the duties of a curator'.[76] Given the task of sorting objects, he had little leisure, but pressed by Liversidge, he drew up plans for a permanent library, a display space, and a chemical laboratory 'for the prosecution of original chemical and physical research, with especial reference to the products of NSW and Australia generally'.[77] The laboratory was designed by Liversidge to take up subjects he could not pursue at the University, including the chemical classification of Australian flora. For a time, it was to been the laboratory he had never had.

Impressed by the Museum's growth, Parkes backed the project fully, and in May 1881, promised another £500 for a lecture program,[78] and instructed Charles Wilkinson to lend eight assistants from the Mines Department to help with geological labels. On 19 September 1882, Alfred Roberts proudly reported that the 'Museum is rapidly assuming a definite form'. In addition to the fossils and minerals were some 8000 other objects, including magnificent specimens of pottery and glass, all carefully installed in 130 showcases.[79] Of the fifteen sections outlined in Liversidge's report, six were already in an 'advanced state of preparation'.[80] In scarcely two years, Liversidge's Museum had become a reality.

Pride in achievement was manifest all round. But dismay was to follow. Three days after Roberts' optimistic report, disaster struck the Domain. On 22 September 1882, the building caught fire – set alight by an unknown arsonist – and literally exploded.[81] Pieces of iron and wood were found as far away as Double Bay. The search for the villain was unsuccessful. Rumour had it that the Government had stored convict records, as well as objects of virtue, in its transepts,[82] which might have attracted someone seeing no advantage in admitting to a convict past. The press preferred to moralise, blaming the fire on the *hubris* of a palace-proud city. Whatever the case, the Exhibition's embers glowed for days, and virtually everything it contained was lost.

The building was valued at £11,000, but the contents were beyond price. Even so, it was calculated that £10,900 worth of gifts disappeared – books, cases, maps, specimens and exhibits – in addition to nearly £9000 worth of purchases. When a final list was compiled, it included W.B. Clarke's unique collection of mineral specimens, sent by Wilkinson's Mining and Geological Museum; the Linnean Society's entire library and collections (largely gifts of William John Macleay); the Exhibition Committee's minute books; the Government's surveys for new railway lines; and, to historians' unending grief, the colonial census records for 1881. Ironically, the Trustees had 'for security' reasons moved to the wooden palace many items from the safe sandstone of College Street, warning the Minister that they had no space for collections which ran 'great risk of being destroyed'.[83]

The 'museum that never was', which had meant so much to so many, had ceased to exist. If its loss was felt heavily by the city, it bore more heavily still on Liversidge, whose work had literally gone up in smoke. He was desolate. For many months, he even gave up travel. But he drew consolation from the fact that in March, his Faculty of Science had at last been approved. By coincidence, the same month that the Garden Palace burned, the Royal Society of London elected him to its Fellowship. At least two of his dreams had come true. In November, in their ponderous way, the Trustees of the Australian Museum conveyed thanks to him and his committee for their 'zeal and industry', along with their regret at seeing the 'fruits of their labours ... swept away in the general ruin'.[84]

Not until December did Liversidge, preoccupied with university affairs, resume his attendance at Museum committees – but by then, Hunt and Roberts had already begun to pick up the pieces. Roberts, fresh from opening the Royal Prince Alfred Hospital, found accommodation for surviving objects in the so-called Agricultural Hall of the Exhibition. This was admittedly a 'wretched tin shed', in Maiden's phrase, erected next to the Sydney Infirmary for the exhibition of cattle, some two hundred yards away from the Botanic Garden. But it was at least a home, 'pending acquisition of more convenient and suitable premises'.[85]

The new year brought hope of renewal. Over 200 newspapers at home and abroad had reported the great fire. Now it was time to rebuild. Liversidge and Maiden sent over 3000 letters to exhibitors, asking for gifts and donations. Trade books, photographs and chemicals were ordered, and products were bought or borrowed.[86] To restock the ethnological sections, Roberts, with an enthusiasm exceeding his authority, ordered Edward Ramsay to 'board all vessels from the north, New Guinea and the Pacific Islands, to secure specimens and to otherwise use every exertion to re-form this Department'.[87] Their campaign was remarkably successful. Although over 2000 items were lost in the fire, by the end of 1883, the Museum had regained that number and more. Most were to remain in the Australian Museum, and not go to its offspring, but many did, creating a collection of such size that the 'tin shed' was soon full. Liversidge hired carpenters to build display cases even before the government approved their employment; and, well before the 'tin sheds' opened to the public on 15 December, he had begun to look for a more permanent space.

In preparing the case for a technological museum, Liversidge found it useful to call on Victorian expertise, and in August 1883, sent Maiden on a fact-finding mission to Melbourne, where he visited that city's Technological Museum, interviewed Robert Ellery (then doubling as Victorian's Government Astronomer and chairman of its' Technological Museum Committee), and inspected the famous Schools of Mines at Ballarat and Sandhurst. Maiden's first contact with Victoria's scientists opened his eyes to the wisdom of intercolonial contact. Such intelligence as he brought back was useful to Sydney's Museum and to Liversidge's plans.[88]

Amidst the tumult of these months, Liversidge found in Maiden an especially welcome colleague, friend and protégé. Thanks to Liversidge, Maiden was elected to the Linnean Society and to the Royal Society, where he soon became a frequent contributor, officer and, eventually, historian.[89] By 1882, Maiden was well on the way to becoming, in Lionel Gilbert's phrase, a 'well-assimilated colonist', when he capped the year by marrying in Melbourne, and beginning a family in Sydney. Within ten years, his five children, including four girls – inevitably known as the 'beautiful Maidens' – made Liversidge an honorary uncle. Liversidge also encouraged him to enroll at the University, where he took lectures before illness obliged him to stop.[90] During the next few years, they met officially about once a month, sometimes less, but Maiden always served his mentor's interests, and as time went by, they conspired to associate the Museum, the University, the Mines Department, and the Technical College, in projects ranging from the printing of specimen labels, to the sharing of handouts, minerals and models of nuggets.

Liversidge especially welcomed his help in building up a journals collection and beginning experiments on eucalyptus oils.[91] These, Liversidge explained to Henry Armstrong in London, he had no time to perform himself, as 'the little spare time I get is taken up with minerals'. Too few people were chasing too much work. 'I am trying through our local society here to induce people to take up researches, but the difficulties are almost insurmountable in the colonies', he added.[92] In 1883, with the help of Charles Moore of the Botanic Gardens, Maiden began a herbarium in the Museum, and, with Liversidge's encouragement, launched a lifetime of research into the chemical composition of natural products.

Meanwhile, the quest for permanent accommodation continued unabated.[93] The 'tin sheds' were exploding, with 4000 new specimens arriving each year.[94] In his role as Trustee of the Australian Museum and active in its affairs,[95] Liversidge devoted great care to its struggling progeny. Between 11 December 1882 and 28 August 1886, he made fifteen separate donations of books, minerals, wood and fabric specimens, and chemical samples. In 1886, this steady stream turned into a cascade of meteorites and minerals, dyes and wine testing apparatus.[96] With growing interest in Indigenous Australian artifacts, Liversidge sent many specimens, to which he added minerals from the New England district, where miners had begun to work tin, molybdenum, wolfram and antimony, and from the area around Broken Hill, where there was commercial reluctance to announce the discovery of new deposits.[97] Liversidge had become one of the principal mineral collectors in NSW.[98] But there were others, including Wilkinson, who kept Sydney on the rocks. Indeed, so extensive became the combined holdings of the Australian Museum, the Mines Department, the Technological Museum and the University that, by 1886, Liversidge was concerned by the 'danger of the collections becoming universally duplicated'.[99]

To absorb this tide of objects, a share of any building was better than a shed in the Domain, and in March 1886, Liversidge tried tactful persuasion. Writing from his rooms in the Union Club, he asked Maiden whether there had been any movement on their request in 1884 to obtain the use of the 1870 Exhibition Buildings in Prince Alfred Park. 'I often see the present Minister,' Liversidge observed – referring to the medical Dr Renwick, sometime ally on Senate and, at the time, Minister of Public Instruction[100] – 'and should like to be in a position to say something definite about that particular building or any other which might be suitable. He, too, would probably be glad to have the whole of the Agricultural Hall for his hospital.'[101] In August, Liversidge invited the Governor and his wife, Lord and Lady Carrington, to visit the 'tin sheds'. They dutifully came the short distance from Government House, but the effect on their aristocratic sensibilities can only be imagined. Liversidge and the Governor met once more, and then with George Dibbs, the Premier.[102]

However, not even vice-regal flag-waving could conjure resources from a factional government – not even one having an annual budget of £8,000,000, and spending that year £817,767 on public education alone. Less than £34,000 was spent on the colony's university, public library, art gallery and museums, and Liversidge could not get a tenth of that.[103] Liversidge's politics of persuasion took a turn for the worse when, after the famous meeting with the Governor, he and Hunt sought out Arthur Renwick, and came away confident that they had persuaded the Minister to 'turn the Immigration Department into the sheds and put [Maiden] in the main building'.[104] But Dibbs pleaded pressure of business, and when, with the summer recess, his coalition with Jennings fell, the whole parliamentary saga had to begin anew.

In fact, the mid-1880s were not a propitious time for new buildings. From 1883, land revenues began to fall;[105] and in the election of 1887, for the first time, distinct political parties arose. Although they replaced the faction system, they failed to diminish party strife,[106] and less urgent matters – including cultural institutions – were pushed to the sidelines. The temporary Museum in the Domain, never intended for its purpose, was a casualty of the political system; its floor, built on timber piles, subsided under the weight of exhibits, its galvanised iron roof leaked, and drainage, always poor, became pathological. Dr J. Ashburton Thompson, the city's Medical Officer, reported matter-of-factly that, with the growth of fungi and bad smells, the 'sanitary condition of the building renders it unfit'.[107] It did not escape notice that a public institution, ostensibly given to the exhibition of 'sanitary' apparatus, was operating in a most unsanitary manner.

Liversidge's departure on overseas leave in 1887 spared him from these sordid affairs; but in 1888, he returned to the fray. He had visited China and Japan and sent several cases of objects to Sydney – materials which form the basis of the Powerhouse's holdings today – and continued to collect indefatigably wherever he went, sometimes against his curator's better judgement. 'Will you let me know when I can send for any specimens from the Melbourne Exhibition,' he cheerfully asked Maiden, who had to store all these new things. 'Will you please call or send to the [Union] Club for a package?'[108] Liversidge enquired, concerning a collection of artifacts valued at £300 which he had persuaded fellow Union Club member and banker Robert H.D. White to exhibit.[109]

There were also musical instruments which an Indian potentate offered to send by way of exchange, and the Boulton and Watt beam engine that Liversidge had obtained from Whitbread's in London – for all its historical value, a white elephant, impossible to display. And when a shipment of American and Mexican pottery arrived unannounced from Washington, a by-product of Liversidge's rail trip across the United States – 'It had slipped my memory', Liversidge noted, without pausing to apologise[110] – Maiden's patience was sorely tried. In reply, he complained that after six years, the Museum was in peril, not because it lacked objects, but because it lacked space.

By 1888, it was clear that unremitting 'space wars' had exhausted both the Committee and their curator. In the otherwise celebratory year of the colony's Centennial, Maiden wrote despairingly in his annual report:

> The question of the proper housing of ... this Museum has formed the subject of a voluminous correspondence with your department during the last six years. Matters are, however, now approaching a crisis. The Museum is crowded almost beyond endurance ... And the public ... attend in steadily diminishing numbers. The Officers

of the Museum cannot properly attend to the specimens while the efficiency of each employee is diminished by reason of the difficulties which beset him.[111]

Such 'difficulties' included the danger of fire from cigar-smoking pedestrians walking in the Domain, and the macabre sounds and odours of post-mortem inquests taking place on the Sydney Hospital side of the partition between the two buildings – 'the circumstances of which', the curator delicately observed, 'unfavourably affects the attendance of visitors'.[112] Visitor numbers reinforced his argument. Only 26,121 hardy souls visited the Museum on weekdays in 1888, and 312,493 on Sundays, with an average daily attendance of only eighty-three (but 240 on Sundays) – as contrasted with the Art Gallery's weekday attendance of 112,092 (Sundays 82,958), and the Australian Museum's 89,028 (Sundays 39,337). [113]

As with the University and the Royal Society, the story was familiar. The frequent rise and fall of ministries, left cultural institutions low on the ladder of priorities, whatever the size of the colonial budget. Ironically, in late 1889, the situation improved, when political circumstances produced the right result for the wrong reason. Liversidge's determination to get a new building for the Museum was overtaken by rumours of a take-over. Parkes, who returned to power in March 1889, announced that 'the system of Technical Education in this Colony is undergoing a thorough re-organisation'.[114] If the Government took control of the Board of Technical Education, the administration of the Museum by a committee of the Australian Museum's Trustees would be an anomaly. To a meeting of the Trustees on 3 September 1889, Alfred Roberts read a report drafted by Liversidge, outlining the history of the Museum and its achievements; and concluded that without a decent building, 'with due regard to their own self-respect and their duty to the public', the Technological Museum Committee could 'no longer carry on'. As long as there was any hope that the object of benefiting the industrial classes would ultimately be realized, they felt the greatest pleasure in their work; but as it was, 'with the keenest disappointment and regret', they offered their resignations *en bloc.*

It was a pyrrhic gesture, but the Trustees accepted its Committee's resignation, and threatened to close the Museum. However, a fortnight later, sensing ruin, Roberts and Hunt asked that their decision be postponed until they received a reaction from the minister. Behind the scenes, Roberts (and probably Liversidge) saw Parkes, who in mid-September agreed, as part of his plan for technical education, to finance a new building for the Museum; and sent his new Minister of Public Instruction, Joseph Carruthers, a Sydney University graduate (and later Premier), to ask the Trustees to be 'so good as to suggest ... any suitable building in Sydney which it is possible to secure for Technological purposes'.[115] However, in line with his plan, Parkes decided 'that the educational interests of the community will be best served by the amalgamation of the Museum and the Technical College'.[116] In November, the Museum and its curator therefore passed from the jurisdiction of the Australian Museum into a new Technical Education Branch of the Ministry of Public Instruction. On 30 December 1889, with Hunt and Liversidge present, the Trustees capitulated. At a stroke, the Museum had won a building and lost its independence.

It was left to Liversidge to write his Committee's obituary, and to wind up the work of a decade. Although he did not resign from the Trustees, his despair at their inertia tinged his correspondence. 'Sir', he wrote Ramsay brusquely, five months later, 'will you please return the loan objects which you have of mine? Yours truly, A. Liversidge'.[117] Several years would pass before tempers cooled. All the same, he was pleased to be relieved of a responsibility for which he, let alone Roberts and Hunt, could no longer be held accountable.

For the next half-decade, as Liversidge moved from the picture, the Museum was to be Maiden's story. Among their last acts as 'managers', Roberts and Liversidge instructed Maiden to inspect the Exhibition Building in Prince Alfred Park as an alternative site. But the Technological Museum was destined to remain in the 'tin sheds' for another four years – storing its 30,000 objects and specimens in Augean squalor.[118] It was a welcome relief when, eventually, the government agreed to a new place for the collections – on the grounds of Surgeon Harris' colonial home, on the Ultimo street of his name, five minutes from Central Station. There, finally, in August 1893, and in the presence of the Governor, Sir Robert Duff, a fine new Museum was opened to the public – a monument in brick and terracotta, next to the Technical College, and costing £20,000.[119]

Designed in a Romanesque style, and finished with renderings of stone Australian animals and plants, the new Museum was altogether 'adapted', as Richard Baker put it, 'to the necessities and materials of the present day'.[120] Visitors were greeted by a photograph of Liversidge, mounted above the main entrance, bearing the caption, 'through whose special efforts the Technological Museum was founded in 1880'.[121] Today, his memory (without the photograph) genially pervades the 'Powerhouse',[122] and in the atrium where the Boulton and Watt engine delights tourists who have quite forgotten, or have never learned, whose legacy they enjoy.

## 4. ON THE ROAD TO ULTIMO:
### TECHNICAL EDUCATION AND ITS DISCONTENTS

If the Technological Museum, in its glory, was one of Liversidge's pillars of wisdom, a second must surely have been his program for technical education which unfolded at about the same time. In his report of 1880, Liversidge foreshadowed two different dimensions of scientific training: that needed by professional men who looked to the University; and that needed by artisans and craftsmen. Education imparted 'hand and eye' skills to the many, and training of 'mind' to the few. As David Layton has memorably put it, there would be one physics for princes, and another for paupers.[123]

The distinction was hardly new, nor was Liversidge's approach to it. Moving from working-class London had not distanced him from the needs of workingmen; and the experience of Cambridge had made him no apologist for elitism in Sydney. But where others saw political advantage, Liversidge saw a point of principle. Technical education in Britain – from the mechanics' institutes of the 1840s – had borne the seeds of its own contradictions: sponsored by middle-class interests for middle-class purposes, it had been sold as serving the interests of workingmen, and the political economy of crafts and trades, but often had failed to deliver. Moreover, the entire field was dominated by indecision between those who valued education as character forming and those who saw it as training for life.[124] Liversidge was determined to have both, combined if possible in a single system.

Colonial Australia had from its earliest decades shown an interest in technical education of many kinds. In 1833, when Bourke was Governor, Alexander Macleay was Colonial Secretary, and convicts still formed a third of the population, the establishment of the Sydney Mechanics' School of Arts (SMSA) conveyed a compelling vision of colonial education.[125] Its meetings became a platform for Charles Nicholson, Henry Carmichael and John Woolley;[126] and on the School's behalf, and no doubt with future University admissions in mind, Charles Badham rode throughout the colony, 'urging the claims of learning upon the successful rich, and inspiring the poor ... in their efforts towards the attainment of knowledge'.[127]

However, by the middle of the century, the School's star had begun to wane. Its membership of 296 in 1851 was a faint shadow of its 800 a decade earlier. Critics complained that the School, like many of its kind in Britain, failed to offer practical benefits to workingmen; and indeed few were attracted by the annual subscription of twelve shillings – a week's wages for a good cook. Although the 1850s revelled in the rhetoric of training 'hand, mind and eye', the School's library fell into neglect, and its program fell away. Speaking to the Hobart Town Working Men's Club in 1865, Francis Nixon, Bishop of Tasmania, admitted to the heresy that 'mechanics' institutes' had foundered in England because their 'benevolent founders' had failed to 'ascertain the conscious wants of the working classes'.[128] The same held true for Sydney. When Liversidge arrived, the School of Arts in Pitt Street was thirty-five years old, but struggling to survive. The early Victorian fashion for acquiring 'useful knowledge' had expired; what was now needed, he decided, was 'technical' knowledge.

Following the Exhibition of 1851, Britain had given legislative approval to a new generation of 'schools of design', some sustained by government through the Science and Art Department, others supported by guilds and trade associations. Their 'excellent results', in the words of Timothy Coghlan, the NSW Government Statistician, 'could not fail to attract attention in the colonies, where a sound and practical knowledge of the manual arts is of paramount necessity'.[129] A similar message was implicit in the first 'Industrial Exhibition' of NSW, held in 1861, and in the Melbourne Intercolonial Exhibition in 1866, well before the 1870 Centennial Exhibition in Alfred Park and the Garden Palace in the Domain.

Overall, NSW was slower than Victoria to recognise the need for state-supported technical education. Victorian mining interests, appalled at the waste of capital in gold operations, and anxious to apply science to deep-level mining, demanded better education.[130] In 1869, the Victorian government appointed a Commission for Promoting Technological Industrial Instruction to chart a strategy.[131] Schools of Mines, established at Ballarat in 1871 and in Bendigo in 1873, set models for other colonies to follow.[132] When the Sydney School of Arts was repeatedly criticised for its neglect of 'useful' instruction, it began classes in mechanical drawing in 1865, followed by mineralogy and geology in 1869, design in 1870, and chemistry in 1871.[133] By the end of the 1870s, NSW boasted some eighty mechanics' institutes, and the SMSA, aided by a government grant of £200, grew to 2000 members. It was time, however, for Australians to explore ways of introducing technical classes of relevance to workingmen, free from the middle-class burden of 'improvement' that travelled with British institutions. Both management and labour had to work within a system that, as with church schools, often found it easier to duplicate effort than to confront differences.

With the inauguration of the Engineering Association in 1870, and the NSW Trades and Labour Council in 1871, came an interest in finding a solution that had eluded others.[134] Some reformers, like the engineer Norman Selfe, saw government subsidies and practical skills as the key issues.[135] Others, including Edward Dowling, a public servant keenly interested in workingmen's education, preferred to see private support along a broader educational front.[136] On the day before Liversidge's second famous letter to the *Herald* in February 1873, the Sydney Mechanics' School of Arts, led by Dowling, had set up a committee to establish a Working Men's College, along the lines recently made fashionable in London by Frederick Denison Maurice.[137] This was a frankly middle-class push, devoted to elementary education of a literary and scientific character. But it drew support from the conviction that technical education was bound to come. Indeed, in the draft Mining Bill that Parkes had just laid before the Legislative Assembly, there was specific mention of a School of Mines, and lectures in geology, mineralogy and chemistry. To Liversidge's dismay, the Act was silent about where the school should be, or who should pay its costs.[138] But at least the principle was before parliament.

From his experience in London, Liversidge had some sympathy with the idea of a Working Men's College. However, he saw the future of technical education as firmly part of the University's brief. He had his supporters. Robert Abbott, Parkes' first Minister of Mines, opposed making the School of Arts a place of mining education, saying that a School of Mines 'would not be so widely useful [there] as it might be made' if established at the University.[139] Parkes also preferred to give the job to the University, if by so doing the University were better enabled to give 'young men a practical education'. To 'popularise the University by doing this ... would be a great step in the right direction', said Francis Suttor, pastoralist, MLA for Bathurst, and Parkes' Minister of Justice and Public Instruction in May 1876, speaking in reply to a deputation from the Engineering Association, the Builders and Contractors' Association, and the Trades and Labour Council. These gentlemen had asked the Council of Education to establish a program of technical courses under government jurisdiction; but that government, at least, saw it as a university responsibility – if only Manning were prepared to accept it.

For the next two years (1876–78), the issue remained unsettled, as Dowling gathered information from Europe, monitored the experience of British provincial cities, and kept up pressure on government. In 1877, when Parkes returned to office, his government voted £2000 per annum towards the establishment of a 'Technical or Working Men's College'. But who would administer the college? This was the proximate cause of Manning's instruction to Liversidge, then in England, to report on developments in technical education overseas, so as to prepare a plan for Sydney. However, within a year, the advocates of the College had stolen a march on Manning and, with the Government's blessing, it was to be the School of Arts, not the University, that would take over technical education for the colony.

The first lectures of the new Technical College at the corner of Kent and Sussex Streets were not well attended, but plans were laid for an appeal to working men (and women) of all trades. Early in 1879, the College convened an Honorary Council of Advice – a cabinet of 'all the talents' – which included the new Minister of Mines (E.A. Baker), the Mayor, Professors Smith and Liversidge, Edward Combes, Charles Moore, G.P. Ramsay, the Headmaster of Sydney Grammar (A.B. Weigall), and John Sutherland (Farnell's Secretary of Public Works and Vice-President of the School). In May 1879, the School's main building was enlarged, its courses were advertised in the *Illustrated Sydney News*, and its lecturers prepared for business. Student numbers increased from 280, in the 1878 series, to over 1000 in 1880, of whom 400 were manual workers.[140]

The College needed support. A chance to win headlines came in October 1879, one month after the Garden Palace opened, when Dowling called a meeting at the School of Arts to which he invited the Trades Union Congress, then sitting in Sydney, together with visitors from the UK and Germany who were attending the Exhibition. The meeting resolved to offer 'further instruction in the sciences relating to the arts and manufactures in the educational institutions of the colony' – without suggesting precisely how this might be done.[141] Was there to be a single, comprehensive system, of which technical education formed just one part? Or would the colony prefer a binary system, with separate institutions?

The answer was not obvious. In Britain, educators spoke of the need for technical education to meet industrial competition from the Continent, the United States, and Japan.[142] Many preferred to retain the tradition of technical training in apprenticeships, as Liversidge's family had known. However, T.H. Huxley, Lyon Playfair and William Armstrong, and Liversidge by extension, wanted technical education to embrace vocational training (including languages), as well as education in the principles of knowledge, technologies, and 'practice'. At the opening of Finsbury Technical College,[143] Sir Philip Magnus spoke for the same liberal caucus when he said that Finsbury's main purpose was 'not to make scientific men, nor to train scientists, as the Americans call them, but to educate *technikers* as the Germans say' – in other words, to 'explain to those preparing for industrial work, or already engaged in it, the principles that have a direct bearing upon their occupation, so that they may be enabled to think back from the processes they see to the causes underlying them, and thus substitute scientific method for mere rule of thumb'. These were to be not the 'High Priests', as Huxley called them, but the soldiers of science.[144]

Huxley's rhetoric was familiar. But the circumstances of 'settler capitalism', with its political economy based on primary industries, a small manufacturing base and a large service sector, might demand a different solution. Colonial life celebrated the 'practical man'.[145] How to make him also a soldier of science? In October 1880, six months after Liversidge's report was published and the Exhibition closed – but while its flag still flew over the Domain – Dowling and the Technical College Committee joined with the Trades and Labour Council in a two-day 'Technological Conference' at the School of Arts. Manning and Windeyer represented the University, and Renwick and Combes (since 1879, MLA for East Macquarie), the Legislative Assembly. Together with representatives from rural mechanics' institutes, John Sutherland represented the School of Arts, and Norman Selfe, the Engineering Association.[146]

Sydney's first Technical College on Kent and Sussex Street.
Window signs advertise the courses taught.

Parkes gave a spirited address, endorsing 'citizen virtue' in an 'object so laudable, so unexceptionable, and so far reaching'.[147] The same month, Liversidge celebrated the endowment of a new Technological Museum. There seemed hope that the main lines of his report would be realised. Even as he wrote, however, it was clear that there was no unanimity on the central issues. The conference revealed three contrasting positions. The first, advocated by Norman Selfe, argued that technical education should be left to technical men, not university men. His argument was shared by 'practical men' like W.H. Humphreys, President of the Engineering Association, who deplored the fact that 'Professor Liversidge in his report',

> [would] seem to imply that the only way such institutions could be
> made advantageous would be by the establishment of Professorships
> or Schools of Science in connection with the Sydney University.

The Politics of Practical Men

This might help sons whose parents could afford university education. 'But I do not see', Humphreys continued,

> how the University can possibly undertake the tuition of the many hundreds of young Mechanics, even if such ... were willing to trust themselves within such aristocratic enclosures.[148]

The engineers were not interested in subsidising the middle classes, but in improving the lot of apprentices.

A compromise was offered by Edward Combes, who argued that that policy did not require the denial of one side to benefit the other. 'By withdrawing chairs at the University', he said, 'science was not being popularised'. What was wanted, he continued, were more technical museums, more drawing classes in Schools of Art, and more trade schools in the mining areas. With this Selfe agreed, but argued that the 'high class' scientific training that Liversidge's report advocated 'could be better obtained in the older centres of Europe'. What the colony really wanted was, he said, 'just as much science as an ordinary man required, and not the technicalities and theories of science'.[149]

As the conference wore on, Manning saw that 'more' for the College might well mean 'less' for the University. Speaking last, he ventured well beyond any public commitment he had ever made, suggesting that the University would support technical education, and even take 'the technical college into affiliation'.

It fell to the Minister, Joseph Carruthers, to defuse the debate by proposing a 'third way'. There had been, he observed, some talk of a sort of rivalry between the Technical College and the University. But there were no grounds. They were two distinct institutions:

> The one was to instruct the youth of the colony in the higher branches of education, and the other was to instruct the working classes in their particular business.

To this politician, the issue was not a matter of party, but of practice: 'Technical education meant such a knowledge of science or sciences as would enable the working man to do his work more efficiently than he would if he did not possess that knowledge'.[150]

The conference passed three resolutions. The first, proposed by Combes, endorsed the Technical College's program. The second, proposed by Selfe, called upon government to institute a 'proper system of Technical Education in Science and Art', to afford instruction 'principally of a practical character'. On this, Selfe and Manning concurred. Windeyer, speaking for the Technical College, favoured having separate trade schools and a link with the University. The third resolution, proposed by the Trades and Labour Council, asked for evening classes in 'every prosperous locality'. The meeting ended with a deputation to the Minister.

Remarkably, Liversidge missed the October conference. Why is not clear, although that month did find him immersed in research. His report had set out the basic blueprint, and the 'architects' and builders were left to settle the details. He may have seen, in January 1881, the destinies of the Technical College and the University tending to converge. But the conditions of convergence had yet to be agreed. His report, encouraging evening classes and courses for artisans, emphasised the importance of reaching out to 'practical' fields such as agriculture, commerce and architecture. The University would retain its position at the apex of an educational pyramid, with 'practical' subjects becoming more 'scientific' as one climbed upwards. The School of Arts seemed prepared to take the middle ground, willing even to affiliate the Trades Council with the University, provided there were bursaries from the College to Camperdown.[151]

Ann Hone has reminded us that the question of technical education posed a real dilemma for the Government of NSW.[152] Because the issue was not exclusively, or even essentially, one of finance, its resolution was all the more difficult.[153] If Henry Parkes favoured 'high instruction' as the way for the working classes to make themselves equal to the best of their fellows, then the claims of technical education were secondary, even subsidiary to those of the University; and less important than those of the new high schools whose curricula were being directed towards academic knowledge. Parkes conceded the principle of government aid, but on the grounds that it formed part of what Hugh Robson, Inspector of Public Charities, called in 1880, the 'Government scheme for public intellectual improvement'.[154] Knowledge of principles, according to the Huxleyan creed, would lead to practical application; and science would reciprocally prosper by its association with the 'practical'. This Liversidge believed.[155] But would it satisfy the practical men?

Liversidge wished to avoid competing frames of mind. While yearning for a School of Mines, he was opposed to the view, expressed by Parke's Minister for Mines, that the University and the Technical College were two 'entirely distinct institutions' – representing the classical division between the 'higher' branches of knowledge and 'business' education, one way of knowing for the middle class, another for workers. Such class distinctions were too high a price to bear. Early in 1881, Parkes asked the Technical College Committee to report on its plans.[156] Dowling's recommendations were delivered to John Sutherland, a confirmed 'practical man' and politician, but also a Vice-President of the School of Arts, who reported to Parkes in August 1881.[157] Liversidge's recommendations had almost entirely disappeared. Instead, Dowling and Sutherland proposed that the new Technical College should become Sydney's central institution for mining and secondary education. Sutherland also rejected Liversidge's suggestion of 'payment by results',[158] and his suggestion of a 'Normal School' for teachers at the University. He found Liversidge's reservations about evening classes over-cautious, and argued that the future of the working man lay in expanding this sector.[159]

For the Technical College, the Committee asked for an endowment – on terms analogous to the University's – of £2500 per annum; and for a subsidy of £1000 to the School of Arts; for an increased matching grant for the mechanics institutes; and for concessionary fares for students. Contrary to Selfe's wishes, the College as a legal entity was to be left in private hands. The Parkes government, which had recently approved increases to the Royal Society and the University, accepted the College's report in principle. But, as so often, no immediate action ensued.

The destruction of the Garden Palace in September 1882, which postponed hopes for a technological museum, also dampened the prospects for a technical college, which were not quickened when, in January 1883, Parkes again fell from power and was replaced by Alexander Stuart, a Scottish businessman. Stuart, who had served as a NSW

Commissioner in Philadelphia in 1876 favoured state aid to church schools, but had less sympathy with delegating technical education to private hands. In June 1883, he instructed George (later Sir George) Reid, the MLA for East Sydney and his minister of public instruction,[160] to ask the School of Arts to surrender the College to the administration of a government board, in return for a secure annual grant. After the difficulties of the previous four years, it was an offer the School could not refuse.

With vivid memories of 1880, the government tried to play down differences between factions. Edward Combes, the great facilitator, was appointed President of the new Board of Technical Education, along with nineteen men (increased the following year to twenty-four) representing as many different interests. Several were members of the Royal Society and of the School of Arts, including John Sutherland. George Reid, not a university man, but an advocate of education, backed Liversidge, and welcomed closer links between the government, the College and the University. Accordingly, on 25 June 1883, both Liversidge and Russell were invited to join the new Board; and Russell was asked to be Vice-President. Both accepted.

'It will be a work most congenial with my feelings,' Russell wrote the next day, 'and one in which I have been most anxious to take a part.'[161] Windeyer, President of the School of Arts, and also a member of the University's Senate, came on board, although wishing that the Technical College could be affiliated to the University to form a 'national school of learning'.[162] Alfred Roberts declined, but Wilkinson accepted, and so belatedly did John Smith: 'Perhaps a little Scotch caution may be useful to mingle with the enthusiasm of the other members', he observed.[163] In August 1883, the Board of Technical Education was established, and in October, Sydney's Technical College was transferred to its management.[164] The Board's budget for 1884 was set at £17,000; the University's endowment that year was £15,500. Manning knew that an opportunity to score a major goal for the University had been missed.

Over the next six years, the Board developed classes throughout the city and its suburbs. [165] In curricular terms, it was a victory for both Selfe and Dowling. Indeed, Dowling was seconded from the Government Printing Office to be its first Secretary. The School of Arts, cut off from one of its major functions, gradually subsided from an educational institution to a literary body and a library, paralleling the demise of the mechanics institutes in Britain.[166]

Although the 'practical men' had apparently got their way, relationships with the University remained strong, and technical students were actually to be examined by the academics. In June, Windeyer, wearing one hat, wrote to himself, wearing another – as from the School of Arts, to ask the University's Board of Studies to oversee the Technical College examinations, and therefore its curriculum.[167] This was agreed, and the Board also accepted Liversidge's proposal that student teachers enrolled at the College should take courses at the University. As for lectures in physiology, whatever Anderson Stuart might say, Sydney's newly formed Medical School, although still in only its first year, and with only four students, was obviously the best venue.[168]

Thereafter, the Technical College grew rapidly, enrolling 598 students in the first quarter of 1884 and 1000 in the fourth. Of 590 who took its first examinations, 350 passed, a result considered satisfactory. Overall, its courses reached sections of the community earlier ventures had missed. College courses appealed to carpenters, teachers, engineers, and stonemasons alike. Some lecturers sent to country towns were reportedly received 'with some degree of indifference', but others were welcomed as 'productive of much good'.[169]

In October, the College leased a hall, chemical laboratory, and art room from the School of Arts in Pitt Street, while another building was rented in Sussex Street, and a room for cooking classes was found in the Royal Arcade. Dispersion was unavoidable, and inevitably proved a burden. But all were proved right about the evening lectures which, by the end of the first year, had attracted 34,298 people, or an average of 183 each time. In science, as Liversidge feared, the outcome was not brilliant; by far the most popular subjects were commercial studies, drawing and painting. Against the 166 enrolled in the first and 96 in the second, there were only eighteen in practical electricity, and only one in practical chemistry.[170]

## 5. Dreamers and Problem Solvers

As the reform of technical education simmered through the early 1880s, Liversidge moved to centre stage in University affairs. In March 1882, Senate agreed to establish a Faculty of Science, and he was appointed its first Dean. At the same time, to rationalise responsibilities, John Smith's chair was divided, with Smith keeping the chair of Experimental Physics, and Liversidge becoming Professor of Chemistry and Mineralogy. Geology fell from his title, and was privately offered to William Stephens along with the title of Professor of Natural History and Hovell Lecturer.[171] Six months later, Stephens was joined by William Haswell, a demonstrator in comparative anatomy and physiology, and between them, a new Department of Natural History was born. Stephens was more a friend of the Linnean than of the Royal, but became a loyal ally of Liversidge in University matters. On 30 August 1883, Liversidge led him and Gurney in a new campaign, in recommending to Senate new lectureships in architecture, metallurgy, mining and German.

As Dean of Science – and, since June, a Fellow of the Royal Society of London – Liversidge at last spoke with Charles Badham, Dean of Arts, on equal terms. His authority grew in March 1884 with the appointment of William Warren, a former student of the RSM and Owens College, as Professor of Engineering.[172] As Dean of Science, Liversidge played both sides of Broadway, knowing that appointments made to the Technical College could also serve his purposes at the University. When there was no new money for junior staff, he used his patronage to find College jobs for W.A. Dixon in chemistry; for Angus Mackay in agriculture; and for S.H. Cox in geology, mineralogy and mining.[173] All three were Fellows of the Chemical Society of London, and their coming to Sydney was a bonus to the city.

To other College posts, Liversidge sent former students, including John Kinloch (later the University's Yeoman Bedell) in mathematics and Edward Rennie in physics.[174] In 1885, Rennie became the first Professor of Chemistry at the new University of Adelaide, and the first of Liversidge's *protégés* to hold a university chair.[175] The Technical College might have been seen as an example of 'jobs for the boys'. However, no one seemed to object. Even Anderson Stuart accepted an appointment as a College lecturer in physiology.[176] Liversidge never took an appointment himself, but throughout the early 1880s, the Board

The Technology Museum under construction with the College on the right, about 1893.

of Technical Education felt his touch. On his motion, educational collections of minerals and fossils were purchased and lent to country institutes;[177] and quarterly exams were instituted, as more convenient for working men.[178] To Liversidge, the Board referred course advertisements, the collection of statistical returns from teachers, and the distribution of literary works to country districts.[179] Records are incomplete, but he attended eighteen of the Board's twenty fortnightly meetings in 1885, and typically responded punctually to requests for advice on the salaries and duties of technical staff.[180] When he could not attend in person, he made recommendations by letter.

In early 1884, when the Board of Technical Education began meeting regularly, Liversidge's diary was packed with competing obligations – lectures and meetings every morning, laboratories two afternoons a week and committee meetings on a third; meetings

of the Faculty of Science, the Board of Studies, and Senate; and all the extra work involved in preparing new by-laws for medicine and engineering. It is hardly surprising that between August 1883 and February 1885, he attended only twenty-three of the Board's fifty meetings, slightly fewer than the average attendance (twenty-six) of its members.[181] Behind the scenes, however, he remained influential and rarely missed an important vote.

Through the late 1880s, Liversidge faced continued opposition from those who saw the Technical College as a separate space, a competing alternative to university education. W.A. Dixon, for example, opposed giving scholarships to Technical College students to attend the University . 'It should be', he said, 'the aim of the Institution not to divert the current of a man's life into some new channel'; scholarships would tend 'to unsettle young men in their proper pursuits'.[182] But in many cases, Liversidge's wishes prevailed. On his recommendation, for example, scholarships for College students, and College scholarships for country children, were proposed; and, while delayed, were ultimately sanctioned.[183] As a member of the University's Building Committee, Liversidge set up a parallel building committee for the Board of the Technical College, and led consultations with the Government Architect, having the interests of both institutions clearly in view. On Liversidge's recommendation, Henry Russell joined a special inter-institutional committee set up to examine 'how best existing agencies could be used' (Russell, as usual, became chairman); meanwhile, Liversidge arranged for College classes to take place in the University's Great Hall.[184]

Liversidge's greatest intellectual contribution to the Technical College lay in the field of mining education. Mineral discoveries in New England, the Barrier Ranges, and the coal districts north and south of Sydney were prompting fresh interest in a trade that employed over 20,000 people and produced (in 1886) mineral products worth £2.9 million.[185] Neither the University nor the Technical College had made much progress in the direction of mining education. Yet, from its earliest weeks, the Board of Technical Education met entreaties from Schools of Arts in Goulburn, Mudgee, Bathurst and Newcastle, pleading for at least peripatetic lecturers in mining subjects.[186]

In early 1886, John (later Sir John) Shepherd, MLA for East Macquarie, moved parliament to consider a fresh proposal to establish a School of Mines.[187] Local politics had a certain effect, as did the prospect of increasing output from the lead-silver-zinc mines at Broken Hill. Following additional representations from Dowling and the Board of Technical Education, Patrick Jennings' government approved Shepherd's resolution, and in June 1886, Dr Renwick requested a vote of £17,000 to encourage mining and agricultural instruction in country districts. Subsequently, six 'mining' members were added to the Board of Technical Education, who turned to Liversidge, busy with the start of term, for advice.

Liversidge's recommendations, drawing heavily upon his proposals of 1873 and 1880, were few and familiar, but to the point. He rejected a proliferation of country sites, as in Victoria, as any one would cost £500 to build, another £500 to equip, and another £1100 annually to operate. He preferred to build on the two existing 'schools' in Sydney; if the Technical College and the University were fully equipped, 'the wants of Sydney should be amply supplied'.[188] These would also teach subjects other than mining, and should be known as 'Schools of Science and Mining'. The Board accepted his report in principle.[189] But his recommendations were blocked in supply by James Fletcher, Secretary for Mines, and a former coal mine manager, who saw no reason to divert public funds to an educational object.[190] For the rest of Jennings' government – and until the early 1890s – the idea of a mining school was left in abeyance.[191]

When disappointed, Liversidge was rarely at a loss. Taking into account the new Technological Museum, the Technical College, and his Faculty of Science, Liversidge placed an each-way bet. The Centennial celebrations of NSW in 1888 would surely warrant a new exhibition building, which could in due course become the site of a combined Technical College, Technical Museum, Mining and Agricultural Museum, and Patent Office rolled into one – why not, along lines he had proposed in 1880? The vision was taken up by George Reid, then in opposition, who argued that a colony which had 'spent hundreds of thousands of pounds in building palatial edifices in which young men could be educated for the learned professions' could appropriately devote some attention to its 'industrial youth'. Sadly, the Dibbs government disagreed, and the proposal lapsed.[192]

At one level, this was just another setback. However, as he reflected on the buildings in South Kensington recently occupied by the new 'Central Institution' of the City and Guilds in London,[193] Liversidge could see a growing contrast between what England was doing, and what Australia had yet to begin. Whilst parliamentary procrastination held up developments in NSW, the City Parochial Charities Act of 1883 had created an endowment of £59,000 which was working its way into schools and evening classes throughout England. The same year, the City and Guilds Institute of London, which was already spending £25,000 per annum, opened a 'trades school' at Finsbury Technical College, at a cost of over £35,000.[194]

Given the relative differences in size and wealth, comparisons with Britain are unfair, but Liversidge unfailingly looked 'Home' for guidance. In 1885, the NSW government spent about £17,000 on technical education, and about £18,000 on the University and its colleges. On a per capita basis, this translated into about £100 per university student, but only £17 per Technical College student.[195] The wealthy barons of beer, beef and business had not yet made education a priority. There were practical reasons. Of the 2364 students attending the Technical College in 1885, some 327 were clerks, followed by carpenters, scholars, teachers and engineers. Few came from other trades; whether many benefited vocationally, was difficult to tell.

Liversidge faced other battles with 'practical men'. Selfe's concept of a separate 'Technical University' was not what Liversidge (or, so far as is known, Russell) wanted to see.[196] But Selfe's manner, alternately cloying and abrasive, caused resentment within the Board, and ultimately between the Board and the government. This would not have been fatal, had Selfe not begun to draft the Board's *Annual Reports* that appeared as appendices to the Department's annual statements. These were intended to be statistical, rather than interpretative – indeed, this was Combes' custom, and Russell's after him – but during Russell's absence in 1887, when Selfe became Acting President, their tone became more assertive. Behind the cloak of quasi-anonymity, Selfe used these ostensibly 'objective' pages to air differences within the Board; to introduce his own views; and to make public his case for greater independence in the management of its affairs. In 1887, Selfe summarised for the reading public what he found deplorable in 'abstract', modern, university education, alongside quotations from British authors on the moral benefits of technical instruction as means of arresting 'questionable amusements' among the young.

There cannot have been much love lost between Liversidge and a man who asserted, in the Board's *Annual Report*, that 'it is quite possible for university training to unfit young men for the very profession it is intended to lead to.'[197] Whatever the consequences, Liversidge believed that curricula lay in the hands of the College's teachers, and was disinclined to debate the issue further.[198] But if such prosaic matters divided the Board, far more worrying was the growing estrangement between the Board and the Department of Public Instruction to which it reported.

In 1887, James Inglis, journalist, publisher, tea-merchant, and Parkes' Minister of Public Instruction questioned certain decisions made by the Board concerning the examination and certification of teachers,[199] and added his private doubts about the professionalism of some Board members.[200] Inglis opposed Selfe's vision of 'two universities' – one for academic and one for vocational education – and censored passages in the Board's report that were critical of the Department and the Government. Above all, however, he distrusted the Board's independence. Trustees acting as unpaid administrators could not always be relied upon to manage a public activity; or, having received public money, to welcome administrative control.[201]

All this unpleasantness came to a head in May 1888. That month, Liversidge received a request from Inglis for advice on mining instruction, which suggested that the Minister was totally unacquainted with the history of the subject. This was bad enough. Worse, came news that Parkes had decided, as part of his plan for colonial education, to end the Board, with its voluntary members, and to bring the Technical College and Technical Museum into the public service. Henceforth, he who paid the piper would call the tune.

On 27 June, seeing little future for a Board he had joined with such enthusiasm, Liversidge submitted his resignation – the only time he ever resigned from a public appointment. The next day, Russell said he saw 'no prospect of again taking a share in the work of the Board', and also quit.[202] In November 1889, Joseph Carruthers, replacing Inglis in Parkes last ministry, announced that the Board's functions would be transferred to a new Technical Education Branch within the Department of Public Instruction. Carruthers, a graduate of the University, wanted a centralised system that combined independent schools and technical schools in a pyramid with the University as its apex.[203] In this, he was not far from Liversidge; but he failed to see the consequences of the takeover he introduced. He did announce the establishment of a teachers' college at the University, and appointed a former teacher, Frederick Bridges, as the first Superintendent of Technical Education; but otherwise, Liversidge's influence was at an end.

In 1894, Joseph Maiden took Bridges' place, and quietly kept Liversidge's spirit alive, but for the moment, the fighting ceased. Dowling retired, and Selfe departed the scene. Under state direction, academic and technical education in NSW continued to follow parallel tracks, rather than a single road.[204]

In 1983, the official history of technical education in NSW referred to the period between 1883 and 1889 as the 'Days of the Gentlemen Dreamers'. Those dreams ended when, in the author's ominous phrase, 'The Branch Takes Over'.[205] For Liversidge, the period had seen a 'seven years' war' from which there had been no obvious victors. There was still no clear concensus on pedagogical direction. In 1892, new buildings of Sydney Technical College were opened to the public. But at the opening, its founders were nowhere to be seen. Today, it stands – with the Technological Museum next door, opened the following year – as a monument to those like Liversidge who pressed for its realization.[206]

What would be its relationship with the University? The institutions were to be separate and unequal. In describing the difference, Norman Selfe's last address was purer than gold:

Let us have each institution in its own special sphere, exerting its energies as a power in our midst. The one to bring forth the classical scholars, natural philosophers, physicians, lawyers, musicians, historians, linguists, mathematicians, and the more highly cultured classes of our people if you will. But the other to cultivate the directors and workers who will till our lands, rear our flocks and herds, dig for our minerals, weave our clothing, prepare our food, construct our means of locomotion, tunnel our holes, bridge our valleys, build our houses and furnish all the necessities and material wants of life. Instead of this increasing 'class prejudice', let each of these bodies cultivate respect and confidence in the teaching and ultimate objects of the other one.[207]

Ironically, just a century later, the end of the 'binary divide', and the beginning of new groupings in higher education, would give his views a powerful relevance.

# CHAPTER 8

# Dean and Doctor

Commemoration in 1880.

I n May 1876, in what would be the last Anniversary Address from the pulpit of his presidency, W.B. Clarke urged fellow members of the Royal Society to keep their eyes fixed firmly on the ground before them. Their 'true position', as he put it, was to remain that of 'pioneers, sowers, foundation layers', their purpose simply a 'flourishing association of men … [who] have at least a better aim, and a more useful and nobler object for the employment of their leisure'.[1]

Clarke's view of steady progress was conditioned by nearly forty years of hardship and struggle, redeemed by an improving knowledge of the land and its resources. Clarke inspired the science of Australian field geology, then taking shape amidst the dust, rain and cold of the gold fields and mineral ranges. The new world of laboratory science he saw coming, but was not equipped to manage. Yet, within twenty years, colonial science was to move from the domain of gentlemen collectors to the world of professionals. That much he knew; how it would come about, was another question. Clarke's generation had begun a slow revolution in the public recognition of science; he was to be its Marat, not its Robespierre. Liversidge's arrival gave him fresh hope, and a promise of things to come.

## 1. The 'Elizabeth Street Conspiracy'

During the years 1878–79, while Liversidge was in Europe, the work of the Royal Society simmered gently. The Council's most urgent task was to get a building. Liversidge's appeal had been successful, if not overly so: some seventy-three of its 350 members gave a total of £500, which brought them within range of the Government's matching offer of £500. To this was added a gift of £500 from Thomas Walker, the wealthy grazier of Yaralla. Patronage had its privileges: Walker was duly elected an Honorary Member, joining a list of fourteen distinguished men of science, including Darwin, Huxley, Owen and Hector. Thereafter, Council revealed the genius of property-owning Sydneysiders.

For some time, the Society rented premises at 4 Elizabeth Street from the Academy of Art, which had in turn leased them at £250 per annum from their owner. But when, in 1878, the owner sought to raise his rent to £300, the Academy, reluctant to stay and unable to buy, looked to its lodger for a solution.[2] Following negotiations between Henry Russell and E.L. Montefiore – art dealer, Trustee of the Academy, member of the Society, and Liversidge's friend[3] – the arrangement was turned on its head. Where the Academy had previously been the host, now it became the guest, promising to lease its own premises for £200 if the Society would buy them. With this guarantee, and a Savings Bank mortgage of £2000 at six per cent, a purchase was possible.[4] The Society instructed its solicitors (also, happily, members of the Society) to buy the building.[5] As the Society was not incorporated, Smith, Russell and Leibius acted as Trustees for its purchase, at a cost of £3525. The Society had found a 'home' at last.

Although he had made the play possible, the final curtain did not fall until Liversidge was away from Sydney; and so he was when Clarke died in June 1878. On 28 May 1879, with Leibius still overseas, and Liversidge but two days returned, Smith gave the Society's Anniversary Address, consisting almost wholly of obituary tributes to his friends – M.B. Pell, Dr John Dunmore Lang, and most extensively, to the dear Rev. Clarke. Liversidge shared in the mourning. A week later, on 4 June, he reported on his travels. He and Russell were the only speakers.[6] The Society needed their leadership.

Within the coming months, Liversidge and Russell confounded the Cassandras, convinced that the winds of change would not last. For a start, the creation of specialist sections quickened the tempo of meetings. Not all benefited equally, but membership topped 430, and the Government renewed the Society's annual grant without comment. From 1876, the Society held its deficit in check, and by 1879, it was showing a small profit. Even the library that Liversidge assembled was given an insured value of £1000.[7]

In October, the Society celebrated by holding one of the 'most successful and brilliant gatherings' of the season, its first *conversazione* in the Great Hall of the University. According to the *Australian Town and Country Journal*, Hector's photographs, shown by the aid of a hydrogen lamp, Cracknell's telegraph and thermometers, and Russell's clocks captivated an audience of over 800.[8] Smith displayed his favourite half-prism spectroscope, and he and Liversidge showed their new 'Cailletet apparatus' for the liquefaction of gases. Thanks to Liversidge's recent foray in Europe, there was a new microscope, which he displayed alongside 150 books, and new photographs of Venice.[9] Charles Moore decorated the Great Hall with palm leaves, ferns and evergreens from the Botanic Garden, and tables were laid with crimson cloth. The cost was fearful.[10] But such was the price of publicity.

The Society's governance was fully in the hands of its secretaries. In 1880, they persuaded the Council that officers should be elected by ballot only. The Governor of NSW, hitherto the Honorary President, was replaced by a 'working president'.[11] The first, elected for 1880–81, was John Smith. But Smith gave Liversidge *carte blanche*. In March 1880, Liversidge brought James Cox and C.S. Wilkinson onto the Council. This not only improved lines of communication with the Australian Museum and the Department of Mines, but gave Liversidge more votes when he needed them.[12] The effects of his improved 'housekeeping' were immediate. Correspondence was listed before meetings; section chairmen were appointed without delay; and the hall at Elizabeth Street was hired out, frequently to the University's Senate, further helping to pay the mortgage.

Smith's first annual report spoke to a new-found optimism. A sense of history pulsed through the Society, when the Council commissioned portraits of distinguished colonial scientists since Brisbane, and ordered from London a set of decorative plaster busts celebrating British genius from Bacon to Murchison, with Queen Victoria added for good measure.[13] In early 1880, the Society's membership exceeded 450. Applications from women had hitherto been politely declined,[14] but with women entering the University from 1881, they could no longer be confined to the *conversazioni*, nor would Liversidge allow it.[15] Liversidge even moved an amendment (in what might have seemed the most arrant *hubris*) actually to *limit* the Society's membership to a ceiling of 500![16] News of the Society's activity spread afar. In a sense of turnabout as fair play, Robert Etheridge Sr, then at the British Museum, even approached the Council for a grant to publish his catalogue of Australian palaentology – possibly the first time a British scholar looked to Australia for a grant.

'The tide of prosperity', Smith said, we owe to the 'enlightened zeal and indefatigable labours' of Liversidge and Leibius.[17] Fresh from England, Liversidge was determined to elevate the Society's profile. After Clarke's death, the Council raised funds to endow an annual Clarke Medal, to be awarded 'to men of science who have made valuable contributions to our knowledge of the Geology, Mineralogy or Natural History of Australia'. With its rules written by Liversidge, the prize was to be open to all; its value was to be enhanced by appealing to the universe of science: bestowing credit both upon those who gave, and those who received.

The first three medals went to Englishmen – Richard Owen, who despite his anti-Darwinian views, had greatly encouraged natural history in Australia and New Zealand; George Bentham, author of the *Flora Australiensis*; and T.H. Huxley, universally famous, who had made a special point of sending copies of the *Philosophical Transactions* from London to the 'embryo society in Sydney ... which', Liversidge said, 'I hope will eventually grow sufficiently to be recognised as a not unworthy daughter'.[18] With the medal to Huxley went Liversidge's thanks to his teacher. With honorary memberships, the Society displayed similar *politesse*. In 1879, when the Council added to its rolls both Darwin and Huxley, they also included Owen. Owen's pleasure at receiving the award was, as he put it, all 'the more encouraging, as coming from the colony from which I have received some of the most interesting subjects of [my] labours'.[19]

For some time, the Society had procrastinated about a charter. In November, Liversidge drafted a Bill of incorporation, which was eventually passed by parliament in 1881. Next, he persuaded Parkes (then leading a Parkes-Robertson coalition) to cease the annual bickering about money, by setting the government grant permanently at a rate increased from £250 to £350. The government delayed, sought advice,[20] but finally relented, and eventually increased its bounty to £400 in 1882.[21] Before 1881 was out, the Society had paid off another £500 of its mortgage.

Confident, the Society now reached out in other directions. Visibility required communication. Mindful of the benefits, Liversidge applied to Sydney the practice Joseph Henry had encouraged in Washington, DC. Moreover, he appointed a committee to abstract and reprint articles about Australia for publication by foreign societies, along the lines of the American *Popular Science Monthly*. Soon, the Society had acquired an 'instant library' of foreign journals, and within the next two years, a network exchanging 1013 issues of the Society's journal with 284 institutions in 116 cities worldwide.[22] This was assisted by a government grant in aid of postage, a compliment Liversidge returned by offering to distribute 100 copies of the Council of Education's *Annual Report* with his mailings to 'Scientific Societies in Europe, America and the Colonies'.[23]

To keep the Society, its journal and its library flourishing was one thing; but to keep the membership vigorous, even with the specialist sections, was a challenge. In 1881, members heard twenty-eight papers given by thirteen speakers; but of these, Liversidge himself gave nine, and Russell, another five. Their professional work had its own momentum but, beyond giving papers, editing the journal and chairing meetings he had few means of encouraging research. Accordingly, Liversidge persuaded the Council to establish a set of prizes (three at £25 annually) for the best essays submitted on set topics (and, with luck, publishable in the Society's journal).[24] Such incentives were unlikely to set the Tank Stream on fire, but they were made welcome.

In 1881–82, it was Russell's turn to be President. A *conversazione* in August attracted 600 guests, including the Governor, Lord Loftus, and his Lady, to the Great Hall of the University – continuing a tradition that prospered, with rare interruptions, until the 1920s. The event, the *Herald* remarked, 'had another aspect than that of being merely for the amusement of the visitors, for no one could walk round the rooms without

Elizabeth House, 5 Elizabeth Street: the home of the Royal Society of NSW, 1875 to 1927.

being impressed with the achievements of science'.[25] Among the assembled company were William Manning, the Chancellor, and Justice Windeyer, the Senior Fellow of Senate, upon whom, as we have seen, fell the task of managing Liversidge's ambitions. The Great Hall gave an opportunity to impress, and thanks to Liversidge, the press were supportive. The *Herald*, possibly enlarging upon the truth, reported that the Society 'had [for years] done nothing to distinguish itself, and its papers were, with few exceptions, of no value to science or to the world'. But now all had changed. A generous library, displayed 'with care'; a hall adorned with portraits; a reading room with the latest serials, and, within a few months, an official seal as well – to all these improvements, the name of Liversidge was attached.[26] To his 'zeal and industry', 'never sparing leisure or labour', the Society owed its gratitude. With 'taste and judgement' as well as hard work, 'He has brought all his singular talent for organisation to bear'.[27]

The following year, reviewing the Society's journal for 1881, the *Herald* said as much again: 'Everyone must see ... how rapidly Professor Liversidge is succeeding in raising the Royal Society into what a scientific institution of its character ought to be.'

Archibald Liversidge

A Royal Society *conversazione* in the Great Hall, 1879.

As the editor mused:

> It is not easy to do this in a country where the real science workers are numerically small. At present, his efforts have apparently been devoted to give solidity to the organisation, and make it a valuable centre to aid now and hereafter those who are engaged in original research. It requires something more than zeal – a real genius for method must be added, with a determination that there shall be no makeshifts, but that what is done is done well and permanently done, so as to be of lasting value.[28]

The *Herald* was perceptive. With his self-effacing tact, Russell's diplomacy and Leibius' devotion, Liversidge had welded his 'Elizabeth Street conspiracy' into a reforming lobby. Its enemy was not ideology but inertia. The *Herald* captured an essential point: 'popular sympathy is with the highest work of science to a far greater degree than is ordinarily supposed', its editor observed; and the reformed Royal Society, 'desires no less to popularize than to foster true science'. By so doing, its name was now 'on more lips ... than ever before in its history'.[29] In 1872, the Society was a quiet colonial sideshow; within five years, Liversidge had turned it into a main event.

With a population a tenth its size, Sydney could never hope to embrace the full range of London's cultural enterprise. Yet, by 1886, Sydney boasted several learned societies, a natural history museum, a free public library, an art gallery, an astronomical observatory, a technical college and museum, and a botanical garden. Of these, Liversidge was a trustee of three, a founder of two, an official of one and a friend to all. He had learned to make the most of the city's intersecting élites. To have concentrated his efforts on achieving a single great library, or laboratory, or museum, or even a single great piece of research, would have failed to excite the broad spectrum of support upon which colonial science depended.

In reviewing new books by William Macleay and W.A. Haswell, the *Herald* saw a bright future for Sydney's small scientific community. 'A few years ago, we would no more have thought of producing such works than we now think of fitting out an Arctic expedition. We had not the time, we had not the means, we had not the talent and ability. If we wanted information on subjects lying at our own doors, we had to send to Europe for it.' Now, educated men could read the world's press, and contribute to a learned journal that was part of the world's literature. What had happened required a suspension of disbelief. As was well known, 'young colonies do not attract men of studious habits and scientific attainments; the only wonder is that it is different now'.[30]

From a wider perspective, these changes were not yet decisive, or permanent. Liversidge's reforms were still isolated, individual achievements. But attitudes were changing, and where colonial research might in Clarke's day have been difficult to defend, now Australians were expecting it of themselves. As the pieces fell into place, Liversidge could now turn to an even greater struggle, fostering science not only as a cultural agency, but as a culture in itself.

## 2. Sydney and its Faculties: Inaugurating the Faculty of Science

If, in the history of the University of Sydney, the 1880s would be forever known as the Challis Decade, the year 1881 was its *annus mirabilis*. John Henry Challis, a prosperous Sydney merchant and landowner, with no previous academic connection, had bequeathed the University a legacy that would amount eventually to over £250,000. Among the largest benefactions ever given to an Australian university, this gift suddenly vaulted Sydney ahead of Melbourne, and gave scope for developments long deferred. And that was not all the good news. In December 1881, on receiving Sir William Manning's promise to introduce 'modern subjects', Henry Parkes had agreed to double the University's grant from £5000 to £10,000, and to add £1000 for new lecturers in the expectation of the Challis gift 'falling in'. The same year, moreover, saw the beginning of coeducation. The University, whose by-laws had never excluded women from degrees, now formally admitted them – half a dozen years after London, but a decade before Scotland, and half a century before Oxford and Cambridge would do the same.

These three events were to transform Sydney University. The Challis bequest, the doubled endowment, and the admission of women foreshadowed changes that would modernize and draw the University closer to the 'general community'.[31] Moreover, they foreshadowed the coming of a culture of research. In America, Johns Hopkins University had since 1876 pioneered the introduction of German-inspired seminars and graduate degrees. To Hopkins, Huxley had nominated (indeed, accompanied on his maiden journey to Baltimore) Liversidge's Cambridge contemporary, Henry Newell Martin, who was destined to launch a generation in experimental biology, just as Ira Remsen had done in chemistry.[32] In Britain, the 'endowment of research' had become a catch-cry of reformers, and government grants to science, once more a matter of hope, were now a matter of expectation.[33] A dramatic expansion of higher education had begun in provincial England, with new university colleges at Leeds (1874) and Liverpool (1881), linked to the Victoria University of Manchester, adding 'modern studies' to the repertoire of liberal learning. It was fitting for Liversidge to transmit the same impulse to Sydney.

For the colony, the 1880s were especially momentous for public education.[34] Following Parkes' *Public Instruction Act* of 1880, primary education became 'secular and compulsory', if not yet free; and by 1885, the government was spending over £1 million each year on education, nearly £700,000 of which went to some 2000 public schools, reformatories, asylums for destitute children and orphan schools, serving over 180,000 pupils. In late 1883, George Reid, then Stuart's Minister of Public Instruction, set up the first eight 'high schools' in the colony, so beginning a state secondary system; and introduced the Bill that opened the way to courses for evening students at the University.[35] Reform was by no means smooth, and teachers were few and poorly prepared, but enrolments grew, and in due course, the University benefited. To Manning's delight, student numbers increased eightfold, more than doubling from seventy-six in 1880 to 203 in 1884, and then trebling to 647 by 1888. Women, who first entered the University in 1881, rose in number from 121 in 1887 to 211 in 1888.[36] They, with the evening classes, heralded the new 'intellectual commonwealth' for which first Badham and then Liversidge had long campaigned.[37]

With more students, and more money to teach them, the University debated which new directions to take. In 1881, Senate prudently deferred the making of new appointments pending a settlement of the Challis bequest, but decided in the meantime to press government for more capital funds. More space was needed for the women who now graced the uncomfortable benches of the Blacket Building. And if Liversidge's report of 1880 had given Manning the expansionary brief he needed – if the applied sciences were to be the engine that would drive forward all other 'modern' subjects, and bring wealth to the university at large – Senate had to alter its by-laws. This Manning accepted. But he had to move quickly lest momentum be lost.

Senate had not forgotten the unsuccessful University Bill of 1878, and on 4 January 1882, at an unusual summer meeting, set up a committee to prepare a set of new by-laws 'which may be rendered expedient or necessary, by the changed condition of the University, arising from the largely increased sphere of teaching now proposed in it'.[38] To this new 'By-Laws and Curriculum Committee', Manning, as Chancellor and the Rev. Canon Robert Allwood (Vice-Chancellor) appointed three of the five professors (Badham, Gurney and Liversidge); and five other members of Senate (Arthur Renwick, H.C. Russell,

Alexander Oliver, Edmund Barton and William Macleay). Renwick, a member of Senate since 1877, was then Secretary of Mines in the Parkes-Robertson government, and later (1885–87) Minister of Public Instruction, and was deeply committed to reform.[39]

The Committee met first in April, and on 17 May, recommended the reorganisation of the University into four faculties – Arts, Science, Medicine and Law – with the amendment of by-laws to require all undergraduates, after passing their first year in Arts, to 'elect which of the faculties they pleased'. The following day, the Committee framed regulations for a new Faculty of Science, and for new BSc and DSc degrees.[40]

As Liversidge predicted, money did not buy unanimity. At the meeting of 17 May, it was resolved that candidates for matriculation in Arts were to be examined in English grammar and composition, Latin, arithmetic, algebra and geometry and in *any one* of Greek, French, German, elementary chemistry and elementary physics. However, Charles Badham, who had consistently opposed the introduction of applied subjects – and would continue to do so to his death – dissented from the resolution that established the four faculties, and opposed the inclusion of chemistry and physics at matriculation.

Badham's views were well known, having been heard over the years in several Commemoration Addresses and countless Senate meetings.[41] In April 1882, he took the opportunity Commemoration always presented to let the press again know his objection to the view that the University, in determining its curricula, should 'pay deference to public opinion, as if public opinion were something more entitled to respect than public utility'. The rich capital of New South Wales presented to his methuselan eyes (he was then aged sixty-nine) the spectacle of a modern Babylon – with wealthy men 'utterly illiterate'; learned professions 'with very little learning to divide amongst them'; and mercantile classes, 'well below the standard of their *confrères* in Western Europe' in literary taste. The University's salvation lay in adhering to, and not departing from, the 'traditional functions of a university', by making 'classical study the instrument of culture.'[42]

Badham bore no grudge against Liversidge. On the contrary, as he once put it, in navigating the polluted seas of colonial intrigue, 'one good analyst like Professor Liversidge keeps in check a great many unauthorised but enterprising persons who would be glad to undertake the same office for a trifling consideration'.[43] But technical education, suitable for technical people, was not the central concern of higher education. Above all, what set the University apart was its responsibility to teach 'men and women to think'. All else was rhetoric, or worse. 'I do not say', he concluded,

> that you should teach nothing else: teach girls cookery if you please, and teach lads chemistry or geology, we shall want cooks, chemists and geologists; but do not omit to educate them as well, that is make them conscious of every operation of their mind, and able to distinguish every thought and sentiment that arises within them.[44]

Badham's views had their defenders, both in the schools and on Senate. But he spoke to a disappearing world. Even Oxford and Cambridge had successfully assimilated the natural sciences into new schools and triposes, and both were offering liberal education and professional training. Badham was the past; Liversidge, the future.

Still, it was ironic that Liversidge, with his cherished memories of Cambridge, should have become Badham's nemesis. To the values of liberal education, Liversidge was by no means opposed.[45] On the contrary, by the new regulations, to which Liversidge agreed, admission to the new BSc degree (and to medicine) required all candidates to have first passed the first year of Arts – to ensure they had acquired what Manning called a 'substratum of general higher education'.[46] However, Liversidge insisted, in return, all Arts students should be required to take – in addition to Latin, mathematics and one language (either Greek, French or German) – a compulsory (and examinable) course in elementary chemistry and natural philosophy. By way of compromise, he agreed to delete science from the Arts second year, which was given over entirely to Latin, ancient history, mathematics and languages. However, the third year retained courses in geography and geology, zoology and botany, in addition to Latin or Greek, mathematics, either French or German, and philosophy for pass students. Arts students reading for honours could reduce their science obligations by concentrating on classics and mathematics in the second and third years.

On 22 May 1882, Senate accepted this settlement, and also framed proposals for new lectureships in engineering, modern history, modern languages and law. Changes in the by-laws and regulations were given expression in the *University Extension Act* of 1884.[47] The 'moderns' had become the new patricians, or so it seemed. Victory was far from complete, and the 'battle of the by-laws' merely went into suspension. But even as arguments were sharpened for the next round of debate, from England came news of an accolade to Liversidge that none of his elders could match.

Two years earlier, Liversidge had been nominated for election to the 'queen of all learned societies' – the Royal Society of London. His sponsors formed an unusually long and impressive list of fifteen FRSs, eminent in chemistry, physics, physiology, geology, and palaeontology. They included W. Boyd-Dawkins, Henry Roscoe and Carl Schorlemmer of Manchester; H.C. Sorby, the pioneer petrologist and mineralogical chemist of Birmingham; and three of Liversidge's teachers from the RSM, Edward Frankland, Sir Warington Smyth and A.C. Ramsay. From Cambridge, came George Humphry, Michael Foster, and F.M. Balfour; and from London generally, the influential astronomer (and later PRS) William (later Sir William) Huggins; G. Carey Foster, the physicist of UCL; and Henry Woodward and Robert Etheridge Sr, from the British Museum. The candidates' book in Burlington House cited Liversidge's nineteen scientific papers; his ten mineral reports for the NSW government; his work as a Commissioner to Paris; and his report on technical education – 'Good work,' as his nomination read, 'done in various ways for the advancement of natural knowledge in Australia.'[48]

Since its age of reform in the 1840s, and through the early 1860s, the Royal Society of London annually received an average of fifty nominations, producing a long queue of eligible candidates for its fifteen coveted places.[49] Some candidates were nominated time and again. Most who were elected had been put up several times. So it was all the more gratifying to Liversidge that he had waited only two years, and that his election had been a virtual 'shoe-in'.[50] At the age of thirty-six, Liversidge had joined the most exclusive club in the English-speaking scientific world.[51]

Liversidge's preferment instantly marked him out as one of the promising men of his generation. He was only the tenth man to be elected on the basis of work done in Australia, and only the fourth since Ferdinand von Mueller in 1861 – following R.L. Ellery (1873), W.B. Clarke (1876) and Frederick McCoy (1880). For years to come, he would be one of only two (and between 1882 and 1886, the only) FRSs in NSW. He confidently acknowledged the congratulations of Sir George Stokes, senior secretary of the Society, and requested the customary extension of time before taking up his admission, as he was 'not likely to be Home for a year or two'.[52] His subscription of £15 per year would be handled as usual through Trübner and Co. of Ludgate Hill.

The news of his election happily coincided with both the inauguration of the Faculty of Science and the appearance of Liversidge's only book, *The Minerals of New South Wales*. Published first by the Department of Mines, and welcomed by the *Sydney Morning Herald* as a guide 'such as no one has previously given to Australia',[53] the book, with an iconic illustration of gold crystals gracing its cover, was destined to become the standard reference on Australian minerals for a generation.

With John Smith a spent force (he died in 1885), it was inevitable that Liversidge would be 'elected' the 'foundation' Dean of Science. He celebrated by successfully proposing his faithful Leibius for an honorary MA in 1882, the first time an officer of the Mint had been thus recognised. With the division of Smith's chair – Liversidge becoming Professor of Chemistry and Mineralogy, and the professorship of geology held in abeyance – Liversidge could devote his entire attention to his own subject – in what time, of course, he could spare from lecturing to the first-year Arts students whom he was now obliged to teach, and from the BSc students whom he hoped to attract.

Sydney's first BSc students entered in 1883, and the first two graduated in 1885.[54] No one suggested that the Faculty's future would be easy. Grand visions could not flourish alongside such small numbers. But until secondary schools multiplied, one could not expect more. Even private schools that offered science did not encourage it greatly. Since 1876, the matriculation examination set questions in elementary chemistry, physics and geology, but only as options. In 1883, following the compromises of 1881, the structure was again altered, but hardly improved, as Greek, French, German, chemistry and physics all formed a single class of subjects, from which only one could be chosen. The fact that science was optional, discouraged pupils from taking it at all. As coming years would show, this not only limited potential students for the BSc, but also narrowed the expectations that could be held of the few who did.

Still, thanks to the compromise of 1882, there was a small crowd of first-year students having to take chemistry.[55] Finding a building in which to teach them was more problematic. A solution was not made easier by the appointment in 1882 of Sydney's first professor of physiology, Thomas Anderson Stuart, a brilliant twenty-six year old graduate of Edinburgh. Arriving in the opening months of 1883, Anderson Stuart confronted, as had Liversidge at the same age, a huge task – in his case, building a medical faculty and a medical school.[56] He began with six students, and a tiny cottage. Gurney met the new professor and his wife when they arrived from Britain via Melbourne at Redfern station; and Liversidge made them welcome. The professoriate was now six; and if one followed the politics of faction, it was weighted five to one in favour of science.

University staff and third-year students in 1881. Front row from left: Professors Gurney and Liversidge, Vice-Chancellor Canon Allwood, and Professors Badham and Smith.

The expansion of the University necessarily brought with it a redistribution of power and a realignment of influence. At the beginning of the academic year 1883, Manning asked Senate to introduce two major innovations, in both of which Liversidge was destined to play a major role. The first was the Buildings Committee (technically, a subcommittee of Senate), which Manning set up to consider proposals for new structures, and to supervise their construction. With an irony that must have pleased Liversidge, the new committee held its inaugural meeting in Elizabeth Street, and in the rooms he had acquired for the Royal Society. That session of 8 March 1883, chaired by Sir William Windeyer (Manning's Vice-Chancellor), included Russell and Renwick, representing Senate; Gurney and Liversidge, as professorial members, and Anderson Stuart and Stephens, by invitation. The scientists agreed to furnish proposals for Senate, and within a fortnight, Liversidge, Stephens, and Anderson Stuart had sketched requirements for chemistry, natural history and medicine. These, which were to be 'given to the Colonial Architect, and explained to him by the professors',[57] outlined four new buildings, to be erected at the 'southern end of and parallel to the present [Blacket] buildings'. They were to take the form of three pavilions, the one closest to the Blacket Building for chemistry; the middle one for natural history, geology and natural philosophy; and the third for medicine.[58]

Successful proposals for new accommodation and curricula required not only detailed negotiation, but also a united front. In April, while the Buildings Committee saw to matters of sandstone, Senate reconstituted the Professorial Board that had been set up in 1853, and which in 1876 had been rebranded as the Board of Studies. This Board, which included all six professors, was to review proposals that would otherwise consume the time and patience of busy senators. It was an important act of devolution for which the first professors had fought dearly; twenty-five years later, far from being opposed, it was welcomed. However, Senate conceded that Badham was outnumbered by Smith, Gurney, Stephens, Liversidge, and Anderson Stuart. On any issue of substance, the Principal was in a minority of one.

In July 1883, with all six professors present, the Board of Studies began to ruminate again over the by-laws. The old Arts/Science faultline quickly surfaced, and Badham, Dean of Arts, confronted Liversidge, Dean of Science, over the requirement of compulsory science. Four of the six votes went predictably to Liversidge, but on referral to Senate, the conservatives supported Badham, and a compromise was proposed: natural history was to be accepted as a 'foundation subject' in Arts, on condition that the first-year course in elementary chemistry and natural philosophy be dropped. However, the scientists stood fast, and eventually another compromise emerged, by which first-year Arts students would take a single lecture course over three terms, one each devoted to elementary physics, experimental chemistry and natural history. In exchange for the prospect of attracting BA students to honours in science, the scientists agreed to move the courses in geology, zoology and botany from the Arts third year to the second year, and to make it a non-examinable course of lectures.

In acceding to this compromise, the scientists had bowed to Badham.[59] But Liversidge took comfort in compromise, and in the following months, quietly took possession of the Board, as he had the By-Laws Committee, and as he would the Buildings Committee. As he became adept at pushing initiatives and skirting difficulties, the Board's minutes recorded his victories.[60] It was Liversidge who proposed 'Charles Darwin' as the topic of the next Prize Essay for English verse;[61] it was Liversidge who welcomed William Windeyer, wearing his hat as representative of the Sydney Mechanics' School of Arts, to discuss the future of University-Technical College cooperation; and it was Liversidge who persuaded his colleagues to take responsibility for examining Technical College candidates. It was Liversidge who organised the professors into a deputation to the Minister of Public Instruction;[62] and Liversidge who insisted that the Board recommend new lectureships in German, mining and metallurgy.[63] It was also Liversidge who took the initiative in amalgamating the University's Proctorial Board with the Board of Studies,[64] and later, in recommending to Senate ways of reducing its workload.[65] During the next twenty years, many initiatives bore his hand;[66] and in dozens of ways, from fixing 'common weeks' to providing students with examination books,[67] it was to Liversidge that the Board looked for decisions.

Indeed, the Board of Studies (or the Professorial Board, as it became after 1886) was more effective than many other committees on which Liversidge sat. Its revolving chairmanship and weekly term-time meetings imposed a discipline, which in turn contributed to a semblance of consensus. For Manning, the Board meetings gave an opportunity to gauge professorial opinion before Senate debates.[68] It was to the Board, for example, that Manning turned to discuss the evening classes that Badham proposed. Senate favoured moving into the 'evening market', and the idea was popular in the community. But the evening was not to be for Arts alone, and the scientists asked for better equipment and accommodation to meet an expected demand.[69]

Badham's death in February 1884 was a time for great public mourning. The colony had lost a dedicated scholar and educator; the University had lost a singular mind. Yet, no one denied that, with his demise, the Board of Studies entered a new phase, favourable in many respects to many of the reforms that he had resisted. John Smith became the new Dean of Arts (and Stephens doubled up as *pro tem* lecturer in classics), and for a time the Board's meetings in Elizabeth Street came as close to the proceedings of a scientific cabal as the University had ever seen. A Venetian Doge and Council of Ten could not have acted in greater harmony.

Over the next ten years years, the superiority of numbers played directly into Liversidge's hands. In effect, he served three roles simultaneously – *ex-officio* member of Senate, resident 'lawyer' of the Board, and Dean of Science – quite apart from his normal lecturing duties. Attendance at Senate gave him working access to deal-making, while the Board looked to him for detail. As Dean, re-elected again and again for six terms every three years, he became almost the *secrétaire perpetuel* of science. When, in 1886, George Reid approved £1000 for evening classes, of which £200 was to buy apparatus for chemistry and physics,[70] it fell to Liversidge to decide how the money would be spent. Equally, he decided how to divide the government grant for scientific apparatus, typically £500 a year, among chemistry, physics, engineering and natural history.[71] His reputation for even-handedness fostered within the Faculty a degree of harmony difficult to find in the world outside.

His influence extended well beyond Science. In September 1884, Liversidge adroitly navigated through the Board a proposal for a lectureship in architecture, a position that had lain vacant despite its prescription in the engineering by-laws (which Liversidge had drafted).[72] This involved new expenditure, and even without a Badham, Senate managed to resist. In ways fundamentally unchanged since the 1850s, the advancement of any new discipline, especially one where 'professional training' was concerned, had to contend with laymen who wished as often as not to keep things as they were.

The Board saw that in union there was strength. On a motion by Smith in May 1884,[73] approved the following year, the professors took another step towards union, and restyled the Board of Studies as the Professorial Board.[74] In the short term, their creation of a Faculty of Science had done little to allay public criticism. During the second reading of the University Extension Bill in 1884, John Stewart claimed 'that while the University's costs had been exorbitant, its graduates were fit for very little'.[75] The *Echo* hoped the University would 'fling open the gates' and let in not the favoured few, but 'all who desire and deserve it'.[76] The *Daily Telegraph* deplored its reluctance to be 'shaped by local causes': 'Those who would seek to make it a dead imitation of Oxford or Cambridge would make it dead indeed.'[77]

Familiar criticisms received familiar replies. John Smith, a member of the Legislative Council since 1874, countered that, so far as the curriculum was concerned, Sydney students were actually taught 'many things above mere words. It comprised the three branches of languages, mathematics and physical science, each of which appealed to its own particular class of faculties. These branches provided a sufficient foundation for a general education.' It was 'a proper curriculum,' and the University was adding to it.[78] Indeed, the University, had responded to criticism with its evening lectures (which, despite their enthusiastic reception in 1884, had produced only a modest six graduates by 1889), and with extension lectures, modelled on those of Cambridge, which had spread to country towns in 1886. Their purpose – in the words of Walter Scott, Badham's successor as Professor of Classics – was to take knowledge to the country as 'no would-be student should be excluded by poverty.'[79]

At the same time, Liversidge sought to ensure that no would-be student would be excluded from science. The compromise struck in the Arts debate in 1883 was unsatisfactory. It had proved nearly impossible to attract first-year Arts students to the BSc. When five BA students applied to transfer, Badham actually discouraged them. But where there is death there is hope, and three months after Badham's demise, Liversidge proposed amendments in the by-laws to permit third-year Arts students to attend lectures and take examinations in any two of six science subjects (together with their required Latin, other languages and mathematics).[80] The Professorial Board naturally approved the proposal; but on referral, Senate rejected it. The following December, Senate similarly rejected Smith's appeal to reinstate the first-year lectures in physics and chemistry, which had been swept away by the curriculum changes of 1882. Writing from London, in what would be the year before his death, Smith pleaded with Manning:

> I hear through Professor Liversidge that the lectures on chemistry and experimental physics are to be much curtailed. I am very sorry for this, and feel sure that in this country and on the continent of Europe such a policy would greatly be condemned as retrograde. Physical science has undergone an enormous expansion of late years. It is more and more realized as lying at the foundation of national prosperity; and it has thoroughly indicated its right to form a large portion of any education claiming to be liberal, or in fact anything more than elementary ... it is to be regretted that the Sydney University could recede from its former position, and cut down its science teaching.[81]

Smith's protests were to no avail. Nor was he, as Dean of Arts, more successful in loosening regulations for students wishing to transfer to science. The Faculty of Science was forced to recruit all its BSc students directly from matriculation. Whatever the educational arguments, this policy was sure to limit numbers. During the next decade, fewer than three students a year completed the BSc.[82] Whether because of the shades of Badham, or the bias of Senate, science remained, like Prometheus, chained to its rock.

Battles for physical space paralleled the scientists' struggle for curricular space. Rarely were these 'builders' stories' either simple or short. Late in 1883, George Reid, sanctioning funds for education, included an additional £20,000 in the University's vote for 1884 specifically for new buildings for science and medicine. After years of fasting, such a feast beggared belief. But James Barnet, the colonial architect who planned carvings on the Pitt Street frontage of the General Post Office to portray 'selected arts, sciences and customs of the day', was busy with other projects, which threatened to delay the project indefinitely. Fearing the loss of their vote in the year-end parliamentary darkness, the Building Committee – then comprising Badham, Anderson Stuart, Russell, Stephen and Liversidge – asked Senate to 'wait upon the authorities'.[83] This, one of Badham's last meetings, found the elderly classicist unable to oppose the future.

Sandstone carving by Tommaso Sani on the Pitt Street façade of Sydney's General Post Office, said to represent Liversidge.

One sees Liversidge's hand at work when, on 5 December 1883, the Vice-Chancellor, the Registrar, and Anderson Stuart met with Reid.[84] Their reception by a fellow member of the Union Club went well. Reid promised to see the Colonial Architect personally and to sanction contracts due by 31 March 1885, which would enable the University project to proceed. He added, the *Herald* reported, 'that he should be always happy to see any gentlemen on University subjects'.[85]

His word was taken as his bond, but five months later, no further response was received; so on 17 April 1884, the Building Committee, this time with Liversidge in the chair, asked Senate to send another deputation to the Minister.[86] This meeting produced results – sketch plans were prized from the Colonial Architect; and, by the end of May, on Liversidge's motion, the Building Committee – now Liversidge, Anderson Stuart and Russell, with the ailing Smith and Warren co-opted – submitted to Senate a four-point plan providing for chemical laboratories and lecture rooms to be erected 'as speedily as possible before any other part of the permanent buildings'. The dangers of fire and overcrowding in the Blacket Building were obvious, but Liversidge repeated them just the same. Money for building would first go to the Medical School, after which it would be the turn of natural history, engineering and chemistry.

The plan and its priorities were agreed – but problems emerged. Owing to government delays, nothing from Reid's £20,000 was available for spending in 1884. This put pressure on the Senate to review priorities for 1885. There was no debate about the needs of natural history and engineering;[87] but when faced with a choice between medicine and chemistry, Liversidge knew he had lost when Arthur Renwick produced a plea from Anderson Stuart to put the Medical School first. 'In the public interest,' Renwick argued, 'and that of the University, the Medical School should be built without delay.' To this end, another deputation went to Macquarie Street, this time comprising the Chancellor, the Vice-Chancellor, and the Dean of Medicine (Anderson Stuart).[88] Liversidge's chemical laboratory would have to wait.

James Kerr credits Anderson Stuart with 'tenacity, acumen (and friendship with Liversidge)' in bringing about this tactical success for Medicine.[89] Certainly the debate must have tested that friendship. Still comrades in arms – and sometime housemates – the two were now divided by circumstances that would bring for one the prospect of professional expansion, and for the other, years of continued struggle. Liversidge's setback was made all the worse by knowing that, in December, the Legislative Assembly voted £10,000 to house William Macleay's natural history collections, in what would become the Macleay Museum.[90] Chemistry, the mainstay of the curriculum, seemed to be the Cinderella of the Faculty.

If Liversidge was disappointed, he did not show it. On 3 December 1884, even as the Government was placing £15,000 for the Medical School on the estimates for 1885, Liversidge renewed his appeals to Senate. Scarcely had the new year begun before he wrote again, in the humid heat of 4 February, and again in March, at the beginning of the academic year, begging Senate to appropriate funds for chemistry from surplus accounts, in advance of receiving an earmarked government grant.[91]

On 21 March 1885, a special meeting of Senate was convened to visit the site chosen for Anderson Stuart's Medical School, and to discuss the site of the new Macleay Museum. At Liversidge's insistence, Senate was embarrassed into agreeing that chemistry should eventually have a site 'at the southern end of the main University buildings, at a distance of eighty feet from them and in line with the general front' – that is, between the Blacket Building and the Medical School. But when would it come?

At the end of Trinity Term, 1885, Liversidge renewed his campaign. In May, when the Government was sent the University's budget for 1886, Liversidge asked again for £10,000 for a new laboratory.[92] The government declined, but this time, the medical lobby rallied to his support, and on a motion by Renwick, Senate voted £1000 from reserves for a temporary laboratory – a sum which it hoped to recoup by a refund from Government.[93] This, at least, as Manning put it, 'removed operations from their confined and dangerous position' in the Blacket Building, against the day when there could be 'a laboratory of superior class and greater dimensions ... for the Department of Chemistry, or for Science generally'.[94] Liversidge took what he was given, sent out plans for tender, and within months, had a tall brick building, costing only £737, built at the south-eastern end of the (eventual) quadrangle.[95]

There it would stand, until overtaken by the Fisher Library (now MacLaurin Hall) in 1917, its skylighted contours and laboratory furnishings memorable to generations of students. After 1889, it was to be remade into the Women's Common Room, and to become a focal point for undergraduate life.[96] As in 1881, the interests of women and the demands of science seemed fated to coincide. On Senate, Russell – whose daughter, Jane, was then an undergraduate – seemed at pains to insist upon the building's 'temporary character'.[97] But for three decades or more, its very existence spoke in eloquent reproach of the University's parsimony, one of several architectural blemishes upon its sandstone complexion.

When we consider what motivated medical men in their ambition for a new building, we reflect that Anderson Stuart need not have loved chemistry less, to have loved medicine more. Given the fragile and elusive character of government promises, self-interest came first. That year, expenditure on railways, schools and other public works was made difficult by a drought and fall in wool prices. In September 1885, Alexander Stuart's government was threatened with a dissolution and William Trickett, briefly the new minister of public instruction, was far less sympathetic than Reid. In 1884, Trickett said he would support increased funds for the University only where it made 'social refinement and intellectual advancement run hand in hand with material prosperity'.[98] To Senate's modest request, he told the University to draw against its Challis funds. Liversidge, drafting Senate's reply, observed that the Challis bequest was for educational purposes and the endowment of chairs, and 'not [intended] as a direct relief to the public treasury'.[99] But his plea fell on ears that could not hear: the Stuart administration disappeared in October; and its short-lived successor, led by George Dibbs, expired in December, leaving a deficit of £1 million.[100] An equally tenuous Robertson ministry came and went by February 1886, and the refund never came.

Without a research laboratory, Liversidge was limited to desk analyses and editing. His *Minerals of New South Wales*, a vast cataloguing effort, appeared in a second edition in 1882, but he published nothing else between 1881 and 1883. In 1884, he reported only two further analyses, one of which was for his friend F.W. Hutton of Christchurch.[101] Reform within the Royal Society and the work of the Australian Museum distracted his attention; while the work of the Board of Technical Education made increasing demands.

Worse still, at the beginning of 1885, John Smith, aged sixty-four, fell gravely ill. The long years had taken their toll. Following Manning's report to Senate in April, Smith had suspended his lectures for two terms, but in the end was to give no more.[102] After months of slow decline, he died on 12 October 1885. All classes in physics, as well as his own in chemistry and geology, now fell upon Liversidge.

It was fitting that Liversidge was asked to give the eulogy. Smith's death severed the University's last link with its' foundation professors. Significantly, however, Liversidge's remarks were directed not so much to the University, but to the work of public science on which he and Smith had laboured for thirteen years. Liversidge spoke movingly of the Aberdonian's contribution to the water question and to the furtherance of secondary education. Smith had enemies, he had faults, yet he had brought about changes in the same 'quiet, unobtrusive, conscientious manner' that Liversidge commended to his fellow-members and imitated himself.[103]

In his praise of famous men, Liversidge could say little about Smith's zeal for the lecture room, still less of his interest in practical chemistry, and a great deal less about committee work. For nearly a decade, Liversidge had borne the burden of extending science within the University and the wider community. Although he held the chair of experimental physics, Smith had taught no physics for thirty years. In fairness, he had tried to get an additional appointment, and in July 1884, as his health failed, urged the University to choose as his successor a 'professor of natural philosophy', trained along Scottish lines, a man of standing comparable to P.G. Tait of Edinburgh and William Thomson of Glasgow. He suggested this title, as he explained to Manning, because:

> [T]he term Experimental Physics is becoming obsolete. That branch is now like other branches, subjected to profound Mathematical reasoning, and there is consequently no such line of demarcation between it and other portions of Natural Philosophy as formerly existed.

Moreover, he added:

> No-one should be appointed to the Chair of Natural Philosophy who is not comparatively young, of high Mathematical ability, well trained in the use of apparatus and fresh from the centres of progress.[104]

This amounted to Smith's last will and testament to science at Sydney.

Given Liversidge's plans for chemistry, Manning took no immediate action on Smith's proposal for physics. But with Smith's death, the Board of Studies did set out the terms and conditions for his successor, and put these to a saddened Senate.[105] The Board followed Smith in recommending a chair to which should be assigned a 'wider and different range of teaching in Physical Science, including portions of the duties before discharged by the Professor of Mathematics as Professor also of Natural Philosophy'.[106] Liversidge also argued for the appointment to be made in experimental physics, rather than in mathematically-based natural philosophy. In particular, he spoke of the practical physics as taught at Cambridge, and embraced in the Natural Sciences Tripos, as distinct

MISTRY CLASS 1886. N°40 RUBBISH BOX

Students in a chemistry lecture, 1886.

from the 'Newtonian physics' taught in the Mathematical Tripos. In the end, his arguments prevailed, and Senate moved to select Australia's first experimental physicist. The Board might have well have had forebodings that the small rooms Smith occupied would prove wholly inadequate for an energetic successor.

The coming year, 1886, was a turning point for physics in Australia, with the arrival of two Cambridge men – the twenty-four year old W.H. Bragg of Trinity to the inaugural chair of physics at Adelaide,[107] and Richard Threlfall, aged twenty-five, of Caius College and J.J. Thomson's demonstrator at the Cavendish – to Sydney.[108] Threlfall came highly recommended – with a double first in the Natural Sciences Tripos and successes in the laboratory and lecture room. He had, his referees said, 'quite an unusual amount of energy'.[109] Chosen from a field of twenty candidates, Threlfall won largely on Thomson's recommendation.[110] In May, this robust, rugby-playing experimentalist descended upon Sydney, full of all the enthusiasm that Liversidge once knew.

With some foreknowledge of the Sydney situation, Threlfall took the precaution of bringing with him a 'sufficient quantity of instruments, including tools and materials for the manufacture of apparatus to suit a proper laboratory'.[111] Receipts for this unauthorised expenditure, which he presented to Senate on arrival, amounted to an alarming £2000. More worrying still, the young professor greeted the university on 7 June 1886 with a simple demand for a 'Physical Laboratory', 'without which no adequate instruction in the subject could be given'. He could not make do with Smith's modest accommodation in the Blacket Building – a lecture room and office, plus 'the room partitioned off at the top of the staircase, [and] if required, the top room in the tower, hitherto used by the Demonstrator in Chemistry'.[112] 'The fools have appointed me', he is reported to have said;[113] and Threlfall was never one to suffer fools gladly.

For the first time in Liversidge's memory, youthful temerity triumphed over tired forbearance. Historians have marvelled at the speed with which Senate – 'at once', this time – met Threlfall's demands. Manning sent a deputation to the new Premier, Patrick Jennings, requesting that a vote for £8000 for Threlfall's building be placed on the estimates for 1887; and that instructions be given to the Colonial Architect for the 'immediate production' of plans for a building to be put near 'Natural History' (the Macleay Museum then under construction, 100 yards northwest of the main building).[114] The hard-pressed Jennings Government, wrestling with increases in tariffs, crown rents and loans, declined the proposal.[115] But Threlfall was not to be deterred. On 25 October, he sent Senate a third letter, calling attention to the fact that, as the lecture room had not been adapted for teaching optics, he would prefer that Gurney take those lectures instead; and that, for want of a physics laboratory, he would be quite 'unable to give the course of Practical Physics prescribed in the by-laws'. A fourth letter asked why Jennings had struck out Senate's proposal, without even submitting it to parliament.[116]

These were all good questions; but such rough play was unprecedented. To his credit, Manning went in to bat for his new professor. When the beleaguered Jennings failed to meet his demands, Manning sent a second appeal, and this time a promise that £8000 could be had in 1887 was made by Jennings himself, who was well known for his 'princely liberality' towards St John's College, and was a patron of the University. But faced in December 1886 with the likelihood that he would soon fall from power, Senate decided to hedge, 'to take the responsibility of providing the necessary funds in the meantime' (from what source, it is not said),[117] and to build on the site that Threlfall had chosen – in open space west of the Main Building, looking out upon grazing paddocks, on the line of what would later become Science Road.

Years later, Sir J.J. Thomson, embellishing the story, claimed that Threlfall had personally lobbied the Jennings Government; and had won his case just as Jennings, beaten on a division, surrendered to its opposition (Parkes' ministry of 1887–89).[118] That Threlfall could have made an intervention of this kind is doubtful; but Jennings' Minister of Public Instruction was Arthur Renwick who, as a member of Senate, had piloted Anderson Stuart's Medical School through Parliament, and who was responsive to Threlfall's situation as well. In any case, Threlfall claimed a personal victory. As he wrote to Lord Rayleigh, Thomson's successor at the Cavendish, in January 1887:

> I only got the money at last by letting it be generally known that I would go back if I did not get it; rather a bold stroke I took as I dare say many of them would only be too glad.[119]

In May 1887, before any new money arrived, the surplus from the balance of £20,000 granted by the Government in 1883 was assigned to the construction of the physics building and the Medical School, 'as those objects appear to coincide most nearly with the understanding which seems to have existed when the vote was made'.[120] Manning was deeply concerned about rising costs, at a time when Medicine alone – with its demands for £15,000 – threatened to consume the University's entire vote. He mused that the Medical School might become financially independent.[121] In the event, this did not happen, but £9000 was allocated to Physics – 'simple utility being thus all that was aimed at, the building will be of very moderate cost'.[122] Chemistry was given a temporary laboratory of wood and iron, erected nearby at a cost of £1000 from funds that Parliament had so far refused to grant. At least, it removed Liversidge's students from their 'confined and dangerous position' in the Blacket Building. 'It can be made to suffice for some years', Manning hoped.[123]

In matters of bricks and mortar, Threlfall had trumped his Dean. Liversidge's reaction can only be imagined. But publicly he welcomed, where he might have privately opposed. There were favours that could be repaid. There was the friendship of an older Cambridge man for a younger. In any case, Liversidge rose above infighting. Thanks to him, they had a Science Faculty and a science degree. In May 1886, he chose his presidential address to the Royal Society of NSW to point out that:

> the science and professional student is now, after many a hard struggle, emancipated from most of the old classical fetters in cases where he has not the time or inclination to proceed with such studies.[124]

Progress was inevitably slow. There was 'almost an entire absence of real instruction in even the most elementary science' in schools, and virtually no practical experiments, and for graduates, the openings 'are but few, and usually not well paid'. Even in a colony that owed much of its prosperity to mineral wealth: 'The necessity of having well-trained scientific managers to mines, metallurgical works, and manufactories, is hardly yet recognised.'[125] The remedy would take years.

During the remainder of the decade, Liversidge dedicated much of his time to Faculty, Professorial Board and Senate meetings. Indeed, he could make rods for his own back. Owing to his efforts, for example, new by-laws required incoming BSc students to take lectures and examinations in French and German. When this strained a course already heavy with practical classes, he tried to transfer the language requirement from the second year to matriculation, but having to master a modern language did not make a science degree any easier, or more attractive.[126] Nor was it seen as such in the community.

Following Badham's death, and until Walter Scott found his footing, the sciences held the upper hand. Sydney was weak in many Arts subjects in which English universities had chairs.[127] In his Chancellor's Address that year, Manning drew attention to the situation – all the more grievous, as these subjects were 'of the very essence of University education, and the chief source of "culture" and preparation for the world's higher work'.[128] In 1886, when Scott became by rotation chairman of the Professorial Board,[129] Liversidge supported him, and in November, the six professors asked the Jennings government for £6150 to appoint lecturers in history, philosophy, political economy and English. At the same time, however, they requested laboratory materials for advanced students. The Government's response was instructive. Scott's appeal for a chair of modern history – 'to bring teaching in that subject to the degree of completeness already reached by the Department of Natural Science'[130] – was declined; instead, the University was given £1000 for new apparatus.[131]

This episode threw the power of science into bold relief. However, Liversidge and Scott made common cause on many issues. One concerned the rising demand for advanced courses. The phenomenon was hardly new, or unique to Australia – specialisation was a fact of modern life. But this had serious implications for staffing. As Andrew Garran – a graduate both of London (BA 1845, MA 1848) and of Sydney (LLB 1868, LLD 1870), editor of the *Sydney Morning Herald* and parent of a Sydney student, later Sir Robert Garran – pointed out in a letter to the *Sydney Mail* – no one could expect students to embrace all subjects, nor could they embrace even one, without being given more specialised instruction than they were receiving.[132] This, coming from a Trustee of Sydney Grammar School, who treasured the classics, but who also criticised Badham's attitude towards the 'bread and butter sciences'.[133]

In August 1886, five years after Cambridge accepted specialisation by dividing the Tripos into Parts I and II,[134] Sydney debated the mix of courses required of honours students.[135] Interest focused on the second and third years of the Arts degree. Liversidge wanted Sydney students to choose any four of eleven courses in the second year, and then specialise in any three of eleven courses in the third year. About half the courses in each list were to be in the sciences.[136] With Scott's support, Liversidge's recommendations were accepted by the Board and Senate in December 1886. From 1887, following a generalist first year, Arts students could go on to more specialised courses in their second and third years. This was a promising step: but at the close of 1886, the Faculty of Arts was far from able to realize the promise. The University's endowment, raised by Parkes and Reid from £10,000 to £12,000 in 1883, remained static until 1890. For over a decade, only the scientific apparatus and special purposes vote improved, rising majestically from £1000 in 1881 to £7100 in 1888.[137] In the meantime, any new spending had to be met from the Challis bequest.

The year 1886, which had opened so brilliantly, thus closed with unfinished business. Liversidge's laboratory remained a dream on paper, and new research had been postponed. He had written only a few papers that year, all for the Royal Society of NSW, and all about mineral specimens sent to him from New Guinea and the Pacific, with some notes on a new meteorite from Queensland.[138] But not all his frustrations could be laid at the door of Faculty and Senate. On the contrary, he would later look back on 1886 as a tipping point – if not for the University, then certainly for the history of Australian science.

## 3. Science Mobilised: Intercolonial Association

In the years between 1880 and 1886, Liversidge had brought about a kind of vertical integration in the culture of science in New South Wales. On paper at least, the vision embodied in his report of 1880 was now largely realised. There was a vibrant Royal Society, a Technological Museum, a Technical College and a Faculty of Science. Now he could move beyond NSW, and act on his most ambitious project – the federation of Australasian science.

When he returned to Sydney from Paris in 1879, Liversidge had raised a metaphorical Eureka flag among his co-conspirators in Elizabeth Street. In reporting on the geological congress, he admitted that it conflicted with the BAAS meeting (that year, in Dublin); and acknowledged that 'many persons were doubtless surfeited with [such] scientific picnics'.[139] Nonetheless, the value of scientific congresses was well established and, in Australia, they could be copied with profit. Sydney's first International Exhibition, scheduled for 1879, would highlight the colony's scientific achievements as evidence of its cultural vitality. 'I hardly like to propose', Liversidge told his Sydney friends, who doubtless wondered what other ambitious schemes their Honorary Secretary had brought back, 'that a Geological Congress should be held [in Sydney] because the number who could attend would be such a small one'. Instead, he added prophetically,

> the Royal Society of NSW might, perhaps, with advantage, join with the other scientific societies to hold some special meetings, at which papers could be read and discussed, after the model of the British Association.[140]

Suggested by John Smith in 1866, and by the philosophers of Van Dieman's Land much earlier, the idea of an 'intercolonial BAAS' was not new. But with the passing years, the arguments grew more persuasive. In 1871, when Robert Ellery, Victoria's Government Astronomer, enlisted the help of Henry Russell in making the Australian Eclipse Expedition to Cape Sidmouth, the stage was set for what Michael Hoare has called the 'first real attempt at formal, intercolonial scientific cooperation on any scale'.[141]

Intercolonial cooperation of any kind was propelled by the new technologies of the day, notably the railway and the telegraph. In 1872, Russell conducted intercolonial studies of Australian meteorology, based on data gathered by all six colonies, and relayed by wire among them. The NSW government gave Russell £1000 for his part of the project, and Charles Todd received similar promises from South Australia. The transit observations of 1874 similarly underlined the importance of cooperation, and confirmed official readiness to support it. In 1875–76, Ellery used a trip to Europe to enquire about international meteorological programs, and on his return in May 1877, outlined to the Royal Society of Victoria a plan to transmit astronomical observations across the continent. The same year, the astronomers, with the help of the colonial telegraph departments, began an system of 'weather telegraphy' that linked South Australia, Victoria and NSW.[142]

By 1879, acts of cooperation were not a question of principle, but of political will. During his visit in 1873, Anthony Trollope sensed there was 'little tendency' among Australians to 'that combination which seems to me ... essential to their future greatness'. Yet, he confided to his readers, 'that they will at some time combine themselves I look upon as certain'.[143] Liversidge brought that certainty to life. Could there not be a scheme for regular communication, uniting all the colonies?

What Liversidge wanted was collective action to overcome the triple tyrannies of separation, specialisation and scale. Travel could be made easier by subsidised rail and coastal shipping. Specialisation could be reduced by meeting together. Scale was more difficult. If the scientific community was defined by the number of professional men in the three universities (Sydney, Melbourne and Adelaide), two metropolitan museums, six departments of mines, and a half-dozen technical colleges and mining schools, all would fit into one of his lecture rooms. Adding contingents from New Zealand would not greatly increase their discomfort. But 'professionals' could not afford to ignore the different Royal Societies and the other learned societies, with their aggregate membership of between 2000 and 3000 – mostly amateurs and enthusiasts, but all of whom had much to contribute. Nor should Australia neglect the model of the New Zealand Institute, established in 1867 by the 178 members of the eight provincial societies of that colony. Since the 1870s, Sir James Hector and New Zealand had been an inspiration to Liversidge, and had shown what cooperation could accomplish. Now, with Henry Russell at his side, Liversidge began to see the statue in the marble.

Planning began as early as the Garden Palace in 1879, when Liversidge tried to interest Australian geologists in matters arising from the Paris conference of 1878, which were to feature at their next congress at Bologna in 1881. Few responded, and his efforts failed. Charles Moore mourned that 'it was … impossible for the geologists of each colony to meet together'.[144] But it was common knowledge, as Charles Wilkinson put it, that the geology of each colony must be understood by every other: 'Geological science not only compels a union of workers in the different provinces of Australia, but throughout the world.'[145] Eight years on, with more experience of institution-building behind him, Liversidge determined to try again.

Several factors favoured a fresh attempt. Across the continent, the colonial scientific societies shared a history of genteel poverty, and had at best a fragile hold on life. The press liked to poke fun at what it called their elitism. In Melbourne, what Hoare has called the 'privilege-conscious' Royal Society of Victoria kept potential members at bay. In 1883, Robert Ellery, for many years President of the Royal Society of Victoria, claimed that Australia was 'not yet large enough to maintain, in an effective state, a number of scientific societies'. However, many small interest groups flourished in astronomy and natural history, and there was certainly a market for more. In 1885, Robert Litton, editor of the short-lived *Australasian Scientific Journal* (1885), succeeded in launching a Geological Society of Australasia, with Frederick McCoy and Ferdinand von Mueller as Vice-Presidents. The Society grew slowly, but it attracted members from Victoria, NSW and New Zealand, and showed what might be done.[146] In any case, whether one welcomed diversity, or saw 'fragmentation' as dangerous, it was clear that in 'Unity is strength'. Ellery set about improving cooperation, first with NSW and then beyond.[147]

In fact, relations between the sister societies of NSW and Victoria were already quite warm. In 1878, Nicolai Miklouho-Maclay had played off Victoria against NSW in seeking support for a marine zoological station, which belatedly went to Sydney. But where the societies acted together, benefits multiplied. When, in 1881, the Royal Society of NSW made Frederick McCoy the first Australian recipient of its Clarke Medal, it was not seeking to win favour from of its sister Society. But it was an astute gesture.[148] Similar moves followed. In 1882, Victorians rallied to Sydney when Liversidge and the Royal Society of NSW appealed for help in replacing books and specimens lost in the Garden Palace fire.[149] Cooperation could also lend strength in a range of directions, including Antarctic exploration, in which Victorians and Tasmanians were involved, and in surveys of the tropical north.[150]

In 1874, when William Macleay 'broke off' from the Royal Society of NSW to form the Linnean Society, many feared that the colony's small scientific community would not survive. Liversidge's reforms were a calculated response, and as we have seen, won a reprieve. Even so, the Royal Society entered the 1880s unsure whether the specialist sections of engineering and medicine were going to follow the Linneans, and break away. To preserve solidarity at home, the Society needed to reach abroad.

These arguments were played out against a growing tide of colonial nationalism. In November and December 1883, Sydney hosted an intercolonial conference to discuss the future of Papua and New Guinea. To forestall the expansionist interests of imperial Germany, Queensland's Premier had raised the Union Jack at Port Moresby in February, and in April announced its 'annexation'. The British government of Lord Derby at first repudiated the act, but in October 1884 changed its mind and formally claimed the southern coast of New Guinea, ten days before Bismarck informed Britain that the north eastern quarter of the island was already a German protectorate.[151] In Sydney and Melbourne, proceedings focused on a place that had been of scientific and strategic interest since W.J. Macleay's *Chevert* expedition in 1875.

Not for the first time, political events captured scientific interest. In April 1883, Edmond Morin La Meslée, a member of the Paris Geological Society then living in Sydney, called a public meeting at which he proposed the creation of a 'Federal Geographical Society of Australasia'. Rejecting ties with either the Royal Society of NSW or Britain's Royal Geographical Society, La Meslée argued for an association 'independent and national', not colonial: 'Geography is a science that cannot wait,' he said, 'as our very future depends upon the more or less perfect acquaintance which is gained of the natural resources of the country.'[152] A new society was needed to serve the 'information and benefit of the people of Australasia', he said, commanding the 'commercial, political and natural sciences', across an arc stretching from New Guinea to Antarctica. The idea, he pointed out, was not before its time. Through *Petermann's Mittheilungen*, he claimed, German geographers knew more about Australia than Britons did.

The significance of these events was lost neither upon Liversidge nor the Royal Society of NSW. Although safe in its 'castle' on Elizabeth Street, the Society was always grateful to be alive. In 1883, its President, Christopher Rolleston – veteran public servant and councillor of the Society since Denison's day – greeted the Society's survival with near-disbelief: 'I think it not exaggerating,' he told the membership, 'when I say the Society is acquiring such a station in the public estimation that we may, without presumption, look forward to the time when its advice and assistance on questions of public interest involving scientific enquiry may be sought by the Government of the Country.' Such a happy outcome would see the Society – 'respectfully', in the steps of what Rolleston called our 'great English prototype', the Royal Society of London – become an unofficial 'department of science'.[153]

This uplifting forecast, and the upturn in membership revenue on which it was based, spoke well for Liversidge's reforms. But Rolleston exaggerated the Society's vitality. Although its members were many, its leaders were few. Since relieving the Governor of the day from the automatic courtesy title of president, the Society had elected a 'scientific' president, so following its venerable parent society in London.[154] But only six men served as president through the 1880s, and these rotated in a game of musical chairs, with Rolleston, Russell and Liversidge each serving twice. The same few faces dominated its program. Of the 500 members in 1886, only thirty-six contributed papers, and most of these papers were the work of only seven or eight.[155] Liversidge himself delivered thirty-three papers, making him easily the Society's single most productive member.

Nonetheless, by 1886, the Society had money in the bank, and due to a timely gift from the Fairfax family, negotiated by Liversidge, had repaid its mortgage. Thanks to the Jennings' government, subsequently confirmed by James Inglis, Parkes' Minister of Public Instruction, its endowment was doubled by a grant of matching funds, pound for pound based on subscription income. In 1887, this brought in £400 in public funds, and with it new growth. By 1890, the library, housed in handsome cases, was beginning to burst at the seams. But prosperity bred caution. No anniversary address failed to recall the hardship of earlier years. In 1888, the ageing Sir Alfred Roberts supposed it might be 'difficult to measure the exact amount of good which has been accomplished up to the present time'. But he hoped that, with its 'primary difficulties' now surmounted, at last the Society could become 'increasingly productive of practical good'– particularly in what he called 'Nature's own great laboratory', the Antarctic, where there were real prospects for intercolonial cooperation.[156]

In 1885, following an appeal by Baron von Mueller, an Australian Antarctic Exploration Committee was set up in Melbourne, convened jointly by the Royal Society of Victoria and the Victorian Branch of the Royal Geographical Society. In April 1887, Captain Crawford Pasco, RN, and H.K. Rusden, representing the Committee, asked the Royal Society of NSW to mobilise the NSW Branch of the Royal Geographical Society and the Linnean Society in asking the NSW government to support an intercolonial expedition.[157] This was a foretaste of things to come. So too, was the approach, in August that year, from the Royal Society of South Australia, which asked the Society's help in persuading the NSW Government to offer concessionary rail fares to doctors attending the first Intercolonial Medical Congress in Adelaide. The Society agreed. Throughout this correspondence, the hand of Liversidge (bracketed with Leibius) was always in evidence. 'Indefatigable' was the sobriquet Smith applied to both, a term his successors ritually copied. Given their energy, it is not accidental that the Royal Society of NSW became the host of the first pan-Australasian gathering of science.

The immediate inspiration for this meeting arose in September 1884; the occasion, the annual Congress of the British Association in distant Montreal. That year, for the first time in its fifty-three-year history, the BA held its annual meeting outside Britain. The sea voyage was intended to give new life to a body showing signs of senility, and an imperial boost to British science in a corner of Empire where ties of kinship were being weakened by American enterprise.[158] Towards costs, the Canadian Government voted $25,000, the city of Montreal gave $5000, and local citizens, a further $10,000, producing $40,000 (or £8000) which covered the travel of British participants. It was to the Biology Section of this meeting that Liversidge relayed a telegram he had just received from William Hay Caldwell, a 'Balfour Student' of Caius College, Cambridge, then working on the Burnett River in Queensland.[159] 'Monotremes oviparous, ovum meroblastic', read the famous lines in which Caldwell announced his discovery of the oviparous nature of the platypus. For H.N. Moseley of Oxford, President of the Section, who knew Sydney from the visit of HMS *Challenger* in 1874, 'no more important telegram in a scientific sense had ever passed through the submarine cables'.[160]

Behind that terse message, lay a story of Liversidgean influence at its most typical – understated, private, persevering, definitive – focusing an international light upon a uniquely Australian object. Liversidge's contact with Caldwell dated from Cambridge days, but drew upon his interest in Australian caves and rivers. During his visit to England in 1878, Liversidge met Professor W. Boyd Dawkins, of Owens College Manchester, who was accumulating evidence from Europe and Asia to argue that early man lived at the same time as the extinct marsupials that Richard Owen had described.[161] Liversidge's reading of the *Diprotodon* and other fossil discoveries – quickened by visits to collections in New Zealand and Paris – induced him to help Dawkins, and he proposed to study a number of caves in NSW. William Forster, Agent-General for NSW in London, had sent a similar request in 1876 at the behest of Richard Owen and Sir George Macleay.[162] When, however, the governments of South Australia, Tasmania, Victoria, Queensland, and New Zealand were invited to join, all declined. There was perhaps little reason to support a study likely to interest only one colony.

Henry Parkes agreed to sponsor a survey of caves by Edward Ramsay and the Australian Museum, but his government fell before it could get a vote (estimated at £340) into its budget for 1878. When Parkes returned again to power in December 1879, he still found the venture promising.[163] Even so, two years elapsed before money and opportunity combined. In 1881, Liversidge again put the question, this time through the Trustees of the Australian Museum. Funds were granted, and at his request, the Museum constituted a Committee to Manage the Exploration of Caves and Rivers, its first scientific expedition since Alexander Thomson's near-fatal experience of 1867. With James Cox and C.S. Wilkinson, Liversidge was authorised 'to take all necessary steps for the expenditure of the money voted for the purpose', and between June 1881 and March 1882, supervised four surveyors and taxidermists at a cost of £1200 to search the Wellington Caves, the caves near Yass, the rivers to the west and north, and other deserving sites.

Acting as the expedition's manager, Liversidge logged thousands of specimens that Ramsay's team sent back to the Museum – including 'an almost perfect *ramus* of *Thylacoleo*, with the articulating condile so anxiously looked for by Professor Owen', together with the teeth of a *Diprotodon*, the toe-bones of a large Echidna, and the pelvis of a giant kangaroo.[164] From the Wellington Caves came 10,000 specimens, with fragments of thirty species of mammals, birds and reptiles; while Alexander Morton came back from the Burdekin and Mary Rivers of Queensland with 2000 specimens, some of which William Macleay pronounced 'undescribed' and certainly 'new to the Sydney Museum'.

Not surprisingly, the Caves and Rivers expedition was voted the 'most important work carried out by the Trustees in 1882'.[165] Two consequences followed. First, Harrie Wood, Minister of Mines and Liversidge's friend, announced that the government would conserve the Wellington Caves by setting aside 129 acres and appointing a keeper to collect fossils for the Geological and Australian Museums.[166] This was not the first step towards environmental conservation in NSW; credit for this went to John Lucas and Parkes' establishment of the [Royal] National Park in 1879. But it signified a continuing trend that Liversidge would do his best to maintain.[167]

Second, it contributed to the realisation of Liversidge's plans for intercolonial cooperation. The story began in 1882, when Alexander Morton had canvassed the rivers

'The Platypus', *The Australian Sketcher with Pen and Pencil*, 2 December 1886.

of Queensland, looking for specimens of ganoid and dipnoid fishes.[168] He found one specimen of the dipnoi, the *Ceratodus*, a species internationally famous since its discovery and description by Krefft in 1869. It was likely that anyone seeking further evidence of this 'living fossil' would follow Morton to the Burnett River, and the following year, Morton was followed by William Caldwell, a student of Michael Foster's *protégé* (and Liversidge's contemporary), the brilliant embryologist, F.M. (Frank) Balfour.[169]

For a long moment in 1882, British biology mourned the death of its twin suns, Charles Darwin and Frank Balfour – the one of age, the other by accident. In tribute to Balfour, Foster and George Humphry established a scholarship for research in experimental biology. Scientists at home and overseas, including Liversidge, contributed to this 'Balfour Memorial',[170] and in 1883, Caldwell became the first 'Balfour Student', with a stipend of £200 a year for three years.[171]

Caldwell was a dedicated biologist and skilled instrument maker who, with Threlfall and Dew-Smith, had invented a precision microtome for preparing tissue sections.[172] For three years, sharing Balfour's interest in unusual and possibly transitional species, Caldwell had worked on the embryology of Australian fauna, a subject long the preserve of Richard Owen. Like Owen, he was interested in two questions – the reproductive mechanism of the marsupial, and the proper taxonomic classification, by reproductive evidence, of the monotremes.

For decades, studies of the monotremes (that is, the platypus and the echidna) had made them metaphorical metronomes, oscillating between rival theories.[173] Some naturalists believed they were oviparous, laying eggs outside their bodies, like birds and reptiles. Owen and Bennett, however, believed they were ovoviviparous, hatching their eggs within their bodies. Since 1833, Owen's work on 'On the Structure, Generation and Development of Living Australian Montremes and Marsupials' dominated the field. Thanks to Liversidge, George Bennett and Gerard Krefft, many specimens were sent to England and several to Owen – 'some few even alive', as the *Herald* charitably put it.[174] But Owen's interpretation remained untested. No one had seen proof in nature. 'To form a sound basis of taxonomy', in Huxley's phrase,[175] it was necessary to study *in situ* these bizarre creatures that fascinated science.

Caldwell arrived in Australia in September 1883, and made Sydney his forward base. William Macleay lent him a temporary laboratory until the government, at Liversidge's request, found him premises in Macquarie Street to store his goods and specimens. He then spent several months riding through the colony, collecting and observing marsupials, and in April, rode north to the Burnett District to look for *Ceratodus* and monotremes during their breeding season. Assisted by as many as thirty Aboriginal trackers, he spent months collecting echidna eggs, until finally, in August 1884, he had the luck to encounter – and the misfortune to shoot – a platypus that had just laid an egg in the river bank. Caldwell's observation formed the basis of the memorable message, relayed by a neighbouring cattle station, to Liversidge in Sydney, thence to the BA in Montreal.[176] He brought back to Sydney a large collection of echidna and platypus eggs – 'quite easy to get', he wrote Liversidge confidently, 'I cannot understand how they have not been got before'[177] – and a number of *Ceratodus*, which he studied for the next six months.

Predictably, Caldwell's reading of the platypus was queried by Bennett, who was close to Owen, and also by Ramsay, who described himself as a 'doubting Thomas'.[178] But by the time Caldwell's account appeared in the *Philosophical Transactions of the Royal Society*,[179] the scientific world had accepted his conclusions, together with their insight into the possible evolutionary relationships between egg-laying monotremes, birds, reptiles and amphibians. His discovery, which confirmed the prediction of Geoffroy-Saint-Hilaire, made necessary a fundamental correction to the vertebrate taxonomies that were codified by Georges Cuvier in the 1820s, and taught by Owen ever since.[180]

Then and later, Liversidge properly disclaimed any credit for Caldwell's success, but it was widely known that his encouragement had made it possible.[181] Both the Australian and the British press, beginning with *The Times*, fumbled in reporting the news.[182] Coincidentally, the same day, Wilhelm Haacke, Director of the South Australian Museum, announced a similar discovery to his colleagues in Adelaide.[183] But all ears were turned overseas, and British Empire scientists meeting at Montreal became the first to hear the news.[184] 'The honours of the occasion,' exclaimed the *Australian Town and Country Journal*, 'have been carried off by the duck-billed platypus – an ornament to the zoological world which has covered all the curiosities of Canada with the shadow of a great eclipse.'[185] The colonial press rejoiced at the international reception given to Australian fauna; while *Melbourne Punch* satirised the *naïveté* of British scientists discovering Australian 'natives'.[186] Few papers gave a passing thought to the Aboriginal trackers who had presumably known this for ever. Any suggestion that Australia had been merely a suitable piece of real estate, plundered by a British researcher, was dispelled. James Service, Premier of Victoria, saw an opportunity to put Australia on the map. Within days, he telegraphed an invitation to the BAAS, suggesting that it hold one of its future congresses in Melbourne.

Service's invitation was timely. The BAAS, warmed by its experience of 'social imperialism' in Canada, was interested in a voyage to the antipodes.[187] So was Liversidge. On 16 September 1884, in a letter to the *Herald,* subsequently reproduced in England, Liversidge seized the initiative from Victoria, and gave it a twist. Whilst agreeing that Australia always merited a visit, he calculated that, given the distance, fewer than fifty of the BA's 2000 subscribers would make the trip, as against the 400 to 500 who normally attended a congress in Britain. Instead, therefore, of having a British Association meeting in Australia, he proposed that Australasia establish its own Association – a 'federation or union of the members of the various scientific societies in Australia, Tasmania and New Zealand' – with its first meeting in 1888, timed to coincide with the centennial of British settlement in Sydney. 'I am sure that such an Association must come sooner or later if we are to hold our own,' he wrote. '[It] would not only do a great deal for the advancement of science in other Colonies, but would also favour their progress in other ways.'[188] The aim of Caldwell's morphology, as the *Herald* put it, was to show 'the common unity underlying living structures';[189] no simile could more neatly convey Liversidge's plan for the unification of Australian science.

Liversidge's proposal was aired at a reception given by the Royal Society to mark Caldwell's triumphant return to Sydney in November 1884.[190] But after a brief flurry of interest, it was shelved. Larger issues of federalism distracted the political and popular press.[191] Proposals for 'empire federation' had been canvassed in Britain since the 1840s, but a fresh impulse came in November 1884 with the Imperial Federation League and its vision of a single parliament, linking all the British colonies to Britain, 'for Britain's better security'. Although the League eventually failed for want of party support, daughter branches survived in Canada, New Zealand, and in the capital cities of Australia, where local sections of the pro-Empire 'Round Table' recruited Liversidge and several of his colleagues and students.[192]

While an empire federalism dominated by London had its advocates, some – like Richard Jebb and Liversidge – preferred a federal structure linked to colonial self-government. An Australian federation, uniting the six colonies into one, followed by an alliance of 'equals' between Australia and the United Kingdom, spoke to the spirit of an emerging Australian nationalism that had many supporters in England.[193]

In the meantime, technology was coming to the assistance of colonial integration. In June 1883, the railway brought Melbourne closer to Sydney; in January, 1887, Adelaide came to Melbourne, and in January, 1888, Sydney met Brisbane – steps that confronted the contradictions of inconsistent gauges, and what the Victorian Premier, Sir James Patterson, called the 'barbarism of borderism'. Political federation was on its way. The question was whether science would join the train.

For many, the message was clear enough. In 1886, Clement Wragge, the controversial meteorologist who made his name by giving female names to cyclones, caught the sentiment exactly, in his opening address to the newly founded Meteorological Society of Australasia at Adelaide. 'As in politics,' he said, 'so in science, we desire federation.'[194] While politicians wrestled with constitutional niceties, scientists saw federation as a practical solution to the tyrannies of distance, isolation and fragmentation.

Meanwhile, on the other side of the world, British interests conspired with these colonial aspirations. In November 1885, during his Presidential Address to the Royal Society of London, T.H. Huxley – who had since become a Liberal Unionist and anti-Home Ruler as well – proposed the creation of a union of English-speaking men of science, one which would co-opt the United States and the 'settler colonies' into a single professional community, united by language, culture and tradition. Huxley had returned from a visit to the Philadelphia Centennial Exhibition in 1876 impressed by the potential of American science. Now, as Britain was daily bombarded by Germany's successes in applied science and industry, an Anglo-American alliance seemed more than ever vital to imperial interests. Touching on imperial politics, Huxley went on to say, 'Whatever may be the practicality of political federation for more or fewer of the rapidly growing English-speaking peoples of the globe, some sort of scientific federation should surely be possible.' The solution must not offend the *internationale* of science. 'Nothing is baser than scientific Chauvinism', he said, but added, thoughtfully, 'blood is thicker than water'.[195]

Huxley's thoughts struck an harmonious chord with Liversidge who, not yet forty years old, had just been elected President of the Royal Society's colonial counterpart. After a decade on Council, this seemed an ideal moment to revisit the question of scientific federation, which had been incubating since well before his return from Europe in 1879, and on which he had spoken in 1884. Federation would place Australia securely in the context of 'Greater Britain' – and give Australian science the status of a partner among equals.

It was therefore fitting that, as Huxley's *protégé*, Liversidge devoted his Presidential Address of 5 May 1886 to a variation of Huxley's message. To promote the fraternal spirit of Australian and New Zealand science, he suggested that 'arrangements should be made for holding a meeting of those who wished to form an Australasian Association in 1888'.[196] Recasting his ideas of 1879 and 1884, Liversidge proposed to draw on the thirty-eight 'recognized scientific societies' of Australia and New Zealand, comprising on paper between 2000 and 3000 members. He drafted a constitution virtually identical to that of the British Association – with provision for a general committee (or council), local committees to prepare congresses, and specialist sections to promote research – but with the addition of a significant new feature, to appoint delegates from the colonial societies to the larger body. One day, perhaps, the metropolitan BAAS might visit Australia; but until it did, an intercolonial association would speak for itself. Its congresses and committees would have an immediate effect, Liversidge added, by raising the 'high-water mark of thought' in all the colonies. Politically, 'it would tend to stimulate all classes, and disseminate a taste for all branches of knowledge'. Personally, its success would shine like a jewel in the crown of his presidency.

In July 1886, with the backing of his Council, Liversidge wrote to all the learned societies in Australia and New Zealand, inviting a number of delegates (one for each 150 members) to a constitutional convention to be held in Sydney on 10 November. Their purpose: to plan for an inaugural congress in 1888 to 'take stock' of 'all scientific matters ... concerned with Australasia'. The Royal Society of NSW would play host; the University of Sydney would supply the venue.

It may not have surprised Sir Patrick Jennings, then Premier and Treasurer, to receive two urgent appeals for aid to science that August. Just as Threlfall was delivering his *ultimata* to the University's Senate, and they, to the government, Liversidge asked Jennings for assistance – 'as is done elsewhere', he said – towards the expenses of the Association's first congress. The proposal was discussed, and amidst parliamentary confusion, almost defeated.[197] But it was rescued by a combination of parsimony and pride. On the one hand, it was too expensive to bring the British Association to Australia. On the other hand, a distinctly Australasian Association would give 'lectures on the general principles of science, for the general public; [but also] encourage a closer examination of [Australia as] a scientific continent'.[198] Neither object could be achieved by a British body that would merely 'come and go', conceded Sydney's *Daily Telegraph*, 'leaving no better organisation than there is at present among colonial scientists'. A visit from the BAAS would become 'more an opportunity for Australian hospitality than Australian science'.[199]

Against this background, the Agent-General of NSW in London quietly withdrew the tentative invitation made to the BAAS, whilst leaving the door ajar, ready to be reopened at a later date.[200] In Sydney, the *Herald* and the *Telegraph*, and in Melbourne, the *Argus* and the *Age,* reported Liversidge's plan, and Henry Parkes endorsed a 'scheme which will be confined to the Australian Colonies, and which will be acceptable to all classes of this country'. 'It should be the central idea', his backbench colleague, Thomas Garrett added, 'to keep it a purely Australian matter'.[201]

And so an 'Australian matter' it became – albeit with the addition of New Zealand. Soon, political circumstances changed – an election in January 1887 turned Jennings out of office; the Sydney Centenary Exhibition planned for 1888 fell through; and Melbourne set about having an international exhibition instead. But Liversidge went ahead with his preparatory meeting on 10 November 1886. The difficulties were not easily denied. Of the twenty-seven delegates nominated by the thirty-eight societies, only sixteen actually turned up. But these – seven from NSW, two from New Zealand, two from Queensland, and five from Victoria – were of like mind, and agreed upon a constitution. Adopting Liversidge's reading of the BAAS, they resolved to meet and hold elections for officers sixteen months later, on 7 March 1888. The first Congress was scheduled for September. The Victorians, led by Professor W.C. Kernot of Melbourne University, offered Liversidge their warm support, assuring him that his sister colony did not begrudge him 'the honour of initiating' the Association. 'No one,' Kernot said, 'had any shadow of misgivings to the thing being good, and the sooner it was done the better.'[202]

In six short months, Liversidge had swept to a consensus that had been six years in the making. In London, *Nature* summarised the sequence of events, while regretting that the British Association could not 'see their way to visit Australia during the Centennial year'. What British readers made of this is not clear, but it seemed that colonials were to be independent *malgré eux*. According to *Nature's* correspondent, theirs would be an Association 'thoroughly Australian in character'.[203] However, imperial loyalties were not to be diminished. On the contrary, the new body was to be, in Liversidge's words, not an independent agency, but an 'Australian offshoot'.[204] The metaphor of family, the sense of organic unity that Huxley preached, Liversidge warmly embraced.

# 4. Cambridge Revisited

If Liversidge's proposals were father to the deed, their birth was destined to be delayed until he returned. On 1 November 1886, the day that Threlfall wrung approval for his laboratory, and less than a fortnight before the intercolonial delegates were to meet in Elizabeth Street, Senate granted Liversidge overseas leave for the academic year 1887. In the six months since May, the combined tasks of managing the Royal Society, the Technical College and the Museum, organising the Association, and fighting for his laboratory, had taken a heavy toll. Moreover, it had been seven years since his last trip 'Home'. On Christmas Day 1886, leaving Threlfall to transfer his version of Cavendish physics from Bene't Street to Botany Bay, he sailed for Europe, to 'make himself more practically acquainted with the changes which have been taking place in the great centres of scientific work and thought, and more particularly the radical changes which were being made in the methods of teaching practical chemistry'.[205] This time he chose to go via the Pacific, and to see China, Japan and the United States.

Praise followed him to the quayside. Writing from Wellington, James Hector, thanking him for the award of the Clarke Medal, applauded 'the well organised efforts which the Royal Society have instituted for the purpose of developing scientific research in these colonies'.[206] Leaving Port Jackson with a half dozen passengers on the small coastal steamer *Birksgate*, he landed first at Brisbane, where he was collegially elected a corresponding member of the Royal Society of Queensland. From there he steamed east and north, visiting Fiji and Java, before going on to China and Japan.[207] In February, he was ensconced in the Tokio (*sic*) Club, where he presented his credentials to M. Watanabe, President of the Imperial University of Tokyo (and effectively Imperial Minister for Education), who welcomed the idea of contact with Australia. At Liversidge's request, Maiden sent Watanabe lists of Australian materials and catalogues from the Technological Museum;[208] and also sent Liversidge £100 to buy Japanese ceramics, carvings and minerals.[209] Two cases were shipped back to the Sydney Museum on the *Tehran,* thanks to the kindness of William Gowland, his fellow student at the RSM, and now Master of the Imperial Mint at Osaka, who was seemingly in control of 'the whole of the coinage of the Japanese Empire'. [210]

Leaving Japan, Liversidge sailed to San Francisco, where he visited the State's Geological Museum, newly established since John Smith's last trip; and took the Union Pacific for a seven-day journey across mountains, deserts and plains to the East Coast. Along the way, he visited geological surveys and museums (and possibly, at least one scientific society, in Indiana, which elected him a corresponding member). In Washington, DC, he visited the Geological Survey and met colleagues at the Smithsonian Institution. He had received the Smithsonian's *Annual Reports* since his first contact with Joseph Henry in 1875. In 1885, he had sent Washington a consignment of ninety Australian minerals,[211] and he now promised meteorite fragments as well, in exchange for American ethnographic material.[212] Sailing a fortnight later from New York to Southampton, Liversidge duly arrived in London, and finally came to rest in late April at the Savile Club's new premises in Half Moon Street, Piccadilly.[213]

Sad news greeted his return to England. Just a few weeks earlier, his father had died at his cottage in Kent. Liversidge mourned with his brothers and his sister, Caroline, who had nursed their father for many years.[214] In the ensuing settlement, the family coach-building works was left to his elder brother, Jarrett. William had by now a flourishing business, and gladly looked after their sister. Details took months to resolve, and while they seem not to have affected Liversidge financially, they touched him deeply. Moved, perhaps, by intimations of mortality, he spent several days at the British Museum, examining the Harleian MSS, and tracing the Liversidge family from the fifteenth century and the reign of Edward IV. Compiling genealogies and charting the family's branches from Cheshire to Somerset, led him to the township of Liversidge, on the outskirts of modern Bradford, which he visited for the first time.

It was not until the summer that Liversidge returned to work, and began what became his usual round of visits to university friends and chemical laboratories around the kingdom. At Manchester, Henry Roscoe – who had years before been offered a job at the Sydney Mint[215] – was especially hospitable, as was Liversidge's teacher at Cambridge, George Liveing, who had been elected FRS in 1879, just three years before him. In Liverpool and Glasgow, Liversidge took careful notes on laboratories, and began to draft sketches of what he hoped one day would become his own buildings in Sydney. In London, he bought chemical apparatus for the University,[216] and specimens for the Australian Museum.[217] In August, he read with amusement Nature's belated account of the 'withdrawal' of the Agent-General's invitation to the British Association. It was odd to see Australian affairs through British eyes. Nature had interpreted the matter simply as a 'party question'. The truth of colonial politics was never so pure, nor as simple.

As in 1878, so in 1887, Liversidge's happiest hours in England were spent in Cambridge, greeting familiar streets and the rhythms of college life. That fine autumn, the newly-planted horse-chestnut trees on King's Parade gave scant sign of their future fullness, as he made his way from the railway station, first to see his old rooms in Christ's, then through Market Square to Free School Lane.

To the visitor's eye, Cambridge's landmarks remained unaltered, but in the last nine years, much had changed. Fellows of colleges could now marry, and substantial family houses had begun to appear along Grange Road and Madingly Road, to the west, and north of the town, and along Brookside to the south. Colleges for women had been established at Newnham and Girton, and women could now sit Tripos examinations (although they could not yet take degrees). New Triposes had begun in history and law. A decade after the tests were abolished, the dominance of the Church of England on college life was slowly but steadily receding.

Perhaps the most conspicuous changes were in the accommodation afforded science. On the New Museums site – between Pembroke Street and the Corn Exchange – the last tree in the Old Botanic Garden had given way to the advance of new laboratories. The Cavendish – since 1884, under Threlfall's teacher, J.J.Thomson – was flourishing as never before.[218] In chemistry – the only subject that offered laboratory work in Liversidge's day – Liveing was showing a strong (if belated) interest in research, sometimes aided by James (later Sir James) Dewar. The chemical laboratories for which

Liveing had schemed for fifteen years had at last become a reality. At a cost of £30,000 – over five times the expense of Liversidge's ambitions for Sydney – they formed the southern bastion of the New Museums site, between the Cavendish on the north and east, and the Anatomy School on the west.

In 1883, the Samuelson Commissioners reported to Parliament that 'the energy and activity which is being displayed [in the teaching of science] at Cambridge is very remarkable'. This was an understatement. The Natural Sciences Tripos had overtaken the Classical and Mathematics Triposes in popularity, and 'Naturals' now comprised nearly half of all Cambridge undergraduates.[219] The NST's curriculum was said to contain 'in almost every branch of science at once the most advanced and the most practical character'.[220] Over 185 undergraduates were reading chemistry; and although a third of these were medical (and few, according to Liveing, had 'time or inclination for original research'), their presence gave promise of things to come.[221]

Cambridge science had prospered in other ways as well. A new class of university demonstrators had emerged, whose work now complemented college tutorials. Symbolised by the Balfour Students, new opportunities were opening across the sciences.[222] In 1881, five years before Sydney debated the idea of specialised courses for second- and third-year Arts students, Cambridge had divided its Triposes, so that more specialised work was available. More colleges were giving fellowships in science, and there was better coordination between teachers and examiners. Professors – previously 'freelance, each doing the best he could for his own subject'[223] – were now working together on syndicates and committees. At what was already the famous Cambridge Scientific Instrument Company, which traced its roots to the early days of Foster's first team at Trinity, Dew-Smith and Horace Darwin were hard at work. Their success augured well for Cambridge's first 'science park'.[224]

Walking from Christ's to the Philosophical Society, Liversidge met members of Michael Foster's prosperous Physiology School, now numbering over 120 students. Their buildings had cost less than £12,000, but Cambridge had a history of putting its money in people, rather than structures. Foster and Balfour had produced brilliant progeny – in botany, with Liversidge's contemporary, Sydney Vines, and in comparative morphology, with Adam Sedgwick. Anatomy was never better before taught by Alexander MacAllister, and geology by Thomas McKenny Hughes.[225] The mineralogical collections, built up by W.H. Miller, were better housed and displayed. All in all, there was much to envy.

But pleasure, rather than envy, was Liversidge's response, when the prodigal found that his friends had not forgotten him. On the recommendation of Foster (who had since become a professor himself), and seconded by Coutts Trotter, the University awarded Liversidge the degree of Master of Arts (*honoris causa*). It was an unusual step, and perhaps belated recognition of the degree that Liversidge never received, but might have taken had he stayed in Cambridge and sat the Tripos.[226] Foster, ever thoughtful, was nowhere more generous than with former students.

In ways unforeseen, the honorary degree marked a turning point in life, and brought another moment of reflection. Liversidge's years in Sydney had not been badly spent. He had caught the baton from Smith, and created a culture of science where little had existed before. He had transformed the Royal Society into a body with an international presence. As Carl Leibius was fond of saying 'we never got a move on until Liversidge came'.[227] He had helped open the country to development. He had proposed a plan that would unite the scientific community of the continent. He had published thirty-three papers, and a major work of reference, soon to appear in a third edition. He had been elected a Fellow of the Royal Society well before his contemporaries. His salary was greater than that of his teachers, and his standard of living was comparable, if not better. New South Wales seemed in some ways more optimistic than England: the boom of the 1880s and the Challis endowment had opened fresh prospects for expansion. If Sydney's chemistry, botany, mining and engineering were still limited by a lack of space, at least physics was now in the hands of a capable Cambridge man; a modern Medical School was taking shape; and graduates were moving into treble figures and into the highest echelons of colonial society.

Were there reasons for Liversidge to regret his choice, and even to return to England? If so, could he? There was little work to be had in the British chemical and mining industries; members of the Royal School of Mines Association tended to find employment overseas, rather than in Britain. Although there were new chairs in chemistry in the north of England, these tended to favour organic and physical chemistry, not chemical mineralogy. A new chair of chemistry at Dundee might have interested him, but there was nothing immediately open in Scotland or Ireland. Owens and Liverpool had sealed their chemical fates for a generation. Cambridge, while Liveing lived, was much the same. Oxford, with no chemistry school to speak of, was not a possibility. In London, few academics ever moved; at Finsbury, no one looked like dying, and at UCL and King's, for every one who did, there were a dozen able candidates waiting patiently in the wings.

Such calculations must have left Liversidge in no doubt that he had little choice. In any case, how could he compare the flat East Anglian fenland with the majesty of the Blue Mountains? Or the dust and congestion of London with the gardens of Sydney? That year, visiting Cambridge, Liversidge passed a point of no return. Fifteen years lay behind him, twenty stretched ahead to retirement. The best strategy must be, surely, to make the most of Australia; if possible, to select *one* research subject on which to concentrate; to make the most of the niche he had created in the study of Australian minerals; and to devote himself to the organization of science, where his talents had been recognised. Australia had given him a mission. With a house and a laboratory of his own, he could still call Australia home. Such might have been his thoughts as he took the train from Cambridge to Liverpool Street, that first week of November 1887, as a winter chill settled on England, and the warm fires of the Savile Club greeted his return.

The London winter season proved pleasantly busy. Meetings, errands, visits to friends and museums vied with the learned societies, and shopping for new equipment. As usual, Liversidge played the role of 'commercial traveller', exchanging artifacts and specimens for news and views. At his instruction, *Waldheimia* were sent to the Zoological Museum in Cambridge, as were *Ceratodus* in 1879 and various cetaceans in 1883.[228] In November, he arranged for models of the Bingera meteorite to be sent to Lazaras Fletcher, the mineralogist at the Natural History Museum.[229] Finally in December, came the *coup de resistance*, for which he would be forever remembered in Sydney.

Earlier in the year, revisiting family history in the East End, he discovered that the brewing company, Samuel Whitbread and Co. was disposing of a 'very fine if not the finest example' of an original Boulton and Watt beam engine, built in 1785 and working well, but being replaced by the owners with more modern technology. The engine was of great historical interest – Samuel Smiles described it as the 'greatest invention of modern times'.[230] In the words of Robert Thurston, the great engineer, nothing better conveyed the image of science as the promoter of civilization.[231] Liversidge invited Whitbread's to present it to the Technological Museum, and before leaving London, arranged for the engineer, E.T. Cowper, to superintend its shipment to Sydney 'for the benefit of the public'.[232] Liversidge was delighted. Today, it ranks as the world's oldest surviving rotative steam engine. 'Everybody thinks we are very fortunate,' he wrote Alfred Roberts, 'to obtain so interesting an historical engine and the owners would only present it on the understanding that it is set up, exhibited and well cared for.' In 2008, it is just that.[233]

All this contributed to a happy leave-taking from London. Liversidge spent Christmas in Venice, a city for which – as the prints on his bedroom walls testify – he formed a great fondness. But come the New Year, it was time to leave. On 13 January, Liversidge sailed from Venice on the famous RMS *Chusan* (4496 tons), which had inaugurated the P&O steam mail service between the England and Australia in 1882. On the way to Sydney, via Malta and Port Said, Suez and Colombo, Adelaide and Melbourne. Liversidge boarded with Sir Saul Samuel, the Agent-General, and his family. 'Entertainments and amusements, fancy dress balls etc. were carried on with spirit throughout the voyage', in the course of which, the *Herald* solemnly recorded, the 'usual monotony of a sea voyage was reduced to a minimum'.[234]

After five weeks at sea, it must have seemed like returning to a small world when Anderson Stuart joined the ship in Melbourne. All arrived safely in Sydney on 2 March 1888. Thanks to the Samuels, the passengers were greeted at Bradley's Head by a deputation of colonial dignitaries.[235] It was a happy omen. For Liversidge, it marked a return to business. The year 'away' had begun in discovery, continued in introspection, and ended in determination. It would now require all his energy to keep science in the forefront of university policy, and to breathe life into the 'federative spirit'.

# CHAPTER 9

# Science and Solidarity

The main corridor of the Chemistry Building.

Borrowing an appealing phrase, Sir Mungo MacCallum, linguist, dedicated servant of Empire, and Sydney's first full-time Vice-Chancellor, once referred to the 'practical idealism' that sustained 'transplanted Britons' through personal discomfort, even professional loss, in the service of colonial scholarship.[1] It is an expression that fits many of Liversidge's generation, and none more than himself. It exemplified an approach that adapted the 'borrowed culture' of Britain to the circumstances of colonial life. It contained, at base, a view of nature and society that blended individualism and cooperative enterprise. The result was not so much a coherent philosophy as an attitude of mind, a philosophy *sans doctrines*.

Practical idealism was inspired by the bush and the outback, as much as the city.[2] But in the cities, it had a particular resonance. By the late 1880s, the city-states of Australia had become 'moving *metropoleia*', their public architecture recalling the capitals of Europe; their literature, mercurial, critical, even abrasive.[3] Looking to the wealth generated by wool, wheat and minerals, the methods of science had achieved a bourgeois respectability, the worth of which, if not overvalued, was at least no longer questioned. Liversidge, who had known defeat in many forms, found himself on the threshold of success, as the University's emphasis changed from pioneering to pragmatic. The decade was becoming, in Whitehead's phrase, the age of the professional.[4]

In June 1888, Liversidge resigned from the Board of Technical Education, and in September, from the Trustees of the Technological Museum. He was happy to devolve the Museum to Joseph Maiden and Henry Smith.[5] In 1889, he resigned the honorary secretaryship of the Royal Society, and although he agreed to serve as president again in 1889–90, he returned to the backbenches of the Council for the rest of the decade. While he remained a Trustee of the Australian Museum, he sought time for other things. These years would be dominated by administration, organisation and laboratory life, but also by making a home for himself, amidst the saga of university reform.

## 1. The AAAS and All That

When Liversidge returned from England in early 1888, he had before him the prospect of struggle on three fronts – technical education, the Technological Museum, and the University. The year proved a turning point in the history of intercolonial cooperation through the establishment of the Australasian Association for the Advancement of Science. The AAAS (later known as ANZAAS) was a monument to his genius. But the idea had not sprung fully armed from the meetings of November 1886.[6] Like many manifestations of genius, it was an idea whose time had come. At that meeting, the delegates of the colonial societies had agreed to delay their deliberations until his return to Australia, in January 1888. No one questioned his right to lead; no one offered to take his place.

During his visit to England in 1887, Liversidge had seen Empire unity manifested in many ways. Celebrating Queen Victoria's Golden Jubilee, the Prince of Wales had opened the new Imperial Institute in South Kensington, under the direction of Liversidge's friend, W.R. Dunstan. Its stated purpose – assisting the productive application of imperial resources – could be combined with the coordination of colonial science.[7] The first Colonial Conference, held in London in 1887, fostered similar sentiments of imperial federation, directed 'from the top'.

In Australia, a professional variation on the imperial theme had been promoted by the British Medical Association, which established branches in the different colonies. In 1887, the South Australian branch encouraged medicos from all the colonies to attend an inaugural Intercolonial Medical Congress held in Adelaide, to coincide with the fiftieth year of that colony and the opening of Adelaide's International Exhibition. The Congress was warmly welcomed, and endorsed the idea of an organisation like the British Association. But it did not go as far, or as fast, as Liversidge's plan for an Association under Australasian management.[8]

Liversidge was away so long that colleagues feared he might have given up the idea,[9] but on 7 March 1888, only five days after his return, he convened a meeting of the organising committee at the Royal Society's rooms in Elizabeth Street. There was no point in suggesting an imperial federation run from overseas: that promised only delay and division. That is, there would be no Australian branch of the BAAS. Unlike their medical cousins, the scientists would link, without supplanting, the learned societies of Australia and New Zealand, and would focus their energies upon a common cause. Federation was a means, not an end in itself. The end was to achieve an Australian identity.

To this end, it was appropriate that the first president would be an Australian. However, of the Association's first six officers, the only native-born Australian was Henry Russell, the astronomer. In 1886, Russell was the first Sydney graduate to be elected an FRS. He was a past President of the Royal Society (in 1877 and 1885), a Fellow of the University Senate, Vice-President of the Board of Technical Education, a pioneer of meteorology, and Liversidge's closest friend. He was easily elected the first President of the AAAS. To link the traditional and the modern, Liversidge – then senior Honorary Secretary of the Royal Society – was elected a joint Honorary Secretary together with the elderly George Bennett, Secretary of the Linnean Society, while the ageing Edward Strickland, of the Geographical Society, became Honorary Treasurer. Thus the Royal Society of NSW closed ranks with its sister societies. When Strickland died the following year, Russell assumed his duties, reconstituting the 'duo' of Russell and Liversidge, and giving them leave to apply to the Association the dynamism that had produced such success elsewhere.

Initially, the whole burden of organisation fell upon Liversidge, and on his slim resources, but by 12 March, he and Bennett had written to all thirty learned societies in Australia and New Zealand, reporting the recommendations of the preliminary meeting, and requesting their membership rolls, in order to write to each of some 2000 *savants* living in Australia. Meeting first fortnightly, then weekly between April and August 1888, Liversidge set up a secretariat. With Bennett and Maiden as his amanuenses, he recruited local secretaries to represent the five eastern colonies and New Zealand (whether Western Australia would send representatives remained unclear). Steamship and railway companies were dunned for concessionary fares, and the government of NSW was asked for a pound-for-pound grant in aid of expenses. Recalling the BA's early practice, Liversidge tactfully invited the Premier and other leading government and University officers to be Vice-Presidents of the Association, and extended similar invitations to the Presidents of the Royal Societies of Victoria, South Australia, Queensland, Tasmania, and NSW, and the New Zealand Institute.

Liversidge photographed in 1885.

To flavour this salad of mixed interests, Liversidge added flattery and praise. Thus, to Sir James Hector of Wellington, invited to be a Vice-President:

> I hope you will be able to make our first meeting in September next a success by coming over and giving a paper or papers; if you could give an account of the Geological and Geographical exploration of New Zealand since settlement, it would be much appreciated and no one could do it as well as yourself. I think we shall have a very successful meeting. Will you also kindly draw attention to the meeting through the New Zealand Institute?[10]

Refusal was impossible. Liversidge nearly always won his way.

Established thirty-seven years after its American sister,[11] and a decade after its French cousin,[12] the AAAS became the fifth Association of its kind in the world.[13] Like the BAAS, its objects were to give a 'strong impulse and a more systematic direction to scientific enquiry; to promote the intercourse of those who cultivate science in different parts of the Australasian colonies and in other countries; and to obtain a more general

INSOMNIA - ITS
CAUSES AND
Cure.
An Essay.

SCIENCE

The AUSTRALASIAN SOCIETY of
MUTUAL BORES have been
reading no end of papers to each
other lately and nobody any the wiser

The AAAS, or 'The Australasian Society of Mutual Bores', *Bulletin*, 8 September 1888.

attention to the objects of science, and a removal of any disadvantages of a public kind which may impede its progress'.[14] In May, it was decided to hold the inaugural Congress in Sydney a week earlier than planned, to accommodate the University of Melbourne, where it would otherwise have clashed with its academic calendar. So began the tradition of 'conference week'. The committee also timed the conference to avoid conflict with the opening, on 1 August, of the Melbourne Centennial Exhibition, which would otherwise have diminished attendance. Support snowballed: eighty-six subscribed in April, 212 by May, and 501 by early August. When the new Council met in the Royal Society's rooms on 27 August, it was appropriate that Liversidge was thanked by acclamation for 'initiating the movement which had led to the formation of the AAAS'.[15]

It had been largely thanks to his initiative that, between 27 August and 8 September 1888, a decade of hope culminated in a festival of reason, as the first Congress opened in the University's Great Hall. The English businessman and imperial federationist William Westgarth, praised the Association for being 'happily and successfully inaugurated ... and destined, we cannot doubt, to a future of universal distinction in a land of such scientific wonders and novelties as Australia'.[16] Overall, results exceeded expectations. Over 850 people from the six colonies and New Zealand subscribed, and over 730 turned up – the largest meeting of scientists ever seen in Australasia, and one of the largest public gatherings seen in Sydney since the Garden Palace. Representation was worryingly uneven – whilst the largest contingent (560) was predictably from NSW, only thirteen came from Victoria. From South Australia came fifty-nine, from Queensland, forty-six, and from tiny Tasmania, eighteen. Forty came

Officers of the AAAS at the Melbourne Congress, 1890.

from New Zealand, and only a few from WA. But the several sessions introduced men (and a few women) who had previously known each another only by name.

What was rightly heralded as a triumph of the intercolonial principle – and later seen as a dress rehearsal for federation – revealed much about the men responsible. Liversidge was the producer and director, but Russell wrote the script. His inaugural Presidential Address remains one of the most lucid statements of the needs of Australian science ever made. Placing his remarks in context, the new Association, Russell said, was not to be the hobby of a few individuals, 'or of one colony', but of 'scientific men and lovers of science' throughout the continent. Its purpose was to 'work up the facts known in every branch of Australian science', and to advance the 'culture of science', in what he called its most 'comprehensive sense'. It would bring 'to the front many men ... now scattered through the country who have ability and genius ... whose daily work is of another kind', to grapple with material 'questions in chemistry, physics and geology; in mining, mineralogy and engineering; in meteorology, water conservation and irrigation, and every other subject that may promote our national advancement'. The Association was to stand outside politics, rejecting factions. 'Science stands or falls as a whole,' he said, 'if we limit it to certain purposes or persons it ceases to be science.'[17] The Association would speak for a consensus, transcending class and politics, devoted to systematic study of the continent, its people and its resources.

The Royal Society of NSW welcomed the President's speech, and their own implicit role in the proceedings. University men were also pleased to hear Russell, a member of Senate, assure them that 'This Association stands as a protest against the shortsighted and utilitarian policy of those who would cultivate only what they characteristically call the bread and butter sciences.' The range of subjects the Association included in its charter – language, literature and the arts – said as much.

Unhappily, Russell's concessions failed to impress Sir William Manning, who, in moving a vote of thanks, managed to voice an ill-judged criticism of science as an enemy of general education. Liversidge suffered Manning's remarks in silence, but Sir James Hector did not, resorting to language, which at a stroke did more to unify the assembly than all the noble sentiments Russell had just expressed.[18] But Russell's project for the advancement of 'all the sciences' spoke exactly to Liversidge's purpose. On this the officers agreed: the Association was to be not a body of visionaries, but of 'practical idealists', with a program of useful ends.

None of this meant that the Congress had to be all work and no play. The whole week, Sydneysiders and visitors progressed through a majestic range of garden parties, exhibitions and excursions that showed Sydney at its best. The sessions were diverse and dramatic. Hector spoke on the recent Tarawera eruption in New Zealand, and R.M. Johnston, the Government Statistician at Hobart, talked about the novelties of Tasmanian fossils. Dr E.C. Stirling of Adelaide announced the discovery of the marsupial mole, and G.S. Griffiths preached on the duty of Australians to explore Antarctica, 'to secure to this colony, universal attention, and the approbation of the entire civilized globe'.[19]

Liversidge predictably led the Chemistry Section, giving two papers plus a plan of his new laboratory; while Russell, Threlfall and Threlfall's demonstrator, J.A. Pollock, led Physics. Speaking to Section E (Geography), the Hon. John Forrest from Western Australia dwelt upon the importance of preserving the environment, conserving water, surveying minerals, and making common cause where 'colonial history and colonial enterprise' were involved.[20] The infectious spirit of federation was summed up by the young professor of physics at Adelaide, W.H. Bragg, who wrote excitedly to his *fiancée*:

> I think this Association is going to do us a lot of good, especially such as, like me, are willing to work, but don't quite know where to begin. Contact with other and more experienced workers will start us off on the right track.[21]

His sentiments were echoed by von Mueller of Melbourne: 'Ours is a kind of scientific federation full of soul. Everyone can help.'[22]

Guided by Liversidge, the first AAAS Council included Edward Rennie, graduate of both Sydney and London, and now Professor of Chemistry at Adelaide; George Bennett and J.C. Cox, representing 'learned leisure'; and W.C. Kernot, an engineer from Melbourne, whose encouragement had greatly sustained Liversidge in his months of planning. Over four days in September, they and a dozen new friends from the learned societies and government departments agreed on the way forward. Following a suggestion of Anderson Stuart and Dr Alan Campbell of Adelaide, the Council agreed to hold a congress every other year, and to coincide, where possible, with the Intercolonial Medical Congresses. Local Committees were to prepare the Congresses, effectively uniting the local scientific societies in a common cause. At Hector's suggestion, Sydney's Royal Society was to be the Association's corporate headquarters.

Following the practice of the BAAS, the Council set up a series of Research Committees of Investigation, initially thirteen in all. Forrest suggested Antarctic exploration; Ellery of Adelaide proposed the 'state of meteorology'; R.L. Jack of Brisbane, the geological record; and Professor J.G. Black of Otago, the 'state of chemistry, with reference to gold and silver'. W.A. Haswell of Sydney proposed the endowment of a marine biological station; Baldwin Spencer of Melbourne, a bibliography of Australian biology and a committee to review the protection of native birds and mammals. James Wilson, Sydney's Professor of Anatomy, suggested the 'construction and hygiene requirements of places of amusement in Sydney'. Not to be outdone, Liversidge proposed no fewer than six, to include a census of Australian minerals and a bibliography of Aboriginal Australia and Polynesia.

'The great need at present', agreed J. Steel Robertson in the *Centennial Magazine*, is for 'more extended original research'.[23] Yet, the AAAS had to demonstrate its practical worth. After all, learned societies had, in the words of the *Australasian*, 'too often condescended, in [their] dearth of valuable papers ... to the most trivial and ignorant discussions of rabbit killing or other topics of the day'. But, it added hopefully, the Association promised to be an 'institution of a different sort', one which 'represents an important step in the intellectual development of the colonies'. The 'writers who appear before it will be kept up to the mark by the certainty of vigilant and competent criticism'. Above all, the AAAS would directly promote the 'public regard' in which science was held.[24]

Two months later, London joined the refrain. Cautious praise from *Chemical News* and *Nature* was outrun by the excitable prose of the weeklies.[25] 'Our colonies are rapidly coming abreast of us', breathed the *Saturday Review*: 'There is a freshness and a breadth about the work of Australian science which, alas!, we rarely find now in that of the old country.'[26] Amidst bleak reports of economic recession and the German menace, any such good news was welcome. In two years' time, when the Association met in Melbourne, the venerable Baron von Mueller pledged to England the imperial loyalty of the Association, which fostered by its works the 'union of the Empire.'[27] Australian science had confirmed its place in the imperial sphere.

## 2. LABORATORY LIFE: BUILDING CHEMISTRY

If Liversidge's return inspired the first meeting of the AAAS, his return to university life proved no less eventful. In England, debates between old and new methods of instruction, then murmuring in the heavy quarterlies, confirmed him in the conviction that, as he put it to H.E. Armstrong, 'there is much room for improvement'.[28] He might even apply Armstrong's heuristic methods, had he the laboratory and the means. On 5 March, scarcely three days off the ship, Liversidge gave notice of motion in Senate of proposals for a new curriculum and new accommodation. On 19 March, he delivered to Senate plans for a new chemical laboratory, with an understanding that drawings of a new School of Mines would not be far behind.[29]

If anyone was surprised by this speed, it was not Sir William Manning, who had been waiting for Liversidge to return before approaching Parkes on matters of expenditure. The Chancellor, along with the Challis bequest, has been credited with the 'rebirth' of the University in the 1880s.[30]A lawyer who reflected a liberal stream of thought flowing from John Stuart Mill and the founders of University College London,[31] Manning espoused many liberal causes. He pressed for the admission of women to Sydney University in 1881, and helped in the establishment of Women's College. Liversidge had joined with him, and looked to him as a progressive ally on Senate. On 28 March, Liversidge joined a deputation led by Manning, and including the Vice-Chancellor, Sir John Hay and Henry Russell, to meet James Inglis, Parkes' Minister of Public Instruction.[32] To their disappointment, Inglis refused to give funds for a chemistry building, and referred the question back to the University.[33] From the government's point of view, the Challis bequest should be sufficient to meet such demands.

The terms of the Challis bequest had been debated before, and the University steadfastly refused to use it for capital works. Liversidge duly resumed his lectures and managed the AAAS, and it was June before he returned to the question. This time, however, he chose to play the government's card, and asked Senate to lift its restrictions on the use of Challis funds to give £7500 towards a new chemistry building, if the government could be persuaded to contribute an equal amount. At least, a compromise might work. As a tribute to the University's greatest benefactor, Liversidge proposed that his new chemistry building be made a memorial to Challis and bear his name.[34]

Since the Challis bequest had arrived, no one had proposed to commemorate its donor in brick or stone – at least, not beyond a pious marble statue in the Great Hall. A 'Challis Chemistry Building' would be a living monument, at a price the University could accept and the government not refuse. But before a decision could be taken, a new factor entered the equation. In January, Manning won a legal victory of 'Spycatcher' proportions over the British Commissioners of Inland Revenue, by which a considerable sum thought to be owing to Britain in legacy duty was declared to fall under Australian jurisdiction. This the government restored to the University,[35] producing an unexpected windfall of over £12,000. When details reached Senate on 18 June 1888, it was agreed, on a motion by Dr Renwick, to grant Liversidge the figure he had asked, on the matching conditions that he had specified, and on the understanding that the funds be repaid from interest on the Challis income over the next twenty-five years.[36]

This opened the way, and in July, with a new laboratory already designed, and at last in prospect, Liversidge began to look for ways of 'furnishing' it with staff. The assumption that a laboratory would tend itself, or that with a building the rest would follow was, if short-sighted, not an uncommon failing of universities. The history of Cambridge records that, when in October 1870, appeals to university and college endowments failed, the seventh Duke of Devonshire gave £6300 to create the famous physics laboratory that was to bear his family name. But the gift brought with it no funds to pay the salaries (at first, amounting to £660) of the Cavendish's director and staff. For a time, it looked like Cambridge might have a laboratory without a professor.[37] Moreover, the Cavendish Laboratory was not completed until the autumn of 1873, until which time its first Director, James Clerk Maxwell had to teach on a variety of

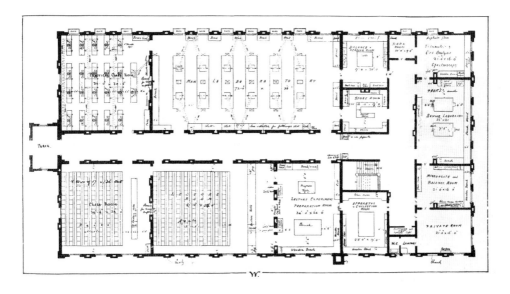

Plan of the new Chemistry Laboratory at Sydney University.

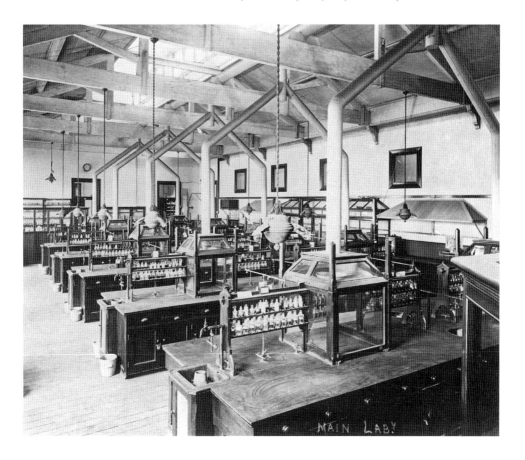

The main Chemistry Laboratory, 1893.

Liversidge's office showing his retort and filter stand.

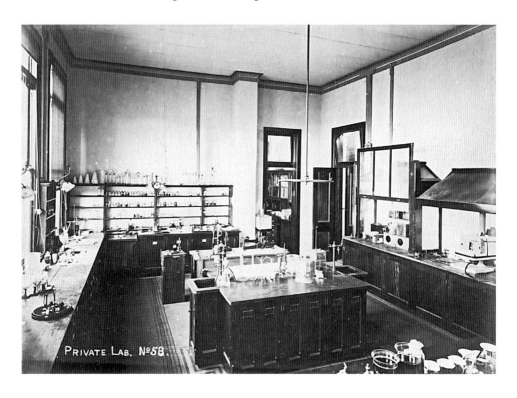

PRIVATE LAB. Nº 58.

Liversidge's private Laboratory.

sites. He had, as he put it 'no place to erect my chair, but move about like the cuckoo, depositing my notions in the Chemical Lecture Room 1st term; in the Botanical in Lent; and in Comparative Anatomy in Easter'.[38]

The lesson was not lost upon Cambridge undergraduates. On coming to Sydney, Richard Threlfall knew enough to insist that a laboratory must come with his chair. A decade earlier, Liversidge had been less well placed to make such demands. But now, twenty years later, his chemistry mentor at Cambridge was getting a fine new building – on Pembroke Street, next to the Cavendish – and if Sydney had kept its professor without a proper laboratory for nearly as long as Cambridge, now it, too, had no excuse to delay any further.

The break point fell to Sir Patrick Jennings, now in opposition in the Legislative Assembly. Member of Senate, friend of Threlfall, and supporter of the AAAS, Jennings tabled a motion requesting Parkes to place £7500 for Liversidge's laboratory on the estimates for 1889. Meanwhile, Liversidge went ahead. On 16 July 1888, he submitted to Senate copies of plans he had sent to the Colonial Architect. A site between the Blacket Building and Threlfall's Physics was chosen for a structure to be completed and fitted out for no more than £15,000.[39] On 6 August 1888, weeks before the 730 delegates gathered for the inaugural AAAS, Senate approved.[40]

When the Congress arrived, it was too early for Liversidge to cheer. It was, of course, Threlfall, and not Liversidge, who could point proudly to a science building, nearly finished, on pastureland along a dusty path two minutes' walk from the Great Hall.[41] And the Macleay Museum, under construction since 1886, had just opened, to the greater glory of its benefactor.[42] Having no visible results yet to show, Liversidge modestly outlined his plans on paper to Section B (Chemistry). Although on the eve of victory, he cautiously styled his lecture, 'The Proposed Chemical Laboratory'. As far back as 1873, he told his audience, 'it was almost as near being an accomplished fact as at the present time'. In the event, the iron and brick structure – begun in 1889 and costing £13,000 – would not be finished in time for March 1890, when the first students took up their numbered places in its lecture theatre,[43] and the laboratories would not be fitted out until the following year. But, when at last it was ready, Liversidge found he had become the envy of the Empire.[44]

As if to caution *hubris*, the administration counted the cost. Behind Manning's embarrassing comments at the AAAS Congress lay budgetary considerations, as well as bias. Wherever new disciplines were admitted and new buildings built, new staff must follow. In August 1888, the Registrar, H.E. Barff, asked Liversidge whether the new curricula he planned for science and engineering 'will necessarily or practically involve any additional cost for Lecturers, Demonstrators, etc'. The Senate, Barff rather unnecessarily explained, wished to understand the 'financial as well as the educational merits of the scheme'.[45] The warning lights went up when Liversidge mentioned that, 'as a contingency', he would require another demonstrator for chemistry, and one or more lecturers in engineering.[46] He offered Barff little comfort, by adding that these changes would,

> not in themselves necessitate any additional expenditure over and above what is required by the present By-Laws, unless the changes should bring a much increased number of students into the Science Departments ...

To deal with potential numbers, Liversidge's list of appointments was limited to four: lecturers in metallurgy, mining and surveying, and an assistant demonstrator in chemistry – total cost, £750, as agreed by the science professors in June 1886. He entered only one plea: 'I have to teach Mineralogy as well as Chemistry', he observed, 'and as a chemist I have necessarily treated Mineralogy for its chemical rather than for its geological side'. When Senate made its new appointments under the Challis bequest, Liversidge asked that his chair be divided, relieving him of mineralogy, and coupling that subject with mining, 'as a portion of Economic Geology'.[47] If Providence had given Liversidge three wishes – a faculty, a laboratory and a School of Mines – this would be his fourth. In 1891, he gave up mineralogy to geology, and became Professor of Chemistry alone.

The close of 1880 was a time of land boom-led expansion, and the University shared in the glow. In January 1890, Liversidge attended the first meeting of professors from all the Australasian universities, held in Melbourne at the invitation of that University, and convened by Professor Edward Morris, head of its English Department. Morris was an enthusiastic supporter of the AAAS's Literature and Fine Arts Section, and welcomed Liversidge as a kindred spirit. The meeting was made pleasant by excursions to the Centennial building, and doubly profitable for the medicos, whose Australasian Medical Congress conveniently coincided.

Back at Sydney, buoyed by Liversidge's success with Senate, his fellow science professors leapt onto his bandwagon. In September 1888, William Stephens and William Haswell, who had moved into the temporary, four-roomed cottage used in 1883 for Anderson Stuart's Medical School, asked the Chancellor for a building devoted uniquely to their subjects of natural history, geology, zoology and botany. They wanted a site just west of the newly built Macleay Museum, then occupied by the temporary Medical School. As it was, they had only one small laboratory and a lecture room, inconveniently far from the Macleay, but not far enough from the smells of the dissection room.[48] They did not fail to mention that Melbourne University had appropriated £15,000 for its new biology building.

The professors' proposal was considered, but deferred for two years, at which point Liversidge pushed it through. Liversidge worked closely with Haswell to strengthen his subject, including the re-launching of marine biology. In 1888, Liversidge also resurrected his proposal of 1884 for University-financed research fellowships.[49] These he proposed should be offered from 1890, at £150 a year, as part of a strategy to encourage 'promising students' to take up research in science or literature, so as to 'qualify themselves to take part in University teaching and examining'.[50] The larger colleges of Oxford and Cambridge were doing the same. At Sydney, however, the BSc degree 'experiment' was inconclusive. Judging from the graduates of 1885, including the celebrated (later Vice-Chancellor and QC) Frank Leverrier, of the few who had taken the BSc, fewer had found employment in science. As in 1885, only two graduated in 1888 with BSc degrees, Archibald Fletcher and Fanny Hunt. Fletcher's destiny is unknown. Miss Hunt, possibly Australia's second woman science graduate, became a headmistress in Queensland.[51] Whether either would have continued in science, if research fellowships had been available, remains an open question.

Such considerations were relevant if the University's new wealth was to be used wisely. The interest from the Challis bequest virtually doubled the University's income, making Sydney not only the richest university in Australia but also, in *per capita* student terms, better off than most provincial British universities. With £182,000 invested at four per cent, the endowment yielded £7200 per annum, or enough for seven new chairs. In fact, five were created by May 1889, bringing the staff to twelve professors in 1890, by which time the windfall had risen to over £250,000.[52] In June 1889, Manning, rightly fearful that the government might one day consider the bequest 'a mere relief to the public Treasury',[53] invited the professors to submit proposals for its use.

Their responses were thoughtful. Walter Scott, the Oxford-educated Professor of Classics and Dean of Arts who succeeded Badham – a progressive who supported modern studies, women's education, teacher's training, and university extension, and whom Barff credits with 'a singleness and earnestness of purpose which he applied with great force'[54] – recommended the creation of five new chairs: in modern history, modern literature, classics, philosophy and political economy. The last of these had been recommended thirty years earlier, and all had their advocates. As for political economy, 'the close connection' of this discipline with questions of 'immediate practical interest' demanded that:

> On the one hand, the promotion of a more general and systematic study of Political Economy is of vital importance to the community at large; on the other hand, the nature of the subject makes it essential that it should be treated academically – that is to say ... free from the direct pressure of political and commercial interests.[55]

Only if these appointments were made, Scott added, would the 'Humanities' reach the 'same degree of completeness as has already been attained in the department of Natural Science'. Mungo MacCallum, newly arrived in 1887 to teach modern languages and literature, agreed with his Arts colleague, but this left the scientists to agree among themselves.

Liversidge, actually the first by date to reply to Manning's letter, set out a case for three new science chairs, in botany, zoology and geology – to separate these subjects, as he put it, from their 'present very involved and unusual combination' – together with a lecturership in mining and metallurgy and another demonstratorship in chemistry. He took care to say that, without them, 'it is impossible to properly carry out the [new] By-laws in Science and Engineering'.[56] However, he also supported the recommendation of new chairs in modern history, music, painting and sculpture.

Haswell supported Liversidge's three chairs in natural history – which would make science at Sydney as complete as at 'any other University in the Colonies', and equal to the larger universities at home;[57] and Warren underlined the need for a lecturer in mining and metallurgy. Unexpectedly, Stephens recommended that Manning delete from his 'shopping list' the subjects of political economy, modern (but not ancient) history, philosophy and theology – domains which, he considered, were 'reasonably or unreasonably looked on with suspicion or dislike by any section of the community, and which, as such, were for prudential motives excluded at the very outset ... of the University'. More radically, Stephens proposed not to create new chairs at all, but rather to transfer to the Challis bequest the salaries of the existing five professors, so as to release the University's state endowment 'to the enlargement and improvement of ... University teaching'.[58] What this implied was not clear. But it was completely out of step with reality. Anderson Stuart did not mince words. Three new medical professorships and two surgical tutors were urgently needed.

To accept all these proposals was impossible, so Manning took a middle course. For all their protests, by comparison with Arts, the scientists had been well served. Most of the University's students were in Arts, but most of its professors were in Science. To balance the picture, Manning recommended to Senate that six new chairs be established. Political economy and a second chair in classical languages were left out, but law, philosophy, anatomy and modern history were brought in. To appease the scientists, a chair in engineering was included, while for Liversidge, there was to be a new chair of geology combined with mineralogy. But no lesser appointments were to be made from the Challis funds; and all new Challis professors were to receive set salaries of £900. Unlike the foundation professors, and Anderson Stuart, Threlfall and Liversidge, none of the new 'junior' professors was to receive lecture fees.[59]

All this was as much as Liversidge could expect. And in 1888 and 1889, he had much to do. In October 1888, explaining to British colleagues the slow progress of his laboratory, Liversidge chose to blame the makeshifts and not the men – but neither did he curse the Badhamites, who had delayed its progress for a decade, nor the politicians of all parties, who had postponed responsibility from year to year.[60] Nor did he mention his colleagues, notably Anderson Stuart and Threlfall, who had overridden his claims with their chariots of self-interest.

By 1890, however, Liversidge had scored a victory over 'procrastination', as Sir William Tilden put it, citing Sydney's new chemical laboratory as an exemplar for the Empire.[61] In February, Liversidge moved his laboratory equipment from the 'interim' building into his new premises fifty yards away. Not surprisingly, beyond a set of analyses of hot spring waters, and four papers for the second AAAS Congress, meeting in Melbourne in January 1890, his research had ground to a halt.[62] 'My own time,' he wrote Thiselton-Dyer in London, 'is now almost completely taken up by the details of fittings, etc. in my new laboratory. The vacations for two years have thus been mopped up. One practically has to be architect and clerk of the works.'[63] Similarly to Lazarus Fletcher, of the Mineralogy Department of the British Museum, who was anxious to obtain a supply of the Barratta meteorite, he moaned: 'I have been moving into my new laboratory and spending my days with carpenters and gasfitters; in consequence everything has been more or less upset and inaccessible.'[64]

At last, however, by the end of 1890, the laboratory was complete, and at beginning of 1891, Liversidge took full possession. What Sydney lacked, in what Tilden called 'architectural pretensions' – in Liversidge's words, it was 'a plain and unpretending building'[65] – it made up for 'in practical convenience'. In suitability and proximity to other science buildings and museums, it was equal to Manchester and superior to Glasgow. In size, it almost rivaled South Kensington. Perhaps there was just a hint of envy in Liversidge's recalling that 'at the Central Institute, South Kensington, the staircase is built of glazed terracotta, and the pillars are of the same; and at the Liverpool College of Chemistry the walls are lined with cream and

Liversidge's lecture on crystallography, 1890.

The preparation room in the Chemistry Laboratory.

First year medical students in the Junior Laboratory, 1893.

gold-coloured glazed bricks'.[66] These might be 'clean, effective and artistic' for England, but they were out of the question for Sydney. Economy rather than eloquence was the colonial theme; but space had been prudently left, the *Herald* approvingly noted, for additions and improvements, so that 'as the city grows from a third to a half ... and even a million of people, the laboratory can be extended with equal steps'.[67]

Thanks to Liversidge's overseas comparisons, his building combined the best features of laboratories for Henry Roscoe at Owens in 1872, H.E. Armstrong at Finsbury in 1882, and by Waterhouse for Liverpool in 1887. Gratifyingly, he was as well situated as Liveing at Cambridge. Over 180 students could be accommodated in one lecture room, and 120 in a second; and several experiments could be kept underway at once. There was a Junior Laboratory for forty students, and a Senior, with benches for sixty. The building had all the features – gas, water, compressed oxygen, hydrogen and coal gas – that money could buy or usage commend.[68]

On the lower ground floor was a mineralogical museum, with models of nuggets and specimens of meteorites, and heavy machinery for the liquefaction of gases, with which Liversidge later entertained his visitors. On the first floor were classrooms and lecture rooms, conveniently near rooms for gas analysis, polariscopy, spectroscopy and photography. Laboratory fittings – paraffin-wax for sealing workbenches against acids, India rubber to prevent leaks from draught cupboards – were 'state of the art'. There was even a room to house Liversidge's collections of minerals and 'old forms of apparatus, etc.' which may be of 'historical interest'.[69] At the end of the building, there was a balance room – something Frankland had lacked in London – and a 'professor's laboratory' which gave Liversidge a private space with cedar furniture that was the envy of his colleagues.

28.—PRACTICAL COURSES.

A.—INTRODUCTORY COURSE FOR JUNIOR AND MEDICAL STUDENTS.

1. Glass working.—Rounding the ends of rods and tubes, drawing, bending and joining tubes, blowing bulbs, mending test tubes.

2. The preparation and properties of gases, *e.g.*, hydrogen, oxygen, carbon, monoxide, carbon dioxide, the oxides of nitrogen and sulphur, chlorine, hydrochloric acid, hydrofluoric acid, ammonia, &c.

3. The structure of flame, flame re-actions, use of blowpipe, reduction of metals on charcoal, residues coloured by cobalt nitrate, incrustations, films, &c., borax and microcosmic salt beads.

4. Spectroscopic reactions.

5. Reactions of Reagents.

6. Qualitative Analysis by wet and dry processes.

7. Reactions and processes for the detection of the alkaloids, sugars, starch, glycerine, alcohol, fusil oil, carbolic acid and similar common substances.

B.—QUANTITATIVE COURSES.

Candidates for the B.Sc. degree in Chemistry, and B.E. degree in Mining and Metallurgy, are required to make correct determinations of the following substances :—

1. Verification of weights. 2. Determination of ash in filter paper. 3. Copper Sulphate. 4. Potassium dichromate. 5. Calcite. 6. Sodium chloride. 7. Rochelle Salt. 8. Ammonio-ferrous sulphate. 9. Lead Nitrate. 10. Siderite. 11. Dolomite. 12. Apatite. 13. Orthoclase. 14. Niccolite (kupfernickel). 15. Smaltite (Co. Ni. and As.). 16. Copper pyrites. 17. Topaz.

And certain of the following :—

18. Blende. 19. Zinc silicate. 20. Pyrolusite. 21. Chromite. 22. Wolfram. 23. Mispickel. 24. Fahlore. 25. Petalite. 26. Beryl. 27. Strontianite. 28. Cinnabar. 29. Coinage—bronze. 30. Lead, tin, bismuth, cadmium alloy. 31. Ilmenite. White lead and pigments. Cements. Iron Ores. Iron and Steel. Fireclay. Oils. Mineral Oils—including flashing points. Coal Gas. Furnace Gases. Coal, including ash and calorific power. Coke. Water for domestic and manufacturing purposes Manures.

Also the following volumetric estimations :—

1. Chlorine. 2. Silver. 3. Potassium and sodium. 4. Sodium hydrate 5. Iron by permanganate and dichromate solutions. 6. Bleaching powder. 7. Nitric acid. 8. Chloric acid. 9. Ammonia.

And the following determinations of organic substances :—
1. Exercises in the purification of substances, including fractional crystallisation and distillation. 2. Boiling and melting points ; specific gravity. 3. Ultimate analyses. 4. Vapour density. 5. Molecular weights of acids. 6. Use of polariscope. 7. Preparation of carbon compounds.

The chemistry curriculum, 1898–99.

The new chemistry building was also a major gain for Arts. As Nature abhors a vacuum, so in 1891 Arts subjects occupied the lecture rooms in the southern half of the Blacket Building that Liversidge had vacated. Forty years of shared dominion ended, as Latin resumed the room that once housed the Mineralogy Museum, and classics, the old chemistry lecture room. When the first Challis Professor of History, George Arnold Wood, arrived from Oxford in 1891, he lectured in the room previously used for chemistry. How lucky he was. For him, a history department with its own rooms was many years away. As for the generation before him, 'institution-building' was no mere figure of speech.

## 3. Gentleman and Scholar: Life at the 'Octagon'

Just as his laboratory building grew and took shape, Liversidge looked to his domestic arrangements. When he returned from England in 1888, he briefly resumed his rooms at the Union Club in Phillip Street. Over the years, the club had become a prominent landmark. But club life, however pleasant, had a provisional quality, and something more permanent was needed. Marriage was not on the horizon, but property was surely within his grasp. During his seventeen years at Sydney, he had saved almost £1000. Annual rent at the club was about £250; to keep this up indefinitely was not sound strategy.[70] To buy a house was common sense.

There were other considerations. Liversidge had no space outside his laboratory office for his burgeoning library, not to mention his *objets trouvés* from Europe, Japan and the Pacific. Possibly even before leaving for Europe in 1887, he had made a decision. Now, luck and circumstance took him to the eastern suburbs, still on the fringes of the city, along a dark, dangerous and dusty New South Head Road, but where a pleasing prospect awaited him.

As early as the 1830s, following Sir Thomas Mitchell's surveys, a few private houses had begun to appear along the muddy foreshores that lined Rushcutters Bay. For years, the district remained a rural woodland, but by the 1870s, hotels and several fine houses had appeared, appropriately called 'villas', with gardens cascading to the harbour shore. Anthony Trollope, visiting Darling Point in 1873, declared it 'perfect', 'inexpressibly lovely'.[71] To Nehemiah Bartley, it recalled the tranquility of Jersey.[72] Holtermann took photographs to show at the Philadelphia Exhibition in 1876 and the Paris Exhibition in 1878.

By the 1880s, a horse-bus connected the neighbouhood of Darling Point (what is now Edgecliff) to the city, and by 1894, a cable tram made the journey brief.[73] Sir James Martin, Robert Lucas-Tooth and others eminent in government and trade moved into the area. As a local historian has put it, 'There must have been occasions when a quorum for the Legislative Council could be found on the Darling Point omnibus.'[74] Near to St Mark's Church, built in 1854, whose towering spire was visible for miles, on a ridge marking the highest ground of 'Mrs Darling's Point', stood a sandstone blockhouse, called, after its shape, 'The Octagon'.

This unusual building, thought to be the oldest surviving European structure on Darling Point, was built in 1832 with convict labour, and may have served originally as a guardhouse, or more likely a signal station, as it had uninterrupted views, east to South Head and west to the observatory above Dawes Point. A grant of twenty-nine acres including the tower was made in 1835 to Thomas Smith, who had interests in the Commercial Bank, Australian Gas Light, and other companies, and who called the larger property Glenrock. In 1842, the site was subdivided, and the Octagon was occupied by Smith's brother, Henry (later known as the founder of Manly), who called it his 'bachelor tower'. The tower is said to have been visited by T.H. Huxley during the memorable stay of HMS *Rattlesnake* in 1847.[75] In the early 1850s, two wings were added, and the guardhouse became a family home. In 1859, house and grounds were bought by Albert Cheeke, a justice of the Supreme Court, for £3000.

The 'Octagon' from the south, 1889.

By the time Liversidge arrived in Sydney, the site was owned by the family of T.S. Mort, the wealthy merchant, who around 1883 subdivided the estate, giving the Octagon to Mort's brother, Edward. On his instructions, Cyril Blacket, son of the University's architect, made weatherboard additions to the sandstone building, and rendered the whole into a fine structure, surrounded by a vast, landscaped garden, valued by estate agents Richardson & Wrench at £2000.[76] In 1884, Edward Mort mortgaged the house to a partnership of his brother Henry and the former Auditor-General, Christopher Rolleston (past President of the Royal Society, a member of the University Senate, and a close ally of Liversidge). This partnership in turn leased the house to George Montefiore, cousin of E.L. Montefiore, a member of the Royal Society of NSW (since 1875) and Liversidge's companion at the Paris Exhibition.

Like his more famous brother, 'E.L.' was both a wealthy businessman and a painter, and a patron of the NSW Academy of Art (in 1880, the National Gallery of NSW), who years earlier had helped Liversidge acquire Elizabeth Street for the 'Royal'.[77] Blessed with such friends, Liversidge was also blessed by good luck – in this case, by a turn of events that crossed the life of a bachelor and the plight of a widow.

In July 1889, George Montefiore died at sea. His wife Dorothy – the pioneering feminist, better known as Dora Barrow – needed capital, and was obliged to move her family 'downhill' to socially inferior Paddington.[78] Her house was offered, possibly thanks to Rolleston, to Liversidge, first to rent and then, when Dora left Sydney in 1890, to buy for £2400. Liversidge paid a deposit of £900 and took a mortgage of £1500 at four per cent – so joining the fewer than fifty per cent of Sydney householders who were owner-occupiers.[79] By the terms of sale, he became 'tenant from year to year', paying charges of £132. At least, this halved his club rent, absorbed less than a sixth of his salary, and gave him a place for his books and objects.[80] He took up residence in 1888, and Hufton took a set of photographs to prove it.

Between the Octagon and the laboratory, Liversidge was now a man of property. It was a permanent change for a man approaching forty-five. From his verandah, he enjoyed gardens sweeping to the foreshore, and breathtaking views in all directions. A generous 'partners' desk' dominated a comfortable library, its walls neatly lined with rows of journals and prints. His drawing room, its walls finely papered and stencilled, was furnished with deep plush sofas and carpets, Japanese screens and Chinese miniatures. His dining room held an elegant table and cedar dressers, and an ornate cast iron fireplace. On the walls of the four-poster bedroom were an engraving of the Piazza San Marco in Venice, and *Vanity Fair*'s 'Spy' caricature of Darwin.

To share his good fortune, and perhaps his mortgage, Liversidge was joined by Anderson Stuart, who had been widowed since the death of his first wife in 1886. Rivals in the Faculty, the two had similar tastes, and Anderson Stuart remained at the Octagon from 1891 until he remarried in 1894. The men shared what Stuart affectionately described as a 'menage', looked after, he tells us (and the 1891 census confirms) by two Goanese houseboys.[81] 'The cooking was excellent', he recalled; but possibly the experience confirmed his preference for married life, and Liversidge's desire for a temple of peace. 'After my marriage, Liversidge paid me a sort of compliment by not taking anybody else',[82] Stuart mused, and indeed, for reasons we can only surmise, Liversidge thereafter lived alone (with his cook and servants). But friendship stood the test. Stuart may have recalled Liversidge's kindness, when he named his first son (also to become a medical doctor) Archibald.[83]

Whatever the nature of his 'menage', Liversidge shared Darling Point with the *gratiné*. Next door, at Greenoaks, lived Major General Hutton, commanding officer of the army at Victoria Barracks, and his wife, a granddaughter of the Marquis of Winchester. Other neighbours included the Morts, the Knoxes, and James Norton, horticulturist and trustee of the Zoological Station at Watson's Bay. Much as the Macleays of Elizabeth Bay a generation earlier, Liversidge mixed science with the gentry. At the same time, he put himself far from the newer generation of professors arriving at the University. His new Arts colleague Mungo MacCallum lived in Dulwich Hill; and Francis Anderson, the educational philosopher, and George Arnold Wood, the historian, who arrived in 1891, lived in the 'same bachelor establishment in Glebe'.[84] Few academics lived in Edgecliffe or Vaucluse.

Given the social and geographical distances, Liversidge went out of his way to acquire a reputation for hospitality. Many remembered his house with affection, including the bachelor mathematician H.S. Carslaw, Gurney's successor in the chair, who arrived from Glasgow in December 1903 just in time to look after the Octagon during Liversidge's overseas leave.[86]

## 4. The 'Battle of the By-laws': Curriculum Reform and Retreat

Liversidge's circumstances at home and work now easily matched those of his successful brothers in Clapton and his father in Kent. Not least in this respect, the colony had been good to him. Liversidge had a fortunate life in what seemed a lucky country. Yet, as dark clouds of recession gathered, nothing tested his practical idealism more severely than the curricular debates that emerged again in the 1890s. As Dean of Science, he was caught between two opposing tendencies. On the one hand, there was his commitment to science as a cornerstone of liberal education; on the other, there were growing demands from the several disciplines, for professional specialisation, education and training. In his attempts to resolve these forces, he met disillusionment, amounting sometimes almost to despair.

By the third quarter of the century, what Foucault has called the 'disciplinary apparatus' of specialised knowledge had transformed expectations in ways that were not unique to Sydney, or to Australia.[87] From Massachusetts to Midlothian, Toronto to Melbourne, disciplines were struggling for autonomy.[88] The liberal arts curriculum had set the subjects to be taught, the mix of skills required, and the balance between breadth and depth.[89] Now, new subjects vied for status. In some places, notably at Harvard and Yale, Oxford and Cambridge, tradition survived, but elsewhere, and notably in Australia, the prospect of profitable learning was, as it had always been, the most important justification of a university degree.

When such secular pressures cast all in doubt, the politics of knowledge played for time. For twenty years since Badham's day, and again from July 1883 and the creation of the Science Faculty, the Faculty of Arts had pressed Senate to reduce the compulsory science taught in the Arts degree. The question revolved not around numbers to be taught – fewer than 200 students were enrolled in all three years in 1887 – but around the allocation of 'student time'. The scientists had fought the issue to a deadlock in 1886. But on returning in 1888, Liversidge discovered that the compromise he thought had been worked out had unravelled. Indeed, no sooner had he left, than William Macleay persuaded Senate, on 21 February 1887, to rescind the proposal it had accepted the previous December, and to refer matters of reform to further discussion. To this, Senate agreed, constituting an Arts Curriculum Committee to consider the question of 'over pressure' in the first year.

This committee consisted of Manning, Edmund Barton, George Knox (an outspoken critic of the Science Faculty), Macleay, MacLaurin and Renwick. The professors were to attend by invitation only; and while they could make submissions, they could not vote.[90] Their deliberations brought Liversidge one of his most convincing victories, and one of his most resounding defeats.

The arts curriculum had changed little in thirty years. The first year stipulated 150 lectures in Latin and in either Greek, French or German, together with 150 lectures in mathematics and 120 lectures divided between chemistry and physics. The second year consisted half of literary studies and half of mathematics (150 lectures each), plus the sixty non-examinable lectures in natural history that Stephens had introduced in 1884. After 1884, third-year Arts consisted entirely of literary subjects and mathematics, in about equal proportion. The curriculum compromise that obliged Arts students to do a natural science course in their first year, removed the requirement of their being examined in it.

James Arthur Pollock
(1865–1922)

Mungo W. MacCallum
(1854–1942)

James T. Wilson
(1861–1945)

William A. Haswell
(1854–1925)

Thomas Anderson Stuart
(1856–1920)

Archibald Liversidge
(1846–1927)

William H. Warren
(1852–1926)

Theodore T. Gurney
(1849–1918)

George Arnold Wood
(1865–1928)

Richard Threlfall
(1861–1932)

T.W. Edgeworth David
(1858–1934)

Liversidge and his Colleagues - Sydney University Professors, about 1890.

Given this poor showing, few Arts students chose to take science courses in their second and third years, where examinations were compulsory. 'There is a very general feeling', observed Walter Scott in March 1887, 'that these regulations are not satisfactory'. That was an understatement. Students arrived with considerable differences in preparation, capacity and motivation. 'A properly arranged Arts curriculum should be sufficiently elastic to allow for these differences', he said.

With Liversidge, Scott objected to a narrow humanities curriculum that fell 'far below the standard, not merely of old-established Universities, such as Oxford or Cambridge, or of the sister University of Melbourne, but of much more unpretentious institutions, such as are springing up in second and third-rate country towns in many parts of England and Wales'.[91] Scott recommended that the curriculum be redefined, whereby the first year would be devoted to 'general preparatory studies', divided 'among the chief branches of knowledge'. These would be mathematics and classics, as before. But in the second and third years, students would be given a choice between literary subjects and science subjects – along the lines that Liversidge had proposed (and the Board of Studies had accepted) in December 1886.[92] Such proposals came dangerously close to destroying the traditional curriculum, even if retaining a 'core'.[93] But in an age of increasing specialisation, who was left to defend an empty castle?

As 'rival' submissions soon revealed, there was little consensus. Scott's proposals were welcomed by MacCallum and Gurney,[94] but W.J. Stephens, anxious to make a distinction between pass and honours students – along lines that Cambridge was then encouraging – proposed giving undergraduates a more restricted choice. In order to increase numbers reading science, without setting science as an obstacle to entry, Stephens proposed to overlook the requirement of chemistry and physics at matriculation, and to require, as before, only class (and not final, recorded) examinations in science in the first year.[95]

Threlfall, on the other hand, supported Scott's second and third year plan, 'conceiving that a man is ever best educated who is educated by the work he loves'; but insisted, against Stephens, on the extension, rather than the dilution of science in the first year. Taking a bead on Senate conservatives, he pointedly objected that, 'Though there may still be people who consider Science unnecessary for scholarship, this is not the opinion of the old world, and it is in distinct opposition to that experience which has carried all the great schools of England to provide adequate scientific teaching.' In any case, given the poor standard of school science in NSW, 'it would be idiotic to require [the student] to learn less of that which he already knows least'.[96] Anderson Stuart joined Threlfall in opposing a reduction of science in the first year: 'unless a medical student gets his Natural Philosophy in his first year he cannot get it at all'.[97] Warren went further still, proposing to substitute mathematics, chemistry or physics for Latin in the first year, and to introduce advanced science in the second or third.[98]

On 27 June 1887, the Arts Curriculum Committee reported to Senate. Its recommendations followed Scott, specifying the omission of required chemistry and physics from the matriculation examination, and the reduction of science in the Arts degree. This fuelled an angry response from the Professorial Board, where five professors – Gurney, Stuart, Smith, Threlfall, and newcomer Mungo MacCallum, a product of Glasgow, Leipzig, and Berlin, just arrived from Wales – rejected the Committee's recommendation, and underlined the need for standards of proficiency in each of the disciplines taught. Threlfall also urged the

Board to ask Senate to reconsider the BSc degree, the unpopularity of which, he believed, had reduced to absurdity arguments for decreasing science in the Arts degree.[99]

To some extent, the professors' protest was successful. The Arts Curriculum Committee went back to Senate on 19 August 1887, with a new set of proposals, and a complex compromise. As the *Herald* observed, this was a 'committee decision': 'the friends and patrons of various branches of learning being unable to agree as to which branches are best adapted for a liberal training, finally agreed to impose them all upon the unfortunate student'.[100] First-year Arts students were thenceforth to take English, Latin and a third language, together with mathematics, and a single course of science lectures. In the second year, they were to continue two languages, mathematics and either a third language or a science subject. The required forty-five lectures in physics and forty-five lectures in chemistry were to be reduced to thirty in each subject, to make space for thirty lectures in natural history. In the third year, students were to take one language, plus any two subjects from a list of nine that included two other languages, logic and mental philosophy and five sciences, each with required, though non-examined, laboratory practice. Astonishingly, it was agreed that first-year Arts students 'who shall have given satisfactory proof to the lecturer of their intelligent attention to the lectures shall not be required to pass the annual examination in these subjects'.[101]

It was at this point that the working relationship between Scott, Dean of Arts and Liversidge, Dean of Science, began to falter, and with its failure came hard feelings that continued until Scott's resignation and Liversidge's retirement twenty years later. The Arts men held the high ground, while the Science men, whatever their intentions, were forced onto the offensive as their only means of defence. While Senate debated and Liversidge fumed, the press viewed the compromise as an inevitable consequence of specialisation. In this form, the Arts curriculum was to remain virtually unchanged for the rest of the century.

There was little hope, however, that the battle would end, and indeed it continued the following year. Nowhere in England had the central problem – of competition between subjects, at different levels, for tightly controlled 'academic space' – generated more light than heat, and Australia could hardly expect otherwise. In the event, it was ironic that the most searching review of the University's educational philosophy would take place when Liversidge was overseas, and unable to participate. Yet, on 12 April 1888, a month after his return, Liversidge shared with Senate several reflections arising from his recent visit to Europe and America.

In the beginning was matriculation. The University was attracting more applicants, as Manning had wished, but the low level of preparation at the public secondary schools posed a serious problem. The University had engineered the worst of both worlds, setting matriculation examinations that restricted access, but setting them at such a low standard that they failed to guarantee a student's success, even in first year. When the late 1880s saw increasing failures at matriculation, and in first-year courses as well,[102] the *Herald* intelligently observed that 'this question is more easily asked than answered'.[103] Contemporary comparisons with London, Manchester, Melbourne, Adelaide and New Zealand suggested that Sydney was asking too little and expecting too much. In science, the position was particularly grave. Of the 121 candidates who sat the Arts matriculation examination in 1887, only twenty attempted the chemistry questions.[104] Adept at circular reasoning, Senate was asked to exclude compulsory science simply because so little science was taught. In the meantime, Liversidge could expect his entering students, whether in Arts or Science, to be, as he put it, 'entirely ignorant'.[105]

Liversidge attempted to reduce the problem to first principles by moving Senate to abolish the matriculation examination altogether. It was, he argued, of such a low standard that it could be a useful measure of neither ability nor preparation. A 'Graduate' writing to the *Herald* concurred: 'In a free country, and in an age that asks a reason for everything, people naturally ask why they should learn Latin and Greek.'[106] Indeed, Anderson Stuart and Liversidge did not wish a curriculum that stood in the way of either modern languages or science. In its insistence upon Latin, and either Greek or another language, the University effectively denied access to hundreds of pupils whose parents could be excused for thinking the University, in Liversidge's phrase, 'exclusive and even aristocratic'. In reality, it was nothing of the kind. But its image was set by its regulations.

Looking overseas, Liversidge had been impressed by reports from across the Tasman. Sydney University's students numbered fewer than 260. New Zealand, with half the population of NSW, had 800. Blame for this sorry ratio Liversidge placed squarely on the 'barrier which the matriculation examination presents'. Such an examination was found unnecessary in America. Why in Australia could not a similar vision prevail? 'Instead of trying to ascertain' by a written test, he asked – in a colony which fed on the myths of individual enterprise and the 'fair go' –

> whether would-be students are *likely* to profit by attendance on the lectures, would it not be much better to let them come to the University, and then see whether they *do* profit [by them]?[107]

The same point had been made and resisted by Senate for decades. It seemed clear, judging by countries that had the freedom to 'select and adopt the best characteristics of institutions existing in all parts of the world', that Sydney would attract more enterprising students if it simply abandoned its entrance requirement. Perhaps, Liversidge added, if admission were made easier, 'the public would also gradually learn that the instruction it affords is very much more practical and useful than is commonly imagined'.[108]

If such an appeal to American practice constituted an attack upon Australia's flagship, Liversidge's next proposals shot at its mainmast. First, he recommended that the science, engineering and medicine curricula should be extended, as European and American universities were now doing. With the confidence of one who had just returned from overseas, he announced that 'it is utterly impossible to do the work [of science] properly in two years: [but that] if three years were necessary', he was willing to accept that the rules requiring science students to spend their first year in Arts 'should be dispensed with'.

Was Sydney's last hostage to general education to be thus abandoned? On the contrary, Liversidge argued, the sciences occupied an increasingly specialised domain, but those few who read science should not suffer the alienation of a separate degree. In October 1887, *Hermes* asked, reasonably enough, whether Sydney's science courses in Arts served any purpose when 'its uses and functions are usurped by the rival degree'.[109] In answer, Liversidge mooted the abolition of the BSc degree itself.

Threlfall's Physics Building (on the left), with Liversidge's Chemistry Building, 1893.

If Liversidge's proposal to abolish matriculation had shocked Arts, his proposal to end the BSc astonished Science. His alternative plan for a degree combining first-year general courses with later specialisation was a reversion to Cambridge. Everyone agreed that it was necessary to retain a core, while allowing room for specialisation. It was supremely ironic to find the father of Sydney's BSc degree citing the Devonshire Commission in defence of the idea that a degree in science alone was 'unnecessary and undesirable'. However, his logic had a point. Whilst in England in 1887, he had learned that opinion at Oxford and Cambridge had hardened against distinguishing 'between students who had in each case received [an] equally liberal and non-professional education'.[110] Instead, Science dons were increasingly called upon to produce, in Gerrylynn Robert's phrase, 'liberally-educated chemists'.[111] If the single Arts degree at Cambridge could embrace the biological, mathematical and physical sciences, perhaps Sydney's specialised BSc was an unnecessary, divisive distraction. After all, at Cambridge

'the BSc', as Liversidge candidly admitted, 'is completely unknown, and consequently is of less value in after life than the older and better known degree of BA'.

Voting on 21 May 1888, the Professorial Board (formerly the Board of Studies), which represented the professors in all four faculties, with the Chancellor and Vice-Chancellor, decided that it could not recommend either the banishment of matriculation, or – despite the reservations of five of its members – the death of the BSc.[112] Both stayed in place. Indeed, instead of abolishing the BSc the Board went to the other extreme, and recommended keeping it while abolishing compulsory first-year Arts for science, engineering and medical students.[113] Thus, the same year that saw the advent of the AAAS, and heard the Vice-Chancellor warn that education was becoming hostage to science, the University contemplated a final divorce between the 'two cultures'. If the Board's proposal was accepted, the assumption that science should companion the classics would become a thing of the past.

Against this proposal, Liversidge was determined to resist. Between March and August 1888, even while preparing the AAAS Congress, Liversidge recast the requirements for science and engineering. If the University wished to retain a BSc, Liversidge preferred that it be a more professional course, and one that built upon an Arts background.[114] Accordingly, from 1889, entering BSc students were to have either first graduated in Arts; or to have passed first-year Arts; or to have passed the Senior Public Examination in Latin, together with either Greek, French or German, and three subjects in mathematics, astronomy and mechanics. After two votes, a motion to apply the same rigorous standard to medical students was also narrowly passed.[115] With the agreement of the Science Faculty, physiography in the first year of the BSc was to be replaced by lectures in biology, chemistry, mathematics and physics.[116] Liversidge had been the architect of the original BSc; now he was the author of its first major reform.[117]

For a time, peace might have descended on the battlefield. The University agreed that science should kept keep its own degree, and set its own rules. But the question of compulsory science in first-year Arts still festered, and relations between Liversidge and Scott came under even greater strain. July 1889 saw a fresh exchange of fire when Scott again tried to reduce the science commitment (from which, in any case, exemption could now be gained) from three courses to one. Liversidge countered with the argument that, far from eliminating physics and chemistry from the Arts degree, these subjects should be made compulsory: and that, if natural history be omitted, compulsory English should be declared optional![118]

A more divisive meeting of the Professorial Board than that of 30 July 1889 can hardly be imagined. Among the eight professors, smoldering sentiments burst into flame, and such were the inflammatory resolutions that day, that all were voted down.[119] The result was a draw. The year that saw Matthew Arnold and T.H. Huxley debating the place of science in higher education was not a time for Sydney's Senate to abandon one of the University's fundamental tenets. When, in November, the Faculty of Arts again asked Senate to reduce the science content of first-year Arts, Senate conspicuously declined.[120] In December 1889, Manning added a rider to Liversidge's motion, to ask whether 'any and what branch of Arts could be added with advantage as an optional substitute for Science in that year'.[121] But this, too, failed. Things were to remain as they were.

From all this committee work, Liversidge came out wearily resigned to a war of attrition. Over the next ten years, the Professorial Board heard repeated proposals for curricular change, almost all in the direction of more specialised studies. Before matters reached the Board, they were now carefully prepared by the deans, and the Board became a court of appeal, in which Liversidge was of course the leading advocate for science. In September 1889, Liversidge set out new rules for the conduct of Faculty business and for submitting proposals to Senate.[122]

The new curriculum for the BSc came into effect in March 1890. Chemistry, physics, biology and mathematics remained as first-year courses, but the third-term course in physiography was thought too elementary to be compulsory, and physical geography and geology were introduced instead.[123] In the second year, a student selected three from six subjects – the full Cambridge spectrum, including botany and zoology, chemistry, geology, mathematics, physics and physiology. The end of 'Natural History' came much to the dismay of Manning who, as he put it, was sorry to see the specialised sciences superseding the 'greatest and noblest of all subjects for any man to study'.[124]

But while the science degree was made more rigorous, it remained unattractive. In 1890, 156 students entered Sydney to read Arts (bringing the total enrolment to 284), but only twelve matriculated in all other subjects, including science. Looking at the University handbook, one might see why. BSc students were required to spend between thirteen and sixteen hours a week in lectures, just as their Arts contemporaries, but also between nine and ten hours in practicals. Medical students spent seven to sixteen hours in lectures, and between four and ten in practicals. Engineers spent between eleven and fourteen hours in lectures, and between eighteen and twenty-one in practicals and drawing classes. Students of biology, geology and physics, were expected to attend fifty lectures each year; and students of mineralogy and palaeontology, ninety.[125]

In December 1889, and again in February and March 1890, shortly after the new Science by-laws went into force, Liversidge moved in Senate that a report be prepared on the amount of work required of first-year students in the various faculties at Sydney and other universities, and on the position of science in other universities and grammar schools. The answers were not encouraging. Liversidge's reforms were desirable, but self-defeating. Some way had to be found to make the science degree popular.

Liversidge faced a paradox: if the timetable was overloaded, it was because he could assume little or no prior knowledge. Given the poor state of science in the secondary schools, students were doomed to spend more hours in elementary courses. There was equal misery on both sides of the lecture bench. Liversidge, Threlfall and Stephens were locked into serving up material for beginners, few of whom had any hope of reaching advanced work within the prescribed time.

By the deadlock agreement of 1889, first-year Arts students had still to take science. But no one was satisfied with this, least of all the scientists who had seen the requirement wither and recede. While first- and second-year Arts students in 1870 had received some 300 lectures in science over three years, by 1887 the science lectures had fallen to ninety, and all were compressed into the first year. Overall, Arts students were now spending less than fifteen per cent of their time in science, and the Arts professors wanted even less. Despite the advances in chemistry and physics since 1850, Sydney University entered the 1890s giving its students less formal attention to these subjects than at any time in the University's history.

By the middle of 1890, the fragile armistice between the faculties was crumbling, and in his *Annual Report* of 1 September, Walter Scott again threw down the gauntlet. Apparently in retaliation for the exclusion of compulsory Arts from the first-year BSc in science (including medicine and engineering), he recommended that the surviving science courses in first-year Arts be further reduced from ninety to thirty lectures, and from three courses to one. Liversidge and Threlfall were furious. In October, they delivered their replies – Liversidge's in seven printed pages, Threlfall in six – both vibrating with anger.

Threlfall, never one to suffer fools, was uncompromising. The Arts proposals, he said, were founded on mere prejudice. Besides, he added, 'Have we not gone on tinkering with the Course to such a degree that no student can tell what he has to expect?'[126] Liversidge began more diplomatically. This argument had already been fought

CHEMICAL LAB. No.39.A.

Chemistry Laboratory, about 1890.

out and settled in England, he said; there was little point in repeating it. Siding with T.H. Huxley, he did not 'wish to see Literature and Mathematics thrust on one side for Science'; yet, with Huxley, he felt the University would fail in its duty if it let students escape without 'a fair grounding in Experimental and Natural Science'. No student should graduate unable to understand 'reference to scientific matters met with in every day reading'. The Arts Faculty was demanding that less than twenty-five hours in a student's career be 'allotted to the study of the great truths of the Universe, and of the conditions under which we live and have our being'. This was manifestly absurd.

In the debates of 1888, Liversidge had brought himself to compromise with Scott. In 1890, he was uncharacteristically blunt. 'The present Faculty of Arts is not,' he demanded, 'from the special nature of its constitution, in a position to give an authoritative opinion upon what branches of science should be included in a liberal education. Languages and mathematics alone do not constitute such at the present day.' Indeed, that was not all. 'The ancient teachers were revolutionists', he declared, 'compared with many of [the] nineteenth century; they tried to teach the best and newest science and philosophy of their day. We apparently must confine ourselves mainly to those matters which have become encrusted with, and hallowed by, the dust of centuries.' The recommendations of the Faculty of Arts, he warned would, if carried out:

... have very much the same tendency as the efforts of many in the cause of religion, who are well aware that it is a great point gained if possession can be got of the immature, to train them according to their own particular precepts, before they are of an age and have sufficient knowledge to judge for themselves.[127]

Liversidge would not be moved. And the Arts Faculty retreated.

For the next thirty years, with varying degrees of protest, Arts students continued to endure the requirement of science courses in their first year. Struggling with Liversidge on chemistry and Threlfall on electricity, they next met Edgeworth David lecturing on the interior of Australia, followed by the business of history and literature with George Arnold Wood and Mungo MacCallum before (or perhaps during) diversions for tea, tennis, and horseplay with flying caps and trenchers.[128] But Liversidge's victory was pyrrhic. Sydney's Arts curriculum had already ceased to serve the humanistic vision so dear to Nicholson and Merewether, and had come to be a preparatory course for lawyers, public servants and school teachers, with little commitment to the traditional values of liberal education.

Soon after Liversidge's last philippic, the 'battle of the by-laws' ground to a halt. Hostilities were not to be resumed for another decade. Liversidge had fought hard for a Faculty of Science, with its own professional degree, but also for science as a component of liberal education. It was ironic that someone who had given his life to practical studies, should find himself fighting against the severely subject-oriented, pre-professional curriculum that Arts had become. After forty years, science was still pedagogically thin. But Liversidge's stubborn resistance had at least maintained the *status quo*. For the next seventeen years, all first-year Arts students, willing or not, met him for chemistry in their first two terms, five days a week at 11 a.m., sitting next to medical and engineering students in lecture rooms designed for 200, and spending afternoons in laboratories that were uncomfortably full with half their number.

With the curriculum battles of 1888 and 1890 behind him, Liversidge could do no more. But it was also clear that the niceties of debate had exhausted him. In 1892, he mentally withdrew behind his new battlements on Science Road. From there, he planned to build the memorial to his discipline that had eluded him since 1872 – a professional School of Mines for New South Wales.

# CHAPTER 10

# Depression and Optimism

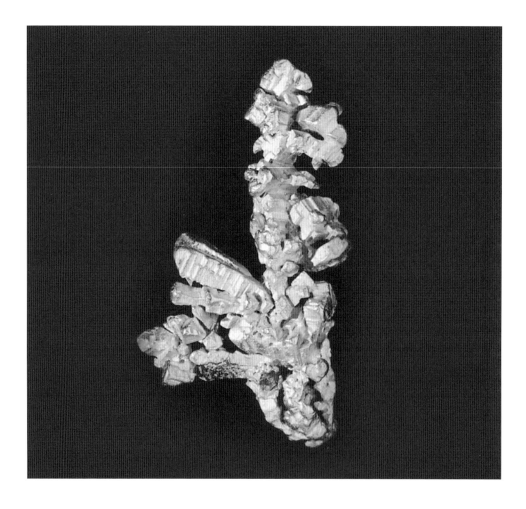

A gold crystal in the shape of Great Britain.
This featured on the cover of Liversidge's *Minerals of New South Wales.*

In 1890, a global recession hit world markets, forcing a drop in prices and a collapse in foreign investment. Australia's small, vulnerable and over-indebted economy faltered. As banks failed, small manufacturers, tradesmen, and professional people were ruined. The effects were felt worst in Melbourne, which was crippled by shipping strikes, and in Queensland, where strikes among the shearers touched every aspect of life. The streets of Sydney saw marches of the unemployed, union demonstrations, and in 1891, a riot at Circular Quay. The country's grief was prolonged through the 1890s by one of the worst droughts on record. It came as no surprise that public works and higher education were caught in a vice, and had little room to manoeuvre.[1]

Against this despairing background, the 1890s also saw the colonies march uncertainly towards 'nationhood'.[2] And the years between 1888, when the AAAS first met in Sydney, and 1898, when it returned, marked a turning point in the public perception of science. Fears of population decline and the 'yellow peril' were heightened by cries for 'national efficiency', calling for the application of scientific methods to enhance life, health and economic growth.[3]

At the University, Liversidge continued to support practical solutions, but was repeatedly baffled by colleagues who chose not to modernise the curriculum, but to defend it behind philosophical barricades that even Oxford and Cambridge were in the process of dismantling. In 1893, Francis Adams, that keen observer of Australia's 'Pacific Slope' and 'Eastern Interior', described the Australian universities as, simply, a 'failure': 'Too timid to boldly make themselves samples of the modern educational theory, they have limited their appeal to the exhausted Anglo-Australian tradition.' For Adams, this was all the more damaging when, after all the debates of the preceding decades, chancellors and senates construed science not as an ally of classical education, but as a threatening alternative, the success of the one to be gained only at the cost of the other. 'If the universities had had the intelligence', Adams wrote, 'to perceive the peculiar character necessary to vitalise their teaching, and the courage to make themselves the mouthpiece of the vague educational aspirations of the community, it might have been otherwise':

> As it is, they are paying and will continue to pay for their want of courage and insight by their impotence, and will have Culture forced upon them from without instead of evolving it themselves from within.[4]

From the outside, Adams' description of the University of Sydney seemed accurate. Inertia within looked like apathy without. Yet his verdict gave scant credit to the struggles of Liversidge and his colleagues, or to Sydney's considerable successes in the classical tradition. Yet, his criticisms were to stick and be amplified, as economic crisis and diminishing resources came to dominate the most demanding decade in Liversidge's career.

For the next ten years, much of Liversidge's life continued to play out in the lecture room, the Professorial Board and Senate. An acknowledged 'boffin of the by-laws', he saw the fortunes of education ebb and flow. His influence grew on key committees, notably the Senate's Buildings, Grounds and Improvements Committee,[5] a body whose influence mimicked that of the Museums and Lecture Rooms Syndicate of Cambridge, the reforming example of which he closely followed.[6] He spearheaded a campaign that led to new faculties, began a School of Mines, and made mining engineering an attractive option.[7] But few of these victories pleased him as much as his research. Only in his mid-forties, he was creating, almost single-handedly, the discipline of geochemistry in Australia. In fitting tribute, the decade of his middle-age ended with honours – at home, the presidency of the AAAS; abroad, the award of an honorary degree. By the end of the 1890s, Liversidge had become one of Australia's statesmen of science.

## 1. ASSAYING VICTORY: THE SYDNEY SCHOOL OF MINES

'There is no doubt', James Gormley observed drily to his colleagues in the Legislative Assembly of NSW on 21 July 1903, 'that the mineral resources of New South Wales have not had the attention which they should have received'.[8] As a pioneering pastoralist, former mayor of Wagga Wagga, and representative of the country party of 'practical men', Gormley spoke for many who deplored the waste and inefficiency of unimproved property, and the vacillation of thirty years on the part of governments towards the mining districts, prospectors and mining companies. The dimensions of minerals policy in NSW were in part legal, in part technical, and in part educational. Practical miners, convinced of their worldly wisdom, were openly sceptical of geological advice, which the growth of a 'mining bureaucracy' at first did little to overcome.

Whether systematic training in the science and engineering of mining had commercial benefits, was as much a matter of conviction as expectation. Given the cost and professional scepticism, mining education in Australia was slow to expand, and schools of mines often had to diversify in order to survive.[9] In these circumstances, Liversidge laboured long and hard to persuade 'practical men' that the advantages of 'scientific mining', already clear in Europe, Africa and the Americas, could, if given a chance, advantage Australia as well.

In pursuit of his plan, Liversidge looked back upon the crest of a wave that he had nearly missed. New South Wales had for years suffered by comparison with Victoria. By 1891 there were no fewer than twelve mining schools, often associated with mechanics' institutes, in mining districts throughout the colony.[10] In South Australia, schools of mines had been established at Gawler, Moonta and Kapunda, followed by others at Peterborough and Port Pirie; and while these were deliberately located near mining operations, all were affiliated with the South Australian School of Mines and Industries, established in 1889 on a site adjacent to the University of Adelaide. That University, which did not yet offer engineering, sent BSc students to its School of Mines for a postgraduate fellowship in mining and metallurgy. Liversidge favoured a similar arrangement for Sydney, with mining laboratories adjacent to science, and with mining engineering forming a strand in the BSc degree.

For over twenty years, Sydney had seen proposals for a school of mines come and go, and for years, no happier fate greeted Liversidge. The *NSW Mines Act* of 1874, which had sanctioned a school of mines, had not specified where it should be, or who should pay for it, and for nearly two decades nothing had been done. In the late 1880s, however, as alluvial beds were worked out, and gold production declined, miners needed geological advice to find less accessible veins, and mining engineers to tap them. Increasing demand for coal prompted a need for geologists to locate fresh seams north and south of

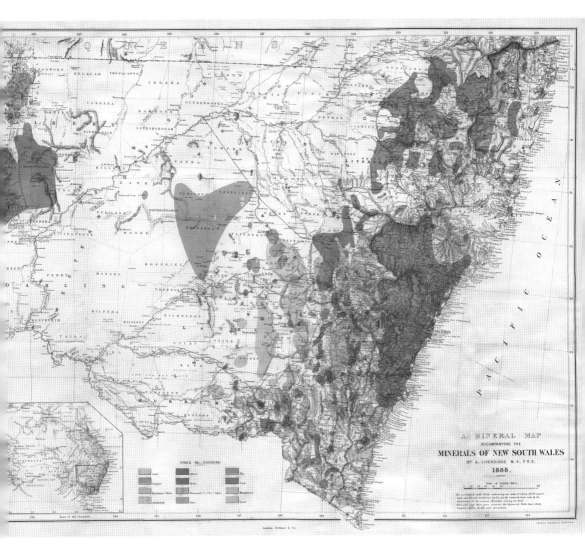

Liversidge's 1888 mineralogical map of New South Wales, adapted from that of W.B. Clarke.

Sydney; whilst the discovery of new silver, copper and tin deposits in the Barrier ranges raised prospectors' hopes.[11] By the 1880s, mining had moved well beyond individuals working shallow claims, and had become big business dominated by public companies, which either hired engineers from overseas, or did without. The industry needed not only educated miners, such as Ballarat produced, but also mining engineers, and men of science with practical experience.

When the Royal Society of NSW held its monthly meeting in July 1884, Leibius and Liversidge gave the leading papers. The Deputy Assayer of the Sydney Mint outlined a history of the new Mount Morgan gold mine in Queensland, and urged the establishment of a colonial mining laboratory, both to examine ores, and for 'extracting by the most approved methods their metalliferous treasures on a large scale'.

Leibius described to members his 1880 tour of European laboratories, during which he visited smelting works at Clausthal, Freiberg and Eisleben. Australia needed a similar establishment, he said, to train a generation of managers and metallurgists equipped to exploit the country's mineral wealth.[12]

All this was music to Liversidge's ears. Indeed, to produce 'scientific' mining engineers had been, in 1873, his maiden ambition, and it remained among his cherished goals. Before leaving for Europe in 1887, he had proposed courses in mining engineering for both the Technical College and the University and, with Warren, had even framed by-laws to install the subject in a new BSc degree in the Faculty of Science. Neither the College nor the University had attracted many students to the subject, Liversidge admitted to the Minister of Lands in 1886. But the demand remained, and he would create the supply.

In April 1888, whilst planning the inaugural meeting of the AAAS and the fitments of his new laboratory, Liversidge reminded Senate of its unfinished business in the matter of a school of mines. He went so far as to suggest the names of possible lecturers. But Senate agreed only to take the matter up with Macquarie Street, and, as ever, government declined to act. Possibly, looming deficits left the government feeling strained.[13] Ministers heard the argument that mining education was best left to private enterprise, and there the matter ended.

However, Liversidge refused to let it die. In 1888, as we have seen, Sir William Manning, the same Chancellor who had commissioned Liversidge to write his Technical Education Report, and who had championed 'the tendencies of modern opinion' – including the admission of women – unexpectedly reverted to his earlier fears that 'applied' subjects might come to dominate the University. Such doubts he aired publicly, if undiplomatically, at the AAAS Congress in August. Liversidge, still wincing from that encounter in the Great Hall, despaired to see the University ossify by denying science and engineering the status they enjoyed in Britain. In any case, with the humanities well represented in new appointments, there was surely no need to fear that Science would ever overwhelm Arts.

In April 1889, Liversidge, his laboratory half built, went to Manning with another proposal for a school of mines. Manning remained sceptical of 'the propriety of the University teaching such subjects as mining and metallurgy'. However, Liversidge icily observed, 'similar institutions, which are certainly not inferior in character and standing, do not think it beneath their dignity or improper to teach these and allied subjects, neither should we lose caste by so doing'.[14] He accompanied his proposal with a list of fifteen analogous departments and courses in Britain, Canada, New Zealand, and Japan. MIT had such a school in Boston, and Columbia University had begun a mining school in New York. Liversidge's School of Mines was not to be a mere 'mechanics institute', but an 'École des Mines', like that of Paris and the Royal School of Mines in London, where lectures were not intended for 'ordinary artisans and working miners any more than University courses of instruction in Architecture are intended for bricklayers and hodmen'. A glimpse of the staff list at Jermyn Street (now removed to South Kensington) revealed men 'of highest eminence' in chemistry, geology, mineralogy and metallurgy, probably the 'very first in their respective subjects'. It was from these that Liversidge would recruit his staff.[15] Manning need not worry: the reputation of the University was safe.

For Sydney, the RSM model was both convenient and appropriate. No British university – nor for that matter, the School of Mines in Cornwall – had succeeded in producing the ideal 'professional scientific mining engineer' suited for imperial service. This class the colony had to create for itself, and for its creation, the University had a mandate from Nature. 'The mineral deposits of the colony', Liversidge reminded the Chancellor, 'are one of the most important sources of wealth to the community; and as the University is indirectly supported in part by the mineral products of the country, it should assist in the scientific development of these resources'. If it did not, 'Those branches of education such as Architecture, Mining and Agriculture, which should tend to support the material condition of the people, and add to the resources of the colony [will be] almost entirely neglected'.[16] Manning delayed his reply. With new law and medical schools in place, and a lawyer himself, he could hardly protest against the importance of professional education. Even so, Cambridge had not rushed into establishing a course in engineering; and Sydney's Chancellor preferred to 'wait and see'.

Depressed by the 'by-laws battles' of 1888–90, and by the delayed construction of his laboratory, Liversidge ended 1889 in a minor key. He had done little scientific work for some time. He had managed to finish a note on the hot springs of the D'Entrecasteaux Islands from samples sent him by Sir William MacGregor, and on chalk and flints from the Solomons; and he wrote a note on the chemistry of amniotic fluids in specimens of the dog-fish, sent him by T. Jeffery Parker in Otago – a reminder of his Cambridge days in physiological chemistry. But these looked to the past. More forward looking, were several papers he began on Australian meteorites and features of gold. These he decided to work up for the second AAAS Congress, which was to be held in Melbourne in January 1890.

The first meeting had raised expectations. Fortunately, the Melbourne meeting was a huge success, with over 1000 papers to be read.[17] Liversidge greeted Baron von Mueller and, for the first time, met the young David Orme Masson, who had migrated from Bristol in 1886, at the age of twenty-eight, to become Melbourne's Professor of Chemistry.[18] Inevitably, Masson was asked to be on the local organising committee, as well as Secretary of Section B (Chemistry). Although he was Liversidge's chemical counterpart, there was a strange silence between them. Not once had Masson corresponded with Liversidge; and it took the AAAS, rather than shared chemical interests, to bring the two together. In fact, they had less in common than might be supposed. Masson was working on the rapidly moving research front in physical chemistry, and was conversant with theoretical questions being set in Germany and America, while Liversidge remained locked in his Anglocentric, descriptive tradition, a man of facts, not of theories.[19] Still, if destined not to be close, the two men were at least complementary. Both were to play key roles in shaping Australian science – Masson, much the younger man, for much longer.[20]

As expected, Liversidge's three papers to Section B were delivered with enthusiasm, and when he returned to Sydney, he spent the rest of the summer preparing his first Presidential Address for the Royal Society. In May, he announced to the Society's members (now stable at 471), the deaths of Julian Tenison-Woods and Felix Ratte, the assistant in mineralogy whom he had helped appoint to the Australian Museum.[21] He also reported the decision to increase the size of the Council, so better to 'further the aims and objects of the Society'; as well as to set prizes for papers on Australian Aborigines, tin deposits and marine biology – 'The right elucidations of [which] might be of incalculable value and benefit to the Colony'.[22] Advertisements followed for Antarctic exploration and the AAAS, through which, 'the general public have an opportunity of learning that there is no great mystery either about scientific workers or their methods of work'. His appeal was practical. 'Of course', he added:

> in some cases long and laborious training is necessary for certain investigations, but a very great deal of a very useful kind can be done by those who have had no special training whatever, provided that they are of ordinary intelligence and use and cultivate their powers of observation, and work diligently upon some selected and definite question.

The results, 'usually depend more upon the willingness to take pains, than what is termed genius or brilliancy of intellect'.[23]

At the University, his tenacity began to pay off when in March 1889, Liversidge was given two slices of the Challis income – £7500 towards his chemistry laboratory, and £800 for a new chair in 'geology and palaeontology with mineralogy'.[24] The latter was the keystone in the arch of a mining school. Politically, it was justified by the curriculum agreed with Arts. Physiography, one of the three science courses surviving the cutbacks of 1887, was still required of all first-year Arts students. Liversidge and Stephens had shared the burden of lectures, but with a class numbering now over forty, there was a case for an additional post. When the first Challis appointments were made in March 1890, Stephens' Chair in Natural History was divided. W.A. Haswell was made Challis Professor of Biology, and Stephens became Professor of Geology and Palaeontology. Some protested that all new chairs should be openly advertised, but Stephens' preferment was a *fait accompli*. In the event, his untimely death only six months later made the appointment of a professional geologist even more urgent.

To cover Stephens' geology lectures, beginning in March 1891, Liversidge turned to T.W. Edgeworth David, a brilliant young Welshman, Oxford graduate and Associate (1878–79) of the RSM. Since coming to Australia in 1882, David had been employed by Wilkinson and the NSW Department of Mines, where he had won a reputation for surveys ranging from the Vegetable Creek Tin Fields in the New England district to the coal seams of the Hunter Valley.[25] He was an heroic figure in Maitland (where two of his children were born), and his discovery of the Greta Coal Seam at Deep Creek proved to be one of the 'richest ever made in eastern Australia'.[26] Liversidge invited David to be an examiner in geology, and in December 1890, asked him to speak to the Royal Society on his coal discoveries. He had a high opinion of the young man's abilities,[27] and was pleased to learn that he was willing, at short notice, and in addition to his official duties, to deliver the lectures to Stephens' students.

When David agreed to provide a temporary solution, Liversidge advertised the geology chair. In December 1890, Senate appointed an interviewing committee in England, as usual co-opting the NSW Agent-General in London.[28] At Liversidge's insistence, the scope of the chair was broadened to 'Geology and Physical Geography'.[29] Among the applicants was W.J. Sollas, Liversidge's laboratory companion at the RSM and Cambridge. For Liversidge, an old friend vied with a new one, in a competition that was to be decisive for Sydney.

At the close of 1890, Liversidge sailed for the third time across the Tasman to attend the third Congress of the AAAS, the first to be held in the sister colony.[30] It was an *Australasian* Association, after all, as Sir James Hector reminded his guests ' ... that aims at, has already achieved, the creation of a common interest in the scientific work of the colonies'.[31] Arriving in Christchurch, Liversidge was welcomed by the Otago Institute, where some 200 New Zealanders – many of whom had views on the federation debate – heard Hector proclaim how cooperation in science offered 'the first truly effective step towards Federation which has yet been achieved'.[32] He paid special tribute to the trans-Tasman friendship begun by Liversidge's first visit in 1877:

> Politicians should take this well to heart. Let them continue to aid all efforts that will tend to bring scientific accumulations in these colonies into a common store, so that each may discover for what purpose it has been best adapted by Nature, and [by which] each may prosper to the full extent of its natural advantages.[33]

Liversidge thanked Hector for organising a meeting 'infinitely better' than Sydney's;[34] and the results seemed to speak for themselves, when the Premier of New Zealand, suitably hectored, granted the Association £750 for expenses, free printing, travel passes, government assistance with maps and books, and leave for government officers to attend the meetings.

For Liversidge, the Congress was a personal success. The New Zealand Institute elected him an honorary member.[35] He was deeply moved. 'I value my membership with you', he later wrote, 'more than anything else I possess except the FRS, although I ought perhaps to rate it even higher than that, since there are 500 of the one and only twenty of the other'.[36] Liversidge was in his element with the naturalist F.W. Hutton, and the two spent hours talking in the Christchurch Club.[37] Unusually, Liversidge delivered no scientific papers – a measure, perhaps, of his near-total absorption in university administration.

In its scientific program, the AAAS was less successful. Poor communication between the colonies made cooperation difficult, and although 550 people attended, six of the committees of investigation set up in 1888 were deemed to have lapsed. Still, Liversidge refused to admit defeat, and pressed the Council to set up more new committees – this time, on sanitation, glaciers, and the chemical composition of mineral waters. There was discussion of Antarctic research and the commitment of the AAAS to polar exploration. And if David Masson did not himself come from Melbourne, he had read on his behalf a brilliant paper on the theory of solutions, in which he anticipated the concept of 'solution temperature'.[38]

In February 1891, Liversidge returned to Sydney, determined to help the AAAS keep its early start. Working closely with Edgeworth David, he grew increasingly to depend upon the younger man. David's charisma and energy were attributes that Liversidge envied and cultivated. In March 1891, Liversidge brought him onto the Council of the Royal Society, where the Welshman soon charmed even the most querulous *savants*. In May, David was offered the chair of geology.

As Liversidge might have expected, the news stunned London. The appointment of a local over a British candidate, *after due advertisement* in England, was almost unprecedented.[39] The Committee had unanimously recommended Sollas for the job.[40] But Liversidge stood his ground, and argued that 'an advanced acquaintance with, and research into the Geology of Australia, together with a special readiness and aptitude for the study of Field Geology, of which the development is so obviously demanded in a new country', decided the issue.[41] David's brief covered the many in Arts as well as the few in Science, and his Oxford education and general knowledge, combined with his lucidity and lecturing skill, made him the ideal man for the job.

With David's appointment, Liversidge's School of Mines came a step closer to realisation. But David's timetable was soon full, so Liversidge renewed his requests of 1889 for junior appointments – including a lecturer in mining and a lecturer/demonstrator in metallurgy and assaying.[42] To show goodwill, he asked Senate to appoint them both on an annual and renewable basis, 'until the success of the course is assured'.[43] When months passed, and nothing happened, he again approached Senate, returning to the larger vision, which he had put to Manning in April 1889 – and asked Senate for £1500 per annum to establish new lectureships in not one, but three new subjects – namely, architecture, mining and agriculture.[44]

At a stroke, Liversidge lifted the particular question of mining education to a broader level of professional training. Since their resignation from the Board of Technical Education in 1888, both he and Russell (who had since become Vice-Chancellor) had tried to claim a role for the University in higher technical education. Now, he, Russell, and Barff went as a delegation to Joseph (later Sir Joseph) Carruthers,[45] Minister for Public Instruction in Parkes's last government, who had abolished the Board of Technical Education, and brought its activities under his department. At first, despite his sympathy for the University (from which he had graduated, BA 1876, MA 1878), and his membership of Senate, Carruthers resisted the idea. Then aged only thirty-three, but destined to have a long and controversial career in colonial and state politics, Carruthers wanted 'a centralised system, to include the independent schools and technical education, with the university at the apex'.[46] He had taken courses from Liversidge, and knew the men now asking his help. However, he warned them that the University risked duplicating the efforts of the Technical College. Officially, he said that his ministry would find money for mining education only if the University would 'harmonise' its operations, assist in the teaching of applied subjects at the College, and pursue 'original research'.[47]

The Minister's response was not unreasonable. But what Liversidge had proposed to Manning in 1889 was a professional school, not an extension of the Technical College. As he saw it, the University had a duty to teach the whole domain of science to managers and engineers; to offer technical instruction suitable for foreman miners and shift captains was quite another matter, which the University might encourage but could not greatly assist.[48] However, Liversidge welcomed Carruthers' encouragement of research. In physics and biology, thanks to Threlfall and Haswell, research was well underway at Sydney. But in mining, Liversidge could do little without more resources. Just managing the chemistry laboratory, which had opened that year, was a full-time job.

In August 1889, during the curriculum debates, Liversidge had gone back to Senate for more help from the Challis bequest, proposing a Challis Lectureship in Mining and a Demonstratorship in Metallurgy, based (as he carefully minuted) in the Chemistry

Department.[49] He lost the first, but carried the second by a Senate vote of eight to five. This did help, but also propelled Liversidge into the personal problems of managing a growing staff. His first demonstrator in chemistry, Dr Albert Helms, resigned in December 1888, amidst some acrimony, after eight years' service.[50] Helms' successor, Edgar Hall, FCS, arrived in 1888, but when he left in July 1889 to work for a mining company,[51] Liversidge had quickly to find a replacement. Fortunately, he found F.B. Guthrie, son of his old friend Frederick Guthrie, Professor of Physics in London. Guthrie senior had nominated Liversidge to the Physical Society of London in 1876.[52] Guthrie junior followed in his father's footsteps, taking a BSc at University College London, and spending a year in Germany, before returning to teach at the Royal College of Science. In Sydney, he was a welcome addition to Sydney's staff, and Liversidge enrolled him in the Royal Society as soon as he arrived.

Alas, in September 1891, after only two years, Guthrie left Liversidge to become the first analytical chemist appointed by the NSW Department of Agriculture. There he began to work on wheat rusts.[53] In 1898, William Farrer, grandly titled 'Wheat Experimentalist', joined the same department, and together Guthrie and Farrer produced the revolutionary 'Federation' variety of wheat. Released for commercial cultivation in 1902, this remarkable product of applied science transformed Australian agriculture and the world wheat market.[54] It also emphasised the commercial value of research,[55] a lesson Liversidge was quick to press upon whoever would hear. He refused to let Guthrie go completely, and called upon him repeatedly over the next eighteen years.[56] But in the short term, Guthrie's departure dealt a blow to his ambitions to set up a research laboratory, with a research group that he could call his own.

In 1891, Liversidge published only one short paper, and that just an addition to his collection of mineral descriptions.[57] Although he had given up the honorary secretaryship of the Royal Society in 1890, his successors passed many routine matters to him, 'not all of ... a pleasant character'.[58] Moreover, the administration of the AAAS, and preparations for the fourth Congress at Hobart, made heavy going. As he wrote to Thiselton-Dyer in London, 'I wish we could have a permanent *Assistant* Secretary'. But a permanent officer would have to sit in Sydney, raising inter-colonial jealousies that the AAAS had not solved. 'I do not think the other colonies would agree to it', he said; and he was probably right.[59]

If Liversidge, as Dean, were to encourage research, he needed energetic workers. In chemistry, to replace Guthrie, he appointed James Schofield, a promising scholar recruited from the RSM in 1892. In geology, Edgeworth David turned out to be exactly what he hoped. By August, even before taking up his chair, David took a leaf from Threlfall's book, and submitted to Senate a comprehensive statement of his needs. David assured Senate that the colony's economy and the University's interest would jointly benefit from increasing expenditure on geology and mineralogy. He reasoned that Liversidge could not be expected to teach the whole of the University's mining courses by himself – 'at considerable personal inconvenience', to put it mildly.[60] David won this, his first battle. On a motion from the ever-helpful Russell, Senate agreed to consolidate geology teaching in his hands. 'Mineralogy' was dropped from Liversidge's title, and went to David; from 1891, Liversidge was 'Professor of Chemistry' alone.[61] While hardly foregoing his interest in minerals, Liversidge accepted a separation of powers that he knew was critical to the future of a mining school.

In the coming months, David learned much from his master tactician, and enjoyed Liversidge's close support. Shortly after his appointment, David was given approval to appoint a Demonstrator in Geology – Smith in earlier times had to wait seventeen years for a demonstrator, and Liversidge, eight – but now, remarkably quickly, came W.F. Smeeth, fresh in 1893 from London and the RSM. Thanks to Liversidge, the Faculty of Science was soon bursting with South Kensington men. The first lecturer in metallurgy, assaying and chemistry (S.J. Speak, appointed in July 1892) was nominated by W.C. Roberts-Austen of the RSM and T.E. Thorpe of the Royal College of Science.[62] Another acquisition was Edward Pittman, Melbourne-born, but RSM trained, with a helpful year of experience at the School of Mines in Bendigo. Both to save money, and to get the benefit of practical expertise, Pittman – since 1882, Wilkinson's successor as Government Geologist – came on board as a temporary lecturer in mining from the beginning of 1893.[63] Their team assembled, Liversidge and David lacked only a building in which to teach.

Manning, despite his liberal credentials, and his profession of liberal causes, could not always be expected to help. Around the same time that he approved the establishment of Women's College, he expressed reservations about mining. In 1892, the University had a staff of forty-four and 598 students, of whom ninety-nine were women. Nearly three-quarters of all students were enrolled in Arts, and the next largest group was Medicine (sixteen). Science enrolled only three per cent.[64] The case for a new building was not overwhelming.

But Liversidge was persuasive, and chance intervened. The order of events is unclear, but towards December 1892, Manning's attention was drawn to a sum of £10,000 lying unexpended on the government's vote for the Mines Department. This item had been carried forward to provide for assays of minerals and the establishment of schools of mines throughout the colony. However, no one had acted upon the proposal, and the vote would lapse if left unspent. In the circumstances of 1893, Manning could not let such an opportunity slip, and moved Senate to apply for the unexpended vote – £1300 for models and diagrams, and £8000 to accommodate the new Professor of Geology and Mineralogy and his lecturer in mining.[65]

Manning's proposal went to the desk of Carruthers, still Minister of Public Instruction. Carruthers looked to Melbourne where, pressured by the Victorian Government, the Ballarat School of Mines and the University of Melbourne had reluctantly entered into an affiliation three years earlier. This was to last only until 1894, but in 1893 the model seemed a good one.[66] Carruthers countered Manning's submission with a proposal to create a similar amalgamation between the University and the Technical College. As this meant a rejection of the idea of regional centres, the way seemed open for Liversidge's preferred solution of a central mining school in Sydney.[67]

In late 1892, Liversidge had asked his old friend Harrie Wood, Undersecretary of the Mines Department, to intervene. Wood summoned an impromptu committee, consisting of himself, David, Edwin Johnston (Undersecretary for Public Instruction), Frederick Bridges (Carruther's Superintendent of Technical Education), and Edward Pittman (representing the Geological Survey). The five recommended that the unspent money go without delay towards establishing a School of Mines in association with the University, with new buildings and courses organised 'on such lines as not to overlap the

'The Mineral and Fossil Collection, Australian Museum',
*The Australian Field*, 29 July 1899.

PLATE II.

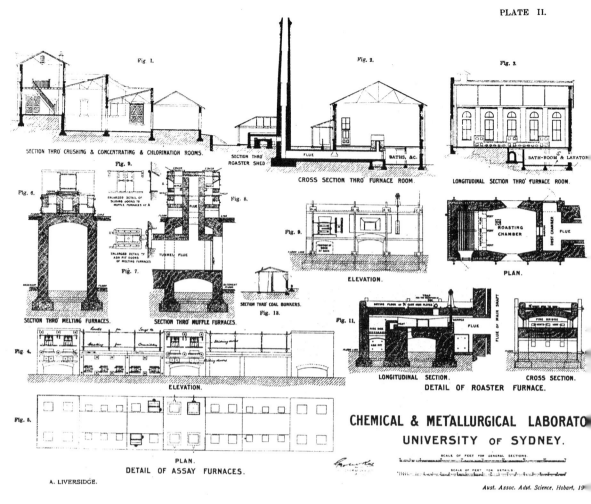

Fig. 1.

SECTION THRO' CRUSHING & CONCENTRATING & CHLORINATION ROOMS.

SECTION THRO' ROASTER SHED

Fig. 2.

FLUE

BATHS, &c.

CROSS SECTION THRO' FURNACE ROOM.

Fig. 3.

BATH-ROOM & LAVATORY

LONGITUDINAL SECTION THRO' FURNACE ROOM.

Fig. 6.

Fig. 9.

ENLARGED DETAIL OF SLIDING DOORS TO MUFFLE FURNACES AT A

Fig. 8.

FLUE

ENLARGED DETAIL OF ASH PIT DOORS OF MELTING FURNACES

Fig. 7.

TUNNEL FLUE

Fig. 9.

ELEVATION.

ROASTING CHAMBER

FLUE

PLAN.

SECTION THRO' MELTING FURNACES.

SECTION THRO' MUFFLE FURNACES.

SECTION THRO' COAL BUNKERS.

Fig. 13.

Fig. 11.

Fig. 4.

ELEVATION.

LONGITUDINAL SECTION.

DETAIL OF ROASTER FURNACE.

CROSS SECTION

Fig. 5.

PLAN.

DETAIL OF ASSAY FURNACES.

A. LIVERSIDGE.

CHEMICAL & METALLURGICAL LABORATO

UNIVERSITY OF SYDNEY.

SCALE OF FEET FOR GENERAL SECTIONS

SCALE OF FEET FOR DETAILS

Aust. Assoc. Advt. Science, Hobart, 19

Plan of the chemical and metallurgical laboratories, about 1902.

teaching imparted at the Technical College, but [to focus upon] the higher branches of the subject at the points where the Technical College teaching left off.[68] Wood's offer was conditional only upon the University agreeing to waive its customary matriculation rules, so as to let Technical College students attend its courses.[69] This perfectly suited Liversidge, who, in his capacity as Dean, manoeuvred the proposal through Senate.[70] The University's waiver was subsequently extended to all students who could afford the annual fee of £130.[71] At last, for at least this minority, the matriculation hurdle had disappeared.

The next step was to build. Calling upon European and American examples, Liversidge submitted plans to Senate's Building Committee and the Government Architect for two new buildings, a furnace room and a milling building.[72] In October 1893, tenders were accepted for completion in twelve months at a cost of £4600.[73] It had all come together just in time. Earlier in the year, the Great Depression had begun – the colonial economy was moving towards a state of panic, as banks closed, revenues collapsed, and T.M. Slatterly, Minister of Mines, found his departmental vote cut by £62,000. Miraculously, Liversidge's building went ahead. In May 1893, the Minister reported proudly that 'the School of Mines was an accomplished fact ... within the University of Sydney'.[74]

In 1894, Liversidge began the first complete course of lectures in the School of Mines – just over twenty years since his famous letters to the *Herald*.[75] Classes began in the Macleay Museum, but by the end of the year, the Department of Geology opened its new building (the 'Old Geology' of today) on the site of the former medical school, facing Liversidge's chemistry building. These were complemented by two new furnace and milling rooms, located across the road, next door to chemistry. Here, Liversidge brought in heavy equipment to crush, concentrate and treat quantities of ores for analysis, and to demonstrate processes for the extraction of gold and silver from their ores.[76] The result was what the Department of Mines was pleased to call a 'well-graded system of scientific education for ... mining engineers and mining managers',[77] whose contribution could not 'fail to effect improvements in our methods of mining and treatment of ores'.[78] Its fame soon spread to England. At last, reported *Nature*, 'the colony [could] offer as complete and effective a course of training in mining and metallurgy as can be obtained in Great Britain'.[79] Sydney students finally had workshop training comparable to that available to Victorians since the 1870s.[80]

The summary effect was indeed impressive: along what was beginning to be called 'Science Road' was a set of buildings serving several disciplines, unique in Australia and on a par with the best universities in the world. Liversidge proudly described the 'complex' to the AAAS at Hobart in 1902.[81] Siting the Mining School next to chemistry not only saved money, but also forged an easy link between the two. Students taking the new Bachelor of Mining Engineering degree received lectures from Liversidge, David and Pittman; those working for a master's degree went on to proof of practical field training. Conversely, students from the Technical College, who had already taken two years in practical mining, could take science subjects at the university. The Technical College, the university and the government thus solved a problem that had resisted solution for twenty years – offering what the Minister called 'a maximum of efficiency ... attained at a minimum cost'.[82]

Liversidge's victory should have been complete. But it was not, or not quite. Almost at once, the new School of Mines came to represent the triumph of commerce over culture. Large private companies, which continued to recruit mining experts from overseas, now had an Australian alternative. As late as 1901, William Knox, president of the Victorian Chamber of Mines, complained to the Royal Commission on Technical Education in Melbourne that Australian mining schools were not producing suitable graduates: 'it is', he said, 'a disgrace to our country that we have not the means of securing these positions for our own men'.[83] But Sydney had already risen to the challenge. Of the fifty who had graduated in engineering by 1900, twenty were in mining engineering, and most went to jobs paying between £750 and £1000 a year – earning as much or more than those who taught them.[84]

This was all well and good, but it was not the solution for science. By the early nineties, Liversidge began to resign himself to the prospect that students in pure science would never (or at least within his lifetime) be as numerous as he wished. Students who took chemistry and geology for the BEng degree soon outnumbered those who took the same subjects (and much else) for the BSc.[85] With Anderson Stuart, Liversidge was obliged to agree that 'the Science Degrees, except in the Department of Engineering ... have been failures practically. People do not understand them sufficiently; they apparently lead to nothing; and from a worldly point of view are, therefore, worth nothing'.[86] The plight of the BSc was a sad commentary on colonial circumstances. Still, thanks to the popularity of engineering and mining, Liversidge could now offer composite degrees in geology, mineralogy, chemistry and metallurgy – a curriculum almost as broad as he had followed at Jermyn Street, and then available at the Royal College of Science. And from this benefits would flow. In 1899, the School of Mines made a successful bid for a new demonstrator in assaying and chemistry, and the following year, won two additional demonstrators in chemistry, who could also help teach pure science.[87]

For the next decade, however, and until the First World War, mining and its cognate subjects were the prominent features of Sydney's scientific image. At best, the School of Mines gave a fuller dimension to Liversidge's network, and made Sydney a minor mining metropolis.[88] 'Mining men' and collectors now sent their specimens to Sydney, just as Murchison's men had earlier sent their specimens to London. One such was Charles Marsh, who, fossicking at the fabulous mines of Broken Hill, sent Liversidge a specimen of the first mineral new to science to be found in NSW – a copper oxide, which Liversidge promptly named *Marshite*, in honour of its finder's 'zeal and diligence'.[89] Some specimens Liversidge sent overseas, notably to the Natural History Museum in London and the Mineralogical Museum in Cambridge, but most he kept for himself.[90] In 1893 and 1894, while working in his laboratory, Edward Pittman 'struck it rich' scientifically, in producing a description of the second new mineral species to be discovered at Broken Hill – a sulphantimonide of cobalt and nickel called *willyamite*.[91] A new tradition of scientific mining and metallurgy had begun.

At the same time, the School of Mines completed Liversidge's building ambitions. From his office in chemistry, across the pastures leading to Newtown, he could survey the fine crenellated 'castles' that he and his 'scientific knights' had built, extending north and south of the narrow path west of the Great Hall. Side by side sat Liversidge's chemistry and mining, Threlfall's physics, David's geology, and Warren's engineering. Haswell had still to find a permanent building for biology, but had use of the imposing Macleay Museum. If, in England, the Royal College of Science, now embracing the RCC and the RSM, was still the model, Sydney had become a respectable and not much reduced version. And in technical education, if South Kensington had come to Ultimo, a little of London had come to Glebe. If a visitor required a monument, *circumspice!*

For Manning, the achievements of science were a mixed blessing. On the one hand, the University had grown. The eighty-one students of 1882, when the Faculty of Science began, had become 376 in 1891. The Faculty may have enrolled only three per cent of all students, but its teaching affected everyone, and its buildings now dominated the site. Perceptibly, the University's centre of gravity had shifted away from Arts. Despairing voices from classics and languages could be heard. On the other hand, Manning cannot have been displeased to see the University getting a favourable press. Somehow, a combination of Liversidge's foresight and Manning's caution had produced an outcome that few in government and society chose to question.

Given this hopeful prospect, the early days of 1894 proved to be doubly cruel. In the year of its birth, the School of Mines, and much else at the university would be placed in grave jeopardy, as the colony reeled before the worst economic depression in living memory.

## 2. The 'Nervous Nineties': University and Colony in Crisis

If the 1880s and early 1890s saw unprecedented prosperity – signalled in Melbourne and Sydney by the great public buildings in Swanston Street and Bridge Street – the mid-nineties saw the reverse. A cyclical world recession in Europe and America burst the bubble of investment, and sent the Australian commodity and financial markets into deep decline. Metropolitan Sydney, once apostrophised by Dilke, refound the squalor of Dickens. Jules Archibald's *Bulletin* became an oracle of aggressive republicanism and civic socialism. Just as the School of Mines opened, Liversidge entered a decade of disharmony. For him, the crisis was both personal and professional.

In January 1890, when the AAAS held its second Congress in Melbourne, there were signs of growing unease. Two years earlier, the Council had set up a committee to enquire into the conditions of labour, with special reference to strikes, and 'to make suggestions for their remedy'.[92] But no remedies had come forward by 1890, or even by 1893, and the Association retreated from further economic debate. Domestically, its finances were insecure, and Liversidge had to wait nearly a year before the NSW Government paid its promised £500 to publish the Congress's first proceedings.[93] In the hard times ahead, the survival of these 'festive occasions' – or what the *Australasian* called 'scientific gaieties'[94] – would depend upon their success in attracting larger audiences. In 1890, the Melbourne *Argus* thought the Association still an 'experiment'; it remained to be seen 'how far its proceedings would stimulate professional students of science, and enlist the sympathy of the general public'.[95] By the mid-1890s, the situation was little better.

Whatever future awaited the AAAS, the Royal Society of NSW was in equally parlous circumstances. In May 1889, scarcely recovered from the inaugural AAAS, Liversidge agreed to be the Society's President for the second time. His 'election' gave him an excuse to resign from the secretaryship, and from its considerable day-to-day correspondence. Thanks to his stewardship, the Society had prospered, and its freehold, well-furnished premises now boasted a library of 1600 books and 200 journals. Its reserves stood at £500. But these good times would not last. In 1890, the Government ended its open-ended commitment to free publication, and substituted a grant, which increased from £400 to £500, but only on condition that the Society match it with subscriptions.

This was a matter of concern. Membership had slipped from a peak of 494 in 1885, to 461 in 1890. In hard times, people were less likely to join, or to renew their subscriptions. And while it was pleasing that a larger proportion of the Society now held scientific credentials (rising from forty-one per cent in 1880 to fifty-three per cent in 1890), this was offset by the departure of amateurs on whose subscriptions the Society depended.[96] Within the small world of colonial science, there were large egos. It was said – with what truth, it is difficult to say – that Sir William Macleay's 'overbearing purse' had divided the community, distorting its priorities, and forcing members to choose between the Royal and the Linnean. 'His feelings are not particularly kindly to me or to any one connected with the Royal Society of New South Wales', Liversidge wrote Thiselton-Dyer, who knew them both.[97]

The Society's financial condition spoke also to its reputation. After all was said and done, Liversidge admitted, the Royal had yet to make a name for itself. 'Much that we have to publish', he confided to Thiselton-Dyer, 'is very poor stuff'.[98] But he never said this in public. On the contrary, his messages to the troops were ever optimistic. 'It may be thought by a few', he announced in his Presidential Address in 1890, 'that we ought now to rest and be thankful; but we are not banded together for that purpose. The objects of the Society will not be advanced if we fold our hands'.[99] To maintain momentum, he pressed the Society to take on a set of projects, some new, some of older vintage – to build, for example, a new Marine Biological Station at Watson's Bay, to sponsor an Antarctic expedition, and to commission a study of Pacific coral atolls and chalk beds.

To share expenses, Liversidge proposed a comprehensive solution – a proper building – a 'modest edition of Burlington House, Piccadilly', as he put it – where both the Royal and the Linnean could have rooms, so gaining economies of scale and rooms to rent. The same space could also house the secretariat of the AAAS, and (as we shall see) international projects for which Sydney would become a regional base. Liversidge's plan proved far-sighted, and at least until today, over ambitious.[100] But at least a conversation had begun. In May 1890, when Manning expressed a desire to have Senate meet in the city, Liversidge proposed an agreement between the university and the Society, which would provide facilities for the one, and rent for the other. Although Manning declined, possibly sensing another Liversidge conspiracy, the idea remained on the table for some time.[101]

Meanwhile, the university's accounts were taking on a Micawberish aspect. Given its private donations, it was a wealthy institution. In 1889, its annual income stood at £26,384, of which £18,800 came from government, including £12,000 for recurrent expenditure and £6800 for special purposes (with £1900 for scientific apparatus). Income from the Challis fund, whence the six new chairs came, amounted to a comfortable £7000 a year from 1890 onwards.[102] In December 1889, the Chancellor was sufficiently optimistic to propose salary increases for professors with over five year's service.[103] But by 1893, fears of recession ended all this. Manning was obliged to hesitate – as it turned out, with good reason.[104]

As early as March 1889, not long after the professors had submitted their 'Challis requests', Senate's Finance Committee, under the careful Scottish eyes of Normand Maclaurin, had produced a list of possible economies, notably in physics, medicine, literature, engineering and anatomy. There was a premonition that cuts might be needed; and these it was hoped, might at least save the University £900 a year.[105] The recommendations were not acted upon, but the following year, Manning set up a fresh Finance Committee, consisting of five Fellows of Senate, who were to be elected annually and meet monthly, and have a mandate to monitor expenditure and review estimates at the end of each academic year. This committee was to be a powerful centralising instrument in his hands, not least in offsetting the centrifugal tendencies of the Science Faculty.[106]

In late 1890, in a series of articles appearing in the *Sydney Morning Herald*, Manning presented a progressive image of the University – a venerable seat of learning, busy and productive, and accessible to growing numbers of men and women who were 'fortunate enough' to have opportunities not open to the colony's pioneers.[107] Such efforts at public relations the professors widely supported. In August 1891 the *Herald* toured

Cyanide vats in the Leaching and Chlorination Room, School of Mines.

the Chemistry Laboratory, at Liversidge's invitation, and heard his evening lecture on carbon chemistry. Normal lectures were rarely, if ever, reviewed in the daily press, but his now featured; and the *Herald* left impressed by the expertise of 'one who has his subject at his fingertips'. There was a special pleasure, the journalist added, in seeing:

> the pleasant way in which life could be spent by a person in love with the subject, and possessed of sufficient means to fit up a laboratory where original research might be uninterruptedly carried on.[108]

In October, the *Herald* returned to do a column on the laboratories themselves, this time praising the industry of Liversidge's 'practical men', his 'aproned and shirt-sleeved students … at work questioning Madame Nature on their own account'.[109]

Good reviews chuffed the Faculty of Science – now comprising the five professors, if we include Anderson Stuart of medicine and Warren of engineering – Faculty meetings were on the whole agreeable occasions. Buildings aside, there were few issues to divide them. The annual vote of £2000 for scientific apparatus in 1890 was divided equally among the four departments of chemistry, physics, natural history (biology and geology) and engineering.[110] Liversidge had a reputation for even-handedness, and within limits, all prospered together.[111]

Partly for this reason, the economic crisis, when it came, was so damaging. The Faculty, if dented by its compromises with Arts, had been consolidated by its victory with the School of Mines. Moreover, there were prospects for research. Liversidge had followed trends led in America by Johns Hopkins in establishing research scholarships. In 1883, the University of Toronto reacted to the exodus of students to the United States by launching a fellowship program of its own. As Toronto did not (until 1897) offer a doctorate degree, these came in the form of teaching assistantships. This was a decided improvement on British practice, and suggested to Liversidge a useful idea. In 1881, he had failed to persuade the University to start a scheme of research fellowships, but now he proposed a scheme of junior demonstratorships to fill this role. It was as much as he could do, against competition from overseas.

In 1889, Sir Lyon (later Lord) Playfair, who had served Prince Albert in the organisation of the Great Exhibition, was called upon to advise the Royal Commission for the Exhibition of 1851, which had been set up to administer the substantial profits arising from the Exhibition. By the early 1880s, the Commissioners had assisted institutions; now, they began to give scholarships to students, under thirty years of age, who demonstrated a special aptitude for research, and who wished to extend their scientific studies 'with a view to promoting scientific culture in the manufacturing districts'.[112] At first, these awards were to be limited to students in British universities, and to 'experimental sciences bearing upon industry'; but at the suggestion of Sir Henry Roscoe, their range was widened to embrace the colonies, and to include fundamental research. Twenty awards were to be made each year, including six to the Empire outside Britain. At £150 p.a. for two years, they were destined to help scores of men and women 'otherwise driven from research', and to 'raise standards of college and university science teaching' throughout the Empire.[113]

The significance of these studentships was quickly recognised.[114] Two could be awarded annually to Australians, and Sydney nominated its brightest and best. In 1892, twenty-one year old Henry (later Sir Henry) Barraclough, one of Warren's third-year engineers, became the first Sydney undergraduate to receive an 1851 Exhibition award, and between 1893 and 1921, this Midas touched another seventeen Sydney graduates (including one woman), of whom ten went on to distinguished careers. In principle, the scholarships were tenable at any university in the world. Some Canadians, Australians, and Englishmen chose destinations in America and Germany in preference to Britain. Barraclough, for example, went to Cornell.[115] The rising chemist, Charles Fawsitt of Edinburgh, went to Germany, where his supervisors were reportedly pleased with the 'patience and skill' of the British and imperial students they received. Indeed they should have been, Henry Armstrong, one of the Commission's advisors, wrote testily in July 1899: 'they have the use of our best material'.[116]

By and large, the tendency to look to the leading centres of Germany was resisted in favour of imperial and (by extension) trans-Atlantic solidarity. 'The idea still seems to persist', Thiselton-Dyer, Director of Kew, observed in 1902, that 'no training in research can be obtained except in Germany. This may still be debatable in exceptional cases', he added:

but as a general principle should not be encouraged by the Commissioners. Competent professors and well-equipped laboratories are increasing in this country, and it is undesirable to withhold the stimulus of research work from them. In the national interest we should encourage research at home, and discourage the notion that it is only 'made in Germany'.[117]

It remained to be seen where young Australians would choose to go. For Liversidge, the 1851 Exhibition award was a mixed blessing. Given its support, bright students could flourish; but they were swept away from their home countries, many never to return. If Australians did not return to Australia – and there were few positions to offer them – they could not do the research that Australia needed. In 1892, W.H. Hamlet, the NSW Government Analyst, repeated before the Chemistry Section of the AAAS Congress at Hobart a sentiment many preferred not to hear. 'Australian chemists inevitably occupy places in the rear guard of the advancing army of science', he said. Preoccupied with assaying, agriculture, sanitation, and what he called 'criminal investigations incidental to our rapidly growing centres of population', they were like soldiers 'at the outposts skirmishing on the frontier of the knowable'.[118]

Apart from draining Australian talent, the 1851 Exhibitions put on report the reputations of teachers as well as students. Both confronted a formidable committee that included Henry Roscoe, T.H. Huxley, Lord Kelvin and Lord Rayleigh, as well as Liversidge's friends, Norman Lockyer and H.E. Armstrong. As to the promise of Sydney's first 1851 scholars, the verdict was 'not proved'. No one was quite prepared to say whether three years of science and engineering at Sydney had fitted Barraclough for research; rather, his award was given him 'specifically to extend his education in branches of science for which there was no provision [in Australia]'.[119] Another Sydney man, Thomas Strickland went to McGill in 1897 to read engineering and fared well; but in 1898, when Joseph Durack, who took a first in physics in his year, was sent by Threlfall to work at the Cavendish on electromagnetic radiation, J.J. Thomson found in him 'no evidence ... either that he is capable of setting a problem to himself or that he has any critical power in comparing his work with other observers'. Two years later, Thompson reported, Strickland's work formed 'a record of lost opportunities'.[120]

It was painfully clear that Sydney might have a few good men, but not enough to lose. In January 1892, whilst in London on leave, Anderson Stuart was invited to lecture to the Royal Colonial Institute on the 'Australian Universities'. He explained the burdensome routines of laboratory – and museum – building required of all Australian professors, and admitted that not many had produced original research. They were unable 'to keep up with the march of discovery by reading only ... particularly in scientific matters. Much time is wasted by not knowing that what one is busily engaged in investigating has all been done before'.[121] Conditions that were once less than satisfactory in W.B. Clarke's generation were much less so when applied to modern physics and chemistry. The solution, if there was a solution, lay partly in regular study leaves, and closer contact with Europe, and in the encouragement of able Australians with an aptitude for research. Liversidge, Warren, Stuart and Threlfall had quarried the stone, and built the buildings; they had now to give them life.[122]

In the Royal Society, at the Australian Museum, and through the AAAS, Liversidge did what he could to promote research in subjects that were, if not unique to Australia, at least likely to be dominated by Australians. In the early 1890s, he tried to apply a similar strategy to the University. The archives record his impatience. In November 1892, flushed with the completion of his laboratories, he proceeded to have them outfitted, at a cost of £2600, only afterwards remembering that he should have asked the Vice-Chancellor's permission (and signature) first.[123] His new curriculum, and the prospect of having advanced students, brought in their train a demand for teachers, who in turn would need space and equipment.[124] Submitting invoices amounting to £1300 to buy new models and diagrams for the Mining School was his forthright response. And sometimes it worked. In December 1892, the government gave him the £1300 he requested, 'until a separate School of Mines under the control of this Department shall be established'; and only in February 1893, was this caveat withdrawn, when Liversidge showed that he could mount a mining course far less expensively than any government department acting on its own.[125]

In the Faculty of Science, where he was re-elected Dean every three years until 1903, Liversidge kept a steady hand on the tiller, rationalising the Faculty's teaching (requiring, for example, BSc, medical, mining and engineering students to take the same courses),[126] and introducing 'interstate' external examiners for undergraduate degrees (a practice since sadly discontinued).[127] With Threlfall, he strengthened third-year science, and encouraged honours students to specialise. His professors he urged to share the burden of lecturing.[128] Laboratory workloads were increased – to nine hours a week for second-year students, and to fifteen hours for third-years;[129] and Senate was asked (in March 1893 and again in August 1894) to require all science students to study German.[130] Senate declined, but Liversidge was unapologetic. Such steps were necessary if Australia wished to foster an élite, upon whom not only the '1851 Commissioners' but also the rest of the scientific world could draw.

It was ironic that, at the beginning of 1894, just as his campaign gathered momentum, Liversidge's hopes were dashed. There had been warning signs, and the same fears that had hastened the coming of the School of Mines, now quickened retrenchment. In 1893, the colonial treasurer had somehow managed to omit from the University's budget the scientific apparatus vote, amounting to some £2700. This could have been a mistake. But as 'Eighties' optimism' disappeared, it was a portent.[131] Owing, the Chancellor said – with sublime understatement – to the 'present and prospective depression in mercantile affairs and the reduction of the estimates by the Government',[132] there were to be no salary increases for professors. Far from it, Senate appointed a committee to review and cut the library's list of periodicals;[133] new appointments of assistants in bacteriology and pathology were postponed; and the Science Faculty was told to suspend Haswell's plans for the expansion of biology.[134] At a time when physics lacked adequate laboratory space for all its first-year Arts students, Threlfall's requests for research space met summary rejection. Physics lectures had to double up in the new Geology lecture room.[135]

Worse was to come. In 1892, the grant to the University peaked at £25,541, including £7200 for buildings and £3200 for the porters' lodges, gates and the cast iron railings – Manning's pet project – that surrounded the embattled institution. Now, the basic budget of 1893 (£13,500) was axed, and the proposed budget for 1894 (£14,700) was slashed to £8000, and again to £6000.[136] From this tragedy came farce. For 1894, £2900 was wiped from the additional endowment, with only £1700 allowed for repairs; and nothing was given for scientific apparatus at all.[137] The scientific apparatus vote, cut from £2400 to £950 in 1892, was completely suspended – indeed, it would not be reinstated until 1902 – and examiners' fees were abolished.[138] Liversidge and his colleagues were helpless to object when Senate asked its Finance Committee to consider ten ways of saving at least £1100 a year.

Manning created and chaired a 'Retrenchment Committee' that presented various apocalyptic futures. Overall, staff cooperated. For example, Liversidge's new lecturer in surveying, G.H. (later Sir George) Knibbs, who had arrived in 1891, offered to work for free 'in view of the University's financial difficulties'.[139] Several in Arts, led by their Dean, Walter Scott – who had already given generously to women's education – accepted voluntary pay cuts.[140] But this was not enough. In June 1893, Manning proposed (and Senate agreed) to put all teaching staff other than professors on six months' notice to quit.[141] Execution of sentence was stayed only by the Premier's decision to hold an independent enquiry into the University's finances.[142] When the acting Chancellor, Judge Alfred Backhouse, and the Registrar, H.E. Barff, called anxiously upon F.B. (later Sir Francis) Suttor – sheep breeder, banker, and Dibbs's Minister for Public Instruction – to protest the University's plight, they learned to their horror that the government was considering whether to suspend the University's entire grant.[143]

Desperate times made for desperate measures. We have seen that in the 1880s, Manning opposed a school of mines; in 1893, he actively supported one. Given the situation, one can appreciate his change of heart. Even while he was negotiating with the Department of Mines, the 'razor gang' of the Retrenchment Committee was contemplating reductions in the laboratory vote, the sacking of laboratory attendants, and the suspension of lectures in certain subjects in alternate years. Judge Backhouse piled Ossa on Pelion by proposing that holders of scholarships and bursaries should be bound by a promise to repay![144] Despair deepened when, despite the opposition of Liversidge, Senate approved.[145] To modern eyes, the Chancellor's only agreeable economy came with the elimination of 'unnecessary' Senate meetings.[146] Quite what savings this produced, the archives do not say.

As Dean, Liversidge was in the thick of the fighting, defending his Faculty as strenuously as he had chemistry.[147] Through July and August 1893, he consulted regularly with the other science professors.[148] As the crisis deepened, his other activities were also hit. Between 1881 and 1891, the Australian Museum's budget had varied between £7000 and £8000 a year, rising to a peak of £11,000 in 1892. But in 1893, this was cut to less than £4000; and nearly a decade would pass before this difference was restored.[149] In the meantime, its staff of thirty-four was reduced to twenty-three. The lost friends compounded Liversidge's grief, made worse by the death of Carl Leibius, from a heart attack, in June 1893, and Robert Hunt, a few months later. The middle months of 1893 saw Liversidge lending a hand at the Australian Museum, where he described new minerals,[150] edited the *Records*,[151] and helped Edward Ramsay write vain appeals to the government. But not even Liversidge could be everywhere at once, and he was pleased that, when the University Extension Board was established in December 1893 – to capture a larger constituency of students in the country districts – Edgeworth David's name was on the list, and his was not.[152]

Edward and Hannah Hufton.

In September 1893, it was a relief to escape by coastal steamer for a fortnight to the fifth AAAS Congress, held for the first time in Adelaide. This proved to be the smallest Congress ever. Overall attendance was not bad, at 488, but despite the strenuous efforts of Rennie and Bragg, only ninety visitors came, but seventy papers were read, featuring reports on ancient fossils and on the meteorological network that now linked the colonies.[153] It was as well that Liversidge came prepared, and dominated the proceedings, giving seven papers, five of which he had already presented to the Royal Society. Three were on analytical techniques, reflecting his new role as a laboratory director.[154] With the other four, he began a series on the geochemistry of gold.[155]

Returning to Sydney, Liversidge's beleaguered colleagues needed more alchemy than chemistry. By the turn of the year, the University's situation, never before so bleak, looked even worse. When Threlfall applied for a typewriter for his department, Senate had to go into committee to weigh the financial implications.[156] Francis Suttor was quoted as saying that the University 'had been liberally and leniently treated by the Government who, without having any voice in its expenditure, had handed over to it large sums of money.[157] Stunned, Senate appointed a committee to consider the implications. 'As you know', Liversidge wrote to Baldwin Spencer in Melbourne, 'times are very hard here'.[158] Senate faced a deficit of over £1300, with little prospect of savings.[159]

Liversidge kept his hand on the tiller. Throughout 1894, he fought to keep his School of Mines open, backing David and his two young RSM recruits, W.F. Smeeth and S.J. Speak. Some rationalisation of stores and apparatus was possible,[160] third-year practicals could be reduced,[161] and when Speak resigned and left for the gold mines in the Transvaal, the duties of his lecturership in metallurgy, were combined with Smeeth's demonstratorship in geology. This produced some savings; but little else could go without cutting into the core.[162] Liversidge persuaded Manning to reappoint E.F. Pittman, the Government Geologist, as a temporary lecturer on an annual basis,[163] and to part with £100 for a junior demonstrator in chemistry, a post he had sought since 1885.[164] Above all, he kept safe the job of Edward Hufton, his laboratory attendant, who had come to him from England in 1881, and who would serve Sydney faithfully for thirty-six years. His sons and grandsons would continue in three generations of service to the University.[165]

Liversidge also kept his hand in the affairs of the Technical College, and at the Technological Museum, where Henry G. Smith, FCS, was given a laboratory to work on the chemical classification of eucalypts and their 'essential oils'. In August 1893, Liversidge watched the Technological Museum move into its new brick building in Ultimo, and rejoiced with J.H. Maiden, who had been appointed its first Curator; three years later, he helped Maiden succeed Charles Moore as Director of the Royal Botanic Gardens.[166]

In early 1894, in an effort to save money, Senate appointed four new sub-committees.[167] After Finance, the two most powerful were Buildings and Grounds and the Calendar Committee – and Liversidge was made chairman of both.[168] His integrity was not contested, but other professors moved to increase their influence.[169] On Senate, Sydney's professors – now twelve in total – demanded a greater role in University decision-making.[170] The professoriate was not without influence; to his credit, Manning, backed by Russell and Renwick, represented its interests on Senate. But professors could still be out-voted. Liversidge, who had been a Fellow of Senate in his own right from 1879 to 1892, was thereafter and until 1902 a Fellow only by the rotational selection of his subject. He was thus only incidentally a member of Senate in the critical 1890s. In September 1894, a lobby from all four Faculties asked that by-laws be altered so that the new professors of mathematics, law and physiology be admitted *ex-officio*.[171] The proposal was at first defeated, but was pursued until it was won in 1898.[172] The chair of chemistry was added to the list only in 1902.

Through 1894, the depression lingered like an overcast sky. In June 1894, Manning persuaded the Premier, Sir George Dibbs that closing the University was unthinkable, and just before the government fell in August 1894, Dibbs introduced a bill to reinstate the University's endowment for 1895.[173] In the event, the bill failed to reach a second reading; a permanent increase would not come until 1902;[174] and the University again deferred all new projects indefinitely.[175] In October 1894, Vice-Chancellor Judge Backhouse, began negotiations all over again, this time with George Reid's Minister for Public Instruction, Jacob Garrard, a prominent teetotaller and trade unionist, remembered today as 'one of the best equipped members [of parliament] to respond to colonial technical, industrial and administrative change'.[176] He was also keenly interested in technical education, which as Liversidge knew, did not spell automatic support for the University. Despite Reid's support, Backhouse met only news of further reductions.[177] The University's endowment for 1895 was to be cut by a further £1500. To meet the salaries bill, all salaries over £300 (which meant all professorial salaries) were to be reduced by five per cent, and fees were increased.[178]

Battered by these misfortunes, the University was dealt a still heavier blow in February 1895, with Manning's death, at the age of eighty-four. As Chancellor for seventeen years, it was not too much to suppose that the strain in later years had helped kill him. On Russell's motion, Sir William Charles Windeyer, one of the first graduates of Sydney University, and a judge of the Supreme Court of NSW, was immediately elected to succeed him.[179] Windeyer offered continuity, and given his support for liberal causes, some sympathy for science. But few could match Manning's sagacity in making savings.[180]

Weighed down by the University's plight, Liversidge took refuge in looking overseas. On the advice of the Australian Museum's palaeontologist, Robert Etheridge Jr., Liversidge sent his minerals collection in 1894 to the World's Fair (the 'Columbian Exposition') at Chicago, winning a prize that he shared with the Museum.[181] His spirits were lifted in May 1895, with the announcement of a third 1851 Exhibition award to a Sydney candidate, and the first to one of his nominees, twenty-six year old John Alexander Watt.[182] But gloom returned in July 1895, with threats from Macquarie Street of a further reduction of £1000 in the University's grant for 1896, giving a budget of only £9000, a free fall from the £15,000 of 1891. With its genius for outrage, the student magazine, *Hermes*, proposed the simple expedient of closing whole Schools altogether – beginning with Mines, which might save very little, but also Law, which would save much more.[183] To their credit, the Chancellor and the Registrar ignored the sanguinary advice. On the basis of 'equal suffering', all staff, regardless of status, were awarded a continuing five per cent pay cut.[184] The cuts stayed in place through 1897.

In September 1895, Senate vainly pleaded with George Reid to restore the endowment, and even to add another £6000 a year, as Dibbs had recommended.[185] Instead, it lost an additional £400 in special funds.[186] But Reid held the line as best he could, and the budget for 1895–96, set to plummet further, stopped when it reached £9000 and £800 for repairs. The £2000 annually allocated for evening lectures was never touched, but in every other respect the crash impinged upon the community, as fees rose and enrolments fell. Some 182 undergraduates matriculated in 1891, but there were only 118 in 1894, beginning what would become a steady decline until the end of the decade. The Faculty of Science, which had only twenty students in 1892, shrank to four in 1894. Engineering struggled to get between fourteen and thirty; Law held its own, but with fewer than forty-five; only Medicine showed an increase, and this was unsure from year to year. As the evening money remained intact, Liversidge proposed a reduction of general fees for evening students;[187] and to salvage numbers Anderson Stuart prophetically suggested spending £10 to advertise the University's professional schools to the 'overseas student' market in India.[188] Both proposals, for opposite reasons, Senate declined.

With financial austerity came a new puritanism, focused on the particularly senseless forms of indiscipline that had become rife among undergraduates at Sydney (just as, observers noted, at Cambridge and Oxford). From the late 1880's, the Professorial Board had issued anti-larrikin rules against student demonstrations, and during Commemoration, but it was a losing battle.[189] At the 1894 Commemoration, Sir Robert Duff, the Governor, made such a poor speech that the professors 'trembled and figetted lest the students should get too unruly and uproarious as at times they threatened to do'.[190] To absorb youthful energies, the University offered sports, and the grounds of Victoria Park had been improved in 1886, and the playing fields extended for the benefit of 'mainly exercise and recreation'. But football (like tennis and golf) was too easily played next to the stained glass of the Great Hall and the Blacket Building. The Professorial Board rarely viewed bad behaviour with good humour, and Senate often handed down fines and punishments.[191]

For Liversidge, whose stammer and temper grew worse with age, violent 'rags' were appalling.[192] His record on student affairs had once been conciliatory. He had, for example, welcomed the coming of a student union in 1874, recalling that he had helped establish the London Union Society in 1869.[193] But little good this did him in the lecture room. In 1892, Liversidge reported 'breaches of discipline' in his first-year class. One J.A. Curran had thrown 'certain things' around the lecture room, one of which struck Liversidge in the face. The 'thing' turned out to be a 'berry', aimed at another student. Liversidge was an unintended target, but he was outraged. Walter Scott for once agreed, and on his motion, the Professorial Board suspended the hapless marksman for a month.[194] Liversidge took such things personally, even when he was not the target, and introduced a system of fines and penalties that the Board used for the next twenty years. To ensure 'discipline and good order', the Board set up a committee, of which Liversidge was a member.[195] In May 1897, following a particularly rowdy commemoration, his committee and the Board resolved to exclude all students from the events except those actually receiving degrees.

By the middle of 1895, Liversidge – and many others – were ready to take leave. Possibly the only bright news the University received that year came in November, when Peter Nicol Russell announced the first of two donations of £50,000 to the School of Engineering. That was a happy day for engineering.[196] But for the Faculties of Science, Arts, Medicine and Law, dependent on a government that ruled with an iron whim, the year ended not with a bang, but a 'bust'.

## 3. Glittering Prizes: The Geochemistry of Gold

Once asked to reflect on the career of Lord Curzon, Churchill described the Viceroy's morning as golden, his noontime as bronze, and his evening, as lead. Liversidge was a metallic metaphor in reverse. His interests during the 1870s in the location of coal and industrial minerals such as tin, iron, and nickel, slowly transmuted into a preoccupation with gold. By the 1890s, his new laboratory gave him a welcome escape from the turbulence around him, and he returned to theories of ore genesis and mineral structure, and the problems of solutions – in this case, 'mineral solutions' – that had occupied him as a young man. It is one of the few pleasant ironies of 1895 that, just as Liversidge found less money at his disposal, so his interests turned to gold.

From his days at the Royal School of Mines, Liversidge's life had been interwoven by threads of gold. An early interest in London, possibly nurtured by Daintree, blossomed in Sydney, when he met W.B. Clarke, and learned about gold in Australia. Liversidge collected references to Australian 'finds' with bibliophilic intensity, and presented his results to the AAAS at Brisbane in January 1895.[197] With Leibius at the Royal Mint, he discussed methods of ore refining. Of the three ways of extracting gold from its ores, the ancient and expensive method of amalgamation had long been replaced by the Wilson method of chlorination, a process which eliminated traces of silver, but left other impurities, the detection of which challenged laboratory techniques.[198] In 1888, experiments with the new Macarthur-Forrest cyanide process at Ravenswood in Queensland, conducted in secrecy by the Cassel Company of Glasgow, were followed by operations at Cassel sites in New Zealand.[199] By 1891, this process had come into general use in South Africa, retrieving gold with higher efficiency, and from ores once thought uncommercial.[200] In the mid-1890s, the technique passed to Australia, and companies that resisted the innovation faced severe competition.[201]

Finding better ways to retrieve gold from its ores was of growing international economic importance. For well over a decade, the discovery of new reserves had become a matter of global concern. The Australian colonies, which had been producing the greater part of the world's gold since the 1850s, were being overtaken by Russia, the United States, and eventually South Africa. Victoria had reported no new finds in ten years; output from NSW and South Australia had fallen; and only Queensland, with Mount Morgan, was increasing its production. In 1886, when the total annual value of Australian gold production was about £5 million, the value of American production reached $35,000,000 (£7 million). This proportion would change following massive discoveries in Western Australia; but in the late 1880s, the British government was deeply worried.

In 1887, William Topley, of the Geological Survey of Great Britain, warned the Chemistry Section of the BAAS Congress in Manchester that the decline of gold production in the Empire had important consequences for the world as well as for Britain. A Royal Commission on the Depression of Trade, reporting in 1888, confirmed his fears. Alluvial deposits were becoming exhausted, and most of the early diggings in NSW had been worked out. Future supplies must, Topley believed, be sought in 'the mining and metallurgical treatment of ores containing [only] small amounts of gold and silver'.[202] Mining companies had turned towards deep quartz veins and reefs. But their practices were often wasteful and ill-informed. The location of deep veins depended on methods of survey and exploration little changed from the days of Murchison and Clarke.

Liversidge's analyses of mine tailings for the Mines Department suggested rich rewards awaited the applications of science. His interests were partly technical and partly philosophical. The chemical laboratory had come to be viewed as an all-powerful instrument of commercial progress. It was not unreasonable to think that knowledge of the ways by which ore bodies are formed would bring better ways of finding them. But such knowledge could also contribute to a better understanding of the earth and its morphogenetic history.

His interests converged on two related questions: how was gold deposited, and how did it come to be where men found it? Not far beneath lay a larger question: what evidence did chemistry afford of evolutionary change? At Liversidge's suggestion, the subject was repeatedly debated at the Royal Society of NSW. In 1887, an unknown author,

Students at work in the Metallurgical Laboratory.

Jonathon Seaver, won the £25 essay prize given by the Society that year for his paper on the 'Origin and Mode of Occurrence of Gold-Bearing Veins and of the Associated Minerals'. The age-old questions remained. Had gold been carried in solution and deposited in veins from circulating meteoric waters heated by volcanic or metamorphic activity, or from aqueous fluids derived from magnetism; or had it been thrown up and concentrated by volcanic or plutonic action from deep within the earth's crust? Liversidge looked for evidence from structure and solubilities. As a contemporary American put it, 'by solution ... the chemist is able ... to imitate the chemistry of ore deposits as carried out in nature'.[203]

From 1889, Liversidge devoted part of his new laboratory to a series of experiments on gold chemistry, directed mainly at processes he felt relevant to the earth's surface, rather than those associated with ore formation. The first of his 'gold' papers – 'On the Removal of Gold from Suspension and Solution by Fungoid Growths' – was based on experiments begun as early as 1881. It was presented initially to the Royal Society of NSW in September 1889, and subsequently to the second Congress of the AAAS in Melbourne in January 1890. In this, he discussed the possible role of 'mould' in precipitating gold from suspension or even from solution by reduction. He experimented with a number of moulds, including *Penicillium*, and other organic substrates, including bread, carefully detailing his observations, and anticipating modern interest in the role of bacteria and other biota in precipitating gold.[204] The following year, he considered the composition of gold appearing in association with other minerals, analysing specimens of auriferous haematite found at the Mount Morgan Mine in Queensland. His discovery that gold was more closely associated with the skeletal silica than with the haematite, may have raised questions about the 'geyser theory' of ore deposition. But, on balance, Liversidge preferred theories of thermal origin.[205]

With Charles Wilkinson's death in 1891, Edward Pittman took over the Geological Survey, and brought his analytical work to Liversidge's laboratory when he was appointed a temporary lecturer in the School of Mines. In 1892, Liversidge conducted experiments to determine the occurrence of gold in the Hawkesbury sandstone and shale surrounding Sydney. W.B. Clarke had seen minute specks of gold in quartz and pebbles on the North Shore, but he had considered them uncommercial. Liversidge hypothesised that, as sandstone was analogous to a gold-bearing sea beach, recoverable gold in commercial quantities could exist in sands there, too. Indeed, from Hawkesbury Rocks, taken from a trench dug in the University grounds, he found sandstone samples yielding 23.8 grains of gold per ton, or the equivalent of 12/-d. per ton in value. When using the cyanide process enabled mining companies to treat ores at a cost of 3/-d. per ton, the very rocks on which the University was built could repay attention. However, this earthshaking proposition was never put to the test, and was not even published until 1894.[206]

Of more immediate application were the results which Liversidge presented in a series of five papers to the Royal Society in early September 1893, and which he repeated in Adelaide at the AAAS Congress later that month. All were reported in England by June 1894.[207] These dealt with (1) the origins of 'moss gold', (2) the condition of gold in quartz and calcite veins, (3) the origin of gold nuggets, (4) the crystallisation of gold in hexagonal forms, and (5) the appearance of gold in 'moiré-métallique'. In each case, Liversidge sought evidence for qualified and sometimes 'falsified' established conclusions.

Today, some of these papers seem more curious than controversial: his matter-of-fact prose does not easily convey his larger purpose. Nor have some of his questions yet been answered. The mechanisms that form moss gold remain a mystery, and the origins of gold nuggets have remained controversial until recent times. Many 'fanciful theories' attempted to explain how gold nuggets had been formed; but were they simply the residual products of aeons of weathering and erosion, or had they formed in place, the result of accretion through chemical action?[208]

In 1864, A.R.C. Selwyn, then with the Victorian Geological Survey, suggested that gold disseminated in rocks and drifts had been dissolved from solution, and re-precipitated around new 'centres' as nuggets. This idea was taken up by Richard Daintree, then Selwyn's assistant, who noticed that in the course of developing photographs, his solutions of gold had 'plated', possibly owing to the accidental presence of an organic substance, such as a piece of cork. These observations launched several enquiries, notably by C.S. Wilkinson in Victoria, and William Skey, analyst of the Geological Survey of New Zealand, and finally, Liversidge, to establish whether nuggets could indeed 'grow' in solution. In 1866, Wilkinson, Daintree's successor, succeeded in coating a number of minerals with gold, from which it appeared to follow that organic substances could act as precipitating agents, depositing gold from solution. In 1870 and 1872, however, Skey demonstrated that organic matter was not necessary to cause precipitation, leaving open the causal role of precipitating agents altogether.

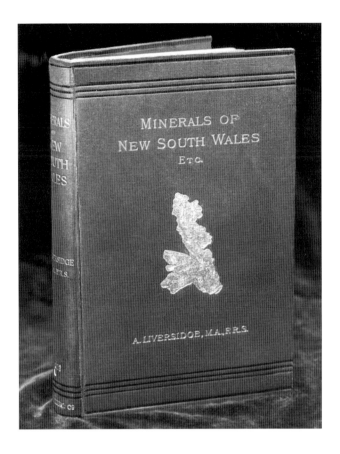

Liversidge's *The Minerals of New South Wales*, published in London, 1888.

Liversidge entered this discussion in his third paper, where he reported a set of lengthy experiments, some lasting for several years – again, the fruits of his new laboratory – investigating the ability of a range of organic materials and minerals to precipitate gold from solution. He found that gold was precipitated by nearly all experiments, either electrolytically or by reduction. He concluded that Daintree had been misled: gold had combined with the glass of his bottle, giving a precipitate.[209] Except where wood or other organic matter was present, the gold nucleus he had placed in each solution increased in weight, confirming that nuggets could 'grow' as long as they were immersed in gold solution. However, it did not follow that, as Wilkinson and Daintree had argued, large nuggets formed by accretion. They seemed instead to be set free from veins, in Rose's phrase, 'almost invariably' by the 'erosion of older auriferous deposits'.[210] This seemingly countered the arguments supporting the accretion hypothesis. Such 'mechanical' explanations reveal how complex the problem of interpretation had become.

To this problem, Liversidge applied his knowledge of crystallography. The crystal lattice structure of a mineral could contain clues to its origin and growth, and as early as July 1884, following a paper by Leibius on recent gold discoveries, Liversidge exhibited to the Royal Society at Elizabeth House a model of gold crystals, prepared from an original by Professor Archer, director of the Museum of Science and Art in Edinburgh (now the National Museum of Scotland) – a model which became his attribute, celebrated first on the cover of his centennial *Minerals of New South Wales* in 1888, and eventually, on his formal portrait, now hanging in the Great Hall of Sydney University. By the late 1880s, he had built an extensive collection of crystallographic models, many of which featured prominently in his lectures.[211]

In September and October 1894, the year following the publication of his chemical experiments, Liversidge delivered two further papers to the Royal Society on the structure of gold nuggets. The first, presented to the Brisbane Congress of the AAAS in January, was summarised in *Nature* two months later; the second was not published until 1897 when it appeared in both the *Proceedings of the Royal Society* and in the *Journal of Chemistry*.[212] In these, he used photography to record his observations on the internal structures of gold nuggets and ingots. His intention, as he explained, was to throw light on the origin of nuggets, and to determine why gold became brittle and unfit for coinage – a line of work that happily combined philosophy with profit.[213] The nuggets were cut and etched with chlorine and other reagents to reveal their crystal structure, and were shown to be evenly crystalline throughout. Ingots showed a similar crystallinity, with the crystals being elongated when the ingot had been rolled.[214]

Liversidge maintained that nuggets had been set free from disintegrating veins, and were not precipitated *in situ*. He continued his work for the next ten years, including specimens from the Klondyke, New Zealand and New Guinea, and extended it to platinum, silver and copper nuggets. The New Guinea nuggets, described in his last paper on the subject in 1906, were unusual, because they were zoned and seemingly not crystalline, and he was uncertain of their origin – perhaps these *were* formed by accretion? Recent studies, however, have shown that these nuggets, too, are crystalline – but at five to fifty micrometres, they are only visible using an electron microscope – and they contain high temperature minerals (tellurides of mercury, gold and silver).[215]

Contemporary opinion ran against Liversidge's conclusions, but history left an open verdict.[216] In 1933, the head of the Mineral Resources Department of the Imperial Institute, writing in one of the first histories of petrology, admitted that 'modern authors have not added very much of importance to the general conception set forth by early workers'.[217] Indeed, Liversidge's work was lost to the geological community, possibly because the *Journal of the Royal Society of New South Wales* was not widely read and the *Journal of the Chemical Society* was read rarely by geologists. Years later, the processes of gold mineralisation remain imperfectly understood. [218] Today, many believe that gold nuggets are formed in the alluvial gravels or soils in which they are found, citing the same arguments as the 'accretionists' of a century or so ago, unaware of Liversidge's observations to the contrary.[219] A re-examination of the topic, repeating and extending his observations and applying modern electron-optical techniques, has confirmed his findings, although whether the matter will rest there is uncertain.[220]

In his experiments, Liversidge had noted that the slow reduction of $AuCl_4$ sometimes produced crystals of different habits including six-rayed stars, hexagons, triangles and octahedrals, and he discussed this in his fourth paper in 1893. Although he apparently did not see them himself, such crystals had been widely observed in nature.

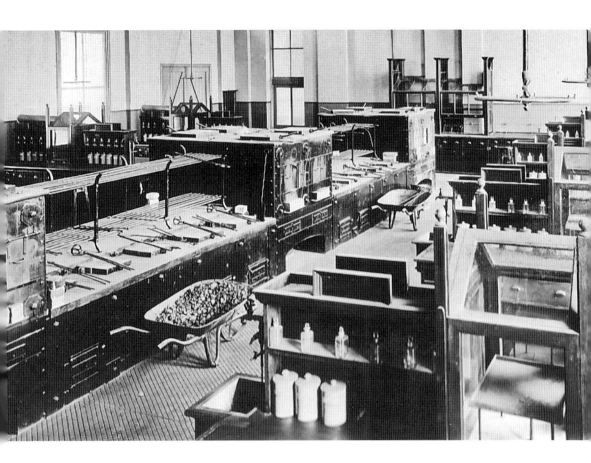

The Assay Rooms in the School of Mines.

They are the product of the gradual solution of gold as organic or chloride complexes by ground and surface water and subsequent precipitation in soils and alluvial sediments. Triangular and hexagonal gold nanocrystals are now being studied for use as catalysts, superconductors and drug delivery systems.

Liversidge's conclusion that nuggets were derived from the decomposition of quartz and other veins begged the question as to how they, and gold deposits as a whole, were formed. Even today, scientists differ. Some contend that many quartz-gold systems in faults and shears were deposited from meteoric groundwaters, heated by metamorphism and circulated through rocks from which they leached gold (the old 'lateral secretion' model). But others argue that gold is deposited by precipitation from magmatic fluids brought to the surface of the earth by volcanic activity. The association between gold and sulphides precipitated in hot springs, from Liversidge's own observations in New Zealand, inclined him towards the latter view, but left the question, as he found it, unresolved.

By the spring of 1895, Liversidge realised that the 'nugget question', and the alleged role of solutions in the theory of ore deposition, were likely to remain unsettled. He therefore turned to a fresh series of experiments which, on 2 October, he presented to the Royal Society in the form of two papers – on 'the Removal of Silver and Gold from Sea Water by Muntz

Metal Sheathing' and on the 'Amount of Gold and Silver in Sea Water'.[221] He followed these papers with a third, 'On the Presence of Gold in Natural Saline Deposits and Marine Plants', in February 1897.[222] The first and third were eventually to have an importance far greater than he anticipated; the third would preoccupy him to the end of his days.

In the first paper, Liversidge examined a specimen of so-called 'Muntz metal' taken from the hull of a coastal trader, and showed that sea water was capable of depositing gold upon it.[223] Today, following research by Australia's Commonwealth Scientific and Industrial Research Organisation (CSIRO), the mining industry is investing heavily in the south-western Pacific to locate undersea volcanic systems which are actively depositing gold-bearing metal sulphides (the so-called 'black smokers').[224] In some sense, the modern era begins with Liversidge. However, the French chemist Joseph Proust is credited with being the first to write on the subject;[225] and in 1822 Alexander Marcet, the English physician and chemist, suggested 'ocean minerals' that might have therapeutic value.[226] Indeed, iodine was detected in seawater as early as 1825, and analysed between 1852 and 1865, as was silver, detected in 1850, and arsenic, in 1857.[227]

In reviewing this literature, Liversidge read the work of Edward Sonstadt, an amateur chemist, patent agent and journalist living in the north of England, and Emil Münster, Professor of Metallurgy at Christiania (Oslo) University.[228] Sonstadt was a keen student of neglected questions, well known for his many papers in *Chemical News*. In 1865, he experimented on the extraction of iodine from kelp, and improved methods then in use.[229] Some years later, he found in the seawater of Ramsay Bay, near the Isle of Man, traces of gold, noticeable in quantity, if 'certainly less than one grain in the ton' (ca 60 mg/tonne). The amount was tiny; even so, at current values, it represented twice as much gold per weight as the lowest grade of land deposit then thought profitable to operate.[230]

Sonstadt used three different procedures, based on assay procedures involving either lead cupellation or the very delicate colour reaction with tin (II) chloride, which led to the colloidal solution known as 'Purple-of-Cassius'. This is still a convenient qualitative test for the presence of gold. His report, published in *Chemical News*, was picked up by the *Athenaeum* in London and by the *Popular Science Monthly* in the United States.[231] A quarter-century later, looking at Sonstadt's modest claims, as reported in the second supplement of Henry Watt's *Dictionary of Chemistry* published in 1875, it was not difficult to believe that, if what Sonstadt had hinted (but not actually promised) were true, there was indeed a case for exploitation. In 1892, Emil Münster of Norway arrived at a much lower result for water in the Baltic.[232] But if such variation existed, why not test further?

In 1866, Richard Daintree suggested that quartz reefs received their gold veins by deposition from the oceans. It followed that the oceans would still have 'supplies' of gold, and other precious minerals, if anyone had the patience to 'mine' them.[233] In 1870, William Skey of the Geological Survey in New Zealand published in *Chemical News* a bromine-iodine method for the detection of minute traces of gold in minerals.[234] Liversidge knew of this work, and made the logical inference that if halogens could be used (by oxidation) to detect gold in minerals, then minerals might be detected in the largest solution of all – the ocean. This could have an important bearing on the mechanism of ore deposition and on prospects for commercial mining, as Daintree had believed.[235]

In 1895, Liversidge set up six Muntz metal 'sites' along the NSW coast to collect samples of seawater. He reproduced Sonstadt's experiments, and his results corroborated Sonstadt's estimates. This news he presented to the Royal Society on 2 October 1895. Liversidge calculated that, if these results were correct, the oceans would contain 75–100 billion tons of gold. A lengthy notice appeared in London in the form of three successive articles in *Chemical News* (with an unusually long delay of a year) during September and October 1896.[236] Despite the delay, Liversidge's conjectures were widely available. But his results, like Sonstadt's, met silence. No one wished to risk his reputation by testing, much less confirming, such calculations.

In early 1897, during his study leave in England, Liversidge sent the Chemical Society of London two papers he had earlier presented in Sydney – one on the crystal structure of gold nuggets, and another on the presence of gold in saline deposits.[237] The second was the more provocative. Following a description of experiments on marine substances – including rock-salt, bittern, seaweed, kelp, and oyster shells – he urged his readers to apply such 'delicate processes' as he had used in analysing seawater, to the determination of gold and silver in minerals and rocks, in the hope thereby of 'throwing light on the origin of gold in veins and similar deposits'.[238] It was clear enough that, at prevailing prices, the oceans might be worth mining.

The paper appeared in the *Proceedings of the Chemical Society* and later in the *Journal of the Chemical Society*; and was reprinted in *Chemical News*. The following year, there appeared only two reports of Liversidge's work of 1895 and 1897, and both were aimed at mining engineers, not chemists.[239] But slowly the idea caught on. Liversidge's estimate of 0.5–1 grains per ton (or about 30–60 mg/tonne) was read by colleagues at the Royal School of Mines, and published in T.K. Rose's *The Metallurgy of Gold*. Although Liversidge was careful to note that his determination was only valid 'in certain cases', not everyone read him so closely.[240]

Prospecting the oceans may have seemed fanciful, but between 1895 and 1903, at least five British patents were taken out for processes to obtain gold from seawater.[241] In 1903, even Svante Arrhenius, the distinguished Swedish chemist,[242] whose influence over the award of Nobel Prizes was legendary, took up the question.[243] However, his experiments confirmed Münster's results, not Sonstadt's, and reduced the world's grand hopes to lesser expectations. The expected 'find' fell tenfold to only 6 mg/tonne.[244] Still, Arrhenius' results remained interesting. One who found them so was Fritz Haber, one of twentieth-century Germany's best-known chemists, famous for his role in the fixation of atmospheric nitrogen.[245] Legend has it that Arrhenius first put the idea to Haber.[246] Surely he, who had won nitrogen from the air, could win gold from the sea? As we shall see, the answer was to seal discussion of the question for the next fifty years.

Long after he published his first papers, Liversidge remained deeply puzzled. Why do gold concentrations vary so widely? And what, to return to his deeper question, does this say about the earth's history?[247] Some years later, he spoke to the Royal Society of NSW on the presence of gold in meteoric dusts – possibly giving clues to the extra-terrestrial origin of the heavy metals.[248] Following his return to England in 1907, he would continue to 'prospect', this time in the soot of Piccadilly. If gold could be found in the ocean, why not in the streets of London? But these thoughts were far distant from the southern winter of

1895, as Liversidge, weary of the 'by-law battles', retreated to his library at Darling Point. He was anxious to finish the papers he would give to the Royal Society, before again sailing to rejoin the community of science at 'Home'.

## 4. THE INTERNATIONAL CATALOGUE AND AN IMPERIAL LLD

In July 1895, Liversidge asked Senate for twelve months' overseas leave. Eight years had passed since his last trip to Europe, and much had happened in his discipline that had, he feared, passed him by. On 2 September 1895, Senate agreed that Russell (now Vice-Chancellor) could serve as Dean of Science to make it possible for Liversidge to go abroad.[249] Indeed, he received leave on Russell's motion, and just before a Senate committee was to report, for the first time, 'on the general principles upon which leave of absence should be granted'.[250] Given the depressing circumstances of 1893–95, his reasons for leaving were understandable. But there was a 'pull' as well as a 'push'.

In May 1895, the Royal Society of London issued a circular invitation to all the world's recognised foreign academies, announcing a meeting to be held in London in July 1896 to consider the future of the International Catalogue of Scientific Literature. This enterprise traced its origins to the Royal Society's *Catalogue of Scientific Papers*, conceived by Joseph Henry of the Smithsonian and debated at the British Association at Glasgow in 1855.[251] The Royal Society had assumed the task, and in the early 1890s, plans were made for a fourth edition, to cover the period 1884–1900, but the effort had proved too great a strain on its resources.[252] In 1893, prompted by Henry Armstrong, the Royal Society proposed that the next catalogue of current scientific literature be prepared cooperatively by a committee of delegates from all the leading countries of the scientific world.[253] The new International Catalogue was to begin in 1901, with the twentieth century, and was to be 'the final word in all branches of pure science'.[254]

In June 1895, the invitation reached Liversidge in his capacity as Honorary Secretary of the AAAS. As the Association had no official standing, Liversidge duly referred it to each of the six colonial societies; but by December, no volunteers had come forward. Given Senate's permission, no one objected to Liversidge representing the Royal Society of NSW during his visit to England.[255] However, as the invitation from London was construed as one for the colonial government, it became necessary for the Agent-General of NSW, Sir Saul Samuel, to become involved. Eventually, in this roundabout manner, diplomatic proprieties were observed, and he received an official letter appointing him Sydney's man.[256]

The hard years of 1893–95 looked like having a bleak successor in 1896, as Liversidge prepared to sail. Both P&O and Orient had mail contracts, and fortnightly sailings, so that one departed from Sydney each week, and took five weeks. In December, he arranged for a relative of a friend from the Royal School of Mines, H.H. Voss from South Africa, to look after the Octagon. It was bad luck that the steamship *Australien* (Messageries Maritimes), on which Voss was to arrive from Marseilles, was quarantined with smallpox the day it reached Sydney. 'I cannot get away by my boat [the *Oceania*]', Liversidge complained to Baldwin Spencer on Christmas Eve, 'for my friend ... is a prisoner'.[257] In missing the *Oceania*, he also missed Mark Twain and his family, returning

India by elephant: Liversidge on a visit to Amber, with Richard Baker and family, 1896.

from Sydney to Europe after a three-month tour of Australasia.[258] But finally his friend's ship 'escaped' from quarantine, and Liversidge left Sydney on the P&O's *Orient* on 14 January 1896.[259] F.B. Guthrie, his former demonstrator, now working for the Department of Agriculture, became an 'acting professor', and delivered Liversidge's inorganic chemistry lectures, for which he was paid from the professor's salary, £300 plus fees.[260]

    This trip was to mark what the *Herald* called 'an extensive tour' – 'wherever there was a scientific institution, college, or technical school, it was visited by the professor'.[261] During February and March, by train and elephant, the forty-nine year old Liversidge traversed India – from Calcutta to Agra, to Bombay – *en route* visiting friends from the RSM now holding senior positions in the Geological Survey. Leaving India, he made passage via Aden and Egypt to Naples. With a train to Paris, and a boat to London, it was not until April that he took up residence at the newly established (1894) Royal Societies Club in St James' Street. At once, as the *Herald* put it, he 'ploughed into that scientific vortex', dividing his time between meetings of the Royal Society, the Geological Society and the Chemical Society, and dining in the Savile Club's new premises at 107 Piccadilly.[262] Refreshed, by his own account, he then embarked on a whistle-stop tour of 'all the principal towns in England where there is a university, technical college or mining school'.[263]

The pace of change in England seemed to catch him by surprise, and made the distance all the more palpable. As he told an interviewer,

> it is only ... when you come into contact with gentlemen who have practically demonstrated their discoveries, or continued those of other people by actual experiments, that you learn that an hour's conversation is worth columns of reading.[264]

The same could easily be said of W.H. Bragg, who would give up Adelaide for Leeds; and Richard Threlfall, who would give up Sydney before another year was out.[265]

London was the centre of Liversidge's universe. 'I went to most of the colleges in India,' Liversidge wrote Baldwin Spencer, 'and want to do the same in Egypt, but I do not care to leave London in the thick of the Society meetings.'[266] With Spencer, he discussed the prospect of electing more Australians to the Royal Society. Both Etheridge and Haswell had gone forward, but he advised Spencer to be patient. Until those two were elected, he said, 'another Australian could not have much of a chance'.[267] As for the rest, life in England was no holiday: 'it is a great pleasure to be home once more and to see old friends, but there is so much to do and see that at times it is very like hard work'.[268]

In London, Liversidge visited Sir William Ramsay's laboratory at University College, the site of Rayleigh's discovery of argon, and the discovery of terrestrial helium. At King's College, he watched Herbert (later Sir Herbert) Jackson experiment with the new cathode-ray tube. At the Royal Institution, he saw James (later Sir James) Dewar demonstrate the liquefaction of air. The apparatus for doing low-temperature work, he noted 'has been considerably simplified and its cost reduced', but as the science vote at Sydney had been so badly cut since 1893, he knew 'we cannot afford to secure it and other improved appliances'.[269] If Liversidge had gone to Cambridge, he would have met the newly appointed (1894) Cavendish Professor of Experimental Physics, J.J. (later Sir J.J.) Thomson, FRS, whose discovery of the electron (in April 1897) was soon to transform chemistry and physics. There, too, he would have met several new '1851' scholars, conducting research with resources an Australian professor could only envy.

With the coming of summer, Liversidge's search for 'scientific spoils' was pleasantly interrupted by a letter from H.E. Barff, Sydney's Registrar, conveying a request from the Senate that Liversidge represent the University at a ceremony honouring Lord Kelvin's Jubilee, which was to be held at the University of Glasgow.[270] Kelvin had often expressed goodwill towards the colonial universities, and of course the Scottish connection with Sydney was one to be celebrated at every opportunity.[271] This act of homage brought Liversidge an unexpected bonus. Kelvin was to receive an honorary Doctor of Laws degree, and on 20 May, when the University asked which 'Colonials, U.S. Americans and Foreigners' should also be honoured, Kelvin said that 'our visitor from New South Wales, Liversidge, should also be on our list'. Indeed, he added, 'I find he is author of many papers which seem very important, on chemistry and mineralogy'.[272]

So it was that in June, Liversidge entrained from St. Pancras for Scotland to receive his first and only honorary doctorate. There he joined a procession of over sixty distinguished men, representing universities from Calcutta to Cambridge.[273] Liversidge accepted the degree not only for Sydney and NSW, but for all Australia. 'Distinguished as a traveller, a chemist and a geologist' who has 'conducted researches of vital importance to the colony', the public orator announced, 'he is a man of wide knowledge, and equally wide sympathy, who has done service to other departments of science'.[274] Liversidge delivered a congratulatory address in return for his 'unexpected honour'.[275] Critics, even Liversidge himself, might say the tribute was intended 'more as a compliment to the Colony of New South Wales and to the University of Sydney than as a personal one'.[276] All the same, it was good to read that one was being recognised, in *Nature*'s stirring words, as 'a typical representative in science of the English colonising spirit'.[277]

That 'spirit' came to the fore when, on 13 July, Liversidge was greeted in London by his Cambridge mentor, Sir Michael Foster, now Biological Secretary of the Royal Society and one of the most powerful men in British science. Foster had organised a meeting of the English-speaking delegations who were attending a week's conference on the International Catalogue of Scientific Literature.[278] Their task was to establish guidelines for collecting and classifying scientific papers, and to agree upon a division of labour in listing and collation. Early differences as to which language should be used were settled diplomatically: the French representative preferred English to German, and the German representative preferred English to French, so English became the international language.

Within a week, these delegates had agreed on the publication of a subject and author catalogue, and set up a Central Bureau in London, with rented offices on Southampton Street. The director of the Central Bureau, Dr H. Forster Morley, was made responsible to an Executive Committee of the Royal Society, chaired by H.E. Armstrong, and including the Biological and Physical Secretaries, Sir Michael Foster and Dr (later Sir) Joseph Larmor.[279] The committee marked a first step towards a global system of scientific information, with its promise of worldwide access. Regional bureaux were to be set up – and being the only Australian present, Liversidge won the Australian post for the Royal Society of New South Wales. Plans were made for a second preparatory conference, to be held in London in October 1898; a third, also in London, would take place in June 1900. Delegates left England confident they had made a contribution of great importance.[280]

If Liversidge's presence in London was fortuitous, he always turned it to advantage. In August, he heard one of the greatest chemists of Europe, Marcellin Berthelot, tell at the International Congress of Applied Chemistry, in London, that the future belonged to the country with the most laboratories and the greatest number of chemists. Chemistry would 'discover processes of commercial success', he said, and 'render nugatory the efforts of those who have secured a monopoly of the accidental gifts of nature'. The message was encouraging to Australians, not least when Berthelot added, 'gold occurs in almost every rock, in every soil of the whole world; to extract it from the streets of Paris or from the plains of North Germany is merely a chemical problem to be resolved in laboratories'.[281] To achieve this was an intriguing possibility, which ten years later brought Liversidge back to the streets of London.

During the late summer and autumn of 1896, Liversidge's diplomacy was called into play during a controversy surrounding the Royal Society's expeditions to the coral atoll of Funafuti, which were sent to test Darwin's theory of coral-atoll formation.[282] Towards the end of the year, Charles Hedley and Robert Etheridge Jr, both on the staff of the Australian Museum, had the temerity to publish independently, and in Australia, some of the biological materials they had collected during W.J. Sollas's unsuccessful British expedition in 1894. Their action flew in the face of the long-established convention that the Royal Society had first claim on any publication ensuing from any expedition it sponsored. As a Trustee of the Museum, and as Michael Foster's friend, Liversidge helped smooth ruffled dignities. Hedley was let off with a mild rebuke, but Etheridge's arrogance so greatly displeased the Royal Society that, in spite of Liversidge's repeated annual nominations between 1892 and 1898, he was never elected a Fellow. 'I am sorry for him', Liversidge confided to Lazarus Fletcher at the Natural History Museum, 'but he is apparently bound to be at cross purposes with some one'.[283] Eventually, the Hedley-Etheridge affair died down; and, on Liversidge's advice, the Royal Society let Edgeworth David take charge of a second expedition.[284] After Liversidge left England in 1897, Foster used his influence with the Admiralty to deploy a Royal Navy hydrographic ship for David's expedition.[285] The resulting second and third expeditions to Funafuti have passed into legend. Although they failed to bring theoretical closure, they marked a turning point in the history of Anglo-Australian scientific relations. By way of thanks, David sent Liversidge bore samples for his studies on gold in seawater.[286]

Towards the end of 1896, as he wrote up for the Chemical Society his papers on the origins of nuggets and gold in seawater, Liversidge obtained a number of European mineral specimens for the Australian Museum;[287] and arranged an exchange of Australian mineral, gemstone and meteorite specimens with Cambridge and South Kensington – where the Mineralogy Department of the Natural History Museum has today several on display.[288] In November, these informal exchanges were put on a regular footing, under terms similar to those he had proposed in 1887.[289] Following a request from Sir Saul Samuel, the Agent-General, Liversidge inventoried the NSW mineral collections of the Imperial Institute in South Kensington, while deploring the poor labelling and unsatisfactory conditions in which the collections were kept.[290]

On St Andrew's Day, Liversidge revelled in the anniversary meeting of the Royal Society, which that year honoured the American Ambassador and Lord Salisbury's nephew, A.J. Balfour, then First Lord of the Treasury, later Prime Minister, and a good friend to British science.[291] At dinner, Liversidge sat next to William Gowland (his old friend from the RSM and the London University Union), whom he had last seen in Japan in 1887, and near Arthur Dew-Smith and Michael Foster, to whom he owed so much.[292] On 25 January 1897, just before his leave came to an end, Liversidge again visited Cambridge, where he dined at Christ's with Arthur (later Sir Arthur) Shipley, the college's biology don (and later, Master). With Shipley, also a bachelor, and a spiritual cousin of Michael Foster's group, he became in later years the closest of friends. Through Shipley's good offices, Liversidge was elected a fellow of the Cambridge Philosophical Society.[293] It was a nice honour and a parting gift.

At the end of January 1897, Liversidge sailed from Plymouth. He arrived in Sydney on 1 March, narrowly in time for the start of the new academic year. His life in Britain had been full of recognition and reward. He knew that the affairs of the Royal Society of NSW, the AAAS and the Faculty, were in suspended animation, awaiting his return, but he intended to keep the world better informed about Australia, and about his own work, by sending regular reports to *Nature* and the international journals. He knew he would have to work harder than ever to create a research school at Sydney. He had seen what energy and idealism could achieve. Yet, even as the *Arcadia* drew into Port Jackson, the world itself seemed to slow as Sydney's statesman of science returned to work.

# CHAPTER 11

# Statesman of Science

Council members of the Australasian Association for the Advancement of Science, Sydney, 1898.

It seems but yesterday I saw at dawn
The faint line of the soft Australian shores …
A land, they said, of golden air, where scents
Of sweetest flowers float, and where the grapes
In honied clusters droop, a Paradise –
Of glowing blue and tranquil loveliness.[1]

Thus Australia welcomed Hallam, Lord Tennyson, son of the Poet Laureate, as he steamed past the Fleurieu Peninsula on his way to become the Governor of South Australia. A similar vision greeted Liversidge and the P&O *Arcadia* as they entered Port Jackson on the last day of February 1897.

Liversidge, aged fifty, was in his prime. An honorary LLD had added cubits to his stature. He had met Rayleigh and Ramsay, and seen Röntgen's X-rays. He had met J.J. Thomson, who would soon announce his discovery of the electron, and had seen James Dewar probing the behaviour of gases at such low temperatures that Henry Adams feared for the 'death of matter'.[2] Physical chemistry and radioactivity had entered the curriculum, and mineralogical chemistry and chemical mineralogy were being recast by European crystallographers looking for the physical principles that underlie mineral structures. All this excitement, Liversidge left behind him.

Even so, as he returned to Sydney, Australian scientists were establishing their own traditions. Inventions and patents were multiplying, and agriculture and mining were flourishing. Thanks to the AAAS, the fledgling community of Australasian science had acquired a seemingly permanent, if ambulatory, presence, combining convivial gatherings and reports on local activities with the announcement of new discoveries, mostly from overseas. Elsewhere, there was the usual talk of drought and depression, but more of South Africa, of free trade and protectionism, and the prospect of federation. Authors and artists toyed with European and American fashions, philosophers argued about vitalism and pragmatism, and those who feared for Australia's future joined debates about imperial efficiency and scientific management.[3]

In much of this, Liversidge kept a passing interest; in some, he played a modest part. Overall, he returned to a community grateful to have him back; and within a year, as foreshadowed, he was honoured by the AAAS, by being elected its President. But his ceremonial crowning, *imperium in imperio,* was paralleled by a growing sense of disenchantment, of a kind only a dedicated servant of the university, city and colony could know.

Returning to Sydney, retrieving the reins of office, were acts of near self-denial. Responsibility weighed heavily upon him. In Britain, as in much of Europe and North America, the days of the 'encyclopaedic' professor, single-handedly responsible for an entire discipline, were fast coming to an end. Yet, at Sydney, the tradition remained. Liversidge had to cover the horizon, with what help he could find and afford. Research opportunities were few. At best, he could encourage others, notably through junior appointments, when he could get them. And after years of drought, students came in floods – in 1901, no fewer than 583, by far the majority in Arts, making demands on facilities that were adequate at best for the 297 who had matriculated ten years earlier. Their chemcial education, which he defended as a point of principle (and, perhaps, personal profit), left little time for personal research. Advancing age also took its toll. Colleagues were courteous, but first years were less so; and few seemed to appreciate selflessness. Within five years, a Senate that had thrice given him overseas leave would be willing to let him leave altogether.

## 1. *Fin de Siècle:* The 'Two Cultures' and Their Servants

The last years of the century saw the University climbing slowly out of the trough of depression. Science and Arts, now joined by Medicine and Law, had become Faculties, which gave at most lip-service to the concept of the 'well-rounded scholar'. Badham had not lived to see his deepest fears materialise; but Liversidge saw his ideals wither on the vine. Medicine was now part of an international network. But in chemistry Sydney was standing still. Teaching methods and texts remained static from year to year. Meanwhile, the University's policy towards research was clear: when staff wished to advance scholarship, they did so at their own expense. This was as true in Arts as in Science, but the sciences typically made more noise about it. Thus, the press celebrated the discoveries of J.P. Hill, Haswell's Demonstrator in Biology, who in 1895 located a placenta in the bandicoot. Edgeworth David, with pardonable pride, loudly proclaimed this as 'perhaps one of the most important, if not the most important, scientific discoveries as yet made in Australia'.[4] The news required taxonomers to revise textbook tables of Order Marsupialia. Yet, when Hill applied for two terms' leave to further his research in Edinburgh, he had to do so without pay.[5]

Discouragement followed. In 1898, the redoubtable Beatrice Webb visited Sydney, and recorded her conversations with its professors. They, she said, were depressed by the 'utter indifference of well-to-do Australians for learning of any kind'.[6] Their situation could not easily improve, in a society she found obsessed with racing horses and making money.[7] Australia had, of course, dedicated philanthropists and public libraries, art and theatre, music and literature. But 'the real work of manhood in Australia', sighed the *Herald*, was mainly 'directed towards a practical end'. And even practicality met bureaucracy. Threlfall became restless when Senate denied his application to act as a consulting engineer;[8] and when he learned that Warren had done the same, without asking permission,[9] his mind began to turn elsewhere. After twelve years, his days in Sydney were numbered.

Within the Science Faculty, the engineers alone seemed content, as well they might be, following a gift in June 1896 of £50,000 from P.N. Russell, since returned to England.[10] Founder of one of the colony's pioneering engineering works, Russell is also remembered in the annals of labour unrest. But after leaving Sydney, he was much cultivated (and rewarded, in 1904, with a knighthood). His gift enabled the University to create a new School of Engineering, including a four-year course in mechanical and electrical engineering. However, disagreements between Threlfall and Warren – on such weighty questions as who was to teach physics to electrical engineers – marred Liversidge's decanal peace.[11] Consensus was vital. Students in mining, for example, were technically in engineering; yet their principal instruction was delivered by the chemistry and geology departments.

During Liversidge's absence, Senate had witnessed several revolutions but no changes. The University had a strong helmsman in Sir William Windeyer, but in 1896, he retired to England (and died in Bologna the following September), and was succeeded by Sir Normand MacLaurin, then Vice-Chancellor and Member of the Legislative

Council.[12] A Fellow of Senate since 1883, MacLaurin was not an unknown quantity. Edinburgh-educated surgeon, physician and medical administrator; a director of the Colonial Sugar Refining Company; and a member of both the Royal and the Linnean Societies, he was in many ways sympathetic to the interests of science – especially when these did not conflict with the aspirations of the Medical School. A shrewd investor, and a dedicated federalist, he strongly defended a policy of 'unification' in the curriculum.

MacLaurin's preferences became better known when curriculum battles, thought to be long over, were declared unfinished, and broke out again between Arts and Science. In 1897, Senate appointed another committee to review the Arts curriculum,[13] and to consider again the science requirement for first-years.[14] For Liversidge, David, and Threlfall's *locum*, J. Arthur Pollock, this requirement still specified one course throughout the year, with the three terms divided equally between chemistry, physics, and natural history.[15] By the end of the year, after much debate, the threat passed, and the requirement remained, but Senate, on a motion by MacLaurin, seconded by Renwick, gently reminded the scientists that, by Senate's resolution ten years earlier, their lectures must 'embody a general view of the whole subject with particular reference to the elucidation of general principles'.[16] It was not MacLaurin's fault if such an appeal failed to produce courses that Arts students found interesting.

As the century ended, Liversidge remained Dean of Science, his days measured by the hour-glass of administrative detail – fine-tuning course requirements in mining and engineering;[17] strengthening conditions for higher degrees;[18] and minimising the expensive duplication of laboratory materials.[19] As the economy recovered, so spirits lifted. Regulations were drawn up for a new postgraduate degree in mining engineering,[20] and laboratory practicals were instituted for first-years in all science subjects.[21] In April 1897, Liversidge supported a Senate motion to introduce a degree in agriculture.[22] Happiest when there was money to spend, he was put on the P.N. Russell Fund Committee, and with Threlfall and Warren, had the unaccustomed pleasure of suggesting how the new scholarships should be awarded.[23] He was fittingly the one to propose that a portrait of Russell be commissioned to honour their engineer-benefactor.[24]

From 1898 onwards, as chairman of Senate's Building Committee, Liversidge sought to remedy biology's need for laboratory space, and the need of physics and engineering for additional buildings.[25] A surplus on the Challis account helped him do so, and as from 1898, he and his colleagues were no longer required to suffer the five per cent salary 'tax' imposed during the depression.[26] 'Retrenchment' was still the favourite word of the Finance Committee, but immediate pressures eased.[27] A vacant junior demonstratorship in chemistry was filled;[28] and in 1900, a second post was created.[29] In December 1897, Haswell at last got the demonstratorship in biology, to which he progressively, but unsuccessfully, tried to appoint a woman (Miss Marion Horton, BSc).[30] Laboratory fees for first-year science students were reduced from three to two guineas a term,[31] and eventually the wages of laboratory assistants – including those of the indefatigable Edward Hufton – were increased.[32]

In his beloved School of Mines, Liversidge put the turn to good advantage. With help from the Russell Fund, he divided the single Demonstratorship in Geology/ Lecturership in Metallurgy, and two young men (James Taylor, from the Royal School of

Mines, and W.G. Woolnough, one of David's students) were appointed.[33] With their help, Liversidge kept up some private work – examining, for example, minerals sent to him by mining companies.[34] It was perhaps something to be known by the General Manager of the Broken Hill Proprietary Coal Mines as 'the foremost mineralogist of Australia'.[35] In 1899, the NSW government asked his help in rationalising the several collections of mineral specimens dotting the city. An enquiry was held, and a decision made, by which the Mines Department established a museum to display the collections begun by Wilkinson and developed by Pittman.[36] No solution was found to the problem of overlapping; but at least, the collections of the Australian Museum, the Geological Museum and the University bore testimony to the range of material available for the scholar, the teacher and the 'practical man'.

In many other ways, Liversidge continued his work of public service. Having helped arrange the Australian expedition to Funafuti, Liversidge navigated David's leave of absence through the shoals of Senate, so that an Australian geologist could lead the expedition; and Liversidge hired Edward Pittman to take David's geology lectures and William Dun, Etheridge's assistant in the Geological Survey, to take his palaeontology class.[37] When a Royal Commission on Mining in Queensland recommended that students be sent to Sydney from the 'deep north', Liversidge made haste to welcome them – with open scholarships, where he could[38] – and arranged for mining managers from BHP to visit his mining building, to watch his students at work.[39] The Mining School, after the Medical School, became the principal point of scientific contact between the University and the community, and the University treated its claims generously.[40] In 1899, Senate authorized a new demonstratorship in assaying and chemistry, and another Associate from the Royal School of Mines, Arthur Jarman, was appointed. This gave Liversidge scope to let part-time students who were taking day-time geology, chemistry and mining classes at the Technical College, to attend advanced classes at the University in the evening.[41] The result was to improve, even further, the good relationship between the two institutions.

Following his visit to England, Liversidge returned more interested than ever in the chemical boundaries between inorganic and organic matter,[42] and tried to keep alive his work on gold. He published (first presenting, as customary, to the Royal Society) some new work on the crystalline structure of nuggets,[43] and was pleased to find it noticed in the United States, as well as in England. He had followed Dewar's popular demonstrations of the liquefaction of gases, and these he repeated to appreciative audiences in Sydney; and to at least one appreciative visitor, the Governor's wife, Lady Tennyson, for whom touching 'liquid air' proved a sensational experience.[44] Turning to commercial questions, Liversidge reported the discovery in Europe that 'frozen air' reduced electrical resistance, and speculated on the 'payable consequences' of having the means to increase conductivity in power transmission. As the 'electrification' of NSW was just beginning, his audience voted him special thanks.[45]

Although Liversidge had resigned from the honorary secretaryship of the Royal Society in 1884, he retained an interest in its library, and made the Society a base for the AAAS, as well as for the International Catalogue of Scientific Literature. As such, the Society became a post office and clearing house, as he channelled requests from London to the forty-odd learned societies in Australia.[46] Gathering such information took more time than organising the AAAS. Yet, he knew that, as he confided to Baldwin Spencer in June 1897, 'If the work can be carried out it will be a great boon to everyone out here'.[47] By the end of

1898, he felt under pressure to retreat where he could – he resigned, for example, from the Publications Committee of the Australian Museum[48] – but he continued to conduct from the Octagon and the University a correspondence exceeding dozens of letters a week.[49]

It was deeply ironic that, just as the finances of the University were improving, Richard Threlfall decided to leave Sydney, taking overseas leave in the first instance, but resigning permanently as from 31 October 1898.[50] He led thereafter a highly successful professional life in England; and although Liversidge would see him again – indeed, would work with him during the Great War – his departure was an enormous loss to the small community of science at the University, the AAAS, and the Royal Society. J.A. Pollock, who had understudied Threlfall, and who had begun to make a name for himself in the science of interferometry, was chosen as his successor as Professor of Physics, becoming the second graduate of the University to be elected to a Sydney chair.[51]

Pollock's appointment was welcomed by Liversidge, who as Dean did what he could to strengthen job prospects for young Australians. In 1900, he similarly celebrated the appointment of George Harker, the first junior demonstrator to be appointed from the 'ranks' of his own undergraduates. After Harker did well in his BSc in December 1899, Liversidge put him to work on the chemistry of topaz and on methods of using fluorine as an analytical indicator. But to rejoice one year, was to grieve the next, when Harker won an 1851 Exhibition to the City and Guilds Institution in London.[52] Liversidge inevitably gave his support. 'Although such work' as he had done, Liversidge observed, 'is not brilliant like the discovery of a new element, yet it involves much hard work and thought ... He is very keen to become a chemist in time and I hope that he will, although he has a lot to learn.'[53]

Harker eventually returned to Sydney to offer a much-needed course in organic chemistry. But every such parting was a blow to Liversidge's wider ambitions for his field and for the Faculty. It was good to welcome H.S. Carslaw, a researcher from Glasgow via Cambridge, who promised to strengthen mathematical physics, replacing the unfortunate Gurney in 1903. The University's research record in mathematics, according to Senate, had become 'absolutely non-existent' because 'mentally equipped with every gift except ambition, he [Gurney] ... never published a line'.[54] Despite the University's advances in mining and medicine, as late as the Jubilee in 1902, it was hard to see a bright future for the small band of brothers, struggling to keep science in the Arts curriculum, with students who took early opportunities to leave, and with a BSc course that refused to grow.

## 2. COMPLETING THE CIRCLE OF THE SCIENCES: THE AAAS AND ITS ODYSSEY

In January 1898, as Threlfall contemplated his return to England, Sydney played host to the seventh Congress of the AAAS, now completing its first cycle of capital cities, and returning to Sydney after an absence of ten years. Its record in that decade had been remarkable. Collegiality had overcome cynicism, and hard work had answered faint praise. Behind the scenes, Liversidge's influence had been felt everywhere. Thanks largely to him, six volumes of proceedings had appeared; in addition, over twenty research committees had reported, a program of research grants had begun, and colonial legislatures had been spurred to act on issues ranging from the notification of infectious diseases to the creation of nature reserves.

Admittedly, there were shortcomings. The technical character of its proceedings had distanced some speakers from the lay public, and many research committees had found it difficult to consult. Given the distances, meetings of Council were infrequent, and despite vice-regal patronage, the Association lacked financial security. Publication and other expenses were managed year by year, and congress by congress. In a decade dominated by talk of federation, the Association's federal role remained unresolved. If the several colonial societies were to continue to show allegiance to a central body based in Sydney, formidable tact and continuing enterprise would be required from its honorary secretaries.

It was with mixed feelings that Liversidge greeted his continuing commitment to the association. With six congresses safely behind them, the Association's Council, meeting at Brisbane in 1895, had elected him its next president. No one else in NSW matched his seniority. Because he was the logical choice, and because he was to be overseas in 1897, the Council postponed the seventh Congress until his return. Liversidge actually opposed this, fearing that an intermission of three years would cost the Association its momentum, if not its life; but he was overruled. Plans went ahead for 1898. But the demands were Herculean. 'It takes up a great deal of time', Liversidge admitted wearily, in June 1897, 'and I shall be very glad when it is over'.[55]

In the event, the Sydney meeting of 1898 proved a high-watermark in the Association's history. Close press coverage ensured it a warm reception. The opening ceremony in the Great Hall saw the Premier of NSW, the Chief Justice, the Speaker of the Legislative Assembly, the heads of government departments, and a flock of correspondents pay homage to science, and to its scientific version of Federation. The Mayor of Sydney, who was otherwise engaged, was criticised for his absence. The occasion was deemed as having national importance. Indeed, the *Herald* saw the Association's 'federal constitution' as the 'keynote of its success'. Comparisons with the mother country received full play, as usual, but with an important difference: with survival had come parity. The British Association, no longer 'parent', had become merely an 'elder brother'.[56]

The *savants* arriving in Sydney, proclaimed the *Herald*, were 'not men of local reputation merely, but [men who] have ... done work that is known wherever the English language is spoken, and beyond it'.[57] Over 820 participants and 200 papers testified to the vitality of an organisation which, in the *Herald's* phrase, 'exploited the present, and looked to the future' of science in Australasia.[58] The chemistry and physics sections heard papers on many local phenomena and practical problems, while anthropology and ethnology were attended by 'men who have risked disease and violence in their efforts to obtain a thorough knowledge of the subject'.[59] The *Herald* rejoiced that the Congress 'epitomised in a few days the intellectual life of Australasia.'[60] W.M. Hamlet's founding Congress *Handbook* of 1888 was deemed not only out of print but also, in the best sense, 'out of date'.[61]

For Liversidge, whatever his earlier misgivings, the event was a personal triumph. Manning's undiplomatic remarks in the Great Hall ten years earlier were nowhere mentioned. When Ralph Tate of Adelaide introduced Liversidge to the podium, it was as the 'father of this great scientific movement'.[62] In his Presidential Address, Liversidge

Liversidge, giving his Presidential Address in the Great Hall of Sydney University.

spoke nervously, and at some length, of the advancing research front of chemistry, and of Australia's participation in international science. He was never at ease in large gatherings, and the acoustics in the Great Hall were (and are) seldom to a speaker's advantage. But his words radiated through the scientific and popular press. In London, his address was reprinted by *Nature* and quoted in Parliament.[63] On Baldwin Spencer's initiative, a photograph of the officers was taken, giving the President – looking tired, holding papers – pride of place.

Liversidge's work by no means ended with the Congress's gala farewell. The continuing demands of organisation cost him his summer holiday,[64] and five months afterwards, he was still lobbying the NSW government for publication funds. The jovial

George Reid, his old friend, by now Premier, and normally a reliable supporter, left Sydney at a critical juncture, and only in May did the government agree to the cost.[65] 'I hope it won't take too many years to bring out', Liversidge wearily wrote Baldwin Spencer,[66] and did all he could to produce the proceedings; but without the help of Edgeworth David – preoccupied with his last expedition to Funafuti – he would have been hard-pressed to finish the job.[67] It also fell to him to send out some 8000 circulars, advertising the volume – 'to all the large landowners as well as to *all* [his italics] professional men or officials who were of sufficient standing'.[68]

Nor were his duties finished when the volumes were despatched. As honorary secretary, he had to propel the Association onwards to its next Congress, scheduled for Melbourne in January 1900. The last months of his presidency (1898–99) saw his secretarial correspondence reach Himalayan proportions. 'Good Friday has its uses', he observed in March 1899, 'like Sunday for working up arrears'.[69] Sydney's reception had pleased him, but he hoped for still better things from Melbourne: 'I am glad your Mayor is willing to do so much', he wrote Baldwin Spencer, who had assumed the job of organising secretary for the next Congress, 'it would do the Sydney mayor good to attend the Melbourne series'.[70]

In sometimes delicate matters of Sydney-Melbourne diplomacy, Liversidge always followed strict protocol. But with Baldwin Spencer he was at ease. Over the years, they had found mutual interests in ethnography and photography, and their fascination for collecting transcended the prosaic trials of academic life. Writing upon the fine AAAS notepaper that Liversidge had printed, they appointed Section presidents for the Melbourne Congress – including Ellery, Hutton, and Knibbs.[71] Liversidge insisted on having Spencer's signature on all circulars to ensure that he would get proper 'credit for his work',[72] and cheerfully helped him get details of hotel accommodation, railway concessions and newspaper coverage.[73] Such small but courteous gestures remained Liversidge's trademark. When he listed the names of retiring presidents on the masthead – frankly in hopes of 'inducing more of them to attend' – he shared credit with them all.

If this kind of collegiality was courteous, it was also common sense. In 1900, the professional community upon which Liversidge could draw for support, including academia and government, numbered fewer than 200; and by no means all of these could be relied upon for regular contributions. In the first forty years of the Association, only sixty-seven people delivered five or more papers, and although speakers somehow came forward, Liversidge lived constantly with the fear of a sudden shortfall. On the research side, Liversidge was responsible for running five of the fourteen committees of investigation appointed in 1888; and fewer than twenty scholars or scientists joined any of them.[74] By 1900, over twenty research reports were published, on subjects ranging from municipal sanitation to the prevention of wheat rust – but in most cases, these were the work of the same few hands.

Given its thin base, the Association had an uncertain future. Despite a short list of science-based innovations, such as those of Farrer, the encouragement of research was never easy, nor was it improved by the pretensions of those whom H.C.L. Anderson, NSW's first Director of Agriculture, liked to call the 'city theorist' and the 'clerical agriculturist', who 'live on the farmer, not on the soil'.[75] Even where there were practical papers, the *Herald* feared that the 'general reader' struggled to keep up with the professionals.[76] The AAAS

## THE PICNIC OF THE SEASON.

" An excursion will be made to this delightful and geologically interesting locality. The geological interest attaching to the district is due to the occurrence of elæolite-syenite, phonolite, nosean or haüyne trachyte, tinguaitic and other rocks of the nepheline series. The occurrence of gold in association with the trachytes likewise merits investigation. Heavy hammers are requisite."—Vide official guide.

The AAAS Congress at Hobart in 1902.

could not yet offer up 'those authoritative summings up of the progress of knowledge which one so eagerly looked for in England'. [77] At 'Home', the British Association held annual meetings, at which major discoveries were announced. In Australia, the AAAS met only every two or three years, and although much was said, most had a local bearing. For every 'general' paper given between 1892 and 1895, there were two 'local' papers, and for every 'theoretical' paper, ten 'empirical' ones.[78] This was valuable, but not enough to win the case for science. In his Presidential Address, Liversidge insisted that the 'opening meeting of the Session is almost the only opportunity … to review what we have accomplished and to consider what we might do for the advancement of science'.[79] Section presidents could do as much, but most did not.

Finally, there remained the tyrannies of distance. Melbourne's *Argus* sympathised with the *savants*, 'forced to pursue their solitary labours in separate cities', and therefore apt to become 'monotonous and one-sided'. Many more congresses were needed to stimulate the 'flagging scientific worker – broaden his vision, and give science a higher place in public esteem'.[80] In the meantime, no wonder the public felt isolated from the pulpit.

Following Sydney's successful start, Liversidge reasoned with the worst that might befall. In 1900, the government of Victoria actually declined to publish the AAAS's proceedings, and Spencer contemplated sending them to England for publication. 'New Zealand and South Australia', Liversidge mourned, 'will probably follow suit and perhaps NSW will then find it convenient to do the same, so that after Tasmania [scheduled for 1902], it looks like the beginning of bad times for the AAAS'.[81] As Australia's 'march to nationhood' drew nearer to Federation, ties with New Zealand were loosened, and the AAAS crossed the Tasman only once more (Dunedin, in 1904) during the next quarter century. Meanwhile, despite a growing population, membership of the 'Royals' in the five colonies/states remained static. Some pointed to the splintering effect of new professional societies; others blamed 'suburban spread' away from the capital cities.[82] Between 1900 and 1913, AAAS attendances oscillated between 335 and 820, and did not exceed Melbourne's legendary 1162 of 1890 until 1926.

If the AAAS's future remained uncertain, at least Liversidge's reputation was secure. His last paper in Queen Victoria's century was given to the Royal Society's final meeting of 1900, on the chemistry of the famous Boogaldi meteorite.[83] If he had become the country's expert on the 'metals of the heavens', he was also an expert in more earthly pursuits, taking a keen interest, as we have seen, in promoting the interests of Richard Threlfall, Edgeworth David and Baldwin Spencer to the Royal Society of London. Threlfall, by then returned to England, was elected FRS in 1899 and the following year, David and Spencer were among the 'select fifteen'.[84] In the latter case, Liversidge had carefully timed his support – waiting until Spencer's Horn Expedition gave him a 'good lift'. In the meantime, he proposed Spencer as an Honorary Member of the Royal Society of NSW, to 'show local recognition of his claims'.[85] If, a generation earlier, worthy colonials might have failed to win the 'blue ribbon of British science', at the end of the century, 'I don't think', Liversidge wrote, 'there is anybody left out now in Australia on the biological or geological side'.[86]

In many cases, Liversidge was the principal sponsor; but he also asked colleagues to help where, as he once told Spencer, 'I can't well do it myself'. Reaching across the Tasman, his intervention with Michael Foster, Biological Secretary of the Royal Society, helped secure

The AAAS Congress members in Hobart in 1902. Liversidge is in the front row, fifth from the left.

F.W. Hutton's election. In return, Parker in Otago nominated Haswell in Sydney. Never overstating his own importance, Liversidge was rarely turned down, nor were his colleagues in Victoria. Foster seems to have taken every opportunity to promote Australian candidates, and so confirm the 'organic unity' of imperial science. Others shared his view. Thus, when supporting David Orme Masson in 1901, Sir William Ramsay of UCL (soon to win the Nobel Prize in 1904), commented that it was 'well to elect a colonial now and then, especially, [as he added] as Masson ranks first I should say in our Australian colonies'.[87]

Between 1900 and 1903, five Melbourne professors became FRSs. Sydney, by 1903, with a science staff of seven, had three; and a fourth (J.T.Wilson) was soon to follow; while even tiny Adelaide had one (E.C. Stirling) and by 1907, a second (W.H. Bragg). The Royal Society's imperial outlook transcended Australia. Between 1890 and 1910, fifteen new FRSs were also elected from Canada and South Africa – helping, as Rod Home has aptly put it, to lock the dominions into the imperial network.[88] At the turn of the century, Australian FRSs were more numerous in proportion to the number of scientists in the Australian universities, than were FRSs in the universities and university colleges of Britain. London cemented the loyalty of Australia well beyond Federation.

## 3. 'Federation Follies': University Science and Politics

The Commonwealth of Australia was born on 1 January 1901. Long drawn-out debates gave way to an agreement that bound the six colonies into a single nation, under a British act of parliament. At the same time, the Empire was at war, and caught up in a patriotic tide of imperial sentiment, the new nation reflected not so much the springtime of youth, as the uncertainties of adolescence. Ever since the first military contingents left Sydney for South Africa in October 1899, news from the veldt had dominated the headlines: 'War news', Liversidge sadly agreed with Baldwin Spencer, was 'more attractive' than the affairs of science.[89]

The war in South Africa created deep divisions at the University. The relief of Mafeking in May 1900 was a cause for rejoicing, and pro-British sympathies, religiously voiced by Edgeworth David and Anderson Stuart, were quietly echoed by Liversidge. But the pro-Boer sentiments of the historian George Arnold Wood ruffled the British Empire League and the University's Senate, and in 1902, almost cost him his job.[90] Perhaps his discipline was inevitably political, and those who had attempted to exclude history from the curriculum in the 1850s – and even in the high water days of the 1880s – knew what this meant.

The coming of Federation found Liversidge still at the helm of his Faculty. Dean for nineteen years, interrupted only by overseas leave, he remained, after Gurney's retirement in 1903, the University's oldest link with the generation of the 1870s.[91] Federation had no immediate effect on the universities.[92] However, as the new states were relieved of certain charges, there was the prospect of increasing endowments, and when Sir William Lyne's government showed signs of ending the seven lean years of the last colonial administration, Liversidge began a campaign for his share.

In August 1900 – writing as if the crisis of 1893–96 had never taken place (or, indeed, because it had) – Liversidge sent Senate a request for £5000 to 'remedy existing deficiencies in apparatus', and for £2000 to purchase more.[93] Medicine was asking for twice as much money as any of the other science departments, and for chemistry, Liversidge managed to get only £254.[94] But he needed at least £1845 to keep the Faculty in business. For 1901, his maintenance vote reached pre-depression parity at £1250, and for 1902, John See's government raised the scientific apparatus vote to £2000,[95] which came on top of £855 arising from balances on current account.[96] This kept science going. Yet, relative to increasing costs, what had been adequate ten years earlier, was no longer so.

Ironically, the University was beginning to suffer from its own success. 'At one time', Theodore Gurney told the first Colonial Universities Conference, meeting in London in 1903, 'Sydney University was looked upon as an institution belonging to the lettered class, the wealthy class. Now it is recognised as belonging to everyone, poor and rich. Everyone is encouraged to come, and the good are encouraged to stay'.[97] This was not quite true. Not 'everyone' came, and few of the 'good' actually stayed in academic life, as there were limited prospects for an academic career. What was true for men, was much truer for women.[98] Not until 1908 was a woman employed on the academic staff at Sydney, and she was a junior demonstrator.[99] Even so, Sydney (Melbourne was about the same) now boasted some 600 undergraduates, including 200 in Arts and 200 in Medicine. The University's buildings were bursting at the seams. No new structures had gone up since the improvements to the Medical School, and with increasing numbers using the Blacket Building, everyone felt the congestion.[100] Liversidge pressed for a new biological laboratory and for extensions to physics, chemistry, and the outgrown engineering building. In 1901, he also supported the construction of a new University Union, combining the men's and women's common rooms,[101] although with little expectation that they would be sufficient.[102] To create more space, some on Senate envisaged the now familiar 'quadrangle', with the Blacket Building forming its eastern side, and the new Fisher Library (completed in 1910) forming its southwestern corner.

Summing up the state of the University in 1905, the *Herald* refused to judge harshly an institution, which, 'in anything like the form we now know ... hardly dates back earlier than 1885'.[103] This view applied equally to the Royal Society of NSW. In 1900, Liversidge was elected the Society's President for the third time. He spent a week in the Blue Mountains preparing his Presidential Address, which he delivered in May 1901.[104] His first address in 1886 had effectively launched the AAAS. His second, in 1890, had proposed a range of new projects – a Marine Biological Station, an expedition to the Antarctic, a program of research in the Pacific, and a 'Burlington House' to unite all the learned societies of Sydney under one roof. All this would keep the Society thinking for a decade, and much of it would come to pass. This time, however, he ran up proposals that were even more far-reaching.

Following Queen Victoria's death and funeral, the Society's annual meeting for 1901 was inevitably pitched, in the *Herald's* phrase, in 'a minor key'. But not only the Queen was mourned that evening in Elizabeth Street, when Liversidge surveyed the state of the Society. The Society was in jeopardy. Membership had drifted downwards from 457 in 1890 to 368 in 1901, and was now lower than at any time since 1885. Ironically, the extended suburban railways had not helped. Members living far from the city were reluctant to come 'on a hot summer day or on a cold wet evening'. Those that did could hear an average of twenty-one papers a year, but 'most of them', Liversidge admitted, were 'of a highly technical character'. The problem was familar: 'there are', Liversidge added, 'comparatively few who have a profound knowledge of a subject, and ... at the same time ... the gift of ... imparting their knowledge in a popular style'.[105] His solution, as always, was to have more, not less – more receptions and *conversazioni,* larger annual dinners, and a fresh campaign to recruit new members, notably from the state and private schools.

On behalf of the professional few, he appealed again to the middling many. Liversidge well knew that, with the rise of new professional associations, the Royal's scientific 'sections' had passed their zenith. Still, there were many things they could do, and Liversidge ended his address with three simple proposals. First, just as colonial scientists had been federated, now the scientific societies of the six states must also be federated, by the creation of a new Australian 'Academy' of Science for the Commonwealth. Second, Liversidge urged the Society to campaign for the introduction of the metric system to bring the country into line with the global marketplace.[106] Third, he asked the Society to promote the training of teachers; without more science teachers, schools could never send enough pupils to read science at university. The audience applauded. Liversidge's idealism was transparent. His program was a fitting start for a new decade.

Once again, Liversidge was to be disappointed. His proposed changes were a half-century ahead of their time.[107] However, thanks to his influence, the Society remained an open house for reforming ideas. From its podium, William Warren and Anderson Stuart appealed to the interests of 'national efficiency' and Francis Anderson, the Glaswegian moralist and professor of philosophy, paraded proposals for educational reform. In 1901, Anderson used the Society's forum to criticise the public school system for imposing rigid curricula and cramming, and forsaking the sciences.[108] Public outcry led to demands for a full enquiry, enabling the Royal Society to take credit for the decision of John (later Sir John) See's government to ask one of its members, George (later Sir George) Knibbs, to prepare the first study of primary, secondary and technical education in NSW since

Liversidge's report of 1880. This proved to be also the first comparative study of American as well as European educational institutions commissioned by an Australian government.[109]

It was particularly fitting that credit for the report was shared with J.W. Turner, Liversidge's *protégé*, and his successor-editor at the Royal Society, who was also a member of his Faculty. The Knibbs-Turner report, which appeared in 1906, largely endorsed Anderson's criticisms, and foreshadowed sweeping changes in the secondary school system of NSW. The report also launched Knibbs' public career, and led to his becoming NSW Superintendent of Technical Education in 1905, the first Commonwealth Statistician in 1906, and the first Director of the Commonwealth Institute of Science and Industry in 1921.[110]

It was a measure of the bipartisan support accorded Knibbs' appointment to the enquiry of 1902 that the launching party, held by courtesy of the Lord Mayor in the Town Hall, brought together such contrasting figures as J.H. Carruthers, H.C. Russell, Sir Arthur Renwick, and professors Gurney, Butler, Anderson Stuart and MacCallum. Liversidge could not be present, but asked Anderson Stuart, in giving his vote of thanks, to call for the endowment of research and education, in the same spirit as their colleagues in England had done. If NSW did not do as much or better, he argued, 'they would find themselves very much behind even the old country in a very short time'. In its policy towards science and technical education, Knibbs agreed, 'lay the secret of the future of the Commonwealth'.[111] And if they were to have a future, the universities must provide more of what the Chancellor dismissed as 'bread and butter studies'. The new 'practical' culture would, the *Herald* reported, combine Arts and Science with 'the interests of the community at large'.[112]

The sentiment met an uncertain response. The University was preoccupied with reforms yet incomplete, and skirmishing continued on three well-entrenched fronts: the role of science at matriculation; the science content of the Arts degree; and the sequence and content of subjects in science, engineering and medical degrees. The last of these issues was the easiest, or the least controversial, as the sciences had compacted into a semblance of routine. But the first two were matters on which anyone could have an opinion, and many did. In 1901, Henry Anderson, Fellow of Senate and newly-styled Undersecretary of the Department of Agriculture, urged Senate to admit elementary science as a compulsory subject at matriculation for the Arts degree. His proposal was referred to the Professorial Board. Walter Scott's retirement in 1900 had removed a major opponent, and made a move in this direction possible. Predictably, Anderson's proposal was supported by the science professors. Less predictably, he failed to get their agreement to setting a compulsory science subject in place of a modern language.[113]

In the ensuing debate, Liversidge and David proposed an alternative strategy – a voluntary matriculation examination in chemistry and physics which, if passed, could exempt an Arts student from science courses thereafter.[114] This ingenious plan passed the Board, but not Senate, and thereafter lapsed. But neither Liversidge nor David was prepared to let science disappear altogether from the Arts degree. In 1903, with support from Medicine (including J.T. Wilson and Anderson Stuart), and over the objections of George Arnold Wood, they narrowly retained the existing stipulation of a 'full science course, with practical work', compulsory at some point during the Arts degree.[115] This requirement continued until the First World War.

These compromises were, however, patching up a curriculum that had fragmented almost beyond repair. By 1910, in Australia, as in America and Europe, the 'circle of the sciences' had come apart, and Science and Arts lived in separate, if neighbouring camps. For those few who matriculated in science, Liversidge maintained the possibility of their receiving a liberal education – but only in the sciences.[116] In their first and second years, BSc students received a broad exposure to chemistry, physics, biology, geology, mathematics and physiology. In their third year, they concentrated on two of these subjects. In breadth and depth, the BSc degree was roughly equivalent to the curriculum at Manchester and Birmingham; it was perhaps less demanding than the Natural Sciences Tripos at Cambridge, but gave greater choice than London or Oxford. Indeed, Sydney's BSc resembled its counterparts at Yale and Harvard, where proportionately few took a similar degree. At least, Sydney avoided the all-encompassing 'examination-mania', dictated elsewhere by the quest for honours. Honours in science at Sydney were given for aggregate marks of distinction, and awarded at graduation.[117]

It remains debatable whether University rules or secondary schools' neglect, was the more limiting factor in recruiting students to science. But an undisputed shortage of BSc students did not mean an absence of students to be taught science. As long as the first-year Arts requirement remained, so Liversidge's teaching grew.[118] Indeed, owing to the structure of first-year Arts (and perhaps, to the fact that he was the last professor to receive part of his salary in per capita fees), Liversidge lectured to virtually every student entering the University. Growing numbers spilled over into a demand for more evening courses in physiography and chemistry.[119] In 1899, he supported the promotion of James Schofield, from demonstrator in chemistry and metallurgy to evening lecturer in chemistry to help deal with classes that now exceeded 100.[120]

The early years of Federation gave Liversidge few moments of reflection of the kind that Manning would have relished. However, a visit from the Duke and Duchess of Cornwall brought the University to international notice in 1901, and the University's Jubilee in October 1902 gave Senate a chance to boast its achievements. For the occasion, the Registrar produced a *Short Historical Account*.[121] Fortunately, the University had news to celebrate. With the *University Endowment Amendment Act* of 1902, the government restored the grant to £10,000 which, with the income generated by the Challis bequest, raised hopes of making desperately needed new appointments. The professor of modern languages still held responsibility for all teaching in English, French and German, and there were as yet no faculties for education, veterinary medicine, architecture or agriculture. On the other hand, the University could (and did) point to its success in mining engineering and medicine; and as if to underline the point, in February 1902, approved Edgeworth David's requests for new lecturers in mineralogy and petrology and palaeontology.[122] The University's newest medic, D.A. Welsh, Professor of Pathology, arrived in March 1902, and by May the 'science strength' of the University was looking respectable.[123]

Of course, in respect to research, the story was different. When it came time in the Jubilee celebrations to speak of science, the Dean wisely gave the task to the mellifluous David. Undaunted by vice-regal ceremony, David seized the opportunity to air the University's grievances. 'We have', David said, speaking in the Great Hall to an audience of visiting dignitaries, 'maintained a thoroughly open-door policy in our science teaching

... the science classes and laboratories of the University are open to all comers. Here, truly there is in our University what was Huxley's ideal – a ladder reaching from the gutter to the University'. But what the country needed in research, had yet to be found – 'an ever-vigorous, active and constantly renovated aristocracy, welcoming every intelligent and noble mind'. The explanation was part of a wider problem. Australian science, David said, suffered under a disability from which Europe and America were exempt: 'there is ... little or no scientific opinion in the people of Australia'.[124] How best to foster 'scientific opinion' in the University, and in Australia generally, remained a task for the coming generation.

Between 1852 and 1904, Sydney had awarded 1104 degrees in Arts, but only forty-seven in Science and no research doctorates at all. Against this stood the triumphant progress of university science in Germany and America. The only university in a largely agrarian state still lacked a chair in agriculture, and had no experimental farm. Australians had yet to explore fully the heartland of their country, their oceans, or their coastlines.[125] 'The land and fresh water fauna and flora of Australia are matters of almost as great urgency as is the ethnological study of our aborigines [sic]', David argued. The agenda was endless.

Following the Jubilee, and for the remaining months of 1902, David and Liversidge returned to their classrooms and laboratories. In August, they collaborated in a report on the implications of the new *Mines Inspection Act* of 1901, which appeared to disadvantage graduates in mining and metallurgy,[126] and towards the end of the year, Liversidge took up again his old interest in meteorites. He had sent specimens of the Thunda (Queensland) meteorite to Lazarus Fletcher at the British Museum in 1891, and followed this in 1897 with fragments of the Barratta meteorite that Russell had collected in the 1870s.[127] This time, however, Liversidge sent not only mineral specimens, but analyses as well.[128] His laboratory helped extend his range.[129] In the twelve months from September 1902, he published five papers on the presence of rare metals in meteorites, and on their significance to planetary mineralogy. He was among the first to detect meteoritic gold and platinum, and was pleased to find his results confirmed in England.[130]

These were, however, solitary achievements. Sydney needed research students and chairs in fast developing fields in which Melbourne was taking the lead. Liversidge could not compete with Masson's brilliant chemical quartet, N.T.M. Wilsmore, G.W. MacDonald, D.H. Jackson, and Bertram Steele, all of whom later became professors of chemistry. Sydney remained a teaching laboratory, rather than a research school. Still, what talent Liversidge found, he encouraged, and increasingly filled junior positions with Sydney graduates, rather than looking to London. In March 1901, to replace George Harker, he appointed another former student, Thomas Laby, who soon rewarded his 'talent-spotting' in greater measure.[131]

Laby came from a family of modest means, whose education was interrrupted by the death of his father. Although he passed the Senior Public Examination in mathematics and history, he was deemed unqualified to matriculate, and took a job in the taxation department. He came into science, and probably to Liversidge's notice, through the 'second portal' that was opened by the Technical College, where he enrolled in practical chemistry. In 1899, he was recruited by F.B. Guthrie to the chemical laboratory of the NSW Department of Agriculture, where he and Guthrie wrote a joint paper on commercial

Liversidge and Professor J.W. Gregory (centre) walk under the University clock tower during the University's Jubilee, 1902.

fertilizers. On Guthrie's recommendation, Laby came to Liversidge, and took evening classes at the University in chemistry, physics and mathematics.

Liversidge found in Laby a willing, adaptable worker, and soon put him onto the chemical analysis of meteorite samples. There Laby described the 'inadequacy and tediousness' of current methods for separating iron from nickel and cobalt, and, with Liversidge's encouragement, published his first results in *Chemical News*. His skill in analysis led him to the measurement of weight loss in 'chemical transformations', which led in turn to the 'question of the hour', the measurement of radioactive decay. In 1904, Laby wrote a second paper, this time jointly with Douglas Mawson, then a recent graduate in engineering, whom Liversidge had appointed in 1902 to his second junior demonstratorship in chemistry. Mawson made a name for himself in analysis well before departing for his polar adventures. The paper for the Royal Society of NSW by Mawson and Laby, which Edgeworth David delivered, was on the occurrence of radium in Australia. With self-made apparatus – Liversidge's training stood them in good stead – they had analysed pitchblende, and brought closer the emergence of Australia's uranium industry.

During Liversidge's oveseas leave in 1904, Laby was one of several who stood in for him, helping Guthrie as a demonstrator in the chemical laboratory until 1905 when, backed by Liversidge, he was awarded an 1851 Exhibition to England.[132] Laby first set out for Birmingham, where research in radioactivity was led by John Poynting, FRS, Professor of Physics at Mason College, but he was later advised to move to Cambridge, where he worked with J.J. Thomson at the Cavendish. In 1907, Laby took the new Cambridge 'BA by research' degree with a thesis on alpha particle ionization. Then and later, Liversidge proved a good friend,[133] and lived to see Laby launch a distinguished career in physics, beginning at Wellington in 1909, then at Melbourne from 1915.[134]

At Sydney, as in Cambridge, Liversidge's junior demonstratorships became a 'fast track' for talent. Of the fourteen who held these positions between 1895 and 1907, five joined the research community. Their overall record was not always brilliant, but it was one that neither Pollock in physics, nor Haswell in biology, lacking such positions, could match.[135] Given the dearth of permanent appointments, few could remain at Sydney very long, but there were some who put in a good innings. For example, Mawson's successor, Julius Hogarth (appointed in chemistry in 1904), worked at the University until the Great War, after which he taught at Sydney Technical College. On the other hand, Arthur Jarman, Liversidge's demonstrator in assaying and chemistry, stayed only five years before moving to the chair of mining at the University College of Auckland.[136] His successor, Frederick Eastaugh, was, like Jarman, an associate of the RSM, and might well have done the same. However, he remained at Sydney, where he later (1938) became a Professor of Engineering, and a pioneer in chemical engineering.

Temptations to leave the University were hard to resist. Edward Pittman, borrowed by Liversidge from the NSW government as a temporary lecturer in mining, would have made a fine academic career, had he not been promoted in 1902 to become the Undersecretary for Mines and Agriculture.[137] George Knibbs, a foreman's son, became a surveyor, before becoming for six years Liversidge's omnipurpose lecturer in geodesy, astronomy and hydraulics in engineering, until destiny called him to higher things.[138]

In mining, assaying and metallurgy, lecturers from England tended to teach what they had learned in England, and only then made their way into Australian fields. But with Harker, Laby and Mawson, Liversidge began a new tradition – setting specific problems, frequently of a practical kind, sometimes of commercial significance, generally requiring little more than homemade apparatus, and made appropriate to the interests of government, the AAAS, the Royal Society, or all three. Such research usually focused on Australian or Pacific materials, but its results could also be aimed overseas. In this way, he slowly 'reproduced' a small scientific class. Among its members was Sydney's first science doctorate – awarded in geology – to Walter Woolnough (DSc 1904), who studied with Liversidge and David, and who sailed with David's expedition to Funafuti. Another was Liversidge's first chemical doctorate, James Petrie (DSc 1905), a migrant from Edinburgh who, as Liversidge's 'additional demonstrator' in 1895, had analysed mineral oils extracted from torbanites, a subject that had interested Liversidge since 1881.[139] Petrie also did a detailed analysis of piturie, a natural alkaloid of great pharmaceutical interest, which Liversidge himself had first examined in 1880.[140] Liversidge sent a sample of the substance to Foster's physiological laboratory in

Cambridge, where Langley used it in his research on the nervous system.[141] In Sydney, it led to new work in plant physiology and biochemistry, to which Petrie became a distinguished contributor.[142]

Given the breadth of his interests, Liversidge's laboratory produced a succession of what now seem fragmentary studies, canvassing questions for their intrinsic and occasionally practical interest, without any conspicuous reference to an overarching theoretical program. Unlike Liversidge, Anderson Stuart, Wilson and Martin created a research school in physiology that rivalled Britain. Indeed, it is difficult to locate – either in Liversidge's publications or in those of his students – a unifying theme comparable to contemporary Manchester or Melbourne. Liversidge's 'school of science' continued to resemble the RSM more than a modern chemistry department. Like the RSM, it created its own discourse, notably in chemical geology, mineralogical chemistry, chemical and mining engineering.[143] But in other subjects, Liversidge knew he was losing ground.

This was the official reason why, in August 1903, Liversidge applied for study leave overseas.[144] His application cited his desire to 'visit Europe, and if possible, America', in 'connection', as the Registrar put it, 'with the branches of study and learning pertaining to his Chair'.[145] In fact, exhausted by administration and teaching, he wanted 'to ascertain the progress which is being made in chemistry and applied sciences in the United Kingdom and on the Continent' so that he might bring Sydney up to the moving front.[146] In passing, he also wished to attend the foundation meeting of the International Association of Academies, and a meeting of the International Commission on the Catalogue of Scientific Literature, which had been set up following the third London conference on scientific periodicals in 1900.

Liversidge's leave was approved in December, on the usual condition that he pay his replacement. In the event, it took no fewer than four men, whose salaries came to £750 of his composite pay of £1300.[147] As in 1896, he also had to find a replacement dean. Henry Russell, Liversidge's oldest friend, could have been asked, but he was seriously ill, and had in any case begun a year's leave himself, which would be followed by retirement from the Observatory.[148] Fortunately, Liversidge's willing deputy was at hand, and Edgeworth David became Acting Dean, so that Liversidge could in good conscience embark on his fourth voyage 'Home'.[149]

## 4. English Leave

The hot Sydney summer had begun in earnest by 9 January 1904 when, with his 'Octagon' safely looked after by the young H.S. Carslaw, Liversidge sailed for England. Six weeks later, on 21 February, the *Himalaya* docked at Southampton, and the following day, Liversidge again installed himself in the Royal Societies Club in St James's Street. Seven years and an imperial war separated him from his last visit to London. The papers from Sydney brought him news of the turmoil following George Knibbs' proposals for educational reform. But he was to be spared for a year from Australian debates. Even so, he soon became caught up, and not entirely against his will, in the politics of British science.[150]

Liversidge set out to do England at his usual punishing pace. Within days, he had met colleagues in the University of London, in the City, and at the British Museum of Natural History.[151] Within weeks, he began to travel around the country, always returning to London. On at least three occasions he dined by invitation with the Royal Society Club;[152] and at one such meeting, in Burlington House in May, he exhibited photographs of the Narraburra meteorite and described his discovery of gold and platinum traces.[153] That meeting coincided with a *conversazione* of astronomers, including Sir Norman Lockyer, Sir William Abney, and Professor George Ellery Hale of the Yerkes Observatory in Wisconsin, soon to go to Mt Wilson in California. In April, Liversidge sent a congratulatory message to Sir Henry Roscoe, then celebrating his 'graduation jubilee' in Manchester, in the company of many friends, including Sir William Ramsay, Sir Michael Foster, and Baldwin Spencer, then on leave from Melbourne.[154]

These were moments in what, by spring 1904, had already become a very busy year. Since his last visit to England in 1897, the spectacle of blunder and inefficiency in the South African war had shaken the British establishment, and united a broad spectrum of Liberal Imperialists, Tory Unionists, and new Fabians in pressing for reforms in the conduct of domestic and imperial affairs. 'Wake up, England,' the Prince of Wales had told his subjects in 1900, and Beatrice Webb was not alone in asking whether the English gentleman 'with his so-called habit of command ... [was] the equal of the foreign expert with his scientific knowledge'.[155] In a celebrated essay in 1901, Sidney Webb prophesied that:

> No leader will attract the support of the mass of unpolitical citizens ... without expanding his thesis of national efficiency into a comprehensive and definite programme .... He will in fact, lead the English people – eager just now for national efficiency, they care not how – only by becoming a personified programme of national efficiency in every department of life.[156]

Speaking in Glasgow in March 1902, Lord Rosebery, perhaps the leading Liberal Imperialist, defined 'national efficiency' as:

> a condition of national fitness equal to the demands of our Empire – administrative, parliamentary, commercial, educational, physical, moral, naval and military fitness, so that we should make the best of our admirable raw material.[157]

By 1902, reformers viewed with alarm the competitive presence of Germany, Japan and the United States in world affairs.[158] Britain's institutions, it was said, could not be efficient until they were 'scientific', that is, by cultivating the application of 'scientific method'.[159] The methods of science were vital to national survival.

To make better use of Britain's resources, Rosebery, Lord Haldane and Arthur Balfour crossed party lines to increase provision for technical education and research. In 1902, Edward VII instructed the Eighth Duke of Devonshire, together with Lord Rosebery and A.J. Balfour, to draft proposals for a new college in London, organised along lines similar to Berlin's famous Technische Hochschüle in the Charlottenburg district of the Prussian capital. The same year, the London County Council, guided by the Webbs, issued a damning report on the state of British industrial (and technical education) and in 1904, Balfour's Board of Education decided to create a 'Charlottenburg' in South Kensington by consolidating the Royal College of Chemistry, the Central Institution of the City and Guilds, and the Royal School of Mines. This led in July 1907 to the incorporation of the Imperial College of Science and Technology.

With all these events in train, a second front against 'inefficiency' was opened by Sir Norman Lockyer, civil servant, former secretary of the Devonshire Commission, founder and editor of *Nature*, and self-appointed representative of the reform wing. Since the 1870s, Lockyer's editorials had argued that haphazard encouragement of scientific genius was no substitute for systematic support.[160] In 1903, his campaign invited volunteers, and Liversidge was quick to enlist. 'The opening of the twentieth century', Beatrice Webb had written, 'finds us all, to the dismay of the old fashioned individualist, thinking in communities'. In their usual way, the Webbs worked through dining clubs – including one, dubbed by Beatrice 'the Coefficients', that included J.J. Thomson. It was not difficult for Liversidge to make common cause.[161]

In underwriting ties of tradition and culture, and in constituting what Lord Milner called 'organic Empire unity', science had an important part to play. In his foundation week oration on the 'Functions of a University', delivered at University College in June 1901, Sir William Ramsay suggested that if the work of education included the training of 'men and women for the manifold requirements of the Empire', then 'the best way of doing so' was 'to give them the power of advancing knowledge' through science.[162] His message was underlined by the first Allied Colonial Universities Conference, held in July 1902 at the Royal Society of London when sixty delegates represented the seven universities and three university colleges of England, the four universities of Scotland, the four colleges of Ireland, fourteen institutions in Canada, three in Australia, and one in each of New Zealand and South Africa. Richard Threlfall and Theodore Gurney, now resident in England, were asked to represent Sydney. All were asked to turn their eyes 'forward into the future', taking pride 'in the fact that the British race has been chosen to be the instrument of the diffusion of science and learning in distant lands'.[163]

To perform this civilising role, the conference proposed an Imperial Council of Universities, whose task was to link students and teachers to 'mother England', and so nurture a 'common public opinion of the British peoples'. From this would come a division of labour in the world of learning, described by James Bryce, Oxford historian and former British ambassador to the United States, as one in which the 'sphere of the humanities' would reside in England, leaving the applied sciences – especially the 'practical' sciences – to the colonies. Lest this division stigmatise colonials as mere hewers of wood and drawers of water, Bryce laid upon Britain the obligation of recognising its 'young and bold' colonials, and to learn from them, if not by studying their example, then by receiving them into its institutions of higher education. Mutual cooperation, thus defined, would underwrite the 'unity of the British people throughout the world'.[164]

Reports of the Universities Conference reached Sydney, and Liversidge in London, through Gurney and Threlfall. Neither disagreed that scholarships were a good form of imperial cooperation, nor did they dispute whether Australia should send its ablest students to Britain. This might change, but not yet. The Conference distilled its programme into a single resolution: that an organisation be established to facilitate the regular exchange of information, and to ensure 'that special or local advantages for study, and in particular for postgraduate study or research, be made as accessible as possible to students from all parts of the King's Dominions'. The resulting council, the forerunner of the Empire (now Commonwealth) Universities Bureau, was predominantly British. There was no member on its founding Board from either Canada or Australia.

These developments did not fail to register with Liversidge, who closely followed news of the BAAS meeting at Southport in September 1903, a few months before he arrived. On that occasion, Sir Norman Lockyer had delivered one of the Association's more memorable Presidential Addresses, entitled 'The Influence of Brain Power on History'. Over a decade earlier, an American naval Captain, A.J. Mahan, had published a timely and influential survey of maritime strategy, entitled *The Influence of Sea Power upon History, 1660–1783*. Mahan's book went through many editions, and still commands a place on reading lists.[165] In its time, it constituted a plea for naval preparedness; but in 1903, it also suggested both title and text for Lockyer's campaign. Lockyer's address looked to America and the world overseas, and extended an appeal for the 'endowment of research' in Britain and throughout the Empire.

Preliminary steps towards the state support of universities in the United Kingdom had begun in 1889, when the Rosebery administration established a Treasury committee (the precursor of the late University Grants Committee) to allocate an annual vote of £15,000 among the country's universities and university colleges (excluding Oxford and Cambridge).[166] This was little enough – and in any case, Lockyer feared – too late. Britain, which had spent £120 million to modernise its fleet, was spending less than £13 million per annum on education. In the race among nations, victory would go to the clever and the swift, and to 'science and brains' even more than to 'swords and sinews'. The 'school, the university, the laboratory and the workshop' would be the 'battlefields' of the future. Scientists were becoming the new 'samurai', which for H.G. Wells, writing at the time, was more than merely a figure of speech.[167]

These arguments had been voiced many times. 'The reason why' the government had failed to respond, Lockyer argued, was because such 'appeals ... are the appeals of individuals; science has no collective voice in the larger national questions. There is no organised body that formulates her demands'. Sir William Huggins made a similar case in his Presidential Address to the Royal Society on St Andrew's Day 1902, by endorsing an educational partnership between science and the state.[168] Huggins agreed that, beyond the Royal Society, there was no effective science lobby in Westminster or Whitehall. In principle, the British Association might have done what Lockyer wanted; it had done so in the past. But its finances were limited; its impulse, peripatetic; and it could not report quickly on questions of the hour.[169] Specialist meetings held once a year were not the answer. '[O]ur great crying need', Lockyer concluded, was to 'bring about that organisation of men of science and all interested in science, similar to those which prove so effective in other branches of human activity. For the last few years I have dreamed of a Chamber, Guild, League, call it what you will, with a wide and large membership'.[170]

Lockyer first put his idea to a meeting of the British Association's Conference of Delegates of Corresponding Societies, representing 25,000 members throughout the country, an obvious constituency for any future 'league'. But the BA's council declined to have a special interest group in its ranks, particularly one with a political focus. Lockyer's proposals, when reported by nearly 200 newspapers, met surprisingly heavy criticism. Scarcely any two opposed him for the same reason. Possibly, as one of his biographers suggests, Lockyer 'had overemphasised the importance of science in estimating the national position, to the relative neglect of moral influences'. He had unequivocally 'put the case for the encouragement of science on a materialistic basis'.[171] The central proposal – to strengthen the 'knowledge base' in the interests of 'national efficiency' – was in danger of losing traction.

In matters to do with the financing of research, to paraphrase Sir William Harcourt, all scientists are politicians. Some are also statesmen. Lacking support from the BA, but with the backing of the Royal Society – notably, from Sir William Ramsay, Sir William Crookes and others who had taken part in the 'endowment of research' movement of the 1870s[172] – Lockyer decided to launch a new 'lobby'. From an organising committee meeting in June 1904 at Burlington House, emerged the British Science Guild.[173] That the Guild would find need of Empire loyalists was evident within weeks of its birth.[174]

In July 1904, capitalising upon the enthusiasm his proposal had generated, Lockyer convened the largest delegation of scholars ever gathered in Britain to discuss the future of imperial higher education with the Prime Minister, Arthur Balfour. Over 200 came from Oxford, Cambridge, and the other sixteen universities and university colleges of Britain and the seven of Ireland. Montreal's McGill University sent its Principal, Sir William Petersen.[175] From Sydney, and alone from Australia, came Liversidge. A prepared statement, written by Lockyer and published in *Nature*, was signed by Henry Pelham, Camden Professor of Ancient History at Oxford; Frederick Chase, Norrisian Professor of Divinity, President of Queens' College and currently Vice-Chancellor of Cambridge; Sir Oliver Lodge, Principal of Birmingham University; Sir Henry Roscoe from Manchester; and the ubiquitous Sir Michael Foster.[176]

In Lockyer's words, the scholarly community was united in looking for answers. 'It is a question of an important change of front, of finding a new basis of stability for the Empire in face of new conditions'. What should this include? Certainly, the leaders upon whom Britain depended 'must not', Lockyer proclaimed, be 'entirely untrained in the study of nature and causes of the things which surround them or of the forces which have to be utilised in our daily life'. Indeed, 'their training and education in humanities must also have been of the widest'.[177] To secure the 'two sides of a complete education' and to foster research in both, universities needed state support. The 'essentials' were no longer a few small rooms, books, and 'a small number of teachers of a small number of subjects'. Education for empire must be made 'efficient'.

What precisely this might mean – for Britain or for the Empire – Lockyer left unclear. But – given the presence of Joseph Chamberlain, Chancellor of the Exchequer (but speaking as Chancellor of Birmingham University), flanked by Sir William Ramsay, Sir Richard Jebb and Sir Henry Roscoe – Lockyer could not have had a more responsive hearing. Balfour promised more aid to the universities, encouragement to manufacturers who hired scientific talent, and assistance to the 'exceptional man' – an image as talismanic to the 1890s, as the 'practical man' had been to the generation of the 1870s.[178] Balfour might have gone even further. In May, Sir Michael Foster, now Liberal MP for Cambridge, advocated the creation of 'Regius' laboratories of science at the universities.[179] This, Balfour's government did not accept. But help came in other ways: the following year, the Treasury grant to the university colleges was doubled, with the promise of a further doubling in 1905.

For Lockyer and higher education these were heady days. Liversidge, responding to Lockyer's call, enjoyed being at the centre, even if representing the periphery. Liversidge quickly fell into step with James (later Sir James) Barrett, the distinguished Victorian physician, author, educator and politician, who favoured closer links between imperial and 'Home' universities.[180] For their own reasons, so did the representatives of Oxford, Cambridge and London. Some years later, Dr T. Herbert Warren, Vice-Chancellor of Oxford, told a colonial premiers meeting in London that:

> The universities have had a great part in bringing ... men together in their early and impressionable days, in giving them common sentiments and common loyalty and knowledge of each other. May it not be so still more in the future on a wider scale and in a wider way?[181]

This perspective had two important consequences. Whereas Australian universities had once been thought of as serving colonial interests, now they were seen within an imperial compass. Australia's universities were to provide young men (and women) to support British scholarship, much as Australia provided wheat, wool and minerals to the British economy. Some might remain in Britain; but others would return to Australia, the better to seal the imperial accord. 'Imperial efficiency' required Australian support, in both numbers and influence.

Such understandings, conspicuous in Britain, were less visible in NSW. However, Liversidge worked them into a larger conception of international cooperation. Since the Franco-Prussian war, the scientific academies of Europe had found agreement on a range of issues, such as scientific nomenclature, units of measurement, the coordination of observations, and more recently, the cataloguing of scientific journals.[182] In London, such cooperation was of special interest to the Royal Society, and especially to Sir Michael Foster. 'The problems of the future', he observed:

> must be faced by the best men, and why should not these men work together? Why should not the best men be selected – now an Italian, now a German, now a Frenchmen – because they are best to do the work for which they are best fitted? It is only in this way that we can get the best work done in the future.

Above all, he argued, 'what is wanted in science is organisation'.[183] In Liversidge, Foster found his man.

In 1900, a first General Assembly of what would become known as the International Association of Academies (IAA) took place in Paris. Its mission, in the words of the President of the Institut de France, was no less than that of replacing 'the struggles of war by those of

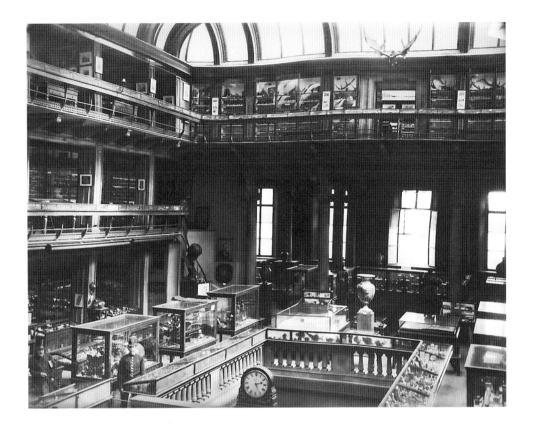

The Geological Museum, London, in 1900.

work'.[184] The vision found ready acceptance, and a second General Assembly was held at the Royal Society on 25–27 May 1904.[185] Foster ensured that Liversidge was invited. As there was not enough time to send an invitation through Colonial Office channels, Foster arranged for him to come as a delegate of the Royal Society instead.[186]

The London meeting of the IAA was a splendid occasion, possibly the most important international and interdisciplinary gathering in first decade of the twentieth century. Scholars from fifteen countries were welcomed by the Lord Mayor at the Mansion House, and received by the King at Windsor.[187] To some delegates, Oxford and Cambridge gave honorary degrees. At a reception to honour them all, Liversidge repeated his 'show' of meteorite photographs. The conference produced plans for international projects in meteorology, geophysics and polar studies, and considered a draft report from the International Council of the International Catalogue of Scientific Literature, to which Liversidge spoke as the representative of Australia.[188] Dinner at the Metropole united him with twelve British colleagues, including W.T. Blanford (RSM, 1852–54), who had retired from the Geological Survey of India in 1882.[189] The IAA's charter, which ultimately enlisted twenty-two countries, did not survive the coming war. But it set a precedent for the International Council of Scientific Unions and other postwar scientific organisations in which Australia would play an important part.

Through many of these proceedings, Liversidge lent a dimension to British plans that would otherwise have been imperial in name only. Not for the last time would an Australian have an influence out of proportion to the size of the country he represented. That summer, in recognition of his imperial service, Liversidge was nominated by George Liveing and John Wesley Judd for the Royal Society's prestigious Royal Medal, of which only two are given annually.[190] This high compliment cited Liversidge's 'investigations of the minerals, meteorites, rocks, and deposits of springs and kindred other substances occurring in Australia and the neighbouring islands, published in a very long series of papers in the "Transactions" of the Royal Society of NSW between 1872 and the present'.[191]

For reasons that remain unclear, his candidature failed; the medals went to the mathematician William Burnside and the pathologist David (later Sir David) Bruce. No report of his nomination survives among his private papers, and of his friends' valiant efforts, he may never have known. But recognition comes in many forms, and in August, still basking in his imperial celebrity, Liversidge was invited to a planning meeting of Lockyer's newly-founded guild. There, he was hailed as Australia's apostle of efficiency, and assured Lockyer that he would carry the message of 'Wake up England!' to the people of New South Wales.

The following week, it was the turn of Cambridge to host Liversidge. That year, the British Association was holding its annual congress in the celestial city. Only 180 delegates were present, but a visit from the Honorary Secretary of the AAAS was a special event. 'For my own part', Liversidge reported, 'I met there more scientific gentlemen than I could possibly have otherwise hunted up in the time at my disposal'.[192] While most stayed in hotels, he stayed at Christ's and dined with Arthur Shipley. The congress over, he returned to London, called on Huggins at the Royal Society,[193] and visited Ramsay at UCL, to discuss current work on radium and radioactivity. 'The great difficulty', Liversidge recalled, with mild understatement, 'is to grasp the substances'; and the greater difficulty, 'that of obtaining quantities of radium of sufficient size'. On all, he gathered notes, so as to explain to Sydney the experiments then being watched so avidly in London and around the world.[194]

Among his chemical colleagues, and in the pages of *Chemical News*, Liversidge found a cordial reception. Only once did he face a sour review. His small volume of *Tables for Qualitative Chemical Analysis*, first published in 1881 to meet the needs of Sydney students, was issued in a second edition by Macmillan that summer.[195] The book's intention was simply to render any student, 'equipped with blowpipe and candle', independent of a laboratory and employable anywhere, whether in town or in the bush. The *British Medical Journal* praised it as welcome if it could prevent the 'slovenly methods of manipulation so well-known to every teacher of chemistry'.[196] That was, indeed, the intention. But it was precisely Liversidge's erudition that annoyed *Nature*'s reviewer, who thought that 'To thrust this work upon a beginner who is not a specialist is almost equivalent to expecting a student of mechanics, who is not to be an engineer, to work a lathe or use a planing machine'.[197] Such criticism was an unintended compliment: his expectations were exactly those that *Nature* intended. Liversidge had to make chemists of men who were unlikely ever to get more instruction than he was able to give them.

As his leave year drew to a close, Liversidge devoted himself to the usual tasks of 'academic housekeeping'. He shopped for new apparatus, and viewed with mixed envy and pleasure the success of South Kensington, to which Balfour had recently promised a new chemical and physical laboratory. In the Midlands, a new school of mines had been proposed for Birmingham, and throughout the kingdom, chemistry was reaching larger numbers of students. Australia, despite its brave efforts in the 1880s, was being outrun.

However, there were ways that Australia could join in Britain's success, and these Liversidge carefully noted. One lay through an extension of the British Science Guild. In October 1905, seven months after Liversidge returned to Sydney, Lockyer formally launched the Guild at a ceremony in the Mansion House. With Lord Haldane as its first president, the Guild was determined to secure the cooperation of 'all occupations and all classes throughout the British Empire', so as to achieve no less than the 'application of science and scientific method in all departments of human endeavour'.[198] By 1907, its membership had reached nearly 600.[199] No corner of the Empire would escape its masonic eye. On 6 December 1905, the Guild's General Committee resolved to form affiliated societies and branches in the provinces and the colonies. And the first of these 'local' committees, Lockyer reported in January 1907, would be established in Australia – 'thanks to the kindness of Professor A. Liversidge', Statesman of Science.[200]

# CHAPTER 12

# Downhill All the Way:
# Reflection and Retirement

Sir George Le Hunte, Governor of South Australia, receiving AAAS members at Government House, Adelaide, 7 January 1907. Liversidge stands to the right of the woman in the white gown.

At the end of 1904, Liversidge dutifully packed his bags, notes, apparatus, and memories of an eventful twelve months. On the second day of the New Year, he gave a farewell dinner at the Royal Societies Club; on 20 January 1905, he sailed for Sydney. After an uneventful passage, trans-shipping in Melbourne, he arrived on the *Marmosa* on 3 March – as usual, not a day too soon for the start of the academic year.

This time, Sydney Harbour was an anticlimax. Liversidge had enjoyed England, and he parted with regret. The British universities were alive with the sounds of building. Interviewed at quayside by the *Daily Telegraph*, he spoke admiringly of the new polytechnics, multiplying throughout the kingdom, doing valuable work 'in some instances as advanced as anything attempted at our own University'.[1] Cambridge, Manchester and London were opening new frontiers in chemistry and physics. The sight of so much activity had been both exhilarating and depressing. Liversidge brought back the £350 worth of apparatus he had been authorised to buy for his department and the Mining School. 'I could have invested double that sum with profit had it been available', he added, with a hint of bitterness.[2] He returned to a Faculty not yet touched by the 'research ethos'.

The *Herald* sympathised with the plight of the professor, that first April of his return: 'the author of an ideal republic of learning would hardly take Sydney as the fabric of his dream'.[3] Speaking at the University's Jubilee in 1902, Sir Samuel Griffith, a Sydney graduate, now Federal Chief Justice, had said: 'The great defect in Australian life is the want of apprehension of the value of knowledge in itself'.[4] The same year, Sir James Barrett, later Chancellor of Melbourne University, cited Sydney's virtues, in recommending to Melbourne a science degree of the kind available there and in America.[5] This he followed in 1912 with a visit to the United States, where he was 'deeply impressed with the universities' energy, drive, liberal support and receptivity to new ideas'.[6] As long as Australia's universities were dominated by elementary teaching, their staff shackled to large and compulsory classes, research could hardly be expected from them. In 1900, there were twelve professors. But of the 291 papers published by University staff between 1852 and 1905, over 100 had been written by Liversidge alone.

## 1. Beyond the Jubilee

With some justice, Sydney's press conveyed a popular image of Australia's universities as pleasant backwaters, on whose tangled banks tradition flourished. Of course, this was an exaggeration.[7] Many staff were demanding flexibility, imagination, and a place for their country on the 'changing surface of knowledge'. Francis Anderson, the philosopher, whose criticisms of Sydney's secondary education had helped launch the Knibbs-Turner report, lent ammunition to Liversidge, when he wrote that 'universities must be tested by the way in which they respond to the ever-growing demands of national life', a sentiment shared by advocates of 'efficiency' everywhere.[8] Speaking at the University's Commemoration Day in 1905, the Chancellor, Sir Normand MacLaurin, agreed that the 'matter in hand is to raise the culture of the nation to a higher level'. And if, as was popularly believed, the University had not 'reacted to the modern spirit so completely as it will have to do', yet it held, 'within itself the full possibility of progress'.[9] A cluster of impressive buildings had arisen along what was becoming known as Science Road. The

University's goal, the *Herald* observed, was to see that 'no affection for tradition limits our eagerness toward, or our capacity for improvement'.[10]

During Liversidge's absence in 1904, many changes had been proposed, and some had come to pass. The coming of the Commonwealth had returned a degree of optimism to university life, and for a short time, a little more money. In December 1904, the Senate revived recommendations for new faculties in commerce, music, agriculture, education and architecture.[11] For these, there was not yet money; but for certain new departments – dental studies, public health and military studies – provision was being made, and the rest was promised.[12] The Professorial Board, after two years' consideration, had reviewed again the matriculation requirements for Arts, and slightly eased the way for those studying science at school to enter university.[13] Senate also approved a new Arts curriculum, which gave a greater choice of subjects, whilst retaining a compulsory science course with practical work. Latin remained compulsory for Arts, and either Latin, Greek, French or German for Science. Despite his absence, Liversidge had at least not lost ground.

The Faculty of Science had introduced several refinements. Liversidge's colleagues had reduced first-year subjects from five to three, requiring everyone to do chemistry and physics, but giving a choice between biology, geology and mathematics. They also extended the degree course in mining and metallurgy to four years (as he had proposed), while his 'ring-ins' took steps in other directions that he had anticipated, but not pursued.[14] In April 1904, for example, Frederick Guthrie began the University's first lectures in the History and Philosophy of Science.[15] Later in the year, James Schofield organised a summer course in laboratory chemistry, which during January and February 1905, packed 200 secondary teachers into a laboratory designed for forty.[16] The Faculty revised its financial arrangements; student fees were henceforth shared between professors, demonstrators and the University, creating useful funds for assistants, and, perhaps less usefully, for administration.[17]

In 1905, the Government acceded to P.N. Russell's gift, for which it granted £25,000 for the construction of new engineering buildings. With the state's budget for that year came £2500 for maintenance; and although Medicine took the lion's share (with chemistry and mineralogy allocated only £555), no one (except perhaps Liversidge) took this badly.[18] The Medical Faculty was now the second largest in the University, and arguably the third largest in the Empire.[19] If Anderson Stuart 'did little [himself] to further the discipline of physiology',[20] he made many brilliant appointments, including Charles (later Sir Charles) Martin (at Sydney, between 1891–97) and J.T. Wilson; Martin was elected FRS in 1901, and Wilson followed in 1909. To those who had forgotten how long it had taken to reach even this state of affairs, the *Herald* reminded its readers that the University's entry into the 'scientific age' dated only from the 1880s, and even then thanks principally to the endowments of its philanthropic foursome – Challis, Fisher, Russell and Macleay. 'When one remembers all these things, the efficiency reached is little less than astonishing'. All the same, the *Herald* warned, 'it would be more than astonishing if those who can accomplish such things as these should be heard to sigh because they had no more worlds to conquer'.[21]

There were indeed many worlds left to conquer. Continuing criticism of the University played on timeless themes. The nineteenth century Australian public had no more sympathy for 'ivory towers' than their twentieth century successors. In any case, the

SYDNEY UNIVERSITY AND
GROUNDS, 1903

1. University Main Building
2. Great Hall
3. Fisher Library
4. Men's Common Room
5. Women's Common Room
6. Medical School
7. Department of Chemistry,
   Metallurgy, Assaying
   and Mining
8. Department of Geology
   and School of Mines
9. Department of Physics
10. Department of Engineering
11. Department of Biology
12. Macleay Museum
13. Gardener's Lodge
14. Messenger's Lodge
15. Caretaker's Lodge
16. Cricket Ground
17. Attendant's Lodge
18. Tennis Courts
19. St Paul's College
20. St John's College
21. St Andrew's College
22. Women's College
23. Prince Alfred Hospital

real point of comparison was becoming America, not England; and in the United States, pontificated the *Herald*, 'manhood is mainly practical and directed towards a practical end'. Australian universities (which had admitted women since 1881) would be 'manly' if they were to be similarly directed, their degrees, 'equipping youth for the duties of life'.[22] Publicly, if cautiously, MacLaurin the Scot went far beyond Manning in embracing the program this ethos implied: 'Every practical art', he said, 'which can be conducted on the whole in the intellectual way may, as I believe, if circumstances permit, be considered to be not unworthy of the attention of the University'. Research would be encouraged, 'so soon as the resources of the place will allow'.[23]

As always, resources remained a problem. But Liversidge's problem ran deeper. The University's governance continued largely in the hands of its conservative Senate. Of its twenty members, four were ex-officio representing the four faculties, while the remaining sixteen were elected by convocation and appointed for life. In 1906, all but three had been appointed before 1890. Of the sixteen most were lawyers – six, in 1905, were judges – who resisted the claims of research, preferring – in Anderson Stuart's phrase– 'that the professors' time ought to be given entirely to teaching'. Sydney's reputation in the arts was beyond reproach (or at least, beyond discussion). Surely, Senate argued, the University's first duty was to the fundamentals of undergraduate education in all its disciplines.

Precisely how that duty should be discharged, however, remained an open question. Outside the gates, the curriculum had few friends and many critics, particularly of the first year. Of Sydney's 800 undergraduates in 1905 (up from 598 in 1892), more than forty per cent read arts and twenty-five per cent read medicine. Science and Engineering still accounted for less than a third. Many Arts undergraduates were teacher-students, seeking degrees to qualify them in their profession. Taken together, a mixed company of this kind would be unlikely to find Sydney University an adventure in ideas. Nevertheless, for seventeen years, the regulations for first-year Arts, in place since the fratricidal days of 1887, had not substantially altered. All first-year students in all four faculties were still required to take English, Latin, another foreign language, mathematics and elementary science.[24] Science was interpreted as one term of chemistry, one term of physics, and one term of natural history.

This definition of science, and the way it was taught, was the principal target of debate. One of the most vocal protests against the curriculum since the University's foundation erupted during Liversidge's absence, at a conference held to review the recommendations of the Knibbs-Turner report. On that occasion, A.O. Black, the NSW Inspector of Public Schools, took the podium to denounce the University as 'steeped in mediaevalism, and hidebound by the traditions of long past ages'. The place, he suggested, needed 'an alpine avalanche, or an eruption like Krakatoa, to sweep away the musty traditions of bygone days' which, he said, 'had grown up round her, stifling all growth, and preventing her from taking her proper place at the head of our national education'.[25] This was sensible, if abrasive, advice and bore equally on all subjects. But Black's special fury was reserved for Liversidge, whom he felt symbolised the past, not the future, of science education. Liversidge's didactic methods, stressing careful procedures and Huxleyan 'object lessons' was, he considered, tantamount to teaching Latin by dictation and mathematics by rote. Flushed with enthusiasm for the progressive 'heurism' that H.E. Armstrong had popularised in England since the 1880s, Black wanted teachers to set simple problems, leaving pupils to find solutions independently. Student projects, along German lines, were said to encourage independent thinking, and to rescue learning from the stultifying habits of set pieces and drills. Such had been the practice, after all, of the Royal College of Chemistry in Liversidge's day, although what was done then was hardly glorified as such.

The irony was oppressive. Having always seen himself in the progressive party, Liversidge was now vilified as a symbol of the past. In fact, he was no more or no less guilty than most of his contemporaries, and his methods were in use until well into the twentieth century. Questions of method and pedagogy sharply divided science educators everywhere. In London, Sir William Ramsay objected to Armstrong's heurism as an expensive fallacy. 'I believe in facts', he said, 'and I believe that the basis of theory is not grasped until the facts on which it rests have been learned'.[26] Content before method was the dominant creed, and one which Liversidge shared; unprogressive, perhaps, but still deeply implanted in the best universities of Britain and the United States.

In Liversidge's absence, it fell to Edgeworth David, as Acting Dean, to reply to Black's criticisms. Defending the University and its methods, he admitted there was always room for improvement. But, he added, given the 'distance of our University from the great teaching centres of the Northern Hemisphere ... science-teaching here, if it is to be judged by its fruits, is as well up to date as the circumstances will admit of'.[27] David then proceeded, slightly wide of his critics, to march past a roll of Sydney's science greats, including W.A. Haswell, who with T. Jeffery Parker had written one of the most widely used zoology texts in the English language;[28] J.P. Hill and James Wilson, who had unravelled the status of the platypus; Richard Threlfall, now long gone, but well remembered as a pioneer in electrical physics, and James Pollock, who succeeded him. Above all, David added, with a note of defiance, it could be said that of the University's eight professors of science, medicine and engineering, three were Fellows of the Royal Society of London, a record equalled by few British universities outside Oxbridge and London.

As for Liversidge's teaching methods – well-schooled lectures, careful dictation of elementary principles, basic laboratory technique in chemical analysis – these, David explained, were recognised as the 'best and most expeditious for working out definite results'. They incorporated a philosophy of undogmatic empiricism that Liversidge (and many others) believed 'best for this country'. Such pedagogical methods were all the more necessary, David added, as long as students came to University with little scientific background, unprepared for independent work. First-year Arts students had, at best, rudimentary exposure to practical work and even science students were asked to begin from first principles. 'It is heartrending', the *Herald* observed, 'to see men being taught to read a barometer, or rushing feverishly through a course of simple analysis'.[29] Liversidge himself warned against stressing analytical work to the extent that it 'degenerated into mere rule-of-thumb use of tables'.[30] But surely Black would appreciate there was little hope for progress until school science improved? Loyally supported by his colleagues, Liversidge's reputation survived the ordeal.[31] Black's views were to prove central in the reform of the state's secondary school system in 1911–12. But it was not a pleasant note on which to welcome Liversidge back to Sydney.

## 2. Prometheus Bound

Unhappily, this note set the tone of Liversidge's last three years at the University. Between 1905 and 1907, he grew steadily more disenchanted. For most of the 150 or so largely uninterested, ill-disciplined seventeen year-olds taking first-year Arts, Liversidge's course in inorganic chemistry was their first exposure to the culture of science. His lectures were delivered in the large hall of his new laboratory (today, the main lecture room of the Pharmacy Building), which for decades afterwards bore the imprint of student numberplates, like police identity tags, on its tiered benches. Meeting three times a week, confronted by Edward Hufton's careful preparations, overburdened with facts, the course sorely tested both teacher and taught. Good behaviour, a rare quality among students at best, evaporated as he spoke.[32] Noisy battles broke out between rival gangs of 'greasers' (engineers) and 'butchers' (medical students), and Liversidge bore the brunt of their ridicule. 'Livy', as he was called, reacted like any teacher approaching sixty might do, with a mixture of fragility, patience, and bad temper.[33] His childhood stammer, carefully controlled for years, re-emerged; and his shyness made matters worse. Writing decades later, Herbert Moran, a well-known Sydney physician, shared his painful memories of a particularly desperate moment during his own first year:

> The old professor looked up at the chemistry class, over his metal-rimmed spectacles, with an air of hopeless irritation. Really, it was too bad. Such a demonstration was most unseemly. A note of defeat was in his voice as he spoke. He would have to dismiss the class and let them get up the subject as best they could from Roscoe's text-book. To this mild threat the only answer was more stamping. The professor seemed at a loss about what to do next.

> He was a mild and humble man of science, something like the caricature of an old-fashioned German lecturer. His hair was iron grey and his vision short-sighted. Worst of all, he had a slight impediment in his speech which the students, with childish glee, were quick to greet each time with stamping. Before each vowel he would place the letter 'n', so that oxygen became 'n ... n ... noxygen'.[34]

Moran's untitled portrait is all too clearly one of Liversidge, in those unhappy days after his return from England. Photographs confirm his wintry appearance, while Edward Hufton, his faithful laboratory assistant, sadly confirmed the discovery of 'noxygen'.

Such moments were regrettably common. Inventive students would spread filter-paper tinctured with ammonia and iodine across the lecture room, interrupting 'Livy's' lectures with slight explosions as the papers dried. Or they would arrive before him, and rearrange the reagents on the lecture bench. On one famous occasion, a student in the crowded back seats, high above the main floor, let loose a pigeon. The lecturer and audience were distracted by the frightened movements of the hapless bird. Legend has it that one Dominic McGlone, a medical student, rose to say 'Professor Livermore [sic], I propose that a Vigilance Committee of ten students be formed to help you keep n ... n ... norder'. The impudence was ignored; Liversidge grasped at the suggestion, and immediately dismissed the class with a caution. The episode entered Sydney folklore.[35]

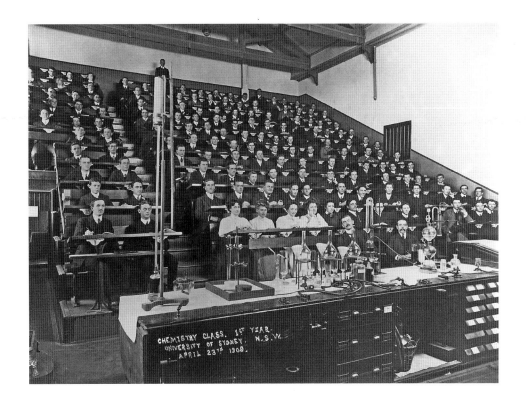

A first-year Sydney chemistry class photographed in 1909.

It could hardly have raised Liversidge's morale, had he known how poorly his meticulously prepared lectures were received. To a highly motivated Arts student like Garnet Portus – later Rhodes Scholar, historian, broadcaster and educator – 'the science courses were a joke'; chemistry, was 'hopeless', its rubrics descending from on high like mosaic tablets, presenting laws to be acquired by heroic acts of memory.[36] In fact, as photographs reveal, Liversidge's lectures were models of articulate, comprehensive, and – in both chemical and pedagogical terms – state-of-the-art practice.[37] He spared no effort or expense to bring to his less than grateful students the benefits of personal contact with the 'new learning' of Europe.

However, it seems that by 1905, Liversidge had let his game slip into a dogmatic style that bored even the most serious student. To Portus, he was just a 'nervous old chap, utterly incapable of keeping order, and pigeons, dogs and crackers used to appear at lectures with us'. To pass the first-year course seemed easy. It seemed that 'a well-kept notebook, with the name of the owner carefully erased with chloride of lime, commanded a fair price in the undergraduate black market', and that mere attendance, 'plus a respectable-looking notebook', seemed to satisfy the examiners.[38]

J.A. Pollock, the physicist, who followed Liversidge in the second term of first year, escaped such severe censure; but even in physics, Portus recalled, 'we learnt very little'. How could they, he protested, 'when we Arts people never did a single experiment in a laboratory?' Only with the third term, and Edgeworth David's physiography, did Portus begin to learn, and then by what he called 'the discipline of interest'. The diligent, pedantic Liversidge could not compete with the effortlessly charismatic David. 'Looking back', Portus recalls, 'it would have been much wiser to have given us callow Arts freshers over to David for a whole year, and let him talk to us about science in general and its place in the scheme of things. Then we might have got some glimmering of what the scientists were getting at. We certainly did not do so by watching experts from afar and mucking up in lectures'.[39] In any case, for Arts students, there was far more to learn from MacCallum's lectures on Tennyson, or from Anderson's classes in philosophy. For medics, there was also the fluent J.T. Wilson and the urbane, if unfocused, Anderson Stuart.

Liversidge and his methods were not without their defenders. F.W. Robinson, a Sydney Arts undergraduate between 1906 and 1909, and later Professor of English at the University of Queensland, numbered him among the Homeric heroes he counted who, with John Smith, had been one of the first willing to admit the 'disturbing element' of women into their lectures. Liversidge 'deserved well of his University', he recalled, 'in his later years ... one feels with some shame that we, with the thoughtless cruelty of youth, made life and lecturing a burden to him'.[40] E.C. Andrews, another Sydney Arts graduate (1891–94), later Government Geologist of NSW, and sometime President of ANZAAS, who fell 'under the spell' of Edgeworth David, knew Liversidge as one who had 'left his mark indelibly on Australian science'. But, he sadly agreed, Liversidge was 'no disciplinarian'. A 'fine type personally', and 'senior students respected him, but freshmen made his lectures a farce'.[41]

## 3. Imperial Networks and Applied Chemistry

If Liversidge was obliged to be stoic with his students, he showed less patience with Senate. In May 1905, the *Herald* – possibly prompted – told the outside world that a division of his chair was urgently required, 'on the ground that [chemistry] cannot possibly be effectively covered by one man'. Chairs were also needed in organic and physical chemistry, subjects that were 'revolutionising the science'.[42] Liversidge knew they were long overdue. That month, he asked the Senate to employ Schofield to give first-year lectures in organic chemistry, to be paid for by deducting from his own salary an appropriate proportion of students' fees;[43] and T.H. Laby was hired to do extra demonstrations for two terms.[44] The same month, he asked Edgeworth David to continue as Dean, and so to act in his place as ex-officio member of Senate.[45] Thereafter, Liversidge resigned from both the deanship and Senate. There were other things he wanted to pursue.[46]

Among these, were links between the scientific community of Sydney and the Society of Chemical Industry (SCI), the Institute of Chemistry, and the Institute of Civil Engineers in Britain. In many respects, all three represented the new professionalism of applied science, combining academic research with commercial application. The first was inspired by Ludwig Mond, Edmund Muspratt and other leading industrialists in the Lancashire chemical trade. Aided by Henry Roscoe, Professor of Chemistry at Owens College in Manchester, their efforts assumed a national character, when a meeting at the Chemical Society in Burlington House led to an organisation that enlisted over 360 members – and by 1906, numbered over 4000.[47] Local sections of the SCI existed in Liverpool and Manchester before Liversidge joined in 1883, but his name appears on the Society's lists as an Original Member. He regularly received its journal, and took an active interest in its affairs. He watched closely when the SCI crossed the Atlantic, to form its first overseas section in New York in July 1894. By the turn of the century, there were five more sections in different cities of Britain and Canada. Liversidge saw an excellent opportunity to add Sydney to the list.

He had cause to act quickly. In August 1900, David Orme Masson of Melbourne, also a member of the British SCI, decided to establish an independent society of chemical industry in Victoria. His reasons were twofold: 'one', he said, 'mainly sentimental, and one eminently practical'. In the first place, the venture was an experiment, which might fail; and if it failed, 'it seemed better in every way that the humiliation of it should be kept quietly and decently within our geographic boundaries'. The second was financial: members of sections of the 'Home' society had to be no fewer than thirty in number, and to pay an annual subscription of thirty shillings over and above local expenses. For their subscription, members received the SCI's journal. Masson thought there would be no difficulty in finding thirty members in Victoria, but it seemed 'pretty certain that the plan ... of forming an independent society with a modest half guinea subscription appeals to and benefits a much larger number'. If his declaration of independence failed, he was willing to risk British displeasure. Others agreed. That month, 105 members enrolled in the Society of the Chemistry Industry of Victoria; no qualifications were required, except 'an interest in industrial chemistry in some way or other'.[48] The Society held its first general meeting in Melbourne on 8 September 1900, opening with a paper on the use of the cyanide process in the extraction of gold.[49]

With this initiative, Masson laid claim to having established the oldest chemical society in Australia. But Liversidge was both bolder and more conservative. He knew his community; if there were a risk of failure, he did not admit it. As for the cost, thirty shillings was a minor disincentive to those – not above fifty, in any event – who were likely to join. More important, a link with the 'Home' society would be professionally useful. Liversidge's motives, we may infer, were threefold. First, his imperial loyalties remained undiminished, and he actively sought British connections. Second, while it was desirable to encourage Australian chemistry, disseminating its findings to a wider world was a different matter. This facility the SCI, through its journal, could more readily supply. The imperial link would also remind capitalists that investment in Australian research could 'ultimately result in increased commercial activity, with profit to Australia, and possible benefit to the world'.[50] Victoria's stronger industrial base was perhaps large enough to support its own society; but NSW's smaller base could benefit by being linked with London. Finally, Liversidge turned Masson's precautionary principle on its head. There was indeed abundant evidence that small scientific societies might fail in the colonies; but if this one were given an imperial mainstay, it would be much more likely to survive.

In November 1901, fifteen months after Masson's meeting in Melbourne, Liversidge organised a meeting at the Royal Society of NSW in Elizabeth Street to consider the proposition. With their agreement, Liversidge approached London, and in July 1903, received permission to form a Sydney Section of the SCI. The Section began with the required membership of thirty, including thirteen from the laboratories of the Colonial Sugar Refining Company, by far the largest single employer of chemists in the state. Founding members included representatives from Commonwealth Portland Cement, as well as veterans from the Sydney Mint, the Department of Agriculture, and the Technological Museum. From the Hawkesbury Agricultural College at Richmond came Henry W. Potts, recently arrived from Victoria to become the College's innovative Principal.[51] From Sydney University, came James Schofield, and from the Linnean Society, Robert Greig-Smith, then holding a Macleay Research Fellowship. Liversidge became the Section's first Chairman, and Thomas Walton, chief chemist of CSR, its Honorary Secretary. As ever, Liversidge was fortunate in his friends. Walton had worked closely with his fellow Scot, Thomas Steel, of the Colonial Sugar Refining Company, in developing a system of chemical monitoring for the sugar refineries. Steel was also widely read in natural history, becoming an officer of the Linnean Society of NSW, and a correspondent of Baldwin Spencer.[52] They made a good team.

The Sydney Section met four times a year, and until September 1907, Liversidge was a regular participant.[53] His methods were those that had guided him at the Royal Society and the AAAS. First, he inspired others; then he ceded control. In every case, his strategy combined Australian patriotism with Empire loyalty. Thomas Walton became to the Sydney Section what Leibius had been to the Royal Society, Schofield to the chemical laboratory, and Maiden to the Technological Museum. None of Liversidge's lieutenants seem to have suffered from their association with him; on the contrary, all prospered, letting him take the lead until the idea was well established.

Walton so well managed the Section's business that, at his death in 1917, he was known in London as the Sydney Section's 'outstanding personality'.[54] The technique that had served Liversidge since Cambridge served him well again; and by gentle persuasion, papers came forward from students and friends – from Robert Greig-Smith on vegetable gums, James Petrie on paraffin oils, George Harker on the fermentation of cane sugars, and Henry Smith on eucalypt oils.[55] To SCI meetings came experiments on saturated solutions by Thomas Steel; on nitroglycerin by C. Napier Hake, the chief inspector of explosives in Victoria; on New Zealand coals by A.M. Wright, of the Colonial Laboratory in Christchurch; and on the tanning properties of Western Australian barks by E.A. Mann, of the Government Laboratory in Perth.[56]

Overall, the SCI Section provided Liversidge with another platform to develop research relevant to the Australian economy. Systematic reporting by the British press could move Australian ideas into a wider imperial marketplace. Similar reasoning underlay his efforts in May 1905 to obtain accreditation for Sydney University courses by the Institute of Chemistry in London.[57] British certification smoothed the entry of Sydney graduates into industrial and commercial circles, and incidentally offered an impressive defence against those who, like Inspector Black, might have criticised Sydney's teaching methods. From 1906, Sydney's graduates could acquire the profitable initials FIC.[58] The same year, at Liversidge's suggestion, Edgeworth David negotiated an analogous agreement with the Institution of Civil Engineers in London, by which the ICE accepted Sydney's engineering degree as a qualification for membership.[59] Liversidge's vision of imperial science was never more confident.

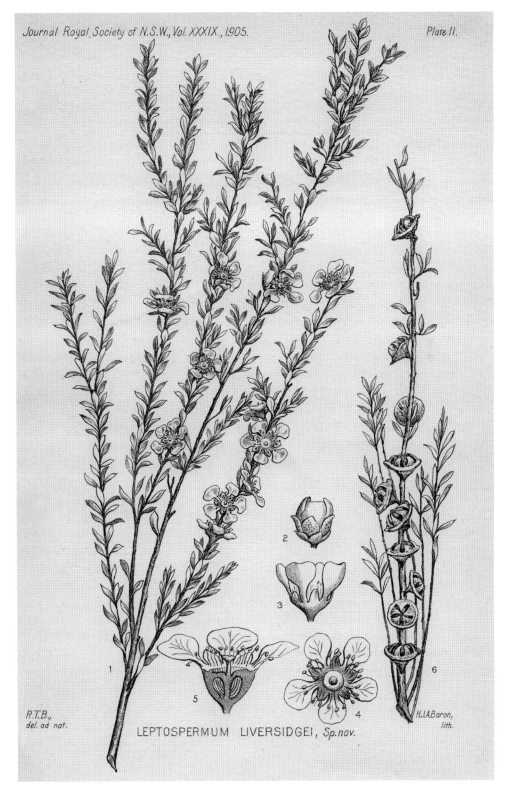

R.T.B.,
del. ad nat.

LEPTOSPERMUM LIVERSIDGEI, Sp.nov.

H.J.A.Baron,
lith.

*Leptospermum liversidgei*: a tea tree dedicated in 1905 to Liversidge and named in his honour by
R.T. Baker and Henry G. Smith.

However, the prospects for making good use of an imperial superstructure turned upon having a robust Australian base. This required a much closer integration of British and Australian effort. In 1886, Liversidge had not opposed a visit of the BAAS to Australia, which might have led to further developments in this direction; but he saw that Australian interests required first the formation of a federated colonial science, large enough in size and substance to offer Britain collectively what it could not easily obtain by itself. Intercolonial cooperation must precede imperial cooperation. Thus, while he encouraged every possible contact with British professional associations, he insisted on Australian science having a first footing. Those who recalled his suggestion to the Royal Society of NSW, in 1901, for a scientific academy to serve the whole of Australia, knew such ideas were ahead of their time. Until technology could overthrow the tyrannies of distance, the vision of a national scientific community remained remote. But Liversidge persevered. His experience as Australia's 'national' representative in England, and his recognition of a flourishing imperial network, especially in medicine and related sciences, encouraged him to pursue Australian science in national terms.

The thought was father to the deed, and in May 1905, Liversidge approached Angus and Robertson, the publishers, with a proposal to begin in Australia 'a monthly magazine of the Nature type'. This *Australian Journal of Science*, as he called it, would have thirty-two pages and cost twelve shillings. With a print-run of 1000, its regular appearance would, he said, unite a community which the AAAS alone could reach at best every two years.[60] He mailed a prospectus to 7000 potential subscribers throughout Australia and New Zealand. Alas, his efforts were to no avail. Geography – and perhaps indifference – worked against him. 'The support received', he conceded, was 'very small'.[61] Angus and Robertson withdrew their interest, and the idea was dropped. Some thirty-four years later, the proposal would be revived, and a journal christened with Liversidge's title, and launched by ANZAAS. For the moment, the opportunity vanished.

But Liversidge had still several cards to play. In 1906, he turned to the British Science Guild (BSG) which, under Norman Lockyer, had begun to appoint specialist subcommittees to apply 'scientific methods to public affairs'. The first two, set up in 1906, were devoted to education and agriculture; and between 1906 and 1914, a further fifteen were set up to study the British chemical industry (1906); prepare the Franco-British Exhibition (1906); and survey the cost of instruments in schools (1907). The Guild went on to apply the 'principles of scientific management' to the coordination of charitable bodies (1908), the synchronisation of clocks (1908), the naming of streets (1908), and the conservation of natural resources (1909). Other committees reviewed national policy on the use of explosives (1909), public ventilation (1909), and the manufacture of technical optics (1909). Deputations went to government on a range of issues, from the amendment of the patent laws to the sponsorship of anthropometric research. It was not difficult to see similar possibilities for Australia. As honorary secretary of the AAAS, Liversidge had sent several approaches to the federal government. In 1907, he helped establish a Sydney branch of the BSG, which attracted a small but loyal following, and promised more in time to come.

In some sense, these organisational successes compensated Liversidge for disappointments at the University. In June 1905, he persuaded Senate to spend £400 on a new balance room next to his private laboratory.[62] But the Faculty's maintenance vote for the following year fell back to £2300 – a large sum by historic standards, but now much less than was needed to keep up with the rest of the world.[63] Deploring Senate's unwillingness to generate new money, he offered (or threatened) to donate £250 of his own salary towards the appointment of permanent staff to teach in organic and physical chemistry.[64] James Schofield gave his required lectures, but was unable to do more. Senate demurred. Liversidge despaired.

Meanwhile, his research had reached a point of diminishing returns. Away from his bench for over a year, he contributed only one exhibition to the Royal Society of New South Wales in 1905, and that year published only two papers, a minor note, and a report on the International Catalogue.[65] During the next two years, he delivered two exploratory papers to the Royal Society, both on the crystal structures of gold, and showed increasing concern to push on with his studies of gold mineralisation and ore genesis.[66] But time was against him. Old friends were dying: for Captain Hutton of Otago, he wrote a moving obituary in 1905, that revealed much of his own regret at being unable to do more.[67] Although only fifty-nine, Liversidge's photographs show a man much older.

With the passage of time, came the recognition that, whilst he had taught hundreds of students, he had produced few 'pupils'. Worse, few of Liversidge's science students had become school teachers – except for the women, of whom there were still but a handful (nine in 1910). The BSc, on which he pinned so many hopes, had failed to fly. By 1905, Sydney had graduated 1148 BAs but only forty-nine BSc degrees – in 1906, only six took the science degree, and most of these were destined for medicine, rather than research.[68] To this rule, there were brilliant exceptions – in 1900, Liversidge could count Douglas Mawson, Oscar Vonwiller, later Sydney's Professor of Physics, Arthur Jarman and Ernest Le Gay Brereton, and Thomas Laby. And south of the Murray, the picture was not much brighter. Melbourne, strong in chemistry, was weak in other fields, and had only twenty-two science students against Sydney's forty-three, and only thirty-eight engineers against Sydney's eighty.[69]

As he neared his sixtieth birthday, Liversidge wanted to leave a legacy not of legions of first years, but of 'men who are working for a degree in pure science ... who will be able to bring reputation to their schools by the original work they do'.[70] Such scholarships and bursaries as Sydney offered were few, and were shared by all the Faculties.[71] The University lacked both the will and the means to encourage scientific research. Of course, there was the larger community to draw upon, and from this Liversidge drew consolation. In NSW, there were perhaps 290 chemists, of whom 143 were analysts and fifty were metallurgists in commercial operations.[72] This community Liversidge had helped create, so inaugurating the 'fellowship' that ran the Mint, the museums, the Department of Mines, the Department of Agriculture, and several firms. But such a community of busy people could never constitute a 'research' community, nor had it much 'learned leisure' to spare.

The prospect of leaving a 'successor' weighed heavily upon him. The need for this was nowhere more evident than in the Royal Society of NSW. The task of promoting the public understanding of science lay with the Society. But when Liversidge's hand left the tiller, its course changed from revolution to routine. Even the *Herald* seemed to give up on its savants, whose monthly papers, it said, had become mechanical, 'purely technical', rarely provoking general discussion.[73] The Society's Council showed little charisma, and with meetings dominated by an ageing generation, attendance fell away. Membership, which had peaked at 494 in 1885, dropped to 374 in 1900 and to 326 in 1905. For one reflecting on the returns from a life's investment, this was depressing news.

## 4. A Time to Leave

At the beginning of 1906, Liversidge signalled Sir Normand MacLaurin, the Chancellor, that unless the Faculty were given more funds, he would resign. In March, he spoke again to MacLaurin about re-organising the Chemistry Department, presumably in the hope of gaining teaching relief so that he could do more research himself.[74] Apparently, he received no satisfaction.[75] Something like a straw broke, when Lawrence Hargrave – who, at his suggestion, had joined the Royal Society in 1877 – offered the University, free of charge, all the models and records of his pioneering flying machines. Senate declined the gift, on the grounds that the University could not afford the cases.[76] This officious response moved Liversidge to fury. Senate seemingly knew much about cost but little about value.

In early April, the *Herald* devoted its Commemoration Day editorial to a sympathetic critique of the University. Drawing a contrast with an outdated view of Oxford and Cambridge, the *Herald* judged that Sydney had 'grown with the country, and developed with its necessities, that it had not been quite oblivious to science, which is advancing every day'. Such faint praise, for someone who actually knew how rapidly Cambridge and even Oxford had recently developed, offered little hope. The editorial Liversidge read appeared alongside another devoted, simply, to 'The Australians'. With his pen, Liversidge marked the lines the editor had reserved for criticising what he considered Australia's unwillingness to accept responsibility for its institutions; its tendency, as he put it, to place 'ease before duty'. That seemed to sum it up. Liversidge could only agree that 'Australians must take themselves seriously – their methods of education, their political powers and privileges, their own self-culture, and the necessity for widening their outlook'. As the editor concluded: 'We cannot afford to remain behind the rest of the world, for ... the period of our isolation has come to an end'.[77]

For over twenty years, Liversidge had tried to bring this realisation to Sydney. Since his return from leave, however, trivial irritations had taken on cumulative weight. The prospect of unending struggle in what seemed a thankless cause, was made worse by the knowledge that there were fewer with whom he could share the fight. Russell was seriously ill. Edgeworth David was overseas. Anderson Stuart was no longer the companion he once had been. Carslaw was agreeable company, but busy in his own right, and the difference of age worked against a close friendship. Even old sparring partners – Badham and Scott – had died or retired, and Liversidge had little in common with many of the 'new' professors arriving since the 1890s. He scarcely knew George Arnold Wood of history, W.J. Woodhouse of Greek, Thomas Butler of Latin, or Francis Anderson of philosophy. In the city, friends at Elizabeth House were disappearing. The membership of the Union Club had changed, and with families moving to the suburbs, many former dining companions no longer lived in the city. There were few collegiate dinners with the Chancellor, such as those Deas Thomson once gave at Barham – or at least, few

Liversidge and Dr A.W. Howitt, AAAS President, at Government House, Adelaide, 7 January 1907.

to which he was invited. Possibly, the lack of a wife and family limited easy access to society. Although he lived in one of the most fashionable parts of Sydney, he seems to have had progressively less contact with his neighbours. Over thirty years separated him from the Macleays of Elizabeth Bay House, the scene of such scientific gaiety, which had by now fallen into genteel decay, its garden overgrown since the death of Lady Macleay in 1903.[78] Liversidge's domestic life revolved around the Octagon, and (judging from his photographs) some happy hours with the Hufton family.

Of course, there were compensations. The Australian Museum retained a warm place in his affections, and he saw Edward Ramsay socially, if less frequently after the 1890s. But the Museum had suffered severe reverses, from which it was slow to recover, and in the absence of funds, there was little that Liversidge, as a Trustee, could do to shape developments. The same was true of the Technological Museum. There, however, the pioneering work of Joseph Maiden on essential oils – along lines that Liversidge had suggested – and the work of Maiden's assistants, Richard Baker and H.G. Smith, on the chemical classification of eucalypts – were producing useful tools for settling questions that botanical chemistry had not resolved.[79] All this, Liversidge found especially satisfying, not least because Baker and Smith were analytical chemists, and Londoners as well.

Baker was born in Woolwich, the son of a blacksmith, and, like Liversidge, took the Science and Arts Certificates from South Kensington before coming to Sydney as a school teacher in 1879. Smith, the son of a plumber, migrated to Sydney for his health, and took evening classes in science under W.A. Dixon at the Technical College. Baker joined Maiden as the Technological Museum's first economic botanist in 1888, and Smith followed, as economic chemist, in 1899. Both became members of the Royal Society of NSW and the AAAS; both, at different times, also lectured at the University. Together, they followed Ferdinand von Mueller and Maiden in pioneering phytochemistry, and helped establish the eucalyptus oil industry.[80] Their collaboration celebrated Liversidge's practical idealism, and Liversidge was touched when they named a species of Leptospermum in his honour.[81]

Baker and Smith also kept Liversidge's memory alive at the Museum in Harris Street, which he visited from time to time – pleased to see the exhibits impressively laid out. At Liversidge's request, the Museum had three storeys, one dedicated to each of the three kingdoms of Nature – animal, vegetable and mineral. In time, the building itself was destined to lose its glamour, and become a proverbial eyesore, before its replacement by the modern Powerhouse in 1988. But in 1906, its message was still compelling – with a thematic 'storyline' that brought chemistry, mineralogy, and botany together with the products of mining, manufacture, agriculture and the industries connected with them – a perfect illustration of South Kensington principle improved by Australian example.[82]

As 1906 drew to a close, Liversidge kept quietly active in his laboratory at the University and in his library at home, advising on the state's new powder magazine,[83] examining crystals sent him by collectors, and presenting a paper to the Royal Society on gold nuggets from New Guinea.[84] This paper he repeated to the Adelaide Congress of the AAAS in January 1907, where some 335 people attended, and which was deemed a great success. In the event, this was to be his last congress. The speeches and garden parties assembled friends from South Australia, Victoria and New Zealand, many of whom he would never see again. On l March 1907, Henry Russell died. Russell had been like an elder brother and Liversidge was devastated.[85] More grief came in June, when news came of Michael Foster's sudden collapse and death – ironically, during a speech he was giving to the British Science Guild in London. Cut off from the past, stymied in the present, Liversidge saw little hope for the future.

In late March 1907, Edgeworth David, who had succeeded Liversidge as Dean in 1904, orchestrated Warren, Anderson Stuart, and Liversidge in writing a joint letter to Senate demanding an increased vote for scientific apparatus. But their demands were not, and could not be met. With expanding numbers, the University needed £15,000 more than its vote in 1907 just for new teaching staff in Arts and the Medical School. The situation was not helped by uncertainties surrounding the allocation of federal grants to the states, which led the Premier of NSW, J.A. Carruthers – the Sydney graduate who had been Minister for Public Instruction under Parkes and Reid during the 1890s – to postpone all discussion of endowments until after the general election of 1907. Between April and May, the *Daily Telegraph* ran a series of thirteen articles on the history and circumstances of the University, all recommending increased funding, but there was no response from Macquarie Street. The federal election that brought Alfred Deakin to power saw much renewed rhetoric about Australia's contribution to 'imperial efficiency'.[86] But there was no immediate prospect of greater aid to university science, or, for that matter, to universities generally.

It is impossible to know precisely Liversidge's state of mind when, on 6 May 1907, he submitted to Senate his resignation, effective that December.[87] His decision seems to have caught the Registrar by surprise. Liversidge was aged sixty, but was not obliged to retire. Indeed, the University had no clear precedents for retirement on grounds other than illness. Woolley had died at sea, Gurney and Pell had been invalided out, Scott and Threlfall had resigned, but Badham and Smith had died in harness, and it seemed that the University expected nothing less from its professors. Threlfall's resignation had found the University quite unprepared to deal with a professorial departure. Indeed, given the column inches belatedly devoted to Threlfall's good works, Sydney's press might be forgiven for thinking he had never left.[88] But if there was praise for famous men once they departed, there was no fixed provision for emeriti. Fortunately, Liversidge had been as prudent in personal finance, as in professional life. He owned the Octagon, free of debt; his savings amounted to perhaps £800 a year for the last twenty years; and possibly the University might contribute something to his retirement after all.[89]

Given these prospects, he had little to gain by staying, and nothing to lose by going. For Sydney, he had great affection, but not the love he had for London. In any case, the city had changed. Darling Point had suffered the misfortunes of suburban subdivision. Liversidge's neighbour, James Norton, had died in 1900, and Sir Edward Knox in 1901, and they were succeeded by new families with whom Liversidge had no contact. On the other hand, the 'pulls' from England were stronger. In his files, he saved a leading article from the *Herald*, which commented on the fate of the 'colonial professor'. 'He is', the *Herald* admitted, inadvertently drawing a portrait that fitted Liversidge exactly, 'to a large extent isolated', in a field in which, 'it is only by associating with the men who are doing the new work that one can understand the work that is being done'. For the serious scholar, 'books are always behind the times, and periodicals are very often a mere Tantalus cup of information'.[90]

On 14 May, writing from the Octagon, Liversidge confided to Fletcher at the British Museum his decision to return to England. He missed, he said, being close to news 'from the inside'. Perhaps conditions in Australia would eventually improve, but as his life passed from midday to evening, there were many things to do before nightfall. 'I am sorry to give up for some reasons', he concluded, 'especially now that I am beginning to get some good students in Chemistry, but I shall be very glad to get back amongst my earlier friends – there is no place like the old home'.[91]

When Liversidge's decision was announced by Senate, at least one response was immediate and predictable. The dykes he had defended in the first-year Arts curriculum met a flood of injured pride, and within weeks, Mungo MacCallum, Dean of Arts, moved to reduce the lecture time allotted to all three science subjects.[92] However, MacCallum failed to get the matter to a vote, and in 1910, the Professorial Board again tried, and again failed, to exempt Arts students from first-year science.[93] Ironically, to maintain science in first-year Arts had become a conservative principle, supported by the Chancellor as a defence of the status quo against specialisation. Nonetheless, the writing was on the wall. After much further debate, Senate finally abandoned the science requirement for Arts in 1915.[94] It seems as if the University were waiting for Liversidge to leave.

Within days of Liversidge's announcement, Edgeworth David took the matter of his succession to Senate; and in June, the University advertised for a replacement at the same salary, less fees, of £900. A search committee was constituted, which included Edward Knox, Jr of CSR (and Darling Point), Frank Leverrier (one of Liversidge's first BSc graduates, and a now a distinguished lawyer), Anderson Stuart, and Edgeworth David. The Chancellor asked Liversidge to advise – apparently, he could not quit even by resigning – and he agreed. [95] The Faculty of Science urged that extensions to the Chemistry Building, for which Liversidge had fought so hard, be made forthwith. If existing facilities could not keep an old professor, they were unlikely to attract a new one.

Following Liversidge's suggestion, and on a motion by Edgeworth David, Senate followed customary practice and appointed a selection committee in Britain. This time, the committee included Liversidge's colleagues Professors Sir William Ramsay of UCL; Alexander Crum Brown of Edinburgh University; W.A. Tilden, of the Royal College of Science (now Imperial College); and H.B. Dixon of Manchester.[96] Three candidates were shortlisted: Bertram D. Steele, a Masson student, then working in Queensland; Gilbert T. Morgan, a graduate of London University and the RSM, a lecturer at the RSM; and J.W. Mellor. All three had DSc degrees, and two were future Fellows of the Royal Society. Subsequently, two internal candidates, Liversidge's assistants Frederic Guthrie and James Schofield, were added to the list.[97] The contest narrowed to Morgan and Steele, and after an unprecedented debate, Morgan was offered the job. The internal candidates knew the ropes, and it was to Schofield that Liversidge left his course and laboratory notes. But in the end, the overseas man won. Gilbert Morgan was a rising star. His appointment augured well for the future of science at Sydney.

With his replacement in sight, Liversidge prepared to leave Sydney. In October, he gave the Australian Museum some of his ethnographic objects, and negotiated with the Museum to buy some of his books and journals.[98] He also offered the Museum his collection of crystals and gemstones at their market value of £400.[99] His extensive holdings of Aboriginal weapons, including New Hebridean arrows collected by missionaries and given to John Smith, were poorly valued, so he kept them, to see if they would be wanted by Cambridge or the British Museum.[100]

In November came Liversidge's last meeting in the Faculty of Science, an occasion marked by his seconding a motion to establish, after years of union with

Liversidge's study at the Octagon.

Japanese urns in the Octagon's ante-room, similar to those Liversidge obtained for
Sydney's Technological Museum.

Science, a separate Faculty of Engineering.[101] This was not to be realised until 1920, when new Faculties in Agriculture, Architecture and Veterinary Science were also established, but he had laid the groundwork. Elsewhere, he tied up loose ends. He continued to work for the Senior Examinations until the very end, and put his staff in line for positions thereafter.[102] He also secured the position of Edward Hufton, who had served him for nearly twenty-six years.[103] In May, on the formal announcement of Liversidge's retirement, the *Daily Telegraph* recorded the community's 'debt of gratitude apart from that which he has earned by his faithful service in the classroom'.[104] In December, Senate entitled him Professor Emeritus in recognition of his thirty-six years' service, and in token of the 'great and beneficial influence which he has exerted in the extension of scientific study and investigation in Australia'.[105] The same week in which Dame Nellie Melba returned to Sydney, Liversidge exchanged contracts on the beautiful Octagon, and booked his passage 'Home'.[106]

At the University, the changes he had witnessed remained both great and incomplete. The solitary professor, reader and two dozen students of 1872 had become five professors, eight lecturers, a dozen demonstrators and over 350 students in each year. Regrettably, all were wanting equipment 'the need for which', recalled the *Herald*, was still 'very far from being realised'.[107] In his time, Liversidge helped oversee some £218,000 in new construction. But bricks and sandstone told less than the full story. Staff numbers had grown, but not in proportion, so that by 1913, for 1600 students there were only fifty-seven full time and fifty-two part time staff in all grades, or a ratio of about eighteen to one, considerably worse than the ratios of British provincial universities (although ridiculously light by current standards).[108] The weight of numbers stifled research and contributed to a 'school-like' discipline.[109] By 1910, the BSc enrolled forty-seven students, but advanced instruction was limited to a few, and the research they did was of a narrowly practical nature or required limited apparatus, or both. In the phrase of one charitable contemporary, this hardly constituted a 'scientific ideal'.[110]

Elsewhere, the picture was little better.[111] Among the six Australian universities, only Melbourne under Masson had so far created a research school in chemistry. What Liversidge proposed in 1904, Melbourne achieved in 1908, creating six science fellowships at £150 each.[112] Everyone, including the editors of the *Herald*, knew that 'given (1) postgraduate students, (2) moderate endowments, (3) reasonable accommodation and equipment, Sydney would not be long in earning for us a great name in science research'.[113] But this lay years away. Of more immediate satisfaction was the University's investment in mining engineering, which put Sydney on a par with the Royal School of Mines, recently (1907) incorporated in the new Imperial College of Science and Technology in South Kensington. 'We give them science: and they must pick up the practical side themselves', Liversidge told the press. And this his students did, working in the mining districts of NSW during their university vacations. Thanks to Liversidge, 'science walked behind the shoulder of the most adventurous prospector', returning wealth from ores once thought unpayable.[114] Although mining students became fewer after 1905, many of the ninety who had graduated by that year made their mark in Asia, South America, South Africa, and New Zealand, as well as in the £30 million-a-year minerals industry of Australia.[115]

The northern aspect of the Octagon, 1907.

Ironically, just as Liversidge pulled up stakes, between 1908 and 1909, a reforming climate, and greater attention to 'national efficiency', brought impressive changes to the University. At last, the endowment was doubled from £10,000 to £20,000, permitting appointments in economics, botany and organic chemistry, and expansion in engineering and medicine. Within the next two years, scholarships were awarded, and evening classes extended. Of all this, Liversidge would have approved.

On Christmas Eve 1907, Liversidge gave an interview to the *Sydney Mail*. Officially, his reasons for leaving were two: he wished, first, to set up a laboratory of his own – specifically for research on the occurrence, deposition and mineralisation of gold; second, he wanted to be in 'touch with men in his own lines of science', with access to facilities 'only to be found in or handy to London'. He promised to return to Australia in 'a few years' time'.[116] It was not a time to cavil. Nothing so well conveyed the goodness of the man at the University, as his parting from it. 'Never', he was heard to say, 'had an appeal for funds for scientific purposes been made to Parliament in vain'.[117] Only one accustomed to taking the long view could have been so generous.

On 2 January 1908, Liversidge was farewelled at the Union Club. Two days later he stood for the last time on the pier at Circular Quay. His ship, the P&O *Mongolia*, was the same on which, in 1872, Jules Verne had launched his immortal English traveller around the world in eighty days.[118] For Liversidge, the passage had taken a lifetime. The *Mongolia* left on the morning tide, and as he steamed past Fort Denison, Liversidge could just glimpse the Octagon and its ornamental gardens, a vestige of tranquillity shimmering in the summer heat. The record does not show whether anyone was there to see him leave, or to say goodbye.

# CHAPTER 13

# Returning Home:
# Efficiency and Sociability

Geological Museum, Jermyn Street, London.

In December 1907, just as Liversidge sailed from Sydney, Edgeworth David left with the Shackleton Expedition to the South Pole. Many remarked on the contrast between the two friends – the shy, short, stammering, retiring bachelor versus the confident Oxford-educated classicist and family man, charismatic, personally courageous, larger than life. But over the years, the two remained close. Not by chance would David be asked one day to write his friend's obituary.

Liversidge reached London in February 1908. No sooner had he arrived, than his legendary good health gave way to an unwelcoming winter. He succumbed to a 'run down', as he called it – and his first six months of retirement were spent confined to an off-season hotel on the South Coast.[1] It grieved him that he had not seen his friends, nor been able 'to do any of the many things I came home to do; the finishing and publishing of papers must be deferred indefinitely; it is a great disappointment to me'.[2]

It was not until June that he made his way to London, not yet fully recovered, and took up lodgings at the United University Club in Suffolk Street, off Pall Mall, just above Trafalgar Square. For a subscription of 95 guineas, and a substantial rent, he enjoyed a grand house recently rebuilt in a style combining Adam and Louis XVI.[3] There he could receive friends and entertain colleagues.[4] From there, he could also launch expeditions to Cambridge, Wales and the South Coast. Indeed, within minutes of the Royal Society and its library, he could even walk to work.

## 1. A COLONIAL RETURNS

Perhaps to his surprise, he found that, if he had in some ways never left England, part of him would never leave Sydney. He read the *Sydney Morning Herald* as assiduously as ever, and his news clipping collection burgeoned. In 1909, John Collier's fine portrait was hung in the Great Hall, showing Liversidge holding his favourite cluster of gold crystals.[5] Thanks to Liversidge and his generation, science in Australia had grown, even if it still had shallow roots. In October 1907, the London *Daily News* carried a particularly negative picture of Australia and of Australians – a place and people 'making no appeal to the imagination'. Australians, it said, had grown too suddenly comfortable and were immersed in material satisfactions, 'wholly indifferent to that richer life which is the basic note of great and enduring states. It has produced no voice, no literature, no art.'[6] Writing in the margin, Liversidge felt moved to agree. But he did not resist the fact that significant changes were underway. Applied science flourished in the first decades of Federation.[7] Sydney Technical College opened new buildings, and in 1912, the Faculty of Science debated ways of taking up 'advanced technological instruction with particular reference to Australian industries'.[8] The University appointed more junior staff,[9] and expanded its range of courses.

Whether by design or necessity, Liversidge's influence in University affairs did not suddenly disappear. Indeed, on at least four occasions during the next four years, he sat on the University's London committees for chair appointments in botany, zoology, and in both inorganic and organic chemistry.[10] In early 1908, despite his intercession, Gilbert Morgan, the committee's choice, declined the chemistry chair offered him,[11] and Schofield and Guthrie stood guard until a successor could be found.[12] The second time around, five British universities – Birmingham, Glasgow, Leeds, Liverpool and Manchester – were asked to nominate candidates, and Liversidge, although scarcely out of his sickbed, was asked to help.[13] Again, loyal Schofield and Steele were passed over,[14] and the committee decided in favour of Charles Fawsitt, a student of Alexander Crum Brown at Edinburgh, and currently Lecturer in Metallurgical Chemistry at Glasgow. Arriving in Sydney in 1909, Fawsitt was to hold office for thirty-seven years, retiring only in 1946.

Research, it was said, 'took little of Fawsitt's personal attention',[15] but not all his junior colleagues were so disinclined, and the coming decade saw several promising developments. In 1911, for example, Senate was asked to provide earmarked funds for research,[16] including library and laboratory assistance, shorthand typists, and sabbatical leave.[17] In 1912, the NSW Parliament endowed six Science Research Scholarships of £150 each – an idea that Liversidge had proposed thirty years earlier.[18] In 1912, the Faculty of Science enrolled its first fourth-year honours students, all embarking upon research theses.[19] By 1914, the Registrar could boast that there were, in all Faculties, eighty-two men and women working for research degrees.[20]

In 1912, responding to demands that Liversidge had voiced for years, the University established a second chair in chemistry, designated specifically for 'organic and applied chemistry'.[21] Liversidge noted that the advertisement spoke of the contribution of chemistry to economic development. The chair went to Robert (later Sir Robert) Robinson – a scion of the distinguished Manchester 'family' of organic chemists – whom Liversidge interviewed in London.[22] Robinson began well, but soon despaired of Sydney's undisciplined first-year medical and engineering students, and returned to England, after only three years, in 1915. Although he had begun important research in Sydney, his decision proved wise. On the basis of what he did once he returned, he was elected an FRS in 1920, knighted in 1939, and received the Nobel Prize in 1947.[23] In 1916, Robinson's successor was selected by another committee, on which Liversidge also served.[24]

Robinson's experience showed that a Sydney chair had ceased to be a once-and-for-all move, and had become a stepping-stone to advancement elsewhere. For a research-oriented man with ambition, Sydney could be a dead-end. The ghost of Threlfall was constantly invoked, and with it, his memorable remark, quoted by David Branagan, that 'I made the greatest mistake of my life [in coming to Sydney]. If I had stayed in Cambridge, I should have been in with J.J. [Thomson] in the discovery of the electron'.[25] W.H. Bragg, who moved to England from Adelaide in 1909, might have sympathised.

Whether Liversidge could, or should, have repatriated to England sooner, we shall never know. What we do know is that, once retired and restored to health, he became 'Australia's man in London' on all matters dealing with science. He was the Australian Museum's representative at the Museums Association's annual conference in 1908,[26] and in 1909 he represented Sydney University at the Darwin Centenary in Cambridge.[27] In 1909 and 1910, *Nature* asked him for a concise history of the AAAS and of the Royal Society of NSW,[28] and the Natural History Museum asked him to analyse Australian minerals.[29] At every AAAS Congress, the Council sent him greetings, addressing him reverently as 'Founder'; and he represented the Australian Association at the British Association's Congresses at Dundee in 1912 and Birmingham in 1913. When he could, he helped Sydney men win imperial recognition. Beginning in 1910, for example, he annually endorsed the candidature of

J.H. Maiden for the Fellowship of the Royal Society,[30] and persuaded the ageing Sir Joseph Hooker and his successor, Sir William Thiselton-Dyer, to back him.[31]

It must have heartened Liversidge to learn from R.T. Baker, Maiden's successor at the Technological Museum, that his photograph still hung in the vestibule.[32] In 1910, he was pleased that the Australian Museum had acquired space to show its Papua-New Guinean collections.[33] But disappointment followed when, after protracted negotiations, the Museum declined to buy his exhibition collection of NSW gems and minerals and his prize-winning collection of minerals from Polynesia, New Zealand and the USA.[34] There was disappointment, too, when he exhibited (and received medals for) his collections at the Japan-British Exhibition of May 1910 and the Empire Exhibition held in honour of King George V's Coronation in 1911,[35] but had to do so in a private capacity, because the NSW government declined to offer its sponsorship.[36] In 1914, at the Franco-British and Anglo-American Exhibitions, he exhibited his papers, gems and minerals for the last time. In 1915, he lent his collection to the Imperial Institute for two years – some of it permanently, at it turned out – simply 'to keep it out of storage'.[37]

## 2. A New Life in London

Liversidge's personal history reveals few features of the prodigal son, but his return found a family ready to receive him. Over the years, the Liversidges had prospered. Of Archibald's eight siblings, only Caroline and his three brothers were still alive, but William and Jarratt had successful businesses and livery company positions in the City, and John George, who had moved to Chiswick, had become a pillar of New Brentford, an overseer of the local Board of Guardians, and a respected churchwarden. Liversidge's brother-in-law (Caroline's husband), J.E. Balfern, now ran the Spring Grove Works of 'Balfern and Liversidge', 'dyers and bleachers and laundrymen, club and hotel contractors', which had prominent premises on Bangor Road, Kew Bridge, just north of Kew Gardens. When, at the age of seventy-two, John George died, the family funeral in Brompton Cemetery on 5 May 1908 was a Galsworthian event. His sons – naval Engineer Commanders John G. and Edward – and their sons, Harold, Leslie, Wilfred and Edgar, and daughters Vera and Jennie – were there, as were his brothers William and Jarratt, and two of John's great-nieces.[38] All were present, except Archibald, still on the coast, convalescing. Gradually, however, he was reunited with his family, and there was no doubting his respected place in its circles. His great-nephews and nieces still remember the cordial, if slightly forbidding geniality of 'The Professor', when taken to his club for tea.

Academic 'housekeeping' aside, it was not until February 1909, that Liversidge returned to laboratory research. From his rooms at the United University Club he could walk easily up Albemarle Street to the Royal Institution, rich in science behind its Palladian façade, and once home to Davy, Faraday and Tyndall. Over the years, he had used its library, but in February 1910, he decided to apply for a research place at its remarkable subsidiary, the Davy Faraday Laboratory. The 'Davy Faraday' dated from 1894, when Ludwig Mond, the wealthy industrial chemist, a man ever mindful of opportunity, gave £45,000 to equip a building and £60,000 to endow research in 'purely scientific chemistry and physical chemistry, that borderland between chemistry and physics from which, in my

The Royal Institution of Great Britain.

opinion, we may hope to learn more about the real nature of things than from any other branch of natural science'.[39] Mond specified that applicants were to be selected according to experience and independence; that admission be free; but that spaces must be re-applied for every year. A new laboratory was fitted up in the basement of no. 21 Albemarle Street, with rooms on the third and fourth floor for photography.[40] *The Illustrated London News* carried an engraving, showing the Prince of Wales at its opening. The first researchers were admitted in January 1897.

The Director of the Royal Institution was James Dewar, the pioneer of low temperature research, whose work on the liquification of gases had won him international fame. Widely disliked for his abrasive temper, Dewar's attitude towards students and assistants was famously condescending.[41] He preferred to work alone, and during his lifetime, the RI failed to produce a significant research group. But this did not deter Liversidge, who had followed Dewar's work since the 1870s. He was particularly excited by Dewar's liquification of hydrogen in 1898, which Liversidge repeatedly demonstrated to enthusiastic Sydney audiences the following year.[42] In London, his fame preceded him. Backed by Dewar, Sir William Crookes, and the RI's Treasurer, Sir James Crichton-Browne, Liversidge was easily accepted into what Gwendolen Caroe, W.H. Bragg's daughter, irreverently called the RI's 'temporary hotel accommodation', where he was to remain a guest for the next fourteen years.[43]

'The Opening of the Davy Faraday Research Laboratory',
*Illustrated London News*, 22 December 1896.

Most of the researchers at the Davy Faraday were young, working for higher degrees, or continuing research begun elsewhere. Neither gender, nationality nor age was a barrier.[44] Of the six who applied in 1910, four were accepted; Liversidge was the only retiree, and also the only analytical chemist.[45] His laboratory companions were generally few – in 1910, only about seven in each term – but the atmosphere was congenial. As the young crystallographer J.D. Bernal later put it, the Davy Faraday created an ambience of innovative, rather than simply unorthodox ideas. The younger men included H.T. (later Sir Henry) Tizard, later FRS, chairman of DSIR and Britain's wartime leader in radar, down from Oxford to study chemical indicators,[46] and Joseph Petavel, who held a PhD from Freiburg, and who later became Director of the National Physical Laboratory. Both were students at Oxford, which did not then have the facilities the RI supplied.[47]

Among the other 'retirees' in the laboratory were Hugo Müller, a German-born, Göttingen PhD and naturalized British subject, who had been chief chemist at De La Rue's paper works for many years, and John Young Buchanan, Scottish chemist, and veteran of the *Challenger* Expedition. They were joined by James Emerson Reynolds, FRS, formerly Professor of Chemistry at Trinity College Dublin, and an expert on organic silicon compounds; and Dr George Senter, a Scottish electrochemist, destined to become Principal of Birkbeck College and Vice-Chancellor of London University. Of these, Liversidge already knew Buchanan, Reynolds and Müller, and came to know them better.

Liversidge saw the RI as a place to carry out work he was unable to do elsewhere.[48] As promised in his application, he supplied his own apparatus and chemicals.[49] Using specimens sent from the Australian Museum, he began to examine marine organisms, tree ash and other organic substances for trace elements.[50] In deducing that tree bark collects traces of silver, he began to develop techniques of 'chemical prospecting' for timbers from Australia and elsewhere.[51] By the end of 1910, he had found what he called 'interesting results'. His sources yielded traces of silver, but before publishing, he needed more evidence.[52] 'I am very much interested in the investigations', he wrote Etheridge, 'and if the additional samples give the same results I think you will be also, but I do not wish to give particulars until I have confirmed them'.[53] He was possibly anxious to avoid any possibility of miscalculations such as he might have made in his earlier work on gold in seawater. But his caution was to prove regrettable; none of his fresh results were destined to appear in print, or even to survive in manuscript. His last Australian paper on gold crystals was reported in *Nature* in January 1908.[54]

## 3. Imperial Chemistry and the British Science Guild

Liversidge was pleased to see some of his papers being taken up in the American and Canadian literature.[55] But he published little in the next few years. In fact, he was distracted by chemical societies that refused to ignore his reputation. As early as May 1910, he was already missing days at the Davy Faraday, and complaining – not, perhaps, without a certain self-indulgence – that 'as in Sydney, my time is being eaten up by committee meetings'.[56] The year before, he had been asked to join a committee of the Institute of Chemistry – which included such long-standing friends as Sir Alexander Pedler,[57] M.O. Forster, FRS, and Percy Frankland, FRS (son of Liversidge's teacher) – to consider ways of strengthening 'chemical relations' in the Empire. The Institute appointed local representatives in India, South Africa, Australia, New Zealand, Canada, Egypt and the West Indies. In June 1910, these became known as 'Hon. Corresponding Secretaries', and thereafter acted as the 'eyes and ears' of the Institute – advising London and supplying 'Home' with news of colonial vacancies. Liversidge championed the view that the organisation was 'British in the best and most modern sense of the term' – that is, imperial in all its ways.[58]

In much the same spirit, Liversidge was elected one of six vice-presidents of the Chemical Society in 1910, and remained in office until 1913. Alongside him were William (later Sir William) Jackson Pope, FRS, who had succeeded Liveing in the chemistry chair at Cambridge; and Gilbert Morgan who, with fine irony, had just declined to succeed him at Sydney. Thus, within two years, Liversidge had gained a place at the centre of 'imperial chemistry', and found himself among men with wide experience of the world. Britain made use of him, and of others like him. When Sir Thomas Holland retired from the Geological Department in Curzon's India, he was invited to be Rector of Imperial College. Important jobs were also found for Henry Dwyer, John Perry, FRS, and W.E. Ayrton, returning from service in Imperial Japan – men who had strong views on what Britain should do to improve its competitive position in applied science and international affairs.[59]

Like others who shared his views, it was only a matter of time before Liversidge was brought back into Norman Lockyer's great reforming engine of 'national efficiency', the British Science Guild.[60] Indeed, as early as 15 January 1908, shortly after returning to London, he was nominated to the Guild's Executive, and was annually re-elected until 1920. As a pressure group without premises, the Guild met in the rooms of the Royal Society, the Royal Society of Arts, the Chemical Society, the Geological Society, the Institution of Electrical Engineers, and the Iron and Steel Institute. How familiar all this must have seemed to the arch-organizer of Elizabeth Street. Even if he could not attend the Guild's annual meeting at the Mansion House in January 1909, Liversidge vigorously pursued its remit – to enquire into and report upon subjects ranging from university endowment to coal smoke abatement, from the improvement of water supply to the synchronisation of its clocks. The Guild not only embraced his skills, but also put him to work on issues he knew about, with people whom he had known for most of his life.

Thus in 1910, Sir Thomas Lauder Brunton – once, a year behind him at Cambridge, now a distinguished physiological chemist – invited Liversidge to join a Guild committee enquiring into the provision being made for medical research and postgraduate studies in London.[61] The committee of fourteen included two old friends from Australia – Sir Almroth Wright, FRS, and Dr C.J.C. (later Sir Charles) Martin, FRS – and two old English friends, Sir Archibald Geikie, FRS, and the chemist A.W. Crossley, FRS. When the committee reported, Liversidge's hand was strongly at work in its recommendation of imperial research fellowships,[62] which was sent to the Royal Commission on University Education in 1911, and which was ultimately accepted by the University of London.[63]

At its prime, the Guild reached into every corner of public life in Britain, and involved many of England's most public scientists. It was, indeed, at a meeting of the Guild in 1907 that Sir Michael Foster spoke for the last time. Foster's dedication was mirrored in Liversidge, who in July 1910, signed a memorial to the Prime Minister, recommending a new program of agricultural research.[64] For two years, he served on the Guild's Technical

Education Committee, chaired by Raphael Meldola, alongside Alexander Pedler, John Perry and George Beilby, whom he knew from the Institute of Chemistry.[65] In 1910, this committee reported on evening classes for artisans, the management of technical institutes, and mining education. Drawing information from the Technical Education Committee of the London County Council and the Board of Education, and from Dr Frederic Rose, formerly the British Consul in Stuttgart, whose reports had alerted Britain to Germany's rising power,[66] Meldola's committee condemned the low level of technical education in universities, the lack of cooperation between universities and polytechnics, and the lack of government funds for industrial training. The Guild's message was clear: unless the government gave greater attention to these issues, Britain could not successfully compete in the race of nations.[67]

The committee's recommendations of bursaries, scholarships, training schemes and national certificates touched off debate in press and parliament. When, during 1912 and 1913, Meldola's committee joined forces with the Guild's Education Committee, Liversidge found himself working with Sidney Webb, Sir Philip Magnus, and a score of others under the chairmanship of Sir William Mather.[68] Their efforts produced a major report, recommending a comprehensive scheme of industrial organisation, a minimum school leaving age of fourteen, continuation classes to the age of seventeen and 'suitable secondary schools for all who can profit by them'. The 'Mather Report' was taken up by the Oxford historian-turned-president of the Board of Education, H.A.L. Fisher, and many of its recommendations featured in the *Education Act* of 1918.

The Guild pressed for better coordination between universities and technical colleges; increased grants to the universities; reduced fees; improvements in teachers' conditions; and a shift of expenditure to the national government.[69] Liversidge had long sought these goals, and it was satisfying to see that their time had come – in England, if not yet in Australia. Between 1910 and 1914, with a membership not exceeding 750 in Britain, 210 in South Africa and sixty in Canada, the Guild kept up a flow of enquiries into the work of the Home Office, the Post Office, the Board of Trade, the Board of Agriculture and Fisheries, the Board of Education, the Post Office and the Colonial Office. Science was 'being applied to the purposes of government'. Prophetically, the Guild also launched reviews of imperial strategic mineral, coal and petroleum reserves, and other sources of energy.[70] A 'Science and State' Committee, set up under Lockyer in May 1914, began a chain of enquiries that eventually led to the creation of the 'scientific civil service'.

To suggest that these sweeping changes were exclusively the result of the Guild's work, would be to exaggerate its importance. But Liversidge and his circle did prepare the way. In January 1914, the Guild's Executive decided to create a General Purposes Committee of six, to sift all matters not specifically relevant to any other. To this body, Liversidge was appointed, and on it he sat continuously, preparing reports on government policy concerning venereal disease, radium therapy and technical education, while at the same time, attending the Guild's Medical Committee, monitoring the Royal Commission on London University, and preparing testimony for the employment and school attendance bill then before parliament. Three decades spent in meetings in Sydney well prepared him for the task.

Fortunately, the Guild was not to consume all his time, nor did the Davy Faraday, which he visited at most once or twice a week. In the first five years of his 'troisième age', his greatest enjoyment came with club life and its scientific conversations. In 1912,

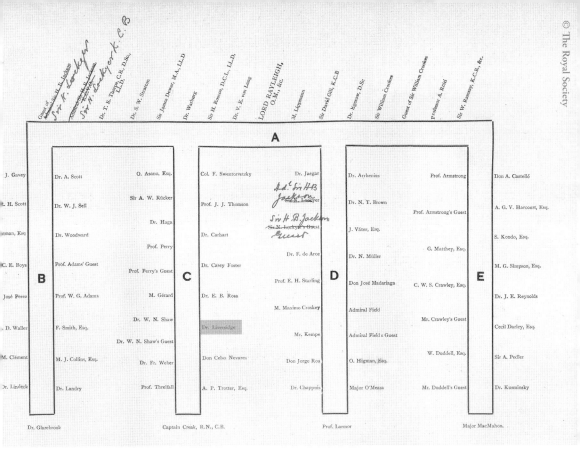

Seating plan for the Royal Society Club dinner in the Empire Hall, Trocadero Restaurant, London, 12 October 1906, to welcome delegates to the International Conference on Electric Units.

he was nominated to the Athenaeum by William Boyd Dawkins, FRS, and Henry Roscoe, FRS (by then a Knight, a Privy Councillor, and a member of the Athenaeum's Committee). By preference, however, Liversidge remained faithful to the Savile Club, now removed from near Piccadilly to upper Mayfair. At the Savile, he became especially close to fellow bachelor Arthur Shipley, whom he had known in Cambridge, with whom he had served on Sydney's London committee, and who was a member of the Savile's Club Committee from 1907 to 1910. When, in 1910, Shipley became Master of Christ's College, Cambridge, Liversidge found reason to visit his old college more often. The two men agreed on much, from national characteristics to motor cars.[71] Both gravitated towards the politics of the Athenaeum, but above all, both were active in the Royal Society Club (RSC), the 'social wing' of the Royal Society.

## 4. The Royal Society and Sociable Science

Liversidge was first introduced to the Royal Society Club as early as 1887, but was not proposed for full membership until 1905, when his sights began to turn towards London. There were few vacancies, but supported by Sir William Huggins (then President of the Royal Society), and seconded by Sir William Crookes, he was eventually elected in June

1908.[72] The event coincided with his return from convalescence, and gave a new dimension to his life. If its Signature Books are a reliable guide, for at least as long as he resided at the United Universities Club, the RSC became almost a second home. Its dinners he attended faithfully – monthly at first in 1908, then weekly from 1909 until November 1913. Once again, his administrative skills were not overlooked: the RSC elected him its junior treasurer in June 1909, second treasurer in 1910, and senior treasurer the following year. Between 1907 and 1910, the archives, neatly kept in his own precise hand, show him as the Club's de facto secretary.[73]

If the Royal Institution gave Liversidge a research base, and the Guild, a public presence, the RSC gave him continuing access to Australian science. Visiting London in March 1896, Anderson Stuart spoke to the Club on the history of Sydney University, and the Club heard news of Edgeworth David's success at Funafuti, as it did from David himself, when he came to London in February 1900. In June 1911, on David's return from Shackleton's 1908–10 Antarctic Expedition, and again in 1914, he was Liversidge's guest, and more than once Liversidge introduced Australian subjects into the Club's agenda. On 14 March 1912, for example, the Club had an 'Australian evening'. Anderson Stuart – as Liversidge's guest – spoke on the prospect of white settlement in the tropical north, after which Liversidge demonstrated the optical properties of Ceylon's sapphires and recounted his search for traces of silver and gold in land and marine plants, in the soot of Albemarle Street, and in the dust of Piccadilly.

A vision of London paved with flecks of gold, within a nugget's throw of Jermyn Street, may have failed to excite the six members present that evening. But for Liversidge, these dinners were seminars. When Henry Armstrong brought an American chemist, a Dr Anderson, to talk about methods of using colloidal graphite as a lubricant, Liversidge took careful notes. If exotic discoveries were to one's taste, there were Smith Woodward on Pleistocene man, C.V. Boys on soap bubbles, and H.E. Armstrong on pigmentation. Whether they welcomed Flinders Petrie on recent discoveries in Eygpt, or Sir William Ramsay on the spectroscopic wavelengths of neon and krypton gas, dinners at the RSC were a delight that Liversidge fully enjoyed.

In July 1909, after living a year in clubland, Liversidge needed a proper residence, and moved from Piccadilly to a rented house in Hornton Street, W8.[74] This was a large, pleasant 'dower' cottage, built for the housekeeper of William Phillmore's vast Kensington Estate, close to the Underground and Hyde Park, a short walk from the museums and libraries of South Kensington and the Royal College of Science, and an omnibus ride from the Royal Institution. There he took a forty-five year lease at £180 a year from Reginald Acland, Registrar of Oxford – a 'quaint sort of cottage', as Liversidge described it, 'which has escaped destruction in spite of the rebuilding which has gone on around it. It even had a studio, which I thought of turning into a laboratory'. [75] The area suited him well; as it did his near-neighbour, Herbert Hoover, the future American president, recently arrived from the mining districts of Western Australia.

In some sense, he had come full circle. Hornton Street was but two miles from his birthplace in Turnham Green, and perhaps a half-hour's walk from his family's graves in Brompton Cemetery. Nephews and nieces visited him, and the rent was manageable. But even this could not be a permanent home. His bookshelves groaned with reports sent by the Royal Society of NSW, the Technological Museum, and the Department of Mines,[76] and so many journals that he began to redirect them to the Australian Museum.[77] Among his books, were copies of his *Minerals of New South Wales* of 1888, interleaved with blank pages, awaiting the day when he could add the descriptions that he, and subsequently Pittman, had recorded. Meanwhile, he had continued to collect works on 'White Australia' and Aboriginal ethnology.[78] All this needed space.[79] He also needed a laboratory of his own. Kensington soot, even if traced with gold-dust, was no substitute for the clean air he once had in Woollahra. Liversidge wanted a house in the country.

For many years, Liversidge had prudently saved much of his salary. In March 1905, after fifteen years, he had paid off the mortgage on the Octagon, which he sold in 1907 for over £4000.[80] This he took to England, together with savings of £15,000, in addition to his pension of half-salary, or £450 per annum. In twenty years' time, he would leave an estate valued at over £46,000 – in today's money, perhaps £4,500,000. In the sense familiar to civil servants of Macaulay's India, the Empire had paid him well. In 1914, it was time to receive his reward.

Readers of Galsworthy may recall the day that Soames and Bosinney walked to inspect Soames' new house near the suburban village of Kingston-upon-Thames, just on the edge of the Surrey countryside. Here, in 1914, into the prosperous domain of the Edwardian middle classes, in an area called Coombe Warren on Kingston Hill, along Coombe Ridge, bordered by Warren and George Roads – right next door, it is said, to Galsworthy's father – Liversidge became again a 'man of property' – a house then called Fieldhead, and now called Hampton Spring.

Fieldhead occupied a site near a Tudor conduit built by Cardinal Wolsey in 1514 to supply fresh water to nearby Hampton Court.[81] Around 1907, two dwellings were built in its grounds. One, with a rateable value of £185, was probably the least of the large establishments rising along George Road.[82] Coombe Court, once owned by the Marquis of Ripon, was far more grand. In this fine house, with ten bedrooms, a stately garage, and gardens with breathtaking views from a long verandah, Liversidge could almost think himself at Darling Point. Liversidge joined the main house to a dependent cottage, built a library, and refitted a large room (some say, the former billiard room) with a cement floor for use as a laboratory. Kingston is not far from central London by bus or train, but Liversidge travelled by car, with a driver, and hired three servants to run his home. What was less than an acre in Surrey, was a principality in Utopia.

Liversidge's contentment would not outlive the year. By a tragic coincidence, following two years of careful preparation,[83] the British Association – the 'travelling Palladium of British science' – embarked for Australia, and arrived in Sydney the week that Britain declared war on Germany. Liversidge was not among them. 'Owing to unforeseen circumstances', he wrote Etheridge, 'I am not able to carry out my intention of going out with the BA'.[84] What those 'circumstances' were is not clear. He may have had work arising from the Franco-British and Anglo-American Exhibitions, where he exhibited his papers, gems and minerals for the last time. But his health, his house, his life in Britain, all militated against another long voyage, even to celebrate 'Australia's scientific coming of age'. In any case, during the first week of August 1914, all thoughts of congresses disappeared, as Liversidge and the nation turned to finding ways of serving King and Empire.

# CHAPTER 14

# Gentlemen to Arms

The skeleton of Zeppelin L33, which crashed and burned near West Mersea in Essex,
September 1916.

In his Final Despatch to the Government, published in the *London Gazette* on 21 March 1919, Field Marshal Sir Douglas Haig departed briefly from a lengthy catalogue of explanations for his conduct of the war, to acknowledge that superiority in what he called 'mechanical contrivances', which had helped bring the Allies victory. 'In this respect,' he said, 'the Army owes a great debt to science, and to the distinguished scientific men who placed their learning at the disposal of their country'.[1] In these few words, much was implied. Behind such obvious 'contrivances' as the tank and the aeroplane, the submarine and poison gas, there lay a host of innovations that bore directly on the outcome of the war, both at home and at the Front. The contribution made by British scientists, often shrouded in secrecy, was central to this 'war of invention', and many Australians found service in their company.[2]

## 1. THE SCIENTISTS GO TO WAR

Britain's scientific mobilisation began in August 1914, but gathered serious momentum in October following the infamous 'Manifesto' of ninety-three German professors, which transgressed international conventions, and wedded German science to the war aims of the Fatherland.[3] Until then, British scientists, including J.J. Thomson and William Ramsay, who had many German friends, shared the hope that the crisis would pass, but by November, such hopes had vanished. That month, the Royal Society resolved itself into a War Committee, and a Conjoint Board of Scientific Societies – a 'War Cabinet', based at the Royal Society – was mustered to coordinate Britain's twenty-seven scientific and professional societies.

According to Threlfall, some of these were 'well-known names, some pure scientific merino, others men of technical eminence'.[4] Their deliberations began slowly, but started to bear fruit only after April 1915 when, with the 'shell crisis' and the advent of chemical weapons in Flanders, and the declaration of unrestricted submarine warfare in the Atlantic, the government fully awakened to the need for 'brain power', and prepared to fight on the 'scientific front' as deadly a war as that being waged in France. During the next twelve months, between June 1915 and July 1916, Britain transformed the organisation and means of waging scientific war; by the end of 1917, those means were put to effect; and by the summer of 1918, aided by American industry, Britain and the Empire were preparing to fight well into the future.[5]

The immediate response of scientists to the war, in Britain and elsewhere, was a rush to the colours. The universities emptied, and the National Physical Laboratory lost a quarter of its staff. The Royal Navy had a formidable scientific reputation, but the War Office had far less experience in using men with scientific training. Early in the war, the Parliamentary Recruiting Committee issued a general list of men wanted for the Front that included 'navvies, tunnellers, and chemists'. Science graduates were allowed to volunteer without reference (or prejudice) to their technical skills, and many were lost in the first months of fighting.[6] Early attempts to volunteer technical expertise were met with scepticism, until loud protests from the press forced the government to review its policies. Only in early 1915 were attempts made to reserve the skills of the country's academic and industrial chemists.[7] When, in March 1915, no chemist was appointed to the board of British Dyes Limited, newly formed to run factories essential to explosives manufacture, the profession rebelled.[8]

By mid-1915, this picture had dramatically changed. In rapid succession, the British government was forced to respond to a catastrophic shell crisis, the advent of chemical warfare, and the appeals of British science and industry. In Cambridge, it had become impossible, as the young 1851 Exhibition Scholar from Sydney, Lancelot Harrison wrote, 'to find any outlet for patriotic energies, there being such a horde of unemployed "dons" and "tutors", who have first call on anything in the nature of war work here'.[9] But news of the death at Gallipoli of the brilliant young physicist, H.G.J. Moseley, poignantly recalled by Rutherford,[10] signalled a change in the tide of public opinion that swelled with the casualty lists. In December 1915, the Conference of Headmasters confidently proclaimed the struggle to be 'eminently a war of science', deplored the 'unwisdom' of employing young students of 'mathematical and scientific ability in line battalions', and prophesied an 'industrial armageddon' if such men were not husbanded for laboratory or other technical work.[11]

In May 1915, following the Germans' use of chlorine gas at Ypres, the Institute of Chemistry and the Registrars of the universities received a request from the War Office for men with a knowledge of chemistry, to join the 'body of corporals' that became the Special Brigade (for gas warfare) of the Royal Engineers.[12] The same month, Arthur Henderson announced the formation of a Privy Council Committee on Scientific and Industrial Research, with a Scientific Advisory Council (ACSIR); within a year, this would lead to the creation of the Department of Scientific and Industrial Research (DSIR).[13]

In July 1915, A.J. Balfour, succeeding Churchill at the Admiralty, created a Board of Invention and Research (BIR), which set up several research panels,[14] and shortly thereafter, the Ministry of Munitions began recruiting chemists and engineers for its Inventions Branch and newly-established explosives factories. In the same month, following overtures from the Royal Society and the Chemical Society, a Committee of the Privy Council for Scientific and Industrial Research was created, followed within a year by the new DSIR.[15] Shortly afterwards, an Experimental Committee for chemical warfare research was formed at General Headquarters behind the lines in France.

On 23 June 1915, Harold Tennant, MP, Asquith's Under-Secretary of State for War, reported to Parliament that practically all the laboratories in the country were now at the disposal of the War Office.[16] Several large chemical firms and university departments – notably, at Cambridge, Manchester, UCL and St Andrews – dedicated themselves to military projects.[17] The war soon embraced all the disciplines. Chemists and chemical engineers produced explosives and propellants, aerial photographs, smoke shells and poison gases. Geographers and surveyors produced artillery maps. Mathematicians calculated anti-aircraft trajectories. Physicists devised apparatus for sound-ranging, counter-mining and submarine detection. Physiologists devised gas masks. Bacteriologists worked on trench sanitation, the prevention of disease, the preparation of rations, and the treatment of wounds. Biologists developed techniques of camouflage, and many other scientists and engineers joined businessmen on committees of the Board of Trade and the Board of Agriculture and Fisheries, supervising wartime production of textiles, electrical goods, ships, aeroplanes and foodstuffs.[18]

In mobilising science, Australia followed Britain. Federal and State Munitions Committees were established, and academics recruited to them. T.R. Lyle, Professor of Natural Philosophy at Melbourne, was on the eight-man Federal Munitions Committee, while Professor Henry Payne, Dean of the Faculty of Engineering, became one of six members of the Federal Arsenal Committee. In Adelaide, Edward Rennie (Chemistry) and Robert Chapman (Engineering) became members of their State Munitions Committees, while Theodore Osborne (Botany) worked on the application of yacca gum to the synthesis of picric acid.[19] Both Sydney and Melbourne universities assisted with chemical and physical tests of steel, and universities and technical colleges alike sent men into munitions manufacture in Victoria and New South Wales, aided by returning chemists with experience of British factories.[20]

The cry for enlistment, which met an enthusiastic response in the population generally, was not immediately evident among university staff. As in Britain, and the United States, many academics regretted the loss of contact with German colleagues. On 10 August 1914, a week after war was declared, the British Association's Australian visit was in full swing. Three German scientists were given honorary degrees. 'The cheers for these', the Melbourne *Argus* reported, 'were warmer than those for their British *confrères*', and Sir Oliver Lodge made a 'fine and dignified speech, deploring the war and hoping for peace'.[21] Of the seven German scientists visiting Australia for the BAAS Congress, only two – Professor Peter Pringsheim, a physicist of Berlin, and Dr Fritz Graebner, an ethnographer from Köln – were interned, and these only because they refused to promise not to serve in the German army.[22]

By October 1914, as in Britain, the 'internationalism' of August had disappeared. Although no spies were arrested, anti-German hysteria swept the country.[23] Melbourne University lost the services of the only two Germans – lecturers in music and German – on its staff. As in England, not all Australian scientists immediately succeeded in getting war work. Medical staff and engineers had no problem.[24] But others found it more difficult. By December 1914, only eight of Sydney's 165 academic staff had joined the AIF. A.C. Rothera of Melbourne, Australia's first lecturer in biochemistry, met a particularly harsh fate. As W.A. Osborne recalled: 'All the military authorities could think of in the way of utilising his exceptional gifts was to give him the task of inspecting latrines', and in the course of his duties, he contracted pneumonia and died.[25] Eventually, several Sydney University staff – including Associate Professor George Nicholson of English, and Dr Henry Lovell, later Australia's first Professor of Psychology – joined Professor J.T. Wilson of Anatomy in the Military Intelligence and Censor's Department, which was directed from Melbourne.[26] Professor J. Douglas Hewart of Sydney's Veterinary Science School was placed in charge of the Australian Army Veterinary Service.[27]

In April 1915, J.P.V. Madsen, the radio physics pioneer, then a lecturer and later Professor in Electrical Engineering, was commissioned a captain and appointed Instructor of Field Engineers and Officer Commanding the Engineer Officers' Training School in Sydney. And when an Australian Mining Battalion was formed in November 1915, James Pollock – Threlfall's successor at Sydney – began working on acoustic geophones. In early 1916, Pollock, then aged fifty-one, was commissioned a captain, and with Edgeworth David, aged fifty-eight, went with the Australian Mining Corps to the Western Front. David ultimately became the principal geologist of the BEF, attached to GHQ in France.

Australia's scientific contribution was brilliantly recorded in the field of applied chemistry. In early 1915, the British munitions industry faced a desperate situation. The surging demand for high explosives had drained the country's industrial capacity. But to fill its factories Britain needed applied chemists. As J.A. Pease, President of the Board of Education, told the House of Commons in May, there were four German dyestuffs firms that employed 1000 chemists, while Britain had only 1500 chemists in the whole of her industry.[28] More were desperately needed to direct the great new factories being built at Gretna and Queensferry and a hundred other sites.

In September 1916, a first group of Australian chemists left to join Moulton's Department of Explosives Supply. They included five members of Sydney Science Faculty – F.A. Eastaugh, lecturer in metallurgy and assaying; J.W. Hogarth and J.M. King, demonstrators in chemistry; Cecil E. Tilley, demonstrator in geology, and D.A. Pritchard. Once in England, they met former pupils of Melbourne's David Orme Masson, including Professor Bertram Steele from Brisbane, and Professor N.T.M. Wilsmore from Perth, as well as Melbournians A.C. Cumming, then a lecturer in chemistry at Edinburgh, who became manager of H.M. Factory at Cragleith, G.S. Walpole, a chemical adviser to the Aeronautical Department of the Admiralty and J.I.O. Masson, Orme Masson's son, then a chemistry lecturer at UCL, who became chemist-in-charge of propellants at the Royal Arsenal, Woolwich. In 1917, with a second group of chemists went A.C.D. Rivett, lecturer in chemistry at Melbourne and organiser of the BA's meeting in Australia, who first joined Brunner Mond, then became a process manager at H.M. Factory, Swindon, where he was responsible for the manufacture of amatol. He did theoretical work of great importance to physical chemistry, and gained organisational experience that would prove invaluable to postwar Australia. [29]

The war united the Australian academic community in Britain. The University of Queensland sent to munitions work both its professor of chemistry, B.D. Steele, and its lecturer in organic chemistry, T.G.H. Jones. A.J. Gibson was a captain in the Australian Intelligence Corps before he became Queensland's foundation professor of engineering.[30] From Western Australia came E.A. Weston (veterinary science) and M. Aurousseau (geology), W.G. Townsend (mining and engineering) and D.C.J. Hill (geology). Edward Shann, Professor of History and Economics, then aged thirty-three, enlisted, but was returned by the Army when the University had difficulty replacing him in the lecture room.[31] Instead, he served as chairman of a State Royal Commission on wartime living costs. Professor H.E. Whitfield (mining and engineering) became an inspector of steel manufacture, working first for the British Government in the United States, and then for the Ministry of Munitions in England.[32]

By the end of 1915, over thirty-six Sydney and forty-seven Melbourne staff were on active military service.[33] Others stayed and taught. In 1915, Anderson Stuart lectured to new recruits on 'Egypt', and on 'How to Keep Fit'. Mungo MacCallum offered 'Some Reflections on the War'; George Arnold Wood weighed 'The Immediate Responsibility of the War'; and D.A. Welsh, Professor of Pathology, spoke of 'The Great Opportunity' the war had given Australians to serve the British Empire, 'the mightiest instrument under God for the uplifting of the human race'.[34] Robert Irvine, Sydney's first professor of economics, spoke on 'National Organisation and Efficiency', a theme that resonated with Liversidge. Between

June and September 1915, Melbourne University sponsored a similar series, including 'Chemistry and the War', by Orme Masson, and 'The Dominions and the War' by Thomas Laby, Liversidge's pupil and Melbourne's newly appointed professor of natural philosophy.[34]

## 2. Mr Britling Sees it Through

Like the literary character of H.G. Wells, and the student of Huxley whom he sometimes resembled, the outbreak of war found Liversidge at home in London – quietly combining research at the Royal Institution with learned society meetings and visitors in Kingston.[35] The pace of life at the RI had slowed considerably, owing to the senescence of James Dewar, who refused to retire.[36] Ludwig Mond had planned to accommodate at least twenty-two research workers, but by 1911, their number, which had averaged seven or eight when Liversidge arrived, had dropped to five, and before 1914, never exceeded eleven. At the outbreak of the war, there were only eight; and by the end of the war, three, of whom Liversidge was one. When Dewar grudgingly made his laboratory available for war service, Liversidge looked elsewhere.

It was to be a fraught event, overlaid by the catastrophe that was soon to engulf the world. As the British Expeditionary Force fell back before Mons, Liversidge celebrated his sixty-eighth birthday. He could have avoided military service. But this was not destined to be a war only for young men. In November 1914, Lord (John Fletcher) Moulton, QC, PC, FRS, senior wrangler, Fellow of Christ's, and two years older than Liversidge, was appointed to chair the War Office's Committee on Explosives. W.J. Reader took it as 'an elegant comment on the place of science in English life', that when the government sought a man of scientific attainments for a post, exposed to all the strain of wartime, 'their choice fell on a Lord of Appeal nearly seventy years old'.[37] But as events were to show, it was an inspired choice, as thanks to Moulton, Britain rapidly made up lost ground in munitions production.[38]

Moulton was not an isolated case. If the wartime President of the Royal Society, Sir William Crookes, was eighty-two, age could be no bar to 'young' men like Liversidge not to Sir Ernest Rutherford, forty-four, of New Zealand, who became Chairman of the Admiralty's Board of Invention and Research; let along to (Major) Sir Douglas Mawson, thirty-two, Liversidge's student and Antarctic explorer, who oversaw munitions shipments to Russia.[39] Professor (Colonel) S.H. (later Sir Henry) Barraclough, forty-five, of Sydney's Engineering Department, arrived in September 1916 with 3000 Australian munitions workers.[40] Edgeworth David was a hero when, at fifty-eight, he donned uniform. 'Your Excellency may be interested to hear,' he wrote the Governor-General, 'that at the present moment both the Admiralty and the War Office are using the services of scientific men on a scale that would never have been dreamed of before the war ... You will not see anything in the press about the doings of our men ... but they are all putting in really solid useful work.'[41]

Liversidge – a member of the Volunteer Rifles at Cambridge, and a Vice-President of the University Scouts at Sydney – was keen to serve. He had at least two nephews in the Royal Navy. Appreciatively, he noted *Sydney Morning Herald* reports that Canada and the United States had followed Britain in introducing conscription. To his mind, Australia was not doing enough.[42] By December 1914, only thirty-eight Sydney University students had joined the AIF,[43] and by September 1915, only sixty-three of the 700 male non-medical students at Melbourne had enlisted.[44] However, the pace gradually picked up, and by early 1916, over 200 Sydney students were on active service.[45]

Overall, perhaps thirty-five per cent of Australian university students enlisted, compared with fifty per cent or more students from the British universities. But by the end of 1915, some three-quarters of all eligible Sydney medical students were on active service.[46] Of the twenty-two second-years in Engineering in 1915, eighteen enlisted together, and one left for munitions work. Of the thirty-one third-year engineers, twenty-two joined up. From fifty-three science students came forty soldiers, three of whom won MCs and a quarter of whom were killed.[47] By the end of the war, Sydney had sent some 1800 students and staff, and Melbourne, 1723. Of these, 199 and 251 (ten per cent and fifteen per cent) were killed, and a further fifteen per cent were wounded.[48]

In the spirit of collective service, Liversidge conscripted himself. He was in good company. Richard Threlfall and W.H. Bragg, both in their fifties, had found jobs in the Admiralty, and Bragg's son, W.L.Bragg, was in the Army. (Both would receive the Nobel Prize in Physics in 1915, and both later returned to distinguished careers in England.)[49] Threlfall's career was accelerated by the war. After his return to England in 1898, he had become Director of Research at Albright and Wilson, chemical manufacturers, in Birmingham. He was asked to advise the Admiralty on chemical questions and in October 1914, received reports, subsequently shown to be unfounded, that the Germans were using a non-inflammable gas in their new zeppelins. On searching the literature, he found mention of a gas well near Hamburg producing a small but significant percentage of helium. When he told this to the Admiralty, he was asked to locate, through the American branch of his firm, a source accessible to the Allies.

Enlisting Professor (later Sir John) MacLennan of the physics department at Toronto, Threlfall developed a scheme to produce helium on an industrial scale.[50] In July 1915, following this success, he was recruited by his Cambridge mentor, Sir J.J. Thomson, to the Admiralty's Board of Invention and Research (BIR).[51] Thereafter, he became a founding member of the Advisory Council on Scientific and Industrial Research, and, in 1916, chairman of two DSIR research boards. He combined his 'committee war' with laboratory experiments at Oldbury, which led to the development of phosphorus bombs and smoke screens. For all this, he was knighted in 1917.[52]

'Threlfall's war' was scientific warfare at its most spectacular. By contrast, Liversidge's war was fought not on the battlefront, nor even in the laboratory, but in quiet conferences, corridors and committees. The Davy Faraday Laboratory at the Royal Institution was emptied, and Liversidge was one of only three workers left; but he kept on and tried, as he wrote William Haswell at Sydney, 'to do a little work'.[53] However, his skills were eventually put to good use in areas where he was almost uniquely qualified to serve. Between 1916 and 1918, he served in four capacities – first, in helping young chemists find military and munitions jobs (a role not unlike that given C.P. Snow in the Second World War); second, in galvanising the British Science Guild for wartime work; third, in developing proposals for a strategic minerals survey of the Empire; and finally, in consulting on alloys for the Advisory Committee for Aeronautics and the BIR's Airships Sub-Committee.

As a Vice-President of the Chemical Society, and as a member of the Overseas Branches Committee of the Institute of Chemistry, Liversidge helped recruit British, Australian and other imperial chemists for war work. Working alongside M.O. Forster, Percy Frankland, Alexander Pedler and Arthur Smithells,[54] he was joined in 1917–18 by A.E. Leighton, representing the Australian Arsenal, who worked with Moulton at the Ministry of Munitions, before returning to organise Australia's emerging defence industry.[55]

Next, Liversidge focused his talents on the British Science Guild, which he had so keenly endorsed in Australia a decade before. It was an idea whose time had come. On 21 December 1915, as Australians offered thanks for the successful ANZAC withdrawal from Gallipoli, the press praised Henderson's scheme for directing science towards industry, and in Melbourne, the Commonwealth Government – prompted by William Osborne, Professor of Physiology, and David Orme Masson, Professor of Chemistry – proposed adopting the British model DSIR.[56] Prime Minister 'Billy' Hughes, was a ready sponsor, persuaded that science possessed an 'immense store of knowledge ready for instant application'.[57] The six states had already pooled research on wheat and other foodstuffs, and the mining industry was contributing to the imperial effort,[58] but Hughes 'caught fire at the vision of a whole nation inspired by the scientific spirit',[59] and on 22 December 1915 committed the federal government to spending £500,000 to help 'science and business to further the ends of industry'.[60]

In January 1916, NSW and Victoria separately urged the British Government to extend its new DSIR to the entire Empire, with a common fund, 'supported by contributions from the United Kingdom and Overseas Dominions', for the 'pooling or consolidation of the resources of the Empire for the purposes of scientific research' – a principle that Lloyd George readily accepted in a memorandum issued that March.[61] As a preliminary to imperial cooperation, in May 1916, Canada established a National Research Council along British lines, followed the next year by South Africa and, eventually, by New Zealand.[62] However, Hughes decided to proceed on independent lines, and in January 1916, established an Advisory Council of Science and Industry, which drafted proposals that led in September to the creation of an Institute of Science and Industry – the forerunner (in 1926) of the Council for Scientific and Industrial Research (CSIR), and today's CSIRO.[63]

## 3. A Community of Scholars

In May 1916, Andrew Fisher, formerly Australia's Labor Prime Minister and now High Commissioner in London, devoted special mention to Australia's scientific war at the Guild's annual meeting. This was a 'red letter' day, as Sir Alfred Keogh put it: the first occasion on which a Dominion had preached to the British on the need to support science and industry. Fisher was a suitable evangel. It was he who, as Prime Minister, had welcomed the British Association on its celebrated visit to Australia in August 1914; it was he who had, at the outbreak of war, promised Australia's aid to 'the last man and the last shilling'. To the BSG's approving audience, he reported on the meetings held by the new Prime Minister five months earlier in Melbourne. 'In the world of Commerce', Fisher warned, 'we have pitted ourselves against a nation whose Science has been the handmaiden of Industry'.[64] Australia's answer was to establish a federal institution which, going beyond the remit of Britain's proposed Advisory Council for Scientific and Industrial Research, would mobilise science in cooperation with the learned societies, the universities, the museums, and the Departments of Mines and Agriculture in the different states.

In the elastic rhetoric typical of such occasions, Fisher urged his British listeners that 'There must be an end of "muddling along"'. Primary and manufacturing industries, 'the foundation of the national structure, must be aided by legislative enactment, rendered practicable by scientific research'. In this way, he said, 'the laboratory will be linked to the Factory and the Mine, and to the soil, the basis of all wealth'.[65] Joining his remarks, James MacDougall, a Melbourne industrialist, believed this would strengthen imperial cooperation and lessen the distance between Australian and British interests. 'We have seen the German push his way in. Almost every place that is raised for the purpose of refining metals ... is controlled by a German', he observed. Now Australia, heralded by her fighting men, would be bound fast to Britain by her minerals and industry – by blood and ore, no less.[66] In so doing, freed of the weight of shackling history, Australia would create new horizons of her own. 'If [traditions] are retarding your progress', Fisher told his British listeners, 'cut the throats of the traditions'. 'Laughter and applause' followed.[67]

Liversidge sent his apologies that day.[68] But during the next two years, he was active on the Guild's Committees for General Purposes, Medicine, and Education. The latter was chaired first by Sir William Mather and later by Sir John Cockburn; its membership of thirty included Sir Philip Magnus, MP, the founding director of the City and Guilds of London Institute and veteran educational reformer;[69] A.T. Pollard and Sidney Webb, both with experience of the City of London College; and Professor John Perry, of Finsbury. Soon, they were joined by Professor T. Percy Nunn, the philosopher of science from London University's Institute of Education. In November 1916, Liversidge's committee issued a report on National Education, which had been completed two years earlier, but which now bore directly on the work of Asquith's Reconstruction Committee and on the Consultative Committee for the Board of Education, launched under Sir J.J. Thomson, which would set the agenda for science education in Britain after the war.

Liversidge sat with the Education Committee during 1916 and 1917, drafting a memorandum on science in British universities,[70] and submitting evidence to the Royal Commission on University Education in Wales. He also prepared commentaries on H.A.L. Fisher's Education Bill, on scholarships for higher education, and on the teaching of science in secondary schools.[71] In 1918, the committee began a study of industrial research and the supply of trained workers; its report, published in 1919, foreshadowed the DSIR's postgraduate research grants scheme, along lines similar to those that Liversidge had recommended to Sydney twenty years earlier.

Oscillating between the Chemical Society, the Institute of Chemistry, and the Guild, Liversidge made good use of the Royal Society Club. The 'wartime' Club engaged a dining room in a hotel near Piccadilly, which functioned as an unofficial annexe for Dominion and Allied scientists when duty brought them to London. Three times, Richard Threlfall entertained Professor (now Captain) James Pollock, his friend and successor at Sydney, who had sailed with Edgeworth David to France, and who was now directing the Second Army Mining School at Proven. In 1917, Sir William McCormick, first chairman of the DSIR, dined with the Deputy Controller of the Air Board (Captain Groves, RN), the Director of Chemical Warfare (Major Thuillier), the Chairman of the Wheat Commission (Sir John Beale), and the Director of Naval Recruiting (Captain White, RNVR). Other 'scientific soldiers' among the guests included Douglas Mawson, Colonel Arthur Smithells (professor of chemistry at Leeds); and Professor (Major) Walter B. Cannon, the distinguished American physiologist who came to deliver the Royal Society's Croonian Lecture in 1918. In November 1918, came the young Major W.L. Bragg, just back from his sound-ranging successes at the Front.[72]

War work was regularly discussed at these dinners with the American science attachés – Professor Henry Bumstead and Dr Durand – posted by the National Research Council in Washington to London and Paris:[73] on one occasion, the conversation turned to metal fatigue; on another, the use of benzene and alcohol as petrol substitutes.[74] The Club pursued its even tenor to an enviable degree. Certainly, rationing never cast a shadow: on 21 March 1918, the day that von Ludendorff launched his last great offensive on the Western Front, members enjoyed a typical nine-course meal at Prince's Hotel.[75] For Liversidge, the Club became a second home. The only surviving dinner table plan – that for the Club's Anniversary on 28 June 1917 – is in his hand. On it, we see him seated between W.H.M. (later Sir William) Christie, the astronomer, and Sir Alexander Pedler, of the RSM and BSG, with Sir J.J. Thomson in the chair, and not far from such old friends as H.E. Armstrong, Sir Jethro Teall, the geologist, and A.S. (later Sir Arthur) Woodward, the palaeontologist of Piltdown fame. With him that evening were also colleagues from Sydney (Threlfall), the Royal School of Mines (Sir William Watts) and London University (Sir William Tilden). Few images so well convey the unity of imperial science, mobilised for Britain's war effort.[76]

Such informal enterprise suited Liversidge well. But it was as an adviser that he wanted to serve his country; and while his contributions were modest, and incomplete, they nonetheless reflected his uncanny ability to select the significant, and to foreshadow developments that were well in advance of their time.

## 4. Airships and the Admiralty

On 19 January 1915, the first zeppelin to cross the North Sea dropped its bombs on the Eastern counties of England.[77] Although the Kaiser proved reluctant to allow the bombing of British cities until the summer – London was not raided until 1 June – air raids added a new dimension of terror to the food and material shortages already being caused by submarine warfare. Fears, which had flickered sporadically in the press since the 'air scare' of 1909, became a reality.[78] Bombing continued intermittently until the autumn of 1918. Casualties were relatively light,[79] but there was substantial damage and disruption; and more than twelve squadrons with 110 aeroplanes and 2200 men had to be kept from France for air defence.[80] Britain's air defences in 1915 had been 'rudimentary in the extreme' and forty-fifty raids took place that year.[81] But by the summer of 1916, British pilots had mastered the art of destroying airships by tracer bullets, focused in the convergent beams of searchlights.[82] This was, of course, just the beginning. Worse was to follow when, in May 1917, Germany sent fixed-wing bombers in the place of dirigibles, and opened a new era of fear.[83]

Several zeppelins flew over Kingston during 1915, but Liversidge's surviving correspondence makes no reference to them. We do know, however, that in early 1916, he sought information from one of his naval nephews, Ted (Edward, the second son of John George), later an Engineer Admiral. In March, Ted sent him pieces of an airship (the Zeppelin L15, he believed) that had been shot down over Kent.[84] Liversidge recognised the materials, and made contact with the Admiralty. As early as May 1909, the government had formed an Advisory Committee for Aeronautics (ACA), under Lord Rayleigh, supported by experimental facilities at the National Physical Laboratory (NPL).[85] But the ACA was caught off guard by the zeppelin attacks, and in 1915 and 1916 were still desperately seeking information on German airship manufacture.

In early 1916, Liversidge sent the Admiralty a proposal to shoot down airships by firing shells with a 'sticky flaming substance' to ignite the hydrogen envelopes.[86] His suggestion was passed over; as was, apparently, a reminder to the Director of Air Services, Rear Admiral C.L. Vaughan-Lee, of the advantages of helium over hydrogen as a 'balloon gas'.[87] But Threlfall had been concerned with helium for over a year, and encouraged Liversidge to pursue his ideas.[88] Over lunch at the Athenaeum, he discussed a range of problems with Horace Darwin, FRS, director of Cambridge Instruments Ltd.; Sir Alexander Kennedy, FRS, an old friend and now Professor of Marine Engineering at University College London; and Sir George Beilby, FRS, the industrial chemist and first director of the DSIR's Fuel Research Board, who had earlier worked for the Admiralty and the NPL on the surface structure of metals.[89] All were interested, and each was able to help. Darwin was on the BIR's Airships Subcommittee,[90] and Beilby was a member of the BIR's main committee.[91] It was, however, Threlfall who took up Liversidge's ideas, and brought them to the board's attention.

Following his helium work, and thanks to J.J. Thomson, Threlfall remained on the Airships Subcommittee, where he was asked to coordinate research useful to naval (or, in principle, to all military) aviation.[92] In June 1916, he wrote Liversidge that 'the very matter we talked of was up for discussion', and the Admiralty wished to have 'all existing information on light alloys in a form easily accessible'.[93]

In Threlfall's view, the Admiralty tended to rely upon what the NPL or Vickers, the munitions firm, 'likes to tell them'. To produce an independent assessment, he promised, was no 'light and easy job'; yet it would certainly 'do the country a good turn'.[94] Liversidge agreed, and on Threlfall's recommendation, the BIR asked him to review the literature on aluminium/magnesium alloys, and to report on which alloys, already in use by the Germans, could be usefully adopted by the British. At about the same time, Threlfall invited him to join the Executive Committee of the Conjoint Board of Scientific Societies, chaired by Sir J.J. Thomson, which included many well-known figures, such as Sir Herbert Jackson, Sir Robert Hadfield, Sir Ray Lankester, and Professor William Watts of the Royal School of Mines. 'What is wanted,' Threlfall observed, 'is men who have the knowledge and experience necessary to run such a business. Now there is nobody in England who has your experience.'[95]

For reasons that are unclear, Liversidge declined to join the Conjoint Board, but he did agree to advise the Admiralty. Between June, when the two men lunched at the Savile Club, and early August 1916, Liversidge completed a draft report for the BIR. Its fate could have been that of many chronicled by the Inventions Departments of the Ministry of Munitions and the Admiralty, and so memorably caricatured by Heath Robinson.[96] Most of these inventions were destined for oblivion. As a BIR historian wearily put it, by the end of the war, 'very nearly every idea which emanates from the public has either already been considered by the Admiralty, or is hopelessly impractical. The numbers vary with the frequency of air raids and the sinking of ships by submarines.'[97]

However, Liversidge's ideas were different. Summarising the literature, he outlined the value of magnesium alloys in reducing weight and producing savings in aircraft design.[98] Their durability was well known. Liversidge recalled that a magnesium ingot, obtained from the Great Exhibition of 1851, and kept in his laboratory at Sydney had scarcely oxidised at all in thirty-five years. Given its advantages, he recommended a set of experiments on beryllium-magnesium alloys, and asked the Imperial Institute in London to ascertain where in Canada beryl might be found. 'It may be urged', Liversidge added, 'that if magnesium and its alloys were of value for aircraft, etc., the Germans would have used them, and that we therefore need not trouble about them'. But, he reasoned, 'that does not necessarily follow, for they may have been unable to do so'; possibly, he mischievously noted, because 'the Germans have not often been discoverers or inventors, but they show great aptitude in making use of the discoveries and inventions of others'.[99]

Beginning in July 1916, Liversidge attended the monthly meetings of the BIR's Airships Subcommittee, then led by Threlfall and R.J. Strutt (later fourth Baron Rayleigh, Professor of Physics at Imperial College). At first, the BIR declined to pursue the studies that he recommended, as lying beyond its remit; but Threlfall persevered, and got him a grant of £250 to employ a seaman assistant from the Royal Navy's Airship Station at Watney Island, Barrow-in-Furness. That autumn, Liversidge got a special break. Following intense German air attacks during the first week of September ('Zepp Sunday', 2–3 September saw a massive attack on London),[100] the Royal Flying Corps improved its defences; and on the night of 24 September, the L33 was forced down in Essex, giving British scientists the opportunity to study a relatively undamaged enemy airship at first hand.[101] It was discovered that the frames of the envelope were made of an aluminium alloy, and the gondola of aluminium sheeting that 'looked like burnished steel'.[102] Its petrol tanks were well protected, apparently self-sealing, and made of an unusual substance. Liversidge went to Colchester to examine the Zeppelin, and sent a memorandum on the self-sealing petrol tanks to the RN Airship Station at Hoo, near Rochester, where his nephew was stationed, recommending them for British use. The Airships Subcommittee agreed that his experiments should be referred to the National Physical Laboratory (NPL) at Teddington. If the NPL declined, the Admiralty would sponsor him at the Royal School of Mines.

Thus encouraged, Liversidge spent October working on alloys, using samples of magnesium sent him by William Wardle and Co. of Liverpool, and Johnson Matthey and Co. of London. He took advice and laboratory space from H.C.H. (later Sir Harold) Carpenter, FRS, another chemist turned metallurgist (at the time, Professor of Metallurgy at the Royal School of Mines); and conducted experiments on tensile strength with samples sent from the Magnesium Metal Co. in London.[103] By January 1917, he was confident that beryllium, magnesium and copper alloys had a critical role to play in aeroplane engine and petrol tank design, and recommended full scale tests.[104] In March, with the Subcommittee's approval, Liversidge turned the matter over to the NPL, and to Dr Walter Rosenhain, a German-born Melbourne graduate who had gone to Cambridge as an 1851 Exhibition Scholar in 1897. In 1901, Rosenhain married Louisa Monash, sister of John (later General Sir John) Monash, and remained attached to Australia ever after. Accompanying the British Association on its visit to Australia in 1914, he did much to spur the development of Australian defence technology.[105]

In 1906, Rosenhain had become the NPL's first Superintendent of Metallurgy and Metallurgical Chemistry, where he helped to launch the discipline of materials science.[106] By 1916, his department was deeply involved in practical projects, such as determining the effects of heat and pressure on the properties of metals, and helping to manufacture substitutes for German chemical glassware, porcelain and optical glass. There was also in his department a section devoted to fabrics for balloons and airships,[107] and it was presumably this section that took up Liversidge's proposals.[108]

The archives do not tell us how far these plans progressed, or how far the NPL took up his suggestions. The use of alloys was of obvious industrial interest. As early as 1902, the Institution of Mechanical Engineers established an Alloys Research Committee (ARC), and persuaded the NPL to take up the subject; and since at least 1908, Rosenhain had been investigating a number of copper, aluminium and magnesium alloys, contributing along the way to the theory of 'specific strength'.[109] In February 1917, the Advisory Committee for Aeronautics, possibly prompted by Liversidge's work for the rival Airships Subcommittee, set up a Light Alloys Subcommittee, under Henry Fowler, Superintendent of the Royal Aircraft Factory. This – along with Rosenhain and the Aeronautical Inspection Department of Naval Construction – was charged to 'institute research ... into alloys and their development' and to 'assist in the removal of difficulties which may arise in their production and use'.[110] They met at the Royal Society, and relied upon the NPL for support. Under this subcommittee's auspices, alloys were developed for aeroplane engines and shell fuses; little was published until years later, but then many of these alloys were used in aircraft and automobile manufacture.[111]

Liversidge's project was but one of many and, by June 1917, was soon absorbed into the NPL's program. Indeed, so quickly did his proposals pass into the mainstream, that the subject soon vanished from the agenda of the Airships Subcommittee.[112] Although as late as November 1916, the German literature carried many descriptions of magnesium alloys, Rosenhain believed that they had no value 'for ordinary purposes'. However, one such alloy, developed during the war, turned out to be a precursor of the alloy used in the airframe of the Concorde.[113] Moreover, in September 1918, a specimen from the gondola of Zeppelin 149, brought down over France, revealed traces of beryllium in the Duralumin used by the Germans, as Liversidge had predicted. This information was used to improve the design and structure of the 'R' series of airships that were built after the war. On both counts, Liversidge had again seen the future, and it worked.

# 5. Minerals and Men

Whilst still working on alloys, Liversidge gave time to several other questions. In November 1916, Threlfall asked him to help the Conjoint Board revive an 'ancient proposal' for a mineralogical 'census' of the Empire.[114] Munitions manufacturers and the metals industries had awakened sharply to the need to be better informed about the distribution of imperial resources. This was not a simple task, and one beyond Britain's Advisory Council of Scientific and Industrial Research, as it involved dealing with India, the colonies, and the dominions. But it was a natural job for Liversidge. 'This information must be got before any steps can be taken and you are the most fitted to do it if you will',[115] Threlfall said. Liversidge could not resist the invitaton. But the situation he found would have daunted a younger man, and a wiser one.

The scene confronting Britain's use of raw materials was one of confusion and waste. Britain entered the war in 1914 with no minerals policy, and within two months, reserves were already severely stretched. An industrial nation, built upon a mountain of coal, surrounded by a sea of oil and natural gas, accustomed to buying raw materials in the free market, Britain had neglected the requirements of a long war. Even in peacetime, the country was not self-sufficient in many essential materials, and, while the fact of foreign dependence was well known, no policy existed for the wartime mobilisation or control of Empire resources. For example, before 1914, the wolfram (calcium tungstanate) deposits of Burma were worked mainly by British companies. However, most of the ore went to Germany for the production of tungsten, which then went to Sheffield for use in manufacturing high-speed tool steel.[116] In the pre-war world economy, this was a plausible practice; but with the advent of massive demands for iron, steel and other metals, it was not.

By the middle of 1915, industrial shortages put the 'minerals war' on a strategic footing. When Germany was deprived of tungsten ore, she substituted molybdenum from Norway, but Britain countered by buying up the whole Norwegian output, and by threatening Norway's imports of food.[117] Similarly, the Allies attempted to reduce Swedish iron and steel shipments to Germany by threatening Sweden's oil and coal supplies. Far-sighted advocates of 'strategic minerals policy' saw the Empire acting as a whole, a single resource unit, as well as a political and military unit, capable of exerting vast leverage upon the world's economy.[118] But this required centralised, frequent, comparative and systematic information. In Whitehall, while many departments took a 'more or less desultory interest in minerals production', 'no one department had a special duty to watch over the development and proper utilisation of our mineral resources'.[119]

As with nearly every technical innovation in the Great War, the way forward was pioneered by interested professionals. The idea of producing 'minerals intelligence' on a global basis was not new. In 1910, the International Geological Congress, held in Stockholm, launched a comprehensive, two-volume survey of the world's iron ore deposits, which for the first time presented statistics for the British Empire and elsewhere. But this was soon outdated. And whilst the Imperial Institute and the Geological Survey of Great Britain regularly produced reference notes on specific mineral resources, they were too static, too slow and, of course, too revealing, to meet the needs of a nation at war.

Government and industry wanted up-to-date details of mineral reserves delivered in secret, at weekly, even daily, and not yearly intervals.[120]

The issue had been discussed in Britain as early as 1911, when an Imperial Conference signalled the need for 'promoting imperial development on scientific lines'.[121] In 1913, Sir Robert Borden, Prime Minister of Canada, concerned at growing United States domination of Canadian resources, proposed to the Privy Council the formation of a 'Mining and Metallurgical Industries Bureau', to foster the development of imperial resources, and to regulate their foreign control.[122] Borden's proposal was referred to the Board of Trade and the Home Office, and to the Imperial Institute, where the Director, Liversidge's friend Wyndham (later Sir Wyndham) Dunstan, set up a committee to consider it. However, the Institute, handicapped from birth by a charter of ill-defined goals, was the least suitable body to run with the idea. Predictably, as cynics observed, nothing came of it.[123]

In 1915, Dr (later Sir) Aubrey Strahan, FRS, Director of the Geological Survey, issued a series of Special Reports on the Mineral Resources of Great Britain. But these took no account of the Empire, or of its relevance to war-time needs. The Survey was itself hard-pressed to supply reports on water supply for the Army in France.[124] What was needed, wrote Henry Louis, Professor of Mining at Armstrong College, Newcastle, was a 'Ministry of Production' to calculate the resources of the Empire as a whole, and to render it impossible that 'the control of any portion of the Empire's mineral production should ever pass into alien hands'.[125]

In September 1916, as Britain counted the cost of the Somme offensive, representatives of the Iron and Steel Institute, the Institute of Metals, the Institution of Mining Engineers and the Institution of Mining and Metallurgy asked Sir William McCormick, Chairman of the ACSIR, to coordinate Britain's minerals war. Allied and Dominion surveys lacked a point of reference in London. 'It cannot be doubted', they added:

> that if a properly organised and efficiently conducted Department of Minerals and Metals had been in existence, much valuable time, many lives and vast sums of money would have been saved to the Nation ... and much of the cost and inconvenience to British industries depending largely for their raw materials on mineral products could have been saved.[126]

When nothing happened, in December, the Society of Engineers began to compile records of the occurrence and distribution of minerals of imperial concern,[127] and early the following year, G.C. Lloyd, Secretary of the Iron and Steel Institute, sent a report on iron and metalliferous ores to the DSIR. This soon went into a second edition, as the iron masters of England struggled for information.

Not until the close of 1916 did the question of preparing an imperial minerals and metals policy at last appear on the agenda of the Conjoint Board of Scientific Societies. At Threlfall's suggestion, the Board's Secretary, William Watts, asked Liversidge to begin by finding out what the Geological Survey, the Board of Trade, and the Imperial Institute had already done. In January 1917, his report drew a picture of disarray. Geological Survey reports did not venture beyond the British Isles. The Board of Trade similarly neglected the Empire. And while the Imperial Institute had prepared forty-eight reports on minerals, and published regularly in its quarterly *Bulletin* on ore reserves in the colonies and overseas, little of this information routinely reached the Ministry of Munitions, let alone the captains of British industry.

Liversidge saw the problem as one of specialisation and access. Much information was published routinely in the *Mining Journal*, the *Mining Magazine*, and in *Mining Studies* (of Australia), and by no fewer than eight British and colonial mining, mineralogical, metallurgical and engineering societies. India and the Crown Colonies also published reports and surveys of their own. But these were 'far too numerous and lengthy for the average man of business to make use of'. Moreover, Liversidge claimed, 'comparatively few people know of these publications, and still fewer know how to get access to them'. Issued independently and 'at places thousands of miles apart', most were 'not easily procurable in London, and quite unavailable outside the respective government offices'. 'No library', Liversidge said, 'except perhaps that of the British Museum, contains the whole of them, and it is very unlikely that any of the other libraries would be able to spare the necessary space'.[128]

In Liversidge's view, there was nothing in Britain to compare with G.A. Roush's *Mineral Industry: Its Statistics and Trade*, published in New York, which had gone through twenty-four editions since its appearance in 1891. Even so, British industry 'ought not to remain dependent upon foreign countries' for such works of reference, which formed a prerequisite for an efficient minerals policy. He recommended the establishment of a new journal, along the lines of the American magazine *Mineral Industry*; greater cooperation between the relevant scientific societies and professional bodies; and an imperial census of mineral resources, illustrated by maps and diagrams, pointing to areas of the world likely to repay future exploitation.

In March 1917, the Conjoint Board accepted Liversidge's report, and instructed him to interview representatives of the minerals and mining establishment, with a view to obtaining their cooperation. In the meantime, however, there was a counter-suggestion from the Royal Commission on the National Resources of the Dominions, which had sat since 1912 under Sir Edgar Vincent (later Lord D'Abernon), and which reported the same month. The Commission pointed to Britain's dependence on Canada's and Australia's molybdenite, Canada's nickel and New Zealand's scheelite. Yet, trade in these commodities was not recorded in official British returns, nor was there 'a systematic or scientific attempt ever made to survey the foodstuffs and raw materials required by the Empire'.[129] The Commission insisted on having an imperial policy to 'resist the pressure of foreign powers in controlling raw materials'. In advance of such a policy, Britain needed a 'preliminary survey' of the Empire's capacity to produce for the 'sustenance of people, the maintenance of industry and munitions of war'.

In its Final Report, the Commission recommended a 'Coordinated Mineral Survey of the Empire' to give urgent attention to particular metals – notably, tungsten and zinc. The Commission was cautious in its praise of the Imperial Institute; and believed its proposals of September 1916 would encounter 'constitutional and administrative difficulties'.[130] In preference, they wished to see the creation of a new Imperial Development Board.

These recommendations posed a threat to many interests. For one, Dunstan demanded that the Imperial Institute be given funds to begin its own bureau of mineral intelligence. Even *Nature* agreed that this would be to the nation's advantage, but the journal also urged the creation of a new department of minerals and metals to coordinate scattered efforts, and to 'deal with the whole subject from the point of view of a great imperial industry'.[131] To make matters more confusing, the following month, Christopher Addison's Ministry of Munitions stole a march on everyone, by establishing a Mineral Resources Development Committee (later to become the wartime Department of Mineral Resources) under Sir Lionel Phillips, whose brief was to deal with the 'examination and development of such mineral properties (other than coal or iron ore) in the United Kingdom as are considered likely to be of special value for the purposes of the war'.[132] Advising Phillips was a group of mining experts, including Professor F.W. Harboard, Edgar Taylor and Aubrey Strahan.[133] The Committee took rooms in the Ministry's premises at the Victoria Hotel in Northumberland Avenue.

With unpardonable neglect had come massive duplication, which, if continued, would threaten the entire basis of Britain's war effort. In June and July 1917, as Liversidge made his way through this minefield, the scene dramatically changed. First, the United States entered the war and the US government created a War Industries Board to organise production and coordinate purchasing. In the rush to mobilise, competing agencies were created in Washington, as in London, with, as the US Bureau of Mines later admitted, 'grave consequences' for the war effort. Disaster was 'averted only by the adoption of a hastily conceived national programme which ultimately proved to be absurdly extravagant both in management and money'.[134]

Even so, the summer months of 1917 and early 1918 saw an impressive application of American ingenuity to munitions production, invoking the close cooperation of government and industry.[135] Several federal departments – including the US Geological Survey, the Bureau of Foreign and Domestic Commerce, and the Bureau of Mines – which had been accumulating data for years, began to meet demands, both Allied and domestic, for information on availability and price, treatment and costs, risks of shipment, and methods of production. Statistics, routinely published on an annual, or at best monthly, basis, were produced at ten-day intervals. Above all, the semi-academic program of the Bureau of Mines and the Geological Survey was suspended, and staff were tasked to investigate reserves of petroleum and critical minerals – notably manganese, chromite, pyrite, sulphur and platinum. In February 1918, the US government established a Joint Information Board of Minerals and Derivatives, with representatives of thirty-two departments, the War Industries Board and the Geological Survey, and a War Minerals Committee, consisting of academic and government geologists, which came forward to compile and present information.[136]

This rush of activity was not lost upon London. In May 1917, Lord Balfour of Burleigh, chairman of a Committee on Commercial and Industrial Policy established by Asquith's government in July 1916, wrote to Lloyd George, urging the creation of a British coordinating Intelligence and Advisory Bureau.[137] The committee agreed that 'we rely, and shall, we think, continue hereafter to rely, even more than in the past, upon imported raw materials'.[138] In the event, the government preferred to follow the Dominions Commission's recommendation and to establish an Imperial Development Board for scientific research and trade.[139] But nothing was done for minerals policy until June 1917, when the War Cabinet, having just survived the worst months of U-boat attacks, and facing increasing munitions demands in France, put the issue to the Imperial War Conference. Encouraged by Christopher Addison, Minister of Munitions, the conference reprised the ideas of 1916, and the creation of an inter-departmental committee 'to prepare a scheme for the establishment ... of an Imperial Mineral Resources Bureau'; to collect information in regard to the mineral resources and metal requirements of the Empire; and to advise what action, if any, may appear desirable to enable such resources to be developed and made available to meet requirements.

This time, pressured from Canada and Australia, Lloyd George accepted the recommendation – the first attempt, according to *The Times*, to 'give concrete shape to any resolution of the [Imperial] Conference'. The committee was chaired by Sir James Stevenson, Director-General of the Department of Inspections at the Ministry of Munitions (until March 1916, Deputy Director-General of the Department of Explosives Supply), and included the High Commissioners of Canada, New Zealand and South Africa, and the Under Secretary of State for India. Australia was represented by the industrialist W.S. Robinson, manager and director of Broken Hill Associated Smelters. Wyndham Dunstan attended for the Imperial Institute. The committee began sittings from its headquarters in the Hotel Metropole, on Northumberland Avenue.

When Liversidge entered the scene in June and July 1917, the picture was not pretty. The Americans had set up agencies which had as yet little direct connection with their British counterparts. The Dominions Commission and the Balfour Committee had fallen out, and the Board of Trade, the Ministry of Munitions and the Imperial Institute had so far produced only another committee. Confusion was the worse compounded by the uncertain reception given the recommendations of the mining institutes.

Thus, the Institute of Mining and Metallurgy welcomed Addison's idea of an Imperial Mineral Resources Bureau. But Sir Lionel Phillips, for the Ministry of Munitions, suggested that the whole question should be referred to the Ministry of Reconstruction. G.C. Lloyd, Secretary of the Iron and Steel Institute, preferred a commercial model to anything he had seen. And G. Shaw Scott, Secretary of the Institute of Metals, welcomed the idea of a census, but withheld giving support until the government decided on the direction of its minerals policy in general. L.T. O'Shea, Secretary of the Institution of Mining Engineers, pointed out that they had already asked the DSIR for a Department of Mines and Minerals, and could take no further action until the DSIR replied. Meanwhile, none of the professional institutions seemed to have time for the Imperial Institute, where Dunstan promised 'merely to proceed as circumstances will allow'.[140]

For three months in the summer of 1917, Liversidge was perhaps the one man in London who knew the range of alternatives confronting the British government in this field. As a representative of the Royal Society, he was also well placed to influence any scheme that might emerge. Regrettably, by the end of the summer, the opportunity had passed.[141] The idea of embarking upon an up-to-date mineral census, without which a strategic policy was unattainable, was left to drift. Phillips' Department of Mineral Resources in the Ministry of Munitions declined to deal with imperial minerals. Stevenson's Imperial Mineral Resources Bureau limited itself to preparing a scheme for Imperial governments

Liversidge photographed in 1918.

to consider. And, to the dismay of the mining institutions, the DSIR, approached as early as September 1916, had not replied by August 1917. The Imperial Institute, already over-committed to an enormous range of vegetable and mineral surveys, was thus left, *faute de mieux*, with a responsibility it could not discharge.

In September 1917, Henry Louis reviewed the state of affairs in *Nature*.[142] The government recognised the need for closer cooperation in minerals information and intelligence. However, far from looking to imperial cooperation, minerals policy had become ensnared in ideological debate. Policy was to be driven not by strategic concerns, but by a political decision to leave mineral production to market forces. The outlook for intervention in minerals – as in iron and coal – was bleak. By the end of the war, received wisdom, concentrated in the final report of Phillips' department, held that governments, far from acting more intrusively, should 'withdraw from mining matters in all respects'.[143]

When Curzon spoke of Allied fleets floating to victory on seas of oil, he could have had in mind the whole domain of natural resources. Britain depended upon imperial (and latterly, American) mines and quarries. Liversidge saw the problem and knew its solution. But, as so often before, his ideas came before their time, and fell prey to the pressures of the moment. A minerals census was postponed indefinitely, as was coordinated imperial action, and 'minerals intelligence' would remain mere record-keeping for the next twenty years. The war ended with two 'mineral regimes' in Britain – the Imperial Mineral Resources Bureau and the Imperial Institute. In July 1918, over a year after it was first proposed, the Bureau was established by the Ministry of Reconstruction, in consultation with the Secretaries for the dominions and India. The mining engineer, Sir Richard Redmayne was elected chairman.[144] Dominion cooperation was endorsed by the Imperial Economic Conference held later that year, but not immediately acted upon. As regards the Imperial Institute, little was expected, and little was received. The Institute struggled until its closure in 1956.[145]

For Liversidge, the outcome of his war work was frustrating. 'I wish', he wrote Haswell in April 1917, 'I could have been of more direct use, like David, Pollock, Welch, Wilson and others'.[146] His regrets resonated with Douglas Mawson, to whom active service was also denied.[147] But whereas the younger Mawson had at least a uniform, Liversidge had none. His photograph, taken by William Stoneman for the National Portrait Gallery in 1917, shows a tired civilian, burdened with his wartime occupations. Typically, he kept his own counsel, and when the Armistice came, Liversidge was again working on his own, at the home that had become his headquarters.

# CHAPTER 15

# Last Rites

'Fieldhead' in Kingston-on-Thames, Liversidge's home from 1913 to 1927.

In mid-1914, just before the outbreak of the war, Liversidge took up residence at Fieldhead, in suburban Kingston-upon-Thames, in a district known as Coombe Warren, a world of fine houses and gardens. If the sweeping views of the Thames valley, the ornamental garden surrounding Cardinal Wolsey's conduit, the well-kept orchard and domestic staff, the motor car in the carriage driveway – if all these seemed hostages to fortune, then at least his private laboratory and library recalled memories of Darling Point and the Australian life he had left behind.

Liversidge's war was fought mostly from Fieldhead, where his dining room had become a perpetual 'open house' for visiting Australian and American scientists. The approaching end of the war found him ready for new challenges. In August 1918, Sir Richard Gregory, Lockyer's successor as editor of *Nature*, and Liversidge's co-conspirator in the British Science Guild, invited him to organise a BSG exhibition of scientific products and appliances manufactured by British wartime industries.[1] 'The shock of war', as *The Times* put it, had 'modified the attitudes of the devotee of pure science to industrial problems, [while] the manufacturer has had proof that the head of the research worker is not always in the clouds'.[2] Liversidge was to be present, as so often, at the moment of reconciliation.

Preparing objects for an exhibition always showed Liversidge at his best. This one was sponsored by the Ministry of Munitions and the Board of Trade, and represented over 300 firms where wartime 'science had come down to business', displaying objects and instruments previously available only from enemy countries.[3] The exhibition opened in early January 1919, at King's College London, and ran for a month. Among its 30,000 visitors, many heard lectures on prospective civilian applications of wartime technology. The section of the exhibition on non-ferrous metallurgy and light alloys, showing materials newly acquired by the aircraft industry, brought Liversidge particular pleasure. Unfairly, when a dinner was held on 15 January to celebrate the Guild's 'victory' and the exhibition's close, Liversidge was ill, and sent his regrets.[4]

In the official reckoning that followed the Armistice, Liversidge's war service was never formally recognised. In Sydney University's *Gazette*, he read of students and colleagues who were honoured in different ways. In 1920, S.H. Barraclough received a knighthood, as did their good friend, W.H. Bragg. Douglas Mawson, already knighted for his Antarctic endeavours, received a military OBE. In 1927, Richard Threlfall, already KBE, was promoted to GBE. Edgeworth David, a front-line military hero, was deservedly knighted before he returned to Sydney. But for Liversidge, there were no mentions in despatches, no public thanks. His disappointment can only be imagined; it was carefully concealed, and his feelings survive only in inferences from letters to friends.

Towards the end of the war, reports from the front poisoned his usually circumspect pen. He read, as ever, the *Sydney Morning Herald*; but an article in April 1918 on 'What Australians have done' – subtitled, 'Triumphs in the Art of War' – brought the acid comment: 'What has Australia got from the war? They have been kept well in the limelight'.[5] Writing from a holiday house in Derbyshire in August, just as Australian divisions were mounting final operations against the Hindenberg Line, he warned W.A. Haswell, who had no discernible intention of returning to England, 'not to come home while the war is on, for conditions here are not so easy as they are in Australia, where

The Liversidge family coach-building works in 1926.

the war seems to be hardly felt'.[6] A few months later, in 1919, he added: 'Now that the Armistice is on, things are beginning to improve ... and we are beginning also to get a little more freedom, but we are still rationed ... I don't complain nor do other people except the munitions toilers who really do not know how to spend their enormous wages'.[7] Miraculously, his extended family escaped the war's worst injuries: his naval nephews survived four years of relative inaction, and none of his more distant relatives served, or suffered. For his part, there were only the wounds of frustrated enterprise. And these, at the age of seventy-two, were slow to heal.

## 1. *APRÈS LA GUERRE*

In 1919, Liversidge began to pick up threads of his prewar research by returning to the Davy Faraday Laboratory, and resuming work on trace elements. It must have brought him pleasure to learn that Fritz Haber, the distinguished German chemist, who had once 'brought nitrogen from the air', was now being asked to bring 'gold from the sea'. The Treaty of Versailles committed Germany to the payment of reparations amounting to US $33 billion or, as Haber put it, 50,000 tons of gold at $32 an ounce.[8] Given Germany's financial situation, Haber was asked to increase his country's reserves by 'mining' the oceans. He had before him nine published determinations, among which Liversidge's were the most promising. Perhaps his far-sighted proposal of 1895 would at last be taken seriously.

Between 1920 and 1928, Haber conducted a long series of experiments, ultimately involving some 2000 samples of sea water.[9] At first – until 1923 – his analyses gave results similar to those of Liversidge. Assuming these to be correct, he then turned to the question of extraction. If ambient values were as high as 5 mg per tonne, the process might produce a profit. Financial houses in Germany backed his research, and in 1923, the Hamburg-American Line installed a laboratory and extraction plant on the *Hansa,* and Haber made a series of measurements on transatlantic voyages to New York.[10]

As Haber traversed the seas, however, he found that concentrations of gold vary considerably by region. Ten times as much appeared in a given volume in the North Atlantic as in the South Atlantic. Taking over a hundred samples from offshore waters near the Californian gold fields, he found that even tidal changes greatly affected the result. Moreover, it seemed that when methods satisfactory for high concentrations were used with water containing low concentrations, measurements reflected the presence of gold in the actual containers being used. Such contamination must have affected Liversidge's results as well, although no one queried his findings because of it. Haber accepted Liversidge's values. The zeal to *extract* had overtaken the need to *test*. Eventually, however, Haber was forced to conclude that Liversidge was wrong.[11] Modern measurements yield about .004 mg per tonne.[12] Liversidge was mistaken by a factor of ten.[4]

Ironically, Haber published an account of his work the same year that Liversidge died.[13] His report divided blame between Sonstadt, who was undoubtedly deceived by reagent contamination (and who, in 1892, seemed to admit as much),[14] and Liversidge, whom he faulted on technical grounds. Liversidge had used methods that required extremely sensitive techniques of extraction. Alas, in Haber's words *diese vertrautheit hat Liversidge nicht besessen.*[15]

In retrospect, it is remarkable that one of Germany's leading chemists took seven years of careful study to reach this conclusion. Still, as his biographer notes, to find a 'profitable concentration', in chemical terms, was not unlike finding a school of herring, in terms of acoustics.[16] Liversidge's results, based on experiments lasting less than a year, were undoubtedly flawed, a fact for which the limitations of his apparatus were easily to blame.[17] In the event, however, the techniques that Haber developed – capable of detecting $10^{-8}$ g per litre, or about $10^{-2}$ mg per tonne – proved of lasting benefit to the commercial extraction of bromine and magnesium. Good results came of Liversidge's false start, even if extracting 'gold from the sea' remained elusive.

Despite these discouraging results, Haber's report did not bring 'chemical prospecting' to an end. Following the Bolshevik revolution, the Soviet government sent geochemists to Siberia to test river water, in hopes of retrieving gold with which to run the country. Their results were inconclusive, but speculation continued. Between 1935 and 1938, William Caldwell in the United States attempted to settle the matter, and produced results suggesting that Haber's outcomes were at best an upper limit.[18] But if gold were present in sea water – and it clearly is, along with many other minerals – its concentration could not be reliably determined by contemporary technology. J.E. Coates, writing in 1939, believed that all analyses made before 1928 were both inaccurate and incommensurable, partly because of wide differences in technique and the absence of reliable rules for replication.[19]

Advertisement for Glover, Webb & Liversidge.

Liversidge in London left the question of trace elements in abeyance, and returned to crystallography, where there were many at the Royal Institution to encourage him. In these years, the RI was becoming a central actor in the X-ray analyses of crystal structures, developing techniques that would prepare the way for the coming of molecular biology. But from this work, Liversidge was by age and background largely excluded. There were many testimonies to his laboratory skills – in isolating crystals, for example – and his reputation as a mineralogist was secure.[20] But mineral chemistry had become a static subject, and in crystallography, chemistry was being overtaken by physics, of which he knew little. It was also unfortunate, if inevitable, that his descriptive *Minerals of New South Wales*, which had endured for a generation, gave way to a flood of new information. By 1920, Edward Pittman's *Minerals* had become the standard reference. These developments Liversidge faced stoically, but with pardonable regret. Pittman's volume was better written, perhaps, but omitted any acknowledgement of his pioneering professor, sometime employer, and founder of Sydney's School of Mines.

Liversidge's last years were destined not to see the completion of his research. The Davy Faraday Laboratory, with its tiny budget, was spared the worst of wartime retrenchment, but the economic recession of 1923 and the 'Geddes Axe' hit its work severely. Throughout Britain, scientists recoiled, proclaiming the 'frustration' of research.[21] The situation was no better in Australia. Returning to Sydney, Edgeworth David mourned that, despite the experience of the war, 'neither in the Old Country nor in this country did the people properly appreciate the importance of science for national progress'.[22] Accepting the Royal Society's Clarke Medal from Liversidge's successor, Charles Fawsitt, David urged Australians to take up 'propaganda' on behalf of science. Public support was sorely needed. To the *Herald* in 1920, H.S. Carslaw wrote angrily that Sydney University's 'new chairs in botany and organic chemistry, just created by the University, could not compensate for 'hopelessly inadequate laboratory accommodation, no lecture rooms, and practically no rooms for research'. 'The sober truth', he added, 'is that the University of Sydney is far behind what it ought to be as a place of scientific work and research of importance'. Instead, it was looked upon 'simply as an institution for manufacturing doctors, lawyers, engineers, teachers ... [and] I wish I could add ... statesmen'.[23]

Sydney University's financial situation, which had seen glimmers of improvement before the war, suffered the predictable effects of rapid postwar expansion. In 1914, the University admitted 1736 students; in 1919, with returning soldiers, there were 2764, and the next year, 3356.[24] Thereafter, admissions fell slightly, so that in 1923, the University actually had nine fewer students than in 1919. But unplanned growth brought untimely stress. The 'industrial' image Liversidge feared had become a fact of life in 1922 when, with the onset of recession, the University plunged £8000 into deficit.[25] There was money for some things, but not for others: for the completion of the Gothic quadrangle, but not for promised chemistry and physics buildings.[26] Not until 1923 did the new Botany School arise at the northern elbow of the Sydney campus, linking the quad with the Macleay Museum. The University did expand its faculties from four to eight – to include Agriculture, Engineering, Architecture, and Veterinary Science. But departmental votes were reduced 'to the lowest possible limits'.

Through the late-1920s, the University's plight worsened, with staff working in poor accommodation and lecturing without tutorial support. At the same time, the 'core curriculum' and the tradition of liberal learning, fashioned with such care in the 1850s, and debated so fervently in the 1880s, became casualties to increasing specialisation. Australian students, G.V. Portus told a British readership, were being 'professionalised rather than educated'.[27] Perhaps it was true, as Portus claimed, that Sydney University was coming to look like a cross between a 'university kindergarten' and a 'bustle of specialists', its teachers reduced to 'factual hens with their beaks close to a chalk line on the ground, self-hypnotised by the process of collecting more and more data'.[28]

Worse, it was said that in the teaching of chemistry, there was, despite the presence of fine minds, an almost anti-theoretical bias. Perhaps Liversidge could see himself in the mirror of his student:

> We seem to love pulling things apart rather than putting them together. We look into the microscope but never see the microcosm. We are too busy with facts to bother about ideas.[29]

If true, this was not Liversidge's fault, nor was it a condition unique to Sydney. Melbourne, in the words of its Vice-Chancellor, was little better off – its 'congeries of professional schools' becoming by the 1920s, a 'sorry substitute for higher education'.[30] Years before, Anderson Stuart had claimed that Sydney had ceased to be a place for 'polishing schoolboys', but only by becoming a place where students were seeking 'degrees ... with money value'.[31] Despite his efforts, the University retained into the postwar years an aura of aloofness, remaining as much as ever a community unto itself. Thanks to teachers' scholarships and bursaries, the doors of higher education were no longer closed to the working classes; yet, as late as 1937, Melbourne calculated that sixty per cent of its students came from middle-class families with an income over £300,[32] and Sydney's situation was broadly the same. Despite the efforts of Liversidge and his generation, the universities still seemed exotic in a country that had not yet 'naturalised' higher education.[33]

Mercifully, Liversidge did not live to see the Great Depression. Nor, however, would he see Australia's slow recovery, beginning in the late 1930s, and the growing spirit of enterprise among its educators during and after the Second World War. Sydney University was to draw fresh breath, and extend its influence through the new University College at Armidale, its first offspring in New South Wales, and a harbinger of things to come. Even so, state governments did little to improve the conditions of learning. As in Liversidge's day, funds were earmarked for specific purposes, which left less for common services. Sydney perhaps managed a little better than Melbourne, and much better than Adelaide, because its endowment was greater. Still, in 1937, faced by what Vice-Chancellor Stephen Roberts called the 'tyranny of numbers', Sydney's 3000 undergraduates were taught by only 237 staff; Melbourne's 3071 students were a little better off, but neither university had ratios that compared favourably with their British counterparts.[34] By the 1940s, Sydney was simply too big for its budget – having as many students as Oxford and, if not as many as Cambridge (6000), Harvard or California, only a fraction of their resources.[35]

Although the universities' absolute income grew in the 1930s, their financial problems remained. It seemed that if Australia were ever to become a 'clever country', it would be in spite, and not because of its universities, cramped as they were in outdated buildings, and drawing upon community support far less than their sister institutions in Canada and the United States. As late as 1939, Australia had a lower participation rate of students per population than any English-speaking country outside Britain; and although it could rightly be said that its graduates were 'absorbed' into the workforce – Melbourne prided herself on the first hundred graduates she sent into agriculture – the national economy and the public service were not to feel their benefit for another twenty years.[36] Liversidge, in

William and Sophia Liversidge, about 1908.

the phrase of John Dunmore Lang, had 'hammered at the gates of futurity'.[37] Tragically, he was not to see the revolution in higher education that, delayed by the Second World War, followed the Murray Commission in 1957, and that led at last in the 1960s and 1970s to a period of unprecedented expansion in buildings, staffing and research.[38]

All this is to look ahead. Liversidge was fated to know an enduring reality, rather than a hopeful prospect. But he could take some joy when, in June 1926, S.M. Bruce, Prime Minister of Australia, opened the long-awaited Council for Scientific and Industrial Research (CSIR). The culture of imperial science had long kept Australia in continuing interaction with the wider world. Bruce recommended that Australian research be placed 'on a par in importance with imperial cooperation in defence';[39] and indeed, in the 1920s, the prospect of imperial scientific cooperation never looked better. The Imperial Geophysical Survey helped strengthen collegial ties,[40] and Australian expatriates featured prominently in British scientific life.[41] Britain returned the favour, continuing to welcome Australians to its halls of honour. In 1925, John Kenner, Sydney's fourth professor of organic chemistry, was elected a Fellow of the Royal Society of London,[42] in the same year in which three of the fifteen newlyelected FRSs were resident in Australia.[43] Liversidge lived to see this, and from it took pleasure.

# 2. Last Days at Fieldhead

The late 1920s saw Liversidge withdraw from society. He attended few BSG meetings, and gradually withdrew from the circuit of the Savile, the Athenaeum, and the RSC. Visitors spoke warmly of the hospitality they received in Kingston and every year the Royal Society of NSW sent fraternal greetings. But Liversidge distanced himself from Australian obligations, and in 1922, declined to be the Society's delegate at the London conference of the *International Catalogue of Scientific Literature*.[44] In June 1923, illness prevented his accepting an invitation from Edgeworth David and the new Australian National Research Council to represent Australia at the meeting of the International Union of Pure and Applied Chemistry in Cambridge.[45] This, at least, he must have regretted: over 100 delegates from twenty-three countries attended, and the program embraced all his interests.[46]

Liversidge's association with the RI came to an end in 1924, a year after James Dewar died. It seemed 'the vitality of the Institution had ebbed away',[47] and even Ernest Rutherford declined to move from Manchester to become its Director. In June 1923, however, the directorship and professorship were given to W.H. Bragg, who brought fresh energy to Albermarle Street. Along with J.D. Bernal – the brilliant young Cambridge crystallographer, who would return to Cambridge in 1927, and do much to launch the field of molecular biology – Liversidge was granted readmission as a visitor.[48] Bragg set about creating a new school of crystallography, which might have appealed to a mineral chemist. But Liversidge, aged seventy-six, decided to call it a day.

With the passage of years, conferences were missed, and Sydney friends disappeared. In 1920, Robert Etheridge died, followed in 1922 by Joseph Carne, the NSW Government Geologist. In 1924, Edgeworth David and W.H. Warren retired, and Joseph Maiden – Director of the Royal Botanic Gardens since 1896 – died the following year. It fell to David to recall happier days when he sailed in 1925 to complete his *Geology of the Commonwealth of Australia* in the relative detachment of England.[49] Liversidge warmly received him in London. Loyally, if accidentally, David was to be with him at the end.

As his research slowed, Liversidge took pleasure in the company of close friends. One was Arthur Shipley – nearly fourteen years his junior, and Master of Christ's College, Cambridge – with whom Liversidge found an interest in motoring through the East Anglian countryside. The two bachelors were also keen microscopists, and in the heroic days of Michael Foster and F.M. Balfour, both had known the excitement of establishing new disciplines. Both were also familiar with the inconveniences of travel. Where Liversidge circumnavigated the globe, Shipley knew America well, and visited Princeton, which gave him an honorary degree.

Liversidge saw in Shipley certain qualities that he lacked, and perhaps certain opportunities that he had foregone. Shipley had stayed in Cambridge with the best research men of his generation – whose company Liversidge enjoyed only fleetingly, every seven years. Shipley had continued to pursue mainstream interests in zoology, while Liversidge had missed a tide in biological chemistry that, if taken at the flood, might have led him to greater fortune. Liversidge's style was ever technical and dry, while Shipley wrote with a lucidity that his friend frankly envied. Shipley's articles for *Discovery*, like his wartime account of British science, are still classics.[50] And there was the inevitable

Liversidge's grave in Putney Vale Cemetery.

question of recognition. As head of the British universities' wartime mission to the United States, Shipley had been knighted, and promoted GBE in 1920.[51] Liversidge ended his life without public honours. It is unsurprising – but who shall say unjust? – that until 2004, we would find Shipley, and not Liversidge, in the *Dictionary of National Biography*.

In August 1926, Liversidge published his last technical paper, describing the chemical nomenclature he thought should apply to the newlydiscovered substances called 'vitamins', whose properties were attracting growing interest at the RI.[52] But his eyes were now turning more to the past than to the future. His obituary of Sir Lazarus Fletcher, written for the Royal Society, was a tribute to the few who had quietly tended the gardens of classical mineralogy. In what was to be his last paper, Liversidge recalled his correspondence with missionaries in Fiji in 1876.[53] In his laboratory at Kingston, he pottered with trace elements. But the work went slowly, and little was said of it. In what proved a half-hearted attempt to revive his social life, Liversidge joined the Carlton Club in 1925. But central London seemed increasingly distant. He missed the RSC's annual dinner in 1926, and managed only one the following year, at the Hotel Jules, on 7 July. That summer marked his last trip to town.

Early September 1927 was unseasonably cold in southern England, and the garden at 'Fieldhead' was fast losing its summer splendour. Liversidge's relatives were scattered the breadth of London, Kent and Surrey. Their family firm had prospered. As Glover Webb and

Liversidge, with premises in the Old Kent Road, it expanded to produce large commercial vehicles, including refuse trucks 'especially favoured by some of the central London boroughs'.[54] But Liversidge's contact with his brothers suffered from infrequent repair, and daily life was increasingly in the hands of his housemaid and chauffeur. His health, which had rarely slowed him, finally gave way. Charles Fawsitt remembered seeing Liversidge a few years earlier, when he looked much as he did in the official portrait of 1908.[55] But his eyesight, strained by years of reading and microscopy, began to fail. Signing the candidates' book at the Carlton Club, his once firm hand was visibly unsteady. At last, his heart betrayed his constitution. Liversidge died peacefully in his canopied bed at Fieldhead on Monday, 26 September 1927. His devoted sister, Caroline Sophie Balfern, came up from Surrey to be with him at the end. It was she, who had helped bury their father, who now erected his headstone, 'in loving memory', in Putney Vale Cemetery, a few miles away.

### 3. REMAINS OF THE DAY

'I don't believe in cleverness but [in] tenacity of purpose', T.H. Huxley had once said. 'If you fail in anything – never mind; [ask] what's the next thing to come?'[56] To this he might have added Galsworthy's admonition to the English – 'let us never give ourselves away'. In death, Liversidge was their countryman. In an early morning drizzle three days later, to his grave by the cemetery roadside, came members of his family and a small but impressive imperial delegation.

There was Edgeworth David, a loyal friend for forty years, representing the University of Sydney, the Royal Society of New South Wales, and the Australian National Research Council; David was to write the official obituary for the Royal Society. With him in the cortège walked Dr George Prior, Keeper of the Mineralogy Department of the Natural History Museum, and Dr G.F. Herbert Smith, the Museum's Assistant Secretary. Arthur Shipley had died the previous fortnight; in his absence, Christ's College was represented by the Rev. Valentine Richards. These were joined by Liversidge's former colleague, E. Kilburn Scott, then on the staff of the Institution of Electrical Engineers, and representing the British Science Guild; Arthur Hutchinson, Professor of Mineralogy at Cambridge and Master of Pembroke College, representing the Mineralogical Society; G.S. Carr, representing the Chemical Society; and Dr A.B. Walkom, representing the AAAS and the Linnean Society of NSW. Finally, there was the anatomist Professor J.T. Wilson, late of Sydney and now of Cambridge, a friend since the 1890s, representing the Australian universities in the Bureau of the Universities of the British Empire.

Liversidge's will, read in December, conveyed the imperial reach of his legacy. From his net estate of £39,197, he left funds to endow three scholarships – £2000 at Sydney University, £1000 at Christ's College, Cambridge, and £1000 at the Royal School of Mines – as well as £1500 to endow four Research Lectureships, to be awarded by Sydney University, the Royal Society of NSW, the AAAS, and the Chemical Society of London. A further £500 was given for a Research Lectureship at Christ's, to be awarded alternately in chemistry and in the 'English language'. Similar sums went to the Marine Society (for the equipment and instruction of poor boys for the Royal Navy and Merchant Service), the Church of England Homes for Waifs and Strays, and the *Arethusa* Training Ship. Smaller sums were left to his nieces and nephews, and to his housemaid, housekeeper, gardener, and chauffeur at Fieldhead.[57]

In 1922, Liversidge had donated to the Australian Museum his fine collection of crystals, rich in Indian and Ceylonese minerals, which it had refused to buy in 1908.[58] Now, to the British Museum's Department of Mineralogy, he left the rest of his vast collection – 3000 specimens, mainly from Australia, including a sixty-five pound section of the Thunda meteorite, forty sections of gold nuggets cut to exhibit their internal structure, forty gemstones, and a collection of lantern and microscope slides. His saddest gift was the sum of £100 he left to the Chemical Society to ensure the publication of his last papers on trace elements, which were destined never to appear in print. Medals, diplomas, news clippings, books, pictures, photographs and maps of Australia he willed to the Royal Society of New South Wales; some would later find their way to the Mitchell Library in Sydney. His library, including beautifully bound sets of his offprints, was bequeathed to Christ's College, Cambridge, where it remains today in the crepuscular corridors of the Old Library. Across St Andrew's Street from Christ's, in the deconsecrated church of that name, there stands a memorial to James Cook's son, who studied at Christ's but, unlike Liversidge, died in Cambridge without seeing the world.

Tributes flowed from many directions, some promptly, others not until the 1930s. *The Times* led with a simple obituary: 'Professor Liversidge: Scientific Work in Australia'. The *Morning Post* followed, with a tribute to 'An Eminent Scientist'. England, it said, claiming him for 'Home', had lost from 'the ranks of chemists a personality who left his mark on the science both as an original worker and as an administrator'. Over twenty other notices appeared, in papers ranging from the *Sydney Morning Herald* to the *Surrey Comet*.[59] The learned societies were of one voice. The Geological Society of London remembered him as 'hospitable and friendly', and recalled his papers on minerals, meteorites and gold. Sir Ernest Rutherford, in his Presidential Address to the Royal Society of London in November, spoke of him as 'one of the pioneers of scientific education in one of our great Dominions ... an ardent collector of minerals and meteorites' and an exponent of experimental mineralogy.[60] Equally lofty tributes came from some who scarcely knew him. 'His influence on the development of scientific education in the Dominion', as Sir John Hammerton put it, 'was deep and lasting'.[61]

In Sydney, he was duly remembered. 'Largely through his activities,' *Hermes* reported, 'science was elevated to the rank of a separate faculty; under his care, [it] grew from childhood to a sturdy adolescence ... He will long be remembered as one of the Fathers of our University.'[62] The Sydney press was polite, even restrained. Liversidge had, after all, been absent from Australia for twenty years. Only slowly would news of his death stir others to recall the skills, virtues and qualities that shone through what Edgeworth David called his 'splendid striving' in teaching, administration and research. He was a founding father, and left his hand on science in the University of Sydney. But his greatest memorial, David wisely recalled, was at once more simple and more profound: the intangible gift of fraternity he had conveyed from his native to his adopted land.[63] In the end, it was for this, his ennobling vision of the imperial scientific community, no less than for his services to knowledge, that Liversidge and his legacy would be long and best remembered.

# CONCLUSION

# Measures and Meanings

In 1895, on the death of T.H. Huxley, Michael Foster added a 'few words' to the many that flowed in tribute to his famous teacher and friend. Most scientific men, Foster said, kept two minds – one devoted to science, the other to religion. Not so Huxley: for him, Foster said, science was 'all in all'.[1] For Liversidge as well, science was 'all in all'. And where he did not pioneer discovery himself, he showed others how to do so, in ways that that shaped the expectations of Australian science.

In the Great Hall of Sydney University hangs a formal portrait of Liversidge, painted by the Hon. John Collier. The artist shows him in academic dress, lecturing, rather than working behind a desk or in a laboratory. Behind him appear a blackboard and a map. To one side, rests a flask of chemicals. The portrait is apparently of someone wishing to be known as a teacher, rather than as a researcher, administrator, or dean. So much is true to life; but there is also modest invention. Liversidge appears tall, but was in fact short; he looks cool and debonair, but was at the time deeply frayed; he stands calm in front of a class, which in later years he rarely was.

In its appeal to convention, the portrait invites us to consider Liversidge as a man of contrasts – appropriately, an 'Archibald Paradox'.[2] As we have seen, he is remembered as astringent in conversation, but also a genial host; a stammering lecturer, yet a keen teacher; an academic progressive, supporting the admission of women and 'modern studies', yet also a campus conservative, endorsing an *élite*. He was an assiduous administrator who disliked meetings; an adroit politician who avoided controversy; a bachelor don who enjoyed family life; a man who lived alone by choice, yet was 'club-able' to a fault. As a patriotic 'Greater Briton', Liversidge was both an Australian federationist and an Empire loyalist; a secular Darwinian, who numbered churchmen among his friends; a man careful with his purse, but also a man of property, rising from artisan London to the comforts of Woollahra and Kingston-on-Thames, and leaving at his death a small fortune in legacies. In many ways, Liversidge remains, as Harvey Beecher has said of William Whewell, an 'unresolved nebula'. Accounts of him, as of Whewell, remain 'testimony to the difficulty of placing a man with such wide-ranging interests into a fathomable pattern'.[3]

Liversidge's life is the story of a modest man who seized the advantages of position and timing, and made the most of an opportunity to join an expanding world of science, commerce and trade. His career coincided neatly with a practical idealism then working its way from Balliol to Botany Bay, which in turn coincided neatly with colonial expectations, and with the changing role of the modern British university. He flourished in times of enthusiasm, and coped in times of despair. In ways in which he was not luminous himself, he proved an excellent conductor of light.

However, despite his voluminous papers, he has left us little with which to form a view of the man himself. We know he was at ease in the realm of the gentlemen's club, learned society and first-class saloon; but we do not know whether he drank or smoked, or whether he read fiction or enjoyed art. As a younger son, he had a conventional upbringing; he enjoyed and repaid the affections of his family. But after his father's death in 1887, there is no surviving correspondence with his brothers. His grandnephews and grandnieces remember him only as a kind, distant figure.

Nor is there evidence of romantic love, whether won or lost. Not unlike Mr Featherstone in *Middlemarch* or Penn's uncle in *Pendennis*, Liversidge was a late-Victorian bachelor; but in that, he was far from being alone among professional men married to their work. From Cambridge onwards, Liversidge lived in a masculine society, and arranged his life as carefully as he did his mineralogical cabinet. Like A.A. Lawson, Sydney University's foundation professor of botany, he treated 'women with the detached courtesy as of a celibate priest'. Unlike Lawson, however, there is no evidence that he had 'for the friendship of men ... that genius which the old Greeks have idealized'.[4] Instead, like his sparring partner, Walter Scott, the classicist, and his friend W.C. Kernot of Melbourne – whom Geoffrey Blainey has described as a 'kindly and plump, wearer of the broadcloth'[5] – Liversidge was a bastion of the British male universe. Into this comfortable existence, women (excepting cook and servants) were an intrusion. [6]

In relation to religion and politics, Liversidge's surviving correspondence reveals regrettably little. Anglican by convention, his disposition seems to have been neither conspicuously religious, nor party political. As with many self-made men, an unassertive liberalism of the Union Club and the Savile seems to have effortlessly given way to a liberal conservatism appropriate to a dean, then to a Carlton Club conservatism consistent with a colonial statesman. Possibly the best clue we have to his beliefs survives in an annotation on a letter from J. Beete Jukes, the geologist, who wrote that a scientist 'does not ask what is the nature of God, nor even of spirit': 'I seek not for a solution of mysteries: I only deprecate the use of mystification .... God Himself, if he addressed me, must address me through my reason, since He has given me no other or higher faculty by which to receive His communications'. Two pages later, Jukes added, 'I am no Leveller; political and social equality is a dream. Men must be as various in station and condition as they are in face'.[7] In the margin, Liversidge has simply written, 'agreed'.

Throughout Liversidge's papers, we find little evidence of lonely introspection. Rather, the story is one of self-imposed duty. Happiness lay in ceaseless work – in building, collecting, and in the occasional ruffling of received opinion. The early adjectives of Liversidge's youth were Smilesian in their sobriety:

> Self-reverence, self-knowledge, self-control
> These three alone lead life to sovereign power.

So Tyndall quoted Tennyson to his London students in 1868,[8] lines that Liversidge preserved on the first page of his lecture notes at the RSM in 1869–70, next to a quotation from one of Huxley's prize-day speeches:

> I don't believe in cleverness: but [in] tenacity of purpose. The first rung of the ladder is not to stand upon, but simply a standpoint by which you may raise yourself still higher. In your upward course never mind how many tumbles you have so long as you don't get dirty.
> If you fail in anything, never mind - [just ask] what's the next thing to be done?[9]

The 'next thing to be done' – in the words of the Cambridge anatomist, Sir George Humphry – was equally straightforward, namely, to unfold the plan of the universe and the operation of natural law. To this project, Liversidge harnessed himself energetically, with the goal of withdrawing function and design in nature 'from the mysterious region of life, into the more intelligible domain of science'. But 'so high a point on the hill of knowledge, a point', Humphry admitted, 'I have imagined rather than yet seen, [and] can be but slowly reached. Much labour is required to clear away the thickets and level the ground, lest the springs of genius carry us down rather than up'.[10]

For Humphry's expedition of the mind, the young Liversidge was well prepared. As a composite portrait of his teachers – Huxley, Foster, Smyth and Frankland – Liversidge incorporated the sensibilities of reform in scientific London and Cambridge, with their insistence upon research and publication. In this sense, he represented an intermediate stage in the professionalisation of science, between an early stage dominated by the 'omni-competent specialist' – the 'gentlemen of science', such as W.B. Clarke and members of the early British Association for the Advancement of Science – and the generation of professional scientists who followed him in the 1890s.[11] At a time when the 'circle of the sciences' was fragmenting into separate disciplines, his generation was among the last to profess, or possess, responsibility for an entire field of knowledge. And in the colony of New South Wales, he played many parts with equal ease. Actor in some roles, director in others, he stage-managed the theatre of science in colonial Sydney. For service to Empire, he was well suited. Like his contemporaries (and classmates) in Japan, India, South Africa, Canada, and New Zealand, he lent a particular style to British science overseas. And like his colleagues J.T. Wilson and Edgeworth David, he was an 'Empire man through and through'.[12]

At the University of Sydney, Liversidge shared this calling, if little else, with the classicist Charles Badham. He would have disapproved of the historian George Arnold Wood in opposing British policy in South Africa.[13] On the contrary, his social and political life was well suited to an orderly, imperial Anglo-Celtic world, with imperturbable, unselfconscious views of class, race and gender. There is no evidence that he thought deeply about social conditions, either at 'Home' or abroad. A man born to commerce and trade, he showed little interest in Labour. Judging by the discipline he enforced in his classes, he would have sided with Valerie Desmond's criticism of Australians 'at shirk'.[14] Perhaps unfairly, he might have found a place among the 'little set of professors and professional people' who greeted Beatrice Webb during her visit to Sydney in 1898, and who struck her as crushed between 'money-making and the race track … somewhat soured by the contempt of the rich people and the impotence of the rest of the population towards any attempt to raise the tone'.[15]

Liversidge supported education for women, on the grounds that this was an idea whose time had come; and perhaps also because women could make good science teachers. His interests in the Indigenous peoples of Australia and the Pacific were keen but conventional; his writings and trophies were those of an amateur ethnographer. He shared Baldwin Spencer's fascination with Aboriginal culture, but sought little understanding of Aboriginal life. His views as to the ultimate fate of the Aborigines – given a climate of opinion that assumed the eventual elimination of native peoples – conformed to colonial expectations. Britain had made him what he was, and he did not question his terms of reference.

***

Looking to Liversidge's scientific legacy, we measure his contribution to a view of colonial science to which he virtually gave his name. T.H. Huxley once spoke of science as 'organized common sense'.[16] It is generally accepted that Liversidge was 'the greatest organizer of science that Australia has seen and surely no-one in that country ever worked more unselfishly and with greater singleness of purpose than he to serve science for its own sake'.[17] When he arrived, two dozen scarcely communicating scientific societies were scattered around Australia and New Zealand; the government of New South Wales was inconstant in its support of science; and a city the size of Bristol made no systematic public provision for technical education. By his labours, Sydney gained a revived Royal Society, meeting in its own premises; a Technical College, teaching thousands of clerks and apprentices; a Technological Museum, productively investigating Australian plants; and an Australian Museum having one of the best mineral collections in the southern hemisphere. By his good offices, Sydney also became the 'headquarters' of the Australasian Association for the Advancement of Science, the Australian base for the International Catalogue of Scientific Literature, and a centre for mining education in New South Wales.

Within a space of thirty-six years, Liversidge designed a framework for public science and popular education that walked the distance from the Australian Museum to the Botanic Garden, from the University in Glebe to the Technology Museum in Ultimo. To the critical tradition of Jermyn Street, he added the confidence of Cambridge, not yet free of natural theology, nor of college crammers, but on the threshold of a 'revolution' in teaching and research. The result combined optimism and purpose in 'practical idealism'.

Colonial professors had first to build their buildings before they could teach in them. In a colonial university that borrowed elements of its coat of arms from Oxbridge, his views on liberal education, on technical instruction, on the relationship between pure and applied science, all were based on a view of public utility. 'Give me a laboratory, and I will raise the world' could have been, Bruno Latour muses, the message of Pasteur.[18] Liversidge embraced a similar Archimedean principle. Where Pasteur was the 'master' of the microbe, Liversidge was the master of minerals. Like his colleagues in medicine and engineering, he dismantled barriers between town and gown. Where miners and managers became 'scientific', lives were saved and the economy benefited.[19] From gold fields, tin tailings, and coal measures in NSW, to nickel miners and missionaries in the Pacific, he demonstrated that the interests of science and society were inseparable.

Through his efforts with the Royal Society and the AAAS, Liversidge contributed wealth of a special kind. European exploration and discovery had produced capital in ideas, routinely traded with metropolitan institutions in exchange for validation, recognition, and fame. The generation of Owen, Darwin, Huxley and Hooker all profited from insights sent or gathered from the 'periphery'.[20] The role of the colonies was one of deference. As W.B. Clarke told the Royal Society of NSW, 'We have done well to reflect a borrowed light, rather than aim at shining with an effulgence of our own.'[21] However, Liversidge went beyond deference to offer colonial science an Australian identity. In noticing differences, he questioned theories that were fashionable at 'Home'. This project, as George Humphry put it, required skill at 'grouping and explaining phenomena'. The task suited him well. By the end of the century, thirteen new mineral species had been identified, four in NSW, including one by himself. For most, there was not yet a commercial market, as there was for the silver, lead and zinc of Broken Hill, for the discoveries of Tasmanian copper at Mt Lyell, and for Queensland copper and gold at Mt Morgan and Mt Isa. In ways that might have appealed to his brothers in the City of London, Liversidge helped create an imperial mineral market for Australia. It is not insignificant that many whom he recruited to Sydney were trained at the same 'school of business' in London, the Royal School of Mines.

*** 

'The highest perfection of every object' – reads the motto above the door of the Victoria and Albert Museum in South Kensington – 'lies in the complete fulfillment of its function'. If Liversidge's world was the laboratory, his laboratory was the world itself. At the same time, he echoed the lessons of his teachers, for whom the business of science required curiosity, skill, an orderly mind, and a distrust of generalisation. As he once wrote of Frederick Hutton, so might we also say of him:

> an ardent worker and observer ... ever ready to give a generous support to the efforts of others; a close reasoner, of clear and independent thought, a pleasant companion and a loyal friend. He greatly disliked publicity, he had a soldier's directness and simplicity of purpose and a strong abhorrence of anything in the nature of pretentions or shams.[22]

For Liversidge, virtue lay in the act of collecting, comparing, cataloguing, and exhibiting. His approach was, as F.B. Smith has put it, 'positivist, synoptic, taxonomic', and much of his life was devoted to helping 'ease the difficulties of students educated in that format, but poorly prepared' for it.[23] This was a task that the time required. The image perfectly suited his sense of duty. Duty could be read as deference. But it also spoke to a sense of stewardship. Foucault once used the simile of 'unearthing' the 'archaeology' of science, to see how scientists construe their task, and make assumptions about the world.[24] The conventions that guided Liversidge were common to his culture – and followed the assumption that all knowledge must be empirically based, and valued by use. His approach to geology, chemistry and crystallography, as to the rituals of collecting and exchange, and to the business of exhibition and display, were perfect illustrations of an imperial ideology that emphasised efficiency, order and stability.

To paraphrase George Basalla, Liversidge embodied three roles central to the natural scientist – as observer, traveller and solver of puzzles.[25] Following the teachings of William Smith and Charles Lyell, Liversidge's generation was taught to examine the world, to consult the Book of Nature directly, as the only means of clearing up anomalies (or fathoming 'misprints') in the record of Creation.[26] Natural history had become, as Nicolaas Rupke has put it, the history of nature. The Great Chain of Being had been temporalised into a 'great chain of history'.[27] And as he observed, so Liversidge travelled, as much or more than any scientist then living in Australia. To furnish the lives of his 'Solomon's House', Francis Bacon's plan was to send abroad 'merchants of light' to garner 'books and abstracts and patterns of experiments'. Liversidge was a merchant – a philosopher-traveller transported by European technologies four times across the globe, visiting India and Egypt, New Zealand, Fiji and Japan, and crossing the United States from west to east. In the act of travelling, he galvanized the links that bound imperial science.

Finally, Liversidge was an arch 'puzzle-solver', a detective in an age that marvelled in Nature's curiosities. In 1863, Thomas Huxley told an audience of working men in Jermyn Street that the ideal scientist was just that – a 'detective' who worked to complete the picture-puzzle of a world, whose general outlines were already known. Once you eliminate the impossible, whatever remains, no matter how improbable, must be the truth, said Sherlock Holmes; and for twenty years before Arthur Conan Doyle, the deerstalker dressed the mental picture of the detective-*savant*. The same image – complete with field microscope and blowpipe – especially favoured the colonial scientist, sent around the world to resolve the mysteries of an unfamiliar Nature. If, as Henry James once wrote, it takes an old civilisation to set a novelist in motion, Australia and the Pacific were the perfect locations for a chemical geologist – an 'El Dorado' and 'an Arcady combined, a land of promise for the adventurous and a home of peace and independence for the industrious'.[28] Of course, the perfect Huxleyan detective – and whether Liversidge resembles Inspector Lestrade more than the sage of Baker Street, the reader must judge – soon realised that Nature is not the same in Mittagong as in Middlesex. But if Liversidge at first saw Australia through the eyes of his teachers, he quickly acclimatised himself to the brightness of his new home. He was hardly the first to appreciate its novelties.[29] But it would take all his detective's skill to unravel just some of her secrets.

Never one for grand syntheses, Liversidge had no overarching research plan, no single great question. As Esmond Wright once said of Benjamin Franklin – also the youngest son of an artisan – Liversidge was a utilitarian, not a utopian.[30] As Beaglehole said of James Cook, he was a 'genius of the matter of fact'.[31] And like the Liversidges of London, he was in many ways a businessman who took satisfaction in getting things done. A tactician rather than a strategist, his training prepared him to work on a wide range of subjects, from minerals and meteorites to kerosene shales. It is not surprising that his *protégés* moved into such different fields – Laby and Mawson into radiochemistry, Guthrie into agricultural chemistry, and Maiden into chemical botany.

However, it is in this special combination of qualities – of imperial loyalty and colonial self-interest, of analytical skill and practical idealism that we find the essence of what we may call 'Liversidgean science', a complex of values that became Liversidge's colonial hallmark and imperial legacy. In his lifetime, his achievements were fully recognised. However, for almost a century after his departure from Sydney, and for at least a generation after his death, his legacy was largely neglected. It is instructive to ask why.

***

In writing the history of modern science, historians usually look to significant observations or discoveries, critical experiments, or theoretical generalisations. Some scientists achieve immortality by bestowing their names upon plants and animals, processes or minerals. Others become immortal in the eponymy of 'effects', 'rules' and 'constants'. To these conventions, Liversidge forms an exception, for perhaps three reasons. First, as W.H. Brock has noted, chemistry and geology were in Liversidge's day 'increasingly forced into each other's company'. Chemists and geologists read each other's language, and political economy sanctioned a marriage between them to solve problems of commercial value and scientific importance. But which should be the dominant partner was left unclear. As Brock puts it, 'should geologists approach their rocks … veins … and volcanoes as chemists, or as geologists equipped with chemical tools … and concepts?'[32] But the research front in chemistry was moving fast towards organic and physical chemistry, and by the 1890s, few chemists read geology as intently as their predecessors. In the passage of time, chemical geology – and geological chemistry – were relegated to Geological Surveys and their museums.

This separation of powers between chemistry and geology took place at a time when geology was itself turning in other directions. However, the history of this 'turn' has yet to be written. As Roy Porter once observed, 'if the first half of the century is rightly known as the Golden or Heroic age in the history of geology, the period between the *Origin of Species* and the First World War is a curiously Dark Age'.[33] As 'historical sciences', geology and mineralogy have always fired the imagination. However, when we move beyond the fossil disputes, the glacial debates, and the 'Devonian controversies' of the 1840s, that is, beyond the struggles of 'Genesis and geology' and the emergence of the geological profession,[34] and into the last decades of the nineteenth century during which evolutionary theory was assimilated, we find less scholarship. A period in which Geological Surveys were professionalised, and new alliances between academia and commerce were formed, has yet to find its historian. Few of Liversidge's contemporaries have received the accolade of a 'life and letters'.

One explanation for this neglect may be found in the inelegiac nature of much late-Victorian geology. In his account of the Devonian controversy, Martin Rudwick draws attention to what he calls 'characteristic' scientific debate, that is, debate that has resulted in significant additions to reliable knowledge. Besides these, he considers there are debates that are not 'characteristic', in the sense that they have not been resolved, and as such are consequently not regarded as having added (or added yet) to the corpus of knowledge. Inevitably, the latter may prove of interest more to historians than to scientists; and in Australia, where the history of geology has been written largely by geologists, it is not surprising that for many 'unresolved' questions of the kind that Liversidge asked, 'closure' remains remote, and its history, unwritten.[35] As Mott Greene argues, it was not until well after the turn of the twentieth century, with the arrival of glacial theory and theories of continental drift, that geology acquired a dramatically new image.[36]

Much the same can be said of classical mineralogy which, during the third quarter of the nineteenth century, became relegated to the academic sidelines, becoming – like comparative anatomy – a subject for the museum world, a relic of Victorian enthusiasm. Not until well after Liversidge's day did the new disciplines of petrology and materials science rise to take up interests that he had pioneered. It remained for twentieth-century physics, aided by electronic, aerospace and information technology, to develop the mineralogy of the rare earths and make more sense of his work.[37]

If Liversidge has until now escaped most historians of geology, neither has he been remembered as a great chemist. He receives no entry in J.R. Partington's massive *History of Chemistry*, nor in Alexander Findlay's *History of Chemistry*; nor – perhaps more surprising, given his efforts for the Chemical Society and the Institute of Chemistry – in Findlay and Mills' patriotic account of *British Chemists*.[38] Liversidge pioneered the commercial applications of mineral chemistry. However, in 1933, when his near-successor at Sydney, G.T. (later Sir Gilbert) Morgan lectured on modern trends in *Inorganic Chemistry*, his work received no mention.[39] The next generation was writing a history that emphasised certain fields to the exclusion of others. The special nature of his work in inorganic chemistry was overtaken by rapid strides in organic and physical chemistry and their industrial applications. Once the principles of atomic theory were established, and the laws of valency and solution confirmed, the disciplines that he represented passed from the sunlight into the shade. Even the discovery of 'missing elements', so as to 'complete' the periodic table, was taken to be merely a matter of time. Kelvin's famous speculation that the future of physics would consist in the derivation of atomic weights to yet further decimal places, borrowed a view that many also held of chemistry. All this was to change with the discovery of X-rays and radioactivity, and the application of the 'new physics' to theories of matter. But for Liversidge, these came too late.

During his lifetime, Liversidge received cordial recognition, especially in Britain,[40] although it remained for North Americans – especially Whitman Cross, Henry Washington and Joseph Iddings to 'discover' him and cite his findings internationally.[41] Today, we may claim him as the first Australian geochemist, and a pioneering figure in the discipline that was later to be launched by the Americans Frank Wigglesworth Clarke and Henry S. Washington, of the US Geological Survey, and the Russians V.I. Vernadsky and A.E. Fersman.[42] It was not, however, until well after Liversidge's death that Victor Goldschmidt and his colleagues in Oslo and Göttingen, aided by optical and X-ray analysis,[43] transformed classical mineralogy into a branch of modern physical science.[44] Working in the Royal Institution at the same time as W.H. Bragg and J.D. Bernal, Liversidge may have read of their work, but he was by then too advanced in age to contribute to it.

In certain respects, there are intriguing parallels between the ways in which we may see Liversidge and Lord Kelvin, at whose Glasgow Jubilee in 1899 Liversidge received an honorary degree. Towards the end of his life, Kelvin famously claimed that his efforts to produce 'grand theory' had failed. As his recent biographers have noted, Kelvin did not stray far during his lifetime from the research program he sketched out forty years earlier.[45] His intellectual agility won him fame and fortune; yet, he could not accept some of the major shifts then taking place in physics, and few colleagues shared his assessment of the most urgent theoretical problems facing science at the beginning of the twentieth century. By then, Kelvin's Victorian brand of engineering physics looked old-fashioned; he had espoused some views that turned out to be erroneous; and his work seemed far removed from the interests of his younger students. Overall, it is scarcely surprising that his reputation suffered with the coming of relativity and quantum theory.[46]

In much the same way, Liversidge was bound by the limitations of his world view and his discipline. In certain respects, he was slow to assimilate new theory. From 1889, applications of Arrhenius' theory of electrolytic dissociation and van 't Hoff's theory of osmotic pressure had begun to treat the earth's magma as a saturated solution from which mineral species may have crystallised. But these insights do not surface in his scientific papers, although the questions that he set honours students in 1894 show that at least he knew of them.[47]

Even so, Liversidge was possibly more progressive than his contemporaries might have granted, insofar as he was working on the emerging frontiers of planetary theory and geochemistry.[48] Some subjects he touched turned to gold – both figuratively and in fact – even if others required the invention of analytical instruments his generation did not possess.[49] His descriptive work presaged the large-scale application of chemistry to Australian agriculture,[50] and launched the field of 'biogeochemistry'.[51] Years would elapse before atomic absorption spectroscopy, electron microscopy, neutron-activation techniques and proton-microprobes capable of detecting one part per billion would arrive to correct his results.[52] Today, the theory of ore genesis, which he saw as through a glass darkly, is still hotly debated.[53]

As we have seen, theory was a realm which Liversidge carefully refrained from entering. European science had failed to produce a consensus on the origins and structure of the earth. But there was no reason to import unnecessary debate to Australia. In this, Liversidge agreed with W.B. Clarke who, when supporting the change in name of the 'Philosophical Society of NSW' to the 'Royal Society of NSW', emphasised the critical distinction between the two. 'It is one thing', Clarke said, 'to respect the method by which a logical argument is to be maintained, and another to defend the introduction of investigations which are often based on conjecture, and are altogether speculative'.[54] Clarke and Liversidge spoke to a generation with living memories of the intrusions of *naturphilosophie* into a Paleyan natural theological world view of design and purpose. In England, speculation still smacked of radicalism, both political and philosophical. On the 'periphery', radicalism was dangerous. However, there were also practical disadvantages in identifying too closely with speculative views. It was not so much conceptual conservatism as common sense for the Royal Society of New South Wales to endorse William Macleay's dictum that 'we cannot do better in the present state of natural history in Australia than confine our attention to observing, cataloguing and describing'. 'The synthetical work' as Macleay put it, 'may well ... be left for the present to the legion of writers who aspire to what is foolishly called 'high science'.[55]

With this declaration, Liversidge solemnly agreed, and in this practice, we find both his principles and his politics. They informed the methods in which he had been trained, and in which he trained every student who entered Sydney University. If they rarely took him far beyond the lectures notes of his teachers, at least they spared him the worst *ignes fatui*. Writing in 1905, Arthur Hutchinson, Professor of Mineralogy and Master of Pembroke College, Cambridge, might well have had Liversidge in mind when he surmised that the 'calling' of mineralogical chemistry consisted precisely in what Liversidge actually did –'analyses scattered in short papers through the pages of many periodicals ... Work resulting either in the discovery of new principles or in the coordination of established facts is only of rare appearance'.[56]

By this token, Liversidge's lifework was inherently factual, objective, custodial and descriptive. He was a master practitioner of 'inventory science'. The three editions of his *Minerals of New South Wales* were working dictionaries of the Book of Nature. The mineral curators in Harvard's Museum of Comparative Zoology, Oxford's University Museum, the Mineralogical Museum of Cambridge, and the Muséum d'Histoire Naturelle of Paris were his spiritual cousins, who corrected and revised his work. [57] Science for them was about reducing Nature to order. If one looked for a larger justification, that could easily be found. Just as chemical nomenclature was to dispel confusion and conjecture, so classification by source, characteristics and use would lead to consensus. By careful analysis and arbitrage – encapsulated in the jury system of the exhibitions he loved so well – science would replace conflict with cooperation, serve the interests of Empire, and hasten the federation of the world.

This image of Liversidge lives on through the printed text. However, there remains a possibility that, after all is said, we may not have read the last word, and Liversidge's story may still be incomplete. If history does not credit him with a single, great crowning scientific achievement, future historians may find him nearing the threshold of several smaller discoveries.[58] His last, unpublished papers, if ever they are found, may point to significant ideas about the location of ore deposits by the analysis of surface environments, a prospect of great value to miners and environmentalists alike, and one worthy of a proper place in history. And insofar as his work on trace elements and meteorites led him to reflect on the origins of terrestrial life, Liversidge remains a pioneer. It remains to be seen whether he was in fact more far-sighted than his contemporaries dreamed. Certainly, he was more ambitious.

***

During his Australian journey, Liversidge met the same strange country that D.H. Lawrence saw, when he wrote of its 'subtle, remote timeless beauty'. For Liversidge, the study of nature led not to passive acceptance, but to active management. He knew that a complete picture of Australia needed many hands, and many years, and might never be complete. He returned to England with this work unfinished. Eighty years after his death, we see him as a man of his time, if also well ahead of it. But nowhere was his influence felt more deeply than at the University of Sydney, which he served for thirty-six years. For reasons that remain unclear, he failed to receive public honours. It is the task of today's generation to redeem this oversight. Ralph Waldo Emerson once wrote that an institution is the lengthened shadow of a man. Today, Sydney University rivals the best universities in the world, and Liversidge has a secure place in its history. If, by the late twentieth century, memory of his work had dimmed, he has since become a model of the Australian scientist as 'public intellectual'.[59] Certainly, his spirit of adventure and enterprise persists – if we look for it – not only in his reports and papers, and in the name of a street in Canberra, but also in the museums and associations to which he gave life. His vision helped make Sydney – city and University - a 'moving metropolis' of international stature. His memory continues to inspire those who serve science and society under the Southern Cross.

Liversidge's key to the Technological Museum, all that remained after the Garden Palace burned to the ground in September 1882.

# APPENDIX I

# TRIBUTES AND OBITUARIES

When modern biographers disagree, it is rare to find conflicting views on objective dates of birth and death. However, Liversidge's biographers do disagree. Some references published during his lifetime, which he presumably confirmed (and for which he undoubtedly gave information), give the year of his birth as 1847. This was indeed the year first given by his most reliable biographer, Dr. D. P. Mellor, writing in the *Proceedings of the Australian Chemical Institute* in 1957, ostensibly in commemoration of Liversidge's centenary. However, Liversidge's birth certificate in the Registrar General's Office in London gives the year 1846, and this date Mellor uses in his later article on Liversidge in the *Australian Dictionary of Biography*.

Liversidge died on 26 September 1927. Daily newspapers (including the *Sydney Morning Herald* and the *Morning Post*) referred ambiguously to this as being in his 'eightieth year'. Subsequent authors – notably, T.W. Edgeworth David, writing in 1930 in the *Proceedings of the Royal Society of London* – chose 1847, perhaps without confirming the date with the Registrar General. Mistakes in the registration of births were not unknown in the 1840s, and it is possible that 1847 could have been written where 1846 was intended. Several interviews with Liversidge contain errors of fact, which Liversidge did not (or could not) correct. On the balance of evidence, however, I have chosen to follow the date given on his birth certificate, and so have taken 1846 as the year of his birth, as does the *Australian Dictionary of Biography*.

Liversidge's principal obituaries are listed below, in chronological order of their appearance.

## I. CONTEMPORARY TRIBUTES

1.  Anon., 'Professor Archibald Liversidge, M.A., F.R.S., President, Royal Society of New South Wales', *Building and Engineering Journal*, VIII (ns), (18 January 1890), 20.

2.  Anon., 'Our Men of Science' (No. VIII), *Sydney Mail* (12 July 1890), 88.

3.  Mennell, Philip, 'Liversidge, Professor Archibald, MA, FRS', *A Dictionary of Australasian Biography* (London: Hutchinson & Co., 1892), 276-277.

4.  Anon., 'Professor Liversidge', Personal Portraits No. 29, *Illustrated Sydney News*, 30 (45), (11 November 1893), 4.

5.  Anon., 'Professor Liversidge', *Hermes*, 3 (3), (7 July 1897), 3-4.

6.  Poggendorff, J.C. (ed.), 'Liversidge, Archibald', *Biographisch-Literarisches Handwörterbuch* (Leipzig: Verlag von Johann Ambrosius Barth, 1904), IV, 898.

7.  Anon., 'Professor Liversidge: Australia's Premier Scientist', *Sydney Mail* (25 December 1907), 10.

8.  Peile, J.L. and J.A. Venn, 'Liversidge, Archibald', *Christ's College Biographical Reporter* (Cambridge: Cambridge University Press, 1913), II, 615.

9.  Stephenson, H.H. (ed.), 'Liversidge, Archibald, F.R.S.', *Who's Who in Science International, 1914* (London: J & A Churchill, 1914), 387.

10.  Reeks, Margaret, *Royal School of Mines: Register of Old Students, 1851-1920, and History of the School* (London: Royal School of Mines, 1920), 115-116.

## II. OBITUARIES

1.  Anon., 'Professor Liversidge', *The Times*, 28 September 1927, 20.

2.  Anon., 'Professor Liversidge: An Eminent Scientist', *Morning Post*, 29 September 1927.

3.  Anon., 'Professor Liversidge', *Sydney Morning Herald*, 29 September 1927, 12.

4.  Anon., 'Professor Liversidge', *Daily Telegraph* (Sydney), 29 September 1927.

5.  Anon, 'Professor Archibald Liversidge, MA, LLD, FRS', *Surrey Comet*, 1 October 1927, 9.

6. Anon., 'Archibald Liversidge', *Chemical Engineering and Mining Review*, 5 October 1927, XX (1927-28), 4.

7. F[awsitt], Charles E., 'Archibald Liversidge (1847-1927)', *Union Recorder (University of Sydney)*, I (26), (6 October 1927), 201.

8. David, Sir T.W. Edgeworth, 'Prof. A. Liversidge, FRS', *Nature,* 120 (29 October 1927), 625-626.

9. Anon., 'Archibald Liversidge', *Proceedings of the Institute of Chemistry*, Part V, (October 1927), 241.

10. Anon., 'Professor Liversidge', *Hermes*, 33 (3), (Michaelmas Term, 1927), 162.

11. Rutherford, Sir Ernest, 'Presidential Address' (November 1927), *Proceedings of the Royal Society of London*, B. 102 (1927-28), 242-243.

12. Monash, Sir John, 'Resolution in honour of Professor Liversidge' (January 1928) *Report of the Australasian Association for the Advancement of Science*, 19 (Hobart, 1928), xv.

13. Anon., 'Archibald Liversidge, MA, LLD, FRS', *Proceedings of the Royal Society of Edinburgh*, 47 (1928), 388.

14. Anon., 'Archibald Liversidge', *The Australian Museum Magazine*, 3 (5), (1928), 148

15. Aston, B.C., 'Presidential Address', *Transactions and Proceedings of the New Zealand Institute*, 59 (1928), 25.

16. Bather, F.A., 'Archibald Liversidge', *Quarterly Journal of the Geological Society of London*, 84 (1928), lv.

17. Stewart, J. Douglas, 'Archibald Liversidge, MA, LLD, FRS', (2 May 1928) *Journal and Proceedings of the Royal Society of N.S.W.* 62 (1928), 8-10.

18. Anon., 'Professor Archibald Liversidge, FRS', *Christ's College Club Book* (Cambridge: Christ's College, 1929), 96.

19. David, Sir T.W. Edgeworth, 'Archibald Liversidge, 1847-1927', *Proceedings of the Royal Society of London*, A 126 (1930), xii-xiv.

20. Spencer, L.J., 'Liversidge (Archibald) (1847-1927)', *Mineralogical Magazine*, 22 (1930), 397-398.

21. David, Sir T.W. Edgeworth, 'Archibald Liversidge', *Journal of the Chemical Society*, Pt. 1 (1931), 1039-1042.

22. Hammerton, Sir John Alexander, 'Liversidge, Archibald (1847-1927)', *Concise University Biography: A Dictionary of the Famous Men and Women of All Countries and All Times* (London: Educational Book Co., 1935).

23. Anon., 'Liversidge, Archibald, MA', *Who Was Who (1916-1928)* (London: Adam & Charles Black, 1947), II, 635.

24. Mellor, D.P., 'Founders of Australian Chemistry: Archibald Liversidge', *Proceedings of the Royal Australian Chemical Institute*, 24 (August 1957), 415-421.

25. Anon, 'Liversidge, Archibald', *The Australian Encyclopedia* (Sydney: Angus and Robertson, 1958), vol. 5, 343.

26. Mellor, D.P., 'Liversidge, Archibald (1846-1927)', *Australian Dictionary of Biography* (Melbourne: Melbourne University Press, 1974), vol. 5, 93-94.

27. Torney, Kim, 'Liversidge, Archibald (1846-1927)', in Graeme Davison, John Hirst and Stuart Macintyre (eds.), *The Oxford Companion to Australian History* (Melbourne: Oxford University Press, 1988), 395.

28. MacLeod, Roy, 'Liversidge, Archibald (1846-1927)', in H.G. C. Matthew and Brian Harrison (eds.), *Oxford Dictionary of National Biography* (Oxford: Oxford University Press, 2004), vol. 34, 52-53.

# APPENDIX II

# EXHIBITIONS AND COMMISSIONS: MEDALS AND AWARDS

Archibald Liversidge was taken as a child to the Great Exhibition in London, where he began a life-long engagement with the 'exhibition movement'. During his career, he served on six colonial, intercolonial and international exhibitions. As a scholar and collector, he attended two major exhibitions (Paris in 1878 and Melbourne in 1880), and contributed to seven others: Philadelphia in 1876, Amsterdam in 1883, Calcutta in 1883-84, the Colonial and Indian Exhibition in London in 1886, Adelaide in 1887, Melbourne in 1888, London in 1890, and Chicago in 1892. For his services, he received several medals, which today are in the possession of the Royal Society of New South Wales. The Society's collection also includes commemorative medals that Liversidge obtained, honouring Marcelin Berthelot, Henri Moisson and the Coronation of Edward VII, and a copy of the Clarke Medal, which Liversidge commissioned for the Royal Society.

1878     Exposition Universelle Internationale, Paris. One medal with the inscription: 'Archibald Liversidge Esq., Membre de la Commission de la Nouvelle-Galles du Sud', and on case: 'Medaille Commemorative offerte pour services rendus. Archibald Liversedge, Esq., Membre de la Commission de la Nouvelle-Galles du Sud'. There is also a badge marked 'Jury 1878', 'Exposition R.F. Universelle, Paris'.

1879     Sydney International Exhibition. Two medals, with the Inscriptions: 'Archibald Liversidge, Honorary Member of Commission'; 'Awarded to Professor Liversidge for Ethnological Exhibits'.

1879-81   Royal Commission for the Australian International Exhibition. Inscription: 'Professor Liversidge for Services'.

1880-81   Melbourne International Exhibition.

1883     Internationale Koloniale en Uitvoerhandel Tentoonstelling, Amsterdam. Medaille d'Argent. Inscription: 'Professor Liversidge FRS for Work on the Minerals of N.S.W'.

1883-84   Calcutta International Exhibition. Inscription: 'A. Liversidge, Esq., FRS, Member, New South Wales Commission'.

1884     The Royal Society of New South Wales, Sydney. Inscription: 'Presented to Prof. Liversidge, F.R.S. as designer of this medal, 1884'.

1886     Colonial and Indian Exhibition, London.

1887     Adelaide Jubilee International Exhibition, First Order of Merit.

1888     Centennial Exhibition, Melbourne.

1890     International Exhibition of Mining and Metallurgy, London. Inscription: 'Awarded to Professor Archibald Liversidge FRS, Sydney'.

1892-93   World's Columbian Exposition, in Commemoration of the Four

Hundredth Anniversary of the Landings of Columbus, 1892-1893; three medals inscribed 'A. Liversidge', and one inscribed 'Arch. Liversidge, MA, FRS'

# APPENDIX III

# LEARNED SOCIETIES AND SOCIAL CLUBS

Liversidge belonged to the following learned societies and social clubs.

## I. LEARNED SOCIETIES

| SOCIETY | YEAR OF ADMISSION | STATUS |
|---|---|---|
| 1. Chemical Society of London | 1872 | Fellow, VP, 1910-13 |
| 2. Royal Society of New South Wales | 1872 | Fellow, Council, 1873-1907; Hon. Sec., 1875-1885, 1886-1889; Pres., 1885-1886, 1889-1890, 1900-1901; V.P., 1890-1894, 1901-1906. (Honorary Member, 1908-1927) |
| 3. Geological Society of London | 1873 | Fellow |
| 4. Royal Society of Tasmania | 1875 | Corresponding Member |
| 5. Linnean Society of NSW | 1875 | Member (Councillor, 1877-1878) |
| 6. Agricultural Society of New South Wales | 1876 | Councillor (until January 1878) |
| 7. Physical Society of London | 1876 | Member |
| 8. Senckenbergische naturforschende Gesellschaft | 1876 | Corresponding Member |
| 9. British Association for the Advancement of Science | 1876 | Member (V.P., 1897) |
| 10. Mineralogical Society of Great Britain and Ireland | 1876 | Member |
| 11. Société d'acclimatation de l'Ile Maurice | 1876 | Honorary Member |
| 12. Linnean Society of London | 1877 | Fellow (until 1883) |
| 13. Royal Geographical Society | 1877 | Fellow |
| 14. Royal Historical Society | 1877 | Honorary Member |
| 15. Société Mineralogique de France | 1878 | Member |
| 16. Institute of Chemistry of Great Britain and Ireland | 1878 | Fellow (V.P., 1910-1911) |
| 17. Royal Colonial Institute | 1879 | Fellow |

| | | |
|---|---|---|
| 18. Royal Society of London | 1882 | Fellow |
| 19. Society of Chemical Industry | 1883 | Member (V.P., 1909-1912) |
| 20. Royal Society of Queensland | 1887 | Corresponding Member |
| 21. Australasian Association for the Advancement of Science | 1888 | Founding Member Hon. Sec., 1886-1907; Pres. 1897-1898 |
| 22. New Zealand Institute | 1890 | Honorary Member |
| 23. Royal Society of Victoria | 1892 | Honorary Member |
| 24. Geological Society of Edinburgh | 1893 | Honorary Member |
| 25. Kaiserlich Leopoldinisch- Carolinische Deutsche Akademieder Naturforscher zu Halle (now Deutsche Akademie der Naturforscher Leopoldina) | 1894 | Honorary Member |
| 26. Indiana Academy of Sciences | 1896 | Foreign Correspondent |
| 27. Cambridge Philosophical Society | 1897 | Fellow |
| 28. New York Academy of Sciences | 1899 | Corresponding Member |
| 29. Royal Society of Edinburgh | 1900 | Honorary Fellow |

## II. Social Clubs

| CLUB | YEAR OF ADMISSION |
|---|---|
| l. Union Club (Sydney) | 1873 |
| 2. Savile Club (London) | 1879 |
| 3. United University Club (London) | 1901 |
| 4. Athenaeum Club (London) | 1912 |
| 5. Carlton Club (London) | 1925 |

# APPENDIX IV

# MINERAL AND ETHNOGRAPHICAL COLLECTIONS

## MINERAL COLLECTIONS

On arriving in Sydney, Liversidge began to collect and exchange ethnographical, mineral and rock specimens, and over the years, amassed hundreds of items. The history of his collections may be traced through his correspondence and his will, although a clear distinction between what were actually 'his', and what were in fact 'specimens in his keeping' is not always easy to draw.

In 1874, Liversidge obtained the approval of the Ministers of Mines and of Public Works to the collection of rock specimens unearthed in public building, for sharing between the Government's Geological Museum and the University. We do not know how many, if any, specimens came to him by this route. However, during the 1870s, he was in frequent receipt of minerals for analysis by the Mines Department, and many of these would have remained in his possession. He also collected and exchanged many specimens relevant to his own research. We believe, for example, that he sent specimens of mispickel to Henry Roscoe in Manchester, possibly around 1874, coinciding with his emerging interest in moss gold. These specimens are now in the University Museum of Oxford.[1] His 'public acquisitions' were even more important. During his European travels in 1878–79, the Australian Museum commissioned him to acquire £1000 worth of specimens, and the University of Sydney obtained £1000 from the government for him to collect 'geological specimens and physical apparatus'.[2] These efforts produced substantial acquisitions for the Museum and the University.[3]

At the Paris International Exhibition of 1878, Liversidge displayed a large collection of rough gems and gem sands with cut and polished opals from NSW, together with minerals, fossils, precious stones, moss gold, silver, copper, crystallised gold, and microscope sections of meteorites found in Australia, and cobalt ores sent from New Caledonia. In March 1879, at the end of his European visit, Liversidge gave a collection of some forty-six specimens – from tin to gemstones – to the Museum of Practical Geology in London.

From November 1880, Liversidge was in sporadic correspondence with Dr (later Sir) Lazarus Fletcher, Curator of Minerals at the British Museum, who oversaw the move of that institution's collections from Bloomsbury to South Kensington. Liversidge's first gifts were followed in September 1881, by thirteen more specimens – ores, shales and minerals from NSW and New Caledonia. In September 1886, he sent Fletcher a list of gold-bearing stones and other rocks he had collected from the Fairfield goldfields. Another group of eight specimens, given in July 1887, may have followed the Colonial and Indian Exhibition; two more were sent in February 1891, making a total of over fifty.[4] In 1881, he presented the Royal Scottish Museum, in Edinburgh, specimens of zircon sand, pleonaste, tourmaline, Noumeaite and cassiterite. He may also have sent to Edinburgh the gold crystals that are featured on the cover of the centennial edition of his *Minerals of New South Wales* (1888), and in his portrait by the Hon. John Collier.[5]

Over the years, Liversidge supplemented and refined what he considered his personal collection, producing a separate 'exhibition collection', with its own labels.[6] At the Melbourne International Exhibition of 1880, he exhibited gems that he had collected in the New England district of NSW; these gems remained the basis of his Exhibition collection. He sent a larger selection of minerals to the Calcutta International Exhibition; and in 1886, exhibited over 200 specimens at the Colonial and Indian Exhibition in London, in a series of categories ranging from 'Crystallized Gold and other Gold' to 'Diamonds', and including specimens from Berrima, Uralla, and the New England District of NSW. By the Adelaide Jubilee Exhibition in 1887, and the Melbourne Centennial Exhibition in 1888, his gold collection had increased to fifty items, and by the World's Columbian Exhibition in Chicago in 1893, to over ninety. By 1893, his exhibition collection altogether exceeded 500 specimens. His enterprise was recognised by the award of prizes and medals at Melbourne in 1880 and 1888, at Amsterdam in 1883, Calcutta in 1884, London in 1886, Adelaide in 1886 and 1890, and Chicago in 1893 (see Appendix II).

When he retired to England in 1908, Liversidge offered to sell his entire collection to the Australian Museum. At that time, this consisted of four main groups: gems, both cut and uncut; isolated crystals; crystallised gold; and a 'miscellaneous collection' containing minerals from Australia, New Zealand, the

United States, Europe and Polynesia.[7] Liversidge later added another two cases of minerals.[8] The Museum Trustees wished to purchase the crystals, valued at £95, as well as their cabinets and some books, but not the gems or the minerals. Liversidge was unwilling to break up the collection and suggested that the Museum pay for it by instalments. The collection was stored at the Museum while negotiations continued, long after Liversidge had left NSW, but no agreement was reached, and the collection was finally returned to him in England in 1911.[9] Some of it may have been shown at the Coronation Exhibition in London in May 1911, and in later years he continued to display his exhibition collection – at the Franco-British Exhibition in London in 1908; at the Anglo-American Exhibition in May 1914; at the Imperial Institute, London, in 1915; and at the British Empire Exhibition in 1924.

In 1921, Liversidge finally donated to the Australian Museum the crystal collection that the Museum had initially wished to purchase. This included 1185 specimens, mostly of European origin, but also some Asian and American items. Few are from Australia; but some were put on display.[10] In October 1927, following his death, the rest of his impressive collection of 3000 items was left by bequest to the British Museum (Natural History).[11] Some duplicates were sent elsewhere, but most of his gold, mineral, meteorite and gem specimens are today held in South Kensington.

# ETHNOGRAPHICAL COLLECTIONS

In a lifetime of travel, Liversidge collected over 100 ethnological specimens. Some of these he exhibited at the World's Columbian Exhibition at Chicago in 1893, and at the Coronation Exhibition in 1911. His collections began with Fijian and Maori tools, weapons and other implements obtained on his first voyage to New Zealand in 1876–77.[12] His exhibition of greenstone Hei Tikis won him a Bronze Medal at the Sydney International Exhibition in 1879. He also received a silver medal for being a member of the Ethnological Subcommittee of the Exhibition. By 1892, he had collected seventy-four Aboriginal stone implements, on which he gave a descriptive paper to the Royal Society of NSW,[13] and which he sent for display at the World's Fair at Chicago in 1893.[14] In April 1881, he donated to the Australian Museum several casts of Tikis, one of which was destroyed in the Garden Palace fire of September 1882.[15] During the 1890s, he exhibited his ethnographical specimens at his home in Darling Point. These returned with him to England, where twenty were shown at the Coronation Exhibition in 1911. On his death, his collection was given to the British Museum of Natural History, which sent it to the Museum's Department of Ceramics and Ethnography.[16] This collection is now held in the Museum of Mankind in London.[17]

Just before leaving Sydney in 1908, Liversidge donated to the Australian Museum ten poison arrows from the New Hebrides, originally a gift from the Rev. H.A. Robertson to Professor John Smith; the skull of an Aboriginal known as the 'King of Bingera'; and the mandible of the skull of a Maori.[18] The last of these was repatriated to New Zealand in 2000. The arrows and the Aboriginal skull are still in the Museum's possession.

---

[1]  Personal communication, Monica Price, Australian Curator, to author, 18 April 1986.
[2]  *Sydney Morning Herald*, 7 August 1880.
[3]  The 'Liversidge collection' of minerals, now in the possession of the University of Sydney, seems to comprise several smaller collections, some of which can be attributed to Liversidge. The largest section is the Krantz collection, purchased by Senate in 1866. To this Alexander Thomson and Liversidge added specimens. See Senate Minutes, 5 (5 June 1878), 224, and David Branagan and Graham Holland (eds.), *Ever Reaping Something New: A Science Centenary* (Sydney: University of Sydney, Faculty of Science, 1985), 122-124.
[4]  Personal communication, Dr E.A. Jobbins, I.G.S., to author, 11 May 1982.
[5]  Personal communication, Dr Harry Macpherson, Department of Geology, Royal Scottish Museum, to author, 25 March 1983.
[6]  Australian Museum Letterbooks, A.L. to Robert Etheridge Jr., 27 November 1908.
[7]  Australian Museum Letterbooks, A.L. to Etheridge, 1 January 1908.
[8]  Australian Museum Letterbooks, Memorandum by Charles Anderson, 'Professor Liversidge's Collection', attached to Liversidge's letter to Robert Etheridge Jr., 1 January 1908.
[9]  Liversidge Papers, Trustees, Australian Museum, to Liversidge, 5 August 1911.
[10]  Australian Museum Letterbooks, A.L. to the Trustees, 3 May 1920. Liversidge wrote of them that 'the main part of the collection formerly belonged to General Cathcart (in 1846), as I was informed when I purchased it, the remainder consists principally of specimens from Ceylon'.

11  British Museum (Natural History), Department of Mineralogy, Nicolson, Freeland and Shepherd to George Prior, BM (NH), 26 October 1927; Prior to Admiral Edward Liversidge, 2 November 1927; John Liversidge to Prior, 3 November 1927.

12  'Do you remember our visit to the hot lakes [of New Zealand] and the Heitiki you got from a native woman?', as John Webster later recalled, Liversidge Papers, Box 19, f. 15, Webster to Liversidge, 27 February 1910.

13  A.L., 'Notes on some Australian and other Stone Implements', *JPRSNSW*, 28 (1894), 232-245.

14  *Official Catalogue of Exhibits, World's Columbian Exposition*, Department M (Chicago: W.B. Corkcy, 1893), 33.

15  Personal communication, Ms Tessa Corkill, Department of Anthropology, Australian Museum, 5 December 1983.

16  Museum of Mankind (London), George Prior, Minerals Department, British Museum (Natural History), to Mr Smith, Department of Ceramics and Ethnography, 29 November 1927.

17  I am grateful to Ms D.C. Starzecka of the Museum of Mankind for locating these implements for me.

18  Australian Museum Letterbooks, A.L. to Etheridge, 16 December 1907; and J.M. Smith to Liversidge, 10 April 1893, placed with receipt of items, 14 October 1907.

# APPENDIX V

## UNDERGRADUATE CURRICULA IN THE UNIVERSITY OF SYDNEY 1852–1912

### I. THE ARTS CURRICULUM

| YEAR | 1852-1874 | 1875 | 1876-1878 | 1879-1882 | 1883 | 1884-1887 | 1888-1912 |
|------|-----------|------|-----------|-----------|------|-----------|-----------|
| I | Greek, Latin, Logic, Ancient History, Mathematics, Natural Philosophy, Chemistry, Experimental Physics | Greek, Latin, Logic, Ancient History, Mathematics, Natural Philosophy, Chemistry, Experimental Physics | Greek (except when exempted - Clause 72) Latin, Mathematics, Chemistry or Experimental Physics | Greek Language and Literature, Latin Language and Literature, Mathematics, Experimental Physics | Latin, Math. and Nat. Hist., Elementary Chemistry and elements of Nat. Philosophy, One of: Greek, French or German | Latin, Mathematics, Elem. Chem. And elements of Nat. Phil., One of: Greek French or German | English, Latin, One of: Greek, French or German, Mathematics, One term each of: Elem. Physics, Chem. & Nat. Hist. |
| II | Greek, Latin, Logic, Ancient History, Mathematics, Natural Philosophy, Chemistry, Experimental Physics | Greek, Latin, Logic, Ancient History, Mathematics, Natural Philosophy, Chemistry or Experimental Physics | Greek (except as above) Latin, Mathematics and Natural Philosophy, Chemistry or Experimental Physics (2 terms) Geology (1 term) | Greek Language and Literature, Latin Language and Literature, Mathematics, Natural Philosophy, Chemistry, Geology | Latin and Ancient History, Mathematics and Natural History, Two of: Greek, French or German | Latin and Anc. History, Mathematics, Two of: Greek, French or German, Geog. and Geol. Zool. & Botany | Two of: Latin, Greek, English, French or German, Mathematics, Either a third language or one of: Phys., Chem., Nat. Hist. or Physiology |
| III | Greek and Latin, Mathematics and Natural Philosophy, Chemistry, Geology, Experimental Physics, Mineralogy, Physics, Practical Chemistry, Geology, | Greek and Latin, Mathematics and Natural Philosophy, Chemistry, Experimental Physics, Geology, Mineralogy, Geology, Zoology and Botany | Classics, Mathematics, Practical Chemistry, Geology, Mineralogy | Classical-Greek, Latin, Ancient History Mathematical - Mathematics and Natural Philosophy Natural Science -Chemistry, Exp. Comparative Anat. Physiol., Logic and Mineralogy | Latin or Greek Lang. and Literature, Mathematics French or German Lang. and Lit. Ment. Phil. & Logic, Phys. Geog. and | Latin or Greek Lang. and Lit. Mathematics, French or German Lang. and Lit. Ment. Phil. & Logic | One of: Latin, Greek, English, French, German Two of a second or third language, Math., Physics, Chem., Geology, Mental Philosophy |
| Notes: | Any BA candidate who had obtained a 2nd class place in 2nd year was then exempted from that one subject. in that school only; provided no student might be exempted from Classics or Math. unless having obtained a 2nd class place in second year exams. | A new by-law (25.3. 1874 no. 42) stated candidates proficient in any 1 school in 2nd year could be examined | A 2nd class place in Classics and/or Math. in 2nd year then exempted the candidate from that subject. | Students had to attend lectures in all three schools in first and second years; in two schools in third year. take just that subject (at Honours) to pass. | BA students had to pass a, b, and c only. Those on I, II or III Honours list in Class. or Math. could then | | |

450

Archibald Liversidge

## II. THE SCIENCE CURRICULUM

| YEAR | 1883-1885 | 1886-1889 | 1890-1904 | 1905-1912 |
|---|---|---|---|---|
| I | Latin, Mathematics, Elementary Chemistry and elements of Natural Philosophy. One of: Greek, French or German | BSc candidates had to complete 1st year Arts and have passed Matric German and French in 1st year; or vice versa, Latin, Math., Elem. Chem. and elements of Nat. Phil., One of: Greek, French or German | Biology, Chemistry, Mathematics, Physics, Physiography | Chemistry, Physics, Two of: Biology, Geology, Mathematics |
| II | Chemistry, Physics, Natural History, Mathematics, French and German (unless exempted) | Chemistry, Physics, Physical Geography, Geology, Zoology and Botany, Mathematics | Three of: Botany and Zoology, Chemistry, Geology, Mathematics, Physics, Physiology | Three of: Biology, Chemistry, Geology, Mathematics, Physics, Physiology |
| III | Three of: Chemistry, Physics, Mathematics, Mineralogy, Geology and Palaeontology, Zoology and Botany | Any two of: Chemistry, Physics, Mathematics or the three subjects: Mineralogy, Geology and Palaeontology, Botany and Zoology, Comparative Anatomy and Physiology | Two of: Biology, Chemistry, Geology, Mathematics, Physics, Physiology | Two of: Biology, Chemistry, Geology, Mathematics, Physics, Physiology |
| Notes: | For science subjects taught before 1883, see Arts Curriculum. | French and German were removed from the 2nd year curriculum. Honours could be done in one or more subjects in 2nd and 3rd years. | 1st year Arts was no longer compulsory for BSc candidates. | |

# APPENDIX VI

# LIVERSIDGE LECTURESHIPS AND SCHOLARSHIPS

## 1. LIVERSIDGE LECTURESHIPS

In his will, proved on 30 November 1927, Liversidge left legacies for Research Lectures to be given under the auspices of five different institutions. These included a gift of £500 to Christ's College, Cambridge, for the appointment of a Liversidge Lecturer, alternately in chemistry and 'on the English language with special reference to possible improvements in its grammar and spelling'; and four legacies of £500, one each to the Chemical Society of London (now the Royal Society of Chemistry), the Royal Society of New South Wales, the Australasian Association for the Advancement of Science (the AAAS, later ANZAAS), and the Chemistry Department (now the School of Chemistry) of the University of Sydney. The deed of gift specified that the lecturer 'shall not deal with generalities or give a mere review of his subject, or give an instructional lecture suitable for undergraduates, but shall primarily try to encourage research and to stimulate himself and his audience to think, and to acquire new knowledge'.

In 1928, by agreement between the Royal Society of NSW, the AAAS and the University of Sydney, it was decided that the Royal Society and the University should appoint a lecturer in alternate years (beginning with the University in 1930), and that the AAAS (later ANZAAS) should have a Liversidge Lecturer at each of its congresses, beginning in 1930. This tradition has continued at the University and at the Royal Society, and at ANZAAS until recently. At Sydney University, the series had a slow start; by 1945, only three lecturers had been appointed; but after the Second World War, the tempo quickened, and since 1975, the lecture has become a regular event of the University's Chemical Society. At the Royal Chemical Society in London, the most recent Liversidge Lecture was delivered in 2007; and at Christ's College, Cambridge, in 2000.

Many Liversidge Lecturers have been, or have become, internationally recognised. They include at least twelve Nobel Prize winners, and many Fellows of the Royal Society of London and the US National Academy of Sciences, as well as several Fellows of the Australian Academy of Science. A list of the lecturers and their titles has been prepared and donated to the Archives of the University of Sydney. Details have been generously supplied by the staff of the Royal Society of New South Wales; Dr Robert Perrin, currently Secretary of ANZAAS; the staff of the Royal Society of Chemistry (London); Professor Noel Hush, FRS, FAA; Professor Ron Clarke, Dr Graham Holland, Dr Jim Eckart, and Dr Chiara Neto of the School of Chemistry, University of Sydney; Mr Colin Higgins of Christ's College, Cambridge, and by many of the lecturers themselves. A project to reprint the chemical lectures is currently underway, under the editorship of Dr David Collins of Monash University, and will be published by Sydney University Press.

## 2. LIVERSIDGE SCHOLARSHIPS

Liversidge also left two legacies – of £1000 each – to Christ's College, Cambridge, and the Royal School of Mines (now incorporated in Imperial College London); and a further £2000 to the University of Sydney, to endow scholarships 'for proficiency in Chemistry (with a sufficient knowledge of physics)'. These scholarships were to be awarded to students just entering university, and as such were 'intended chiefly to encourage the teaching of science in schools'.

Between 1928 and 1967, the scholarships awarded by the University of Sydney went to the two candidates gaining the highest marks in the NSW School Leaving Certificate examinations. When the Higher School Certificate (HSC) was instituted, this condition was altered. Today, Liversidge Scholarships go to the two candidates receiving the highest marks for Two-Unit Chemistry in the HSC, who also present for physics, and who enter chemistry in their first year. In London, Imperial College awards its Liversidge Scholarships on the basis of proficiency in A-level chemistry (with performance in physics also taken into account) to students entering first year, provided they have not spent more than one year at another university.

# Notes

## Introduction

1 See Ann Moyal, *'A Bright and Savage Land': Scientists in Colonial Australia* (Sydney: Collins, 1986).

2 For the signficance of technology to the expansion of Europe, the *locus classicus* remains Carlo Cipolla's *Guns and Sails in the Early Phase of European Expansion, 1400-1700* (London: Collins, 1965). For the broader context occupied by science in the history of 'Europe overseas', see, among his many works, J.H. Parry, *The Age of Reconnaissance: Discovery, Exploration and Settlement, 1450-1650* (Berkeley: University of California, 1982); also see Peggy Liss, *Atlantic Empires: The Network of Trade and Revolution, 1713-1826* (Baltimore: Johns Hopkins University Press, 1983). Cf. Daniel Headrick, *The Tools of Empire: Technology and European Imperialism in the Nineteenth Century* (New York: Oxford University Press, 1981).

3 For an introduction to the vast North American literature, see R. Stearns, *Science in the British Colonies of North America* (Urbana: University of Illinois Press, 1970); George N. Daniels (ed.), *Nineteenth-Century American Science: A Reappraisal* (Evanston: Northwestern University Press, 1972); Alexandra Oleson and S. Brown (eds.), *The Pursuit of Knowledge in the Early American Republic* (Baltimore: Johns Hopkins University Press, 1976); Nathan Reingold (ed.), *Science in America since 1820* (New York: Science History Publications, 1976), and the recent survey by Robert V. Bruce, *The Launching of American Science, 1846-76* (Ithaca: Cornell University Press, 1987). For Canada, see Trevor Levere and Richard Jarrell (eds.), *A Curious Field Book: Science and Society in Canadian History* (Toronto: Oxford University Press, 1974); B. Sinclair, N.R. Ball and J.O. Petersen (eds.), *Let us be Honest and Modest: Technology and Society in Canadian History* (Toronto: Oxford University Press, 1974).

4 Cf. David Mackay, *In the Wake of Cook: Exploration, Science and Empire, 1780-1801* (Wellington: Victoria University Press, 1985); Nathan Reingold and Marc Rothenberg (eds.), *Scientific Colonialism: A Cross-Cultural Comparison* (Washington, DC: Smithsonian Institution Press, 1987); Richard Grove (ed.), *Imperialism and Conservation* (Cambridge: Cambridge University Press, 1988) and John MacKenzie (ed.), *Imperialism and the Natural World* (Manchester: Manchester University Press, 1989). The world of German imperial influence has found a measure of *Entschlossenheit* in the work of Lewis Pyenson. See his trilogy, *Cultural Imperialism and Exact Sciences: German Expansion Overseas, 1900-1930* (New York: Peter Lang, 1985), *Empire of Reason: Exact Sciences in Indonesia, 1840-1940* (Leiden: E.J. Brill, 1989), and *Civilizing Mission: Exact Sciences and French Overseas Expansion, 1830-1940* (Baltimore: Johns Hopkins University Press, 1993).

5 See the 'Introductions' in both Rod Home (ed.), *Australian Science in the Making* (Sydney: Cambridge University Press, 1988); and Roy MacLeod (ed.), *The Commonwealth of Science: ANZAAS and the Scientific Enterprise in Australasia, 1888-1988* (Melbourne: Oxford University Press, 1988).

6 This argument is by no means restricted to Australia; a similar argument can be made for all settler colonies; and separate but analogous arguments for the settler, plantation and crown colonies of the British Empire, as well as for the colonies of the rival European powers. See Roy MacLeod (ed.), *Nature and Empire: Science and the Colonial Enterprise, Osiris,* vol. 15 (Chicago: University of Chicago Press, 2001).

7 The cultural history of Australian science owes much to the pioneering work of an American, George Nadel, *Australia's Colonial Culture: Ideas, Men and Institutions in mid-Nineteenth Century Australia* (Cambridge, Mass.: Harvard University Press, 1957). Since then, Australian cultural historians have offered many insights into the history of science. See Michael Roe, *Quest for Authority in Eastern Australia, 1835-51* (Melbourne: Melbourne University Press, 1965). For an early appreciation, see Stephen Graubard (ed.), 'Preface' in 'Australia: *Terra Incognita?*', *Daedalus: Journal of the American Academy of Arts and Sciences,* 114 (1), (1985), v-xii.

8    Alfred W. Crosby, Jr., *The Columbian Exchange: Biological and Cultural Consequences of 1492* (Westport, Conn.: Greenwood Publishing Co., 1972). See also his *Ecological Imperialism: The Biological Expansion of Europe, 900-1900* (New York: Cambridge University Press, 1986).

9    This is the enduring message of J.R. Ravetz's pioneering *Scientific Knowledge and its Social Problems* (Oxford: Blackwell, 1971).

10    William H. Goetzmann, 'Paradigm Lost', in Nathan Reingold (ed.), *The Sciences in the American Context: New Perspectives* (Washington, DC: Smithsonian Institution Press, 1979), 22-25.

11    See André Gunder Frank,'The Development of Underdevelopment', *Monthly Review,* 18 (1966), 17-31; Michael Barratt Brown, *After Imperialism* (New York: Humanities Press, 1970). See also Norman Clark, 'Science, Technology and Development' in Roy MacLeod (ed.), *Technology and the Human Prospect* (London: Frances Pinter, 1982) and Clark, *The Political Economy of Science and Technology* (Oxford: Basil Blackwell, 1985).

12    For a preliminary outline of this argument, see Roy MacLeod, 'On Visiting the "Moving Metropolis": Reflections on the Architecture of Imperial Science', *Historical Records of Australian Science,* 5 (3), (1982), 1-16, reprinted in Reingold and Rothenberg (eds.), *op.cit.,* 217-249.

13    J.A. Hobson, *Imperialism: A Study* (New York: J. Pott and Co., 1902); V. I. Lenin, *Imperialism: The Highest Stage of Capitalism. A Popular Outline* (New York: International Publishers, 1939), usefully surveyed in Norman Etherington, *Theories of Imperialism: War, Conquest, and Capital* (London: Croom Helm, 1984).

14    Donald Denoon, *Settler Capitalism: The Dynamics of Dependent Development in the Southern Hemisphere* (New York: Oxford University Press, 1983).

15    Warwick Armstrong 'Land, Class, Colonialism: The Origins of Dominion Capitalism', in W.E. Wilmott (ed.), *New Zealand and the World: Essays in Honour of Wolfgang Rosenberg* (Christchurch: University of Canterbury, 1980), 28-44.

16    Cf. Ian Inkster, 'Scientific Enterprise and the Colonial "Model": Observations on Australian Experience in Historical Context', *Social Studies of Science,* 15 (4), (1985), 677-704.

17    See Ian Inkster 'Support for the Scientific Enterprise 1850 - 1900', in Home (ed.), *op.cit.,* 102-132.

18    Donald Fleming, 'Science in Australia, Canada and the United States: Some Comparative Remarks', *Proceedings of the 10th International Congress of the History of Science,* I (Paris: Hermann, 1962), 179-196.

19    Daniel J. Boorstin, *Hidden History: Exploring our Secret Past* (New York: Harper and Row, 1987), 72-75.

20    Roger G. Kennedy, *Men on the Moving Frontier* (Palo Alto: American West Publication Co., 1969), 2, 5.

21    See N. D. Harper, 'Frontier and Section: A Turner "Myth"', *Historical Studies,* 5, No. 18 (1952), 135-153; 'The Rural and Urban Frontiers', *Historical Studies,* 10, No. 40 (1963), 421.

22    For New Zealand, see Michael Hoare and L.G. Bell (eds.), 'In Search of New Zealand's Scientific Heritage', *Bulletin of the Royal Society of New Zealand,* 21 (Wellington: Royal Society of New Zealand, 1984); Charles A. Fleming, 'Science, Settlers and Scholars: The Centennial History of the Royal Society of New Zealand', *Bulletin of the Royal Society of New Zealand,* 25 (Wellington: Royal Society of New Zealand, 1987).

23    See, for example, Marie Boas Hall, *All Scientists Now* (Cambridge: Cambridge University Press, 1984); and Peter Alter, *The Reluctant Patron: Science and the State in Britain, 1850-1920* (Oxford: Berg, 1987).

24    A.P. Thornton, *The Imperial Idea and its Enemies* (London: Macmillan, 1959); see also his *Doctrines of Imperialism* (New York: John Wiley, 1965); for a helpful linguistic exegesis, see Richard Koebner and Helmut D. Schmidt, *Imperialism: The Story and Significance of a Political Word, 1840-1960* (Cambridge: Cambridge University Press, 1964).

25    For comparative historical scholarship, seeking bridges between the history of science, imperial history and the history of economic development, see Daniel Headrick, *Tentacles of Progress: Technology Transfer in the Age of Imperialism* (New York: Oxford University Press, 1988). See also Roy MacLeod and Milton Lewis (eds.), *Disease, Medicine and Empire: Perspectives on Western Medicine and the Experience of European Expansion* (London: Routledge, 1988); Roy MacLeod and Phillip F. Rehbock (eds.), '*Nature in its Greatest Extent': Western Science in the Pacific* (Honolulu: University of Hawaii Press, 1988); and the suggestive cases that inform Philip Curtin, *Death by Migration: Europe's Encounter with the Tropics* (New York: Cambridge University Press, 1989).

26    A most interesting example, in the case of museums, is afforded by Susan Sheets-Pyenson, *Cathedrals of Science: The Development of Colonial Natural History Museums during the Late Nineteenth Century (*Kingston: McGill-Queen's University Press, 1988), chapters 4 and 5.

27 Lucile Brockway, *Science and Colonial Expansion: The Role of the British Royal Botanic Gardens* (New York: Academic Press, 1979); Richard H. Drayton, *Nature's Government: Science, Imperial Britain and the Modern World* (New Haven: Yale University Press, 2000); see also Roy MacLeod, 'The Ayrton Incident: A Commentary on the Relations of Science and Government in England, 1870-1873', in A. Thackray and E. Mendelsohn (eds.), *Science and Values* (New York: Humanities Press, 1974), 45-78;

28 See Wilfrid Airy, *The Autobiography of Sir George Biddell Airy* (Cambridge: Cambridge University Press, 1896); W. F. Cannon, 'Scientists and Broad Churchmen: An Early Victorian Intellectual Network', *The Journal of British Studies*, 4 (1), (1964), 65-88; James Secord, 'King of Siluria: Roderick Murchison and the Imperial Theme in Nineteenth-Century British Geology', *Victorian Studies*, 25 (4), (1982), 413-442; Robert Stafford, *Scientist of Empire: Sir Roderick Murchison, Scientific Exploration and Victorian Imperialism* (Cambridge: Cambridge University Press, 1989).

29 Trevor H. Levere, 'The History of Science of Canada', *British Journal for the History of Science*, 21 (4), (1988), 419-420. See also Carl Berger, in *The Sense of Power: Studies in the Ideas of Canadian Imperialism, 1867-1914* (Toronto: University of Toronto Press, 1970).

30 George Arnold Wood, *The Discovery of Australia* (London: Macmillan, 1922).

31 J. Steel Robertson, 'Natural Science in Australia', *Centennial Magazine*, 2 (1889-1890), 523-527.

32 Indeed, Linnaeus proposed that the Great South Land, still known as New Holland, should be renamed 'Banksia'. See Moyal, *op.cit.*, 22; J.H. Maiden, *Sir Joseph Banks, The Father of Australia* (Sydney: Gullick Press, 1909).

33 Ernest Scott, 'The History of Australian Science', *Report of the Australasian Association for the Advancement of Science (ANZAAS)*, 24 (Canberra, 1939), 1-16.

34 CF. Alan L. McLeod (ed.), *The Pattern of Australian Culture* (Ithaca: Cornell University Press, 1963).

35 C.M.H. Clark, *A History of Australia* (Melbourne: Melbourne University Press, 1978 and 1981), vols. IV and V.

36 Hartley Grattan, *Introducing Australia* (Sydney: Halstead Press, 1944); George H. Nadel, *Australian Colonial Culture: Ideas, Men and Institutions in mid-Nineteenth Century Eastern Australia* (Melbourne: F.W. Cheshire, 1957).

37 See Ann Moyal, 'The History of Australian Science', *Historical Studies, Australia and New Zealand*, 11, (1963), 258-259; Michael Hoare, 'Science and Scientific Associations in Eastern Australia, 1820-1890' (Unpublished Ph.D. dissertation, Australian National University, 1974); Hoare, 'Light in Our Past: Australian Science in Retrospect', *Search*, 6 (7), (1975), 285-290; T.G. Vallance, 'Origins of Australian Geology', *Proc. Linn. Soc. NSW*, 100 (1975), 13-43; T.G.Vallance and D. F. Branagan, 'New South Wales Geology - Its Origin and Growth', in A.P. Elkin (ed.), *The Royal Society of New South Wales: A Century of Scientific Progress* (Sydney: Royal Society of New South Wales, 1968), 265-279; Branagan, 'Words, Actions, People: 150 Years of Scientific Societies in Australia', *Proc. Roy. Soc. NSW*, 103 (1971), 123-141; Branagan (ed.), *Rocks-Fossils-Profs: Geological Sciences in the University of Sydney, 1866-1973* (Sydney: Science Press, 1973).

38 Price Warung, 'Rev. W.B. Clarke, MA, FRS, The Nestor of Australian Philosophers', *Cosmos*, no. 11 (31 July 1895), 533-539.

39 The model takes familiar form in the writing of A.P. Elkin (ed.), *A Goodly Heritage: Science in New South Wales* (Sydney: ANZAAS, 1962), and *A Century of Scientific Progress: The Centenary Volume of the Royal Society of New South Wales* (Sydney: Royal Society of New South Wales, 1968). It may not be coincidental that Elkin was an anthropologist. See Tigger Wise, *The Self-Made Anthropologist. A Life of A.P. Elkin* (Sydney: Allen and Unwin, 1985). However, there is an interesting analogy in the model used by Rob Pascoe to describe the history of Australian literature, distinguishing what he calls an 'external', colonial period of derivative literary conventionalism between 1788 and the 1880s from what he considers to be an 'internal period' of conscious nationalism evident from the 1890s. See Rob Pascoe, *The Manufacture of Australian History* (Melbourne: Oxford University Press, 1979).

40 See Ernest Scott, 'The History of Australian Science', *Australian Journal of Science*, 1 (1939), 116.

41 Richard E.N. Twopeny, *Town Life in Australia* (London: Elliot Stock, 1883).

42 See Peter Robinson, 'Coming of Age: the British Association in Australia, 1914', *Australian Physicist*, 17 (2), (1980), 24, and Rosaleen Love, 'The Science Show of 1914: The British Association meets in Australia', *This Australia*, 4 (1), (1984-85); 12-16.

[43]  Geoffrey Serle, *From Deserts the Prophets Come: The Creative Spirit in Australia, 1788-1972* (Melbourne: Heinemann, 1973).

[44]  For a discussion of contemporary historiography, see the 'Introduction' in MacLeod (ed.), *The Commonwealth of Science, op.cit.*, 1-16.

[45]  George Basalla, 'The Spread of Western Science', *Science,* 156 (5 May 1967), 611-622.

[46]  A refreshing introduction to the pitfalls of functionalism, in deriving explanations from cultural microcosms, and the shortcomings of evolutionism and diffusionism, in avoiding questions central to political economy, is given by Eric R. Wolf, in *Europe and the People without History* (Berkeley: University of California Press, 1982). His analysis, however, is directed to the relations between Europe and 'subject peoples', and not to the particular relations of Europe and the settler colonies.

[47]  Aptly noted by George Bindon in 'Harold A. Innes: Science and Technology at the Periphery', Conference on 'Nationalism and Internationalism in Science: Australia, America and the World of Science', University of Melbourne, 22-26 May 1988.

[48]  MacLeod, 'Moving Metropolis', *op. cit.*, 4.

[49]  See 'Cultural Imperialism and Exact Science Revisited', *Isis*, 84 (1), (1997), 103-108.

[50]  This change in perspective has been underway for some time. See, for example, Roy MacLeod and Deepak Kumar, 'First Indo-Australian Seminar in the History of Science', *History of Australian Science Newsletter*, 1988, and MacLeod and Kumar (eds.), *Technology and the Raj* (New Delhi: Oxford University Press, 1995). See also Roy MacLeod (ed.), *Nature and Empire: Science and the Colonial Experience, op.cit.*

[51]  Quoted in R.T.M. Pescott, 'The Royal Society of Victoria from Then, 1854 to Now, 1959', *Proceedings of the Royal Society of Victoria*, 73, n.s. (1959), 2.

[52]  E.C. Andrews, 'Heroic Period of Geological Work in Australia', *JPRSNSW,* 76, (1942), 96-97.

[53]  See William J. Lines, *Taming the Great South Land: A History of the Conquest of Nature in Australia* (North Sydney: Allen & Unwin, 1991).

[54]  *Daily Telegraph* (Sydney), 3 February 1898.

[55]  See Nathan Reingold, 'Reflections on 200 Years of Science in the United States', *Nature,* 262 (1976), 9-13; Reingold (ed.), *Science in Nineteenth Century America: A Documentary History* (New York: Hill and Wang, 1964).

[56]  Quoted by Gavin de Beer, *Charles Darwin* (London: Nelson, 1963), 107, and in Ann Mozley Moyal, *Scientists in Nineteenth Century Australia: A Documentary History* (Sydney: Cassell, 1976), 60.

[57]  George Seddon, 'Eurocentrism and Australian Science: Some Examples', *Search,* 12 (1981-82), 446-450.

[58]  An interesting example is Sir John Franklin's establishment of public education and the Royal Society of Van Diemen's Land. See Kathleen Fitzpatrick, *Sir John Franklin in Tasmania, 1837-1843* (Melbourne: Melbourne University Press, 1949).

[59]  This follows closely upon Fleming, *op.cit.*, 179-196.

[60]  Cf. the argument of Nathan Reingold and Arthur Molella, in 'Theorists and Ingenious Mechanics: Joseph Henry Defines Science', *Science Studies,* 3 (4), (1973), 328-351. See also Nathan Reingold, 'American Indifference to Basic Research: A Reappraisal' in George M. Daniels (ed.), *Nineteenth Century American Science: A Reappraisal* (Evanston: Northwestern University Press, 1972), 38-62.

[61]  Robert V. Bruce, *The Launching of American Science, 1846-76* (Ithaca: Cornell University Press, 1987), 87, 11.

[62]  See for example, Ann Moyal, 'Invention and Innovation in Australia: The Historian's Lens', *Prometheus,* 5 (1), (1987), 92-110.

[63]  Roy MacLeod, 'The "Practical Man": Myth and Metaphor in Anglo-Australian Science', *Australian Cultural History,* No. 8 (1989), 24-49; cf. Eric Ashby, 'Universities in Australia', *The Future of Education,* No. 5 (Sydney: Australian Council for Educational Research, 1944), 33 *et passim.*

[64]  David Denholm, *The Colonial Australians* (Harmondsworth: Penguin Books, 1979).

[65]  Cf. Ian Rae, 'Chemists at ANZAAS: Cabbages or Kings', in MacLeod (ed.), *The Commonwealth of Science, op.cit.,* 166-195.

[66]  And not always, as wartime experience later revealed, to Australia's disadvantage. See Roy MacLeod, 'The "Arsenal" in the Strand: Australian Chemists and the British Munitions Effort, 1916-1919', *Annals of Science,* 46 (1), (1989), 45-67; Rod Home, 'Science on Service, 1939-1945', in Home (ed.), 220-251.

[67]  On Brisbane, see Ann Moyal, *Scientists in Nineteenth Century Australia* (Sydney: Cassell, 1976), 33-38;

Shirley Saunders, 'Sir Thomas Brisbane's Legacy to Colonial Science: Colonial Astronomy at the Parramatta Observatory, 1822- 1848', *Historical Records of Australian Science*, 15 (2), (2004), 177-209.

[68] *Sydney Gazette,* 29 April 1824.

[69] Roe, *op.cit.,*157-158.

[70] John Woolley, 'Schools of Art and Colonial Nationality: A Lecture (1861)', in *Lectures Delivered in Australia* (Cambridge: Macmillan, 1862), 5.

[71] Alan Moorehead, *Cooper's Creek* (London: Hamish Hamilton, 1963).

[72] G.L. Fischer, *The University of Sydney, 1850-1975* (Sydney: University of Sydney Press, 1975).

[73] Nathan Reingold, 'Definitions and Speculations: The Professionalization of Science in America in the Nineteenth Century', in A. Oleson and S. Brown (eds.), *The Pursuit of Knowledge in the Early American Republic* (Baltimore: Johns Hopkins University Press, 1976), 33-70.

[74] See the *Australian Era* (vol. I, 1850), cited in Roe, *op.cit.*, 156.

[75] See M.E. Hoare, 'Learned Societies in Australia: The Foundation Years in Victoria, 1850-1860', *Records of the Australian Academy of Science*, 1 (1967), 7-29.

[76] Roy MacLeod (ed.), *University and Community in the Nineteenth Century: Professor John Smith, 1825-1885* (Sydney: University of Sydney History Project, 1988).

[77] 'Education in Science', *Sydney Magazine of Science and Art*, I (1857), 119-120.

[78] Cited in Roe, *op.cit.,* 158.

[79] Such as found in England, and ably described by Sheldon Rothblatt, *Tradition and Change in English Liberal Education* (London: Faber and Faber, 1976).

[80] For the extent to which the interests of material culture and colonial idealism shared a common perspective, see Sally Kohlstedt, 'Natural Heritage: Securing Australian Materials in 19th Century Museums', *Museums Australia* (December 1984), 15-32; and 'Australian Museums of Natural History: Public Priorities and Scientific Initiatives in the 19th Century', *Historical Records of Australian Science*, 6 (1983), 1-29.

[81] Cf. the work of William Barton Rogers at MIT, as described in Robert R. Shrock, *Geology at MIT, 1865-1965: A History of the First Hundred Years of Geology at Massachusetts Institute of Technology* (Cambridge, Mass.: MIT Press, 1977), 99-204.

[82] Brian Fitzpatrick, *The Australian People, 1788-1945* (Melbourne: Melbourne University Press, 1946), 217.

[83] Anthony Trollope, *Australia and New Zealand* (London: Chapman and Hall, 1873).

[84] J. A. Froude, *Oceana* (New York: Scribner and Son., 1887)**,** ed. Geoffrey Blainey (Sydney: Methuen, 1985).

[85] This long-established phrase has been made fashionable by Frank Turner, who defines public science as the 'body of rhetoric, argument and polemic' produced by scientists in order to 'justify their activities to the political powers and other social institutions upon whose good will, patronage and cooperation they depend.' See F.M. Turner, 'Public Science in Britain, 1880-1919', *Isis***,** 71 (1980), 589-608, at 589; see also Roy MacLeod, *Public Science and Public Policy in Victorian Britain* (Aldershot: Variorum, 1995) and *The Creed of Science in Victorian England* (London: Ashgate, 2000).

[86] For linkages with Australian history, see Michael Roe, *Nine Australian Progressives: Vitalism in Bourgeois Social Thought, 1890-1960* (St. Lucia: University of Queensland Press, 1984). For the British movement, with which Liversidge and his circle were in harmony, see Geoffrey Searle, *The Quest for National Efficiency* (Oxford: Blackwell, 1971) and *Eugenics and Politics in Britain, 1900-1914* (Leyden: Noordhoff, 1976). For progressivism and the rhetoric of efficiency in American history, see Samuel Haber, *Efficiency and Uplift: Scientific Management in the Progressive Era* (Chicago: University of Chicago Press, 1964); Samuel P. Hays, *Conservation and the Gospel of Efficiency* (Cambridge. Mass.: Harvard University Press, 1959; New York: Athenaeum, 1972) and Robert Wiebe, *The Search for Order, 1877-1920* (London: Macmillan, 1967).

[87] See Thomas Vallance and David Branagan, 'The Earth Sciences: Searching for Geological Order', in MacLeod (ed.), *The Commonwealth of Science, op.cit.,* 130-146, and J.M. Powell, 'Protracted Reconciliation: Society and the Environment', *idem.,* 249-271.

[88] Ralph Waldo Emerson,'The American Scholar', in Stephen E. Whicher (ed.), *Selections from Ralph Waldo Emerson* (Boston: Houghton Mifflin, 1957), 63-79.

[89] For the effect of the telegraph on intercolonial meteorological and astronomical cooperation, see Michael Hoare, 'The Intercolonial Science Movement in Australasia, 1870-1880', *Records of the Australian Academy of Science*, 3 (2), (1976), 7-28. See also Ann Moyal, *Clear Across Australia: A History of Telecommunications* (Melbourne: Nelson, 1984).

90   John Wade (ed.), *The Sydney International Exhibition, 1879* (Sydney: Museum of Applied Arts and Sciences, 1979); see also Graeme Davison, 'Exhibitions', *Australian Cultural History,* No. 2 (1982-83), 5-21.

91   'The Australian Association for the Advancement of Science', *Saturday Review,* 66 (3 November 1888), 519-520.

92   The Bicentennial helped greatly. See, for example, 'Australia!' a special issue of the *Mineralogical Record,* 19 (6), (1988).

93   This period is well, if briefly, described in Richard White, *Inventing Australia; Images and Identity, 1688-1980* (Sydney: George Allen and Unwin, 1981).

94   See also D.J. Mulvaney and J.H. Calaby, *'So Much that is New': Baldwin Spencer, 1860-1929, A Biography* (Melbourne: Melbourne University Press, 1985); W. L.Weickhardt, *Masson of Melbourne: The Life and Times of David Orme Masson, Professor of Chemistry, University of Melbourne, 1886-1923* (Melbourne: Royal Australian Chemical Institute, 1988).

95   The motto of the Imperial College of Science and Technology, of which the of Mines today remains a part.

96   To borrow the title of W.J. Gardner, *Colonial Cap and Gown: Studies in the Mid-Victorian Universities of Australia (*Christchurch: University of Canterbury, 1979).

# Chapter 1

1    For these early notes, I am indebted to Mr. William Liversidge, of Abingdon, Oxfordshire; to the Guildhall Library and the Southwark Public Library; and to the staff of the East Sussex County Record Office.

2    *Kelly's London Suburban Directory,* 1894.

3    The two eldest, John George and Edward William, joined the Royal Navy. Both became Engineer Rear Admirals during the First World War, and one would send two sons to the Navy in the Second World War. Eventually, both were to become executors of their uncle's estate.

4    One of William's sons became President of the Oxford Union and a successful barrister; and a grandson, a colonial servant, teacher and local historian.

5    Arthur's son, another Jarratt, was educated at St. Paul's, and became an accountant for the family firm, which in 1930 profitably merged with British Oxygen. Percy Balfern remained a director of the company until his retirement in 1937.

6    In the late 1950s, the company was merged into the Charrington Group, and twenty years later, into the Coalite Group. In 1975, their works were removed to a wartime air station at Hamble in Hampshire. In 1983, the Group shortened the company's name to 'Glover, Webb Ltd'. As late as the 1980s, the firm specialised in sophisticated refuse vans, prison vans and police vehicles. I am indebted for this information to A.F. Carlile, formerly Product Sales Manager of Glover, Webb, Ltd.

7    Around 1863, Caroline Sophia met and become engaged to John Edward Balfern, a hotel contractor and provisioner, and who went into partnership with her eldest brother, John George. In 1880, Archie's sister Jane died, and in 1881, Caroline Sophia and John moved to suburban Bexley in Kent, to a small house that was renamed 'Buxted Lodge' after Elizabeth's birthplace, on prosperous Parkhurst Road, a few minutes from the railway station.

8    William's family lived in Turrett Lodge, Victoria Park Road in South Hackney until his wife Eliza died in 1881, when he moved with the children a little north to Rowhill near Hackney Downs in Lower Clapton. In 1900, he retired to 'Flintfields', Whyteleafe, in Surrey, where he lived until his death in 1911.

9    In Parkhurst Road, Bexley, Kent. For assistance in tracing this information, I am indebted to Mr. S.A. Liversidge of Billericay, Essex, and to the Borough of Bexley Local History Museum.

10   *Annual Report, Borough of Hackney Microscopical and Natural History Society* (April 1878). Some years later, Jarratt joined him. *8th Annual Report* (1885).

11   Cf. *3rd Annual Report, Hackney Microscopical and Natural History Society* (April 1880), 15; Mary P. English, *Mordecai Cubitt Cooke: Victorian Naturalist, Mycologist, Teacher and Eccentric* (Bristol: Biopress, 1987).

12   Charles Butler, 'Inaugural Oration' (London, 1816), 30, quoted in J.N. Hays, 'Science in the City: The London Institution, 1819-40', *Brit. J. Hist. Sci.,* 7 (26), (1974), 147.

13   Sidney Webb, *London Education* (London: Longmans, Green, 1904), 133.

14   *Ibid.,* 135.

15   *The Times Higher Education Supplement,* 30 September 1988, 9; Richard D. Altick, 'London's Royal Polytechnic

Institution', *New Scientist,* 79 (1978), 36-38.

16 David Layton, *Science for the People* (London: George Allen and Unwin, 1973).

17 Winifred Abbott and Alan Walker, *The Royal Polytechnic Institution and the Polytechnic, 1838-1870* (London: The Polytechnic of Central London, 1988). The Polytechnic celebrated its sesquicentenary in 1988.

18 The Polytechnic of Central London was founded by an amalgamation of the Regent Street Polytechnic and the Holborn College of Law, Language and Commerce.

19 In 1970, the City of London College joined with Sir John Cass College and the King Edward VII Nautical College to form the City of London Polytechnic, or 'City Poly' as it is popularly known today.

20 Charles Mackenzie, 'Original Prospectus of the Metropolitan Evening Classes, December 1848', in *The Calendar of the City of London College, 1882-3,* 188.

21 *The Calendar of the City of London College, 1861-62,* 29.

22 Walter G. Bell, *A Short History of the Worshipful Company of Tylers and Bricklayers of the City of London* (London: H.G. Montgomery, 1938), 32.

23 In 1883, aided by large donations from City Companies, the City Corporation, the Bank of England and City churches, the College moved to a new building in White Street. From 1891, the London County Council made a grant towards maintenance. As it expanded (to over 4,000 students by 1939), it gradually absorbed other premises. Sussex Hall was destroyed in the Blitz in 1940, after which the College (subsequently, the Polytechnic) moved to its present location in Moorgate.

24 Rev. T.H. Bullock, quoted in L.A. Terry, *The City of London College: An Historical Account of the College from its Foundation in 1848 to the Present Day* (London: The College, 1964, revised by G. Harris, 1963), 9. For archival material, I am grateful to Ms. Maureen Castens, Library and Learning Resources Service, City of London Polytechnic.

25 B.T. Hall, *Our Sixty Years: The Story of the Working Men's Club and Institute Union* (London: Working Men's Club & Institute Union, 1922), 16; cf. Rev. Henry Solly, *Working Men's Social Clubs and Educational Institutes* (London: Simpkin, Marshall & Co., 2nd ed., 1904); Rev. Harry Jones, *East and West London* (London: Smith Elder, 1875); John Taylor, *From Self-Help to Glamour: The Workingman's Club, 1860-1972* (History Workshop Pamphlet no. 7, 1972).

26 *Journal of the Evening Classes for Young Men,* no. 9 (September 1859), 65.

27 A. Abbott, *Education for Industry and Commerce in England* (Cambridge: Cambridge University Press, 1933), 137, Appendix, 223-224. Stuart Maclure, *One Hundred Years of English Education, 1870-1970* (London: Allen Lane, 1970); Sidney Webb, *London Life* (London: Longmans, Green, 1904), 120.

28 Terry, *op.cit.,* 9.

29 *Journal of the Evening Classes for Young Men,* no. 1 (January 1859), 6.

30 For the contemporary enthusiasm for examinations and its critics, see Roy MacLeod (ed.), *Days of Judgement: Science, Examinations and the Codification of Knowledge in Victorian Britain* (Driffield: Nafferton Studies in Education, 1982).

31 *First Report of the Department of Science and Art,* 1854. (1783). xxviii. Appendix A, Sir J.E. Tennant to Treasury, 16 March 1853.

32 *The Calendar of the City of London College* (1862-63), 15.

33 *Ibid.,* 38.

34 Liversidge was to rank among the College's most famous alumni. He was the first and, until joined by J.W. Gregory in 1888, the only Fellow of the Royal Society to have been educated at the College. *The Calendar of the City of London College* (1882-83), Annual Meeting, 27 June 1882, 10; Cf. *Calendar* (1929-30), 132.

35 H. Llewellyn Smith, *Report of the Special Committee on Technical Education, London County Council Report,* No. 57 (11 July 1892), 27.

36 George Fownes, *Manual of Elementary Chemistry, Theoretical and Practical* (London: John Churchill, 1844; 14 editions by 1889); Henry Roscoe, *Lessons in Elementary Chemistry, Inorganic and Organic* (London: Macmillan, 1867, 1869): Thomas Eltoft, *A Systematic Course of Practical Qualitative Analysis* (London: Simpkin, Marshall, 1879). Probably J.A. Stöckhardt's *The Principles of Chemistry* (Cambridge, 1850) was also available. Fownes was also the author of *Chemistry, as Exemplifying the Wisdom and Beneficence of God* (London: John Churchill, 1844), a volume written in the manner of the Bridgewater Treatises of the 1830s. It is unclear whether, or if so to what extent, Liversidge's interests in chemistry were influenced by the deistic, as distinct from the pedagogical arguments advanced by his teachers on its behalf.

37  'Suggestions for the Final Examination in 1868', *Journal of the Royal Society of Arts,* XV (1866-67), 561.

38  Benjamin Kennedy of Shrewsbury, quoted in J.W. Adamson, *English Education, 1789-1902* (Cambridge: Cambridge University Press, 1930), 246, citing the *Report of the Royal Commission on the Public Schools* (the Clarendon Commission), 1864. (3288). xx, *Mins. Evid.* vol. 3, Winchester Evidence, Q. 503; xxi, vol. 4, Shrewsbury Evidence, Q. 700.

39  This speculative encounter between Liversidge and Thomson is based on inference from the remarks made by Sir Edward Deas Thomson, Chancellor of Sydney University, in his Commemoration Address on 5 April 1873, when he welcomed Liversidge to Sydney. See Mitchell Library ML A1531-4, Deas Thomson Papers, Commemoration Address, 1873.

40  F.S.M. de Carteret Bisson, *Our Schools and Colleges* (London: Simpkin, Marshall, 1872), 294, Appendix, xxi; Frederic Boase, *Modern English Biography* (London: Frank Cass, 1965), vol. 5, 403-404; Frank Miles, *King's College School: Alumni, 1831-66* (unpublished). I am grateful to Ms. Patricia Methven, King's College Archivist, for helping me with details of Gibsone's life.

41  B.W. Gibsone, 'Description of an Apparatus for Preventing the Escape of Sulphuretted Hydrogen', *Journal of the Chemical Society,* 20 (1867), 415-417.

42  Adamson, *op.cit.,* 246; *Clarendon Commission, Mins. Evid.* vol. IV, Q. 40.

43  *Fourteenth Report, Department of Science and Art,* 1867. (3853). xxiii. 50.

# CHAPTER 2

1  Cf. J.S. Flett, *The First Hundred Years of the Geological Survey of Great Britain* (London: HMSO, 1937). De la Beche, one of the most important men in English science in the second quarter of the nineteenth century, awaits a full biography. But see P.J. McCartney, *Henry De la Beche: Observations on an Observer* (Cardiff: Friends of the National Museum of Wales, 1977).

2  James Secord, 'The Geological Survey of Great Britain as a Research School, 1839-1855', *History of Science,* XXIV (1986), 223-275.

3  Prince Albert in *The Times,* 13 May 1851, quoted in Theodore Chambers, *Register of the Associates and Old Students of the Royal College of Chemistry, the Royal School of Mines and the Royal College of Science* (London: Hazell, Watson and Viney, 1896), xi.

4  See Martin Guntau, 'The Mining Academy of Freiberg - A Centre of Geoscientific Teaching and Research', *Journal of Mines, Metals and Fuels,* 22 (1974), 223-227; Louis Aguillon, 'L'École des Mines de Paris: Notice Historique', *Annales des Mines,* No. 3 (1889), 433-686; Gabriel Chesneau, *L'École des Mines* (Paris: Association Amicale des Anciens Élèves de l'Ecole Nationale Supérieure des Mines, 1931); André Thépot, *Les Ingénieurs des Mines du XIXème Siècle: Histoire d'un Corps Technique d'Etat, tome 1: 1810-1914* (Paris: Editions ESKA, 1998).

5  The published volume (London: David Bogue) appeared in 1852. The other lecturers included Edward Forbes, on Australian rocks and fossils; Lyon Playfair, on gold chemistry; W.W. Smyth, on the dressing of gold ores; John Percy on assaying; and Robert Hunt, on the statistics of gold mining. As none of these men had ever been to Australia, the book is a tribute to imperial self-confidence. See Robert A. Stafford, 'Preventing the "Curse of California": Advice for English Emigrants to the Australian Goldfields', *Historical Records of Australian Science,* 7 (3), (1989), 215-231.

6  Cf. Robert A. Stafford, 'Geological Surveys, Mineral Discoveries and British Expansion, 1835-71', *Journal of Imperial and Commonwealth History,* XII (1984), 5-32.

7  Chambers, *op.cit.,* xxx.

8  *First Prospectus of the Government School of Mines and of Science Applied to the Arts,* 1852, quoted in Chambers, *op.cit.,* xv.

9  Cf. Sir Thomas Wemyss Reid, *Memoirs and Correspondence of Lyon Playfair, First Lord Playfair of St. Andrews* (London: Cassell and Co., 1899), 49-52.

10  See Gerrylynn K. Roberts, 'The Establishment of the Royal College of Chemistry: An Investigation of the Social Context of Early-Victorian Chemistry', *Historical Studies in the Physical Sciences,* 7 (1976), 437-485.

11  Lyon Playfair, 'Hofmann Memorial Lecture', Royal College of Chemistry, May 1893, quoted in Reid, *op.cit.,* 52.

12  J. Bentley, 'The Chemical Department of the Royal School of Mines: Its Origins and Development under

A.W. Hofmann', *Ambix,* 17 (1970), l53-81.

13  Chambers, *op.cit.,* xiv-xvii.

14  Archibald Geikie, *Life of Sir Roderick I. Murchison* (London: John Murray, 2 vols., 1875); James Secord, 'The King of Siluria: Roderick Murchison and the Imperial Theme in Nineteenth Century British Geology', *Victorian Studies,* 25 (4), (1982), 413-442; Robert Stafford, 'Roderick Murchison and the Structure of Africa: A Geological Prediction and its Consequences for British Expansion', *Annals of Science,* 45 (1), (1988), 1-40; Stafford, *Scientist of the Empire: Sir Roderick Murchison, Scientific Exploration and Victorian Imperialism* (Cambridge: Cambridge University Press, 1989).

15  *Tenth Report of the Department of Science and Art,* 1863. (3143). xvi. Appendix V, 193-194.

16  Between 1857-1861, the institution reverted to its previous title of 'Government [*sic*] School of Mines', and in 1863 became the Royal School of Mines. In 1881, under the leadership of its Dean, T.H. Huxley, the School was restructured, and biology, chemistry, and physics went to the renamed college, the 'Normal School of Science', while the Royal School of Mines retained its title, identity, and the mining disciplines. In 1890, the title of the Normal School of Science was altered to the 'Royal College of Science'. In 1907, both the RCS and the RSM became constituent colleges of the new Imperial College of Science and Technology. The ASRM was (and is still) given to those who specialise in the division of Mining and Metallurgy; the title 'Associate of the Royal College of Science' was given to others. Chambers, *op.cit.,* xli; H.E. Eme léus, 'Early History of the Royal School of Mines and the Royal College of Science, 1837-1857', *The Record,* ser. III., no. 5 (1937), 59-62, and 'Early Development of the Royal College of Science at South Kensington', *The Record,* ser. III, no. 6 (1937), 80-82. See also Gerald J. Whitrow, *Centenary of the Huxley Building: Personalities and Events in the History of Imperial College* (London: Imperial College of Science and Technology, 1972), A. Rupert Hall, *Science for Industry: A Short History of the Imperial College of Science and Technology and its Antecedents* (London: Imperial College, 1982), and Hannah Gay, *The History of Imperial College London, 1907-2007: Higher Education and Research in Science, Technology and Medicine* (London: Imperial College Press, 2007).

17  Department of Science and Art, 'Geological Survey of the United Kingdom: Museum of Practical Geology and Royal School of Mines,' *Report of the Seventeenth Session,* 1867-68, 4. .

18  Calculations from Margaret Reeks, *Register of the Associates and Old Students of the Royal School of Mines and History of the Royal School of Mines* (London: Royal School of Mines (Old Students' Association, 1920), 1-212.

19  'Editorial', *The Royal School of Mines Magazine: A Journal of the Students,* I (1877), i. Indicatively, the first issues of this short-lived magazine for 'old miners' carried articles on 'A Geologist's Life in Kashmir' and the eclipse expedition to Siam.

20  Such was Sir William Denison, RE, who became Governor of NSW. See Denison, *Varieties of Vice-Regal Life* (London: Longmans, Green & Co., 1870). For the Royal Engineers, see Whitworth Porter, *History of the Corps of Royal Engineers* (London: Longmans Green & Co, 1889; 2nd ed. Chatham: Institution of Royal Engineers, 9 vols., 1951-58).

21  *First Report of the Royal Commission on Scientific Instruction and the Advancement of Science* (afterwards, *Devonshire Commission*) 1871, *Mins. Evid.,* Q. 5793.

22  Roy MacLeod, '"Instructed Men" and Mining Engineers: The Associates of the Royal School of Mines and British Imperial Science, 1851-1920', *Minerva,* XXXII (4), (1995), 422-439.

23  *Report of the Select Committee on Scientific Instruction,* 1867-68 (432). xv. *Mins. Evid.* Q. 7958.

24  J. Vargas Eyre, *Henry Edward Armstrong, 1848-1937: the Doyen of British Chemists and Pioneer of Technical Education* (London: Butterworths Scientific Publications, 1958); G.Van Praagh (ed.), *H.E. Armstrong and Science Education* (London: John Murray, 1973); W.H. Brock (ed.), *H.E. Armstrong and the Teaching of Science, 1880-1930* (Cambridge: Cambridge University Press, 1973).

25  Roy MacLeod, 'Sir Alexander Pedler (1849-1918)', *Oxford Dictionary of National Biography,* vol. 34, 390-391.

26  Cf. Martin Wiener's compelling, if contentious, *English Culture and the Decline of the Industrial Spirit, 1850-1980* (London: Macmillan, 1981).

27  *Report of HM Commissioners for the Universal Exhibition of Works of Industry, Agriculture and Fine Art, held at Paris in the Year 1867,* 1868-69. (4195). xxiii, 15. See Eric Ashby's enduring *Technology and the Academics: An Essay on Universities and the Scientific Revolution* (London: Macmillan, 1958).

28  In his widely-shared view, Britain had shown 'little inventiveness and [had] made but little progress in the

peaceful arts of industry since 1862'. *Report relative to Technical Education,* 1867. (3898). xxvi, 6.

29 'The Paris Exhibition and Industrial Education', *Journal of the Society of Arts,* XV (7 June 1867), 477-78; *Correspondence with the Schools Inquiry Commission on Technical Education,* 1867-68. xxviii. Pt.I.

30 *Archives of the British Association for the Advancement of Science, Report of Council to the General Committee,* Dundee, 4 September 1867. In his presidential address to Section D at Nottingham in 1866, Huxley complained of the neglect of science in the schools, attributing this to the lack of encouragement afforded by Oxford and Cambridge. Farrar, speaking at the same meeting, recommended a committee of investigation. In February 1868, this committee's report appeared as a parliamentary paper. See *Report of the Committee appointed by the Council of the British Association for the Advancement of Science to consider the promotion of Scientific Education, 1867-68.* (137). liv. l. See John W. Adamson, *English Education, 1789-1902* (Cambridge: Cambridge University Press, 1930), ch. 4, 'Disseminating a Knowledge of Science', esp. 383-394.

31 'Industrial Progress and the Education of the Industrial Classes in France, Switzerland, Germany, etc.,' *Letter from Bernhard Samuelson, MP, to the Vice President of the Committee of Council on Education concerning Technical Education in Various Countries Abroad.* 1867-68. (13). liv. 67.

32 *Report of the Select Committee on Provisions for Giving Instruction in Theoretical and Applied Science to Industrial Classes* (commonly, the Select Committee on Scientific Instruction; and hereafter, the *Samuelson Committee*), 1867-68. (432). xv. l. This began in March and reported in July 1868. For context, see Roy MacLeod, 'Scientific and Technical Education in the Nineteenth Century', in Gillian Sutherland (ed.), *Critical Commentaries on Education: British Parliamentary Papers* (Dublin: Irish Academic Press, 1977).

33 Letter from B. Samuelson, Esq., MP, to the Vice-President of the Committee of Council on Education concerning Technical Education in various Countries Abroad'. 1867-68 (13). 26 November 1867, 57.

34 *Minutes of the Lords of the Committee of Council on Education relating to Scientific Education,* 1867-68. (193). liv. 17.

35 For course marks, see *Archives, Imperial College,* Examinations' Book, 1866-70.

36 *Fifteenth Report of the Department of Science and Art,* 1867-68. (4049). xxvii. Appendix A, 5.

37 In May 1868, 6,875 pupils sat the exams; in 1869, 12,988. The 'examination revolution' had begun. Cf. *Seventeenth Report of the Department of Science and Art,* 1870 (C.174). xxvi. 45.

38 *Fourteenth Report of the Department of Science and Art,* 1867. (3853), xxiii, 50; *Fifteenth Report,* 1867-68. (4049). xxvii. Appendix B, 56, 64.

39 *Tenth Report of the Department of Science and Art,* 1863. (3143). xvi. Appendix V, 189. These are not to be confused (as several of Liversidge's obituarists have done) with the Royal Scholarships, which were given on the basis of aggregate marks in examination results to students who had already attended either Government school for at least one year. In Liversidge's second year (1868), they were won by Bickerton, Sollas and Gordon Broome (later Professor of Mineralogy at Toronto). See Chambers, *op.cit.,* 20l.

40 Cf. *Report of the Samuelson Committee, Mins. Evid.* Q. 401.

41 Until he was encouraged by a teacher in Painswick, appointed under the Science and Art Department, Alexander Bickerton was 'unaware of any special aptitude for science'. However, by excelling at examinations, he won the attention of the RSM's staff. A brilliant career, which included several school appointments, began soon after leaving London, but ended in disappointment at Canterbury, where he felt his 'proven capacity as an organiser in Science teaching was never utilised'. See Alexander Turnbull Library (Wellington), MS 259, f. 90, Committee of Enquiry into Canterbury College, New Zealand University, December 1894, 2. His colonial experience presents an interesting contrast to that of Liversidge.

42 W.J. Sollas, FRS became a distinguished geologist, and a leading authority on Palaeolithic man – see his *Ancient Hunters and their Modern Representatives* (London: Macmillan, 1911). See also E.A. Vincent, 'W.J. Sollas (1849-1936)', *Oxford Dictionary of National Biography,* vol. 51, 534.

43 Roy MacLeod, 'Imperial Reflections in the Southern Seas: The Funafuti Expeditions, 1898-1904', in Roy MacLeod and P.F. Rehbock (eds.), '*Nature in its Greatest Extent': Western Science in the Pacific* (Honolulu: University of Hawaii Press, 1988), 159-194. Se also 'Tavita' [T.W. Edgeworth David].'The Funafuti Expedition', *Hermes,* III (6), (30 November 1897), 8-11; IV (5), (29 October 1898), 8-11.

44 Unsuitable labs were not unique to London. Henry Roscoe describes his life as a student in Heidelberg in 1853, where the chemical laboratory 'had been an old monastery' and where students 'used spirit lamps, drew ... [their] water from the pump, and threw down [their] useless precipitates on the tombstones under [their] feet'. Henry E. Roscoe, *The Life and Experiences of Sir Henry Enfield Roscoe* (London: Macmillan,

1906), 47-48.

45 A.A.E., 'The Royal College of Chemistry in the Seventies', *The Record*, III series (1936), 27.

46 *First Report of the Normal School of Science* (1881-82), 1.

47 W.J. Sollas, 'The Master', *Nature*, 115 (9 May 1925), 747-748. Sollas recalled the comment ten years later. *Sollas Papers (Imperial College Archives)*, W.J. Sollas to W. Watts, 13 June 1935.

48 *First Report, Devonshire Commission, 1871. Mins. Evid.* Q. 5830.

49 The three were Huxley, Tyndall and Frankland. See Roy MacLeod, 'The X-Club: A Scientific Network in Late-Victorian England', *Notes and Records of the Royal Society*, 24 (2), (1970), 305-322; Ruth Barton, '"An Influential Set of Chaps": The X-Club and Royal Society Politics, 1864-1885', *The British Journal for the History of Science*, 23 (1), (1990), 53-81.

50 In the previous decade, the Royal College of Science for Ireland (earlier known as the Dublin School of Mines) had outpaced the Royal School of Mines in its range of subjects, including mathematics and the sciences relevant to agriculture. A historical comparison of the two institutions is overdue.

51 A[rchibald] L[iversidge], 'The London School of Mines', *Sydney Morning Herald*, 11 December 1872.

52 *Ibid.*

53 Armstrong remembered the mechanical drawing course as 'so lacking in feeling and object that the subject never appealed to my dull intelligence'. H.E. Armstrong, 'Pre-Kensington History of the Royal College of Science and the University Problem', *Address to the Old Students' Association* (London: Royal College of Science, September, 1920), 6. Liversidge agreed, and recorded in this subject the lowest marks – 55 – of his academic career. (*Imperial College Archives*, 'Examination Books, 1866-1870').

54 For Frankland's life, see M.N.W. and S.J.C., *Sketches from the Life of Edward Frankland* (London: Spottiswoode and Co., 1902). This has little material concerning Frankland's teaching, or his influence upon students. For this, see H.E. Armstrong, 'First Frankland Memorial Oration of the Lancastrian Frankland Society', *Chemistry and Industry*, 53 (1934), 459-466; and more recently (and perhaps definitively) by Colin Russell, *Edward Frankland: Chemistry, Controversy and Conspiracy in Victorian England* (Cambridge: Cambridge University Press, 1996).

55 Frankland Archives (Private), Frankland to H. Bence Jones, 10 June 1862, quoted in Russell, *op. cit.*, 218.

56 Robert Bud and Gerrylynn Roberts, *Science versus Practice: Chemistry in Victorian Britain* (Manchester: Manchester University Press, 1984), 150.

57 W.H. Brock, 'Prologue to Heurism', in M.J. Seabourne (ed.), *The Changing Curriculum* (London: Methuen, for the History of Education Society, 1971), 71-85. A British Association committee in 1867 drew an heuristic distinction between scientific *information* (which included natural history, physiology and geology), and scientific *training*, which included experimental physics, chemistry and botany. In this context, chemistry was 'remarkable for the comprehensive character of the training which it affords. Not only does it exercise the memory and the reasoning powers, but it also teaches the student to gather by his own experiments and observations the facts upon which to reason ... it leads to general ideas and conclusions, [while] it checks over-confidence in mere reasoning, and shows the way in which valid extensions of our ideas grow out of a series of more and more rational and accurate observations of external nature'. *Report of a Committee appointed by the Council of the British Association for the Advancement of Science to consider the Best Means for Promoting Scientific Education in Schools*, 1867-68. (137). liv. l.

58 *Report relative to Technical Education*, 1867. (3898). xxvi. 12-13.

59 J.F. Donnelly, 'Representations of Applied Science: Academics and Chemical Industry in late Nineteenth-Century England', *Social Studies of Science*, 16 (1986), 195-234.

60 A table of Frankland's lecture titles is valuably set out in Russell, *op. cit.*, 309.

61 Armstrong, 'Pre-Kensington History', *op.cit.*, 6.

62 A.L., 'The London School of Mines', *op. cit.*

63 Edward Frankland, *Lecture Notes for Chemical Students: Embracing Mineral and Organic Chemistry* (London: John Van Voorst, 1866); cf. his monumental *Experimental Researches in Pure, Applied and Physical Chemistry* (London: John Van Voorst, 1877).

64 H.E. Armstrong, 'Our Need to Honour Huxley's Will', *Huxley Memorial Lecture* (London: Imperial College, 1933), 57.

65 *First Report, Devonshire Commission, 1871, Mins. Evid.*, Q. 5728.

66 Frank A.J.L. James, 'The Practical Problems of New Experimental Science: Spectro-Chemistry and the Search

for Hitherto Unknown Chemical Elements in Britain, 1860-69', *British Journal for the History of Science,* 21 (1988), 181-94. Today, mineralogists also use infra-red spectrography to measure the vibrational frequencies of atomic groups; and differential thermal analysis to measure water and temperature variations.

67 Jim Eckert, 'The Early Years: for Chemistry Alumni', *ChemNews* (Sydney: School of Chemistry, 2006), 7. I am indebted to Dr Eckert of Sydney University for his comments on the teaching of chemistry, yesterday and today.

68 A.A.E., 'The Royal College of Chemistry in the Seventies', *op.cit.,* 127.

69 Herbert McLeod was, as Colin Russell puts it, Frankland's 'right hand man'. (See Russell, *op. cit.* 309); he prepared all his lecture demonstrations. Not content to be a 'lab assistant' only, he wrote several papers. See H. McLeod, 'Apparatus for determining the Quantities of Gases Existing in Solution in Natural Waters', *Journal of the Chemical Society,* 22 (1869), 307-312; and 'On a New Form of Apparatus for Gas Analysis', *idem.,* 313-323. A devoted Anglican, he held strong views on the relations of science and religion. See W.H. Brock and Roy MacLeod, 'The Scientists' Declaration: Reflexions on Science and Belief in the Wake of *Essays and Reviews, 1864-1865,' British Journal of the History of Science,* IX Part 1 (31), (1976), 39-66; and Frank James (ed.)*, Chemistry and Theology in Mid-Victorian London: The Diary of Herbert McLeod, 1860-1870* (London: Mansell, 1987). In 1871, McLeod became Professor of Physical Science (later Chemistry) at the Royal Indian Engineering College at Cooper's Hill, and later, editor of the Royal Society Catalogue of Scientific Papers, in which connection Liversidge would meet him again.

70 Wilhelm George Valentin became well known for his laboratory manuals. His *Course of Qualitative Chemical Analysis* (London: J. and A. Churchill, 1873) progressed through four editions by 1876; through a further five edited by W.R. Hodgkinson by 1890; and, as *Valentin's Practical Chemistry,* a tenth by 1908. His *Introduction to Inorganic Chemistry* (London: J. and A. Churchill, 1872), went to a third edition in 1876. His *Twenty Lessons in Inorganic Chemistry* (London, 1879) remains a classic of laboratory exposition.

71 Friedrich Wöhler, *Handbook of Inorganic Analysis: 122 Examples Illustrating the Most Important Processes for Determining the Elementary Composition of Mineral Substances* (London: Walton and Moberley, 1854).

72 Liversidge was not to see the advent of electron microscopes or instruments to measure magnetic and radioactive properties that today identify chemical 'signatures' in minute traces.

73 Russell, *op. cit.*, 303.

74 Armstrong, 'Pre-Kensington History', *op.cit.,* 4.

75 Frederick Guthrie, 'Fourth Annual Dinner, RSM, 15 December 1876', *The Royal School of Mines Magazine: A Journal of the Students,* I (1877), 95.

76 *Ibid.*

77 Armstrong, 'Pre-Kensington History', *op.cit.,* 4.

78 G.G. Stokes, '4th Annual Dinner, RSM', *The Royal School of Mines Magazine: A Journal of the Students,* I (1877), 97.

79 Alexander W. Bickerton, 'Old Times by an Old Boy', *Phoenix,* 23 (June 1911), 187-189.

80 *Ibid.*

81 Thus his paper on Mineral Chemistry in June 1868: 'What are the chief impurities present in natural waters; to which of them is the quality termed "hardness" due; and what is the meaning of the term "previous sewage contamination"'. Department of Science and Art, Royal School of Mines, *Report of the Eighteenth Session,* 1868-69, 27.

82 Russell, *op. cit,* 303.

83 His signed copy of H. Kopp, *On the Past and Present of Chemistry* (London: W. Clewes, 1868) is in the Liversidge collection of the Royal Society of New South Wales.

84 *Seventeenth Report of the Department of Science and Art,* 1870. (C.174). xxvi. 466.

85 Armstrong, *op.cit.,* 62.

86 *Reports of the Royal Commission on the Pollution of Rivers,* 1870. (C. 37). xl. l and 449; 1871.(C. 347). xxv, 689; 1872.(C. 603). xxxiv, l; 1874. (C. 951). xxxiii, 1.

87 *Report of the Royal Commission on Water Supply,* 1868-69. (4169). xxxiii. *Mins. Evid.* Qq. 6191- 6432; *Reports on the Analysis of the Waters supplied by the Metropolitan Water Companies during 1869, 1870, and 1871,* 1872. (99). xlix. 801.

88 *Royal Commission on Water Supply, op. cit.,* 12.

89 Bickerton, *op.cit.,* 188.

90    For a concise discussion, see Colin Russell, 'Sir Edward Frankland (1825-1899)', *Oxford Dictionary of National Biography*, vol. 20, 766. See also Armstrong, *Huxley Memorial Lecture, op.cit.,* 64. His results were published in E. Frankland and H.E. Armstrong, 'On the Analysis of Potable Waters', *J.Chem.Soc.,* 21 (1868), 77-108. See also his *Water Analysis for Sanitary Purposes, with Hints for the Interpretation of Results* (London: John Van Voorst, 1880) – a perfect *vade mecum* for travellers (and would-be teetotallers). This book appeared too late for Liversidge's immediate use (see chapter 4), but it enjoyed wide acclaim and a second edition in 1890. For the history of water supply in London, see Roy MacLeod, ed. for David Owen, *The Government of Victorian London* (Cambridge, Mass.: Harvard University Press, 1982).

91    *Certificate of the Royal College of Chemistry,* signed by Edward Frankland, 9 November 1868. The original is now with the Royal Society of New South Wales.

92    Archibald Liversidge, 'Detection of Magnesia in the Presence of Manganese (Notes and Queries)', *Chemical News,* 17 (24 January 1868), 49.

93    The Royal School of Naval Architecture and Marine Engineering held its classes in South Kensington from 1864 until it was transferred to Greenwich in 1873.

94    Valentin's untimely death in 1879, at the age of fifty, was greatly mourned. Liversidge Papers, Box 19, f.11. At about this time, Liversidge prudently purchased a copy of Michael Faraday's *Chemical Manipulation: Being Instructions to Students in Chemistry* (London: John Murray, 1830). This survives in his gift of books to the Royal Society of New South Wales.

95    Armstrong went to Leipzig, and wrote a PhD thesis on aromatic compounds with Frankland's teacher, Hermann Kolbe. See Armstrong, 'Pre-Kensington History', *op.cit.,* 7.

96    Imperial College Archives, Sollas Papers, W.J. Sollas to W. Watts, 13 June 1935.

97    'Liversidge', *Journal and Proceedings of the Institute of Chemistry,* Pt. V (1927), 241.

98    'Sulphur in Coal-Gas (Notes and Queries)', *Chemical News,* 21 (29 April 1870), 204.

99    In 1854, when the Professor of Natural Philosophy at the Royal Institution, Tyndall turned down Playfair's invitation to join the staff of the School; but six years later, persuaded by the offer of £200 a year, succeeded Stokes as lecturer in physics.

100   John Tyndall, 'An Address to Students', *Fragments of Science* (London: Longmans, Green and Co., 1879), vol. II, 91.

101   Cf. John Tyndall, *Sound: A Course of Eight Lectures delivered at the Royal Institution* (London: Longmans, Green, 1867). The book went through five editions. Liversidge's copy is in the Rare Books Room of Fisher Library, University of Sydney.

102   Cf. John Tyndall, 'Colour of the Sky, the Polarisation of Skylight, and the Polarisation of Light by Cloudy Matter', *Proc. Roy. Soc.,* 17 (1866), 223-233; 'A New Series of Chemical Reactions produced by Light', *Proc. Roy. Soc.* 17 (1866), 92-103. See also his popular 'On the Relations of Radiant Heat to Chemical Constitution, Colour and Texture', *Fortnightly Review,* 4 (1866), 1-15; and 'On Chemical Rays and the Light of the Sky', *idem.,* 11 (1869), 226-248. The latter two were reprinted in the first edition of his *Fragments of Science.*

103   Observing suspended particles of dust in luminous beams, Tyndall found that much suspended matter was organic, and could be destroyed by passing the air through a red-hot platinum tube. His observations were first given in a lecture at the Royal Institution on 21 January 1870, and later printed as 'On Dust and Disease' in *Fragments of Science for Unscientific People* (London: Longmans, Green, 1871), 277-327. H.C. Bastian replied in *The Modes of Origin of Lowest Animals: including a discussion of the experiments of M. Pasteur, and a reply to some statements by Professors Huxley and Tyndall* (London: Macmillan, 1871). In January 1876, Tyndall presented further research to the Royal Society, which later appeared in his *Essays on the Floating Matter of the Air in relation to Putrefaction and Infection* (London: Longmans, Green, 1881), esp. 28-38. This celebrated controversy culminated between January and March 1878: see Tyndall, 'Spontaneous Generation', *Nineteenth Century,* 3 (1878), 22-47 (reprinted in the second volume of *Fragments of Science: A Series of Detached Essays, Addresses and Reviews* (London: Longmans, Green, 1879), vol. II, 292-336). This was followed by Bastian's reply in the *Nineteenth Century,* 3 (1878), 261-277; and Tyndall's 'Last Word', *Nineteenth Century,* 3 (1878), 497-508. For the debate, see John Farley, *The Spontaneous Generation Controversy from Descartes to Oparin* (Baltimore: Johns Hopkins University Press, 1977). On Tyndall, see A.S. Eve and C.H. Creasey, *Life and Work of John Tyndall* (London: Macmillan, 1945).

104   Unfortunately, Eve and Creasey, Tyndall's principal biographers, offer little insight into his educational work at the School of Mines. For suggestions, see Roy MacLeod, 'John Tyndall', *Dictionary of Scientific*

Biography, XIII (New York: Scribners, 1976), 521-524; and for fuller documentation, Roy MacLeod *et al.*, *John Tyndall, Natural Philosopher, 1820-1893: A Catalogue of Correspondence, Journals and Collected Papers* (London: Mansell, 1974).

[105] John Tyndall, 'Miracles and Special Providences', *Fortnightly Review,* 7, ns I (1866), 645-660, reprinted in the first volume of his *Fragments of Science* (London: Longmans, Green and Co., 1871), 43-70.

[106] John Tyndall, 'Scientific Materialism: Address to the Mathematical and Physical Section', *Report of the British Association for the Advancement of Science* (Norwich, 1868), reprinted in the first edition of *Fragments of Science, op.cit.*, 109-122. See also 'The Scientific Use of the Imagination', *Report of the British Association for the Advancement of Science* (Liverpool, 1870) also reprinted in the first edition of *Fragments.* His views on materialism were later summarised in his famous 'Belfast Address'. See *Report of the British Association for the Advancement of Science* (Belfast, 1874). This was reprinted in the second volume of the reissued *Fragments.*

[107] Tyndall to Murchison, 10 June 1868, quoted in Eve and Creasey, *op.cit.,* 130.

[108] Archibald Liversidge, *Notes from a Course of Lectures by Warington S. Smyth, 1868.* This copy is held by the Royal Society of New South Wales.

[109] Lady (Grace) Prestwich, *Life and Letters of Sir Joseph Prestwich* (Edinburgh: William Blackwood, 1899), 42. Prestwich, later Secretary of the Geological Society, advised Liversidge's generation that 'geology is entirely a science of observation and comprehension; accustom yourself to employ these talents – notice the effects of all you witness, study their causes, and you cannot fail to become a good geologist'. *Idem.,* 20.

[110] Royal School of Mines, *Report of the Seventeenth Session,* 1867-68, 24.

[111] Niels Stenson codified a set of principles in the sixteenth century, but it was not until the invention of the goniometer in 1780 that angles could be measured and symmetries precisely deduced. In 1784, René Just Haüy derived a set of laws of crystal symmetry that appeared applicable to the entire mineral kingdom.

[112] Dana's system reached its sixth edition in 1892, and is now in its ninth. There are today many classificatory systems based on chemical, physical, genetic and other characteristics. Many mineralogists use the system proposed by Hugo Strunz, of Germany, in 1938, as revised. Cf. *Encyclopaedia of Rocks and Minerals* (London: MacDonald, 1977), 52. In 1980, the *Cyclopedia of Minerals* (5th Edition) listed 2,000 mineral species, and about sixty new species are added every year. Of these, fewer than 100 have so far had economic value, but this represents a ten-fold increase over the number known in the age of 'iron and steel'.

[113] John F.W. Herschel, *Preliminary Discourse on the Study of Natural Philosophy* (London: Longman, Rees, 1831), 354.

[114] W.J. Lewis, 'Preface', *A Treatise on Crystallography* (Cambridge: Cambridge University Press, 1899). Lewis, a student of Story-Maskelyne at Oxford, succeeded Miller as Professor of Mineralogy at Cambridge (1881-1926). *Proc. Roy. Soc.,* A 111 (1926), xliv.

[115] Vanda Morton, *Oxford Rebels: The Life and Friends of Nevil Story-Maskelyne, 1823-1911, Pioneer Oxford Scientist, Philosopher and Politician* (Gloucester: Alan Sutton, 1987).

[116] Mineralogical Society (London), Correspondence, Nevil Story-Maskelyne to J.H. Collins, 27 September 1875.

[117] Gustav Rose, *Das Krystallo-chemische Mineralsystem* (Leipzig: Wilhelm Engelmann, 1852), remained the classic introduction to chemical mineralogy for thirty years. See John G. Burke, *Origins of the Science of Crystals* (Berkeley: University of California, 1966).

[118] Cf. William T. Stearn, *The Natural History Museum at South Kensington* (London: Heinemann, 1981), chapter 15.

[119] Archibald Geikie, *Memoir of Sir Andrew Crombie Ramsay* (London: Macmillan, 1895), 355.

[120] *First Report, Devonshire Commission,* (C.318). 1871. *Mins. Evid.* Q. 752.

[121] C. Le Neve Foster, *Mineralogical Memoranda* (Helston: R. Cunnack, 1867). A copy is held in the Liversidge collection, in the Library of the Royal Society of New South Wales.

[122] Sollas Papers*,* W.J. Sollas to W. Watts, 13 June 1935.

[123] James D. Dana, *The System of Mineralogy* (New York: George P. Putnam, 1837; second edition, 1844; third edition, 1850). Dana's text reached a sixth edition (edited by Edward Salisbury Dana) in 1914.

[124] J. Beete Jukes, *The Student's Manual of Geology* (Edinburgh: A. and C. Black, 2nd ed., 1862), 5.

[125] *Second Report, Devonshire Commission,* 1872. [C. 536]. xxv. *Mins. Evid.* Q. 656.

[126] H.H. Bellot, *University College, London, 1826-1926* (London: Athlone Press, 1929), 268-269.

[127] Not until 1871 did the University of Edinburgh create a chair of Geology and Mineralogy, with a legacy from

Murchison. Archibald Geikie (1835-1924) was its first incumbent (1871-1881).

128 Bellot, *op.cit.,* 268.

129 T.G. Bonney, *Memories of a Long Life* (Cambridge: Metcalf, 1921), 42. The income from the chair in the 1840s, depending upon student fees, ranged between £14 and £48 a year. Bellot, *op.cit.,* 268.

130 Cf. Martin J. S. Rudwick, *The Great Devonian Controversy: The Shaping of Scientific Knowledge among Gentlemanly Specialists* (Chicago: University of Chicago Press, 1985).

131 See MacLeod, 'Imperial Reflections in the Southern Seas', *op.cit.*

132 Robert Etheridge and Harry G. Seeley, *Manual of Geology* (London: Charles Griffin, 1885). Seeley was Sedgwick's assistant in Cambridge from 1860 to 1870.

133 Alexander Green, *Geology for Students and General Readers* (London: Daldy, Isbister and Co., 1876), Part I: Physical Geology; Joseph Prestwich, *Geology: Chemical, Physical and Stratigraphical* (Oxford: Oxford University Press, 1888); Archibald Geikie, *Class-Book of Geology* (London: Macmillan, 1886).

134 Jukes, J. Beete. *The Student's Manual of Geology* (Edinburgh: A. and C. Black, 3rd ed., 1872), 2-3.

135 The *locus classicus* of Sorby's petrology is his 'On the Microscopical Structure of Crystals Indicating the Origin of Minerals and Rocks', *Journal of the Geological Society,* 14 (1858), 453-500. See W.H. Wilcockson, 'The Geological Work of Henry Clifton Sorby', *Proceedings of the Yorkshire Geological Society,* 27 (1947), 1022.

136 Kalervo Rankama and T.G. Sahama, *Geochemistry* (Chicago: University of Chicago Press, 1950), 10. This information was communicated to the English-speaking world in a letter from Bunsen to Roscoe, as related in Roscoe's autobiography. I owe this reference to Dr. Graham Holland.

137 B.M. Kedrov, 'Dimitri Mendeleev', *Dictionary of Scientific Biography,* IX (New York: Scribners, 1974), 288-289.

138 William Crookes, 'Preface to the First Edition', *Select Methods in Chemical Analysis* (London: Longmans, Green, 1871, 1886).

139 Most minerals are isomorphs, chemically different yet having the same crystalline structure, but there are also minerals having the same chemical composition but different crystal structures. For example, 'calcite' and 'aragonite' are both pure $CaCO_3$, but have different crystal-lattices. Such 'polymorphs' can be indicative of changing environmental conditions where they are found. From the study of such minerals and 'solid solutions' of minerals – families of minerals with varying chemical compositions, in which different atoms occupy similar structural positions (eg, magnesite and siderite are both (Mg, Fe) $CO_3$) much may be learned about the history of geological change.

140 For the early development of mineralogy and geology, including the Wernerian interpretation, the Huttonian alternative, and attempts at synthesis see Rachel Laudan, *From Mineralogy to Geology: The Foundations of a Science, 1650-1830* (Chicago: University of Chicago Press, 1987). See also Roy Porter, *The Making of Geology: Earth Science in Britain, 1660-1815* (Cambridge: Cambridge University Press, 1977) and Roy Porter, 'The Terraqueous Globe', in George Rousseau and Roy Porter (eds.), *The Ferment of Knowledge* (Cambridge: Cambridge University Press, 1980), 285-326.

141 Mineralogical Society (London) Correspondence, Robert Hunt to J.H. Collins, Truro, 18 September 1875.

142 'Dendritic Spots on Paper. (Preliminary Note)', *Hardwicke's Science Gossip: An Illustrated Medium of Interchange and Gossip for Students and Lovers of Nature,* 5 (1 April 1869), 81. Liversidge returned to the subject in later years: see his 'Dendritic Spots on Paper', *J. Chem. Soc.,* 25 (16 May 1872), 646-648; *Chemical News,* 25 (14 June 1872), 284; correspondence in *Chemical News,* 26 (5 July 1872), 11.

143 Alex Dolby, 'Debates over the Theory of Solutions: A Study of Dissent in Physical Chemistry in the English-speaking World in the late-Nineteenth and early Twentieth-centuries', *Historical Studies in the Physical Sciences,* 7 (1976), 297-404.

144 'Nuclei and Supersaturated Saline Solutions', *Chemical News,* 22 (26 August 1870), 97.

145 'Nuclei and Supersaturated Solutions of Sodic Sulphate', *Chemical News,* 22 (19 August 1870), 90-92; 22 (26 August 1870), 97; 'Experiments upon Supersaturated Solutions of Sodic Sulphate', *Chemical News,* 22 (18 November 1870), 242-244; 22 (25 November 1870), 253-254; 22 (2 December 1870), 245; 22 (16 December 1870), 298.

146 'On Supersaturated Saline Solutions' (20 June 1872), *Proc. Roy. Soc.,* XX (1872), 497-507.

147 These are also now in the Library of the Royal Society of New South Wales. They amplify Smyth's *Lecture on Mining for Teachers,* printed by HMSO in 1862.

148 Percy studied chemistry in Paris with Gay-Lussac and Thenard before taking a medical degree in Edinburgh in

1838. He was elected FRS in 1847, and became famous in 1848 with the invention of an improved process to extract silver from its ores, which found wide application in the American West. His appointment to the RSM followed in 1851. He also influenced Thomas Gilchrist in developing his process for making Bessemer steel. 'John Percy', *Dictionary of National Biography*, XLIV (London: Smith Elder, 1895), 426

[149]  Chambers, *op.cit.*, lvii.

[150]  Liversidge contributed to Percy's collections (which contained, at his death in 1889, over 4,000 specimens) several selected species of moss gold from Australia. See J.F. Blake, *Catalogue of the Collection of Metallurgical Specimens formed by the late John Percy, Esq., MD, FRS, now in the South Kensington Museum* (London: HMSO, 1892), 226.

[151]  Armstrong, *Huxley Memorial Lecture, op.cit.*, 57.

[152]  Bickerton, *op.cit.*, 189.

[153]  Sollas Papers, *op. cit.*, Sollas to Watts, 13 June 1935.

[154]  W.J. Sollas, 'The Master', *Nature* 115 (9 May 1925), 747-748.

[155]  Secord, *op.cit.*, 260.

[156]  For conceptual background, see Martin J.S. Rudwick, *The Meaning of Fossils: Episodes in the History of Palaeontology* (New York: Science History Publications, 1972).

[157]  Imperial College Archives (London), London Union Society, A.L. to F.E.A. Manning, Secretary, University of London Union, 23 April 1924.

[158]  W.A. Haswell and T. Jeffery Parker, *A Textbook of Zoology* (London: Macmillan, 1897). This popular text reached four editions by 1928, and seven by 1962; it also reached a world-wide audience (and, in its 7th edition by A. J. Marshall and W.D.Williams (1972), still does).

[159]  *First Report, Devonshire Commission, Mins. Evid.*, Q. 1210.

[160]  Archibald Geikie, *Memoir of Sir A.C. Ramsay* (London: Macmillan, 1895), 331.

[161]  H.E. Armstrong, 'Autobiography', *The Central* (June 1938), cited in Alexander Findlay and W. Hobson Mills, *British Chemists* (London: Chemical Society of London, 1947), 58-95.

[162]  Bickerton, *op.cit.*, 188.

[163]  In 1873, he accepted the chair of chemistry at Canterbury (salary £ 450). See his 'University Reform: The Inaugural Address for 1881' (Christchurch: G. Tombs, 1881). See R.M. Burdon, *Scholar Errant: A Biography of Professor A.W. Bickerton* (Christchurch: The Pegasus Press, 1956).

[164]  Richard A. Proctor, *Wages and Wants of Science Workers* (London: Smith, Elder, 1876).

[165]  Cf. Roy MacLeod, 'The Support of Victorian Science: The Endowment of Science Movement in Great Britain, 1868-1900', *Minerva*, IX (2), (1971), 197-230; Roy MacLeod, 'The Resources of Science in Victorian England', in Peter Mathias (ed.), *Science and Society* (Cambridge: Cambridge University Press, 1972), 111-166.

[166]  For Danby (1840-1924), see chapter 3.

[167]  Lyon Playfair, 'Universities in their relation to Professional Education', *Address to St. Andrew's Graduates' Association,* 8 February 1873 (Edinburgh: Edmonston and Douglas, 1879).

[168]  Sheldon Rothblatt, *The Revolution of the Dons: Cambridge and Society in Victorian England* (London: Faber, 1968).

# CHAPTER 3

[1]  F.A. Reeve, *Victorian and Edwardian Cambridge from Old Photographs* (London: Batsford, 1978), 3; cf. P. Stocker, *Public Health and the Housing Problem in Cambridge, 1830-1900* (Cambridge: University of Cambridge School of Architecture, 1979).

[2]  Reeve, *op.cit.*, 3.

[3]  *The Times,* 2 December 1884.

[4]  'A Fragment Found in a Lecture-Room', in E.H. Coleridge, *The Poems of Samuel Taylor Coleridge* (London: Humphrey Milford, 1935), 35.

[5]  Cf. Thomas Thornely, *Cambridge Memories* (London: Hamish Hamilton, 1936).

[6]  J.P.D. Dunbabin, 'Oxford and Cambridge College Finances, 1871-1913', *Economic History Review,* 28 (1975), 631-647.

[7]  J. Steegman, *Cambridge* (London: Batsford, 1940), 95.

8    For the wider context, see Roy MacLeod and Russell Moseley, 'Breadth, Depth and Excellence: Sources and Problems in the History of University Science Education in England, 1850-1914', *Studies in Science Education,* 5 (1978), 85-106.

9    The long battle for religious freedom, leading to the final abolition of the 'Tests' in 1871, is related in Winstanley, *op.cit.,* especially Chapter 3.

10   Miss Emily Davies opened a college in Hitchin which moved in 1873 to Waterhouse's buildings at Girton. In 1871 Miss Anne Jemima Clough opened a hostel for women students in Regent Street, which by 1875 moved to Newnham, and became Newnham College. Cf. Rita McWilliams-Tulberg, *Women at Cambridge* (London: Gollancz, 1975).

11   J. Mayor, 'Commemoration Sermon', *The Eagle,* X (1878), 194-195.

12   See his copious account of 'The University of Cambridge' in R.B. Pugh (ed.), *The Victoria History of the Counties of England: A History of Cambridgeshire and the Isle of Ely* (London: Dawsons, 1967), III, 150-311, esp. 235-266. For the administrative history of the University, see D.A. Winstanley, *Later Victorian Cambridge* (Cambridge: Cambridge University Press, 1947); A.I. Tillyard, *A History of University Reform from 1800 AD to the Present Time, with Suggestions towards a Complete Scheme for the University of London* (Cambridge: Cambridge University Press, 1913).

13   Walter F. Cannon, 'Scientists and Broad Churchmen: An Early Victorian Intellectual Network', *Journal of British Studies,* 4 (1964), 56-88 and 'The Role of the Cambridge Movement in Early Nineteenth-Century Science', *Proceedings of the Xth International Congress of the History of Science* (Ithaca, 1964), 317-330; and Susan Faye Cannon, *Science in Culture: The Early Victorian Period* (New York: Science History Publications, 1978).

14   See Christopher N.L. Brooke, (ed.), *A History of the University of Cambridge*, vol. 4, (Cambridge: Cambridge University Press, 1992).

15   On the mathematical tripos, see John Gascoigne, 'Mathematics and Meritocracy: The Emergence of the Cambridge Mathematical Tripos', *Social Studies of Science,* 14 (4), (1984), 547-584; and P.M. Harman (ed.), *Wranglers and Physicists: Studies on Cambridge Physics in the Nineteenth Century* (Manchester: Manchester University Press, 1985).

16   On the history of the Natural Sciences Tripos, see Roy MacLeod and Russell Moseley, 'Breaking the Circle of the Sciences: The Natural Sciences Tripos and the Examination Revolution', in Roy MacLeod (ed.), *Days of Judgement: Science, Examinations and the Codification of Knowledge in Victorian Britain* (Driffield: Nafferton Publishers, 1982), 189-212.

17   See J.F.W. Herschel, *Preliminary Discourse on the Study of Natural Philosophy* (London: Longman, Rees, Orme, Brown and Green, 1831); Adam Sedgwick, *A Discourse on the Studies of the University* (Cambridge: John Parker, 1833) and William Whewell, *Of a Liberal Education in General, and with particular reference to the Leading Studies of the University of Cambridge* (London: Waterlow and Sons, 1848); see also W. Whewell, 'Herschel's Preliminary Discourse', *Quarterly Review,* 45 (July 1831), 374-407. For context, see Robert G. McPherson, *The Theory of Higher Education in Nineteenth-Century England* (Athens, Georgia: University of Georgia Press, 1959), and J.W. Adamson, *English Education, 1789-1902* (Cambridge: Cambridge University Press, 1930).

18   See T.G. Bonney, *A Chapter in the Life History of an Old University; being the Introductory Lecture of the Session 1881-82, delivered to Students of University College London* (Cambridge: Deighton, Bell, & Co., 1882).

19   *Special Report from the Select Committee on the Oxford and Cambridge Universities' Education Bill,* 1867 (497), xiii; Hansard, *Parliamentary Debates,* 3rd Series, 187 (5 June 1867), cols. 1614-45.

20   Mary Archer, and Christopher D. Haley (eds.), *The 1702 Chair of Chemistry at Cambridge: Transformation and Change* (Cambridge: Cambridge University Press, 2005).

21   'The Natural Sciences at Cambridge', *The Lancet,* 1867 (ii) (9 November 1867), 586.

22   *Returns from Universities of Oxford and Cambridge (on Professors), 1870-75,* 1876. (349). lix, 327. At Oxford, the situation was broadly similar: there were forty-four university chairs, but these were principally in theology, law and languages. twelve were in mathematics, but only six were in the natural sciences, and three of these were created as recently as the Honours School of Natural Sciences in 1854. Of Oxford's 370 college fellows, fewer than a dozen were in science; of 107 college scholarships annually offered, only six were open to science. *Third Report, Devonshire Commission,* 1873. (C. 868). xxviii, xliii. Oxford in this period has been well served by A.J. Engel, *From Clergyman to Don: The Rise of the Academic Profession in Nineteenth-Century Oxford* (Oxford:

Clarendon Press, 1983). The picture of science at Oxbridge compared inequitably with that at Manchester and Glasgow, for instance, where even smaller staffs carried larger loads for lower salaries, but still invested in laboratory practice. The Professor of Natural Philosophy at Owens College, Manchester earned £400, and the Professor of Chemistry, only £150; whilst their counterparts at Cambridge (and, for that matter, Melbourne) earned salaries of £500 and £300 respectively. Glasgow paid William Thomson (later Lord Kelvin) £380, which had to cover his assistant; his colleague in chemistry earned £200. However ill-paid and understaffed, the Scottish and provincial universities at least had buildings.

23  *Third Report, Devonshire Commission,* 1873. (C. 868). xxviiiii, xlii.

24  J.G. Crowther, *The Cavendish Laboratory, 1874-1974* (London: Macmillan, 1974); cf. Romualdus Sviedrys, 'The Rise of Physical Science at Victorian Cambridge', *Historical Studies in the Physical Sciences,* 2 (1970), 127-146.

25  For Woodward's influence, see Hugh Torrens, 'Early Collecting in the Field of Geology', in Oliver Impery and Arthur MacGregor (eds.), *The Origins of Museums* (Oxford: Clarendon Press, 1985), 211-212.

26  *The Cambridge University Calendar, 1870* (Cambridge: Deighton, Bell, 1870), 295.

27  The discussion was prompted by Lt. Col. Alexander Strange, a former army officer who had returned from the trigonometrical survey in India to preach the gospel of science. J.G. Crowther, *Statesmen of Science* (London: Cresset, 1965), 237-272

28  *Report of the British Association for the Advancement of Science* (Exeter, 1869), 213-4.

29  A.J. Meadows, *Science and Controversy: A Biography of Sir Norman Lockyer* (London: Macmillan, 1976).

30  R.V. Jones, 'The Domesday Book of British Science', *New Scientist,* 4 March 1971, 481-483.

31  Cf. *Reports of the Royal Commission on Scientific Instruction and the Advancement of Science* (afterwards the Devonshire Commission). First Report, 1871. (C. 318). xxiv; Second, 1872 (C. 536), xxv; Third, 1873 (C. 868), xxviii; Fourth, 1874. (C. 884). xvii; Fifth, 1874. (C. 1087). xxii; Sixth, 1875. (C. 1279). xxviii; Seventh, 1875. (C.1297). xxviii; Eighth, 1875. (C.1298). xxviii.

32  George Liveing, Alfred Newton and Coutts Trotter gave evidence in November 1870. Cf. *Supplementary Report of the Devonshire Commission,* 1872. (C. 536). xxv. *Mins. Evid.* Q. 4537, 4704, and 5034.

33  *Idem., Mins.Evid.* Q. 4548 (George Liveing, 30 November 1870).

34  'Natural Science at Cambridge', *The Lancet,* 1869 (ii), (11 December 1869), 816.

35  *Admissions Book,* Christ's College, Cambridge, 1870; *Christ's College Magazine,* 36, No. 113 (1927), 64-65.

36  H. Rackham, 'Christ's College, Cambridge', in Pugh (ed.), *op.cit.,* III, 429-436.

37  Cf. *Eighteenth Report of the Department of Science and Art,* 1871. (C. 397). xxiv. 461. Liversidge and his friends received 'special mention' in a list of Smyth's former pupils, submitted to the Devonshire Commission in July 1870. See *Supplementary Report, Devonshire Commission,* 1872, xxv. *Mins.Evid.* Q. 2333. In February 1871, Frankland added other names to the list; among them were 'G. Liveing, Professor of Chemistry, Cambridge, and Dr Smith, Professor of Chemistry, University of Sydney'. *Idem., Mins.Evid.* Q. 5681.

38  R.B. Somerset, 'Introduction', *The Student's Guide to the University of Cambridge* (Cambridge: Deighton Bell, 3rd ed., 1875), 25.

39  *Third Report, Devonshire Commission,* 1873.(C. 868). xxviii, xliv.

40  Winstanley, *op.cit.,* 191.

41  'Natural Science at Cambridge', *The Lancet,* 1869 (ii), (21 August 1869), 278.

42  T.G. Bonney, 'Natural Science at the University of Cambridge', *Nature,* 1 (3 March 1870), 452.

43  Charles Darwin, 'Autobiography' in Francis Darwin (ed.), *The Life of Charles Darwin* (London: John Murray, 1892), 23.

44  Although expressing its preference for doing so through the award of professorial fellowships. John Peile, *Christ's College* (London: F.E. Robinson and Co., 1900), Ch. IX, 'Reform', esp. 284-290.

45  *Supplementary Evidence, Devonshire Commission,* 1872. (C. 536). xxv. *Mins. Evid.,* Q. 4537.

46  Christ's College Archives, 'Scholarship Examinations', 1870.

47  Their number continued with T.W. Horscroft Waters, Demonstrator in Physiology, 1882-1887; A.C. Haddon, Reader in Ethnology; Sydney Vines, later Fellow of Magdalen College, Oxford, and Sherardian Professor of Botany, and the redoubtable A.E. (later Sir Arthur) Shipley, FRS, Professor of Zoology, Master of Christ's between 1910 and 1927, and one of Liversidge's closest friends. For Shipley, who helped create the 'high position of British zoology', see *Christ's College Magazine,* 36, no. 113 (1927), 1-8; and 36, no. 114 (1928), 76-79. For Vines, 'who, more than anyone else in Britain, made Botany', see *Christ's College Magazine,* 63, no.

133, (1954), 112-113.

48  J. Peile, *Christ's College Biographical Register, 1505-1905* (Cambridge: Cambridge University Press, 1913), II, 482.

49  Tom Collins, *School and Sport: Recollections of a Busy Life* (London: Elliot Stock, 1905), 13.

50  Cf. John Bickerdyke, *With the Best Intentions: A Tale of Undergraduate Life at Cambridge* (London: W. Swan Sonnenschein, 1884).

51  *The Students' Guide, op.cit.,* 91.

52  *Ibid.,* 86.

53  They included Philip Collins, a twenty-one year-old Londoner, who read mathematics (BA 1871) and became a lawyer; John Vernon (BA, 1872), an Irish classicist, who later read for the Irish Bar; Charles Taylor, a Yorkshireman (BA 1874), who later became a Canon of Manchester; Charles Fitch, twenty (BA 1873), a classicist and later barrister; and Thomas Weston, a Haileybury man who kept eight terms and left without a degree, going into steel smelting in Wales and coffee in Ceylon, before joining Allied Insurance in 1894.

54  Cambridge University Library, MSS Department, Annual Reports, Cambridge Union Society, 1871, 1872.

55  Miller, *St. John's, op.cit.,* 70.

56  Christ's College Archives: Boat Club Minutes, 1867-75, Records of Michaelmas Term, 1870, Lent and Michaelmas Terms, 1871.

57  Christ's College Archives: Athletic Club, Minutes, 1864-1898.

58  *Supplementary Report, Devonshire Commission,* 1872. (C. 536). xxv. Appendix IX, 35. Oxford had a similar number which occasioned no surprise. See Janet Howarth, 'Science Education in late-Victorian Oxford: A Curious Case of Failure', *English Historical Review,* 102 (403), (1987), 352.

59  See Roy MacLeod and Russell Moseley, 'The "Naturals" and Victorian Cambridge: Reflections on the Anatomy of an Elite, 1851-1914', *Oxford Review of Education,* 6 (1980), 177-195.

60  *The Student's Guide, op.cit.,* 26.

61  Bonney, *op.cit.,* 452.

62  *Supplementary Report, Devonshire Commission,* 1872. (C. 536). xxv. *Mins. Evid.,* Q. 1108.

63  *Ibid.,* Q. 1151.

64  For a candidate to do well in the six days' final examination could easily be, as George Liveing admitted, 'quite beyond the powers of any student'.

65  'Don't let this Tripos sit on your back like Sinbad's Old Man of the Sea or pursue you like Frankenstein's Monster', Donald Macalister, later the distinguished medic of St. John's, was warned by his tutor in 1876. The Tripos tested any undergraduate's will to succeed. For an introduction to this aspect of Cambridge, see Edith Macalister, *Sir Donald Macalister of Tarbert* (London: Macmillan, 1935), 31.

66  *The Student's Guide, op.cit.,* 220.

67  Gerrylynn K. Roberts, 'The Liberally-Educated Chemist: Chemistry in the Natural Sciences Tripos, 1851-1914', *Historical Studies in the Physical Sciences,* 11 (1), (1980), 157-183. See also Roy Porter, 'The Natural Sciences Tripos and the Cambridge School of Geology, 1850-1914', *History of Universities,* II (1982), 193-216; David Wilson, 'Experimentalists among the Mathematicians: Physics in the Cambridge Natural Sciences Tripos, 1851-1900', *Historical Studies in the Physical Sciences,* 12 (2), (1982), 325-371. These articles form part of a wider project on the history of the Tripos, convened by Roy MacLeod, *Science with Honours: Natural Science and Liberal Education in late Victorian Cambridge.*

68  The tripos examiners varied slightly each year: in 1871 and 1872, they were F.J.A. Hunt (one of the earliest graduates of the tripos) and Coutts Trotter, both of Trinity; Philip Main of St. John's, and – establishing a precedent for 'external examiners' – W.H. Flower from the Royal College of Surgeons in London. Trotter remained year after year, and gave the examinations continuity.

69  Collins, *School and Sport, op.cit.,* 16.

70  *Cambridge University Reporter,* 15 March 1871, 251.

71  John Venn, *Early Collegiate Life* (Cambridge: W. Heffer and Sons, 1913), 258-259.

72  *Cambridge University Reporter,* 15 March 1871, 252. Hudson (1838-1915), editor of the *Eagle* from 1872, taught in Cambridge from 1862 to 1881, when he became Professor of Mathematics at King's College, London.

73  Typical questions: 'Can Peter's vision be accounted for on the hypothesis of false perception?' 'Show the un-Jewish character of Jesus' teaching.' 'Show the economic disadvantages that would attend the perfect

demonstration of Christianity.' 'A miracle is an effect without a cause: Answer this objection.' Second Prelims Examination Papers, March 1872, *Cambridge University General Almanack and Register for 1873* (Cambridge: H. Willis, 1874), 130-1.

74    Indeed, the year after Liversidge came up (that is, for those matriculating in 1871), Christ's insisted that its successful Exhibitioners in natural sciences complete their Prelims at the earlier time. (*Cambridge University Reporter,* 30 November 1870, 129). It is not clear whether this stipulation existed in 1869 for those matriculating in 1870.

75    *The Correspondence of Charles Darwin, op.cit.,* 76, 101.

76    Barlow, *op.cit.,* 54.

77    *The Cambridge University General Almanack and Register* (Cambridge: H. Wallis, 1871, 1872).

78    That is, five feet, nine and a half inches. *Cambridge University General Almanack and Register for 1872* (Cambridge: H.Wallis, 1873), 290; and *for 1873,* 299.

79    T.H. Huxley, assisted by H.N. Martin, *A Course of Practical Instruction in Elementary Biology* (London: Cambridge, 1875).

80    *Special Report from the Select Committee on the Oxford and Cambridge Universities' Education Bill,* 1867. (497). xiii. *Mins.Evid.* Q. 4204.

81    *Idem.,* Q. 4203

82    *The Cambridge University Calendar, 1870 (*Cambridge: Deighton, Bell, 1870), 294.

83    'A clear lecturer, but a little lazy-minded', George Stokes recalled. Cambridge University Library, Add. MSS, 6582 (314), Rev. E. Atkinson Papers, 3 (1877), George Stokes to Atkinson, 23 February 1877.

84    On Miller, see T.G. Bonney, 'In Memoriam: W.H. Miller', *The Eagle,* No. 63 (1880); J.P. Cooke, 'Memoir of William Hallowes Miller', *Proc. Amer. Acad. Arts and Sciences,* 16 (1881), 460-468: Harriet S. Miller, *Memorial of W.H. Miller* (Cambridge: privately printed, 1881).

85    Geison, *op.cit.,* 121.

86    See A.F.R. Wollaston, *Life of Alfred Newton* (London: John Murray, 1921), 'Dr Shipley's Reminiscences', 104-105. For a more sympathetic reading of his life, see the tributes by J.W. Clark, A.G. Peskett, and S.F. Harmer, 'Alfred Newton', *Cambridge Review,* 28 (1907), 479-480.

87    F.O. Bower, *Sixty Years of Botany in Britain (1875-1935): Impressions of an Eyewitness* (London: Macmillan and Co., 1938), 13. See *Memorials, Journal and Botanical Correspondence of Charles Cardale Babington* (Cambridge: Macmillan and Bowes, 1897).

88    W. Thiselton-Dyer, 'Michael Foster- A Recollection', *Cambridge Review,* XXVIII (30 May 1907), 439-40.

89    John Venn and J.A. Venn (eds.), *Alumni Cantabrigienses: A Biographical List of All Known Students, Graduates and Holders of Office at the University of Cambridge from the Earliest Times to 1900* (Cambridge: Cambridge University Press, 1951), Part II, V, 456.

90    W.H. Miller, *A Treatise on Crystallography (*Cambridge: J. Deighton, 1839).

91    Cf. Story-Maskelyne's evidence to the House of Commons, *Returns from the Universities of Oxford and Cambridge (on Professors), 1870-75,* 1876. (349). lix, 16.

92    *Supplementary Report, Devonshire Commission,* 1872, Appendix IX, 35.

93    In Cambridge, he enjoyed a modest reputation; as professorial successor to the omniscient Whewell, he could well have had less. W.J. Lewis, who followed him in 1881, knew him as a popular lecturer, given to practical work, adept at building models, and respected for independence of mind, enthusiasm, and attention to detail. Bonney, *op.cit.,* 303.

94    Harriet Miller, *op.cit.,* 9, Story-Maskelyne to Mrs Miller, 21 May 1880.

95    'Nuclei and Supersaturated Saline Solutions', *Chemical News,* 22 (19 August 1870), 90-92; 22 (26 August 1870), 97.

96    As long ago as 1845 Liveing, one of Hofmann's first students at the Royal College of Chemistry in London, went to Cambridge to read the Mathematical Tripos. In 1853, he won a fellowship and college lectureship in chemistry at St. John's, where for many years he gave catechetic lectures on Paley. Edward Miller, *Portrait of a College: A History of the College of Saint John the Evangelist in Cambridge* (Cambridge: Cambridge University Press, 1961), ch. IV, 79.

97    Cf. T.R. Glover, *Cambridge Retrospect* (Cambridge: Cambridge University Press, 1943), 87-90.

98    St. John's College Archives, Liveing Letters, f.13, Liveing to W.G. Palmer, 8 August 1921.

99    *Report of the Select Committee on Scientific Instruction, 1867-68,* Appendix 20, 475.

100  *Ibid.*

101  In 1702, the University gave Vigani the title of professor, so creating the oldest continuously occupied chair of chemistry in Britain. But this happy precedent the University failed to match with stipend or apparatus. The subject languished after Vigani's death in 1764, until the University offered a small lecture room (but still no stipend) to Richard Watson, a mathematician and classicist, who knew nothing of chemistry before his election to the chair in 1764. Watson's vigorous efforts to improve himself, and his students, ceased when he became Regius Professor of Divinity in 1771.

102  F.G. Mann, 'Chemistry at Cambridge', *Proc. Chem. Soc.* (July 1957), 190-3.

103  Miller, *Portrait of a College, op.cit.*, ch. IV, esp. 64. Cf. A.C. Crook, *Penrose to Cripps: A Century of Building in the College of St. John the Evangelist, Cambridge* (Cambridge: St. John's College, 1978), 28.

104  Glover, *op.cit.,* 87-90.

105  The manner of his dying was only possible in the Cambridge of the new century: in stepping back from the street to avoid a passing car, he upset a girl riding on a bicycle, and fell himself, breaking a thigh, which developed complications. W.E. Heitland 'George Downing Liveing, 1827-1924', *The Eagle,* xliv (1924-26), 96.

106  *Ibid.*

107  F.A. Reeve, *Cambridge* (London: Batsford, 1964), 106.

108  *Ibid.*

109  St. John's College Archives, College Laboratory Records, f. 82.5, Vouchers and Accounts; Main, *op.cit.*, 191.

110  Cambridge University Library, Add. MSS, 6582 (406), Rev. E. Atkinson Papers, 3 (1877), Liveing to Atkinson, 6 October 1877.

111  Shipley, *op.cit,* 313.

112  Venn and Venn, *op.cit.,* Part II, Vol. IV, 183; Mills, *op.cit.,* 470.

113  This was very gradual. Downing continued its laboratory until 1920, and Girton until 1935.

114  Venn and Venn, *op.cit.,* Part II, Vol. IV, 183.

115  Heitland, *op.cit.,* 97.

116  Mann, *op.cit.,* 192.

117  'Professors' Lectures'. Flysheet by G.D. Liveing (December, 1875). Copy in Cambridge Pamphlets (J.W. Clark collection). Folio Series, Vol. V, Item 73A.

118  Heitland, *op.cit.,* 97.

119  Henry Roscoe, *Lessons in Elementary Chemistry: Inorganic and Organic* (London: Macmillan, 1867). *Chemistry, Inorganic and Organic: with Experiments* (London: John Churchill, 1867); *Laboratory Teaching, or Progressive Exercises in Practical Chemistry* (London: John Churchill, 1869); *Metals: Their Properties and Treatment* (London: Longmans, Green, 1870); he had already used F.A. Abel and C.L. Bloxam, *Handbook of Chemistry, Theoretical, Practical and Technical* (London: John Churchill, 1854). W.A. Miller, *Elements of Chemistry, Theoretical and Practical* (London: J.W. Parker and Son, 1855); *Introduction to the Study of Inorganic Chemistry* (London: Longmans Green, 1871).

120  August W. von Hofmann, *Introduction to Modern Chemistry, Experimental and Theoretic* (London: Walton and Moberley, 1865).

121  William Buckland, *Geology and Mineralogy, considered with reference to Natural Theology, with additions by Professor Phillips* (London: William Pickering, 1836, 2 Vols; reprinted in 1837, 1858, ed. by Frank Buckland of Christ Church, and published by Bell and Daldy, 1869).

122  Charles Adolphe Wurtz, *Leçons de Philosophie Chimique* (Paris, 1864).

123  Marcellin Berthelot, *Chimie Organique fondée sur la Synthèse* (Paris: Mallet-Bachelier, 1860). 2 vols.

124  St. John's College Library, Science Accessions List, *ca.* 1870. I am grateful to the Librarian of St. John's for his assistance in locating these volumes.

125  Not without a battle in the Arts Schools. (*Cambridge University Reporter,* 15 February 1871, 190). His victory followed a successful proposal to hire a laboratory assistant, paid by the University. (Cambridge University Archives, Museums and Lecture Rooms Syndicate, Min. VI. 65, f. 274.)

126  *Supplementary Report of the Devonshire Commission,* 1872. (C. 536). xxv. Appendix IX, 35.

127  *Chemical News,* 22 (16 December 1870), 298. Letter dated Cambridge, 8 December 1870; *Cambridge Chronicle,* 22 June 1867 and 22 April 1871; *The Lancet* (1867) (i) (15 June 1867), 750. *The Lancet* records there were 'several other candidates, some of whom were not members of the University'. *(idem).*

128  T.W. Danby (1840-1924), son of a London businessman, studied at the Royal School of Mines between 1857 and 1860, where he won the Government Prize in 1858 and the Duke of Cornwall's Scholarship, and took the ARSM. He went up to Cambridge in 1860, matriculating at Queens' and Caius, and apparently studied sometime at Heidelberg, before migrating to Downing in 1862. There he held his fellowship between 1867 and 1869, at a stipend of £150. After leaving Cambridge, he became a Fellow of the Geological Society, and published a translation of Carl Wilhelm Fuchs' *Practical Guide to the Determination of Minerals by the Blowpipe* (London: Field and Tuer, 1875). He ultimately became a Chief-Inspector of Schools, for South-Eastern England. See Cambridge University Archives, *CUR*, 39.11, 'Professor of Chemistry'; *Cambridge University Reporter,* 1 February 1871, 171; 'College News', *Cambridge Chronicle,* 29 April 1871; *The Times,* 24 March 1924. In fact, his claim to be 'the first person who has been elected a Cambridge Fellow mainly on account of distinction in the natural sciences' was disputed by G.E. Paget, who recalled earlier lecturers at St. John's, Peterhouse and Downing. *The Lancet,* 1867 (ii), (26 October 1867), 540; (16 November 1867), 625. But Danby's name was remembered. Under regulations then in force, he was first required to pass the Mathematical Tripos, and this he did, becoming Senior Optime in 1864, and proceeded to lead the first class of the NST the same year. See *Alumni Cantabrigienses, op.cit.,* Part II, Vol. II; Chambers, *Register of the Associates and Old Students of the Royal College of Chemistry, the Royal School of Mines and the Royal College of Science* (London: Hazell, Watson and Viney, 1896), 45.

129  *Cambridge University Reporter,* 3 May 1871.

130  Cambridge University Archives, CUR 49, (240). Announcement by the Professor of Chemistry.

131  John Wale Hicks (1840-1899), later Bishop of Bloemfontein, mountaineer, author of tracts on Church doctrine, including *The Connection between the Natural and the Spirit World* (Cambridge, 1890). He served as Liveing's demonstrator, from 10 am to 6 pm each day in term, until 1877, when he published a textbook on inorganic chemistry, and left to enter the Church. See *Who Was Who, I (1897-1915)* (London: A. & C. Black, 1920), 337.

132  'Cambridge and Chemistry', *Nature,* 31 (19 February 1885), 377.

133  'Process for the Estimation of Fluorine', *Chemical News,* 24 (10 November 1871), 226 and discussion; 'Notes on Science' (17 November 1871), *Journal of the Royal Society of Arts,* 20 (1871- 72), 19.

134  Cf. J. N. Langley, 'In Memoriam: Sir Michael Foster', *Journal of Physiology,* XXXV (1907), 233-46; Gerald L. Geison, *Michael Foster and the Cambridge School of Physiology: The Scientific Enterprise in Late Victorian Society* (Princeton: Princeton University Press, 1978).

135  Cf. John Willis Clark, 'Coutts Trotter', in *Old Friends at Cambridge and Elsewhere* (London: Macmillan and Co., 1900), 314-8. After his untimely death, at fifty, in 1887, Venn recorded that it was 'to him more than to any other single person' that 'the indubitable improvements effected in University matters during his short academic career' were due. *Alumni Cantabrigienses, op.cit.,* Part II, VI, 235. Foster called his death a 'calamity', and wrote his obituary: 'Coutts Trotter', *Nature,* 37 (15 December 1887), 153-154.

136  *Supplementary Report, Devonshire Commission,* 1872. (C. 536). xxv. *Mins. Evid.,* Q. 5064.

137  Geison, *op.cit.,* 75-76.

138  W. Gaskell, 'In Memoriam: Sir Michael Foster', *Cambridge Review,* XXVIII (7 February 1907), 221-222.

139  R. Robson, 'Michael Foster and Trinity College', *Trinity Review* (Easter, 1965), 10.

140  D.H.M. Woollam, 'The Cambridge School of Physiology', in A. Rook (ed.), *Cambridge and its Contribution to Medicine* (London: Wellcome, 1971), 43-44.

141  Sir Henry Dale, 'Sir Michael Foster, KCB, FRS, A Secretary of the Royal Society', *Notes and Records of the Royal Society of London,* 19 (1), (1964), 17.

142  Robson, *op.cit.,* 11.

143  For their later relationship, see Chapters 4, 7 and 8. Cf. correspondence between Foster and Liversidge, 1 July 1887, in *JPRSNSW,* 21 (1887), 216-217.

144  See M. Foster (letter) 'Science at Cambridge', *Nature,* 28 (16 August 1883), 374.

145  [M. Foster], 'Science in the Schools', *Quarterly Review,* 123 (1867), 244-258.

146  G.M. Humphry, 'Address in Physiology' (delivered to the BA at Nottingham, 1866), *Journal of Anatomy and Physiology,* 1 (1867), 1. For Humphry, see *Proc. Roy. Soc.,* 75, (1905), 128-130.

147  *Ibid.,* 90.

148  W.H. Gaskell, 'Sir Michael Foster', *Proc. Roy. Soc.,* 80 (1908), lxxi.

149  Langdon-Brown, *op.cit.,* ch. VI.

150  For a discussion of contemporary practices in experimental medicine on the Continent, see William Coleman and Frederic L. Holmes (eds.), *The Investigative Enterprise: Experimental Physiology in Nineteenth-Century Medicine* (Berkeley: University of California Press, 1988).

151  H. Hale Bellot, *University College London, 1826-1926* (London: University of London Press, 1929), 314.

152  Rolleston, *op.cit.,* 90.

153  See Churchill College Archives, Physiological Society Minute Book, 1876-1892.

154  Together they published *A Course of Practical Elementary Physiology and Histology* (London: Macmillan, 1876) which reached seven editions by 1899.

155  Langley, 'In Memoriam', *op.cit.,* 238.

156  Humphry, *op.cit.,* 2.

157  Quoted in a review of C.A. Rolf, *Outlines of Physiological Chemistry, Journal of Anatomy and Physiology,* VII (June 1873), 319-320.

158  A. Crum Brown and T.R. Fraser, 'On the Connection between Chemical Constitution and Physiological Action', *Journal of Anatomy and Physiology,* II (1868), 224.

159  E. Klein *et al., Handbook for the Physiological Laboratory* (London: John Churchill, 2 Vols., 1873).

160  *Ibid.*, 5.

161  Cf. Crum Brown and Fraser, *op.cit.;* James Blake, 'Observations of Physiological Chemistry', *Journal of Anatomy and Physiology,* V (1871), 247-250.

162  See Dale, *op.cit.,* 2l.

163  J.L.W. Thudichum, 'Researches on Kryptophanic Acid, a Normal Ingredient of Human Urine', *Journal of the Chemical Society,* 23 (1870), 116-133. See also J. L.W. Thudichum and J.A. Wanklyn, 'Researches on the Constitution and Reaction of Tyrosine', *Journal of the Chemical Society,* 22 (1869), 277-291, and related papers. For this pioneer of physiological chemistry and chemical pathology, see David L. Drabkin, *Thudichum: Chemist of the Brain* (Philadelphia: University of Pennsylvania Press, 1958). For his government work, see Roy MacLeod, 'The Frustrations of State Medicine, 1880-1899', *Medical History,* XI (1967), 15-40.

164  In 1873, the *Journal of the Chemical Society* began a separate section for physiological chemistry.

165  Archibald Liversidge, 'Note upon Kryptophanic Acid' (April 1872), *Journal of Anatomy and Physiology,* VI (ii) (May 1872), 422-425. Apparently, he was right, as Klein's *Handbook,* published early the following year makes no reference to the subject. Ten years later, cryptophanic (*sic*) acid was listed by Henry Power and Leonard Sedgwick in *The New Sydenham Society's Lexicon of Medicine and the Allied Sciences* (London: The New Sydenham Society, 1882), Vol. II, with the appropriately acidic comment, 'the existence of this body as a distinct chemical body is doubted'.

166  M. Foster, 'Notes on Amylolytic Ferments', *Journal of Anatomy and Physiology,* I (1867), 107-113.

167  J.L.W. Thudichum, *A Manual of Chemical Physiology, including its points of contact with Pathology* (London: Longmans, 1872), reviewed in *The Lancet,* 1872 (i), (8 June 1872), 798.

168  The paper did not appear in print, however, for over a year. See 'On the Amylolytic Ferment of the Pancreas', *Journal of Anatomy and Physiology,* VIII (1), (November 1873), 23-29; 'Reviews', *The Lancet,* 1873 (ii), (20 December 1873), 881.

169  Cambridge University Library: Cambridge Papers J5990, CUNSC, Easter Term, 1863.

170  Their motto: *Omnia quae secundam naturam fluent, sunt habenda in bonis* (Cicero). Cambridge University Library: Cambridge Papers J5990, CUNSC, 1872.

171  The six included J.C. Saunders (Downing, later a Fellow and Anglican vicar), C.T. Whitmell (Trinity and later an HMI), C.F.J. Yule (St. John's), H.M. Martin (Christ's), J.E.H. Gordon (Caius) and P.H. Carpenter (Trinity, later Science Master of Eton).

172  *Nature,* 6 (6 June 1872), 110. The new members were W.S. de Mattos of Trinity and G.G. Paget of Caius.

173  *Nature,* 25 (16 March 1882), 474.

174  Rolleston, *op.cit.,* 24.

175  Shortly to follow him was Arthur Milnes Marshall (Caius, 1882), later Professor of Zoology at Owens College, Manchester.

176  Cf. Geison, *op.cit.,* 183, 187 *et passim.*

177  E. Klein, J. Burdon Sanderson, M. Foster, T. Lauder Brunton, *Handbook for the Physiological Laboratory* (London: John Churchill, 1873), reviewed in *The Lancet,* 1873 (i), (3 May 1873), 631, and in the *Journal of Anatomy and Physiology,* VII (ii), (1873), 322-325; see also Michael Foster, *A Course of Elementary Practical*

*Physiology* (London: Macmillan, 1876), reviewed in *The Lancet,* 1876 (ii), (18 November 1876), 720.

178   Dew-Smith, a keen photographer, knew the South Seas well. He became a close friend of Liversidge and is said to have been the model for the character of Attwater, in Robert Louis Stevenson's and Lloyd Osbourne's *The Ebb-Tide. A Trio and Quartette* (London: W. Heinemann, 1894). See M.J.G. Cattermole and A.F. Wolfe, *Horace Darwin's Shop: A History of the Cambridge Scientific Instrument Company, 1878 to 1968* (Bristol: Adam Hilger, 1987), 9-10 *et passim.*

179   Archibald Liversidge, 'On the Amylolytic Ferment of the Pancreas' (November 1872), *Journal of Anatomy and Physiology,* VIII (1), (November 1873), 23-29; reprinted in M. Foster (ed.), *Studies from the Physiological Laboratory* (Cambridge: Cambridge University Press), I (November 1873), 45-51.

180   For their significance in physiology, see Edward Sharpey Schaefer, 'History of the Physiological Society during its First Fifty Years, 1876-1926', *Supplement to the Journal of Physiology* (London, 1927); and W.F. Bynum, 'A Short History of the Physiological Society, 1927-1976', *Journal of Physiology,* 263 (1976), 23-72; Sir Walter Langdon-Brown, *Some Chapters in Cambridge Medical History* (Cambridge: Cambridge University Press, 1949), ch. VI.

181   Cf. Ruth G. Hodgkinson, 'Medical Education in Cambridge in the Nineteenth Century', in Arthur Rook (ed.), *op.cit.,* 79-106, esp. 99-100.

182   Cf. Michael Foster, *On Medical Education at Cambridge* (London: Macmillan, 1878); Bellot, *op.cit.,* 314-315.

# Chapter 4

1   For the University's early history, see H.E. Barff, *A Short Historical Account of the University of Sydney* (Sydney: Angus and Robertson, 1902) and his successor, Robert A. Dallen, 'Early Days of the University of Sydney', *JRAHS,* 12 (1926), 271-288. The University's official history is set out in Clifford Turney, Ursula Bygott and Peter Chippindale, *Australia's First; A History of the University of Sydney* (Sydney: University of Sydney in association with Hale and Iremonger, 1991) in two volumes. For an overview, see W. J. Gardner, *Colonial Cap and Gown* (Christchurch: University of Canterbury, 1979).

2   Details of the building are to be found in *The Great Hall and the Professorial Board Room: A Brief History and Description* (Sydney: Australasian Publishing Co., ca. 1965), and David Lawton and Jeremy Steele, *The Great Hall Guide* (Sydney: University of Sydney, 1981).

3   Sir Charles Nicholson, 'Commemoration Address', 1859, in *Inaugural Addresses, delivered on the Occasion of the Opening of the University of Sydney by the Vice-Provost and the Professor of Classics, in 1852; also, Report of Addresses at Various Commemorations held in Subsequent Years, and delivered by the Chancellor, Sir Charles Nicholson, Bart.* (Sydney, nd), 47. For details of the Great Hall, see Bertha McKenzie, *Stained Glass and Stone: The Gothic Buildings of the University of Sydney* (Sydney: University of Sydney, 1989).

4   The London committee comprised G. B. (later Sir George) Airy, Astronomer Royal; Sir John Herschel, the distinguished astronomer and Master of the Mint; Professor Henry Malden, who was also consulted about the establishment of new universities in India; and Henry Denison, a Fellow of All Souls, Oxford.

5   See Cecil W. Salier, 'Professor Morris Birkbeck Pell', *JRAHS,* 18 (1932), 246-251; N.J.B. Plomley, *Several Generations* (Sydney: Wentworth Books, 1971).

6   C.M.H. Clark, *A History of Australia* (Melbourne: Melbourne University Press, 1978), vol. 4, 89.

7   See Christopher Rolleston, *Statistical Review of the Progress of New South Wales in the Last Ten Years, 1862-1871* (Sydney: Government Printer, 1872), 122. See also James Davidson, *The Sydney-Melbourne Book* (Sydney: Angus and Robertson, 1986).

8   Sir Charles Nicholson, 'Inaugural Address', 11 October 1852, quoted in R. Therry, *Reminiscences of Thirty Years' Residence in New South Wales and Victoria* (Sydney: Sydney University Press, new ed. 1974), 441.

9   A simplified account is given in Ian Westbury, 'The Sydney and Melbourne Arts Courses, 1852-1861', in E.L. French (ed.), *Melbourne Studies in Education* (1961-62), (Melbourne: Melbourne University Press, 1964), 256-284.

10   Sir Charles Nicholson, 'Inaugural Address', quoted in Dallen, *op.cit.,* 278.

11   Geoffrey Bolton, 'The Idea of a Colonial Gentry', *Historical Studies,* 13 (1968), 320; see Michael Roe, *Quest for Authority in Eastern Australia, 1835-51* (Melbourne: Melbourne University Press, 1965), chapter 2.

12   The Queen's Colleges were brought into existence by the Crown in December 1845, and the Queen's University

incorporated by charter on 3 September 1850

13  R.A. Dallen, 'The Chancellors of the University of Sydney since its Foundation', *Journal and Proceedings of the RAHS*, XIX (iv), (1933), 2; *University of Sydney Calendar, 1852-53*, 15. The initial Senate included graduates of Cambridge, Dublin, Edinburgh and Oxford; three Queen's Counsel, two medical doctors, and three clergymen, the Colonial Secretary, the Auditor-General, and the Speaker of the Legislative Council. For the University's Act, see Turney *et al., op. cit.,* 636.

14  Macmillan, *op.cit.*, 29.

15  For further information, see Turney *et al., op. cit.*

16  The relationship of the University and its affiliated colleges was resolved (except for the issue of the religious certificate) by a compromise arrangement written into the Affiliated Colleges Act, 1854.

17  Clark, *op.cit.,* 89.

18  Dallen, *op.cit.* 277. The Australian discourse commonly referred to English precedents. See, for example, 'The University and the Colleges', *Sydney University Magazine,* No. 1 (January 1855), 98-120.

19  'Report from the Select Committee on the Sydney University', *Papers of the Legislative Council,* 1849, 1, cited in Macmillan, *op.cit.*, 37.

20  See Leonard Huxley, *Life and Letters of Thomas Henry Huxley* (London: Macmillan, 1900), vol. I, 83 and 100, Huxley to W.S. Macleay, 9 November 1851; Jordan Goodman, *The Rattlesnake: A Voyage of Discovery to the Coral Sea* (London: Faber and Faber, 2005), 293.

21  On which, see E.G.W. Bill, *University Reform in Nineteenth-Century Oxford: A Study of Henry Halford Vaughan, 1811-1885* (Oxford: The Clarendon Press, 1973), chapters 8, 10 and 11.

22  John Woolley, 'Inaugural Address' (11 October 1852), in Sir Charles Nicholson (compiler), *Inaugural Addresses delivered on the Occasion of the Opening of the University of Sydney* (Sydney: The University of Sydney, 1852), 15.

23  T. W. Moody and J. C. Beckett, *Queens, Belfast, 1845-1949: The History of a University* (London: Faber & Faber for the Queen's University of Belfast, 1959).

24  Roy MacLeod and Russell Moseley, 'Breaking the Circle of the Sciences: The Natural Sciences Tripos and the Examination Revolution', in Roy MacLeod (ed.), *Days of Judgement* (Driffield: Nafferton, 1982), 189-212.

25  Colleges were opened to Anglicans at St. Paul's from 1852; to Catholics at St. John's from 1857; and to Presbyterians at St. Andrew's from 1867. Relations with the Church of England were clarified by the Affiliated Colleges Act, 1854, and by the repeal of the religious certificate provision of this Act in 1888.

26  Personal Communication, 26 March 1990. See Ken Cable, 'The University of Sydney and its Affiliated Colleges, 1850-1880,' *The Australian University,* 2 (3), (1964), 183-214.

27  *Report from the Select Committee on the Sydney University, Votes and Proceedings of the Legislative Assembly of New South Wales (*hereafter *VPLANSW),* IV (1859-60)**,** 165.

28  *Ibid.*

29  *Ibid.*, Evidence, 79.

30  The University of Sydney Incorporation Act Amendment Act, 1861.

31  Macarthur Papers (Mitchell Library), William Macarthur to James Macarthur, 10 February 1861, cited in Macmillan, *op.cit.*, 58.

32  See K.J. Cable, 'John Woolley (1816-1866)', *Australian Dictionary of Biography,* vol. 6, 435-437, and Woolley, *Lectures Delivered in Australia* (Cambridge: Macmillan, 1862).

33  Turney *et al., op. cit.,* 125**.**

34  Macmillan, *op.cit.*, 53.

35  Macmillan found details of 129 Sydney undergraduates who graduated between 1852 and 1862. Of eighty-four whose backgrounds were known, twenty-one were sons of merchants; the fathers of fourteen were lawyers, twelve were pastoralists, two were ministers of religion, nine were artisans, eight retailers, and three were builders. See Macmillan, *op.cit.,* 58.

36  Geoffrey Blainey, *A Centenary History of the University of Melbourne* (Melbourne: Melbourne University Press, 1957).

37  G. Morris, 'The Development of Science Subjects in Australian Secondary School Education', *Australian Journal of Education,* 8 (1), (1964), 47.

38  Macmillan, *op.cit.,* 56.

39  Alan Barcan, *A Short History of Education in New South Wales* (Sydney: Martindale Press, 1965).

40    G.L. Fischer, *The University of Sydney, 1850-1975* (Sydney: The University, 1975), 18.

41    Senate Minutes, 2 July 1853. Smith was given a housing allowance of £120 pa (Senate Minutes, 20 September 1852). In 1852, the three professors were granted an increase in salary of £150, considered to be temporary, to account for the rise in the cost of living from the time of their appointment until their arrival in the colony (Senate Minutes, 18 July 1852).

42    R.J.W. Le Fèvre, 'The Establishment of Chemistry in Australia', in A.P. Elkin (ed.), *A Century of Scientific Progress: The Centenary Volume of the Royal Society of New South Wales* (Sydney: Royal Society of New South Wales, 1968), 341, 344.

43    The NSW system was established in 1848, and administered from 1848 to 1866 by the Board of Education. Following the Public Instruction Act of 1866, the Board was succeeded by the Council of Education, and this in 1880, by the Department of Education. See Jean Ely, 'The Institutionalisation of an Ideal: Centralisation in Educational Administration in NSW, 1848-1880', in S. Murray-Smith (ed.), *Melbourne Studies in Education, 1973* (Melbourne: Melbourne University Press, 1973), esp. 231-248.

44    In 1876, Smith also received an honorary LLD from the University of Aberdeen. See Michael Hoare and Joan T. Radford, 'John Smith (1821-1885)', *Australian Dictionary of Biography*, vol. 6, 148-150.

45    See Robert H. Kargon, *Science in Manchester: Enterprise and Expertise* (Manchester: Manchester University Press, 1977).

46    'On the Advantages which Science and Commerce derive from each other', *Australian Era*, 1 (6) (February 1851), 88.

47    Cf. John Smith, 'The Action of Sydney Water upon Lead', delivered to the Philosophical Society on 15 August 1856; *SMH*, 15 August 1856; see Max Kelly, 'Picturesque and Pestilential: The Sydney Slum Observed, 1860-1900', in Max Kelly (ed.), *Nineteenth Century Sydney. Studies in Urban and Social History* (Sydney: Sydney University Press, 1978), 66-80.

48    See Lionel Gilbert, *The Royal Botanic Gardens: A History, 1816-1985* (Melbourne: Oxford University Press, 1986); Ronald Strahan, *Rare and Curious Specimens: An Illustrated History of the Australian Museum, 1827-1979* (Sydney: The Australian Museum, 1979).

49    J.H. Maiden, 'A Contribution to a History of the Royal Society of New South Wales', *JPRSNSW*, 52 (1918), 257.

50    Or, to give its title full justice, the Australian Society for the Encouragement of Arts, Science, Commerce and Agriculture. See W.R. Browne and Ida A. Browne, 'Scientific Societies in Australia', *RACI Journal*, 28 (1961), 100-109; but see also David Branagan, 'Words, Actions, People: 150 Years of the Scientific Societies in Australia', *JPRSNSW*, 104 (1972), 123-141.

51    See Alan Barnard, 'Thomas Sutcliffe Mort (1816-1878)', *Australian Dictionary of Biography*, vol. 5, 299-301; C.J. King, 'Charles Moore (1820-1905)', *idem.*, 274-275.

52    Maiden, *op.cit.*, 257.

53    John Smith, 'Presidential Address', *JPRSNSW*, 15 (1881), 3.

54    Ten years later, with Royal Assent, it became known as the Royal Society of New South Wales. See Maiden, *op.cit.*, 215-361. See also C.H. Currey, 'Sir William Thomas Denison (1804-1871)', *Australian Dictionary of Biography*, vol. 4, 46-53.

55    Sir William Denison, *Varieties of Vice Regal Life* (London: Longmans, Green, and Co., 1870), vol. 1, Denison to Sir Roderick Murchison, 25 June 1856.

56    John M. Ward, *Earl Grey and the Australian Colonies, 1846-57: A Study of Self-Government and Self-Interest* (Melbourne: Melbourne University Press, 1958).

57    Denison, *op.cit.*, vol. I, 354, Denison to Murchison, 25 June 1856.

58    *Ibid.*

59    *Historical Records of Australia*, VII (1), 587, cited in T.G. Vallance, 'The Start of Government Science in Australia: A.W.H. Humphrey, His Majesty's Mineralogist in New South Wales, 1803-12', *Proceedings of the Linnean Society of NSW*, 105 (2), (1981), 107-146.

60    Similar appointments were made in Victoria in 1852, Tasmania in 1859, and Queensland in 1868. See D. F. Branagan and T.G. Vallance, 'Samuel Stutchbury (1798-1859)', *Australian Dictionary of Biography*, vol. 6, 216-217.

61    After Denison came to office, Stutchbury fell from vice-regal favour and his appointment was ended. See David Branagan, 'Samuel Stutchbury (1788-1859)', *Oxford Dictionary of National Biography* (Oxford: Oxford

University Press, 2004).

62  W.B. Clarke Papers (Mitchell Library) MSS 139/49. Clarke to Thomson, 27 March 1867, 21 November 1868. Thomson to Clarke, 15 December 1868. See Ann Mozley, 'William Branwhite Clarke (1798-1878)', *Australian Dictionary of Biography*, vol. 3, 420-422.

63  See Ann Moyal (ed.), *The Web of Science: The Scientific Correspondence of the Rev. W. B. Clarke, Australia's Pioneer Geologist* (Kew: Australian Scholarly Publishing, 2003).

64  Institute of Geological Sciences (Geological Museum), G. S. M. 1/6 f. 99/103. Sir Henry Young to Lord Grey, 13 January 1851.

65  Institute of Geological Sciences (Geological Museum), G.S.M. 1/6 f.104. De la Beche to Colonial Office, 16 July 1851.

66  Theodore Chambers, *Register of the Association and Old Students of the Royal College of Chemistry, the Royal School of Mines and the Royal College of Science* (London: Hazell, Watson and Vines, 1896), xxxviii.

67  A.R.C. Selwin was appointed at £500, but his salary was raised by Lt. Governor Latrobe first to £800, and later to £900. See D.F. Branagan and K.A. Townley, 'Alfred Richard Cecil Selwyn (1824-1902)', *Australian Dictionary of Biography*, vol. 6, 102.

68  Jukes had written an extensive account of the goldfields, which Liversidge read before leaving England. See J. Beete Jukes *et al., Lectures on Gold: For the Instruction of Emigrants about to proceed to Australia* (London: David Brogue, 1852), and J. Beete Jukes (edited by his sister), *Letters and Extracts from the Addresses and Occasional Writings* (London: Chapman and Hall, 1871). Copies of these remained in Liversidge's library.

69  Institute of Geological Sciences (Geological Museum), G.S.M. 1/6/9 f. 266. Murchison to H. Merivale, 7 January 1856.

70  See G.C. Fendley, 'Sir Frederick McCoy (1817-1899)', *Australian Dictionary of Biography*, vol. 5, 134-136.

71  The Deas Thomson Scholarship set a valuable precedent. In 1871, the Earl of Belmore granted funds for an annual gold medal to be given to an undergraduate 'proficient in geology and practical chemistry, with special reference to agriculture'. Senate accepted, so to 'promote that important branch of industry'. Senate Minutes, 22 May 1871, 4.

72  'A School of Mines', *Sydney Magazine of Science and Art,* II (1859), 70.

73  He became known for his photographic work using the new 'wet plate' method invented by Archer and Fry in England in 1850. Many photographs of the University under construction, and of other colonial scenes, have been attributed to him; these form 'the most important collection of early photographic material so far discovered in this country, and one of the few collections of such material existent anywhere in the world'. D.S. Macmillan, 'Professor John Smith and the Beginnings of Photography in Australia', *Proceedings of the Royal Australian Chemical Institute,* 26 (1959), 343. See also Catherine Snowden, 'The Strayfaring Professor: John Smith and "Learned Leisure"', in Roy MacLeod (ed.), *University and Community in Nineteenth-Century Sydney: Professor John Smith (1821-1885),* (Sydney: University of Sydney History Project, 1988), 14-30.

74  See Le Fèvre *op.cit.,* 341-354.

75  Alfred P. Backhouse, 'Some Notes from the Pigeon-Holes of My Memory', *JRAHS,* XIX (1), (1933), 55. See also K.J. Cable, 'Alfred Paxton Backhouse (1851-1939)', *Australian Dictionary of Biography*, vol. 7, 127-128.

76  These were collected and published in two volumes: *Wayfaring Notes: Sydney to Southhampton by way of Egypt and Palestine* (Sydney: Sherriff and Downing, 1865) and *Wayfaring Notes (Second Series): A Holiday Tour Round the World* (Aberdeen: A. Brown and Co., 1876). Smith took overseas leave in 1862-63, 1871-72, and 1881-83.

77  *SMH*, 28 January 1876.

78  Smith was at first paid less than his colleagues because he was expected to receive fees from extension lectures. On 20 September 1852, Senate resolved that Smith be 'placed upon a similar footing with the other Professors in the augmentation of salary'. (Senate Minutes, 20 September 1852). When Smith was given leave in 1862, his salary was £950 pa. I am indebted to Dr Peter Chippindale for this information. The salary information given in Bruce R. Williams and David R.V. Wood, *Academic Status and Leadership in the University of Sydney, 1852-1987* (Sydney University Monographs No 6, 1991), 41, is slightly incorrect. For correct details, I am grateful to Mr Wood.

79  But from which, as David Branagan has observed, they had to pay for assistants and horses.

80  T.P. Anderson Stuart, 'The Majority of the Medical School: Being an Account of the Rise and Progress of the School', *Australian Medical Gazette,* 21 (1902), 491-503; J.L. Young *et al.* (eds.), *Centenary Book of the*

*University of Sydney Faculty of Medicine* (Sydney: Sydney University Press, 1984), 39-41.

81    Le Fèvre, *op.cit.*

82    David Branagan (ed.), *Rocks, Fossils, Profs: Geology and Science at the University of Sydney* (Sydney: Science Press, 1973), 3.

83    Institute of Geological Sciences, GSM 1/8/381-2, Murchison to Deas Thomson, 26 January 1866.

84    Mitchell Library, A 1531, Deas Thomson Papers, Edward Deas Thomson, 'Chancellor's Address', 5 April 1873.

85    G.P. Whitley and Martha Rutledge, 'Johann Ludwig (Louis) Gerard Krefft (1830-1881)', *Australian Dictionary of Biography*, vol. 5, 42-44.

86    See Moyal *op. cit;* Clarke Papers, M.L. MSS 139/49. Thomson to Clarke, 2 August 1871.

87    Clarke Papers, M.L. MSS 139/43. James Manning to Clarke, 25 December 1871.

88    Senate Minutes, Special Meeting, 30 November 1871, 16.

89    Charles Watts was paid from Smith's salary, and acted for him in the laboratory for a further £5 per week.

90    See G.C. Bolton, 'Richard Daintree (1832-1878)', *Australian Dictionary of Biography*, vol. 4, 1-2. Daintree merits a full biography. Geoffrey Bolton has given us a tantalising glimpse in *Richard Daintree: A Photographic Memoir* (Brisbane: Jacaranda Press, 1965).

91    Cf. E.J. Dunn and D.J. Mahony, 'Biographical Sketch of the Founders of the Geological Survey of Victoria', *Bulletin of the Victorian Geological Survey*, 23 (1910), 7-48; Thomas A. Darragh, 'The Geological Survey of Victoria under Alfred Selwyn, 1852-1868', *Historical Records of Australian Science*, 7 (1), (1987), 1-25.

92    Clarke Papers M.L. MSS 139/37. Thomson to Clarke, 3 March 1871.

93    Clarke Papers, M.L. MSS 139/37. Daintree to Clarke, 21 January 1872.

94    In 1895, Judd was appointed Dean of the Royal College of Science, succeeding T.H. Huxley. In 1877 he was elected FRS, and in 1895, was gazetted CB. He became widely known, thanks to his popular texts and his monograph on Krakatoa. But he was also distinguished for his microchemical work on bores taken from Funafuti, during the celebrated test of Darwin's theory of coral atoll formation (1896-1904). On this he and Liversidge were again to come into contact. See R.M. MacLeod, 'Imperial Reflections in the Southern Seas: The Funafuti Expeditions, 1898-1904', in Roy MacLeod and P.F. Rehbock (eds.), *'Nature in its Greatest Extent': Western Science in the Pacific* (Honolulu: University of Hawaii Press, 1988), 159-194.

95    Smith, *op.cit.*, second series, 217-222; 230.

96    And in 'choosing new chemical apparatus and ascertaining their uses, so as to keep Sydney University abreast in the March of Science'. Mitchell Library, A 1531-3, Deas Thomson Papers, Minnie Smith to Deas Thomson, 14 January 1875.

97    Senate Minutes, 4 April 1872.

98    Leonard Huxley, *The Life and Letters of Thomas Henry Huxley* (London: Macmillan, 1900), vol.1, 81, T.H. Huxley to Henrietta Heathorn, 5 August 1852.

99    Charles Darwin, *Journal of Researches into the Natural History and Geology of the Countries visited during Voyage of HMS 'Beagle' round the World* (London: John Murray, 1845), 420.

100    For an entertaining mixture of motives, see Alexander Marjoribanks, *Travels in New South Wales* (London: Smith, Elder and Co., 1847), 234.

101    Henry Kingsley, 'Advice to Emigrants', *The* [Edinburgh] *Daily Review*, May 1870, quoted in Kaye Harman (ed.), *Australia Brought to Book: Responses to Australia by Visiting Writers, 1836-1939* (Sydney: Boobooks, 1985), 59.

102    Opportunities were just opening in the provinces. The Yorkshire College of Science, Leeds, was established in 1874, and Mason College in Birmingham in 1881. For a contemporary reading of the employment situation, see J.H. Gladstone, 'Presidential Address', Section B (Chemistry), *Report of the British Associaton for the Advancement of Science*, vol. 42 (Brighton, 1872), 64-69, cited in *Chemical News*, 26 (16 August 1872), 1-9.

103    T.G. Bonney, 'Natural Science at the University of Cambridge', *Nature*, I (1870), 452. See also Roy MacLeod and Russell Moseley, 'The "Naturals" and Victorian Cambridge: Reflections on the Anatomy of an Elite, 1851-1914', *Oxford Review of Education*, 6 (2), (1980), 177-195.

104    M. Foster and Adam Sedgwick (eds.), *The Works of Francis Maitland Balfour* (London: Macmillan, 1885), 22, citing Darwin to Balfour, 6 July 1881.

105    For his posthumous influence, see Roy MacLeod, 'Embryology and Empire: The Balfour Students and the Quest for Mechanism in the Laboratory of the Pacific', in Roy MacLeod and F.R. Rehbock (eds.), *Darwin's*

*Laboratory: Evolutionary Theory and Natural History in the Pacific* (Honolulu: University of Hawaii Press, 1994), 140-165.

106 Sollas (1849-1936) was also elected a Fellow of St. John's, Cambridge in 1882, but went to a chair in Oxford in 1897. For his tribute to Huxley, see W.J. Sollas, 'The Master', *Nature*, 115 (2897), (9 May 1925), 747-748.

107 James Harrison, *Australia Felix Monthly Magazine*, I (June 1849), 41-42.

108 Calculated from the biographical entries in Margaret Reeks, *Register of the Associates and Old Students of the Royal School of Mines and History of the Royal School of Mines* (London: RSM, 1920).

109 This salary afforded a reasonable middle-class living, at a time when wheelwrights and stonemasons, the 'labour aristocracy' in Sydney, earned between £100 and £130 a year, and cooks and servants between £20 and £30. *Australian Almanac for 1873* (Sydney: John Ferguson, 1873), 163.

110 *The Londonium,* 5 (2), (April 1921), 41.

111 Denison, *op.cit.*, I, 309.

112 Leslie Stephen, *Sketches from Cambridge* (London and Cambridge: Macmillan, 1865), 94.

113 See A.L., 'Dendritic Spots on Paper', *J. Chem. Soc.,* 25 (16 May 1872), 646-648.

114 Smith, *Wayfaring Notes,* (Second Series): *A Holiday Tour Round the World, op.cit.,* 261. The *Northumberland*'s saloon passengers numbered forty, including Liversidge and the Smiths, while its second and third class held 105. Cf. *The Argus,* 17 September 1872.

115 The population of New South Wales increased from 198,168 in 1851 to 358,278 (excluding Victoria) in 1861, to 519,182 in 1872. Revenue increased from £406,000 in 1851 to £1.5 million in 1861 and to £2.7 million in 1871. Trade increased from £3.3 million in 1851 to £11.9 million in 1861 to £20.9 million in 1871. See Christopher Rolleston, 'Statistical Review of the Progress of New South Wales in the last 10 Years, 1862-1871', *TPRSNSW,* (1872), 122-123.

116 Roy MacLeod and Milton Lewis, 'A Workingman's Paradise? Reflections on Urban Mortality in Colonial Australia, 1860-1900,' *Medical History,* 31 (4), (1987), 387-402.

117 Pell declined the accommodation that the Senate offered him, and was granted an allowance of £150 in lieu. Senate Minutes, 19 July 1852.

118 Dynever Terrace was then a row of four houses between 12-18 College Street. Liversidge lived there until at least May 1873. Liversidge Papers, Box 10, A.L. to Russell, 22 May 1873. The site is now occupied by the NSW Department of Police.

119 See Michael Hoare and Martha Rutledge, 'Sir William John Macleay (1820-1891)', *Australian Dictionary of Biography,* vol. 5, 185-187; Nancy Gray, 'Alexander Walker Scott (1800-1883)', *Australian Dictionary of Biography,* vol. 6, 93-94.

120 R.F.F. Gillespie, 'Forty Years On', *Draft History of the Australian Club,* 52, quoted by Angel, *ibid.,* 67; see R.H. Goddard, *The Union Club, 1857-1957* (Sydney: Halstead Press, 1957), 1-2.

121 See Martha Rutledge, 'Sir Alfred Stephen (1802-1894)', *Australian Dictionary of Biography,* vol. 6, 180-187; John Atchison, 'Robert Hoddle Driberg White (1838-1900)', *idem.,* 389-390; F.K. Crowley, 'Sir William Charles Windeyer (1834-1897)', *idem.,* 420-422.

122 Gardner, *op.cit.*, 66.

123 Russel Ward, *The Australian Legend* (Melbourne: Oxford University Press, 2nd ed., 1966), 21.

124 Russel Ward, *Australia: A Short History* (Englewood Cliffs, NJ: Prentice-Hall, 1965), 30.

125 *SMH,* 23 April 1869.

126 George Nadel, *Australia's Colonial Culture: Ideas, Men and Institutions in Mid-Nineteenth Century Eastern Australia* (Melbourne: Cheshire, 1957), 119.

127 *Select Committee on the Botanic Gardens, Mins.Evid., VPLCNSW,* 1855, vol. 1, 1176.

128 See Sybil Jack, 'Cultural Transmission: Science and Society to 1850', in Rod Home (ed.), *Australian Science in the Making* (Sydney: Cambridge University Press, 1988), 45-66.

129 The Sydney press regularly noticed itinerant scientific lecturers – like the Mr. Gilchrist of Philip Street, who advertised 'pleasing and interesting experiments on the nature and property of air', and gave discourses similar to Dr. Pepper's demonstrations in London, *Sydney Gazette,* 18 July 1818, as well as on 'tidbits' of scientific new, such as the discovery of meteorites in America, and Sir Humphrey Davy's election as President of the Royal Society. *Sydney Gazette,* 6 November 1819, 16 June 1821.

130 Michael Cannon, *Australia in the Victorian Age: Vol. 3. Life in the Cities* (Melbourne: Nelson, 1975), chapter 8.

131 Quoted in Manning Clark, *A History of Australia* (Melbourne: Melbourne University Press, 1981), vol. 4, 92.

[132] *Empire,* 25 August 1859.

[133] See Elena Grainger, *The Remarkable Reverend Clarke* (Melbourne: Oxford University Press, 1982).

[134] Asa Briggs, *Victorian Things* (London: Batsford, 1988), Introduction.

[135] Graeme Davison, 'Exhibitions', *Australian Cultural History,* 2 (1982/3), 5-21, esp. 6.

[136] 'Exhibition of the Products of New South Wales', *Sydney University Magazine,* no. 2 (April 1855), 146-166.

[137] See Robert W. Rydell, *All the World's A Fair: Visions of Empire at American International Expositions, 1876-1916* (Chicago: University of Chicago Press, 1984), 1-9.

[138] *SMH,* 6 May 1869.

[139] Brian Fletcher, *The Grand Parade: A History of the Royal Agricultural Society of New South Wales* (Sydney: Royal Agricultural Society of NSW, 1988), 68-71.

[140] Harley Wood, 'George Robert Smalley (1822-1870)', *Australian Dictionary of Biography,* vol. 6, 136-137. With Charles Todd, Smalley fixed the boundary between South Australia and NSW.

[141] *Select Committee on the Botanic Gardens, VPLCNSW,* 1855, I, *Mins. Evid.,* Q. 1156 (Charles Moore, 10 August 1855).

[142] In 1927, the functions of the Sydney Mint were transferred to the new Commonwealth Mint in Canberra.

[143] A.W. Ward, 'William Stanley Jevons (1835-481)', *Dictionary of National Biography,* vol. X, 811; see also H.A. Jevons (ed.), *Letters and Journal of W.S.J.* (London: Macmillan, 1886), and R.D. Collison Black and Rosemond Konekamp (eds.), *Papers and Correspondence of William Stanley Jevons* (London: Macmillan, in association with the Royal Economic Society, 1972); see also J.A. La Nauze, 'William Stanley Jevons (1835-1882)', *Australian Dictionary of Biography,* vol. 4, 480-481, and La Nauze, 'Jevons in Sydney', in *Political Economy in Australia: Historical Studies* (Melbourne: Melbourne University Press, 1949), 26-44.

[144] For recent work, see Harro Maas, *William Stanley Jevons and the Making of Modern Economics* (Cambridge: Cambridge University Press, 2005), and Graeme Davison, 'The Unsociable Sociologist- W.S. Jevons and his Survey of Sydney, 1856-8', *Australian Cultural History,* 16 (1997-98), 127-150.

[145] NSW Archives, Sydney Mint Papers*,* 3/1665. Letterbook 1, f. 100, Staff List, 1854.

[146] Reeks, *op.cit.,* 'Robert Hunt', 96; see also William Warren, 'Anniversary Address', *JPRSNSW,* 27(1893), 4. The colonial Hunt is not to be confused with Robert Hunt, FRS (1807-1887), Keeper of Mining Records in Jermyn Street (1845-1878), and author of *Mineral Statistics, 1855-1884,* for whom see *Concise Dictionary of National Biography,* 1965 ed., 662.

[147] G.P. Walsh, 'Charles (Carl) Adolphe Leibius (1833-1893)', *Australian Dictionary of Biography,* vol. 5 (1851-1890), 79; see also *Daily Telegraph,* 20 June 1893, and T.P. Anderson Stuart, obituary notice of Dr. Leibius, *JPRSNSW,* XXVIII (1893), 36.

[148] NSW Archives, Sydney Mint Papers 3/1665. Letterbook 2, 357, Ward to Treasurer, 15 May 1860; 2, 450-451, April 1861. See Francis B. Miller, 'On the Application of Chlorine Gas to the Toughening and Refining of Gold', *Journal of the Chemical Society (of London),* 21 (1868), 506-513.

[149] Charles Moore, 'Anniversary Address', *JPRSNSW,* 14 (1880), 1-2.

[150] A.L., 'The Royal Society of New South Wales', *Nature,* 83 (23 June 1910), 503.

[151] John Smith, 'Anniversary Address', *JPRSNSW,* XIII (1879), l.

[152] For comparative statistics, see Davidson, *op.cit.,* 294-295.

[153] Cf. Arnold Thackray, 'Natural Knowledge in Cultural Context: The Manchester Model', *American Historical Review,* 78 (1974), 672-709.

[154] Report to Council by George R. Smalley and Edward Bedford, 26 July 1865, quoted in Maiden, *op.cit.,* 303.

[155] R.T.M. Pescott, 'The Royal Society of Victoria from Then, 1854 to Now, 1959', *Proceedings of the Royal Society of Victoria,* ns 73 (1961), 1-40. The Society's counterparts were the Royal Society of Tasmania (Van Diemen's Land), begun in 1843 from the Botanical and Horticultural Society of Van Diemen's Land; the Royal Society of Victoria, founded in 1855 as the Philosophical Institute of Victoria by the amalgamation of the Philosophical Society of Victoria and the Victorian Association for the Advancement of Science; this took the title 'Royal' in 1859. There was also the Adelaide Philosophical Society, founded in 1853, which became the Royal Society of South Australia in 1880; the Philosophical Society of Queensland, founded in 1859, which became the Royal Society of Queensland in 1885; and the Royal Society of Western Australia, created in 1914, from the Western Australian Natural History Society.

[156] *Rules of the Royal Society of New South Wales,* adopted 1 November 1865, quoted in Maiden, *op.cit.,* 305.

[157] Quoted in Elkin, *op.cit.,* 17.

158 T.H.S. Escott, *Social Transformations of the Victorian Age* (London: Seeley and Co., 1897), 320.

159 See G.P. Walsh, 'Henry Chamberlain Russell (1836-1907)', *Australian Dictionary of Biography*, vol. 6, 74-75.

160 A.L., 'The Deniliquin or Barratta Meteorite' (18 December 1872), *TRSNSW*, 6 (1872), 97-103.

161 'The University of Sydney', *SMH*, 29 August 1870.

162 Anthony Trollope, *Australia and New Zealand* (Melbourne: George Roberts, 1873), 149.

163 See T.H. Huxley, 'Universities, Actual and Ideal' (1874), in *Science and Education* (London: Macmillan, 1893).

164 Gardner, *op.cit.,* 68.

165 W. Lauder Lindsay, 'The Place and Power of Natural History in Colonisation, with special reference to Otago (New Zealand)', *Edinburgh New Philosophical Journal*, 17 (ns), (1863), 141.

166 Dallen, *op.cit.,* 281.

167 John Woolley, 'Lecture at the Sydney School of Arts', June 1854, in *Lectures Delivered in Australia* (Cambridge: Macmillan, 1862).

168 *Australian Almanack, 1872* (Sydney: John Sheriff, 1872).

169 *Speeches by Sir Hercules Robinson,* (Sydney: Gibbs, Shallard, 1879), 127-128.

170 *Australian Almanack, 1873* (Sydney: John Ferguson, 1873), 163.

171 Of this, £1.7 million arose from taxes, and £497,900 from the sale and lease of Crown land. See Trollope, *op.cit.,* 159.

172 See Wilma Radford, 'Charles Badham (1813-1884)', *Australian Dictionary of Biography*, vol. 3, 68-71.

173 *The Athenaeum*, cited in Thomas Butler, 'Memoir of Professor Badham', in *Speeches and Lectures delivered in Australia by the late Charles Badham, DD* (Sydney: William Dymock,1890), 23. Similar sentiments were expressed by A.E. Housman in 1899. See James Diggle and FRD Goodyear (eds.), *The Classical Papers of A.E. Housman* (Cambridge, 1972), 472. See also '"Some Oxford Scholars": Una conferenza inedta di J.U. Powell', *Eikasmos*, VII (1997), 245-282.

174 Trollope, *ibid,* 135; Sir Robert Randolph Garran, *Prosper the Commonwealth* (Sydney: Angus and Robertson, 1958), 76.

175 Charles Badham, 'Sydney University Commemoration, 1867' in *Speeches and Lectures delivered in Australia by the late Charles Badham, DD* (Sydney: William Dymock, 1890, 1.

176 *Ibid*.

177 Charles Badham, 'Commemoration Address, 1870', in *ibid.*, 27-32.

178 Charles Badham, 'Commemoration Address, 1868', in *ibid.*, 8.

179 The memorable phrase recalled by Sir Robert Garran, *op.cit.,* 76.

180 Charles Badham, 'Commemoration Address, 1868', 11.

181 N.J.B. Plomley, *Several Generations* (Sydney: Wentworth Books, 1971), 84.

182 Senate Minutes, 1 May 1872, 29-30; 2 October 1872.

183 *SMH*, 27 February 1873.

184 Liversidge Papers, Box 18, f. 52. University of Sydney, Lent Term, 1874. Set of Apparatus necessary for the course of Experimental Chemistry and Qualitative Analysis. Liversidge Papers, Box 15, f. 14, 'Chemical Apparatus', *SMH*, 23 February 1874.

185 Senate Minutes, 4 December 1872, 44; 8 January 1873, 46.

186 Senate Minutes, 5 February 1873, 49.

187 Anthony Trollope, *Australia and New Zealand* (Melbourne: George Roberts, 1873), 182.

188 *SMH,* 21 June 1872.

189 'The London School of Mines', *SMH*, 11 December 1872.

190 'School of Mines', *SMH*, 6 February 1873.

191 *Ibid*.

192 The Lawrence Scientific School at Harvard was established in 1847; the Sheffield Science School at Yale was founded in 1854, and renamed in 1861. Lawrence Veysey, *The Emergence of the American University* (Chicago: University of Chicago Press, 1965). See also George Wilson Pierson, *Yale College: An Educational History, 1871-1921* (New Haven: Yale University Press, 1952), 50.

193 'School of Mines', *SMH*, 27 February 1873.

194 *Ibid*.

195 *Ibid*.

196  *Ibid.*

197  *Ibid.*

198  *First Report of the Royal Commission on Scientific Instruction and the Advancement of Science,* 1871. [C. 318]. xxiv, 643. Cf. Chambers, *op. cit.,* xxxviii.

199  *SMH,* 27 February 1873.

200  Senate Minutes, 5 March 1873, 67.

201  Mitchell Library, Deas Thomson Papers, A 1531, Edward Deas Thomson, 'Chancellor's Address', 5 April 1873.

202  Charles Badham, 'Sydney University Commemoration, 1873', in Badham, *op.cit.,* 61-62.

203  Senate Minutes, 1 October 1873, 67-68; 5 November 1873, 69.

204  Senate Minutes, 3 December 1873, 72; 4 February 1874, 77.

205  *Empire,* 7 November 1873.

206  Charles Badham, 'Sydney University Commemoration, 1872', in Badham, *op. cit.,* 50.

207  Barff, *op.cit.,* 93.

208  Senate Minutes, 10 February 1874. His title included the designation of Assistant Lecturer in Chemistry; this was modified in March (Senate Minutes, 4 March 1874, 80) to Demonstrator in Practical Chemistry. In March 1877, his salary was increased to £600 per annum. (University of Sydney Archives, Cash Book, 1877).

209  Australian Museum Letter Book (AMLB), B 10.74.61. A.L. to Secretary, Australian Museum, 7 February 1874.

210  Edward Deas Thomson, 'Commemoration Address', 22 March 1874.

211  Mitchell Library, Macleay Papers, ML MSS 2009/115. Item 3, Diary entry for 28 March 1874.

212  See 'The Macleay Museum', *SMH* 11 April 1874, filed in Liversidge Papers, Box 15. See also Peter Stanbury and Julian Holland, *Mr Macleay's Celebrated Cabinet* (Sydney: The Macleay Museum, 1988).

213  Liversidge Papers, Box 15, f. 14; *SMH,* 30 March 1874.

214  *Empire,* 30 March 1874.

# CHAPTER 5

1  Archibald Liversidge, 'Sydney Water', Letter to the Editor, *SMH,* 30 January 1874; *Union Club Letterbook, 1870-79,* f. 166, F.H. Wilson to Edward Bell, 21 February 1874; R.H. Goddard, *The Union Club, 1857-1957* (Sydney: Halstead Press, 1957).

2  See also 'Purification of Water - A question asked by "A Young Householder"', and A.L., reply, *SMH,* 4 February 1874.

3  F.H. Wilson, Letter to the Editor, *SMH,* 16 March 1874.

4  A.J.C. Mayne, *Fever, Squalor and Vice: Sanitation and Social Policy in Victorian Sydney* (St. Lucia: University of Queensland, 1982); *Second Report of the Commission of Inquiry into the Sanitary State of Large Towns and Populous Places, 1845,* XVIII [C. 602]; cf. *Edwin Chadwick, Report on the Sanitary Condition of the Labouring Population of Great Britain* (ed. M.W. Flinn), (Edinburgh: Edinburgh University Press, 1964); Anthony S. Wohl, *Endangered Lives: Public Health in Victorian Britain* (Cambridge, Mass.: Harvard University Press, 1983).

5  Milton Lewis and Roy MacLeod, 'A Workingman's Paradise? Reflections on Urban Mortality in Colonial Australia, 1860-1900', *Medical History,* 31 (1987), 387-402. For contemporary parallels with 'marvellous Smelbourne', see Graeme Davison, *The Rise and Fall of Marvellous Melbourne* (Melbourne: Melbourne University Press, 1978).

6  Since well before Chadwick made the problem fashionable, the supply of clean water was one of the most important, but also one of the most vexing duties of local government – compounded by legal complexities, administrative ambiguities, and technical disagreements. Cf. R.A. Lewis, *Edwin Chadwick and the Public Health Movement* (London: Longmans, Green, 1952); S.E. Finer, *The Life and Times of Sir Edwin Chadwick* (London: Methuen, 1952).

7  Cf. Shirley Fisher, 'The Pastoral Interest and Sydney's Public Health', *Historical Studies,* 20 (1982), 73-89; Shirley Fitzgerald, *Rising Damp: Sydney, 1870-90* (Melbourne: Oxford University Press, 1987).

8  John Smith, 'The Action of Sydney Water upon Lead', *Phil. Soc. of NSW,* 13 August 1856, reported in the

*SMH,* 15 August 1856.

9  Cf. John Simon, *Reports of the Medical Officer of the Privy Council* (1855-76); Royston Lambert, *Sir John Simon, 1817-1904, and English Social Administration* (London: McGibbon and Kee, 1963). John Eyler, *Victorian Social Medicine: The Ideas and Methods of William Farr* (Baltimore: Johns Hopkins University Press, 1979).

10  Cf. Margaret Pelling, *Cholera, Fever and English Medicine, 1825-1865* (London: Oxford University Press, 1978); see also Christopher Hamlin's instructive, 'Muddling in Bumbledom: On the Enormity of Large Sanitary Improvements in Four British Towns, 1855-1885', *Victorian Studies,* 32 (1), (1988), 55-83.

11  For comparative perspective, see Christopher Hamlin, 'Providence and Putrefaction: Victorian Sanitarians and the Natural Theology of Health and Disease', *Victorian Studies,* 28 (1985), 381-411.

12  John Smith, 'On the Water Supply of Sydney', *TRSNSW,* 2 (1868), 86-96.

13  *Council Minutes,* 6, 14, 21, 27 July; 3, 17 August; 14, 24 September; 5 October 1870.

14  Cf. 'Conference on the Steps to be taken to Ensure Prompt and Efficient Measures for Preventing the Pollution of Rivers', *Journal of the Royal Society of Arts,* 23 (1874-78), 91.

15  Public fears were fueled by a devastating report from Dr. G.F. Dansey, the City Health Officer. Dansey was appointed in 1862, succeeding Dr. Isaac Aaron, the first Medical Officer of Health of the City of Sydney, who had been appointed under the Sydney Corporation Act of 1857, but who had been dismissed from office for criticising those responsible for overcrowding in Sydney's slums. Dansey met a similar fate in 1888. C.J. Cummins, *A History of Medical Administration in New South Wales, 1788-1973* (Sydney: Health Commission of NSW, 1973), 66-67.

16  Cf. J.T. Bunce, *History of the Corporation of Birmingham with a Sketch of the Earlier Government of the Town* (Birmingham: Cornish, 1878, 1885); See Mayne, *op.cit.* For London in the same period, see Roy MacLeod (ed. for David Owen), *The Government of Victorian London, 1855-1889: The Metropolitan Board of Works, the Vestries, and the City Corporation* (Cambridge, Mass.: Harvard University Press/ London: Blackwell, 1982)

17  For Pell, see I.S. Turner, 'Morris Birkbeck Pell (1827-1879)', *Australian Dictionary of Biography,* vol. 5, 428-429.

18  *State Archives,* 4/805/3, Sydney City and Suburban Sewerage and Health Board, Archibald Liversidge to Secretary, 14 May 1875, 6 July 1875.

19  'Professor Liversidge's Analysis of Sydney Water', *SMH,* 9 July 1875; 'The Sydney Water Supply', *SMH,* 11 July 1875; 'Analysis of Water Supply', *Echo,* 15 July 1875; 'Sydney Water', *Echo,* 16 July 1875.

20  *State Archives,* 4/805/3, Sydney City and Suburban Sewerage and Health Board, A.L. to Pell, 22 July 1875.

21  For example, William Corfield's *Digest of Facts relating to the Treatment and Utilisation of Sewage* (London: Macmillan, 2nd ed., 1871), a copy of which he possessed.

22  For example, *The Echo,* 20 August 1875, made much of them.

23  *Liversidge Papers,* Box 15, f. 172. 'Water Pollution Prevention Bill', *SMH,* 16 July 1875. 'The Act gave a great amount of work to plumbers', as Liversidge recalled, 'as cisterns had to be installed in thousands of homes. Further, the new catchment areas on the Nepean and other rivers had to be provided'. (Liversidge margin notes).

24  For background to this epidemic, see Noel Deerr, *Cane Sugar* (London: Norman Rodger, 1921), 'Cane Diseases', and C.G. Hughes and E.V. Abbott (eds.), *Sugar Cane Diseases of the World* (Amsterdam: Elsevier, 1961).

25  'Report upon the Sugar-cane Disease in the Mary River District, Queensland', *Queenslander,* no. I (6 May 1876), 21; no. II (13 May 1876), 20; no. III (20 May 1876), 20; no. IV (27 May 1876), 19; no. V (3 June 1876), 13; no. VI (10 June 1876), 10.

26  Joseph Bancroft, 'Development of Sugar Cane Disease in Queensland', *The Sugar Cane,* IX (l September 1877), 476-480. For further discussion, see 'Report of the Board appointed to enquire into the Causes of Disease affecting Live Stock and Plants', *Votes and Proceedings of the Legislative Assembly of Queensland,* 3 (1876), 14-21; 5 (1878), 711-712. See also M. Josephine Mackerras, 'Joseph Bancroft (1836-1894)', *Australian Dictionary of Biography,* vol. 3, 84-85. For local information, I am grateful to Mr. John Kerr of St. Lucia, and to Mrs. M. Walker, of the Maryborough, Wide Bay and Burnett Historical Society.

27  See D.A. Herbert, 'A Story of Queensland's Scientific Achievement, 1859-1959', *Proc. Roy. Soc. Queensland,* 71 (1959), 8-9.

28  Cf. *The Sugar Cane,* 8 (1 August 1876), 429-435; 8 (1 September 1876), 471-480; 8 (2 October 1876), 539-545; and 8 (1 December 1876), 633-641; also reprinted in the *Moniteur de la Nouvelle-Calédonie, Journal*

*Officiel* (October 1876). A collection of the reports, printed in Sydney, was reviewed in *Chemical News,* 34 (1876), 272.

29  The Société d' Acclimatation de l'Ille Maurice. His report was received in August 1876. See *Transactions of the Royal Society of Arts and Sciences of Mauritius,* X (ns), (1878), 55, 60. For this reference I am grateful to Dr. Ly-Tio Fane, of the Sugar Industry Research Institute, Réduit. Mauritius, a British colony since the Napoleonic wars, was in the nineteenth century the British Empire's single largest producer of cane sugar. Visiting Port Louis in 1847, T.H. Huxley described it as a 'complete paradise'. See Jordan Goodman, *HMS Rattlesnake* (London: Faber and Faber, 2005), 88.

30  Archibald Liversidge, 'Report upon the Disease in the Sugar Cane', *The Sugar Cane,* VIII (1876), 641.

31  Other afflictions include mosaic, gum disease, leaf scald and 'Fiji' disease. See F.S. Earle, *Sugar Cane and its Culture* (New York: John Wiley and Sons, 1928), 109-161.

32  Henry Ling Roth, *A Report on the Sugar Industry in Queensland* (Brisbane: Gordon and Gotch, 1880), 71-72.

33  This contributed ultimately to the establishment of the Bureau of Sugar Experiment Stations, and the Sugar Research Institute in Mackay. See Norman King, 'The Foundation and Development of the Bureau of Sugar Experiment Stations', in *Fifty Years of Scientific Progress* (Brisbane: Government Printer, 1950), 5-15. For technical assistance, I am grateful to Dr. Owen Sturgess, Director of the Bureau.

34  Sydney City and Suburban Sewage and Health Board, *Third Report upon the Quality of the Sydney City and Suburban Water Supply* (Sydney: Government Printer, 1876), 7.

35  NSW State Archives, 4/805/3, Sydney City and Suburban Sewerage and Health Board, A.L. to Pell, 3 August 1876.

36  Sydney City and Suburban Sewage and Health Board, Minutes of Evidence in *Twelfth and Final Report of the Board appointed on the 12 April 1875, to inquire into and Report as to the Best Means of Disposing of the Sewage of the City of Sydney and its Suburbs, as Well as of Protecting the Health of the Inhabitants thereof* (11 May 1877), *VPLANSW,* 3, 1876-1877, *Mins. Evid.* (9 October 1876), Q. 3484-3502.

37  NSW State Archives, 4. 805. 3. A.L. to Chairman, Sydney City and Suburban Sewerage and Health Board, 6 October 1876; 'Dredgings from Farm Cove' (21 October 1875), *VPLANSW,* 5, 1875-1876, 386-387; 'Analyses of Water' (23 October 1875), *VPLANSW,* 4, 1875-1876, 387-388.

38  *The Metropolitan Water Board Act, 1880,* 43 Vict. c.32.

39  David Branagan,'Bricks, Brawn and Brains: Two Centuries of Geology and Engineering in the Sydney Region', *Journal and Proceedings of the Royal Society of New South Wales,* 129 (1996), 1-32; W.V. Aird, *The Water Supply, Water Sewerage and Drainage of Sydney, 1788-1960* (Sydney: Metropolitan Water, Sewerage and Drainage Board, 1961).

40  'Deodorization of Sewage Matters' (Correspondence with John Smith and C. Watt), *ARDM, 1879* (Sydney: Government Printer, 1880), 29-30; *VPLANSW,* 5 (1879-80), 533-536.

41  R.J.W. Le Fèvre, 'The Establishment of Chemistry in Australia', in A.P. Elkin (ed.), *A Century of Scientific Progress* (Sydney: Royal Society of New South Wales, 1968), 354.

42  William E. Ford, 'The Growth of Mineralogy from 1818 to 1918', *American Journal of Science,* 4th ser., 45 (1918), 250.

43  Cf. Aaron Ihde, *The Development of Modern Chemistry* (New York: Harper and Row, 1964), 142; Evan M. Melhado, *Jacob Berzelius: the Emergence of his Chemical System* (Madison: University of Wisconsin, 1981), 308-326.

44  Mineralogical Society (London), Correspondence, Maskelyne to J.H.Collins (Cornwall School of Mines, Truro), 27 September 1875.

45  Mineralogical Society (London), Correspondence, Samuel Haughton to J.H. Collins (Cornwall School of Mines, Truro), 13 January 1876.

46  *Idem.*, R. Hunt to J.H. Collins, 18 September 1875.

47  Worner and Vallance, *op.cit.,* 358; T.G.Vallance, 'John Lhotsky and Geology', in V. Kruta et al., *Dr John Lhotsky, The Turbulent Australian Writer, Naturalist and Explorer* (Melbourne: Australia Felix Literary Club, 1977).

48  Jan Lhotsky, 'Mineralogy of Australia', *New South Wales Magazine,* 1 (1), (August 1833).

49  W.A. Cawthorne, *Menge the Mineralogist: A Sketch of the Life of the Late Johann Menge* (Adelaide: J.D. Shawyer, 1859); B.J. Cooper, *et al.*, 'Johannes Menge – South Australia's First Geologist', *Abstracts of the Geological Society of Australia*, 15 (1986).

50  Howard K. Worner and Thomas G.Vallance, 'Early Australian Mineralogy', *The Mineralogical Record*, 19 (6), (1988), 360.

51  Cf. the work of Robert Hunt and Carl Leibius on simplifying the refining of gold in F.B. Miller, 'On the Application of Chlorine Gas to the Toughening and Refining of Gold', *J. Chem. Soc.*, 21 (1868), 506-513.

52  On Wood, see Thomas A. Darragh, 'Charles Sturtevant Wood, Analyst. His Work and his Correspondence with James Hector', *Historical Records of Australian Science*, 7 (3), (1988), 231- 271.

53  David Branagan, 'A.R.C. Selwyn: Pioneer of the Australian Geological Profession', *Abstracts of the Geological Society of Australia*, 12 (Sydney: 7th Australian Geological Convention, 1984), 76-78.

54  Worner and Vallance, *op.cit.*, 360. Its formula is Au$_2$Bi.

55  *Ibid.*, 361.

56  Edward Salisbury Dana, *The System of Mineralogy of James Dwight Dana: Descriptive Mineralogy* (London: Chapman & Hall, 1914), vii-ix, preface to the third (1850) and fifth (1868) editions.

57  A.M. Thomson, *Guide to Mineral Explorers in Distinguishing Minerals, Ores and Gems* (Sydney: John Sands, 1869). A copy, marked 'from the author with his compliments' is preserved among Liversidge's books in the Library of the Royal Society of New South Wales.

58  W.B. Clarke, 'Annual Address', *TRSNSW*, 2, 1869, 12.

59  Patrick Stirling, *The Australian and Californian Gold Discoveries and their Probable Consequences* (Edinburgh: Oliver and Boyd, 1853); John Calvert, *The Gold Rocks of Great Britain and Ireland* (London: Chapman and Hall, 1853); Edward Hargraves, *Australia and its Gold Fields* (London: H. Ingram, 1853); Paul E. de Strzelecki, *Physical Description of New South Wales and Van Diemen's Land* (London: Longmans, 1845); Simpson Davidson, *The Discovery and Geognosy of Gold Deposits in Australia* (London: Longman, Green, Longman and Roberts, 1860), 90.

60  W.B. Clarke, *Researches in the Southern Gold Fields of New South Wales* (Sydney: Reading and Wellbank, 1860). Liversidge's annotated copy is in Christ's College, Cambridge.

61  See Michael Hoare, 'The "Half-Mad Bureaucrat": Robert Brough Smyth (1830-1889)', *Records of the Australian Academy of Science*, 2 (4), (1973), 25-40; 'Robert Brough Smyth (1830-1889)', *Australian Dictionary of Biography*, vol. 6, 161-163.

62  R. Brough Smyth, *Goldfields and Mineral Districts of Victoria* (Melbourne: John Ferres, 1869), 235.

63  In July 1873, Liversidge obtained Adam Sedgwick's copy of Elie de Beaumont's *Recherches sur Quelques-unes des Revolutions de la Surface du Globe* (1829-30), and was familiar with his work.

64  Cf. John Challinor, *The History of British Geology: A Bibliographical Study* (Newton Abbott: David and Charles, 1971); G.C. Amstutz *et al.*, *Ore Genesis: State of the Art*, No. 2 (New York: Springer Verlag, 1982); A.H.G. Mitchell and M.S. Garson, *Mineral Deposits and Global Tectonic Settings* (New York: Academic Press, 1981); R.L. Stanton, *Ore Petrology* (New York: McGraw Hill, 1972).

65  Thomas Sterry Hunt, 'The Chemistry of the Primeval Earth', *Chemical News*, 15 (1867), 315-317, reprinted in his *Chemical and Geological Essays* (Boston: James R. Osgood and Co., 1875); W.H. Brock, 'Thomas Sterry Hunt', *Dictionary of Scientific Biography*, 6 (1972), 564-566.

66  W. Westgarth, *The Colony of Victoria: Its History, Commerce and Gold Mines* (London: S. Low, 1864), 304.

67  A. Selwyn, *Essays* (Melbourne: Government Printer, 1861), 223, cited in Westgarth, *op.cit.*, 309.

68  W.B. Clarke, *Researches in the Southern Goldfields of New South Wales* (Sydney: Reading and Wellbake, 1860).

69  W.B. Clarke, *On the Progress of Gold Discovery in Australasia from 1860-71* (Sydney: np, 1872), 9-10.

70  Davidson, *op.cit.*, 132.

71  For Daintree, see G.C. Bolton, 'Richard Daintree (1832-1878)', *Australian Dictionary of Biography*, vol. 4, 1-2.

72  Cf. C.S. Wilkinson, 'On the Theory of the Formation of Gold Nuggets in Drift', *Transactions and Proceedings of the Royal Society of Victoria*, 8 (1867), 11; Richard Daintree, *Report of the Geology of the District of Ballan, including Remarks on the Age and Origin of Gold* (Melbourne: Geological Survey, Government Printer, 1866), 9-10.

73  A.R.C. Selwyn and George Ulrich, *Notes on the Physical Geography, Geology and Mineralogy of Victoria* (Melbourne: Government Printer, 1866), 43.

74  J.H. Gladstone, 'Presidential Address', Section B (Chemistry), *Report of the BAAS* (Brighton, 1872), quoted in *Chemical News*, 26 (16 August 1872), 71.

75  Edward Schenck, 'Chemical Science', Presidential Address, *Report of the BAAS*, 57 (Manchester, 1887), 629.

76  Cf. Theodore Ranff, *The Origin and Formation of Auriferous Rocks and Gold* (Sydney: Turner and Henderson, 1878). Ranff presented Liversidge with a copy of the 1889 edition of this book.

77  Geologists still debate whether gold was deposited by a process of ancient volcanic syngenesis, or in 'epigenetic' rocks, formed in more recent times. See Neil Williams, 'Exploring for Magmatic Origins', *Nature*, 321 (26 June 1986), 812.

78  Brough Smyth, *op.cit.,* 235.

79  He shared credit for the identification of *marshite* in 1892. Worner and Vallance, *op.cit.*, 363. See Chapter 6, infra.

80  A.L., 'On the Formation of Moss Gold and Silver' (6 September 1876), *JPRSNSW*, 10 (1876), 125-133, 134.

81  *Ibid.*

82  *Ibid.*, 133.

83  Archibald Liversidge, 'On the Formation of Moss Gold and Silver', *op.cit.*, 134.

84  John Lucas carried the Coal Mines Regulation Act and amendments to the Mining Act under Robertson's ministry. However, the government fell before Lucas could 'persuade his colleagues to spend a surplus of £1,750,000 on such impressive public buildings as the schools of mines and design'. R.W. Rathbone, 'John Lucas (1818-1902)', *Australian Dictionary of Biography*, 5 (1851-1890), 108.

85  A.L., 'Notes on the Iron and Coal Deposits at Wallerawang, and on the Diamond Fields', in *Mines and Mineral Statistics of New South Wales, op. cit.*, 94-103; *Sydney Mail*, 27 February 1875; Liversidge Papers, Box 19, f. 7. Colonial Secretary to Sir James Martin, 9 August 1877.

86  A.L., 'Report upon certain of the Coals, Iron Ores, Limestones and Copper Ores of New South Wales', Part 1, *Annual Report, NSW Department of Mines* (afterwards *ARDM*), 1875 (Sydney: Government Printer, 1876), 127-147.

87  A.L., 'Analyses of Samples of Coal from NSW', *ARDM*, 1877 (Sydney: Government Printer, 1878), 24-29; 'Report on the character and value of copper ores of NSW', *ibid.*, 34-37; 'Report upon the Theoretical Calorific Power of certain Samples of NSW Coal', *ibid.*, 209.

88  ARDM, 1875, op.cit., 151. Reprinted in 'Analyses of auriferous tailings from New South Wales', *ARDM*, 1877, 14-19.

89  A.L., 'Report upon certain of the Auriferous and Stanniferous Tailings, Auriferous and Argentiferous Copper Ores, and Other Minerals of New South Wales', Part II, *ARDM*, 1875, 149-167. Report quoted in *ARDM*, 1877, 31-32, 33.

90  Mitchell Library (Brown Papers), Correspondence and Papers, A 1686-22, vol. 5, 147-157, A.L., 'Report on Geological Specimens from New Britain and New Ireland' (nd, ca. October 1875).

91  A.L., *The Minerals of New South Wales* (Sydney: Government Printer, 1876) was eventually superseded by Edward Pittman's *Mineral Resources of New South Wales* (Sydney: Geological Survey of New South Wales, 1901), complemented by George Smith, *A Contribution to the Mineralogy of New South Wales* (Sydney: Government Printer, 1926); this in turn was overtaken by E.C. Andrews, *The Mineral Industry of New South Wales* (Sydney: Department of Mines, 1928), and this, by N.L. Markham and H. Basden (eds.), *The Mineral Deposits of New South Wales* (Sydney: Geological Survey of New South Wales, 1974).

92  Speech by T.S. Mort at a luncheon celebrating the opening of a Sydney slaughterhouse, *SMH*, 4 September 1875.

93  For English models, see Robert Kargon, *Science in Victorian Manchester: Enterprise and Expertise* (Manchester: Manchester University Press, 1977).

94  'The Naturalist: Marsupials and Montremata', *Australian Town and Country Journal*, 27 September 1884, quoting a letter from A.L. to Mr. W.S. Dowel of Tamworth, NSW.

95  Within the last thirty years, unusual isotopes have been discovered, suggesting the presence of presolar matter; from these, radioactivity measurements have been used to estimate the age of the universe. Thanks to NASA, at least twenty minerals have been found that may have existed in the solar system before the appearance of life on earth.

96  A.L., 'The Deniliquin or Barratta Meteorite' (18 December 1872), *TRSNSW*, 6 (1872), 97-103.

97  A.L., 'Report on the Elsmore Tin Mine' (2 July 1873) in *Australian Town and Country Journal*, 26 July 1873, 110-111, abstract in *London Mining Journal*, 45 (1875), 1096. Tin was later discovered at Tenterfield, NSW, in 1872.

98   *Ibid.*

99   A.L., 'The Bingera Diamond Field' (1 October 1873), *TRSNSW*, 7 (1873), 91-103; reprinted in *Mines and Mineral Statistics of NSW* (Sydney: NSW Intercolonial and Philadelphia International Exhibition, 1875), 104-115; 'Report on the Discovery of Diamonds at Bald Hill, near Hill End' (1 October 1873), *TRSNSW*, 7 (1873), 102-103; and in *Mines and Mineral Statistics of NSW, op.cit.*, 115-116. His results were fully noticed in London. See Liversidge Papers, Box 9, Newspapers, 1 October 1875; 'Notes on the Bingera Diamond field, with Notes on the Mudgee Diamond Field' (24 June 1874), *Quarterly Journal of Geol. Society of London*, 31 (1875), 489-492; *SMH*, 1 October 1875. These diamond fields were worked intermittently until 1904, impeded by a lack of water. See N.L. Markham and H. Basden (eds), *The Mineral Deposits of New South Wales* (Sydney: Department of Mines, 1974), 577.

100  Anthony Trollope, *Australia and New Zealand* (Melbourne: George Roberts, 1873), 192.

101  Review of Thomas Carpenter, 'The Metallurgy of Gold', *Australian Town and Country Journal*, 25 April 1874.

102  A.L., 'Note on a New Mineral from New Caledonia' (7 May 1874), *Journal of the Chemical Society*, 27 (July 1874), 613-615.

103  'Tin Mines of Tasmania', *Supplement to the Mining Journal*, 45 (2 October 1875), 1095-1096; 'On the Stanniferous Deposits of Tasmania', by Mr. Wintle, read by Professor Liversidge, *TRSNSW*, 8 (1875), xxxv; see also W.A. Haswell, *Guide to the Contents of the Australian Museum* (Sydney: Australian Museum, 1883).

104  Senate Minutes, December 1874.

105  R. O. Chalmers, 'History of the Department of Mineralogy, Australian Museum, Part I: 1827-1901', *J. Min. Soc. NSW*, I, (1979), 15-17.

106  NSW State Archives, Department of Mines, 17/2173/74/3891/3698, 3781; Liversidge Papers, Box 19. f.1, Correspondence between Liversidge and Minister for Public Works, 30 September - 20 October 1874, and Minister for Mines, 3-16 November 1874.

107  The collections begun in 1875 rapidly increased after the appointment of the Department's Geological Survey staff in 1876. In 1877, the government bought the mineral collection of W.B. Clarke, which in 1881 was transferred to the Garden Palace Exhibition buildings in the inner Domain. In 1882, some 50,000 specimens, including most of Clarke's collection, were lost in the fire that destroyed the building. (See Chapter 5). In 1886, a new Museum, under the directorship of Mr. J.E. Carne, was begun in the Geological Survey offices in Macquarie Street, and this in 1893 moved to the former Agricultural Hall in the Domain, behind Sydney Hospital and vacated by the removal of the Technological Museum to Harris Street. The Museum moved again, to George Street North, in 1909. E.W. Card was its Curator from 1894 until his retirement in 1927. Cf. Geological Museum Archives, 'The Mining and Geological Museum, Sydney'; 'History of the Mining Museum', *Mineral* (December, 1958), 35; 'The New South Wales Mining Museum: A Useful Institution', *The Australian Mining Standard* (29 December 1909), 690-691.

108  A.L., 'The Minerals of New South Wales' (9 December 1874), *TPRSNSW*, 9 (1875), 153-215; expanded into *The Minerals of New South Wales* (Sydney: Government Printer, 1876); a second edition appeared as *Description of the Minerals of New South Wales, in Mineral Products of New South Wales* (Sydney: Department of Mines, 1882); and a third edition as *The Minerals of New South Wales* (London: Trübner and Co., 1888).

109  For much of the material (but not necessarily the interpretations) in this section, I am indebted to the pioneering work of A. Pigott and Ronald Strahan in Ronald Strahan *et al.*, *Rare and Curious Specimens* (Sydney: The Australian Museum, 1979).

110  Cox, educated in Edinburgh, was a leading figure in Sydney medical circles, an instrumental reformer of the Sydney Infirmary, and a teacher at the Medical School. He was the first Secretary of the Entomological Society in 1862, and later President of its successor, the Linnean Society of NSW (1881-82). He was also first President of the NSW Board of Fisheries (1862) and a Fellow of the Linnean Society of London (1868). He served as a Trustee of the Australian Museum from 1865-1912, perhaps the longest serving member in its history. He published four books of reference on Australian land shells. See Ruth Teale, 'James Charles Cox (1834-1912)', *Australian Dictionary of Biography*, vol. 3, 483.

111  Strahan, *op.cit.*, 37.

112  'Curator, Australian Museum', *VPLANSW*, 4 (1874), 5, Deas Thomson to Colonial Secretary, 5 May 1864.

113  Gerard Krefft, 'The Improvements Effected in Modern Museums in Europe and Australia', *TRSNSW*, 2, (1869), 15-25. See his volumes on *The Snakes of Australia* (Sydney: Government Printer, 1869); *A Short Guide*

to the *Australian Fossil Remains, exhibited by the Trustees of the Australian Museum, and arranged and named by Gerard Krefft* (Sydney: Australian Museum, 1870); and *The Mammals of Australia* (Sydney: Government Printer, 1871).

[114] Mitchell Library, Krefft Papers , MSS 956, f. 39. Agassiz to Krefft, 9 April 1870. Forster, a liberal MLA, and later the Agent-General of NSW in London, was a keen amateur naturalist. He was greatly flattered by the courtesy. See *idem*, f. 49, Forster to Krefft, 26 June 1872. Interest in its discovery has been compared to that stimulated by the *coelacanth* in the twentieth century. Strahan, *op.cit.*, 29.

[115] Mitchell Library, Krefft Papers, MSS 956, Charles Darwin to Krefft, 17 July 1872, 6 December 1876.

[116] Roy MacLeod, 'Evolutionism and Richard Owen 1830-1868: An Episode in Darwin's Century,' *Isis*, 56 (185), (1965), 259-280; Nicolaas Rupke, *Richard Owen, Victorian Naturalist* (New Haven: Yale University Press, 1994).

[117] See Ann Mozley Moyal, 'Sir Richard Owen and his influence on Australian Zoological and Palaeontological Science', *Records of the Australian Academy of Science*, 3 (2), (1975), 41-56.

[118] Barry Butcher, 'Gorilla Warfare in Melbourne: Halford, Huxley and "Man's Place in Nature"', in Rod Home (ed.), *Australian Science in the Making* (Sydney: Cambridge University Press, 1988), 153-169.

[119] See Alan W. Martin, *Henry Parkes: A Biography* (Melbourne: Melbourne University Press, 1980), 240ff.

[120] Mitchell Library, Parkes Papers, vol. 20, f. 348. Krefft to Parkes, 23 September 1873.

[121] Cooper's short parliamentary career ended that year, when he failed to win a seat in the 1874 elections. He left politics, and six years later died broken, bitter and poor. See Bede Nairn, 'Walter Hampson Cooper (1842-1880)', *Australian Dictionary of Biography*, vol. 3, 453-455.

[122] 'Report from the Select Committee on the Sydney Museum', *VPLANSW*, 5 (1873-4), 10, para 5.

[123] *Ibid.*, para. 7.

[124] *Ibid.*, para. 6

[125] The case is summarised in Peter Stanbury and Julian Holland (eds.), *Mr. Macleay's Celebrated Cabinet* (Sydney: Macleay Museum, 1988), 48-49.

[126] 'Curator, Australian Museum', *VPLANSW*, 4 (1875), 15; see Macleay Papers (Mitchell Library) 2009/115, diary entry 16 June 1874.

[127] Teale, *op.cit.*, 483.

[128] Owen seems not to have noticed the insult. In April 1874, he acknowledged receipt of the fossils, in a letter to Bennett that Krefft had printed as a testimonial in his defence. 'My best regards and good wishes to Krefft,' Owen added. 'The Museum will owe much to his energy and devotion. As soon as our new Museum is ready to enable us to sort out our accumulated stores, Sydney shall have my first selection of duplicates'. Sub-committee Report, *Mins. Evid.*, Q. 1214.

[129] *Ibid.*, *Mins. Evid.*, Q. 1547.

[130] British Museum, Owen Papers, Bennett Letters, quoted in Strahan, *op.cit.*, 34.

[131] 'The Science Commission's Museum Report', *Nature*, 9 (26 March 1874), 397-399.

[132] Alleyne, son of a Barbados planter, studied medicine in Edinburgh, fought in the Maori wars, and was in 1852 the first doctor to administer chloroform in NSW. He oversaw the quarantine station, and in 1876 and 1881 won praise for suppressing virulent outbreaks of smallpox in the colony. He was an officer of a number of other benevolent institutions, and was considered a 'progressive'. See Richard Refshauge, 'Haynes Gibbes Alleyne (1815?-1882)', *Australian Dictionary of Biography*, vol. 3, 27.

[133] *Catalogue of the Minerals and Rocks in the Australian Museum* (Sydney: Australian Museum, 1873).

[134] 'Report of the Trustees of the Australian Museum for the Year ending December 1874', *VPLSNSW*, 4 (1875), Appendix 4, 'Report of the Sub-Committee to inquire into certain charges against the Curator' (Report, *Mins. Evid.* (A.W.Scott), Q. 1288.

[135] Mitchell Library, Parkes Papers, vol. 20. f. 402. Krefft to Parkes, 2 March 1874.

[136] Mitchell Library, Parkes Papers, vol. 20, f. 352. Krefft to Parkes, nd. (ca July 1874).

[137] Stephens, educated at Oxford, was encouraged by Sir Charles Nicholson and Benjamin Jowett of Balliol, to come to Sydney as foundation headmaster of Sydney Grammar School in 1856. A friend of Woolley, he campaigned for education in the 1860s, and left Sydney Grammar in 1879 to establish his own private academy, the New School (from 1879, Eaglesfield). He was active in the Philosophical Society from 1857, and in 1862, was a founder of the Entomological Society of NSW. In 1874, he helped form the Linnean Society of NSW. See Cliff Turney, 'William J. Stephens (1829-1890)', *Australian Dictionary of Biography*, vol.

6, 197-198.

138  Mitchell Library, MacLeay Papers, Diary entries, 20 June; 2, 5, 7, 17, 21 July; 6, 13 August 1874.

139  Australian Museum, 'Subcommittee Report', *op.cit.*, 22; Adjourned Monthly Meeting of the Trustees, 30 July 1874.

140  The *20th Annual Report* of the Museum in 1875 apologised for the 'gross and culpable exaggeration' that had reported 239,000 visitors, 'a figure 100,000 in excess of the entire population of Sydney and suburbs'. 'Report of the Trustees of the Australian Museum for the year ending 1873', *VPLANSW*, 4 (1875), 239.

141  Mitchell Library, Macleay Papers, Diary entry, 21 September 1874.

142  Mitchel Library, Macleay Papers, Diary entry, 20 August 1874.

143  And a member of the circle that began, with Macleay, the Entomological Society in 1862. Scott, described as an 'entomologist and entrepreneur', was born in India, educated at Cambridge, and became one of the colony's most successful merchants. Like Macleay, he was a member of the Australian Club. He published on butterflies and moths, and trained his daughters, Harriet and Helena, to become scientific illustrators. (They produced drawings for books by James Cox and, ironically, Gerard Krefft). See Nancy Gray, 'Alexander Walker Scott (1800-1883)', *Australian Dictionary of Biography*, vol. 6, 93.

144  British Museum (Natural History), Owen Papers, Bennett Letters, Bennett to Owen (nd), cited in Strahan, *op.cit.*, 34.

145  Liversidge Papers, Box l4, 'The Museum Doors Nailed Up: The Police in Charge', *Evening News*, 6 July 1874; 'Mr Krefft and the Museum Trustees', *Evening News*, 7 July 1874; 'The Sydney Museum', *The Empire,* 7 July 1874; 'The Australian Museum', *SMH*, 8 July 1874; 'The Museum', *idem*, 9 July; 'The Trustees and Curator of the Museum', *Evening News*, 11 July 1874.

146  'The Australian Museum', *Nature*, 10 (4 June 1874), 82.

147  Mitchell Library, Macleay Papers, Diary entry, 26 November 1874. He subsequently entered the Legislative Council.

48  Legal opinion gave the Trustees unequivocal power to sack him, considering him in the same relation to the Trustees as a professor was to the Senate of the University. See William Dalley, Attorney General, 'Australian Museum: Opinion of Crown Law Officers', *VPLANSW*, 4 (1875), 301-304.

149  Several letters from Krefft to his friends and sometime supporters are held by the British Museum of Natural History. I am indebted to Dr. W. Horning, formerly of the Macleay Museum, for this information.

150  Liversidge Papers, Box l4, Debate on Supply.

151  Sub-Committee Report, *Mins. Evid.*, Q. 1621.

152  Mitchell Library, Parkes Correspondence, vol. 20, f. 360, Krefft to Parkes, 23 July 1874.

153  A.L., 'Iron Ore and Coal Deposits at Wallerawang, New South Wales' (9 December 1874), *TRSNSW*, 8 (1874), 81-91; later reprinted in 'Notes on the Iron and Coal Deposits at Wallerawang, and on the Diamond Fields', in *Mines and Mineral Statistics of New South Wales, op. cit.*, 94-103.

154  A.L., 'Note on a New Mineral from New Caledonia,' (7 May 1874), *J. Chem. Soc.*, 27 (July 1874), 613-615; 'Nickel Minerals from New Caledonia' (9 December 1874), *TRSNSW*, 8 (1874), 75-80.

155  Transits of Venus occur in pairs, eight years apart, at intervals of 105 and 121 years, and only five have been observed scientifically. The first was seen in 1639 by the English curate, Jeremiah Horrocks. The second pair, which Halley predicted for 1761 and 1769, presented European astronomers with an occasion to mount what historians record as the first example of modern international scientific cooperation. The 1874 transit offered a chance to use modern photography to catch the instant of the event. For details, see Eli Maor, *Venus in Transit* (Princeton: Princeton University Press, 2004), 117-123.

156  In 1761, observations in the Northern Hemisphere and the Cape of Good Hope produced a value for the distance, but astronomers decided to renew their observations at the expected transit eight years later.

157  See Geoffrey M. Badger, *The Explorers of the Pacific* (Sydney: Kangaroo Press, 1988), ch 7.

158  P.M. Janiczek, 'Transits of Venus and the American Expedition of 1874', *Sky and Telescope*, 48 (6), (1974), 366. See also Alex Soojung-Kim Pang, 'The Social Event of the Season: Solar Eclipse Expeditions and Victorian Culture', *Isis*, 84 (2), (1993), 252-277, and Pang, *Empire and the Sun: Victorian Solar Eclipse Expeditions* (Stanford: Stanford University Press, 2002).

159  Cf. R.J. Bray, 'Australia and the Transit of Venus', *Proc. Aust. Soc. Astr.*, 4 (1), (1980), 114.

160  K.J. Cable, 'William Scott (1825-1917)', *Australian Dictionary of Biography*, vol. 6, 97-98.

161  For Ellery, see S.C.B. Gascoigne, 'Robert Lewis John Ellery (1827-1908)', *Australian Dictionary of Biography*,

vol. 4, 135-137.

162  H.C. Russell, *Observations of the Transit of Venus, 9 December 1874, made at Stations in NSW, Illustrated with Photographs and Drawings, under the Direction of H.C. Russell, BA, CMG, FRS, etc, Government Astronomer* (Sydney: Government Printer, 1892), v.

163  'Benefits Resulting from Scientific Research', *Australian Town and Country Journal*, 28 June 1873.

164  Russell, Observations, op.cit., vi.

165  'H.C. Russell', *SMH*, 23 February 1907.

166  Hixon was a veteran of several Pacific voyages, on which he showed great 'scientific attainments and curiosity'; and was a member of several colonial commissions on fisheries, defence, and warlike stores. He advocated the building of lighthouses, ran the volunteer naval brigade, and took an interest in the *Vernon*, a training ship for orphan boys. He married a daughter of Francis Lord, and two of his daughters married grandsons of John Fairfax. See Ruth Teale, 'Francis Hixon (1833-1909)', *Australian Dictionary of Biography*, vol. 4, 402.

167  Henry Russell, 'Some of the Results of the Observation of the Transit of Venus in New South Wales', *TRSNSW*, 8 (1874), 93; Cf. Russell, 'Scientific Notes', *TRSNSW*, 9 (1875), 135.

168  Captain Tupman, 'Monthly Notices', *Royal Astronomical Society*, 38 (June 1878), 453.

169  William Scott, 'The Transit of Venus as Observed at Eden', *TRSNSW*, 8 (1874), 113-116.

170  A.L., report in *TRSNSW*, 8 (1874), 99.

171  A.L, 'Observations of the Transit of Venus, 1874, December 8-9', *Memoirs of the Royal Astronomical Society*, XLVII (1882-1883), 74.

172  *Ibid*.

173  Agnes M. Clerke, *A Popular History of Astronomy during the Nineteenth Century* (1885; London: Adam and Charles Black, 3rd ed., 1893), 292.

174  The NSW figure was 0.055 second of arc higher than the correct result. Another opportunity to observe the solar parallax occurred in 1882, and in 1891 Simon Newcomb, the American astronomer, analysing the 'black drop' effect observed in the timing of egress, calculated a value of 8.794 seconds of arc, which produced a distance of 92,956,000 miles. This is close to what we now know to be correct. For the inherent difficulties of measurement, see J. M. Pasachoff, Glenn Schneider, and Leon Golub, 'The Black-Drop Effect Explained,' in *Transits of Venus: New Views of the Solar System and Galaxy, Proceedings of the International Astronomical Union*, Colloquium No. 196 (2004), 1-12.

175  Russell, 'Scientific Notes', *TRSNSW*, 9 (1875), 135.

176  H.C. Russell, 'Results of the Transit of Venus, 1874', *JPRSNSW*, 12 (1878), 244.

177  *SMH*, 6 November 1875.

178  H.C. Russell, 'Anniversary Address,' *JPRSNSW*, 16 (1882), 5.

179  Sydney Observatory Archives, 'Transit of 1874 Papers', Goulburn Report, Liversidge to Russell, 21 November 1875.

180  Russell, 'Scientific Notes', *TRSNSW*, 9 (1875), 135.

181  Russell, *Observations*, *op.cit*.

# CHAPTER 6

1  *SMH*, 4 October 1872.

2  'The Benefits Resulting from Scientific Research', *Australian Town and Country Journal*, 28 June 1873.

3  Charles Badham, *Speeches and Lectures delivered in Australia, by the late Charles Badham, DD, Professor of Classics in Sydney University* (Sydney: William Dymock, 1890), 96.

4  For characteristics of public science in Britain, see Robert Kargon, *Science in Victorian Manchester: Enterprise and Expertise* (Manchester: Manchester University Press, 1977).

5  A.W. Martin, 'Sir Henry Parkes (1815-1896)', *Australian Dictionary of Biography*, vol. 5, 399-402.

6  Cf. R.W. Connell and T.H. Irving, *Class Structure in Australian History* (Melbourne: Melbourne University Press, 1980), 110-111.

7  Ursula Bygott and K.J. Cable, *Pioneer Graduates of the University of Sydney*, University Monographs No. 1 (Sydney: University of Sydney, 1985), 5.

8  *Sydney Mail*, 27 February 1875.

9  Macleay was the first President of the Linnean Society of NSW. Cf. J.J. Fletcher, 'The Society's Heritage from

the MacLeays', *Proc. Linn. Soc. NSW,* XIV (1920), 567-635.

10   Mitchell Library, Macleay Papers, MSS 2009/l15. Diary entries for 14 and 20 March 1875.

11   For Ramsay and Stephens, see A.H. Chisholm, 'Edward Pierson Ramsay (1842-1916)', *Australian Dictionary of Biography,* vol. 6, 3-4; Cliff Turney, 'William John Stephens (1829-1890)', *idem.*, 197-198.

12   'Departure of the *Chevert* for New Guinea,' *SMH,* 19 May 1875; David S. Macmillan, *A Squatter Went to Sea* (Sydney: Currawong Press, 1957).

13   Sir Alfred Roberts, 'Anniversary Address', *JPRSNSW,* 23 (1889), 2.

14   Rev. W. B. Clarke, 'Anniversary Address', *JPRSNSW,* 10 (1876), 2.

15   Charles Moore, 'Anniversary Address', *JPRSNSW,* 14 (1880), 2.

16   *Sydney Mail,* 27 May 1876.

17   W.J.J. Spry, RN, *The Cruise of HMS Challenger, Voyage over Many Seas, Scenes in Many Lands* (London: Sampson Low, 1880), 159. Not all were enamoured of its inhabitants: 'Delightful although it was at Sydney to make so many friends amongst one's own countrymen, after so long a voyage from home, and to enjoy their far-famed hospitality, one could not as a naturalist, help feeling a lurking regret that matters were not still in the same condition as in the days of Captain Cook, and the colonists replaced by the race which they have ousted and destroyed, a race far more interesting and original from an anthropological point of view'. H.N. Moseley, *Notes by a Naturalist: An Account of Observations made during the Voyage of HMS Challenger round the World in the Years 1872-1876* (London: Macmillan, 1879), 276.

18   Union Club Archives, House Committee Minutes (1873-78), f. 63, 17 March 1874.

19   For the *Challenger's* visit to Sydney, see W.H. Hargraves, 'Reminiscences of the Challenger Expedition', *Australian Museum Magazine,* 1 (1923), 212-213.

20   *Challenger*-Briefs von Rudolf v. Willemoes-Suhm (Dr Phil), 1872-1875, *Nach dem tode des Verfassers Herausgaben von seiner Mutter* (Leipzig: Verlag von Wilhelm Englemann, 1877), 96. A copy was given to Liversidge by Professor Rein of Marburg in 1878, and is held by the Royal Society of NSW.

21   W.B. Clarke, 'Anniversary Address', reported in the *SMH,* 13 May 1874.

22   For Jennings and Stuart, see A.E. Cahill, 'Sir Patrick Alfred Jennings (1831-1897)', *Australian Dictionary of Biography,* vol. 4, 477-479; Bede Nairn and Martha Rutledge, 'Sir Alexander Stuart (1824-1886)', *Australian Dictionary of Biography,* vol. 6, 211-214.

23   G.P. Walsh, 'Charles (Carl) Adolph Leibius (1833-1893)', *Australian Dictionary of Biography,* vol. 5, 79.

24   'The Royal Society', *SMH,* 13 May 1875.

25   See Michael Hoare, '"The Half-Mad Bureaucrat": Robert Brough Smyth (1830-1889)', *Records of the Australian Academy of Science,* 2 (4), (1973), 28-29.

26   W.B. Clarke, 'Anniversary Address', *TPRSNSW,* 9 (1875), 1-56, quoted in the *SMH,* 8 July 1875.

27   The Society elected from Melbourne, Frederick McCoy, Baron von Mueller and Ellery; from Brisbane, Louis Bernays, Vice-President of the Acclimatisation Society of Queensland; from South Australia, Dr Richard Schomburgk, Director of Adelaide's Botanic Gardens; from Perth, Frederick Barlee, Colonial Secretary of Western Australia; from Hobart, Dr James Agnew, Hon. Secretary of the Royal Society of Tasmania; and from New Zealand, Dr James Hector, FRS, Head of the New Zealand Geological Department, and Dr Julius von Haast, FRS, Provincial Geologist of Canterbury. 'Royal Society', *SMH,* 8 July 1875. See also J.H.M. Honniball, 'Sir Frederick Palgrave Barlee (1827-1884)', *Australian Dictionary of Biography,* vol.3, 96-99; and F.C. Green, 'Sir James Wilson Agnew (1815-1901)', *idem.*, 18-19.

28   Council Minutes, RSNSW, 4 August 1875, 7 June 1876.

29   Council Minutes, RSNSW, 27 October and 3 November 1875.

30   'On Royal Society' in *Scientific Notes, SMH,* 5 November 1875.

31   'Royal Society', *SMH,* 6 September 1877.

32   See Theodore Gill, 'The Smithsonian System of Exchanges', in *Public Libraries in the United States of America: Their History, Condition and Management,* Part I (Washington, D.C.: U.S. Government Printing Office, 1876), 285-290. I am indebted to Dr. Paul Theerman for this reference.

33   'The Linnean Society', *SMH,* 2 February 1876; William Macleay, 'Presidential Address', *Journal of the Linnean Society of NSW,* I (1875-76), 85.

34   *Ibid.*

35   Calculations from Lists of Members of the Linnean Society (1875) and the Royal Society of NSW (1885).

36   *Papers and Proceedings of the Royal Society of Tasmania,* VI (1871-74), 106.

37  The Senckenberg Society, founded in 1871, was devoted to the study of the 'Humboldtian' sciences. Liversidge was nominated in May 1876 by Professor Paul Reis, with whom he corresponded about technical museums in 1878. The Physical Society was founded at the South Kensington Museum in February 1874, with Professor J.H. Gladstone, the electrical engineer, in the chair. The Society's membership at first drew principally upon Londoners. Liversidge was recommended by Gladstone and by his friends William Garnett, Herbert McLeod (of the Royal College of Chemistry), G.F. Rodwell, G.C. Foster (of UCL), and Frederick Guthrie (his teacher at the RSM).

The Mineralogical Society, whose first President was Henry Clifton Sorby (the 'father of petrology'), was established in London in February 1876; Liversidge was elected in September that year. The Society met variously in the rooms of the Royal Microscopical Society, the Museum of Practical Geology, and (from 1889) the Geological Society in Burlington House. Interestingly, Liversidge was elected before his teacher, George Liveing, and soon after T.H. Bonney, the geologist of St John's, Cambridge, Archibald Geikie, of Edinburgh, and F.W. Rudler of the British Museum. Liversidge remained one of the few colonial members of this predominantly British and Irish society. For an outline of its administrative history, see W. Campbell Smith, 'The Mineralogical Society (1876-1976)', *Mineralogical Magazine,* 40 (313), (March 1976), 429-439.

38  A.L. 'On "Black" and "Ruby" Tin Ore' (8 November 1876), *TPRSNSW,* 10 (1876), 290; 'Fossiliferous Siliceous Deposit from the Richmond River, NSW and the so-called Meerschaum from the Richmond River, NSW' (6 December 1876), *idem.,* 237-240; 'On a Remarkable Example of Contorted Slate' (6 December 1876), *idem.,* 241-242.

39  A.L, 'Note on a Mineral from New South Wales, presumed to be Laumontite' (6 September 1876), *Mineralogical Magazine,* 1 (2), (1876), 54.

40  In 1877, Clarke also received the Murchison Medal of the Geological Society of London for his work on the stratigraphy of New South Wales.

41  E. Docker, 'Joseph Docker (1802-1884)', *Australian Dictionary of Biography*, vol. 4, 79-80.

42  Council Minutes, RSNSW, 27 May 1876, 27 September 1876.

43  *Ibid.*, 2 May 1877.

44  *Ibid.*, 27 September and Monthly Meeting, 13 October 1876.

45  *Ibid.*, Meeting 13 October 1876; Colonial Secretary to A.L., 7 November 1876.

46  *Ibid.*, 28 June 1876.

47  'By laws', Council Minutes, *TPRSNSW,* 9 (1875), x-xi.

48  *Ibid.*

49  *Ibid.*

50  Cf. J.H. Maiden, 'A Contribution to a History of the Royal Society of New South Wales', *JPRSNSW,* 52 (1918), 347.

51  Council Minutes, RSNSW, 27 September 1876.

52  Council Minutes, RSNSW, 6 December 1876.

53  A.L., 'On the Formation of Moss Gold and Silver' (6 September 1876), *JPRSNSW,* 10 (1876), 125-133.

54  The phrase was used by Russell in his Anniversary Address in May 1877 ('The Royal Society', *SMH*, 5 May 1877), and repeated by Liversidge (A.L., 'The Royal Society of NSW', *Nature,* 83 (23 June 1910), 503).

55  Council Minutes, RSNSW, 26 April 1877.

56  *Ibid.*

57  Council Minutes, RSNSW, 12 April 1877.

58  'Notes of Royal Society', *SMH,* 5 May 1877; Annual Address, *JPRSNSW,* 11 (1877), 222.

59  Council Minutes, Royal Society, 4 December 1878; *JPRSNSW,* 12 (1878), 187.

60  Maiden, *op.cit.*, 349. He was to serve as President in 1885-86, 1889-90 and 1900-01.

61  A.P. Elkin, 'The Challenge to Science, 1866: The Challenge to Science, 1966' (28 October 1966), *A Century of Scientific Progress: The Centenary Volume of the Royal Society of New South Wales* (Sydney: Royal Society of NSW, 1968), 23. See also A.P. Elkin, 'Centenary Oration: Part 1: The Challenge to Science, 1866', *JPRSNSW,* 100 (1966), 105-118.

62  Liversidge Papers, Box 13, f. 3. 'Royal Society', *SMH,* 5 October 1877.

63  'Royal Society', *SMH,* 6 September 1877; C. Rolleston, 'Anniversary Address', *JPRNSW,* 12 (1 May 1878), 2.

64  'Royal Society of NSW', *SMH,* 4 October 1877; H. Russell, 'Anniversary Address', *JPRNSW,* 11 (2 May

1877), 5.

65 Russell was also attacked by one of his workmen in 1889. See G.P. Walsh, 'Henry Chamberlain Russell (1836-1907)', *Australian Dictionary of Biography*, vol. 6, 74-75.

66 Council Minutes, RSNSW, 12 December 1877.

67 'Royal Society', *SMH,* 3 May 1878.

68 'Royal Society', *SMH,* 5 May 1877.

69 Badham, *op. cit.*, 81.

70 D.A. Kerr, 'The History of Science Teaching in New South Wales, 1788-1960' (Unpublished M. Ed. thesis, University of Sydney, 1966), 233. For the situation in Victoria see Rosemary Polya, *Australian School Science Text-books, 1850-1939* (Bundoora, Vic.: The Borchardt Library, 1986), 11 *et seq.*

71 Liversidge Papers, Box 20, f. 6, Badham to Chancellor, 30 March 1875.

72 *Ibid.,* Smith to Chancellor, 29 November 1875.

73 *Ibid.,* A.L. to Chancellor, 29 March 1875.

74 Senate Minutes, July 1876; Deas Thomson Papers (Mitchell Library), ML A 1531-3 Microfilm Cy 721, Mrs Smith to Deas Thomson, 14 Jan 1875; Deas Thomson to Mrs Smith, 16 January 1875.

75 Senate Minutes, 7 February 1877, 171. Smith complained he had not received compensation for the afternoon lectures, which had been reduced in number (and in fees) since the curriculum changes of 1874-75. In August, Windeyer proposed that he be given compensation of £150 per annum; Senate reduced this to £30 per annum. Senate Minutes, 5 September 1877, 191.

76 'Rather than give two hours a day to his University duties', Macleay cynically commented, 'for which he gets £1,000 a year, he has resigned and demands a pension on the basis of bad health.' Macleay Papers (Mitchell Library), 2009/115 Diary Entry, 18 October 1876. By the terms of his appointment in 1852, Pell received a pension of half salary for life.

77 Senate Minutes, 4 April 1877, 174.

78 Senate Minutes, 2 December 1874, 95-96; 1 December 1875, 124.

79 Senate Minutes, 1 August 1877, 186.

80 Edward Deas Thomson, *Commemoration Address,* 16 June 1877, 3-4.

81 *Ibid.*

82 William Macleay, 'Presidential Address', *Journal of the Linnean Society of New South Wales,* I (1875-76), 89, reprinted in the *SMH,* 2 February 1876.

83 Rev. G. Brown, 'Notes on the Duke of York Group, New Britain, and New Ireland', *Journal of the Royal Geographical Society,* 47 (1877), 137-150, where Liversidge's analyses are described (140). The results were much later presented to the Royal Society and reported in London. See A.L., 'On the Occurrence of Chalk in the New Britain Group' (4 July 1877), *JPRSNSW,* 11 (1877), 85-91; *Geological Magazine,* 4 (December 1877), 529-534.

84 I.D. Dick, 'Historical Introduction to New Zealand Science', in F.R. Callaghan (ed.), *Science in New Zealand* (Wellington: A.H. and A.W. Reed, 1957), 1-18.

85 Gardner, *op.cit.,* 9-42.

86 Michael Hoare, 'The Reform of Science in New Zealand, 1903-1926', *Second Cook Lecture,* 1976.

87 See R.M. Burdon, *Scholar Errant: A Biography of Professor A.W. Bickerton* (Christchurch: The Pegasus Press, 1956).

88 A.L., 'Water from a Hot Spring, Fiji Islands' (1 September 1880), *JPRSNSW,* 14 (1880), 147-8.

89 A.L., 'Vanishing Customs in the Fiji Islands', *Man,* 21 (81), (September 1921), 133-136; cf. his analyses of the mysteries of chalk, of which earrings were made in New Britain for the Rev. George Brown. His reports appeared – 'On the Occurrence of Chalk in the New Britain Group' (4 July 1877), *JPRSNSW,* 11 (1877), 85-91, but were first reported in G. Brown, 'Notes on the Duke of York Group, New Britain, and New Ireland', *Journal of the Royal Geographical Society,* 47 (1877), 140.

90 See C.F. Gordon Cumming, *At Home in Fiji* (London: William Blackwell, 1882, 1901), which reprints Liversidge's analysis of the healing hot springs. An annotated copy is held by the Royal Society of NSW.

91 'Royal Society of NSW', *SMH,* 24 November 1877.

92 Their friendship culminated in a moving obituary: A. Liversidge, 'Captain Hutton, FRS', *JPRSNSW,* 39 (1905), 139-141. See also G.M. Thomson, 'Biographical Notice of F. W. Hutton', *New Zealand Journal of Science,* 2 (1885), 301-306. For Hutton's work, see John Stenhouse, 'Darwin's Captain: F.W. Hutton and the

Nineteenth Century Darwinian Debates', *Journal of the History of Biology*, 23 (3), (1990), 411-442.

[93] A.L., 'Notes on some of the New Zealand Minerals belonging to the Otago Museum, Dunedin' (Otago Institute, 23 October 1877), *Trans. NZ Institute*, 10 (1877), 490-505; 'Analyses of a Rock Specimen from New Zealand showing the Junction between Granite and Slate' (Otago Institute, 8 November 1877), *Trans. NZ Institute*, 10 (1877), 505-506.

[94] H.F. von Haast, *The Life and Times of Sir Julius von Haast, Explorer, Geologist, Museum Builder* (Wellington: The Author, Avery Press, 1948).

[95] Von Haast Papers (Alexander Turnbull Library, Wellington), MS37. A.L to von Haast, 11 May 1877.

[96] Von Haast Papers, A.L. to 'My Dear Haast', 13 March 1877.

[97] Von Haast Papers, A.L. to von Haast, 12 June, 26 July 1877, and 22 August 1877. Hector spoke to the Royal Society's *conversazione* in the Great Hall of Sydney University on 1 November 1879.

[98] A.L, 'On the Occurrence of Chalk in the New Britain Group' (4 July 1877), *JPRSNSW*, 11 (1877), 85-91.

[99] 'Report upon Mineral Specimens examined for the Mining Department during the year 1877', *ARDM*, 1877 *op.cit.*, 10, 210-212.

[100] Macleay Papers (Mitchell Library), 2009/115. Diary entries, 4 June, 5 August; 15, 16 and 22 October 1876.

[101] *Senate Minutes*, 14 November 1877, 197. For Lackey, see Martha Rutledge, 'John Lackey (1830-1903)', *Australian Dictionary of Biography*, vol. 5, 48-49.

[102] Clifford Turney, Ursula Bygott and Peter Chippindale, *Australia's First: A History of the University of Sydney* (Sydney: University of Sydney in association with Hale and Iremonger, 1991), vol. 1, 157. The total grant for 1882, including apparatus and other expenses, rose from £6,000 to £12,964. See *Statistical Register of NSW for 1888* (Sydney: Government Printer, 1889), 257.

[103] See Appendix III. Holroyd joined the Royal Society of NSW in July 1876.

[104] Royal Botanic Gardens (Kew), Queensland Plant Diseases, f. 82, A.L. to Hooker, 13 April 1877.

[105] His early promise was destined not to be fulfilled. Although a diligent teacher, Gurney retired early in 1902, 'mentally equipped', as one of his colleagues wrote, 'with every gift except ambition'. See I.S. Turner, 'Theodore Gurney (1849-1918)', *Australian Dictionary of Biography*, vol. 4, 309-310. Gurney returned – ill, and aged only fifty-three – to Cambridge, where he met Liversidge many years later. Sydney University made him a professor emeritus in 1905.

[106] Alexander Turnbull Library, Hector Papers, A.L. to Hector, 10 November 1877.

[107] H.C. Russell, 'Anniversary Address', *JPRSNSW*, 9 (1877), 222; *SMH*, 3 May 1877.

[108] Alexander Turnbull Library, Von Haast Papers, A.L. to Julius von Haast, 13 October 1877.

[109] Smith had been overseas in 1862-63, and 1871-72, and would go again in 1881-83. See Hoare and Radford, *op. cit.*, 149.

[110] Senate Minutes, 14 November 1877, 196.

[111] Hoare, *op. cit.*, 22.

[112] Senate Minutes, 3 October 1877, 194. It was forthcoming. See Senate Minutes, 5 June 1878, 227.

[113] Mott Greene, *Geology in the Nineteenth Century: Changing Views of a Changing World* (Ithaca: Cornell University, 1982), 193-194.

[114] These were held in Bologna in 1881, Berlin in 1885, London in 1888, and Chicago in 1892. They did, as Liversidge put it, 'much valuable work in a quiet way'. 'Anniversary Address', *JPRSNSW*, 24 (1890), 37.

[115] *Report of the Executive Commissioners to the Paris Exhibition, 1878*, VPLANSW, 5, 1879-80, 108.

[116] See Graeme Davison, 'Festivals of Nationhood: The International Exhibitions' in S.L. Goldberg and F.B. Smith (eds.), *Australian Cultural History* (Cambridge: Cambridge University Press, 1988), 158-177.

[117] Bede Nairn, 'Edward Combes (1830-1895)', *Australian Dictionary of Biography*, vol. 3, 445.

[118] Martha Rutledge, 'Jules Joubert (1824-1907)', *Australian Dictionary of Biography*, vol. 4, 493-494. For Joubert's remarkable history, see his *Shavings and Scrapes from Many Parts* (Dunedin: J. Wilkie and Co., 1890). For his work as a reforming, but ultimately dismissed Secretary of the Agricultural Society, see Brian Fletcher, *The Grand Parade: A History of the Royal Agricultural Society of NSW* (Sydney: Royal Agricultural Society of NSW, 1988), 57.

[119] Clarke Papers (Mitchell Library) A.L. to Clarke, 13 November 1877.

[120] Liversidge Papers, Box 19, f. 7, Colonial Secretary to A.L., 9 August 1877.

[121] 'Paris Universal Exhibition, 1878', *VPLANSW*, 5, 1879-80, 994.

[122] Von Haast Papers (Alexander Turnbull Library), A.L. to von Haast, 13 October 1877.

123 R.O. Chalmers, 'History of the Department of Mineralogy, Australian Museum, Part I: 1827-1901', *J. Min. Soc. NSW,* 1 (1979), 18.

124 *Council Minutes,* Royal Society of London, 12 December 1877.

125 *Liversidge Papers,* Box 19, f.9, Joubert to A.L., 22 May 1878.

126 On Knox, see A.G.Lowndes (ed.), *South Pacific Enterprise: The Colonial Sugar Refining Company, Ltd.* (Sydney: Angus and Robertson, 1956), 17; A.G. Lowndes, 'Sir Edward Knox, (1819-1901)', *Australian Dictionary of Biography,* vol. 5, 38-39; Helen Rutledge, *My Grandfather's House* (Sydney: Doubleday, 1986).

127 That is, by measuring the degree of rotation of a beam of polarised light passed through the solution. Knox had been commended to Smith by Sir Daniel Cooper, one of CSR's original shareholders, in London. See George Bindon and David Philip Miller's '"Sweetness and Light": Industrial Research in the Colonial Sugar Refining Company, 1855-1900', in Rod Home (ed.), *Australian Science in the Making* (Sydney: Cambridge University Press, 1988), 176.

128 Knox made many voyages to Britain; on this occasion he may have recruited Andrew Fairgrieve, a chemist with 'large experience in refineries in England and France'; and possibly also Thomas U. Walton, BSc of Glasgow, whom Knox met in Greenock, and who migrated to work at CSR in 1881. Graham Holland, 'Thomas Utrick Walton (1852-1917)', *Australian Dictionary of Biography,* vol. 12, 376-377.

129 Bindon and Miller, *op.cit.,* 177, 181.

130 *Liversidge Papers,* Box 19, *Report of the Executive Commissioners to the Paris Exhibition, 1878, VPLANSW,* 5, 1879-80, Appendix II, 79. Combes claimed, and received, over £900 in expenses for his services from the close of the Exhibition until his return to Sydney in July 1879. From Liversidge's marginal comments on his copy of the report, it is clear that he harboured no doubts of Combes' impropriety, but made no public criticism.

131 For Forster, see Bede Nairn, 'William Forster (1818-1882)', *Australian Dictionary of Biography,* vol. 4, 199-201.

132 His qualifications entitled him to become a Fellow without examination – a useful credential, come what may. For context, see Colin Russell *et al.*, *Chemists by Profession: The Origins and Rise of the Royal Institute of Chemistry* (Milton Keynes: Open University Press, 1977).

133 Cf. *Illustrated London News,* 4 May 1878, 402; 11 May, 442; South America, 1 June, 516; Spain, 8 June, 529; Japan, 8 June, 540-541; USA, 15 June, 568.

134 *Liversidge Papers,* Box 19, f. 9, Joubert to A.L., 26 May 1878.

135 *Report of the Executive Commissioners, op.cit.,* 999.

136 *The Times,* 10 December 1878, noted in Liversidge Papers, Box. 10. Henry Roscoe, the Manchester chemist, came to Paris with the British delegation, as a juror for chemical products; why he did not intercede for Liversidge, if indeed he had an opportunity to do so, remains unclear. See H. Roscoe, *The Life and Experiences of Sir Henry Enfield Roscoe* (London: Macmillan, 1906), 163-167.

137 See Robert Fox, 'The *Savant* Confronts his Peers: Scientific Societies in France, 1815-1915', in Robert Fox and George Weisz (eds.), *The Organization of Science and Technology in France, 1808-1914* (Cambridge: Cambridge University Press, 1980), 272.

138 Mitchell Library, Deas Thomson Papers, vol. 3, f. 1531 Badham to Chancellor, (January) 1878.

139 Sir William Manning,'Commemoration Address', 22 June 1878, 2.

140 See Peter Chippindale, 'A Decade of Reform – The University of Sydney 1876-1886', *Sydney University Archives Record* (1998), 7-26. See also Mark Lyons, 'Joseph Leary, (1831-1881)', *Australian Dictionary of Biography,* vol. 5, 73.

141 Sir William Manning, 'Commemoration Address', 22 June 1878, 5.

142 *Liversidge Papers,* Box 19, f.1, Manning to A.L, 31 July 1878. The Minister for Justice and Public Instruction approved the request, to 'enable him to cooperate with E. Combes, Esq. and the Agent-General in procuring information respecting the system for Technical and Industrial Instruction in Europe', *Senate Minutes,* 20 August 1878. Liversidge's papers contain no evidence of co-operation with Combes.

143 'Report upon certain Museums for Technology, Science, and Art, also upon Scientific, Professional and Technical Instruction, and Systems of Evening Classes in Great Britain and on the Continent of Europe, by Archibald Liversidge' (afterwards, *Liversidge Report), VPLANSW,* 3, 1879-80, 787-1059 At the Senate meeting of 20 August 1878, Liversidge was granted an extra two months' leave on full salary (extending his absence until May 1879) to prepare his report.

144 *Liversidge Report,* A.L. to Manning, 10 October 1878.

[145] 'University of Sydney - Endowment of - Correspondence', Sir William Manning to the Minister, 5 August 1878, *VPLANSW,* 1881, 3.

[146] Mitchell Library, Parkes Papers, MSS A1050-2, f. 91, Parkes to Windeyer, 30 August 1876.

[147] Senate Minutes, 13 November 1878.

[148] R.W. Rathbone, 'John Lucas (1818-1902)', *Australian Dictionary of Biography,* vol. 5, 108.

[149] Senate Minutes, 13 November 1878; G1/1/6 'Special Meeting of the Senate held at the Legislative Council on 13th November 1878'.

[150] See S. Murray-Smith, 'Norman Selfe (1839-1911)', *Australian Dictionary of Biography,* vol. 6, 100-101.

[151] Cf. L.A. Mendelson, 'Norman Selfe and the Beginning of Technical Education', in C. Turney (ed.), *Pioneers of Australian Education* (Sydney: Sydney University Press, 1972), vol. 2, 105-148.

[152] *Sydney Mechanics' School of Arts Minutes, 1876-86,* 21 January 1879.

[153] Chalmers, *op.cit.,* 18. See W.A. Haswell, *Guide to the Contents of the Australian Museum* (Sydney: Australian Museum, 1883). They included boron, silicas, garnets, zeolites (aluminum silicates), zincs (from calamine ores), and ores of iron, cobalt, nickel and tin and were put on display in 1880.

[154] 'Nothing ever received for any of this *ceratodus* etc. except a few rubbishy minerals', he minuted years later. Liversidge Papers, Box 19, f. 9, on letter, 14 November 1878.

[155] Museum of Zoology (Cambridge) MSS, A.L. to Willis Clark, 18 October 1878.

[156] J. Caldor, *Robert Louis Stevenson: A Life of Study* (London: Hamish Hamilton, 1980), 88.

[157] Richard D. French, *Antivivisection and Medical Science in Victorian Society* (Princeton: Princeton University Press, 1975). See also Michael Foster, 'Vivisection', *Popular Science Monthly,* 4 (1874), 672-685.

[158] Wellcome Institute for the History of Medicine, London, Schapey-Shäfer Papers, PP/ESS 70, Martin to Schäfer, 24 March 1878.

[159] Martha Rutledge, 'Sir Alfred Roberts (1823-1898)', *Australian Dictionary of Biography,* vol. 6, 34.

[160] Minutes of the Trustees of the Australian Museum, 6 August 1878.

[161] *The Times,* 12 December 1878.

[162] Liversidge Papers, Box 19, f. 12, Archibald Liversidge to Forster, 9 October 1878.

[163] *Liversidge Report,* iv, Archibald Liversidge to Manning, 10 October 1874.

[164] *Liversidge Report,* iv, Fitzpatrick to Archibald Liversidge, 21 September 1878.

[165] Liversidge Papers, Box 19, f. 12, Archibald Liversidge to Captain Jopp, R.E., 5 November 1878.

[166] The vote was not passed until October 1879, when £2,000 was approved. Australian Museum Letterbook (1876-80), f. 518, C. Robinson, Acting Secretary, to Colonial Secretary, 2 October 1879; f. 532, Robinson, Minute, 8 October 1879.

[167] Australian Museum Letterbook (1876-1880), f 343, A.L. to Foster, 5 November 1878; A.L. to Secretary, 12 November 1878.

[168] *Liversidge Report,* A.L. to Minister for Public Instruction, 20 April 1880. Forster's telegram of 16 December 1878 read: 'Technological Museum. Can you now authorise expenditure. Liversidge waiting at great personal sacrifice. Time wasted if necessary purchases not authorised'. Despite telegrams between Forster and the Chief Secretary, the £500 promised in the Secretary's letter of 15 August 1878 was not sanctioned until after Liversidge had left Europe. Liversidge Papers, Box 19, f. 11, A. Jopp to A.L., 20 January 1879.

[169] Liversidge Papers, Box 19, f. 12, A.L. to Forster, 10 March 1879, 24 April 1879. The grant of £500 was finally sanctioned in September 1879, but only £25.17. 2 was spent on books (Messrs Trübner and Co.). The balance of £474.2.10 was paid to Liversidge in December 1880. Charles Buckland, of the Australian Museum estimated that Liversidge's expenses from October 1878 to April 1879 totalled £515.10. 6. See NSW State Archives, Box 10/14284, Charles Buckland, Secretary, Australian Museum, to W. Wilkins, Department of Public Instruction, 18 November 1880.

[170] Alexander Turnbull Library, Hector Papers, Box 1879, A.L. to Hector, 30 January 1879.

[171] Liversidge Papers, Box 19, f. 12, A.L. to Forster, 10 March 1879.

[172] These included: 1. Designs and fittings of certain institutions, notably those of the new Natural History Museum at South Kensington, the South Kensington Museum for Science and Art, the Royal School of Mines Museum, the Museum of Science and Art at Edinburgh, the College of Science at Dublin, the Natural History Museum in Paris, Owens College in Manchester, Mason's College in Birmingham, Yorkshire College in Leeds, the Indian Engineers College at Coopers Hill, and the proposed new Colonial Museum in London; 2. Reproductions of works of art; 3. Economic botany specimens; 4. Fabrics from the weaving school at Leeds;

5. A series of tested specimens for tensile strength; 6. A series of worsted stuffs; and 7. Specimens of pottery. *Liversidge Papers*, Box 19, f. 12, A.L. to Forster, 31 March 1879.

173  *Australian Museum Minute Books, 1879-1883*, 15 January 1880.

174  'Paris Universal Exhibition, 1878', *VPLANSW*, 4, 1879-80, 999.

175  *SMH*, 9 June 1879.

# CHAPTER 7

1  *Liversidge Report, VPLANSW*, 3 (1879-80), 787-1059.

2  See Martin J. Wiener, *English Culture and the Decline of the Industrial Spirit, 1850-1980* (Cambridge: Cambridge University Press, 1981). See also Donald N. McCloskey, 'Did Victorian England Fail?' *Economic History Review*, 2 ser. xxiii (1970), 446-459.

3  See David Layton, *Science for the People: the Origins of the School Science Curriculum in England* (New York: Science History Publications, 1974); see also John W. Adamson, *English Education, 1789-1902* (Cambridge: Cambridge University Press, 1930).

4  See Robert Bud and Gerrylynn Roberts, *Science versus Practice: Chemistry in Victorian Britain* (Manchester: Manchester University Press, 1984).

5  Cf. Roy MacLeod, 'Scientific and Technical Education in the Nineteenth Century', in Gillian Sutherland (ed.), *Critical Commentaries on Education: British Parliamentary Papers* (Dublin: Irish Academy Press, 1977).

6  Cf. Roy MacLeod and Russell Moseley, 'Breadth, Depth and Excellence: Sources and Problems in the History of University Science Education in England, 1850-1914', *Studies in Science Education*, 5 (1978), 85-106.

7  Karl Pearson, *The Grammar of Science* (London: Walter Scott, 1892).

8  For a compelling, if early, reading of a neglected theme in Australian social history, see Michael Roe, *Nine Australian Progressives: Vitalism in Bourgeois Social Thought, 1890-1960* (St Lucia: University of Queensland Press, 1984); for comparative background, see Robert Wiebe, *The Search for Order, 1877-1920* (London: Macmillan, 1967).

9  See Richard Altick, *The Shows of London* (Cambridge, Mass.: Belknap Press, 1978).

10  *Liversidge Report*, ix-xi.

11  *Ibid.*, xii.

12  For a general survey of museum history and administration, see Kenneth Hudson, *Museums of Influence* (Cambridge: Cambridge University Press, 1967).

13  Nicolaas Rupke, 'The Road to Albertopolis: Richard Owen (1804-92) and the Founding of the British Museum of Natural History', in N.A. Rupke (ed.), *Science, Politics and the Public Good* (London: Macmillan, 1988), 63-89.

14  Deborah Campbell, 'A Valuable and Self-Denying Citizen: The Public Life of John Smith', in Roy MacLeod (ed.), *University and Community in Nineteenth Century Sydney: Professor John Smith (1821-1885)* (Sydney: University of Sydney History Monographs No. 3, 1988), 47-59. See also Campbell, 'Culture and the Colonial City: A Study in Ideas, Attitudes and Institutions, Sydney, 1870-90' (Unpublished PhD dissertation, University of New South Wales, 1982).

15  The SMSA's earliest Committee of Management included four officers (President, Vice-President, Secretary and Treasurer), and 'twenty others' of whom thirteen should be operatives. The other committee members (or 'Trustees') were to retire in rotation, five per year. See R.I. Johnson, 'The History of the Sydney Mechanics' School of Arts from its Foundation in 1833 to the 1880's' (unpublished MA thesis, Australian National University, 1967), 60-61.

16  All sides looked for balance between lay interest, professional expertise and financial accountability. Between 1851 and 1881, no fewer than four parliamentary enquiries reviewed the management of the British Museum, without reaching a consistent conclusion. Through the 1860s and 1870s, administrative indecision directly affected plans for the new British Museum (Natural History), which was proposed by Richard Owen. Matters were not helped by a struggle in 1869 and 1870 over the direction of the Royal Botanic Gardens at Kew, which polarized the scientific community and Gladstone's new administration. Eventually, a compromise was reached, by which Kew came under ministerial control, in return for which J.D. Hooker and his successors were allowed to run the Gardens without departmental interference. The experience left unhappy memories. See Roy MacLeod, 'The Ayrton Incident: A Commentary on the Relations of Science and Government in

England, 1870-1873', in Arnold Thackray and Everett Mendelsohn (eds.), *Science and Values* (New York: Humanities Press, 1974), 45-78.

<sup>17</sup> *Report of the Royal Commission on Scientific Instruction and the Advancement of Science* (afterwards, *Devonshire Commission*), Fourth Report, 1874 [C. 884]. xxii. 1.

<sup>18</sup> *Liversidge Report*, xii.

<sup>19</sup> Susan Sheets-Pyenson, *Cathedrals of Science: the Development of Colonial Natural History Museums During the Late Nineteenth Century* (Kingston and Montreal: McGill-Queen's University Press, 1988).

<sup>20</sup> Camille Limoges, 'The Development of the Muséum d' Histoire Naturelle of Paris, c. 1800-1914', in Robert Fox and George Weisz (eds.), *The Organisation of Science and Technology in France, 1808-1914* (Cambridge: Cambridge University Press, 1980), 211-240.

<sup>21</sup> A fine illustration of a the progressive philosophy of machinery and belief in the educational value of its display is to be found in the Christopher Polhem collection, dating from the seventeenth century and today part of the National Museum of Science and Technology in Stockholm.

<sup>22</sup> Ronald Strahan, *et al., Rare and Curious Specimens: An Illustrated History of the Australian Museum, 1827-1979* (Sydney: The Australian Museum, 1979), 38.

<sup>23</sup> *Liversidge Report*, xiv. The Report was read with interest in Victoria. See 'Liversidge on Technology', *Imperial Review,* I (5), (April 1881), 88-91.

<sup>24</sup> *Ibid.*, xv.

<sup>25</sup> *Ibid.*, xvii-xxx., parts of which are lifted almost verbatim from the *Devonshire Commission, Third Report, 1873.* [C. 868]. xxviii.

<sup>26</sup> *Ibid.,* xxix. Cf.T.H. Huxley, presiding at a lecture given by Silvanus P. Thompson, 'Apprenticeship: Scientific and Unscientific', *Journal of the Society of Arts,* XXVIII (December 1879), 33.

<sup>27</sup> *Ibid.*, xxix-xxx.

<sup>28</sup> Richard Yeo, 'Scientific Method and the Image of Science, 1831-1890', in Roy MacLeod and Peter Collins (eds.), *The Parliament of Science: Essays in Honour of the British Association for the Advancement of Science* (London: Science Reviews, 1981), 65-88.

<sup>29</sup> 'President's Address' (28 January 1880), *Proc. Linnean Society NSW,* IV (1879), 471-475. Tenison-Woods was four times an unsuccessful candidate for election to the Royal Society of London between 1883 and 1887.

<sup>30</sup> John Allwood, *The Great Exhibitions* (London: Cassell and Collier Macmillan, 1977), 12.

<sup>31</sup> See Paul Greenhalgh, *Ephemeral Vistas: The 'Expositions Universelles', Great Exhibitions and Worlds' Fairs, 1851-1939 (*Manchester: Manchester University Press, 1989); and Greenhalgh, 'Education, Entertainment and Politics: Lessons from the Great International Exhibitions', in Peter Vergo (ed.), *The New Museology* (London: Reaktion Books, 1989).

<sup>32</sup> Henry Adams, *The Education of Henry Adams* (Boston:Houghton Mifflin, 1907, 1961), 465.

<sup>33</sup> Peter Berger and Thomas Luckman, *The Social Construction of Reality: A Treatise in the Sociology of Knowledge* (New York: Anchor Books, 1967), 92-107; R.Wiebe, *The Search for Order, 1877-1920* (London, Melbourne: Macmillan, 1967); Robert Rydell, *All the World's A Fair: Visions of Empire at American International Exhibitions, 1876-1916* (Chicago: University of Chicago Press, 1984).

<sup>34</sup> See Graeme Davison, 'Festivals of Nationhood: The International Exhibitions', in S.L. Goldberg and F.B. Smith (eds.), *Australian Cultural History* (Cambridge: Cambridge University Press, 1988), 158-177.

<sup>35</sup> 'The World's Columbian Exhibition', *World's Columbian Illustrated ,* I (February 1891), 2, quoted in Rydell, *op.cit.*, 5.

<sup>36</sup> Smithsonian Institution Archives, RU 70. Exposition Records of the SI and US National Museum, 1875-1916, Box 19, [Wilber O. Atwater], US Agriculture Department, 'What the Exposition May and Ought to Be' (nd, ca. 1893), quoted in Rydell, *idem, 7.*

<sup>37</sup> Melbourne was not to steal the stage. In 1877, Jacob Levi Montefiore and his partner Samuel Joseph, two of Sydney's leading businessmen, asked the Society's executive committee to explore the prospect of holding an international exhibition in Sydney. Montefiore, who had organised the art section at previous shows, was a friend of Sir Henry Parkes and a fellow Free Trader. His overseas agent, who had seen exhibitions in Brussels and Paris, suggested that a similar exhibition could be of value to Sydney, by increasing trade and the flow of capital and stimulating immigration. Chambers of Commerce were also interested. If, Montefiore added, the Agricultural Society was unwilling, an approach might be made to Melbourne instead.

<sup>38</sup> Estimates of cost varied. Some members of the Society preferred to lease exhibition buildings at Prince Alfred

Park which, with additions, would cost £18,000. Others preferred a new structure, and proposed a site in the Inner Domain, fronting Macquarie Street, between Bent and Bridge Streets, on the paddocks where Sir Hercules Robinson kept his carriage horses. This was a more attractive but also more expensive prospect, which Robinson estimated would cost £27,000. The Society estimated returns in fees and subscriptions would be only £14,520. A.W. Martin and P.Wardle (eds.), *Members of the Legislative Assembly of New South Wales, 1856-1901, Biographical Notes* (Canberra: ANU Press, 1959), 113-114. No sooner had the Exhibition opened, than Robinson left New South Wales to become Governor of New Zealand. See Bede Nairn, 'Sir Hercules Robinson, 1st Baron Rosmead (1824-1897)', *Australian Dictionary of Biography,* vol. 6, 48-50.

[39] Jennings was keenly interested in education and music, and donated £1,000 to buy an organ for the Great Hall at the University, and much more to the library at St John's College. In recognition of his work for the International Exhibition, he was created CMG in 1879 and KCMG in 1880. See A. E. Cahill, 'Sir Patrick Alfred Jennings (1831-1897)', *Australian Dictionary of Biography,* vol. 4, 477-479.

[40] Graeme Davison, 'Exhibitions', *Australian Cultural History,* 2 (1982/3), 5.

[41] Linda Young, 'The Sydney International Exhibition, 1879: Australia's First Expo', *Heritage Australia,* 7 (2), (1988), 26.

[42] A fine description is given in Renee Free *et al., Sydney International Exhibition, 1879: An Exhibition Celebrating the Centenary of the Sydney International Exhibition* (Sydney: Museum of Applied Arts and Science, 1979), 7-25.

[43] 'Newly risen, how bright you shine'. In 1906, this line was adopted as the official motto of the State of NSW.

[44] Free, *op.cit.*

[45] *Australian Almanac for 1877* (Sydney: P.E.Reynolds, 1877), 204.

[46] E.C. Rowland, 'The International Exhibition of 1879', *JRAHS,* 38 (1952), 289. But see also Bolton, for whom the best day was Australia Day 1880. (Lissant Bolton *et al.*, 'Lost Treasures of the Garden Palace', *Australian Natural History,* 19 (12), (1979), 417.)

[47] *Ibid.*, 414-419.

[48] Liversidge also knew Stephens as a keen amateur naturalist, and as a founder and Council member of the Linnean Society. That busy year, Stephens also helped found the Zoological Society of NSW. Liversidge and he became friends, if not quite allies. See Cliff Turney, 'William J. Stephens (1829-1890)', *Australian Dictionary of Biography,* vol. 6, 197-198.

[49] For a discussion of this classificatory system, from which Dewey's library system was derived, see Rydall, *op. cit.,* 20-21.

[50] 'Australia's Proud Tradition of Expositions', *Bicentenary '88,* 3 (2), (1983), 4.

[51] Young, *op.cit.*, 28.

[52] Hunt was Deputy Master at the Sydney Mint, and an Associate of the Royal School of Mines (1851-53), and had been a member of the Royal Society of NSW since 1878.

[53] *The Bulletin,* 23 April 1881.

[54] See Eugene Ferguson, 'Technical Museums and International Exhibitions', *Technology and Culture,* VI (1), (1965), 30-46.

[55] *Report from the Committee for the Technical College, Sydney Mechanics' School of Arts,* 1881, 5.

[56] On the Board of Trustees, Hunt replaced W.J. Stephens, who served again as a Trustee between 1883 and 1890. Australian Museum, *Council Minutes,* 4 December 1878.

[57] In the event, this was to be nine years.

[58] *Annual Report, Technological, Industrial and Sanitary Museum* (afterwards, *AR, TISM),* 1889.

[59] *Liversidge Report,* 1880, xi.

[60] The adjective 'Sanitary' was added to embrace the idea of the Parkes Museum in London, which pleased Liversidge; and to give scope for medical material, which pleased Roberts.

[61] According to an article in the *Daily Telegraph,* 22 October 1880, the Trustees of the Australian Museum asked the Government to appoint a separate Board of Management for the Museum, 'as the work was growing too great for the Trustees to undertake'. This the Government declined to do.

[62] During his visit to Melbourne, he stayed at the Australian Club. Liversidge Papers, C.R. Buckland to Archibald Liversidge, 15 January 1881.

[63] For MacIntosh, see Martha Rutledge, 'John MacIntosh (1821-1911)', *Australian Dictionary of Biography,* vol. 5, 164-165.

64  *SMH,* 11 January 1881.

65  A.L., 'An Analysis of Moa Eggshell' (Philosophical Institute of Canterbury, 5 August 1880), *Trans. NZI,* 13 (1880), 225-227; *Geological Magazine,* 7 (December 1880), 546-548.

66  'Water from a Hot Spring, New Britain' (1 September 1880), *JPRSNSW,* 14 (1880), 145; *Chemical News,* 42 (31 December 1880), 324; *J. Chem. Soc.,* 40, A (1881), 564, 1019-1020; 'The Action of Sea-water upon Cast-Iron' (1 September 1880), *JPRSNSW,* 14 (1880), 149-154; *Chemical News,* 43 (18 March 1881), 121-123; 'Water from a Hot Spring, Fiji Islands' (1 September 1880), *JPRSNSW,* 14 (1880), 147-148; *Chemical News,* 42 (31 December 1880), 324-325; *J. Chem. Soc.,* 40, A (1881), 564, 1019-1020; 'Notes upon some Minerals from New Caledonia' (1 September 1880), *JPRSNSW,* 14 (1880), 227-246.

67  A.L., 'On the Composition of some Coral Limestones, &c., from the South Sea Islands' (6 October 1880), *JPRSNSW,* 14 (1880), 159-162; abstr. in *J. Chem. Soc.,* 40, (1881) 1011-1012; 'The Alkaloid from Piturie' (3 November 1880), *JPRSNSW,* 14 (1880), 123-132; *Chemical News,* 43 (18 March 1881), 124-126, 138-140; *Moniteur Scientifique,* 23 (1881), 774-780; 'On Some New South Wales Minerals' (3 November 1880), *JPRSNSW,* 14 (1880), 213-225; abstr. in *J. Chem. Soc.,* 40 A (1881), 991-997.

68  A.L., 'On the Composition of Some Wood enclosed in Basalt' (1 December 1880), *JPRSNSW,* 14 (1880), 155-157; 'Upon the Composition of Some New South Wales Coals' (8 December 1880), *JPRSNSW,* 14 (1880), 181-212; abstr. in *J. Chem. Soc.,* 40, A (1881), 980-3; 'On the Bingera Meteorite, New South Wales (Preliminary Notice)' (8 December 1880), *JPRSNSW,* 16 (1882), 35-37; *Jahrb. für Min.,* 2 (1885), abstr. in *J. Chem. Soc.,* 50 A (1886), 133; 'The Deniliquin or Barratta Meteorite (Second Notice)' (8 December 1880), *JPRSNSW,* 16 (1882), 31-33; *Jahrb. für Min.,* 2 (1885), abstr. in *J. Chem. Soc.,* 50 A, (1886), 134; 'On the Chemical Composition of Certain Rocks, New South Wales, &c. (Preliminary Notice)' (8 December 1880), *JPRSNSW,* 16 (1882), 39-46; *Jahrb. für Min.,* 2 (1885), 'Rocks from New Britain and New Ireland (Preliminary Notice)', (8 December 1880), *JPRSNSW,* 16 (1882), 47-51; *Jahrb. für Min.,* 2 (1885), 63.

69  A.L., 'Analyses of Queensland Soils' (20 January 1881), *J. Chem. Soc.,* 39 (1881), 61-63; *Chemical News,* 43 (28 January 1881), 43; 'Stilbite, from Kerguelen's Island' (7 February 1881), *Proc. Roy. Soc. Edinb.,* (1882), 117-119; abstr. in *J. Chem. Soc.,* 40 A (1881), 695.

70  Liversidge Papers, C.R. Buckland to Archibald Liversidge, 14 February 1881. Liversidge's wisdom in keeping his own records became evident when it was revealed that Buckland had been 'milking the accounts' of the Museum of at least £550. Strahan, *op.cit.,* 38.

71  Roy MacLeod, 'Founding: South Kensington to Sydney', in Graeme Davison and Kimberley Webber (eds.), *Yesterday's Tomorrows: The Powerhouse Museum and its Precursors 1880-2005* (Sydney: Powerhouse Publishing in association with UNSW Press, 2005), 42-54.

72  'Frederick Settle Barff', in J.A. Venn, *Alumni Cantabrigienses* (Cambridge: Cambridge University Press, 1940), 148; 'Frederick Settle Barff', in F. Boase (ed.), *Modern English Biography,* 4 (Supplement to vol. I), (London: Frank Cass, 2nd imp., 1965), 263.

73  See Mark Lyons and C.J. Pettigrew, 'Joseph Henry Maiden (1859-1925)', *Australian Dictionary of Biography,* vol. 10, 381-383; Lionel Gilbert, *The Royal Botanic Gardens, Sydney: A History, 1816-1985* (Melbourne: Oxford University Press, 1986), 114-117.

74  Henry M. Franklyn, *A Glance at Australia in 1880* (Melbourne: 1881), 28-31.

75  Robert O. Chalmers, 'History of the Department of Mineralogy, Australian Museum, Part I: 1827-1901', *Journal of the Mineralogical Society of New South Wales,* I (1979), 18, quoting a letter from the firm of Krantz to Secretary, 28 July 1880. August Krantz (1809-1872) studied at the Bergakademie in Freiberg, and supplied much of Europe with mineral specimens. The company still exists (2008) at www.krantz-online.de

76  Gilbert, *op. cit.,* 47.

77  *Annual Report of the Australian Museum for 1881* (Sydney: Government Printer, 1882), 21. A distinction was made between economic and scientific productions, with the former given precedence.

78  *SMH,* 3 May 1881.

79  *SMH,* 15 December 1883.

80  These were animal products, vegetable products, waste products, foods, economic geology and sanitary appliances. *Annual Report of the Australian Museum for 1881* (Sydney: Government Printer, 1882), 22.

81  The Acting Secretary, Charles Buckland, who had recently been dismissed for misappropriating Museum funds, was a key suspect, but disappeared and was never brought to trial. See J.L. Willis, 'From Palace to Power House' (Sydney: Museum of Applied Arts and Sciences, 1982), 28.

82    Chalmers, *op.cit.*, 19.

83    Cited in Strahan, *op. cit.*, 38.

84    *Ibid.*

85    *Ibid.*

86    Museum of Applied Arts and Sciences (afterwards MAAS), MA 2/2, 4 December 1882.

87    MAAS, M 2/2. 2 January 1883; 26 February 1883.

88    By 1884, meetings of the Committee of Management had become 'managerial' in name only. As Liversidge's energies were absorbed by the University and the Royal Society, and as the elderly co-committee members, Roberts and Hunt, faded from the scene, day-to-day administration passed to Maiden. In any event, had not Liversidge's Report of 1880 specifically recommended placing the Museum under a permanent director, assisted by a 'board of advice'? MAAS, Trustees Correspondence, 1883-86. *Maiden Report*, 31 August 1883.

89    In later years, Maiden would be Liversidge's chronicler at the Royal Society of NSW, and his successor in the AAAS. See his 'A Contribution to a History of the Royal Society of New South Wales', *JPRSNSW,* 52 (1918), 215-361. He was elected FRS in 1916. He was also twice president of both the Linnean Society and the Royal Society of NSW. His greatest memorial, outside the National Herbarium in Sydney's Botanic Gardens, are the palm trees and other flora that flourish in the public spaces of modern Sydney.

90    See Lionel Gilbert, *The Little Giant : The Life and Work of Joseph Henry Maiden 1859-1952* (Sydney: Kardoorair Press in association with the Royal Botanic Gardens, 2001).

91    MAAS, Trustees Correspondence, 1883-86; A.L. to Hunt, c.11 August 1885; A.L. to Maiden, 28 May 1885; MAAS, Letterbook, 4/3, f. 695. Maiden to A.L., 22 June 1886.

92    Armstrong Papers (Imperial College), 26A, A.L. to Armstrong, 12 March 1885.

93    The early months of 1884 saw few openings on the accommodation front. The leading possibility at first was a site near the Australian Museum, at the eastern end of King Street, where it was proposed to build a free Public Library, and with it, a Technological Museum, an Industrial College and a School of Design. Liversidge thought well of the plan, and with Roberts assembled specifications for the building, providing for lecture theatres, laboratories and workshops, and 50,000 square feet of exhibition space. If the government did not agree to this, Liversidge supposed that space could be found on land adjoining the Australian Museum. This was not an unattractive solution but it proved unacceptable to the government, which decided there were other uses for that land. Reduced to the role of an estate agent, for which he had little taste, Liversidge grew weary of the business. Then, and later, his appeals fell on deaf ears. Feelings were so low by May 1884, that the elderly Roberts, expressing a personal vote of no confidence in the government, offered his resignation from the Committee. Patiently, Liversidge and Hunt persuaded him to stay, and in good time, as in June came the news that the government had appropriated an increased vote of £4,300 for 1884. TISM Minute Books, 2/1. 5 May 1884; cited in Willis, *op.cit.*, 47.

94    *Annual Report, TISM,* 1889, 15.

95    Prompting, for example, the appointment of the young Frenchman, Felix Ratte, to catalogue its mineral collection. Ratte has been described as the first head of the Museum's Department of Mineralogy (Strahan, *op.cit.,* 38). He was listed as a 'cataloguer' from 1881 until 1887, when he was entitled 'Assistant in Mineralogy'. He completed the *Catalogue of Minerals in the Collection* in 1885 (Chalmers, *op.cit.,* 19). He led a blameless scientific life, and possibly suffered at the hands of the Museum; whatever the cause, he grew depressed and committed suicide in 1890.

96    MAAS, Inward Correspondence, 1886, f. 43, A.L. to Maiden, 22 February 1886; f. 66, A.L. to Maiden, 17 March 1886; f.161, A.L. to Maiden, 21 June 1886; f. 168, A.L. to Maiden, 24 June 1886; f. 193 A.L. to Maiden, 14 July 1886; f. 233, A.L. to Maiden, 9 August 1886.

97    It appears that several minerals were recorded as new discoveries long after their location had become common knowledge – spessartite and rhodonite, for example. See George Smith, *A Contribution to the Mineralogy of New South Wales* (Sydney: Government Printer, 1926), 110.

98    David Branagan (ed.), *Rocks, Fossils, Profs: Geological Sciences in the University of Sydney, 1866-1973* (Sydney: Scientific Press for Department of Geology and Geophysics, 1973).

99    MAAS, Inward Correspondence, A.L. to Maiden, 30 August 1886. A similar point was made in 1888 by Anderson Stuart, who joined the Trustees in 1881, and who noted that the journal collections of Sydney's ten major libraries, which lacked any cooperative plan, had both gaps and duplications. These institutions included the Australian Museum, the Linnean Society, the Observatory, the Royal Society, the University, the Free

Public Library in Bent Street, the Parliamentary Library in Macquarie Street, the Royal Geographical Society of Australasia and the Technological Museum. With £50 from each library, Anderson Stuart produced the first bibliography of scientific literature held in the colony.

100  A graduate of Sydney (BA 1857) and Edinburgh (MB 1860, MD 1861), Arthur Renwick was an important figure in medical education and politics and in NSW. Elected to the Senate of the University in 1877, he entered parliament in 1879 and served in a succession of governments until 1887, and on Senate until 1908. He was also active in commercial life, and made and lost a fortune. A contemporary called him 'a very ambitious little gentleman' with an enlarged idea of his own political worth'. He was several times acting Dean of Medicine, and was knighted in 1894. See Martha Rutledge, 'Sir Arthur Renwick (1837-1908)', *Australian Dictionary of Biography*, vol. 6, 20-21.

101  MAAS. Inward Correspondence, 1886, f. 6, A.L. to Maiden, 17 March 1886.

102  From this, Maiden was excluded. The apparent slight caused the only known rupture in their friendship. The event was even less well received by Dr. Renwick, who was also left out. MAAS. Inward Correspondance, 1886, f. 6, A.L. to Maiden, 30 July 1886.

103  *Statistical Register of NSW for the year 1888* (Sydney: Government Printer, 1889), 181, 257.

104  MAAS. Inward Correspondance, 1886, f. 6, A.L. to Maiden, 13 October 1886.

105  Colonial revenues arose principally from a combination of indirect taxes (customs, excise, licences and stamp duties), land sales and leases, services (including railways, post and telegraph), and rents. In 1882, land revenues peaked at £2.9 million, and fell to £1.6 m in 1886. In 1887, the picture began to improve (to £2.3 million). See *Statistical Register of NSW for the year 1888, op. cit.*, 176, 177.

106  See P. Loveday and A.W. Martin, *Parliament, Factions and Parties: The First Thirty Years of Responsible Government in New South Wales, 1856-1889* (Melbourne: Melbourne University Press, 1966), chapter 6, esp. 129 ff.

107  *AR, Australian Museum,* 1889; supplement, *AR, TISM*, 36-7, quoted in F.R. Morrison, *A Brief Outline of the History and Activities of the Museum of Applied Arts and Sciences* (n.d.).

108  MAAS, Inward Letters, 1889, f. 105, A.L. to Maiden, 12 March 1889; f. 125, A.L. to Maiden, 31 March 1889; f. 172, A.L. memo, 9 May 1889; f. 248, A.L. memo to Curator, 23 July 1889.

109  Robert White, an unlikely combination of popular bank manager, speculator, and member of parliament, was also the NSW Commissioner for exhibitions at Melbourne (1880 and 1888), Calcutta (1883) and Adelaide (1887), as well as for the Colonial and Indian Exhibition in London in 1886. He died in the Hospital for the Insane at Callan Park. See John Atchison, 'Robert Hoddle Driberg White (1838-1900)', *Australian Dictionary of Biography*, vol. 6, 389-390.

110  *Idem.*, f. 336, A.L to Curator, 15 October 1889.

111  Willis, *op.cit.*, 34.

112  *Annual report, TISM*, 1894, 184.

113  *Statistical Register of NSW for the year 1888, op. cit.,* 257.

114  *AR, Australian Museum,* 1889.

115  *AR, Australian Museum,* 1889; supplement, *AR, TISM*, 36-37.

116  *Ibid.*

117  MAAS, Inward Letters, 1890, f. 269, A.L. to Curator, 22 May 1890.

118  *AR, TISM,* 1889.

119  *Report of the Minister of Public Instruction, 1893* (Sydney: Government Printer, 1894), 34.

120  R.T. Baker, *Technological Museum, Sydney* (Sydney: Special Issue for the BAAS, Government Printer, 1914), 7.

121  MAAS, Letterbook, MA 4/32, f. 514. R.T. Baker to A.L., 5 July 1911.

122  Since its creation, the Museum has been renamed thrice. The original Technological, Industrial and Sanitary Museum became in 1888, simply, the Technological Museum; but in 1944, the name was changed to the Museum of Technology and Applied Science; and from 1950, it was officially renamed the Museum of Applied Arts and Sciences. Today, the 'Powerhouse' is its unofficial, if more popular, title. See Graeme Davison and Kimberley Webber (eds.), *op. cit.*

123  David Layton, *Science for the People* (London: George Allen and Unwin, 1973).

124  For Australian versions of these English debates, see *Australian Journal of Education*, 1868 and 1869.

125  In some respects predating British interest in art and design, in a city that still lacked much of either, the School enjoyed vice-regal patronage, and in the beginning at least, the confidence of the colonial bourgeoisie.

At a time when most of the colony remained unexplored, it was fitting that the School's first president should be Major Thomas Mitchell, the Surveyor-General. It was an equally suitable place for those seeking funds for an expedition to search for the lost Leichhardt in 1865. See Jan Todd, 'Colonial Adoption: The Case of Australia and the Sydney Mechanics' School of Arts', in Ian Inkster (ed.), *The Steam Intellect Societies: Essays on Culture, Education and Industry, circa 1820-1914* (Nottingham: University of Nottingham, 1985), 105-130. The SMSA has yet to find its historian, although Jean Riley, 'The Mechanics' School of Arts Movement in New South Wales' (Unpublished essay, University of Sydney, 1983) is a valuable beginning. I am grateful to Ms Riley for her advice.

126 Cf. George Nadel, *Australia's Colonial Culture* (Cambridge, Mass: Harvard University Press, 1957), 143, 182**.**

127 Quoted in Derek Whitelock, *The Great Tradition: A History of Adult Education in Australia* (St. Lucia: University of Queensland Press, 1974), 111.

128 Quoted in *ibid.*, 182.

129 T.A. Coghlan, *The Wealth and Progress of New South Wales, 1889-90* (Sydney: Government Printer, 1891), 533.

130 Stephen Murray-Smith, 'Technical Education in Australia: A Historical Sketch', in E.L. Wheelwright (ed.), *Higher Education in Australia* (Melbourne: F.W. Cheshire, 1965).

131 Murray-Smith, *op.cit.,* 158.

132 See Roy MacLeod, 'Of Mines and Mining Education: The School of Mines at the University of Sydney, 1884-1894', *Earth Sciences History,* 39 (2001), 32-53.

133 L.A. Mandelson, 'Norman Selfe and the Beginnings of Technical Education', in C. Turney (ed.), *Pioneers of Australian Education* (Sydney: Sydney University Press, 1972), vol. 2, 109.

134 P.W.D. Mathews and G.W. Ford (eds.), *Australian Trade Unions: Their Development, Structure and Horizons* (Melbourne: Sun Books, 1968), 20.

135 For a Selfe appreciation, see Mandelson, *op.cit.*, 105-148.

136 W.G. McMinn, 'Edward Dowling (1843-1912)', *Australian Dictionary of Biography,* vol. 8, 329-330. Dowling was also a leading federationist, and author of *Australia and America in 1892: A Contrast* (Sydney: Charles Potter, 1893).

137 'Lectures for Promoting Technical Education ', *VPLANSW,* 5 (1876-7), 810 ff.

138 *Mining Act,* 16 April 1874, clause 8.

139 'Lectures', 10; *SMH,* 31 October 1874.

140 *SMH,* 11 October 1879; Whitelock, *op.cit.,* 113

141 'Technological Education Conference', *SMH,* 11 October 1879.

142 As the colonial debate deepened, the British position became more complex. In 1869, John Scott Russell, the naval engineer, dedicated to Queen Victoria a memorable appeal for state-sponsored Technical Education (J. Scott Russell, *Systematic Technical Education for the English People* (London: Bradbury, Evans and Co., 1869). Cf. George S. Emmerson, *John Scott Russell, A Great Victorian Engineer and Naval Architect* (London: J. Murray, 1977); G.P. Mabon, 'Russell: A Forgotten Champion of Technical Education', *Vocational Aspect,* 17 (1965), 228-229). From the early 1870s, T.H. Huxley and Henry Roscoe took up the same refrain and criticised the wealthy Livery Companies of London for neglecting their historical role in sponsoring technical education for working men. In 1878, their efforts prompted the City Guilds to establish a 'Technical University' for London; and in 1883, the City and Guilds Institute ('City and Guilds of London Institute for the Advancement of Technical Education') opened Finsbury Technical College. In 1884, the City and Guilds established the 'Central Institution' in South Kensington, with professors in chemistry, engineering, mechanics and physics. At Finsbury, Henry Armstrong was joined by John Perry and W.E. Ayrton, fresh from four years' experience of education in Japan; together, these three created a model for technical colleges throughout Britain. (W.H. Brock, 'The Japanese Connexion: Engineering in Tokyo, London and Glasgow at the End of the Nineteenth Century', *British Journal for the History of Science,* 14 (48), (1981), 227-243.) Theirs was to be the equivalent of the Swiss or German polytechnic; its students, prospective teachers and men of trade. In 1907, uniting with the Royal School of Mines, the Central Institution became a part of Imperial College of Science and Technology. See T.H. Huxley, 'Technical Education', *Fortnightly Review,* 29 (January 1878), 48-55; W.G. Armstrong, 'The Vague Cry for Technical Education', *Nineteenth Century,* 24 (July 1888), 45-52; Lyon Playfair, 'Lord Armstrong and Technical Education', *Nineteenth Century,* 24 (September 1888), 325-333; W.G. Armstrong, 'The Cry for Useless Knowledge', *Nineteenth Century,* 24 (November 1888), 653-668.

143 G.G. Walker, 'Finsbury Technical College', *Central,* 30 (1933), 35-48; H.E. Armstrong, 'The Beginnings of Finsbury and the Central', *Central,* 31 (1934), 1-14.

144 *Inaugural Address,* The Finsbury Technical College, 19 February 1883, in Sir Philip Magnus, *Industrial Education* (London: Kegan Paul, Trench and Co., 1888), 256; cf. Frank Foden, *Sir Philip Magnus: Victorian Educational Pioneer* (London: Valentine Mitchell, 1970), 138.

145 For this at times creative tension, see Roy MacLeod, 'The "Practical Man": Myth and Metaphor in Anglo-Australian Science', *Australian Cultural History,* No. 8 (1989), 24-49, and J. Ann Hone,' The Practical Man as Hero? Technical Education in New South Wales in the 1870s and 1880s', *Australian Cultural History,* 8 (1989), 62- 77.

146 See Ann Pugh, 'One Hundred Years Ago: Foundation of the Engineering Association of New South Wales', *Journal of the Engineering Association,* 42 (1970), 83-91.

147 'Technological Conference', *SMH,* 29 October 1880.

148 State Archives, Department of Public Instruction (afterwards, DPI) 10/14282, W.H. Humphreys to E. Dowling, 27 October 1880.

149 'Technological Conference', *SMH,* 29 October 1880.

150 *SMH,* 30 October 1880.

151 *Report from the Committee of the Technical or Working Man's College,* 9 August, 1881 (Sydney: Samuel Lees, 1881), 14.

152 Hone, *op.cit.,* 69.

153 Cf. 'Liversidge on Technology', *op.cit.*

154 Quoted in *ibid.,* 70.

155 Cf. Ruth Barton, 'Scientific Opposition to Technical Education', in M.D. Stephens and G.W. Roderick (eds.), *Essays on Scientific and Technical Education in Early Industrial Britain* (Nottingham: Department of Adult Education, University of Nottingham, 1981), 13-27.

156 Stephen Murray-Smith and Anthony John Dare, *The Tech: A Centenary History of the Royal Melbourne Institute of Technology* (Melbourne: Hyland House Publishing, 1987).

157 See Ken Cable, 'John Sutherland (1816-1889)', *Australian Dictionary of Biography*, vol. 6, 223-224.

158 And by Huxley, who, along with much English educational opinion, turned against the systems that Robert Lowe introduced in the 1850s.

159 *Report from the Committee of the Technical or Working Man's College,* 9 August 1881 (Sydney: S.E. Lees, 1881), 11-17.

160 George (later Sir George) Reid (1845-1918), born in Scotland, son of a Presbyterian minister, came to Australia as a child, worked in trade and entered politics in 1878. He became well known as a political and economic writer, and as an exponent of free trade. He was elected to the Legislative Assembly in 1880, and became Premier of NSW (August 1894 - September 1899). With the coming of Federation, he was elected to the Federal House of Representatives, and served as Prime Minister in 1904-05, and as free trade leader of the Federal Opposition for the rest of the period between 1901 and 1908. Between 1909 and 1916, he was Australia's first High Commissioner to London and a British member of parliament between 1916 and 1918. See Anon., 'Leaders of Technical Education in New South Wales: The Hon. George Houston Reid', *The Australian Technical Journal,* I (2), (30 March 1897), 38-40; W.G. McMinn, *George Reid* (Melbourne: Melbourne University Press, 1989); 'Sir George Houston Reid (1845-1918)', *Australian Dictionary of Biography,* vol. 11, 347-54.

161 NSW State Archives, DPI IU/14282. A.L. to Reid, 26 June 1883; H.C. Russell to Reid, 26 June 1883.

162 Commemorative Address, Fifteenth Anniversary, SMSA, 22 March 1883.

163 NSW State Archives, DPI IU/14282, Smith to Reid, 10 July 1883.

164 For an outline of events, see Joan Ritchie, 'The Development and Significance of Technical Education in New South Wales, 1880-1900', *JRAHS,* 55 (1969), 245-261.

165 Under the Board, Sydney Technical College followed the curriculum of London's City and Guilds Institute, which in 1887 extended its examinations to NSW candidates. Branch technical schools were established at Bathurst, Goulburn and Newcastle along the lines of City and Guilds Institutes in provincial England; while subjects offered were identical to those taught at Finsbury. Instruction was organised in thirteen departments – from agriculture to art, and chemistry to elocution. *Nature* smiled on this fine example of colonial fidelity. *Nature,* 33 (18 May 1886), 462.

166 Whitelock, *op.cit.*, 114. Windeyer resigned from its presidency in 1886, saying 'The tone of the general meetings lately held and the indifference of the great majority of members to the institute's interests, shows me that I would be placing myself in a false position by any longer holding an office to which my repeated election has hitherto been a matter of pride and pleasure'. SMSA Minutes (18 January 1886). Although subscriptions grew from 1,295 in 1876 to 4,372 in 1890, the 'abnormal increase' in lady members accounted for most of the SMSA's growth, and works of fiction had become the mainstay of its library, as they are today.

167 The Board at first declined the invitation on the grounds that there was 'too great a variety of courses with very inadequate instruction in several branches', but on further consideration, accepted, and recommended approval to Senate. Minutes of the Board of Studies, 5th meeting, 18 June 1883; 6th meeting, 23 June 1883.

168 *Report of the Minister of Public Instruction for 1884* (Sydney: Government Printer, 1885), 33.

169 *Ibid.*, 35.

170 *Ibid.*, 33.

171 Stephens' preferment to a chair without prior advertisement did not escape criticism. See, for example, Alexander Oliver, 'On Some Recent Appointments by the Senate', *Sydney University Review,* 2 (April 1882), 89-105.

172 Like Warren, Haswell had been in Australia for some time before joining the university. Whilst Haswell had worked at the Queensland Museum and the Australian Museum, Warren had worked for two years in the Public Works Department, while also teaching at the Technical College. Both were archetypal 'go-betweens', linking basic and applied science, academic and technical education. Haswell became the first Challis Professor of Biology in 1891. See Patricia Morison, 'William Aitchison Haswell (1854-1925)', *Australian Dictionary of Biography,* vol. 9, 226-227. Warren was promoted to a chair of engineering in 1884, and put on the Challis establishment in 1891. See Anon., 'Prof. W.H.Warren. Wh.Sc., M.Inst. C.E. and The P.N. Russell School of Engineering, University of Sydney', *The Australian Technical Journal,* I (11), (30 December 1897), 325-332.

173 State Archives, Board of Technical Education, Minutes, 16 January 1885. S.H. Cox was formerly an Assistant Geologist and Inspector of Mines in the New Zealand Geological Survey, and may have been recruited by Liversidge in New Zealand in January 1883. Board of Technical Education, Minutes, 12 March 1884.

174 *Idem.,* 28 December 1883.

175 See J.M. Barker and D.R. Stranks, 'Edward Henry Rennie (1852-1927)', *Australian Dictionary of Biography,* vol. 11, 361-362. Another was T.H. Laby, who became Professor of Physics at Wellington, then Melbourne, and who was elected FRS in 1931. See Katrina Dean, 'Inscribing Settler Sciences: Earnest Rutherford, Thomas Laby and the Making of Careers in Science', *History of Science,* XLI (2), (2003), 217-240.

176 Board of Technical Education, Minutes, 31 March 1884.

177 Board of Technical Education, Organising Committee Minutes, 19 May 1884.

178 *Ibid.,* 10 October 1884.

179 *Ibid.,* 7 November 1884, 12 July 1886.

180 *Ibid.*, 8 September 1885, 26 July 1886.

181 *Report of the Minister for Public Instruction, 1884* (Sydney: Government Printer, 1885).

182 W.A. Dixon, *Technical Education: Outlines of a Practical Scheme to carry out a system of Technical Education through the Schools of Arts of New South Wales* (Sydney: Colonial Publishing Society Ltd., 1881); *Report of the Board of Technical Education, 1885* (Sydney, Government Printer, 1886), 10.

183 Board of Technical Education, Organising Committee Minutes, 11 May 1885, 23 March 1886, 12 April 1886.

184 Board of Technical Education, Minutes, 12 October 1883.

185 *Statistical Register of NSW for the year 1888* (Sydney: Government Printer, 1889), 235.

186 Cf. Board of Technical Education, Minutes, 29 September 1883, 31 October 1883, 20 February 1884, 12 March 1884, 31 March 1884.

187 *New South Wales Parliamentary Debates,* 18 (1886), iii, 1773 ff.; 20, 2305.

188 State Archives, Department of Public Instruction, 10/14282. 'Recommendations respecting the proposed formation of Technical and Mining Schools in the Principal Mining Centres of New South Wales, as prepared by Professor Liversidge, FRS, 23 June 1886'.

189 His description of mining instruction at the University as 'somewhat limited', and his favoured title, 'Schools of Science and Mining', was deleted. The Board also deleted his suggestion that local residents bear a portion of the cost of mining schools set up in their districts.

190  State Archives, Department of Mines, H. Wood to Fletcher, 23 November 1886.

191  State Archives, Department of Public Instruction, 10/14282. Renwick, Recommendations respecting the proposed formation of Technical and Mining Schools.

192  McMinn, *George Reid, op.cit.*, 48; *New South Wales Parliamentary Debates,* 27 (1887), 2329-30; 2458-64.

193  The Central Institution was known as the Central Technical College from 1893 to 1907, when it became part of Imperial College, and was designated the City and Guilds College.

194  Formally, the 'City and Guilds of London Technical College, Finsbury'. For its history, see W.H. Brock, 'Building England's First Technical College, 1878-1926', in Frank A.J.L. James (ed.), *The Development of the Laboratory: Essays on the Place of Experiment in Industrial Civilisation* (London: Macmillan, 1989), 155-170. See W.H. Brock, *H.E. Armstrong and the Teaching of Science, 1880-1930* (Cambridge: Cambridge University Press, 1973), 34; Cf. P. F. R. Venables, *Technical Education: Its Aims, Organization and Future Development* (London: G. Bell, 1956), 22.

195  In the same period, the colony's secondary schools spent about £5 per pupil. McMinn, *George Reid, op.cit.*, 33.

196  This principle was later embodied in 1948 by the New South Wales University of Technology (now the University of New South Wales), and was to survive in the New South Wales Institute of Technology (now the University of Technology, Sydney).

197  Norman Selfe, 'President's Annual Report', *Report of the Board of Technical Education,* 1887, 18.

198  Cf. *Report of Minister of Public Instruction for 1886* (Sydney: 1887), Appendix VIII, 158.

199  Inglis was the author of *Our Australian Cousins* (London: Macmillan, 1880), which he wrote as a tea-planter visiting from India. The book includes a chapter on Mount Wilson in the Blue Mountains, which he saw as a 'hill station', as did Badham and others of Sydney University. See Martha Rutledge, 'James Inglis (1845-1908)', *Australian Dictionary of Biography,* vol. 4, 457-458.

200  *New South Wales Parliamentary Debates,* 34 ser. 1 (1887-88), 17 July 1888, 6466.

201  Cf. State Archives, Department of Public Instruction, Board of Technical Education, Edward Dowling to Under-Secretary, 28 March 1889, and accompanying documents; Minute, Carruthers, 29 March 1889; 'The Board of Technical Education', *Daily Telegraph,* 30 March 1889.

202  State Archives, Board of Technical Education, Minutes, 11 July 1888.

203  Carruthers was a graduate of Sydney University (BA 1876, MA 1878), who entered politics as a free trader in 1879. He supported Renwick against Edmund Barton for the university seat, and was keenly interested in women's education and technical education. He remained in NSW politics until 1908, becoming Premier in 1904 and a member of federal parliament after Federation. See John M. Ward, 'Sir Joseph Hector McNeil Carruthers (1856-1932)', *Australian Dictionary of Biography,* vol. 7, 574-578.

204  For discussion, see Joan M. Ritchie, 'The Development and Significance of Technical Education in New South Wales, 1880-1900', *JRAHS,* 55 (1969), 245-261; P.R. Proudfoot, 'The Development of Architectural Education in Sydney, 1880-1930', *Historical Studies,* 21 (83), (1984), 197-211.

205  *Spanners, Easels and Microchips: A History of Technical and Further Education in New South Wales, 1883-1983* (Sydney: New South Wales Department of T.A.F.E., 1983), chapters 2 and 3.

206  This debate would be rehearsed again in Liversidge's time by his friend G.H. (later Sir George) Knibbs, in his famous *Report of the Commissioners on Agricultural, Commercial, Industrial and Other Forms of Technical Education, VPLANSW,* 4 (1905), 1-853.

207  State Archives, Board of Technical Education, Minutes, 28 March 1887.

# CHAPTER 8

1  W.B. Clarke, 'Anniversary Address' (17 May 1876), *JPRSNSW,* 10 (1876); quoted by John Smith, 'Anniversary Address' (1879), *JPRSNSW* 13 (1879), 7-8.

2  Council Minutes, 29 April 1878.

3  *Ibid.,* 31 July 1878.

4  *Ibid.,* 26 June 1878.

5  *Ibid.,* 30 October 1878.

6  *SMH,* 5 June 1879.

7  This rose to £2,000 by October 1882, and £3,000 by 1885. Council Minutes, 27 September 1882, 29 April

1885.

8   'Royal Society *Conversazione*', *Australian Town and Country Journal*, 1 November 1879. A copy is preserved in the Liversidge Papers, Box 13.

9   *JPRSNSW*, 12 (1878), 187.

10  '... While they gave great pleasure to the lady friends and relatives of members', Charles Moore morosely reflected, '[they] are nevertheless a heavy tax on our resources ... but I would hope that as long as they do not draw too deeply on our funds that they will be continued'. 'Anniversary Address' (12 May 1880), *JPRSNSW*, 14 (1880), 4.

11  From 1881 to 1900, the Governor was styled 'Honorary Governor'. After Federation in 1901, the title was dropped, and the Governor-General became the Patron, and the Governor, the Vice-Patron. See J.H. Maiden, 'A Contribution to a History of the Royal Society of New South Wales (with Information in regard to other New South Wales Societies)', *JPRSNSW*, 52 (1918), 350.

12  Council Minutes, 24 March 1880.

13  *Ibid.*, 28 April, 25 August 1880.

14  *Ibid.*, 22 December 1880.

15  The first woman was not elected until 1935. W.R. and Ida Browne, 'Scientific Societies in Australia', *RACI*, 5 (28), (1961), 104.

16  Council Minutes, 24 November 1880.

17  John Smith, 'Anniversary Address,' *JPRSNSW*, 15 (1881), 12.

18  Imperial College, Huxley Papers, 21. 227. A.L. to Huxley, 30 July 1881.

19  Annual General Meeting (12 May 1880) *JPRSNSW*, 14 (1880), 300; Owen to Liversidge, 27 October 1879.

20  'Dear Parkes', wrote Robertson, Minister of Public Instruction (May 1880-November 1881), 'I don't know how far you committed the Government on this matter last year. It seems a matter which might be aided, but I am not well informed as to its claims'. NSW State Archives, Department of Public Instruction, 20/12922, f. 80/10309. Liversidge to Minister, 19 November 1880; Minutes, Robertson to Parkes 13 December 1880.

21  Council Minutes, 30 August 1882.

22  Indeed, by 1984, the size of the periodicals library was to prove an embarassment to the Society. In 1985, it was deposited at the University of Newcastle, NSW.

23  State Archives, Department of Public Instruction, 20/12922. Liversidge to Minister of Public Instruction, 19 May 1880.

24  Liversidge prophetically proposed as a topic, 'the embryology and development of the marsupials'. Council Minutes, 31 August, 29 September 1881. The first subjects set were: the Aborigines of New South Wales, the treatment of auriferous pyrites, plants indigenous to New South Wales, and the influence of climate and pasture on wool.

25  'Royal Society's *Conversazione*', *SMH*, 29 September 1881.

26  Council Minutes, 25 October 1882.

27  *SMH*, 5 May 1881.

28  *SMH*, 31 July 1882.

29  *SMH*, 8 October 1884.

30  *SMH*, 7 March 1882, reviewing W. Macleay, *A Descriptive Catalogue of Australian Fishes* (Sydney: S.F. White, 1881), and W.A. Haswell, *Australian Crustaceans* (Sydney: Australian Museum, 1882).

31  Sir William Manning, 'Chancellor's Address', 13 April 1889.

32  J.Z. Fullmer, 'Ira Remsen (1846-1927)', *Dictionary of Scientific Biography*, XI (1975), 370-371.

33  Roy MacLeod, 'The Support of Victorian Science: The Endowment of Science Movement in Great Britain, 1868-1900,' *Minerva*, IX (2), (1971), 85-99.

34  Alan Barcan, *A Short History of Education in New South Wales* (Sydney: Martindale Press, 1965), chapter XIII.

35  W.G. McMinn, Sir George Houstoun Reid (1845-1918)', *Australian Dictionary of Biography*, vol. 11, 347-354.

36  *The Statistical Register of New South Wales*, Annual Volumes, 1881-1888.

37  N.T.K, 'Dr Badham's Proposal', *Sydney Quarterly* Magazine, 1 (October 1883), 87-93; *SMH*, 3 October 1883, 8.

38  Senate Minutes, G 1/1/6, 4 January 1882.

39  Renwick remained on Senate until 1908; he was Vice-Chancellor in 1889-91, 1900-02, and 1906-08, and

several times Acting Dean of Medicine. Liversidge also knew him in connection with his work for international exhibitions. See Martha Rutledge, 'Sir Arthur Renwick (1937-1908)', *Australian Dictionary of Biography*, vol. 6, 20-21.

40 Medicine was referred to the Board of Medical Examiners, and Law to the legal members of Senate.

41 For his commemoration addresses, see Charles Badham, *Speeches and Lectures delivered in Australia* (Sydney: William Dymock, 1890). See also, Wilma Radford, 'Charles Badham (1813-1884)', *Australian Dictionary of Biography*, vol. 3, 68-71.

42 Charles Badham, 'University Studies', *Sydney University Review,* 2 (April 1882), 161-165.

43 *Ibid.*, 164.

44 *Ibid.*, 165.

45 It was, however, bitterly opposed by Anderson Stuart.

46 Sir William Manning, 'Commemoration Address', 29 July 1882, 7.

47 The University of Sydney Extension Act, 1884, permitted the award of degrees in all subjects except divinity.

48 Royal Society Archives*, Candidates Book,* 1880, 7.

49 Roy MacLeod, 'Whigs and Savants: Reflections on the Reform Movement in the Royal Society, 1830-1848', in Ian Inkster and J.B. Morrell (eds), *Metropolis and Province: Science in British Culture, 1780-1850* (London: Hutchinson, 1983), 55-90.

50 With him were elected Valentine Ball (Professor of Mineralogy and Geology at Dublin), William Dittmar (Professor of Chemistry at Glasgow), Walter Gaskell (the Cambridge physiologist), Richard (later Sir Richard) Glazebrook (Demonstrator at the Cavendish and future Director of the National Physical Laboratory), and Francis (later Sir Francis) Darwin, the botanist.

51 Passed over on this occasion, but later admitted, were the notable geometer, George Allan of Galway, and the chemist and science educator, Raphael Meldola, with whom Liversidge would later have much to do.

52 Royal Society Archives, M.C. 12.265 Liversidge to Stokes, 4 August 1882; M.C. 12. 265, Liversidge to Walter White, 4 August 1882.

53 *SMH,* 18 January 1883. This was a revised and extended edition of the paper he first delivered to the Royal Society of New South Wales in 1874 ('The Minerals of New South Wales' (9 December 1874), *TPRSNSW,* 9 (1875), 153-215).

54 Frank Leverrier (later a distinguished QC, inventor, member of Senate, 1907-39, and Vice-Chancellor, 1914-17 and 1921-23) and Ebenezer Clarence Wood were the first graduates. See Martha Rutledge, 'Francis Hewitt Leverrier (1963-1940)', *Australian Dictionary of Biography*, vol.10, 80-81. Of Wood, who also graduated in engineering, little is known. See David Branagan and Graham Holland (eds.), *Ever Reaping Something New* (University of Sydney: Science Centenary Committee, 1985), 5, 111.

55 In 1880, seventy-six students enrolled in the Faculty of Arts. See Turney *et al., Australia's First; A History of the University of Sydney* (Sydney: University of Sydney, 1991), vol. 1, 647.

56 For Anderson Stuart, see William Epps, *Anderson Stuart, MD* (Sydney: Angus and Robertson, 1922); John Atherton Young, Ann Jervie Sefton and Nina Webb (eds.), *Centenary Book of the University of Sydney Faculty of Medicine* (Sydney: Sydney University Press, 1984), 171-198, 199-202, 238-243; See also Patricia Morison, *J.T. Wilson and the Fraternity of Duckmaloi* (Amsterdam: Rodopi, 1997), 62.

57 Minutes of the Building Committee, 19 March 1883.

58 Senate Minutes, 19 May 1884, 31; 'Report of the Building Committee'.

59 *Ibid.*, July 1883.

60 Board of Studies Minutes, July 1883 to December, 1884.

61 Senate Minutes, 11 May 1883.

62 *Ibid.*, 23 June 1883.

63 *Ibid.*, 30 August 1883, September 1883.

64 *Ibid.*, 21 April 1884. The Board of Studies amalgamated with the Proctorial Board in 1886, to form a new Professorial Board. See the University of Sydney website, www.usyd.edu.au/ab/about/history/html.

65 Professorial Board, 29 June 1892.

66 Cf. Professorial Board Minutes, 8 December 1892, when Liversidge revised seven clauses on matters ranging from the length of terms to setting examination periods.

67 *Ibid.*, 23 April 1902.

68 Such as presaged the alteration of By-laws, leading to honours in the Faculty of Science. Board of Studies

Minutes, 23 July, 21 August 1885; Senate Minutes, November 1885.

69   Board of Studies Minutes, 27 July, 20 August, 27 November 1883; Senate Minutes, February 1884.

70   Professorial Board Minutes, 10 August 1886.

71   Senate Minutes, 29 August 1884.

72   Board of Studies Minutes, 19 September 1884; Senate Minutes, December 1884.

73   Senate Minutes, 22 May 1884.

74   So it was formally known from June 1885. Senate Minutes, June 1885, the precursor of the present Academic Board.

75   *VPLCNSW,* 12 (1884), 3046-3047.

76   *Echo,* 1 March 1884.

77   *Daily Telegraph,* 7 May 1885.

78   I am indebted to Dr Peter Chippindale for this information.

79   *SMH,* 15 September 1886.

80   Board of Studies Minutes, 16 May 1884, 45; 10 June 1884, 48-49.

81   Liversidge Papers, Box 20, 108.

82   Branagan and Holland, *op. cit.,* 250.

83   Building Committee Minutes, G.1/11/1, 17 November 1883, 150.

84   Senate Minutes, G.1/1/6, 5 December 1883.

85   Liversidge Papers, Box 15, f.14, SMH, 29 December 1883.

86   Senate Minutes, Report of the Building Committee, 30 May 1884, 155.

87   Senate Minutes, G.1/1/7. 4 June 1884, 30-35.

88   Ibid, 34.

89   James Kerr in Young, Sefton and Webb (eds.), *op.cit.,* 477. Drawings were submitted to Senate in November 1884.

90   Senate Minutes, 3 December 1884, 73. See Peter Stanbury and Julian Holland (eds.), *Mr Macleay's Celebrated Cabinet: The History of the Macleays and their Museum* (Sydney: Macleay Museum, University of Sydney, 1988).

91   Senate Minutes, 3 December 1884, 72, 74-75; 4 February 1885, 89-90; 21 March 1885, 105.

92   Senate Minutes, 18 May 1885, 125.

93   Senate Minutes, 17 August 1885, 155.

94   Sir William Manning, 'Chancellor's Address', 29 May 1886, 16.

95   Senate Minutes, 7 September 1885, 161-162; H.E. Barff, *A Short Historical Account of the University of Sydney* (Sydney: Angus and Robertson, 1902), 102. A photograph of the interior can be found in Branagan and Holland, *op.cit.,* 39.

96   Soon after women were admitted, provision was made for their accommodation 'between lectures' in a small wooden cottage situated at the southern end of what became the Quadrangle. Manning, 'Chancellor's Address', 13 April 1889, 17. There is disagreement concerning the precise positioning of these buildings. I am grateful to Graham Holland for allowing me to examine his photographs of the buildings in the Quadrangle during this period, and for his thoughts on the location of the interim chemistry building and students' common-rooms.

97   Barff, *op.cit.,* 102.

98   Art Society of NSW, *Annual Report* (1884), 25.

99   Senate Minutes, G1/1/7, 7 September 1885, 160; 7 December 1885, 204-205.

100  Hilary Golder, *Politics, Patronage and Public Works: The Administration of New South Wales, vol 1: 1842-1900* (Sydney: University of New South Wales Press, 2005), 213.

101  'On some New South Wales Minerals' (2 July 1884), *JPRSNSW,* 18 (1884), 43-48; and 'Analysis of Slate in Contact with Granite from Preservation Inlet, New Zealand' (with a note by Professor Hutton) (Philosophical Institute of Canterbury, 27 November 1884), *Trans. NZ Institute,* 17 (1884), 340-341.

102  Cf. R.J.W. Le Fèvre, 'The Establishment of Chemistry in Australia', in A.P. Elkin *et al., A Century of Scientific Progress* (Sydney: Royal Society of New South Wales, 1968), 361-362.

103  Cf. A. L., 'Presidential Address' (4 March 1885), *JPRSNSW,* 19 (1885), 151; cf. Roy MacLeod (ed.), *University and Community in Nineteenth Century Sydney: Professor John Smith (1821-1885)* (Sydney: University of Sydney History Project, 1988).

104  Liversidge Papers, Box 1, f. 108, Smith to Sir William Manning, 12 July 1884.

[105] Board of Studies Minutes, 26 October, 30 October, 4 November 1885; *Senate Minutes,* November 1885.

[106] Cf. Branagan and Holland, *op.cit.,* 8.

[107] J.G. Jenkin, 'The Appointment of W.H. Bragg, FRS, to the University of Adelaide', *Notes and Records of the Roy. Soc. of London,* 40 (1985), 75-99.

[108] The Selection Committee in London was convened by Sir Saul Samuel, the NSW Agent-General, and included Robert (later Sir Robert) S. Ball of Greenwich; R.B. Clifton of Bristol; G.C. Foster of University College London; P.G. Tait of Edinburgh, J.J. Thomson of Cambridge, and Sir William Thomson (later Lord Kelvin) of Glasgow. There were twenty candidates, ten from England, five from Scotland, three from Dublin, one from Melbourne and one from Hobart. See Rod Home, 'First Physicist of Australia: Richard Threlfall at the University of Sydney, 1886-1898', *Historical Records of Australian Science,* 6 (3), (1986), 333-357

[109] Sir William Manning, 'Chancellor's Address', 29 May 1886, 9.

[110] *Ibid.*, 9.

[111] Senate Minutes, 7 December 1886, 281; 'Report of the Senate of Sydney University', *VPLCNSW, 1887,* 42 (part 3), 225.

[112] Board of Studies Minutes, 17 December 1885.

[113] Branagan and Holland, *op.cit.*, 8.

[114] 'Report of the Senate of Sydney University', *VPLCNSW, 1887,* 42 (part 3), 225. For the Macleay Museum, see Peter Stanbury and Julian Holland (eds.), *op. cit.*

[115] Golder, *op. cit.,* 230.

[116] Senate Minutes, 25 October 1886, 326.

[117] Report, 225; Senate Minutes, 8 December 1886, 346; 13 December 1886, 349-350.

[118] J.J. Thomson, 'Sir Richard Threlfall, GBE, FRS', *Nature,* 130 (13 August 1932), 229.

[119] Rayleigh Correspondence, US Air Force Geophysics Laboratory, Mass., quoted in Rod Home, *op. cit.*, 339.

[120] Senate Minutes, 2 May 1887, 398.

[121] Sir William Manning, 'Commemoration Address', 14 May 1887, 23.

[122] Senate Minutes, 27 June 1887, 420.

[123] Sir William Manning, 'Chancellor's Address', 29 May 1886, 16.

[124] Presidential Address, *JPRSNSW, 20* (1886), 24.

[125] *Ibid.*, 27.

[126] Cf. R. Threlfall, *Report on the Curriculum of the Faculty of Arts* (n.d., ca. 1887), 2.

[127] Professorial Board Minutes, 20 May 1886.

[128] Sir William Manning, 'Chancellor's Address', 29 May 1886, 6.

[129] Professorial Board Minutes, 18 June 1886. See R. Philps, 'Walter Scott (1855-1925)', *Australian Dictionary of Biography,* vol. 11, 549.

[130] Professorial Board Minutes, 16 November 1886.

[131] Professorial Board Minutes, 28 July 1887. The grant of £1,000 was divided among the departments, but given mostly to natural history and physics. Chemistry, with Liversidge absent overseas, received only £80.

[132] 'Nova Cambria' [Andrew Garran], 'Classical and Modern Education', *Sydney Mail,* 22 November 1884.

[133] See E.K. Bramsted, 'Andrew Garran (1825-1901)', *Australian Dictionary of Biography,* vol. 4., 233-234; Sir Robert Randolph Garran, *Prosper the Commonwealth* (Sydney: Angus and Robertson, 1938), 5, 75-79.

[134] Roy MacLeod and Russell Moseley, 'Breaking the Circle of the Sciences: The Natural Sciences Tripos and the Examination Revolution', in Roy MacLeod (ed.), *Days of Judgement: Science, Examinations and the Codification of Knowledge in Victorian Britain* (Driffield: Nafferton Publishers, 1982), 189-212.

[135] Present were the Chancellor, Walter Scott, Gurney, Threlfall and Liversidge.

[136] Parenthetically, his proposal removed a compulsory language requirement.

[137] T.A. Coghlan, *The Wealth and Progress of New South Wales* (Sydney: Government Printer, 1892), 410.

[138] 'Metallic Meteorite, Queensland (Preliminary Notice)', *JPRSNSW, 20* (1886), 73; 'Notes on Some Rocks and Minerals from New Guinea, etc.', *JPRSNSW, 20* (1886), 227-230; 'Notes on some New South Wales Silver and other Minerals', *JPRSNSW, 20* (1886), 231-233; and 'On the Composition of some Pumice and Lava from the Pacific', *JPRSNSW, 20* (1886), 235-238.

[139] 'The International Congress of Geologists at Paris, 1878', (4 June 1879), *JPRSNSW, 13* (1879), 35-42.

[140] *Ibid.*

[141] Michael Hoare,'The Intercolonial Science Movement in Australia, 1870-1890', *Records of the Australian*

*Academy of Science,* 3 (2), (1976), 9.

142  *Ibid.*

143  Quoted by Kaye Harman, *Australia Brought to Book: Response to Australia by Visiting Writers, 1836-1939* (Sydney: Boobooks, 1985), 67-68.

144  Charles Moore, 'Anniversary Address', *JPRSNSW,* 14 (1880), 6.

145  Charles Wilkinson, *JPRSNSW,* 22 (1888), 27.

146  D. Branagan and T. Vallance, 'The Geological Society of Australasia (1885-1905)', *Journal of the Geological Society of Australasia,* 14 (2), (1967), 349 -351.

147  R.J. Ellery, 'Anniversary Address' (14 September 1883), *TPRSV,* XX (1883), xxvi-xxvii.

148  John Smith, 'Anniversary Address,' *JPRSNSW,* 15 (1881), 16-20.

149  Roy MacLeod, 'Founding: South Kensington to Sydney', in Graeme Davison and Kimberley Webber (eds.), *Yesterday's Tomorrows: The Powerhouse Museum and its Precursors, 1880-2005* (Sydney: Powerhouse Publishing in association with UNSW Press, 2005), 42-54.

150  Hoare, *op.cit.*

151  In 1887, the British Government entrusted Queensland with responsibility for administering the Protectorate of British New Guinea (afterwards, Papua), with financial contributions from NSW and Victoria. In 1888, the Protectorate became a Crown Colony, with a British Governor, administered jointly by Queensland and Great Britain. In 1906, it was annexed to Australia as Papua. In 1920, German New Guinea was assigned to Australia under a League of Nations mandate.

152  'Preliminary Meeting', *Proceedings of the Georgraphical Society of Australasia,* I (1883-84), vii-xvi.

153  Christopher Rolleston, 'Anniversary Address', *JPRSNSW,* 17 (1883), 17.

154  See Roy MacLeod, 'Whigs and Savants: The Royal Society of London and Reform, 1830-1848', in I. Inkster and J. B. Morrell (eds.), *Metropolis and Province* (London: Hutchinson, 1983), 55-90.

155  Ian Inkster and Jan Todd, 'Support for the Scientific Enterprise, 1850-1900', in R. Home (ed.), *Australian Science in the Making* (Cambridge: Cambridge University Press, 1988), 102-132.

156  Roberts, 'Anniversary Address', *JPRSNSW,* 23 (1889), 3.

157  Christopher Rolleston, 'Anniversary Address', *JPRSNSW,* 21 (1887), 3-7; for the Antarctic Expedition Committee, see *idem.,* 13. See also Michael Moore, 'Crawford Atchison Denman Pasco (1818-1898)', *Australian Dictionary of Biography,* vol. 5, 409.

158  On the debatable importance of the British Association to Canada, see Richard A. Jarrell, 'British Scientific Institutions and Canada: The Rhetoric and the Reality', *Trans. Royal Society of Canada,* Series 4, 20 (1982), 533-547; also Roy MacLeod, 'Introduction', and Michael Worboys, 'Social Imperialism and the British Association,' in MacLeod (ed.), *The Parliament of Science: Essays in Honour of the British Association for the Advancement of Science, 1831-1981* (London: Science Reviews, 1981).

159  G.P. Bidder, 'Mr. W.H. Caldwell', *Nature,* 148 (8 November 1941), 557-558.

160  H.N. Moseley, 'On the Ova of Monotremes', *Report of the BAAS ,* 54 (Montreal, 1884), 777. See also Roy MacLeod, 'Embryology and Empire: The Balfour Students and the Quest for Mechanism in the "Laboratory" of the Pacific, 1885-1895', in Roy MacLeod and Philip Rehbock (eds.), *Darwin's Laboratory: Evolutionary Theory and Natural History in the Pacific* (Honolulu: University of Hawaii Press, 1994), 140-165.

161  See W. Boyd Dawkins, *Cave Hunting: Researches on the Evidence of Caves respecting the Early Inhabitants of Europe* (London: Macmillan, 1874); *Early Man in Britain and His Place in the Tertiary Period* (London: Macmillan, 1880).

162  *Exploration of the Caves and Rivers of New South Wales, VPLANSW,* 5 (1882), 18-19 (afterwards, *Caves and Rivers*), Forster to Colonial Secretary, 4 May 1876.

163  *Caves and Rivers,* 20, A.L. to Colonial Secretary, 11 July 1879.

164  *Caves and Rivers,* 44, E.P. Ramsay's Report, 31 December 1881. The evidence supported Owen's theory that the *Thylacoleo,* or 'marsupial lion', was a carnivore, rather than a herbivore, as Krefft had supposed (*Caves and Rivers ,* 34, E.P. Ramsay, 20 September 1881).

165  *Report of the Trustees of the Australian Museum for 1882,* Appendix XI, *VPLANSW,* 2 (1883), 709-724.

166  *Annual Report of the Department of Mines, 1888, VPLANSW,* 4, (1889), 10. James Sibbald was appointed the first Keeper of the Caves.

167  See *Daily Telegraph,* 8 June 1886. This set in train the preservation of the neighbouring Jenolan Caves, the publication of guide books (the first, by the Department of Mines, in 1906), and the opening of the area as a

tourist attraction by the Macquarie Shire Council in 1926.

[168] The *Dipnoi* are fish with gills and lungs and arterial blood. There are three members of this group, one of which is the *Ceratodus*.

[169] His best-known work was *A Treatise on Comparative Embryology* (London: Macmillan, 1881); see also F.M. Balfour (ed.), *Studies from the Morphological Laboratory in the University of Cambridge* (London: Williams and Norgate, 1880), which followed the format of laboratory investigations begun by Michael Foster, and which were later copied by Liversidge in Sydney.

[170] Cambridge University Archives, CUR 100. l and ZOO 10/1- 2. For a brief history, see J.R.Tanner (ed.), *The Historical Register of the University of Cambridge to the Year 1910* (Cambridge: The University Press, 1917), 275-276. See also MacLeod, 'Embryology and Empire', *op.cit.*

[171] Bidder, *op.cit.*

[172] Sir Richard Threlfall, 'The Origin of the Automatic Microtome', *Biological Reviews,* 5 (1930), 357-361.

[173] See the standard accounts in M.E. Hoare, '"All things are Queer and Opposite": Scientific Societies in Tasmania in the 1840s', *Isis,* 60 (2), (1969), 205; A.M. Moyal, *Scientists in Nineteenth Century Australia* (Sydney: Cassell, 1976), 75.

[174] *SMH,* 4 September 1875.

[175] Quoted against the Darwinians by William Macleay at the Linnean Society of NSW, and reported by the *SMH,* 2 February 1876.

[176] As in the discovery of gold, his claim was rapidly followed by others. One month after his report, Dr. William Haacke, Curator of the South Australian Museum, published an eyewitness account of an Echidna laying eggs and hatching them. See *Trans. Roy. Soc. South Australia,* 1884, cited in Moyal, *ibid.,* 375. See also Liversidge Papers, Box 18, reports in *Evening News,* 9 October 1885, and December 1885.

[177] Extract from a letter by Caldwell to A.L., 8 October 1884, *JPRSNSW,* 18 (1884), 138.

[178] See *Town and Country Journal,* 13 September 1884, 549; 20 September 1884, 591.

[179] W.H. Caldwell, 'On the Development of the Monotremes and Ceratodus,' *JPRSNSW,* 18 (1884), 117-122; W.H. Caldwell, 'The Embryology of Monotremata and Marsupalia', *Philosophical Transactions of the Royal Society,* B, 178 (1887), 463-486.

[180] See Harry Burrell, *The Platypus: Its Discovery, Zoological Position, Form and Characteristics, Habits, Life History, etc.* (Sydney: Angus and Robertson, 1927).

[181] 'The Naturalist: Marsupials and Monotremata', *Australian Town and Country Journal,* 27 September 1884.

[182] 'The Royal Society of New South Wales: Mr. Caldwell's Australian Researches', *The Southern Science Record,* 1 (1), NS (January 1885), 16-20. In what must be a classic case of misreporting, *The Times* (4 September 1884), followed by the *Daily Telegraph* (15 October) and the *Herald* (18 December) reported Caldwell as confirming Owen's view that monotremes were 'viviparous', where he had in fact stated the opposite. A running account of the discoveries was carried by the *Australian Town and Country Journal.*

[183] Wilhelm Haacke, 'Meine Entdeckung des Eierlegens der Echidna hystrix', *Zoologischer Anzeiger,* 7 (1884), 647-653.

[184] Kathleen G. Dugan, 'The Zoological Exploration of the Australian Region and its Impact on Biological Theory', in Nathan Reingold and Marc Rothenberg (eds.), *Scientific Colonialism: A Cross-Cultural Comparison* (Washington, D.C.: Smithsonian Institution Press, 1987), 94-95.

[185] 'Science and the Colonies', *Australian Town and Country Journal,* 13 September 1884.

[186] *Melbourne Punch,* 25 September 1884.

[187] See Michael Worboys, 'The British Association and Empire: Science and Social Imperialism, 1880-1940', in Roy MacLeod and Peter Collins (eds.), *The Parliament of Science* (London: Science Reviews, 1981), 170-187.

[188] *SMH,* 17 September 1884.

[189] *SMH,* 18 December 1884.

[190] Special Meeting, *JPRSNSW,* 18 (1884), 141. The meeting was widely reported. See 'The Royal Society of New South Wales: Mr Caldwell's Australian Researches', *The Southern Science Record,* 1 (1), (January 1885), 16-20; 'The Royal Society', *Daily Telegraph,* 18 December, *SMH,* 18 December, *The Echo,* 18 December 1884.

[191] See Henry Parkes, 'Our Growing Australian Empire', *The Nineteenth Century,* 15 (1884), 138-149.

[192] See Leonie Foster, *High Hopes: The Men and Motives of the Australian Round Table* (Melbourne: Melbourne University Press, 1986), chapter 3.

[193] See J.A. Froude, *Oceana: England and her Colonies* (New York: Scribner, 1886).

194 *Meteorological Society of Australasia: History, Rules, Regulations and List of Members* (Adelaide, 1886), quoted in Hoare, 'Intercolonial Science,' *op.cit.*, 12.

195 T.H. Huxley, 'Presidential Address', *Proceedings of the Royal Society of London,* 39 (1885), 278-301.

196 Liversidge's Presidential Address to the RSNSW was reported in London two months later. *Nature,* 34 (29 July 1886), 303.

197 Jennings apparently misunderstood Liversidge's proposal for an Australasian Association and, contemplating his own indecisive arrangements for the colony's centennial, took up Liversidge's letter of 1884, and through the Agent-General in London, peremptorily invited the British Association. The BAAS agreed to consider the invitation at its Birmingham meeting that August, and accepted, provided the costs of British visitors were met. On 8 September, however, Jennings met a two-day barrage in the Legislative Assembly, led by Sir Henry Parkes, which forced him to retreat to a more limited vision of between forty and fifty 'eminent scientific persons', costing perhaps £6,000. Coincidentally, on the same day, the BAAS fixed on Bath for its 1888 Congress. This did not deter rumours of British annoyance with the colony's insistence – 'in somewhat dictatorial terms', as *Nature* put it – on limiting its contribution to the expenses of so few 'of the most eminent representatives of British science'. *Nature,* 34 (2 September 1886), 434.

198 *SMH,* 9 September 1886.

199 *Ibid.; Daily Telegraph,* 3 September 1886. See also *The Argus,* 27 August 1888.

200 *Nature,* 36 (25 August 1887), 398. That door was reopened in 1910, leading to the first (and only) visit of the BAAS to Australia in 1914.

201 *NSW Parliamentary Debates, 1885-1886,* 22, 4700-4704.

202 Liversidge Papers, Box 19 Folder 4, Kernot to A.L., 19 July 1886.

203 *Nature,* 38 (6 September 1888), 437-438.

204 *Royal Botanic Garden* (Kew), *NSW & Victoria Letters,* 1865-1900, vol. 173, A.L. to W. Thiselton-Dyer, 30 September 1888. His use of 'Australian' rather than 'Australasian', presumably accidental, proved to be prophetic.

205 Senate Minutes, 1 November 1887.

206 National Museum of New Zealand Archives, 1883-1886, f. 742, James Hector to Hon. Secretary, Royal Society of NSW, 31 December 1886.

207 The visit to China is mentioned by the *Illustrated Sydney News,* 'Personal Portraits No. 29: Professor Liversidge,' 1 November 1893, 4. However, the Liversidge papers are strangely silent about this visit.

208 MAAS, Inward Correspondence, 1887. Folder 17, A. Liversidge to Hart, 26 February 1887; f. 100 Liversidge to Maiden, 18 March 1887; f. 129, W. Gowland to Maiden, 9 April 1887; f. 149, Gowland to Maiden, 19 April 1887.

209 These fine objects form ten entries in the MAAS Accessions Book, 2 (September 1886-June 1890), ff. 96-104.

210 *Illustrated Sydney News,* 1 November 1893, 4.

211 Smithsonian Institution Archives, RU 305, A.L. to National Museum, Washington, DC, 8 October 1885.

212 The Australian materials arrived shortly; but the Smithsonian's specimens of American and Mexican pottery reached Sydney two years later. See MAAS, Inward Correspondence, 1889, f. 336. A. L. to Curator, 15 October 1889.

213 For the Savile's clientele in this period, see R.D. Thompson, *D'Arcy Wentworth Thompson: The Scholar-Naturalist, 1860-1948* (London: Oxford University Press, 1958).

214 Some time later, Archibald's elder brother Jarrett put up a stained-glass window in Harringay (Hornsey) Church to the memory of their father.

215 Roscoe declined, and William Stanley Jevons took his place. See Harro Maas, *William Stanley Jevons and the Making of Modern Economics* (Cambridge: Cambridge University Press, 2005), 28.

216 *Senate Minutes,* 18 July 1887.

217 Australian Museum Letter Books, 11 July 1887.

218 For an approving commentary on the changes that had taken place in Cambridge, see J.G. Crowther, *The Cavendish Laboratory, 1874-1974* (London: Macmillan, 1974), 110-112.

219 Roy MacLeod and Russell Moseley, 'The 'Naturals' and Victorian Cambridge: Reflections on the Anatomy of an Elite, 1851-1910', *Oxford Review of Education,* 6 (1980), 177-195; Roy MacLeod and Russell Moseley, 'Breaking the Circle of the Sciences: The Natural Sciences Tripos and the Examination Revolution', in Roy

MacLeod (ed.), *Days of Judgement, op.cit.*, 189- 212.

[220] *Second Report of the Royal Commission on Technical Instruction,* 1884. [C.3981]. xxix. 424.

[221] Gerrylynn K. Roberts, 'The Liberally-educated Chemist: Chemistry in the Cambridge Natural Sciences Tripos, 1851-1914', *Historical Studies in the Physical Sciences,* 11 (1), (1980), 157-183.

[222] See MacLeod, 'Embryology and Empire', *op.cit.*

[223] *Ibid.*

[224] M. J. G. Catermole and A.F. Wolfe, *Horace Darwin's Shop: A History of the Cambridge Scientific Instrument Company, 1878-1968* (Bristol: Adam Hilger, 1987).

[225] See Roy Porter, 'The NST and the "Cambridge School of Geology": 1850-1914', *History of the Universities,* II (1982), 193-216.

[226] Grace of the Senate, Cambridge University, 13 October 1887. Liversidge received his honorary degree on 27 October, under University Statute A, chap. II, sec. 18, para. 3, which provided for 'British subjects who are of conspicuous merit, or have done good service to the State or to the University'. I am grateful to Dr. Elizabeth Leadham-Green for this information.

[227] Quoted, *inter alia*, in D.P. Mellor, 'Archibald Liversidge (1846-1927)', *Australian Dictionary of Biography*, vol. 5, 93.

[228] Private Communication, Dr. A.E. Friday, Department of Zoology, Cambridge, 10 July 1984.

[229] BM (Nat. Hist.), Min. Dept. A.L. to Fletcher, 16 November 1887.

[230] Samuel Smiles, *Lives of the Engineers: The Steam Engine, Boulton and Watt* (London: John Murray, 1861-1874, 1904 ed.), vol. 4, 472.

[231] Robert Thurston, *A History of the Growth of the Steam Engine* (London: Kegan Paul, 1878), 2.

[232] MAAS, Inward Correspondence, 1888. f. 5. A. L. to Whitbread, 20 December 1887; cf. correspondence ff. 1-8, 12- 25 December 1887.

[233] MAAS, Inward Correspondence, 1888. f. 45, A.L. to Sir Alfred Roberts, 30 December 1887. In continuous operation for over a century, the engine was valued at £1,000. Liversidge attached wording, to be placed on a plaque, tracing the lineage of the engine from the Earl of Somerset's first steam engine in 1655 to James Watt, who had 'enlarged its powers and uses and brought it to its present state of perfection'. Packed in forty-five crates, weighing thirty-five tons, it arrived in Sydney on board the wool clipper *Patriarch* to a muted greeting from Museum staff suffering from a lack of space. It remained in crates on the wharf for the next five years, and in storage for over thirty years. In 1895, Liversidge reminded Maiden that he had taken 'a great deal of trouble to secure this very valuable and interesting piece of historical machinery'. (MAAS Inward Correspondence, 1895, A.L. to Maiden, 15 August 1895). Ten years later, it was reassembled at the Harris Street Museum, but only in 1988 was it made to work again under steam, and given a suitably prominent place in the newly-rebuilt 'Powerhouse' Museum. For details, see Graeme Davison and Debbie Rudder, 'The Heroic Age of Steam', in Davison and Webber (eds.), *op. cit.*, 201-203.

[234] The *Herald* exceeded its reputation for understatement. Leaving the Mediterranean, the ship met unusually heavy hail and lightning storms off Colombo, and in the Red Sea, monsoons and a cyclone. Escaping with no worse than damaged sails, however, fine weather at last greeted them at Albany. *SMH* 3 March 1888.

[235] *Ibid.*

## CHAPTER 9

[1] See Ken Cable, 'Sir Mungo William MacCallum (1854-1942)', *Australian Dictionary of Biography*, vol. 10, 211-213; Cf. J.P. Hill, 'J.T. Wilson: A Biographical Sketch of his Career', *Journal of Anatomy*, 76 (1941-42), 6. For details of Wilson and MacCallum, see Patricia Morison, *J.T. Wilson and the Fraternity of Duckmaloi* (Amsterdam: Rodopi, 1997), 55-58 *et seq.* For another use of this intriguing expression, see Suzanne Falkiner (ed.), *Leslie Wilkinson, A Practical Idealist* (Woollahra: Valadon Publishing, 1982).

[2] Roy MacLeod, 'The "Practical Man": Myth and Metaphor in Anglo-Australian Science,' *Australian Cultural History*, No. 8, (1989), 24-49.

[3] Of this tendency, the *Bulletin* was an embodiment. See John Docker, *The Nervous Nineties: Australian Cultural Life in the 1890s* (Melbourne: Oxford University Press, 1991).

[4] A.N. Whitehead, *Science in the Modern World* (Harmondsworth: Penguin, 1926), 227-228.

[5] Cf J.H. Maiden, 'Anniversary Address', *JPRSNSW*, 3l (1897), 21-22; SMH, 8 July 1898.

6   For background, see Michael Hoare, 'The Intercolonial Science Movement in Australasia, 1870-1890', *Records of the Australian Academy of Science*, 3 (2), (1976), 7-28; and M. Hoare, 'Learned Societies in Australia: The Foundation Years in Victoria, 1850-1860', *Records of the Australian Academy of Science*, 1 (2), (1969), 7-29, to which I am indebted for some of the material (but not for the interpretation) which follows.

7   H. Mortimer-Franklin, *The Unit of Imperial Federation* (London: Swann Sonnenschein, 1887).

8   *Trans. Intercolonial Medical Congress of Australasia* (Adelaide, 1888).

9   Royal Botanic Gardens (Kew), NSW and Victorian Letters, 1865-1900, Box 173, Liversidge to Thiselton-Dyer, 30 September 1888.

10  NM-NZ Box 1888, f. 242. A.L. to Hector, 21 April 1888.

11  Cf. Sally Kohlstedt, '*Savants* and Professionals: The American Association for the Advancement of Science, 1848-1860', in A. Oleson and S.C. Brown (eds.), *The Pursuit of Knowledge in the Early American Republic* (Baltimore and London: Johns Hopkins University Press, 1976), 299-325; cf. Sally G. Kohlstedt, *The Formation of the American Scientific Community: The American Association for the Advancement of Science, 1848-1860* (Urbana: University of Illinois Press, 1976).

12  Cf. Robert Fox, 'The *Savant* Confronts his Peers: Scientific Societies in France, 1815-1914', in Robert Fox and George Weisz (eds.), *The Organisation of Science and Technology in France, 1808-1914* (Cambridge: Cambridge University Press, 1980).

13  Fifth, that is, within the western European model. The source of the 'Association idea' has been traced to Switzerland in 1815. (Cf. G.V.H. Degen, 'Die Grundungsgeschichte der Gesellschaft deutscher Naturforscher und Ärzte', *Naturwissenschaft*, 11 (1955), 12) and to the better known German association founded in the 1820s. Cf. Charles Babbage, 'Account of the Great Congress of Philosophers at Berlin', *Edinburgh Journal of Science*, 60 (1829), 225-234; J.F.W. Johnston, 'Meeting of the Cultivators of Natural Science and Medicine at Hamburg', *Edinburgh Journal of Science*, 4 (1830-3), 244; H. Duerner and H. Schipperger, *Wege der Naturforschung, 1822-1972* (Berlin: Springer Verlag, 1972). Following the German example, Associations were formed in Italy (1839), Scandinavia (1839), Hungary (1841), Russia (1863), Poland (1869), and Czechoslovakia (1880). Cf. R. von Gizycki, 'The Associations for the Advancement of Science: An International Comparative Study', *Zeitschrift für Sociologie*, 8 (1979): 28-49. These, however, were impermanent assemblies; more permanent ones along the 'Anglo-Saxon' model, begun in Britain, were later founded in the United States (1847), South Africa (1902), Spain (1908), India (1914), and Japan (1925). Since the last war, several similar associations have come into existence in Latin America. Cf. E. Diaz *et al.*, *La Ciencia Periférica* (Caracas: Monte Avila Editores, 1983); H. Vessuri, 'The Social Study of Science in Latin America', Social *Studies of Science*, 17 (3), (1987), 519-554. Since 1986, an African Association has also existed, with general offices in Nairobi.

14  Cf. 'President's Address', *Report of the British Association for the Advancement of Science*, 1 (York, 1831), and *Report of the AAAS*, 1 (Sydney, 1888).

15  ANZAAS Archives (Mitchell Library) MSS. 908/1, f. 125, General Committee Minutes, 27 August 1888.

16  William Westgarth, *Half a Century of Australasian Progress: A Personal Retrospect* (London: Sampson Low, Marston, Searle and Rivington, 1889), 104.

17  H.C. Russell, 'Presidential Address,' *Report of the AAAS*, 1 (Sydney, 1888), 11, 12, 14.

18  'Science and the Arts', *SMH*, 29 August 1888.

19  G.S. Griffiths, 'Australian Exploration - The Duty of Australia', Section E, *Report of the AAAS,* 1 (Sydney, 1888), 413-415.

20  C.A. Fleming, 'Science, Settlers and Scholars: The Centennial History of the Royal Society of New Zealand', *Bulletin of the Royal Society of New Zealand*, 25 (1987), 270; *Report of the AAAS*, 1 (Sydney, 1888), 106-112, 168-183, 303-312, 352-357, 359, 413-415.

21  Quoted in R.W. Home, 'The Problem of Intellectual Isolation in Scientific Life: W.H. Bragg and the Australian Scientific Community, 1886-1909', *Historical Records of Australian Science*, 6 (1), (1984), 19.

22  Ferdinand von Mueller, 'Presidential Address', *Report of the AAAS*, 2 (Melbourne, 1890), 3.

23  J. Steel Robertson, 'Natural Science in Australia', *The Centennial Magazine,* 2 (7), (February 1890), 523-527.

24  *The Australasian*, 11 September 1888, 490-491.

25  'AAAS', *Chemical News*, 58 (1888), 118; *Nature*, 38 (1888), 437-438, 623.

26  *Saturday Review*, 3 November 1888, 519-520.

27  Ferdinand von Mueller, 'Inaugural Address', *Report of the AAAS*, 2 (Melbourne, 1890), 25-26.

28  Armstrong Papers (Imperial College), 26A, A.L. to Armstrong, 12 March 1885.

29  *Senate Minutes*, 19 March, 23 April 1888.

30  Brian Kennedy, *A Passion to Oppose: John Anderson, Philosopher* (Melbourne: Melbourne University Press, 1995), 74.

31  Jeanette Beaumont and W. Vere Hole, *Letters from Louisa: A Woman's View of the 1890s, based on the Letters of Louisa Macdonald, first Principal of the Women's College, University of Sydney* (Sydney: Allen & Unwin, 1996), 31.

32  Liversidge Papers, Box 21, f. 4, Barff to A.L., 26 March 1888.

33  *Senate Minutes*, 5 March 1888; 19 March 1888; 9 April 1888; 21 May 1888.

34  *Senate Minutes*, 4 June 1888.

35  For details of the settlement, see *Report of the Senate, University of Sydney*, 1887, and *VPLCNSW*, 42 (Part 3), (1887) 226.

36  Liversidge Papers, Box 21, f. 4/14, Manning, 'For the Challis Fund Committee', 13 December 1888.

37  It fell to the brilliant William Bateson, Master of St. John's, to wring from a fortuitous reapportionment of local improvement rates, money for the University to meet the salary of James Clerk Maxwell. For background, see J.G. Crowther, *The Cavendish Laboratory, 1874-1974* (London: Macmillan, 1974), 29-32.

38  *Ibid.*, 48.

39  *Senate Minutes*, 16 July 1888; 6 August 1888; 1 October 1888.

40  *Senate Minutes*, 6 August 1888.

41  Rod Home, 'First Physicist of Australia: Richard Threlfall at the University of Sydney, 1886-1898', *Historical Records of Australian Science*, 6 (3), (1986), 333-357.

42  See Peter Stanbury and Julian Holland (eds.), *Mr. Macleay's Celebrated Cabinet: The History of the Macleays and their Museum* (Sydney: Macleay Museum, 1988), 55.

43  *Sydney University Calendar, 1896,* 93.

44  So suggested Sir William Tilden, *Chemical Discovery and Invention in the Twentieth Century* (London: George Routledge and Sons, 1917), 44-46. The NSW Public Works Department, *Statement of Work Done* (1890) gave a figure of £8,754.9.6, but Barff put the final cost, possibly including fittings, at 'about £13,000'. (H.E. Barff, *A Short History of the University of Sydney (*Sydney: Angus and Robertson, 1902), 102). By the turn of the century, additions had brought the cost of the building to £18,000. 'Sydney University', *Daily Telegraph*, 30 April 1907.

45  See K.J. Cable,'Henry Ebenezer Barff (1857-1925)', *Australian Dictionary of Biography,* vol. 7, 173.

46  Liversidge Papers, Box 21, f. 43, Barff to A.L. 8 August 1888.

47  Liversidge Papers, Box 21, f. 46, A.L. to Chancellor, 20 August 1888.

48  See Stanbury and Holland, *op.cit.*, 55.

49  Liversidge Papers, Box 21, f. 55, 56, 57; Senate Minutes, 5 November 1888.

50  Liversidge Papers, Box 21, f. 61, A.L. to Senate, November 1888. These were approved in principle by Senate in September 1889, but no action had been taken by 1892. Professorial Board Minutes*,* 30 November 1892.

51  David Branagan and Graham Holland (eds.), *Ever Reaping Something New: A Science Centenary* (Sydney: University of Sydney, Science Centenary Committee, 1985), 235; Ursula Bygott and Ken Cable, *Pioneer Women Graduates of the University of Sydney, 1881-1921* (Sydney: University of Sydney, 1985).

52  The five new chairs were in Law, History, Anatomy, Biology, and Logic and Mental Philosophy. See Sir William Manning, 'Chancellor's Address', 13 April 1889, 12, and 'Chancellor's Address', 14 April 1890, 4.

53  Liversidge Papers, Box 21, f. 64, W. Manning, 'For the Challis Fund Committee', 13 December 1888.

54  Barff, *op. cit*, 86; see R. Phlps, 'Walter Scott (1855-1925)', *Australian Dictionary of Biography*, vol. 11, 549.

55  Liversidge Papers, Box 21, f. 68, Scott to Committee of Senate, 15 January 1889. Scott had helped establish the Australian Economic Association in 1887, but failed to persuade Senate to establish a chair in political economy.

56  Liversidge Papers, Box 21, f. 61, A.L., 'Memorandum from the Challis Fund Committee', 14 January 1889.

57  Liversidge Papers, Box 21, f. 69, Haswell and Challis Committee, 7 February 1889.

58  Liversidge Papers, Box 2l, f. 71, Stephens to Manning.

59  Liversidge Papers, Box 21, f. 78, Report of the Challis Chairs Committee, 12 March 1889.

60  A.L., 'The Proposed Chemical Laboratory at the University of Sydney' (6 October 1888), *The Building and Engineering Journal*, 5 (213), (1888), 282.

61  Tilden, *op.cit.*, 44-46.

62  A.L., 'Note upon the Hot Spring Waters, Fergusson Island, D'Entrecasteaux Group' (1890), Queensland (8 March 1889), *Ann. Rep. Brit. New Guinea*, (1888-1889), 31-32; *Chemical News*, 62 (28 November 1890), 264-266, repeated in 'Note upon Samples of Hot Spring Waters, Fergusson Island, British New Guinea, procured by Sir William MacGregor, KCMG', in J.P. Thomson, *British New Guinea* (London: George Philip & Son, 1892), 215-217; 'Note on the Foetal Membranes of Mustelus Antarcticus, by T.J. Parker; with an analysis of the Pseudamniotic Fluid, by A. Liversidge' (Otago Institute, 11 June 1889), *Trans. NZI.,* 22 (1889), 331-333; 'Australian Meteorites' (January 1890), *Rep. AAAS*, 2, (Melbourne, 1890), 387-388; 'Notes on some Hot Spring Waters (New Guinea and Solomon Islands)' (January 1890), *Rep. AAAS*, 2, (Melbourne, 1890), 388-394; 'On the Removal of Gold from Suspension and Solution by Fungoid Growths' (January 1890), *Rep. AAAS*, 2, (Melbourne, 1890), 399-407; 'Chalk and Flints from the Solomon Islands' (January 1890), *Rep. AAAS*, 2, (Melbourne, 1890), 417-420.

63  Royal Botanical Gardens, (Kew), NSW and Victoria Letters, 173, f. 136, A.L. To Thisleton-Dyer, 17 September 1890.

64  BMNH, Min. Dept., A.L. to Fletcher, 26 September 1890.

65  A. L., 'The Proposed Chemical Laboratory', *Building and Engineering Journal*, 6 October 1888.

66  *SMH*, quoting Liversidge, 17 October 1891.

67  'The Sydney University - IV', *SMH*, 17 October 1891.

68  E.C. Robins, *Technical School and College Building* (London: Whittaker and Co., 1887).

69  *Annual Report, Dept. of Mines*, 1900, 68.

70  R.W. Burnie (ed.), *Memoirs and Letters of Sidney Gilchrist Thomas, Inventor* (London: John Murray, 1891), 265-266.

71  Anthony Trollope, *Australia and New Zealand* (Melbourne: Robertson, 1871), 137.

72  Nehemiah Bartley, *Australian Pioneers and Reminiscences* (Brisbane: Gordon and Gotch, 1896), 15-16.

73  Cf. Brian and Barbara Kennedy, *Sydney and Suburbs: A History and Description* (Sydney: Reed, 1982), 150.

74  A.H.C. Whitaker, 'Darling Point: The Mayfair of Australia' (M.A. thesis, University of Sydney, 1983), 58. I am grateful to Ms. Whitaker for permission to consult her thesis.

75  G. Nesta Griffiths, *Some Houses and People of New South Wales* (Sydney: Ure Smith, 1949), 120.

76  *Woollahra Rate and Assessment Books*, 1904, 5. For contemporary photographs and estate agents' advertisements, see Eric Russell, *Woollahra: A History in Pictures* (Sydney: John Ferguson, 1980). Cf. James Jervis and Vince Kelly, *The History of Woollahra: A Record of Events from 1788 to 1960 and a Centenary of Local Government* (Sydney: Municipal Council of Woollahra, 1960), 38-39. Caroline Fairfax Simpson *et al.*, *Ascham Remembered, 1886-1986* (Sydney: Fine Arts Press, 1986), 148-149. See Helen Rutledge, *My Grandfather's House* (Sydney: Doubleday, 1986). During both world wars, it was a convalescent home for wounded soldiers. In the 1960s, Ascham School moved to 'Glenrock' and occupied the tower. The 'Octagon' is now home to the school's archives. I am grateful to Ms Gerri Nicholas for this information. See 'The Octagon: Eight Sides, Eight Lives', *The [Ascham] Newsletter*, 1 (3), (2007).

77  J. Arthur Dowling, 'Darling Point, Potts Point and Neighbourhoods', *JRAHS*, 2 (1906), 64.

78  In 1893, she left Australia altogether, and moved to England, where she became a leading suffragette. Judith Allen, 'Dorothy Frances Montefiore (1851-1933)', *Australian Dictionary of Biography*, vol. 10, 556-557. See her memoirs, *From a Victorian to a Modern* (London: E. Archer, 1927).

79  R.V. Jackson, 'Owner-Occupation of Houses in Sydney, 1871 - 1891', *Australian Economic History Review*, 10 (2), (1970), 140-143.

80  His rates and other outgoings were £30 per annum. Woollahra Council, *Rate and Assessment Books*, 1888-9, 83.

81  *NSW Census of 1891*, Darling Point.

82  William Epps, *Anderson Stuart, MD: Physiologist, Teacher, Builder, Organiser, Citizen* (Sydney: Angus and Robertson, 1922), 50, 153.

83  *Ibid.*, 155.

84  R.M. Crawford, *A Bit of a Rebel: The Life and Work of George Arnold Wood* (Sydney: Sydney University Press, 1975), 115-116.

85  Morison, *op. cit.* 77.

86  See Patricia Morison, 'James Thomas Wilson (1861-1945)', *Australian Dictionary of Biography*, vol. 12, 525-527.

87   See Michel Foucault, *Discipline and Punish: the Birth of the Prison* (New York: Pantheon Books, 1977), 172, 176, 193.

88   For similar struggles and contrasting outcomes in Europe, see Konrad H. Jarausch (ed.), *The Transformation of Higher Learning, 1860-1930: Expansion, Diversification, Social Opening and Professionalization in England, Germany, Russia and the United States* (Stuttgart: Kletta-Cotta, 1983).

89   Cf. Roy MacLeod and Russell Moseley, 'Breadth, Depth and Excellence: Sources and Problems in the History of University Science Education in England, 1850-1914', *Studies in Science Education*, 5 (1978), 85-106.

90   Senate Minutes, 14 March 1887.

91   W. Scott, 'Report on the Curriculum of the Second and Third Years in Arts', 5 March 1887, 2; Senate Minutes, 14 March 1887.

92   Senate Minutes, 14 March 1887, W.J. Scott, 'Report on the Curriculum of the Second and Third Years in Arts', 5 March 1887.

93   Senate Minutes, 14 March 1887, 'Paper from the Most Rev. Alfred Barry, on the By-laws in the Faculty of Arts', 14 March 1887.

94   Liversidge Papers, Box 20, f. 232, T. Gurney, 'Memorandum on the Arts Course', April 1887.

95   Liversidge Papers, Box 20, f. 231, W.J. Scott, 1 April 1887.

96   Liversidge Papers, Box 20. R. Threlfall, 'Report on the Curriculum of the Faculty of Arts' (n.d.)

97   Liversidge Papers, Box 20. Anderson Stuart, 'Report on the Curriculum in Arts', 29 March 1887; Liversidge Papers, Box 21, f. 12, 'Report of the Faculty of Medicine', 12 April 1888.

98   Liversidge Papers, Box 21. W.H. Warren, 'Report on the Curriculum in the Faculty of Arts', 9 May, (n.d.) June 1887.

99   Senate Minutes, 27 June 1887, 'Observations of the Professors' (n.d.); Richard Threlfall, 'Report on the Curriculum of the Faculty of Arts' (n.d. but after 5 March).

100   *SMH*, 6 August 1887, 7.

101   See 'Our University', *SMH*, 5 May 1905.

102   Professorial Board Minutes, 27 March 1888.

103   *SMH*, 23 May 1888.

104   Report of the Professorial Board, 9 April 1888.

105   According to a survey conducted by Threlfall, the training schools of NSW in 1886 devoted less than six hours to science of the 130 class hours given by all masters in all subjects. High schools gave science ten hours of 289, and Sydney Grammar School, only eight of 394. Liversidge Papers, Box 21, f. 181. 'Memorandum from Professor Threlfall in reference to the Proposed Changes in the First Year of the Arts Course' [October 1890], 4.

106   *SMH*, Letter from 'A Graduate', 1 October 1888.

107   Liversidge Papers, Box 21, f. 13, Liversidge to Chancellor and Senate, 12 April 1888.

108   *Ibid.*

109   *Hermes*, 2 (3), (6 October 1887), 2.

110   *Ibid.*

111   G.K. Roberts, 'The Liberally-Educated Chemist: Chemistry in the Natural Sciences Tripos, 1851-1914', *Historical Studies in the Physical Sciences*, 11 (1), (1980), 157-183.

112   This decision followed representations from the Headmasters' Association. Liversidge Papers, Box 21, f. 2l, Report of the Professorial Board, 29 April 1888.

113   By this plan, the University substituted a third year of science for students in science and engineering, and established a five-year course for medicine. Liversidge Papers, Box 21, f. 221, Report of the Professorial Board, 8 May 1888; Resolution of the Faculty of Medicine, 12 April 1888.

114   Liversidge Papers, Box 21, f. 106, By-law, 13 December 1888.

115   Chancellor's Address, 13 April 1889; G.1/1/8/17, Senate Minutes, 17 October 1888.

116   Liversidge Papers, Box 21, f. 221. By-law, 13 December 1888.

117   Professorial Board Minutes, 25 June 1888; 31 July 1888; 2 August 1888; 17 August 1888.

118   *Ibid.*, 22 July 1889.

119   *Ibid.*, 30 July 1889.

120   Professorial Board Minutes, G2/1/1, 11 July 1889, 258; 30 July, 262-265, Senate Minutes, G.1/1/8, 16 September 1889.

[121] Liversidge Papers, Box 21, f. 135, Senate Minutes, 16 December 1889.

[122] Liversidge Papers, Box 21, f. 105, By-law Proposals, 8 September 1889.

[123] Senate Minutes, 1 May 1891, 28 July 1891.

[124] Chancellor's Address, 13 April 1889.

[125] Liversidge Papers, Box 21, f. 89, Liversidge notes, 29 April 1889; f. 108, W.J. Stephens to Registrar, 9 September 1889.

[126] Liversidge Papers, 'Threlfall Memorandum', 5.

[127] Liversidge Papers, Box 21, f. 179, 'Memorandum by Professor Liversidge upon the Proposed New Curriculum for the First Year', 10 October 1890.

[128] Cf. G.L. Fischer's recollections of Egerton Barraclough, in 'The University of Sydney in the 1890s: A Student's View', *University of Sydney News*, 4 (1), (9 February 1972), 8-9.

## CHAPTER 10

[1] For the period, see R.A.Gollan, 'Nationalism, the Labour Movement and the Commonwealth', in Gordon Greenwood (ed.), *Australia: A Social and Political History* (New York: Praeger, l955), 145-195; Humphrey McQueen, *A New Britannia* (Ringwood: Penguin, 1975). See also B.K. de Garis, '1890-1900', in F.K. Crowley (ed.), *A New History of Australia* (Melbourne: William Heinemann, 1974), 216 ff; Beverley Kingston, *The Oxford History of Australia, 3, 1860-1900: Glad, Confident Morning* (Melbourne: Oxford University Press, 1988).

[2] For context, see Michael Roe, *Nine Australian Progressives* (St. Lucia: University of Queensland, 1984).

[3] Neville Hicks, *This Sin and Scandal: Australia's Population Debate, 1891-1911* (Canberra: ANU Press, 1978); see also Carol Bacchi, 'Evolution, Eugenics and Women: The Impact of Scientific Theories on Attitudes towards Women, 1870-1920', in Elizabeth Windschuttle (ed.), *Women, Class and History: Feminist Perspectives on Australia, 1788-1978* (Melbourne: Fontana/Collins, 1980); 'The Nature-Nurture Debate in Australia, 1900-1914', *Historical Studies*, 19 (1980), 199-212.

[4] Francis Adams, *The Australians: A Social Sketch* (London: T. Fisher Unwin, 1893), 57-58.

[5] Despite several attempts to become an *ex-officio* member of Senate by virtue of his chair of chemistry, it was not until 1902 that the Professor of Chemistry became a Fellow of Senate in his own right. Senate Minutes, 12 April 1892, 16 August 1894, 11August, 1896, 15 August l898, 16 August 1900, and 19 August 1902.

[6] Arthur E. Shipley, *'J': A Memoir of John Willis Clark* (London: Smith Elder and Co., 1913), esp. Appendix II, 294-341.

[7] Senate Minutes, 21 November 1892.

[8] *NSW Parliamentary Debates*, *LA*, Second series, X (21 July 1903), 843.

[9] Adrian Haas, 'Schools of Mines in Australia, 1870-1920', *JRAHS*, 75 (4), (1990), 287.

[10] Viz, Ararat, Ballarat, Bendigo, Bairnsdale, Stawell, Castlemaine, Creswick, Sale, Harrietville, Kyneton, Maryborough, Daylesford, St. Arnaud and Clunes. See Robert O. Chalmers, *Minerals from Broken Hill,* 34; Haas, *op.cit.,* 282.

[11] Liversidge Papers, Box 10, annotations on 'Mining Intelligence', *SMH*, 16 September 1884.

[12] A. Leibius, 'Notes on Gold', *JPRSNSW,* 18 (1884), 40.

[13] Hilary Golder, *Politics, Patronage and Public Works: The Administration of New South Wales*, vol 1: 1842-1900 (Sydney: University of New South Wales Press, 2005), 229.

[14] Liversidge Papers, Box 21, f. 83. A.L. to Manning, 10 April 1889.

[15] Senate Minutes, 5 August 1889.

[16] Liversidge Papers, Box 21, f. 83. A.L. to 'My dear Mr. Chancellor', 10 April 1889.

[17] This was to be the largest Congress attendance in the first fifty years of the Association's history. See Roy MacLeod (ed.), *The Commonwealth of Science: ANZAAS and the Scientific Enterprise in Australia, 1888-1898* (Melbourne: Oxford University Press, 1988). Chapter 2, 'From Imperial to National Science'.

[18] Len Weickhardt, *Masson of Melbourne: The Life and Times of David Orme Masson, Professor of Chemistry, University of Melbourne, 1886-1923* (Melbourne: Royal Australian Chemical Institute, 1989).

[19] No relationship developed between their respective departments. See Joan Radford, *The Chemistry Department of the University of Melbourne: Its Contribution to Australian Science, 1854-1959* (Melbourne: Hawthorn Press, 1978). According to Dr. Weickhardt, Masson's biographer, no Liversidge letters survive in Masson's

correspondence.

20  Cf. David Orme Masson, 'Presidential Address', *Rep. of the AAAS*, 13 (Sydney, 1911), 1-4.

21  On Tenison-Woods, see D.F. Branagan (ed.), 'The Scientific Work of Tenison Woods: A Symposium', in *JPRSNSW*, 122 (1989), 107 ff; A. Player (ed.), 'J.E. Tenison Woods' Ten Years in the Bush', *Australian Catholic Record*, LXVI (3), (July 1989), 259-278; 'Julian Tenison Woods - Priest and Scientist', *Australian Catholic Record*, LXVI (3), (July 1989), 279-294; 'Julian Tenison Woods: 1832-1889', *Newsletter Eight*, Group for Scientific and Technological Collections (July 1989), 1-2.

22  'Anniversary Address', *JPRSNSW*, 24 (1890), 19.

23  *Ibid.*, 33.

24  Senate Minutes, f. 108, 25 March 1889.

25  David Branagan and H.G. Holland (eds.), *Ever Reaping Something New* (Sydney: University of Sydney, 1985), 14.

26  *Annual Report, Department of Mines, 1890*, 225-229.

27  See David Branagan, *T.W. Edgeworth David, A Life: Geologist, Adventurer, Soldier and 'Knight in the Old Brown Hat'* (Canberra: National Library of Australia, 2005).

28  The Committee enlisted Sir Charles Nicholson, and Liversidge's nominees, Professor John Wesley Judd of the Royal College of Science and the RSM; Professor W. Boyd Dawkins of Manchester University; and Sir Archibald Geikie, Director-General of the Geological Survey.

29  University of Sydney Archives, G.3/46/9, f. 899. Barff to Sir Saul Samuel, 20 December 1890.

30  At the invitation of the New Zealand Institute, Auckland, as the most populous city in the colony, had first been proposed as the venue but, pressed by Captain F.W. Hutton, Liversidge's old friend and a remarkable Darwinian, the Council opted for Christchurch instead. Thanks to Liversidge's careful diplomacy, sensibilities were saved, and the Australians were warmly greeted. See J. Stenhouse, '"The Wretched Gorilla Damnification of Humanity": The "Battle" between Science and Religion over Evolution in Nineteenth Century New Zealand', *The New Zealand Journal of History*, 18 (2), (October, 1984), 143-162.

31  National Museum of New Zealand Letterbook, 1886-1891, f. 547, Hector to Sir Harry Atkinson, 19 March 1890.

32  Sir James Hector, 'Presidential Address', *Report of the AAAS*, 3 (Christchurch, 1891), 4.

33  *Ibid.*, 21.

34  NM NZ, Box 1891, A.L. to Hector, 25 February 1891.

35  NM NZ, Box 189, f. 50/91, A.L. to Secretary, NZI, 20 January 1891.

36  NM NZ, Box 189, f. 57/91 A.L. to Hector, 25 February 1891.

37  In the fullness of time, virtue had its reward. The following year, honouring Hutton's contribution to the success of the AAAS, Liversidge nominated him for election to the Royal Society of London. Hutton became a FRS (1892), and Anderson Stuart did not.

38  Liversidge Papers, Box 19, f. 5. Masson to A.L., 22 April 1892.

39  'I fear your decision will cause unpleasant feeling', the Agent-General cabled, 'and prevent eminent men acting on Selection Committees here. Nicholson and myself feel very strongly on this point'. Liversidge Papers, Box 21, f.209. Samuel to Colonial Secretary, 9 May 1891.

40  Sollas, then Professor of Geology at Trinity College, Dublin, subsequently moved to the chair of geology at Oxford. His defeat at David's hands in 1890 was followed by a professional setback in the coral atoll boring operations at Funafuti in 1896-97. The University Museum at Oxford became his refuge for a generation thereafter. See Roy MacLeod, 'Imperial Reflections in the Southern Seas: The Funafuti Expeditions, 1894-1904', in Roy MacLeod and Fritz Rehbock (eds.), '*Nature in its Furthest Extent': Western Science in the Pacific* (Honolulu: University of Hawaii Press, 1988), 159-194.

41  University of Sydney Archives, G.3/46/10, f. 619. Barff to Sir Saul Samuel, 21 May 1891.

42  Senate Minutes, 19 August 1889.

43  Faculty of Science Minutes, 15 April 1891. The lectureship in metallurgy was to be fixed at £450, and in mining, at £150.

44  Senate Minutes, 15 June 1891. Faculties of Architecture and Agriculture were not created until 1919, when the number of Faculties was increased from four to ten. Branagan and Holland (eds.), *op.cit.*, 19.

45  J.H. (later Sir Joseph) Carruthers later became a member of the National Convention (1897-98) and Liberal Premier of NSW (1904-07).

46    John M. Ward, 'Sir Joseph Hector Carruthers (1856-1932), *Australian Dictionary of Biography*, vol. 7, 574-578.

47    Liversidge Papers, Box 21 f. 208. H.C. Russell, Deputation to Minister for Public Instruction, 11 September 1891.

48    Liversidge Papers Box 21, f. 83. A.L. to Manning, 10 April 1889.

49    Liversidge Papers, Box 21, f. 101, Agenda Papers, Senate Minutes, 19 August 1889.

50    Some years earlier, Helms had complained that he was not given enough freedom of action. His salary was doubled from £200 to £400, but in 1888, this was reduced by £50 in conformity with a new Senate grading system, which withdrew lecture fees to meet the costs of university administration (itself not a happy precedent).

51    Helms did not go quietly. Not long afterwards, he resigned from the Royal Society, of which had been a member since his arrival in Sydney in 1880. And while he lived on in Sydney as a consulting chemist, sometimes working for the Department of Agriculture, his memory remained among Liversidge's least pleasant.

52    Specifically, to the Bald Nob Copper Mine, Emmaville. See *Proc. Roy Soc. NSW,* 31 (1897), xlviii.

53    He had been the first, and for some time, the only assistant to Henry Roscoe in the foundation years of chemistry in Manchester, before moving to a chemistry chair in Mauritius in 1858. See Henry Roscoe, *The Life and Experiences of Sir Henry Enfield Roscoe* (London: Macmillan, 1906), 104-106.

54    In 1890 the Department of Agriculture was set up under the aegis of the Department of Mines and was responsible to the Minister for Mines and Agriculture. A separate ministry of Agriculture was created in 1907. In the agricultural laboratory, Guthrie was the chemist, and N.A.Cobb, the plant physiologist. For Cobb, see Frieda Cobb Blanchard, 'Nathan A. Cobb, Botanist and Zoologist: A Pioneer Scientist in Australia,' *Asa Gray Bulletin*, 3 (2), (1957), 205-272.

55    C.W. Wrigley, 'W.J. Farrer and F.B. Guthrie: The Unique Breeder-Chemist Combination that Pioneered Quality Wheats for Australia', *Records of the Australian Academy of Science,* 4 (1), (1979), 7-25; see also, F.B. Guthrie, 'William Farrer', *Agricultural Gazette of New South Wales*, 17 (1906), 533-536; 'William J. Farrer and the Results of his Work', NSW Department of Agriculture, *Science Bulletin* No. 22, 1926; Archer Russell, *William James Farrer: A Biography* (Melbourne: F.W.Cheshire, 1949).

56    See Mary Cawte, 'William Farrer and the Australian Response to Mendelism', *Historical Records of Australian Science,* 6 (1), 1984), 45-58.

57    Guthrie was Acting Professor during Liversidge's overseas leaves in 1896-97 and 1904-05, and following his retirement in 1908-09. He later became a leading figure in Australian science, as an original member of the Commonwealth Advisory Council for Science and Industry in 1915, and as a member of the Australian National Research Council (ANRC) in 1921. See Appendix V and C.W. Wrigley, 'Frederick Bickell Guthrie (1861-1927)', *Australian Dictionary of Biography* , vol. 9, 143-144.

58    A.L., 'On some New South Wales and other Minerals (Note No. 6)', (2 December 1891), *JPRSNSW,* 25 (1891), 234-241.

59    Royal Botanic Gardens (Kew), NSW and Victoria Letters, 173, f.138, A.L. to Thiselton-Dyer, 17 December 1891.

60    Royal Botanic Gardens (Kew), NSW and Victoria Letters, 173. f. 137, A.L. to Thiselton-Dyer, 26 September 1891.

61    Senate Minutes, David to Senate, 21 December 1891; Senate Minutes, 7 March 1892.

62    Liversidge Papers, Box 21, f. 201; Senate Minutes, 1 June 1891.

63    *Ibid.,* 6 November 1893, 4 December 1893.

64    Branagan and Holland (eds.), *op. cit,* 112.

65    *NSW Statistical Register*, 1898, Tables 39 and 42, 654-655.

66    Senate Minutes. 5 December 1892, 'School of Mines': Manning's Report.

67    Haas, *op.cit.*, 283.

68    Certainly, Sydney, unlike Melbourne, had little competition from the mining districts. Technical classes at Broken Hill and Newcastle had, for a time, boasted 'Schools of Mines', and classes in coalmining, surveying, geology and assaying were indeed offered, as at West Maitland and Wollongong. Cf. New South Wales Department of Technical and Further Education, *Spanners, Easels and Microchips: A History of Technical and Further Education in New South Wales, 1883-1983* (Sydney: New South Wales Council of Technical and Further Education, 1983), 63-67.

69  *Annual Report of the Department of Mines and Agriculture,* 1894 (Sydney: Charles Potter, Government Printer, 1895), 17.

70  Senate Minutes, 5 April 1893. H. Wood to Registrar, 9 May 1893. 'Proposed School of Mines: Report on Mining Education', *SMH*, 6 January 1893.

71  Technical branch students were allowed to substitute a second modern language for a classical language at the matriculation examination. Senate Minutes, 5 April 1893.

72  According to Liversidge, this compared with at least £134 at the Royal School of Mines, £170 at Massachusetts Institute of Technology, and £240 at Columbia University. *Annual Report, Department of Mines, 1900,* 70.

73  Senate Minutes, 6 March 1893.

74  *Ibid.,* 3 October 1893.

75  *SMH,* 10 May 1893.

76  A.L., 'School of Mines', *SMH*, 6 February 1873.

77  Liversidge described his new buildings to the AAAS in 1902. See A.L., 'The Metallurgical Laboratories of the School of Mines, in the University of Sydney (January 1902)', *Rep. AAAS, 1902* (Hobart), 9 (1903), 830-832. The plates are pictured in Branagan and Holland (eds.), *op. cit.,* 114.

78  *Annual Report of the Department of Mines and Agriculture, 1894, op.cit.,* 17.

79  *Annual Report of the Department of Mines and Agriculture, 1893* (Sydney: Government Printer, 1894), 17.

80  *Nature,* 50 (1894), 415. A point for which the Minister of Mines later took credit. See T.M. Slattery, 'The School of Mines', letter to the *SMH,* 24 June 1901.

81  Liversidge Papers, Box 11, Liversidge annotations on the *Daily Telegraph,* 19 September 1906.

82  'The Metallurgical Laboratories of the School of Mines, in the University of Sydney' (January 1902), *Rep. of the AAAS,* 9 (Hobart, 1902), 830-832, with plates.

83  *Annual Report of the Department of Mines and Agriculture,* 1894 *, op.cit.,* 17.

84  Royal Commission on Technical Education, 1901, *Min. Evid.,* 208.

85  *Annual Report, Department of Mines, 1900,* 67.

86  In 1890, there were fourteen engineering students to three in science; in 1894, thirty in engineering to four in science, and by 1899, sixty-eight in engineering to twenty-four in science (University of Sydney, *Annual Reports, 1890-1899*).

87  T. P. Anderson Stuart, 'A Review of University Life in Australia, with its Conditions and Surroundings in 1891', *Proceedings of the Royal Colonial Institute,* XXIII, Part III, (12 January 1892), 93-141.

88  University of Sydney Archives, G.3/46/21. f. 656. Barff to Sir Saul Samuel, Agent-General, 5 September 1899; University of Sydney Archives, G3/46/22. f. 715. Barff to Liversidge, 28 July 1900.

89  Similar to those of Charles Moore in Sydney, von Mueller in Melbourne and Richard Schomburgk in Adelaide. Pauline Payne, '"Science at the Periphery": Dr Schomburgk's Garden', in Roy MacLeod and Fritz Rehbock (eds.), *Darwin's Laboratory: Evolutionary Theory and Natural History in the Pacific* (Honolulu: University of Hawaii Press, 1994), 239-262.

90  C.W. Marsh, 'On Native Copper Iodide (Marshite) and Other Minerals from Broken Hill, NSW', *JPRSNSW,* 26 (1892), 328. In 1897, Marsh described for the first time, linarite, wulfenite and stolzite, specimens of which became part of Liversidge's personal collection. For other minerals at Broken Hill, see O.H. Woodward, *A Review of the Broken Hill Lead-Silver-Zinc Industry* (Melbourne: Australasian Institute of Mining and Metallurgy, 1952).

91  See his 'Nantokite from New South Wales' (10 April 1894), *Mineralogical Magazine,* 10, No. 48 (September 1894), 326-327; 'Boleite, Nantokite, Kerargyrite, and Cuprite from Broken Hill, N.S. Wales' (6 June 1894), *JPRSNSW,* 28 (1894), 94-98.

92  Chalmers, *op.cit.,* 74. Pittman followed Liversidge's *Minerals of New South Wales* of 1888 with a newer and fuller *Minerals Survey* in 1901. The most recent survey is that of N.L. Markham and H. Basden (eds), *The Mineral Deposits of New South Wales* (Sydney: Department of Mines, 1974).

93  'Committees of Investigation appointed at the General Committee of the Sydney Meeting, 31 August 1888: No 1. Conditions of Labour Committee. *Report of the AAAS,* 1 (Sydney, 1888), xxxii.

94  ANZAAS Archives (Mitchell Library) MSS 988/1, f. 153, Publications Committee, 7 September 1888, f. 183, 12 September 1888.

95  *The Australasian,* 18 January 1890.

96  *The Argus,* 15 January 1890.

97  These calculations are drawn from Ian Inkster and Jan Todd, 'Support for the Scientific Enterprise, 1850-1900', in Rod Home (ed.), *Australian Science in the Making* (Sydney: Cambridge University Press, 1988), 114.

98  Royal Botanic Gardens (Kew), NSW and Victoria Letters, 173, f. 136, A.L. to Thiselton-Dyer, 17 September 1890; f. 137, A.L. to Thiselton-Dyer, 26 September 1891. In 1891, at Liversidge's recommendation, Thiselton-Dyer was given the Clarke Medal of the Royal Society of NSW.

99  Royal Botanic Gardens (Kew), NSW and Victoria Letters, 173, f. 137, A.L. to Thiselton-Dyer, 26 September 1891.

100 A.L. 'Anniversary Address', *JPRSNSW*, 24 (1890), 12.

101 A 'Science House' finally came to pass, at the corner of Gloucester and Essex Streets. Land was granted by the State Government in June 1928, and the foundation stone was laid by the Governor in June 1930. See O.U. Vonwiller, 'Presidential Address to the Royal Society of NSW', 6 May 1931, in *JPRSNSW*, 1931. There the learned societies remained until the Government resumed the property in the 1980s. There are hopes of restoring it to its original purpose. I am grateful to David Branagan for this information.

102 University of Sydney Archives, G.3/46/8, f.162, Manning to A.L.,13 May 1890.

103 In July 1889, Senate agreed that of the University's twelve chairs, six would continue to be paid from general endowment, and six from the Challis fund. The first six would attract half fees; the second six (except Modern Literature) would not. In addition, some eleven junior staff, plus 'general staff' would be paid from general funds. The six 'endowment' chairs in classics, mathematics, chemistry, physics, geology and physiology each received £900 plus half fees. Among the six Challis Chairs, history (G.A.Wood), logic and mental philosophy (Francis Anderson) and biology (W.A. Haswell) received £800; anatomy (J.T.Wilson) and engineering, £900 each, and modern literature, £900 with half fees. A separate lecturing fund of £2,000 was given to the Law School, until a chair (P. Cobbett) was added, and all Challis chairs were put on an equal footing. Liversidge Papers, Box 21, f. 190, Senate Minutes, 15 December 1890.

104 Liversidge Papers, Box 21, f. 135, Senate Minutes, 18 December 1989. There was also a gift from the estate of Thomas Fisher, who died in 1884, leaving the University £33,000. With this windfall, Senate proposed to build a proper library, the need for which was self-evident in an institution now numbering over 530 students and forty-three staff, and which was now by a slender margin the largest of the four universities in Australasia. The staff included fourteen professors, four demonstrators, sixteen lecturers, six assistant lecturers and three tutors. The four universities counted together some 2,000 students and 141 staff. See Anderson Stuart, *op. cit*, 99-100, 159. If savings were to come, the staff would be the first and hardest hit. See Liversidge Papers, Box 21, f. 21, Report of a Committee to consider ... salary to be assigned to the several medical lectureships, 5 October 1889.

105 Liversidge Papers, Box 21, f.187, G.Gage to Registrar, 24 September 1890; f. 193, R. Tecce, 3 March 1891.

106 Liversidge Papers, Box 21, f. 76. N. Maclaurin, 2 March 1889.

107 Liversidge Papers, Box 21, f.150. Report of the Committee appointed by the Senate to draw up By-laws for regulating the care and management of the University's finances generally, 28 February 1890. Cf. the Deas Thomson Committee on Finance, 3 March 1851, reprinted in 1890, in Liversidge Papers Box 21, f.186.

108 'The Sydney University', *SMH*, a five part series, beginning 18 October 1890.

109 'The Sydney University: III', *SMH*, 29 August 1891.

110 'The Sydney University: IV', *op. cit*.

111 Faculty of Science Minutes, 22 November 1889. It could be argued that chemistry, long second to physics, was now benefitting at the expense of biology. Where he spotted 'unappropriated interest', he applied it to annual prizes for practical chemistry and physics. See Liversidge Papers, Box 21, f.173, Senate Minutes, 18 August 1890. Liversidge certainly fared better than the young Haswell who, as Challis Professor of Biology, had to give all zoology and botany classes, elementary and advanced, in the three Faculties of Arts, Science and Medicine. See Liversidge Papers, Box 21, f. 102, 27 August 1889. Haswell's thirty-nine students were pressed into a laboratory space inadequate for twenty-five.

112 There seemed good will within the Faculty, and a sense that civility would prevail. In April 1891, he supported Haswell's bid to extend his laboratory and add a Demonstrator to his staff, (see Faculty of Science Minutes, 15 April 1891, 11 November 1891) and in November that year, backed Threlfall's project to start a course in Electrical Engineering, with the implication of £300 in extra equipment expenses. See Faculty of Science Minutes, 6 November 1891.

113 *Eighth Report of the Royal Commission for the Exhibition of 1851*. 1911. [Cd. 5723]. xxi.4; W.H.G. Armytage,

*A.J. Mundella, 1825-1897* (London: Ernest Benn, 1951), 279; Wemyss Reid, *Memoirs and Correspondence of Lyon Playfair* (London: Cassell, 1899), 442-447.

[114] *Eighth Report of the Commission for 1851*. See Roy MacLeod and E.K. Andrews, 'Scientific Careers of 1851 Exhibition Scholars', *Nature,* 218 (1968), 1013-1014.

[115] The scholarships later created by Britain's Department of Scientific and Industrial Research (DSIR) had their precedent in these awards. In the 1930s, the Commission boasted that 'no important university in the Dominions is without its quota of Professors and Lecturers who owe their early training to our awards'. Among the 600 Scholars appointed by 1935, the Commission could count sixteen Vice-Chancellors, 144 Professors, forty Fellows of the Royal Society and two Nobel Prizewinners. (*Ninth Report*, 1935). By 1961, 918 Scholars had produced ninety-nine FRS's and eight Nobel Prizewinners. (*Record of the Science Research Scholars of the Royal Commission for the Exhibition of 1851, 1891-1960* (London: Exhibition of 1851, 1961), Foreword). For their importance in Canada, see Yves Gingras, 'Financial Support for Post-graduate Students and the Development of Scientific Research in Canada', in Paul Axelrod and John G. Reid (eds.), *Youth, University and Canadian Society* (Kingston: McGill-Queen's University Press, 1989), 304-309.

[116] Faculty of Science Minutes, 21 March 1892. Barraclough later returned to a lectureship at Sydney and became Professor of Mechanical Engineering. During the First World War he was given command of Australian munitions workers in England. He was knighted in 1920. J.M. Antill, 'Sir Samuel Henry Barraclough (1871-1958)', *Australian Dictionary of Biography,* vol. 7, 184-185. See Roy MacLeod, 'The Arsenal in the Strand: Australian Chemists and the British Munitions Effort, 1916-19', *Annals of Science,* 46 (1989), 45-67.

[117] Royal Commission on the Exhibition of 1851 (South Kensington) Archives, 'Opinion of Scientific Examiners on Reports of Scholars, 1891-1900', f. 130, July 1899.

[118] *Ibid.*, f. 247, Memorandum from W.T. Thiselton-Dyer, August 1902.

[119] W.H. Hamlet, 'Presidential Address', Section B, *Report of the AAAS,* 4 (Hobart, 1892), 49-50.

[120] See Exhibition of 1851 Archives, f. 33, 'Barraclough', Final Report. William H. Ledger, Sydney's second '1851', and also an engineering student with first-class honours, was nominated not by his professor, but by Judge Backhouse, the Vice-Chancellor, who sensibly declined to express any 'opinion as to his capacity for research'. In fact, although he had shown great promise, Ledger proved unable to define his research project, and ultimately left science. See Senate Minutes, 5 April 1893. 'In this case', the Committee reported, 'it does not seem the objects of the Scholarship have been accomplished'. Royal Commission on the Exhibition of 1851 (South Kensington) Archives, f. 48 'Ledger'. After fourteen years at the South Australian School of Mines, he became General Manager of the Sydney Steel Company.

[121] Royal Commission on the Exhibition of 1851 (South Kensington) Archives, f. 159. Report by J. J. Thomson, 1900. This opinion did not stop Durack from becoming a lecturer in physics at King's College, London, but it probably helped consign him to his later fate as Professor of Physics at Allahabad.

[122] Anderson Stuart, *Journal of the Royal Colonial Institute,* 3 (1891-92), 157.

[123] See Barry Butcher, 'Science and the Imperial Vision: The Imperial Geophysical Experimental Survey, 1928-1930', *Historical Records of Australian Science*, 6 (1), (1984), 42.

[124] Senate Minutes, 21 November 1892.

[125] *Ibid.*, 5 December 1892.

[126] *Ibid.*, 19 December 1892.

[127] *Ibid.*, 28 March 1893.

[128] Including his good friend Kernot of Melbourne. Senate Minutes, 21 November 1892.

[129] *Ibid.*, 18 April 1893, 2 May 1893.

[130] *Ibid.*, 2 May 1893.

[131] Faculty of Science Minutes, 28 March 1893. He later withdrew the motion, but only when its application proved impossible to enforce. *Ibid.,* 10 April 1893, 16 August 1894.

[132] Manning attempted to retrieve the loss by asking for a reduced vote of £1,200 (Senate Minutes, 6 February 1893) but even that proved impossible to get into the Government's Supplementary Estates, and only £800 was eventually forthcoming (5 June 1893). The impact of the shortfall was softened only by news that the £10,000 from unexpended funds for the School of Mines was on its way (20 February 1893). With it came W.F. Smeeth, nominated by Professor J.W. Judd of the Royal School of Mines to be David's first Demonstrator in Geology. (6 February 1893)

[133] Senate Minutes, 20 February 1893.

134 *Ibid.*, 20 March 1893.

135 *Ibid.,* 27 March 1893.

136 *Ibid.,* 5 April 1893.

137 Senate Minutes, 7 August 1893, 14 August 1893, 5 March 1894.

138 Public Accounts, *VPLANSW*, 1891-1895.

139 *Ibid.*, 2 April 1894.

140 *Ibid.,* 15 May 1893. For George (later Sir George) Knibbs, future Executive Director of CSIR, see S. Bambrick, 'Sir George Handley Knibbs (1858-1929)', *Australian Dictionary of Biography*, vol. 9, 620-621.

141 Senate Minutes*,* 7 August 1893, 14 August 1893.

142 *Ibid.*, 5 June 1893.

143 *Ibid.*, 19 June, 3 July 1893.

144 *Ibid.*, 1 May 1893. For Suttor, see Ruth Teale, 'Sir Francis Bathurst Suttor (1839-1915)', *Australian Dictionary of Biography*, vol. 6, 227-228; The Government conveniently argued that the so-called 'additional endowment' of £8,900 over the University's original £5,000 had been only a 'temporary' measure, as it technically was; and that, given the Challis money, which had brought the total endowment to £37,000 in 1891, the University could forgo this public 'addition'. 'Sydney University: Gradual Withdrawal of Government Support: Reasons for this Step', *SMH*, 27 April 1894. This figure represented £5,000 endowment, £8,900 'additional endowment', £7,200 for buildings, £2,400 for apparatus, and £2,000 for evening lectures. (Public Accounts, *VPLANSW,* 1891.) However, the University could forgo nothing without also giving up scientific apparatus, cutting evening lectures, or removing staff. In the event, all these steps and more were contemplated.

145 Senate Minutes., 21 August 1893.

146 *Ibid.*, 3 October 1893.

147 *Ibid.*, 3 October 1893.

148 *Ibid.*, 17 April 1893.

149 That is, Haswell, Threlfall, Anderson Stuart, Warren and David.

150 Not until 1909 did the Museum's budget again reach £11,000. See R. Strahan, *Rare and Curious Specimens: An Illustrated History of the Australian Museum, 1827-1979* (Sydney: The Australian Museum, 1979), 50.

151 Australian Museum Letterbook, 1891, f. 295, Sinclair to A.L., 20 July 1891; f. 216, Hill to A.L., 23 May 1894; Etheridge to A.L., 9 June 1894; Liversidge Papers, Box 19; f. 5, Ramsay to A.L., 13 April 1892.

152 Australian Museum Letterbook, 1892, f. 16, Sinclair to A.L., 8 January 1892.

153 Senate Minutes, 4 December 1893. The 'extension movement' took the University to the suburbs, then to Bathurst and Maitland and beyond, to 'exploit and assist' rural eagerness for information. It did not survive long – as one critic put it, the University's 'out-stretched hand was withdrawn to the shelter of a fur-coated sleeve' – but for a time it brought enormous visibility to those professors who, like David and Wood, participated. (Arthur W. Jose, *The Romantic Nineties* (Sydney: Angus and Robertson, 1933), 60-67). What Liversidge lost by turning his energies inward into the University, David gained by looking outwards, becoming a celebrity in ways that suited, and rewarded, personality and style, and counted in the reckoning for honours.

154 *Report of the AAAS ,*5 (Adelaide, 1893).

155 That is: 1) 'The Specific Gravities of some Gem Stones' (28 September 1893) *Rep. of the AAAS*, 5 (Adelaide, 1893), 404-407; reprinted in *J. Amer. Chem. Soc.* (23 August 1893), 16 (1894), 205-209; (2) 'Schedules for the Use of Students in Mineralogy' (28 September 1893), *Rep. of the AAAS* , 5 (Adelaide, 1893), 408-409; and 'A Combination Laboratory Lamp, Retort, and Filter Stand' (6 September 1893), *JPRSNSW*, 27 (1893), 347-348; (28 September 1893) *Rep. of the AAAS*, 5 (Adelaide, 1893), 325.

156 His four 'gold papers' were: 'On the Condition of Gold in Quartz and Calcite Veins' (6 September 1893), *JPRSNSW*, 27 (1893), 299-303; (28 September 1893), *Rep. of the AAAS*, 5, (Adelaide, 1893), 324; 'On the Origin of Gold Nuggets' (6 September 1893), *JPRSNSW,* 27 (1893), 303-343; (28 September 1893) *Rep. of the AAAS*, 5, (Adelaide, 1893), 324; 'On the Crystallization of Gold in Hexagonal Forms' (6 September 1893), *JPRSNSW,* 27 (1893), 343-346; (28 September 1893), *Rep. of the AAAS*, 5, (Adelaide, 1893), 324; and 'Gold Moiré-Métallique' (6 September 1893), *JPRSNSW*, 27 (1893), 346-347; (28 September 1893), *Rep. of the AAAS*, 5, (Adelaide, 1893), 325. For the publishing history of these papers, see *Liversidge Bibliography.*

157 Senate Minutes, 5 February 1894.

158 *SMH,* 27 April 1894.

159 Mitchell Library, Baldwin Spencer Papers, MSS 29/8, A.L. to Spencer, 26 April 1894.

160 Senate Minutes, 2 April 1894.

161 Senate Minutes, 14 August 1893.

162 *Ibid.*, 3 October 1893.

163 University Archives, G3/46/15. Barff to Speak, 4 December 1894. When Speak resigned in 1894, the salary of his successor as Lecturer in Metallurgy was reduced from £450 to £400. Senate Minutes, 6 August 1894.

164 Senate Minutes, 2 April 1894, 1 April 1895, 2 December 1895. Pittman was annually reappointed until he left the University, at his own wish, in 1902.

165 *Ibid.,* 17 August 1885. The appointment was James Petrie.

166 Senate Minutes, 4 February 1895. See C.E.F. F[awsitt], 'In Memoriam: Edward Hufton', *Hermes,* 23 (1917), 49. See also the memoir by his granddaughter, Clarice Morris (a science graduate of the University), 'The Diary of Edward Hufton, An Assisted Immigrant, 1879', *JRAHS,* 45 (6), (1960), 314-351.

167 Roy MacLeod, 'Founding: South Kensington to Sydney', in Graeme Davison and Kimberley Webber (eds), *Yesterday's Tomorrows: The Powerhouse Museum and its Precursors 1880-2005* (Sydney, Powerhouse Publishing in association with UNSW Press, 2005), 42-54.

168 That is, By-laws; Buildings, Grounds and Improvement; Organisation; and Calendar were added to Finance.

169 Senate Minutes, 2 April 1894.

170 Their representation had been a debating point since the beginning of the University. In 1885, the six *ex officio* seats fought for by Smith and Woolley under the University's amending Act of 1861 just covered the six men in post. But by the early 1890s, there were ten chairs, and likely to be more. In May 1888, Manning referred the question of 'the best mode of making use of the services of the Teaching Staff in the government of the university' to a committee under his own chairmanship, which sat for over a year. It reached, however, no conclusions, beyond the general expectation that Departments and Deans would provide the best form of 'local government', and that the best Senate was one free of departmental in-fighting (Liversidge Papers, Box 21, f. 129, Report of the Select Committee of the Senate, 15 November 1889) .

171 By no means were the professors to have their own way, especially in these austere times. The Convocation of Sydney graduates, far from approving of the representation of professors as by right, proposed there should be no *ex officio* members of Senate at all, and that the professors be heard through elected Deans (Liversidge Papers, Box 21, f. 200, Resolutions of Convocation, 22 May 1891). At the time, a compromise was struck, by which the six professors on the 'general endowment' were admitted to Senate; and the four Faculties each nominated one representative, chosen on a rotating basis among departments, to serve *ex officio* for two years. (Liversidge Papers, Box 21, f. 155, 24 March 1890. Renwick, *Report of Ex-officio Fellows Committee*, 2 March 1891.) In December 1891, however, the four Deans were replaced by representatives of the four 'branches of learning', presumably in rotation.

172 Senate Minutes, 3 September 1894.

173 Senate Minutes, 7 September 1896, 5 September 1898.

174 *Ibid.,* 2 July 1894.

175 By the University Endowment Amendment Act, 1902, the University's base budget was raised to £10,000. For discussion, see *SMH*, 24 April 1905.

176 Senate Minutes, 6 August 1894.

177 Bede Nairn, 'Jacob Garrard (1846-1931)', *Australian Dictionary of Biography*, vol. 4, 234.

178 Senate Minutes, 15 October 1894.

179 A proposal to divert a larger share of fees from the professors' income could have brought £1,550 more into the University's purse, but at the risk of abridging moral (and in some cases, contractual) obligations, and was defeated, albeit narrowly (Senate Minutes, 4 February to 1 April 1895). As a last resort, Senate, in the summer of 1894-95 looked anxiously at the capital of the Challis Fund, now standing at £25,000, on deposit with the Colonial Treasury (Senate Minutes, Special Meeting, 13 February 1895).

180 Dr. H. Normand MacLaurin, the distinguished Scottish naval surgeon and Vice-Chancellor of the University between 1887-89, was again elected Vice-Chancellor. Cf. *Hermes*, 2 (30 October 1896), 2-4.

181 See 'The Late Chancellor', *Hermes*, I (26 May 1895) 5. The cuts exposed deep cracks in the University's finances, for example, Sir William Macleay offered to endow a Lectureship in Bacteriology, but the Faculty of Science had no laboratory resources, and so deferred the gift (*Hermes*, 3 June 1895). Macleay responded by threatening the future of the entire Macleay bequest, and the situation required much patient negotiation (*Hermes*, 1 July 1895; Faculty of Science Minutes, 14 October 1892). Other programmes – including six-year-

old plans to build the proposed Fisher library, were similarly deferred (Senate Minutes, 6 May 1895; 3 June 1895). In its anguish, the University even 'devolved' the cost of its five telephones (£42 in 1894) from general funds to the laboratory vote of the Science Faculty, until it was discovered there was no laboratory vote to pay for them. When, in 1895, a further £1,000 was cut from the University's 'additional endowment', and all but £750 to its repair fund, hard choices had to be made. Indicatively, the four stone turrets on the Tower, costing £664, were paid for; but the science laboratories came to a halt.

182  Liversidge Papers, Box 19, f.7, Executive Committee to Liversidge, 19 May 1894.

183  *Ibid.*, 6 May 1895. Watt's history is emblematic of the chequered careers of men reading science at Sydney in his generation. He matriculated in 1887, took his BA in 1890, then spent two years as an assayer for gold and silver mining companies. In 1892 he re-entered the University, took an MA in Logic and Mental Philosophy, then did the BSc in two years. In 1896, he became field geologist on Baldwin Spencer's Horn Scientific Expedition to the MacDonnell Ranges, and subsequently worked in petrology with Tate at Adelaide. He later took medical qualifications and ended his days as a physician in Tenterfield, NSW. Exhibition of 1851 Archives, f. 79.

184  'Retrenchment', *Hermes*, I (3), (17 July 1895), 1-2.

185  Senate Minutes, 5 August 1895.

186  *Ibid.*, 2 September 1895; 12 October 1896.

187  *Ibid.*, 7 December 1896.

188  *Ibid.*, 4 February 1895.

189  *Ibid.*, 11 November 1895.

190  In 1891, Threlfall, a robust footballer, alleged that student behaviour at Commemoration was responsible for making the event a 'dull and discreditable fiasco'. Professorial Board Minutes, 30 September 1891.

191  Louisa Macdonald (Principal of Women's College) to Eleanor Grove (nd, 1894), quoted in Jeannette Beaumont and W. Vere Hole, *Letters from Louisa* (Sydney: Allen & Unwin, 1996), 153.

192  Professorial Board Minutes, 14 June 1893, 17 August 1893, 5 October 1893, 17 April 1896.

193  The Union did not provide an escape valve for student high spirits, devoting itself entirely, in the words of a later Registrar, 'to the mental culture of its members' by means of debates and lectures (Robert Dallen, *The University of Sydney: Its History and Progress* (Sydney: Sydney University Union, 2nd ed.1925), 52). Facilities for games, and the necessary pastimes of eating and drinking were not to come until 1911. Until then, there were few places for students to enjoy an independent 'culture', and every inducement to fractious behaviour.

194  Imperial College Archives, Liversidge Papers, A.L. to F.E.A. Manning, Secretary, London University Union, 23 April 1924. The 'ULU' in Malet Street, WC. 1, is now a familiar fixture of university life in Bloomsbury.

195  Professorial Board Minutes, 2 May 1892.

196  *Ibid.,* 11 November 1891.

197  See 'Engineering: A Chat with Professor Warren', *Daily Telegraph*, 14 February 1896.

198  'Contributions to the Bibliography of Gold ' (January 1895), *Rep. of the AAAS*, 6 (Brisbane, 1895), 240-256.

199  By which gold is dissolved in an aqueous solution of chlorine gas and then precipitated by sulphuric acid.

200  By the MacArthur process, gold, which is highly soluble in sodium cyanide, is dissolved and concentrated in cyanide and then extracted by electrolysis or precipitation on lead-zinc. This process enabled miners to recover gold economically from low grade and sulphidic ores.

201  M. Eissler, *The Metallurgy of Gold: A Practical Treatise* (London: Crosby Lockwood and Son, 1888, 5th ed. 1900), 515-529.

202  By 1897, the cyanide process was responsible for ten per cent of the world's production. Of this, South Africa contributed half, Australia nineteen per cent, New Zealand, sixteen per cent and the United States, twelve per cent. See Alan Lougheed, 'The Discovery, Development and Diffusion of New Technology: The Cyanide Process for the Extraction of Gold, 1887-1914', *Prometheus*, 7 (1), (1989), 61-74.

203  W. Topley, 'Gold and Silver: Their Geological Distribution and Their Probable Future Production', *Rep. of the BAAS* (Manchester, 1887), 512.

204  Victor Lenher, 'Solubility of Gold in certain Oxidizing Agents', *Journal of the American Chemical Society,* 26 (1904), 550-554.

205  A.L. , 'On the Removal of Gold from Suspension and Solution by Fungoid Growth', *Rep. of the AAAS,* 2 (Melbourne, 1890), 399-407.

206  'On some New South Wales and other Minerals (Note No. 6)' (2 December 1891), *JPRSNSW,* 25 (1891),

234-241; cf. H.C. Russell, 'Presidential Address', *JPRSNSW*, 26 (1892), 30.

[207] 'Preliminary Note on the Occurrence of Gold in the Hawkesbury Rocks about Sydney' (3 October 1894), *JPRSNSW*, 28 (1894), 185-188.

[208] 'On the Origin of Moss Gold' (6 September 1893), *JPRSNSW*, 27 (1893), 287-298; *Chemical News*, 69, (30 March 1894), 152-155; 'On the Condition of Gold in Quartz and Calcite Veins' (6 September 1893), *JPRSNSW*, 27 (1893), 299-303; *Chemical News*, 69 (6 April 1894), 162-163; abstr. in *J. Chem. Soc.*, 66, (2) (1894), A, 354; 'On the Origin of Gold Nuggets' (6 September 1893), *JPRSNSW*, 27 (1893), 303-343; *Chemical News*, 69 (1 June 1894), 260-262; 69 (8 June), 267-268; 69 (15 June), 281-284; 69 (22 June), 296-298; 69 (29 June), 303-304; 70 (6 July 1894), 6-8; 70 (13 July), 21-22; 'On the Crystallization of Gold in Hexagonal Forms' (6 September 1893), *JPRSNSW*, 27 (1893), 343-346; *Chemical News*, 69 (13 April 1894), 172-173; *J. Chem. Soc.*, 66 (2) (1894), A, 353-354; 'Gold Moiré-Métallique' (6 September 1893), *JPRSNSW*, 27 (1893), 346-347; *Chemical News*, 69 (4 May 1894), 210.

[209] Robert O. Chalmers, *Australian Rocks, Minerals, Gemstones* (Sydney: Angus and Robertson, 1967), 288.

[210] Rose, *op.cit.*, 76. For a typical popularisation, see Frederick Danvers Powers, 'Australasian Views on the Origin of Gold Nuggets', *The Australian Mining Standard*, 27 July 1895, 422-433, 434-436.

[211] T.K. Rose and W.A.C. Newman, *The Metallurgy of Gold* (London: Charles Griffin, 7th ed., 1937), 8-9.

[212] Some of these survive in the possession of the Inorganic Chemistry Department at Sydney University.

[213] 'Structure of Gold Nuggets' (5 September 1894), *JPRSNSW*, 28 (1894), 331; (3 October 1894), *JPRSNSW*, 28 (1894), 333; *Chemical News*, 70 (1894), 199; 'On the Internal Structure of Gold Nuggets', (January 1895), *Rep. of the AAAS*, 6 (Brisbane 1895), 240; 'On the Crystalline Structure of Gold and Platinum Nuggets and Gold Ingots' (3 October 1894), *JPRSNSW*, 31 (1897), 70-79; *J. Chem. Soc.*, 71 (1897), Pt. 2, 1125-1131.

[214] Cf. *Council Minutes, RSNSW*, 7 August 1895; *Evening News*, 8 August 1895.

[215] Liversidge bequeathed his gold nugget specimens to the British Museum and they are now housed in the Natural History Museum. They are in the form of double polished sections mounted on slides.

[216] C.R. M. Butt to R. MacLeod, personal communication, December 2007.

[217] John Calvert, *The Gold Rocks of Great Britain and Ireland, and A General Outline of the Gold Regions of the World, with a Treatise on the Geology of Gold* (London: Chapman and Hall, 1892); J. Malcolm MacLaren, *Gold: Its Geological Occurrence and Geographical Distribution* (London: The Mining Journal, 1908). Cf. also T.K. Rose, *The Metallurgy of Gold* (London: Charles Griffin, 5th ed., 1906), 8. By 1937 and the seventh edition of this text, compiled by fellow Associates of the Royal School of Mines, Liversidge's observations were confirmed, described in terms of molecular arrangements, and placed mineralogically in what was known of secondary gold formations; Cf. W. Lindgren, 'Characteristic Features of Californian Gold - Quartz Veins', *Bull. Geo. Soc. America*, 6 (1895), 237; William Blake, 'Gold in Granite and Plutonic Rocks', *Trans. American Institute of Mining Engineers*, XXVI (1896), 288-298; Cf. Victor Lenher, 'Solubility of Gold in Certain Oxidising Agents', *J. Amer.Chem. Soc.*, 26 (1904), 550-554; 'The Transportation and Deposition of Gold in Nature', *Economic Geology*, 12 (1917), 744-750; 'Further Studies on the Deposition of Gold in Nature', *Economic Geology*, 13 (1918), 161-184; MacLaren, *op. cit*, 83.

[218] Thomas Crook, *History of the Theory of Ore Deposits, with a Chapter on the Rise of Petrology* (London: Thomas Murby, 1933), 15.

[219] This is the conclusion reached by Drs. J.E. Glover and D.I. Groves (eds.), *Gold Mineralization Seminar, University of Western Australia 1978, Summary of presented papers* (Perth: University of Western Australia, 1979). I thank the authors for a copy of these proceedings.

[220] A.F. Wilson, 'Origin of Supergene Gold and Nuggets in Laterites and Soils', *Australian Journal of Earth Sciences*, 31 (1984), 303-316.

[221] R.M. Hough, C.R.M Butt, S.M. Reddy and M. Verrall, 'Gold Nuggets: Supergene or Hypogene?', *Australian Journal of Earth Sciences*, 54 (2007), 959-964.

[222] A.L., 'On the Amount of Gold and Silver in Sea-Water', and 'Removal of Silver and Gold from Sea Water by Munz Metal Sheathing', *JPRSNSW*, 29 (1895), 335-366; *Chemical News*, 74 (1896), 146-148, 160-161, 166-168, 182-184, 191-194.

[223] A.L., 'On the Presence of Gold in Natural Saline Deposits and Marine Plants', *Journal of the Chemical Society*, 71 (1897), 298-299.

[224] Muntz metal was a copper-zinc alloy used for cladding roofs and sheathing ships and pier piles.

[225] See Ciaran Murphy, 'CSIRO Leads Sea Hunt for Goldmine', *The Weekend Australian*, 15-16 May 1999;

Richard Macey, 'Miner's Gift of Gold Comes up the Chimney', *Sydney Morning Herald,* 20 December 1997; Murray Hogarth, 'Stacks of Treasures', *Sydney Morning Herald*, 4 June 1999; 'Seafloor Explorers Race to Peg Out the "Ring of Fire"', *Mining News*, 23 July 2007; 'Nautilus Finds Sunken Treasure', *Mining News*, 10 December 2007.

[226] R.W. Boyle, 'The Geochemistry of Gold and its Deposits', *Geological Survey Bulletin*, No. 280, (Washington, DC: US Geological Survey, 1979), 60.

[227] 'Alexander Marcet', in *World's Who's Who in Science* (Chicago: Marquis, 1968), 1108.

[228] Joe B. Rosenbaum, *et al.*, 'Gold in Sea Water- Fact or Fancy?', *The Mines Magazine,* 59 (September 1969), 14-17.

[229] Edward Sonstadt' (Obituary), *Journal of the Society of Chemical Industry,* 17 (1908), 885. For Münster, see J.C. Poggendorff (ed.), *Biographisch-Literarisches Handwörterbuch,* III (Leipzig: Verlag von Johann Ambrosius Barth, 1898), 947.

[230] Edward Sonstadt, 'New Process for Estimating the Iodine in Kelp Liquors, Mineral Waters, etc.', *J.Chem. Soc.,* XXV (1872), 1116; cf. 'Presence of Calcium Iodate in Sea-Water', *J.Chem.Soc.,* XXV (1872), 597-599.

[231] Edward Sonstadt, 'On the Presence of Gold in Sea-Water', *Chemical News,* XXVI (4 October 1872), 159-161; *J. Chem.Soc.,* XXV (1872), 1119-1120. No quantative data which might have indicated the reproducibility of these qualitative tests or their potential application in gravimetric or colorimetric analysis were provided.

[232] *The Athenaeum* (10 October 1872), 472; Edward Sonstadt, 'Gold in Sea-Water', *Popular Science Monthly,* II, (1872-1873), 255.

[233] E. Münster, 'On the Possibility of Extracting the Precious Metals from Sea-Water', *J. Soc. Chem. Ind.,* 11 (1892), 351.

[234] Richard Daintree, *Report on the Geology of the District of Ballan including Remarks on the Age and Origin of Gold,* (Melbourne: Government Printer, 1866).

[235] See C.R. Resanius, *Quantitative Chemical Analysis* (London: J. & A. Churchill, 7th edition, 1900), vol. II, 355.

[236] MacLaren, *op. cit,* 104.

[237] 'On the Amount of Gold and Silver in Sea-Water' (2 October 1895), *JPRSNSW,* 29 (1895), 335-349.

[238] 'On the Crystalline Structure of Gold and Platinum Nuggets and Gold Ingots' (3 October 1894), *JPRSNSW,* 31 (1897), 70-79; 'On the Presence of Gold in Natural Saline Deposits and Marine Plants' (4 February 1897), *Proc. Chem. Soc.,* 13 (174), (1897), 22-23.

[239] 'On the Presence of Gold in Natural Saline Deposits and Marine Plants' (4 February 1897) *Proc. Chem. Soc.,* 13 (174), (1897), 22-23. His method was to add 0.5 to 5 gram of $(Fe^{II})$ sulphate to an unfiltered solution of 100 to 1000 gram of salt water; to scorify the precipitate with lead, and cupel the product. The natural salts were found to contain 1-2 grains of gold per ton (65-130 mg/tonne); kelp and bittern gave 14-20 grains per ton.

[240] J.R. Don, 'The Genesis of Certain Auriferous Lodes', *Trans. Amer. Inst. Min. Engineers,* 27 (1898), 564-668; J.W. Pack, 'Gold From Sea Water', *Mining and Science Press,* 77 (1898), 154-155.

[241] T. Kirke Rose, *The Metallurgy of Gold* (London: Charles Griffin, 5th edition, 1906), 37; see Liversidge's notes on his copy in Christ's College Library.

[242] *J. Soc. Chem. Ind.,* 14 (31 October 1895), 874; 21 (15 March 1902), 349; 21 (31 December 1902), 1536; 22 (31 January 1903), 97; 22 (31 August 1903), 953.

[243] H.A.M. Snelders, 'Svante August Arrhenius', *Dictionary of Scientific Biography,* 1 (1970), 296-302.

[244] Elizabeth Crawford, *The Beginnings of the Nobel Institution: The Science Prizes, 1901-1915* (London: Cambridge University Press, 1984).

[245] Cf. P. De Wilde, 'Sur l'Or contenue dans l'Eau de Mer', *Archiv. Soc. de Phys. et d'Hist. Naturelle,* 5 (1905), 559-580.

[246] Cf. Morris Goran, *The Story of Fritz Haber* (Norman: University of Oklahoma, 1967), 92; M. Goran, 'Fritz Haber', *Dictionary of Scientific Biography,* 5 (1972), 622.

[247] Goran, *op. cit.,* 91.

[248] Boyle, *op.cit.,* 63.

[249] A.L., 'Meteoric Dusts, New South Wales', *JPRSNSW,* 36 (1902), 241-285.

[250] University of Sydney Archives, G3/46/16. f. 883, Barff to Liversidge, 16 August 1895, *Senate Minutes*, 25 September 1895, 3 February 1896.

251 Senate Minutes, 5 August 1895.

252 Henry was eager for Americans – in many respects, still 'scientific colonials' themselves – to have ready and regular access to the world's scientific literature. By adroit diplomacy, he persuaded the Royal Society, as the senior academy in the English-speaking world, to assume the project's direction and control. By 1872, the Catalogue had begun to materialise; and by 1884 there had appeared twelve large volumes in two editions, listing, by author, the scientific papers published by over sixty national journals recognised throughout the world between 1800 and 1883. It did not take long for the enormity of the task to overwhelm the editorial staff in London. The volumes were archival by the time they were published. A subject index, repeatedly proposed, was repeatedly deferred.

253 The first series (1800-1863) was completed in 1872; the second (1864-1873), in 1874; and the third (1874-1883), between 1891 and 1896, all published by Cambridge University Press. Two supplementary volumes appeared in 1902. The fourth series (1884-1900) was published on the eve of war in 1914. See H. Forster Morley (ed.), *International Catalogue of Scientific Literature, First Annual Issue* (London: Harrison and Sons, 1902), v.

254 *Nature,* 54 (16 July 1896), 148.

255 W.W. Bishop, 'The Record of Science', *Science,* LVI (1922), 214.

256 *Council Minutes,* 11 December 1895.

257 If Sydney had failed to confirm his invitation, Foster was prepared to have Liversidge attend as one of the Royal Society of London delegates. *SMH* ,1 March 1897; *Council Minutes,* 26 May 1897.

258 Mitchell Library, Baldwin Spencer Papers, MSS 29/8 Liversidge to Spencer, 23 December 1895.

259 His visit to the 'great and wealthy and enlightened commonwealth of New South Wales' is described in the first half of *Following the Equator* (New York: American Publishing Co., 1897). Cf. *Mark Twain in Australia and New Zealand* (Ringwood, Vic.: Penguin, 1973), 123.

260 Partly as a result of Liversidge's application for a year's leave, Senate fixed rules to govern all academic leave thereafter. These allowed (a) a maximum of two terms at half-pay, after seven years' service, subject to satisfactory teaching arrangements; (b) special consideration after twenty years' service or 'long service leave'; (c) permission to travel outside Australia during the long vacation; (d) limited sick leave. (Senate Minutes, 2 December 1895) Policy for study leave was thus born in an atmosphere of austerity and retrenchment, and not in the spirit of scholarly adventure that Smith and Russell had encouraged.

261 In 1884, Liversidge's salary was set at £900, to which were added £289 in student fees. His fees for public examinations brought him another £178, for a total of £1,365. His income would have been slightly higher in 1895. His emoluments may be contrasted with Smith's highest salary of £1,040, and Badham's £1,833 plus house allowance (or about £2,200 in total). See Basser Library, MS. 15, Anderson Stuart Papers, (1884) Salaries, 13. Senate Minutes, 2 September 1895, 11 November 1895. University Archives, G. 3/46/16f. 925, Barff to Guthrie, 3 September 1895. Guthrie was assisted by J.A. Schofield.

262 'Return of Professor Liversidge: A Scientist on his Travels', *SMH,* 1 March 1897.

263 *Ibid.*

264 *Ibid.*

265 *Ibid.*

266 Rod Home, 'The Problem of Intellectual Isolation in Scientific Life: W.H. Bragg and the Australian Scientific Community, 1886-1909', *Historical Records of Australian Science,* 6 (1), (1984), 19-30.

267 Mitchell Library, Baldwin Spencer Papers, MSS 29/8, f. 257, Liversidge to Spencer, 26 April 1896.

268 *Ibid.* Spencer was elected FRS in 1900.

269 *Ibid.*

270 *SMH,* 1 March 1897.

271 Senate Minutes, 4 May 1896. University of Sydney Archives, G3/46/17, Letters, 1895-96, f. 757, Barff to Liversidge, 5 May 1896.

272 For a definitive life, see C.W. Smith and M.N. Wise, *Energy and Empire: A Biographical Study of Lord Kelvin* (Cambridge: Cambridge University Press, 1989).

273 Glasgow University Archives, ff. 4140a, 4140b, Kelvin to Jack, 20 May 1896; University of Sydney Archives, G3/46/17, Letters, 1895-96, f. 757, Barff to A.L., 5 May 1896.

274 These included Sir George Stokes and J.J. Thomson, Carey Foster and Sir Henry Roscoe. America sent the noted astronomer Simon Newcomb of Johns Hopkins; France, Emile Picard, of Paris; and Germany, Waldemir

Voight, of Göttingen.

275   *Glasgow Herald*, 17 June 1896; Glasgow University Archives, f. 4140b.

276   Senate Minutes, 1896, Liversidge to Senate, 3 August 1896.

277   Glasgow University Archives, f. 4222, A.L. to Rev. Dr. Stewart, 20 June 1896.

278   *Nature*, 54 (4 June 1896), 105. On his return, Liversidge reported the event to the University. (Senate Minutes, 3 August 1896.)

279   Royal Society Archives, New Letterbook, 13, f. 133. R. Harrison to A.L., 6 July 1896.

280   It was also agreed to hold a second congress in London in 1898. An interim meeting of the Preparatory Committee was later held in 1899, and a third international conference in 1900. The first formal meeting of the International Council came in 1900. Publication began in 1902. Cf. First Annual Issue, *International Catalogue of Scientific Literature* (London: Harrison and Sons, 1902), v-ix.

281   See Liversidge, 'Presidential Address', (6 January 1898) *Rep. of the AAAS*, 7 (Sydney, 1898), 1-56. For the technical difficulties, see W. Boyd Rayward, 'The Search for Subject Access to the Catalogue of Scientific Papers, 1800-1900', in Rayward (ed.), *The Variety of Librarianship: Essays in Honour of John Wallace Metcalf* (Canberra: Library Association of Australia, 1976), 146-170.

282   'England in the Age of Science', *Saturday Review*, 82 (1 August 1896),102-103.

283   See MacLeod, 'Imperial Reflections', in MacLeod and Rehbock (eds.), *op. cit.,* 159-194.

284   British Museum (Nat. Hist.) Min. Dept., Liversidge to Fletcher, 9 September 1897.

285   Royal Society Archives, New Letterbook, 14, f. 303. Foster to Liversidge, 6 March 1897.

286   With Liversidge's agreement, David took Poole and Woolnough from the School of Mines with him, and gave them both a unique research experience.

287   *SMH*, 3 June 1897; Liversidge Papers Box 18: Funafuti files.

288   Australian Museum Letterbook, 1896, f. 533, Etheridge to A.L., 8 June 1896.

289   British Museum (Nat. Hist.) Min. Dept., A.L. to Fletcher, 30 October 1896. The specimens sent to London in 1897 by Liversidge and the New South Wales Department of Mines included noumeaite from New Caledonia; glaucophane, topaz and porphyry, from Tasmania; several minerals from Broken Hill; and a fragment of the Barratta meteorite. *NSW Department of Mines, Specimens*, 9 September 1897. These followed fragments of the Thunda meteorite in 1891. See *History of the Collections contained in the Natural History Departments of the British Museum* (London: British Museum, 1904), vol. I, 43l.

290   British Museum (Nat. Hist.) Archives, f. 1768, A.L. to Sir William Flower, 14 November 1896. Australian Museum Archives, A 23.79.1, A.L. to Curator, 21 April 1897.

291   Liversidge Papers, Box 19, f. 1. A.L. to Sir Saul Samuel, 8 January 1897.

292   See Lord Rayleigh, *Lord Balfour in his Relation to Science* (Cambridge: Cambridge University Press, 1930).

293   Liversidge Papers, Box 14, 'Plan of Tables, Anniversary Dinner', Royal Society, 30 November 1896.

294   Cambridge Philosophical Society*, Council Minutes, 1897. Liversidge was elected in February 1897 and remained a member until his death. For this information I am grateful to the Society's Executive Secretary, Ms. Judith Winton-Thomas

# CHAPTER 11

1   Hallam, Lord Tennyson, 'First Sight of Australia, 1899', *United Empire*, 1 (August 1910), 531.

2   'Return of Professor Liversidge', *SMH*, 1 March 1897.

3   See the early chapters of Michael Roe, *Nine Australian Progressives: Vitalism in Bourgeois Social Thought, 1890-1960* (St Lucia: University of Queensland Press, 1984).

4   T.W. Edgeworth David, 'Presidential Address', *JPRSNSW*, 30 (1896), 29.

5   Senate Minutes, 7 December 1896.

6   A.G. Austin (ed.), *The Webbs' Australian Diary* (1898) (Melbourne: Sir Isaac Pitman and Sons, Ltd, 1965), quoted in Kaye Harman (ed.), *Australia Brought to Book* (Sydney: Boobook Publications, 1985), 148.

7   Jeanette Beaumont and W. Vere Hole, *Letters from Louisa: A Woman's View of the 1890s, based on the Letters of Louisa Macdonald, first Principal of the Women's College, University of Sydney* (Sydney: Allen & Unwin, 1996), 15.

8   Senate Minutes, 8 April 1896.

9   *Ibid.*, 1 June 1894.

10   *Ibid.*, 15 June 1896. This was followed in 1904 by a second gift of £50,000, the income from which was to provide scholarships, open to apprentices at engineering works or to those who qualified at the Technical College. The second gift was contingent upon the State Government agreeing to give a further £25,000 towards a building. See 'The Engineering School', *SMH*, 13 February 1904. Russell was knighted in the Coronation Day honours in 1904. See 'Modernising the University', *SMH*, 2 July 1904. See also Arthur Corbett and Ann Pugh, 'Sir Peter Nicol Russell', *Australian Dictionary of Biography*, vol. 6, 76-77.

11   Liversidge Papers, Box 19, f. 1, Threlfall to Registrar, 20 July 1897.

12   The redoubtable Judge Alfred Backhouse was re-elected Vice-Chancellor. Senate Minutes, 12 October, 2 November 1896. See Ann M. Mitchell, 'Sir Henry Normand MacLaurin (1835-1914)', *Australian Dictionary of Biography*, vol. 10, 327-329.

13   Faculty of Science Minutes, 20 May 1897.

14   *Ibid.*, 7 December 1896; 7 June 1897.

15   Born in Cork, a graduate of the Royal University of Ireland, Pollock came to Sydney in 1885 as Russell's astronomical assistant at the Observatory, but soon left to read physics with Threlfall, taking a BSc with medal in 1889. He was appointed demonstrator in physics in 1890 and acting professor when Threlfall resigned in 1898. The following year, he was given the chair. Like David, he was a local candidate, 'at least as strong' as any recommended by the London committee chaired by the Agent-General. See 'Professor J.A. Pollock', *Daily Telegraph,* 13 April 1899. J.B.T. McCaughan, 'James Arthur Pollock (1865-1922)', *Australian Dictionary of Biography*, vol. 11, 253-254.

16   Senate Minutes, Special Meeting 22 November 1897. The Arts Curriculum Committee also made other recommendations – for example, concerning the introduction of tutorial classes, and the reduction of fees – which were later implemented.

17   See Faculty of Science Minutes, 25 October, 1 November 1898; 3 August 1899.

18   Faculty of Science Minutes, 30 April 1900.

19   *Ibid.*, 12 December 1900.

20   *Ibid.*, 30 June 1896.

21   *Ibid.*, 30 June 1896.

22   Senate Minutes, 7 June 1897; 9 August 1897.

23   *Ibid.*, 5 July 1897.

24   *Ibid.*, 6 September 1897.

25   *Ibid.*, 10 November 1898, 5 December 1898, 3 December 1900.

26   *Ibid.*, 6 December 1897.

27   *Ibid.*, 7 February 1898.

28   *Ibid.*, 5 September 1898. Mr. C. Walker was appointed.

29   University of Sydney Archives, G.3/46/22, f. 175, Barff to A.L., 28 July 1900.

30   Senate Minutes, 6 December 1897; Ursula Bygott and K.J. Cable, *Pioneer Women Graduates of the University of Sydney, 1881-1921* (Sydney: University of Sydney, 1985), 47; David Branagan and Graham Holland (eds.), *Ever Reaping Something New: A Science Centenary* (University of Sydney: Science Centenary Committee, 1985), 226; Claire Hooker, *Irresistible Forces: Australian Women in Science* (Melbourne: Melbourne University Press, 2004).

31   Senate Minutes, 14 November 1898.

32   *Ibid.*, 10 April 1899. In 1901, the faithful Hufton's salary was raised from £162 to £175 per annum, *ibid.*, 2 December 1901.

33   *Ibid.*, 6 December 1897, 7 February 1898.

34   A.L., 'Native Silver Accompanying Matte and Artificial Galena' (7 December 1898), *JPRSNSW*, 32 (1898), xli-xlii.

35   British Museum (Natural History), Mineralogy Department, William Poole to A.L., 2 November 1900; George Sweet to A.L., 15 November 1900.

36   Liversidge Papers, Box 15, Public Works Committee; Box 19, f. 1, Charles Lyne to A.L., 22 March 1899. See *SMH*, 30 March 1899, and Report of the Parliamentary Standing Committee on Public Works, Minutes of Evidence, 'Report relating to the proposed Public Offices, Phillip, Bridge and Young Streets, Sydney', *VPLANSW* (3 August 1899) Third Session, 1899, vol. 4, 937-954.

37   Each was paid only £60 from the professor's salary. *Senate Minutes*, 1 March 1897, 8 May 1897. Earlier,

Dun had assisted David in his study of the Hunter River coalfields. When Etheridge went to the Australian Museum, he became the Survey's palaeontologist and librarian. From 1902 to 1934 he continued as a part-time lecturer in palaeontology at the University. See D.F. Branagan and T.G. Vallance, 'William Sutherland Dun (1868-1834)', *Australian Dictionary of Biography*, vol. 8, 364.

38  Senate Minutes, 6 September 1897.

39  Cf. British Museum (Natural History), Mineralogy Department, W. Poole to A.L., 10 February 1901.

40  University of Sydney Archives, G.3/46/23. Barff to A.L., 27 April 1901.

41  Senate Minutes, 14 October 1901.

42  Cf. A.L., 'The Blue Pigment in Coral (Heliopora coerulea) and other Animal Organisms' (7 December 1898), *JPRSNSW*, 32 (1898), 256-268; *Chemical News*, 80 (21 July 1899), 29-31; *idem.*, 80 (28 July 1899) 41-43; abstracted in *J. Chem. Soc.*, 78 (1900), A, (1), 70.

43  A.L., 'On the Crystalline Structure of some Silver and Copper Nuggets' (3 August 1898), *JPRSNSW*, 34 (1900), 255-258; abstracted in *J. Chem. Soc.*, 80 (1901), A, (2), 662; 'On the Crystalline Structure of Some Gold Nuggets from Victoria, New Zealand, and Klondyke' (1 November 1899), *JPRSNSW*, 34 (1900), 259-262; *J. Chem. Soc.*, 80 (1901), A, (2), 662; 'Boogaldi Meteorite' (5 December 1900), *JPRSNSW*, 34 (1900), xlvi-xlvii; reported in *Nature*, 63 (11 April 1901), 579.

44  On 28 April 1903, the Governor-General and his wife visited the laboratory, and Lady Tennyson commented on 'a most beautiful building with every sort of room fitted up for technical teaching. They showed us liquid air - a lovely sort of grey-blue colour, which was owing to most of the hydrogen [i.e., nitrogen] being consumed from it & only oxygen remaining. It was quite a liquid but when thrown on anything, did not wet it and was icy cold to the touch, the professor throwing some on his hand to show us how it dissolved in vapour and then asked me if I would like to feel it, and & I said yes, he having assured me it would not hurt, only feel icy cold - & it was icy cold enough to frostbite my hand & I have the marks still, which was [*sic*] horribly painful for some hours & felt exactly like a bad burn. He was very much surprised at its treating me so & said he had never known it affect anybody like that before, and it must have been from my skin being so tender.' Alexandra Hasluck (ed.), *Audrey Tennyson's Vice-Regal Days. The Australian Letters of Audrey Lady Tennyson to her Mother, Zacyntha Boyle, 1899-1903* (Canberra: National Library of Australia, 1978), 279.

45  'Liquefied Air', *Daily Telegraph*, 17 August 1899.

46  Royal Society of London, Archives, New Letterbook 17. f.109. Assistant Secretary, Royal Society to A.L., 28 July 1898.

47  Mitchell Library, Baldwin Spencer Papers, MSS 29/8, A.L. to Spencer, 4 June 1897.

48  Australian Museum Letterbook, 1898, f. 837, Sinclair to A.L., 3 November 1898.

49  A.L., 'International Catalogue of Scientific Literature' (3 May 1905), *JPRSNSW*, 39 (1905), xi-xv.

50  Senate Minutes, 6 December 1897, 7 March 1898, 2 May 1898, 10 October 1898. R. Home, 'First Physicist of Australia: Richard Threlfall at the University of Sydney, 1886-1898', *Historical Records of Australian Science*, 6 (3), (1986), 333-357.

51  The other being Thomas Butler, Lecturer in Classics from 1880 and Professor of Latin from 1890. Senate Minutes, 10 April 1899.

52  In 1900, Harker served briefly as Acting Demonstrator in the absence of James Schofield, at £3 per week. Taking his D.Sc. in London, Harker returned to Sydney in 1904, becoming a founder member of the Sydney Section of the Society of Chemical Industry, and working for CSR until 1911. In 1914, he became a Lecturer and Demonstrator in Organic Chemistry at the University, where he remained until 1927, when he was employed by the Cancer Research Fund. Senate Minutes, 4 June 1900; *Report of Senate, Sydney University Calendar*, 1927.

53  Royal Commission on the Exhibition of 1851 Archives, f. 174, A.L. to Committee, 1899. Harker's early work was published, with Liversidge's encouragement, in the *JPRSNSW* and in the annual reports of the AAAS.

54  Quoted in David Branagan and Graham Holland (eds.), *Ever Reaping Something New* (Sydney: University of Sydney, Faculty of Science,1985), 74.

55  Mitchell Library, Baldwin Spencer Papers, MSS 29/8, A.L. to Spencer, 4 June 1897.

56  *SMH*, l October 1897.

57  *SMH*, 3 January 1898.

58  *Ibid.*

59  *Ibid.*

60    *SMH*, 6 January 1898.

61    W.M. Hamlet (ed.), *Handbook to the Sydney Congress* (Sydney: AAAS, 1898), Preface.

62    *Daily Telegraph*, 7 January 1989.

63    Liversidge Papers, Box 13, f. 10.

64    'Only five days', he lamented to Baldwin Spencer. See Mitchell Library, Baldwin Spencer Papers, MSS 29/8, A.L. to Spencer, 7 April 1898.

65    *Ibid.*

66    Mitchell Library, Baldwin Spencer Papers, MSS 29/8, A.L. to Spencer, 13 May 1898.

67    Mitchell Library, Baldwin Spencer Papers., MSS 29/5-11, David to Spencer, 1 August 1898.

68    Mitchell Library, Baldwin Spencer Papers., MS 29/8, A.L. to Spencer, 9 September 1898, 9 August 1899.

69    *Ibid.*, A.L. to Spencer, 31 March 1899.

70    *Ibid.* A.L. to Spencer, 26 October 1899.

71    *Ibid.*, A.L. to Spencer, 31 March 1899.

72    *Ibid.*, A.L. to Spencer, 9 August 1899.

73    *Ibid.*, A.L. to Spencer, 26 October 1899.

74    Committees of Investigation, *Rep. of the AAAS*, 1 (Sydney, 1888), xxxii-xxxiv.

75    H.C.L. Anderson, 'Presidential Address, Section G', *Rep. of the AAAS*, 5 (Adelaide, 1893), 164. In February 1890, Anderson, a Sydney graduate, who received the Belmore Medal for agricultural chemistry, was appointed by Parkes to head a department of agriculture within the NSW Department of Mines. In 1891, he established the Hawkesbury Agricultural College on a 'sound, scientific and practical foundation', and in 1892, set up an experimental farm at Wagga Wagga. He was a member of the University's Senate from 1895-1919, and helped to establish Faculties in Agriculture and Veterinary Science in 1910. For his political career, highlighted by his work for the Public Library of NSW and as Director of the newly established Department of Agriculture from January 1908, see C.J. King, 'Henry Charles Lennox Anderson (1853-1924)', *Australian Dictionary of Biography*, vol. 7, 55-56. See also the official history by Dr. Peter Milray, *In the Service of Agriculture* (Sydney: NSW Department of Agriculture, 1990).

76    *SMH*, 6 January 1898.

77    *SMH*, 9 January 1904.

78    Ian Inkster, 'Scientific Enterprise and the Colonial "Model": Observations on Australian Experience in a Historical Context', *Social Studies of Science*, 15 (1985), 694.

79    A.L., 'Presidential Address', *Rep. of the AAAS*, 7 (Sydney, 1898), 5.

80    *Melbourne Argus*, 7 January 1890.

81    Mitchell Library, Baldwin Spencer Papers, MSS 29/8, A.L. to Spencer, 20 June 1900.

82    In 1901, the membership of the Royal Society of NSW fell from a peak of 457 to 368. By 1914, those of Queensland and South Australia had only 100 members; Tasmania, 200; and Victoria, 94. See 'Miscellaneous Notes on Australia, People, Activities', in George Knibbs (ed.), *Federal Handbook for the 84th British Association Meeting* (Canberra: Commonwealth Government Printer, 1914), 596-597.

83    A.L., 'Boogaldi Meteorite' (5 December 1900), *JPRSNSW*, 34 (1900), xlvi-xlvii; reported in *Nature*, 63 (11 April 1901), 579.

84    Royal Society, *Record of the Royal Society* (Edinburgh: Morrison Gibb Ltd., 1940, 4th ed.), 497-498.

85    Mitchell Library, Baldwin Spencer Papers, MSS 29/8. A.L. to Spencer, 26 April, 27 August 1894.

86    *Ibid.*

87    University College London, Ramsay Papers, Ramsay to W.M. Hicks (Professor of Physics, Sheffield, on the Council of the Royal Society), 17 March 1901. Masson was elected FRS in 1903.

88    Rod Home, 'The Royal Society and the Empire: the Colonial and Commonwealth Fellowship, Part 2: After 1947', *Notes and Records of the Royal Society of London*, 5 (1), (2003), 47-84.

89    Mitchell Library, Baldwin Spencer Papers, MSS 29/8, A.L. to Spencer, 9 November 1899.

90    R.M. Crawford, *'A Bit of a Rebel'. The Life and Work of George Arnold Wood* (Sydney: Sydney University Press, 1975), chapter XIII.

91    Senate Minutes, 14 October 1901.

92    See Roy MacLeod, 'Science, Progressivism and "Efficient Imperialism"', in Roy MacLeod and Richard Jarrell (eds.), *Dominions Apart: Reflections on the Culture of Science and Technology in Canada and Australia, 1850-1945, Scientia Canadensis*, 17 (1,2), (1995), 8-25.

93  Senate Minutes, 13 August 1900. Gurney brought back more apparatus with him when he visited Europe on leave in 1884. See I.S. Turner, 'Theodore Thomas Gurney (1849-1918)', *Australian Dictionary of Biography*, vol. 4, 310.

94  Senate Minutes, 3 December 1900.

95  Senate Minutes, 2 December 1901.

96  Senate Minutes, 3 February 1902, 5 May 1902.

97  *Report of the Allied Colonial Universities Conference,* London, 9 July 1903, 93.

98  See Sara Maroske, '"The Whole Great Continent as a Present": Nineteenth-century Australian Women Workers in Science', and Farley Kelly, 'Learning and Teaching Science: Women Making Careers, 1890-1920', in Farley Kelley (ed.), *On the Edge of Discovery* (Melbourne: Text Publishing, 1993) 13-35, 35-76.

99  See Gretchen Poiner and Roberta Burke, *No Primrose Path: Women as Staff at the University of Sydney* (Sydney: University of Sydney, 1988), 6.

100  'Facts and Features of Sydney University', *SMH*, l June 190l.

101  Senate Minutes, 4 November 1901.

102  *Ibid.*, 2 May 1904.

103  'Our University', *SMH*, 28 April 1905.

104  British Museum (Natural History), Mineralogy Department, W. Poole to A.L., 4 March 1901.

105  A.L., 'The Royal Society of New South Wales', *SMH*, 8 May 1901.

106  A.L., 'Presidential Address' (1 May 1901), *JPRSNSW*, 35 (1901), 1-29; reported in *Nature*, 64 (18 July 1901), 296.

107  Milton Lewis, 'The Royal Society of Australia: An Attempt to Establish a National Academy of Science', *Records of the Australian Academy of Science*, 4 (1), (1978), 51-62. The metric system was eventually introduced into Australia by the Metric Conversion Act of 1970.

108  See Francis Anderson, 'Our State School Education', *Daily Telegraph*, 27 December 1901, 1 January 1902. For his career, see W.M. O'Neil, 'Sir Francis Anderson (1858-194l)', *Australian Dictionary of Biography*, vol. 7, 53-55.

109  *Report of the Commission on Primary, Secondary, Technical and Other Branches of Education, VPLANSW,* 2nd Session, 1904; 'The Education Commission,' *Daily Telegraph*, 10 March 1902; 'The Education Commissioners', *SMH*, 11 April 1902.

110  *NSW Commission on Primary Schools and Technical Education.* See Susan Bambrick, 'Sir George Handley Knibbs (1858-1929)', *Australian Dictionary of Biography*, vol. 9, 620-621.

111  George Knibbs, quoted by the *SMH*, 11 April 1902.

112  'Bread and Butter Studies', *SMH*, 3 May 1902.

113  Professorial Board Minutes, 19 August 1902. The compulsory Latin requirement remained until 1910, when it was replaced by either French or German. Professorial Board Minutes, 21 March 1906, 30 June 1910, 28 May 1913.

114  Senate Minutes, 7 May 1901, 10 June 1901, 2 September 1901. Professorial Board Minutes, 2 July 1901.

115  Senate Minutes, 29 October 1903. Pollock was prepared to compromise with the requirement that Arts students do science, by obliging BSc students to graduate first in Arts. The Faculty of Science disapproved before the proposal reached Senate.

116  In 1905, the Faculty altered its regulations so as to require at matriculation three subjects, recommended by the Arts Curriculum Committee, of which one was English; another a language (either Latin, Greek, French or German), and a third, mathematics. A fourth could be taken from a list of options, which included science subjects.

117  Faculty of Science Minutes, 10 November 1903, 17 November 1903.

118  Clifford Turney, Ursula Bygott and Peter Chippindale, *Australia's First; A History of the University of Sydney* (Sydney: University of Sydney in association with Hale and Iremonger, 1991), vol. 1, 280.

119  Senate Minutes, 4 March 1901; Faculty of Science Minutes, 10 October 1901.

120  For further details of his career, see Appendix V and 'James A. Schofield', *Who's Who in Australia, 1922* (Sydney: Angus and Robertson, 1922), 243.

121  H.E. Barff, *A Short Historical Account of the University of Sydney* (Sydney: Angus and Robertson, 1902).

122  Senate Minutes, 3 February 1902. Herbert S. Jevons was appointed. (Senate Minutes, 5 May 1902).

123  Senate Minutes, 3 February 1902.

124 T.W. Edgeworth David, 'University Science Teaching', *Record of the Jubilee Celebrations of the University of Sydney* (Sydney: Brooks and Co., 1903), 117.

125 'Sydney University: Jubilee Celebrations', *SMH*, 4 October 1902. In March, David was given overseas leave for a year, to report on 'systems of technical education in Europe and America'. Senate Minutes, 3 March 1902.

126 Senate Minutes, 11 August 1902.

127 British Museum (Natural History), Department of Minerals, Letterbook 6 (1897-99), 2 (1888-91), A.L. to Fletcher, 9 September 1897.

128 Senate Minutes, 30 December 1902.

129 In cutting a specimen of the Mt. Sterling meteorite, obtained by the Australian Museum, he failed by his own admission to keep accurate weight measurements. See Australian Museum Letterbook, 1903, A.L. to Etheridge, 17 January 1903.

130 A.L., 'Meteoric Dusts, New South Wales' (3 September 1902), *JPRSNSW*, 36 (1902), 241-285; 'Gold in Meteorites' (3 September 1902), *JPRSNSW*, 36 (1902), xxiv; 'On the presence of Platinum and Iridium Metals in Meteorites' (3 September 1902), *JPRSNSW*, 36 (1902), xxix-xl; 'The Boogaldi, Barratta Nos. 2 and 3; Gilgoin Nos. 1 and 2; and Eli Elwah or Hay Meteorites, New South Wales' (5 November 1902), *JPRSNSW*, 36 (1902), 341-359; 'On the Narraburra Meteorite' (2 September 1903), *JPRSNSW*, 37 (1903), 234-242.

131 Katrina Dean, 'Inscribing Settler Sciences: Earnest Rutherford, Thomas Laby and the Making of Careers in Science', *History of Science,* XLI (2003), 217-240.

132 Royal Commission on the Exhibition of 1851 Archives, f. 242, Laby application, 22 February 1905.

133 Melbourne University Archives, Laby Papers, A.L. to Laby, 7 August 1906; Royal Commission on the Exhibition of 1851 Archives, f. 242, A.L. to Laby, 20 May 1904.

134 Laby took his Liversidge-inspired problem to Cambridge, where it formed the basis of his work on the effects of ionizing radiation in gases. He remained in England until 1909, when he was offered the chair of physics at Victoria College, Wellington. In 1915, he took the chair of physics at Melbourne, where he remained for the rest of his life. He became a key advisor to the Commonwealth Government and a leader in the work of ANZAAS. See Roy MacLeod (ed.), *The Commonwealth of Science* (Melbourne: Oxford University Press, 1988), chapter 2. According to Sir Harrie Massey, later Director of the DSIR in England, who knew him well, his work remained, like that of Liversidge, 'inadequately recognised'. *Nature*, 158 (3 August 1946), 157. See also H.S.W. Massey, 'T.H. Laby', The Laby Memorial Lecture, *Australian Physicist,* 17 (December 1980), 181-187; D.K. Picken, 'Thomas Howell Laby (1880-1946)', *Obituary Notices of Fellows of the Royal Society*, 5 (May 1948), 733-755; Cecily Close, 'Thomas Howell Laby (1880-1946)', *Australian Dictionary of Biography,* vol. 9, 640-641.

135 In 1903, Senate again denied Haswell's request for a separate chair or lectureship in botany. Senate Minutes, 1 June 1903. Branagan and Holland, *op.cit.*

136 *Senate Minutes*, 5 March 1906.

137 'The New Under-Secretary for Mines', *Daily Telegraph*, 3 September 1902; *SMH*, 13 October 1902.

138 For George Knibbs, see footnote 110.

139 A.L.,'On the Torbanite or 'Kerosene Shale' of New South Wales', *American Journal of Science*, XXII, No. 127 (July 1881), 32.

140 A.L., 'The Alkaloid from Piturie' (3 November 1880), *JPRSNSW*, 14 (1880), 123-132; see also Edward Rennie, 'Presidential Address', Section A, *Report of the AAAS*, 2, (Melbourne, 1890), 58.

141 Gerald Geison, *Michael Foster and the Cambridge School of Physiology: The Scientific Enterprise in Late Victorian Society* (Princeton: Princeton University Press, 1978), 286.

142 Petrie subsequently became a Macleay Fellow of the Linnean Society between 1907-25, and later a Bosch Fellow in Cancer Research at Sydney University. See *Who's Who in Australia,* (1922), 2178; *SMH*, 31 March 1927; A.B. Walkom, *Linnean Society of New South Wales: Historical Notes of its First Fifty Years* (Sydney: Australasian Medical Publishing Co., 1925).

143 Engineering had burgeoned to six academic positions by 1904. Senate Minutes, 2 May 1904.

144 Senate Minutes, 2 March 1903; University of Sydney Archives, G.3/46/26, f. 235, Registrar to A.L., 27 February 1903.

145 Senate Minutes, 17 August 1903.

146 'Travels of a Scientist', *Daily Telegraph*, 4 March 1905.

147    F.B. Guthrie was again made Acting Professor, and J.A. Schofield, his Demonstrator, was made Assistant Professor, each at £250; Laby was made Acting Demonstrator at £125 and a Junior Demonstrator was paid £100. Senate Minutes, 17 August 1903.

148    'The Government Astronomer', *Daily Telegraph*, 14 December 1903.

149    Senate Minutes, 7 December 1903.

150    See 'The Education Question', *Daily Telegraph*, 15 January 1904; 'Education Conference', *Daily Telegraph*, 14 April 1904.

151    British Museum (Natural History), Mineralogy Department, A.L. to Fletcher, 10 May 1904, 2 June 1905.

152    In April, October and December, 1904: the first, as guest of David Gill, to hear a talk on Antarctic discovery; at the second, as guest of J. Teall; and at the third, with T. Lauder Brunton, the physiologist, and his friend from Cambridge.

153    A century later, Bonham's auction house in London mounted a sale of meteorites, which it hailed as marking a 'coming of age of meteorites as collectable objects'. See 'Meteorites Worth their Weight in Gold', *SMH*, 22 October 2007.

154    Henry Roscoe, *The Life and Experiences of Sir Henry Enfield Roscoe* (London: Macmillan, 1906), 361.

155    *Ibid.*, 194.

156    Sidney Webb, 'Lord Rosebery's Escape from Houndsditch', *Nineteenth Century and After*, CCXCV (September 1901), 374, quoted in Beatrice Webb, *Our Partnership* (ed. by Barbara Drake and Margaret I. Cole) (London: Longmans, Green and Co., 1948), 223.

157    See Bernard Semmel, *Imperialism and Social Reform* (London: George Allen and Unwin, 1960), 63.

158    Cf. D.S.L. Cardwell, *The Organisation of Science in England* (London: Heinemann, 1957, 2nd edn. 1972).

159    'Science and the scientific method can be applied, not to the discovery of a right end, but to a discovery of a right way of getting to any particular ends'. Beatrice Webb, *op.cit.*, 211.

160    Cf. Roy MacLeod, 'The Social Framework of *Nature* in its First Fifty Years', *Nature*, 224 (1 November 1969), 441-461.

161    Semmel, *op.cit.*, 72-82.

162    Sir William Ramsay, *Oration on the Functions of a University* (University College London, June 1901), 13.

163    *Official Report of the Allied Colonial Universities Conference*, 1903, 72.

164    *Ibid.*, 93, 101-102.

165    A.J. Mahan, *The Influence of Sea Power upon History, 1660-1783* (Cambridge, Mass.: John Wilson and Son, 1890; London: Sampson Low and Co, 1890). Twenty-five American editions had appeared by 1925. It was last reissued in New York (American Century) in 1957, and in London (Methuen), in 1965.

166    On the history of the University Grants Committee, see R.O. Berdahl, *British Universities and the State* (Berkeley: University of California Press, 1959), and Michael Shattock and Robert O. Berdahl, 'The British University Grants Committee 1919–83: Changing Relationships with Government and the Universities', *Higher Education*, 13 (5), (1984), 471-499.

167    See H.G. Wells, *A Modern Utopia* (London: Chapman and Hall, 1905).

168    'The Supreme Importance of Science to the Industries of the Country, which can be secured only through making Science an Essential Part of All Education', reprinted in Sir William Huggins, *The Royal Society; or, Science in the State and in the Schools* (London: Methuen, 1906).

169    W.H.G. Armytage says that the British Association was not anxious 'to embark on the task of applying scientific methods to public affairs', *Sir Richard Gregory* (London: Macmillan, 1957), 67. But see Roy MacLeod, 'Introduction', in MacLeod and Collins (eds.), *The Parliament of Science* (London: Science Reviews, 1981).

170    Presidential Address to the British Association, Southport, 3 September 1903, reprinted in J.N. Lockyer, *Education and National Progress* (London: Macmillan, 1906), 182-184.

171    See H. Dingle in T.M. Lockyer and W.L. Lockyer (eds.), *The Life and Work of Sir Norman Lockyer* (London: Macmillan, 1928), 186.

172    Roy MacLeod, 'The Support of Victorian Science: The Endowment of the Research Movement in Great Britain, 1868-1900,' *Minerva*, IV (2), (1971), 197-230, and 'The Resources of Science in Victorian England', in Peter Mathias (ed.), *Science and Society* (Cambridge: Cambridge University Press, 1972), 111-166.

173    Roy MacLeod, 'Science for Imperial Efficiency: Reflections on the British Science Guild, 1905-1936,' *Public Understanding of Science*, 3 (1), (1994), 155-194.

174    Branches were established in NSW and South Australia.

175 See 'Sir William Peterson (1856-1921)', in W. Stewart Wallace (ed.), *Macmillan Dictionary of Canadian Biography* (Toronto: Macmillan, 1963, 3rd ed.), 593.

176 'Universities and the State', *Nature*, 70 (21 July 1904), 271-273.

177 *Ibid.*, 271.

178 Roy MacLeod, 'The "Practical Man": Myth and Metaphor in Anglo-Australian Science', *Australian Cultural History*, 8 (1989), 24-49.

179 Sir Michael Foster, 'The State and Scientific Research', *The Nineteenth Century*, LX (May 1904), 740-751.

180 James Barrett, *The Twin Ideals: An Educated Commonwealth* (London: H.K Lewis, 1918); S. Murray-Smith, 'Sir James Barrett (1862-1945)', *Australian Dictionary of Biography*, vol. 7, 186-189.

181 *Second Annual Report of the British Science Guild* (afterwards, *ARBSG*), 15 January 1908, 12.

182 Brigitte Schroeder-Gudehus, 'Division of Labour and the Common Good: The International Association of Academies, 1899-1914', in Carl Gustaf Bernhard *et al.*, *Science, Technology and Society in the Time of Alfred Nobel* (Oxford: Pergamon Press, 1982), 8-9.

183 Michael Foster, 'The International Medical Congress', *Nature*, 49 (12 April 1894), 563.

184 Comte de Franqueville, *Discours prononcé à l'occasion de la première Assemblée générale de l'Association internationale des Académies* (Paris: Imprimerie National, 1901), 8.

185 *The Times*, 24, 25 May 1904.

186 Cf. Liversidge Papers, Box 14, f. 4. R. Harrison, Assistant Secretary, Royal Society to A.L., 17 May 1904, and subsequent reports.

187 'International Association of Academies', *Nature*, 70 (2 June 1904), 106; Schroeder-Gudehus, *op.cit.*, 3-20.

188 *Report of the Council of the Royal Society*, 1904, 'Report of the Proceedings of the Second General Assembly of the International Association of Academies' (London: Royal Society, 1904); 'International Catalogue of Scientific Literature', *Nature*, 70 (19 May 1904), 54.

189 Liversidge Papers, Box 14, f. 4, 'IAA Report', Liversidge's marginal notes.

190 For the Royal Medals, see Roy MacLeod, 'Of Medals and Men: A Reward System in Victorian Science, 1826-1914', *Notes and Records of the Royal Society*, 26 (1971), 81-105.

191 Royal Society Archives, Council Minutes, 3 (7 July 1904), 93.

192 'Travels of a Scientist', *Daily Telegraph*, 4 March 1905.

193 Sir William Huggins, *The Royal Society; or, Science in the State and in the Schools* (London: Methuen, 1906).

194 *Daily Telegraph*, 4 March 1905.

195 A.L., *Tables for Qualitative Chemical Analysis* (London: Macmillan & Co., 1904).

196 University of Sydney Archives, G.9 School of Chemistry, Review file.

197 J.B.C. 'Chemical Analysis for Beginners', *Nature*, 71 (3 November 1904), 4-5.

198 *1st AR BSG* (1907), 2.

199 *1st AR BSG* (1907), 5.

200 *Ibid*. Liversidge's name appears on the first published list of the General Committee in 1906.

# CHAPTER 12

1 *Daily Telegraph*, 4 March 1905.

2 *Ibid.*

3 *SMH*, 28 April 1905.

4 *Record of the Jubilee Celebrations of the University of Sydney*, 24.

5 James W. Barrett, *The Future of Melbourne University: A Lecture* (Melbourne: Austral Publishing Co., 1902), 11.

6 S. Murray-Smith, 'Sir James Barrett (1862-1945)', *Australian Dictionary of Biography*y, 7, 186-189.

7 'Our University', *SMH*, 28 April 1905.

8 Francis Anderson, 'The University and National Education', *Hermes*, Jubilee Number, (1902), 37-42. See also Francis Anderson, 'Tendencies of Modern Education, with some Proposals of Reform', *Address to the Public School Teachers' Association of NSW*, April 1909 (Sydney: Angus and Robertson, 1909); and 'Organisation of National Education', *SMH*, 2 and 4 October 1909.

9 'Our University', *SMH*, 28 April 1905.

10 *Ibid.*

11  Senate Minutes, 5 December 1904.

12  Senate Minutes, 5 December 1904, 4 September 1905. Professorial Board Minutes, 23 March 1905, 30 March 1905; Senate Minutes, 4 November 1905.

13  Senate Minutes, 23 May 1904, 4 July 1904. According to Professor MacCallum, it was hoped to remedy a situation in which admission to Arts had become too easy, and that to Science, too onerous. 'The University Curriculum', *Daily Telegraph*, 26 June 1905. By the new rules, which came into force in March 1907, all Arts matriculants were required to reach a junior pass in all subjects, and the senior standard in two. Students entering science or the professional schools would do subjects as before, including mathematics, Latin and another language, but in only two of these subjects was a senior pass required. As no science was required at matriculation, the extent to which this made science more attractive remained questionable. See 'Our University: VII: In Transition', *SMH*, 9 June 1905.

14  Senate Minutes, 4 July 1904.

15  *Ibid.*, 11 April 1904.

16  Faculty of Science Minutes, 28 September 1904.

17  Faculty of Science Minutes, 2 November 1904.

18  Senate Minutes, 2 May 1904.

19  Patricia Morison, *J. T. Wilson and the Fraternity of Duckmaloi* (Amsterdam: Rodophi, 1997), 189; the largest may have been Edinburgh, which sent Sydney both Anderson Stuart and Wilson.

20  J. Atherton Young, 'Sir Thomas Peter Anderson Stuart (1856-1920)', *Australian Dictionary of Biography*, vol. 12, 131.

21  'Our University', *SMH*, 28 April 1905.

22  'Modernising the University', *SMH*, 2 July 1904.

23  'Our University', *SMH*, 5 May 1905.

24  The only choice being that of a second language, Greek, French or German.

25  'Education Conference', *Daily Telegraph*, 14 April 1904.

26  Sir William Ramsay, 'The Teaching of Chemistry', *The Educational Times*, 44 (1891), 273.

27  'The University and Science Teaching', *Daily Telegraph*, 15 April 1904.

28  T. Jeffery Parker and William A. Haswell, *A Textbook of Zoology* (London: Macmillan, 1897), 2 vols. The book went through seven editions by 1975.

29  'Our University: IV: The Science Schools', *SMH,* 19 May 1905.

30  A.L., *Tables for Qualitative Chemical Analysis* (London: Macmillan, 1904).

31  'The University and Science Teaching', *Daily Telegraph*, 15 April 1904.

32  See, for example, Anderson Stuart's comments about his 'unruly boys' in William Epps, *Anderson Stuart, MD Physiologist, Teacher, Builder, Organiser, Citizen* (Sydney: Angus & Robertson, 1922), 60.

33  See S. Craddock, 'The Reminiscences of S. Craddock', *Hermes*, 24 (2), (August 1918), 177.

34  Herbert Moran, *Beyond the Hill Lies China: Scenes from a Medical Life in Australia* (London: Peter Davies, 1945), 1-2. Moran, a dedicated Irish-Catholic Australian, went on to captain the first rugby tour of Britain, serve in Mesopotamia during the Great War, and become a fervent admirer of Mussolini's Italy. See Roslyn Pesman Cooper, 'An Australian in Mussolini's Italy: Herbert Michael Moran', *Overland*, 115 (1989), 44-53.

35  Moran, *op.cit.*, 1-2.

36  See G.V. Portus, *Happy Highways* (Melbourne: Melbourne University Press, 1953), 56. For his career, see W.G.K. Duncan, 'Garnet Vere Portus (1883-1954)', *Australian Dictionary of Biography*, vol. 11, 262-264.

37  Some forty-eight glass plate negatives, dating from 1890, and probably made by Edward Hufton, display in detail Liversidge's lectures, and are held in the Historic Photograph Collection of the Macleay Museum.

38  Portus, *op.cit.*, 56.

39  *Ibid.*

40  F.W. Robinson, 'The Great Hall and Voices of the Past', *The Union Book of 1952* (Sydney: The University of Sydney Union, 1953), 82.

41  Australian Academy of Science (Basser Library), E.C. Andrews Papers, Box 2, Biographical notes, 36-37.

42  'Our University: IV: The Science Schools', *SMH*, 19 May 1905.

43  Senate Minutes, 3 April 1905. Schofield received £100 from Liversidge's emoluments. University of Sydney Archives, G.3/46/31, f. 105, Registrar to A.L., 4 September 1906.

44   *Ibid.*, 12 June 1905; for which Laby received £25 per term in addition to his salary.

45   In 1902, Senate finally adopted by-laws making the professors of modern literature, law, physics and chemistry *ex-officio* members in their own right. *Ibid*, 1 September 1902.

46   Faculty of Science Minutes, 18 April 1905, 15 August 1906, 3 September 1906.

47   Henry Roscoe, *The Life and Experiences of Sir Henry Enfield Roscoe* (London: Macmillan, 1906), 173.

48   David Orme Masson, 'Presidential Address, SCIV', *Proceedings of the Society of Chemical Industry of Victoria*, l (1902), 24-25, as quoted in Len Weickhardt, *Masson of Melbourne: The Life and Times of David Orme Masson* (Melbourne: RACI, 1989), 49-50.

49   See 'Australasian News', *The Chemist and Druggist*, 57 (2), (1900), 608.

50   Henry G. Smith, 'Recent Work on the Eucalypts', *Journal of the Society of Chemical Industry*, 27 (15 August 1907), 857.

51   For 'Principal Potts' and the College he made famous 'throughout the world for the combination of practice and science', see H.G. Holland and B.R. Rose, 'Henry William Potts (1855-1931)', *Australian Dictionary of Biography*, vol. 11, 266-267.

52   Steel wrote the key references on *Peripatus* and the turbellarians in the *Australian Encyclopedia*. See H.G. Holland and J.R. Simons, 'Thomas Steel (1858-1925)', *Australian Dictionary of Biography*, vol.12, 59.

53   *Society of Chemical Industry, Sydney Section, Office Bearers, 1905-1916*. Liversidge was succeeded by Greig-Smith as chairman in 1906, and became Vice-Chairman between 1906 and 1908.

54   Jubilee Number, *Journal of the Society of Chemical Industry*, 50 (July 1931), 46.

55   R.Greig-Smith, 'The Bacterial Origin of the Vegetable Gums, Part I', *Journal of the Society of Chemical Industry*, 23 (3), (1904), 105-107; Part II, 972-973; James Petrie, 'The Mineral Oil from the Torbanite of New South Wales', *idem.*, 24 (1905), 996-1002; George Harker, 'The Fermentation of Cane Molasses, and its Bearing on the Estimation of the Sugars Present', *idem.*, 25 (1906), 831-832; Henry G. Smith, 'Recent Work on the Eucalypts', *Journal of the Society of Chemical Industry*, 26 (1907), 851-857.

56   Thomas Steel, 'The Temperature of Solutions heated by Open Steam', *Journal of the Society of Chemical Industry*, 24 (1905), 606-607; C. Napier Hake, 'A Cause of Exudation of Nitroglycerin from 'Gelatin Compounds'', *idem.*, 24 (1905), 915-916; A.M. Wright, 'Analyses of some New Zealand Coals', *idem.*, 24 (1905), 1213-1214; E.A. Mann, 'A Possible New Commercial Source of Alcohol', *idem.*, 25 (1906), 1076-1078.

57   Senate Minutes, 1 May 1905.

58   *Ibid.*, 5 February 1906.

59   *Ibid.*, 3 December 1906.

60   Liversidge Papers, Box 19, f. 3, Angus and Robertson to A.L., 18 May 1905; University of Sydney Archives, G.9, School of Chemistry, Preliminary Announcement.

61   Meeting of the General Council, *Report of the AAAS*, 11 (Adelaide, 1907), xxiv.

62   Senate Minutes, 12 June 1905.

63   *Ibid.*, 14 August 1905.

64   *Ibid.*, 3 July 1905.

65   'On so-called Gold-coated Teeth in Sheep' (7 June 1905), *JPRSNSW*, 39 (1905), 33-4; 'International Catalogue of Scientific Literature' (3 May 1905), *JPRSNSW*, 39 (1905), xi-xv.

66   'Gold Nuggets from New Guinea showing a Concentric Structure' (5 December 1906), *JPRSNSW*, 40 (1906), 161-162; 'On the Internal Structure of some Gold Crystals' (2 October 1907), *JPRSNSW*, 41 (1907), 143-145.

67   'Captain Hutton, FRS' (obit.), *JPRSNSW*, 39 (1905), 139-141.

68   'Our University: IV: The Science Schools', *SMH*, 19 May 1905.

69   'Sydney University: Students' Ways and Means', *Daily Telegraph*, 2 May 1907.

70   *Ibid.*

71   In 1906, the University offered five fellowships, twenty-nine scholarships and seventeen bursaries in all subjects. 'Sydney University: Its Financial Position', *Daily Telegraph*, 30 April 1907.

72   F.B. Guthrie, 'Presidential Address', *JPRSNSW*, 38 (1904), 16; *Daily Telegraph*, 5 May 1904. Guthrie counted only ninety-six chemists in manufacturing industry in New South Wales.

73   *SMH*, 8 August 1907.

74   Senate Minutes, 5 March 1906.

75   *Ibid.*, 3 September 1906.

76   Senate Minutes, 5 March 1906; *Daily Telegraph*, 4 January 1907.

77   'The University', 'The Australians', *SMH*, 7 April 1906.

78   For its more recent, happier history, see Clive Lucas, 'Elizabeth Bay House, Restored', *Art and Australia*, 16 (1), (1978), 76-84.

79   See Richard T. Baker and Henry G. Smith, *A Research on the Eucalpyts, Especially in regard to their Essential Oils* (Sydney: Government Printer, 1902, 2nd ed., 1920). See J.L. Willis, 'Richard Thomas Baker (1854-1941)', *Australian Dictionary of Biography*, vol. 7, 154-155 and H.H.G.McKern, 'Henry George Smith (1852-1924)', *Australian Dictionary of Biography*, vol. 11, 646-647.

**80**  See Roy MacLeod and Kimberley Webber, 'Empowering: Applied Research and the Commercial Museum', in Graeme Davison and Kimberley Webber (eds.), *Yesterday's Tomorrows: The Powerhouse Museum and its Precursors, 1880-2005* (Sydney: Powerhouse Publishing in association with UNSW Press, 2005), 96-109.

81   R.T. Baker and H.G. Smith, 'On an Undescribed Species of Leptospermum and its Essential Oil', *JPRSNSW*, 39 (1905), 124-130. The *Leptospermum liversidgei*, known as the 'mozzie blocker', is a handsome native plant grown from Darwin to Hobart. See Amanda Phelan, 'Dealing with Mozzies Just Takes a Little Common Scent', *Sydney Morning Herald*, 29 December 1997.

82   See the *Yearbook of New South Wales, 1908* (Sydney: Government Printer, 1909), 88.

83   Senate Minutes, 5 November 1906.

84   A.L., 'Gold Nuggets from New Guinea showing a Concentric Structure' (5 December 1906), *JPRSNSW*, 40 (1906), 161-2; *Report AAAS*, 11 (Adelaide, 1907), 365.

85   See David's appreciation, 'The late H.C. Russell', *SMH*, 2 March 1907; G.P. Walsh, 'Henry Chamberlain Russell (1836-1907)', *Australian Dictionary of Biography*, vol. 6, 74-75.

86   Alfred Deakin, 'Science and the Empire', *Nature*, 76 (9 May 1907), 37-38.

87   It was accepted the following day. University of Sydney Archives, GS/46/32, f.135, Barff to A.L., 7 May 1907.

88   See, for example, 'Sydney University: The Science Side', *Daily Telegraph*, 10 May 1907.

89   For discussion of pensions at Sydney University in this period, see Bruce R. Williams and David R.V. Wood, *Academic Status and Leadership in the University of Sydney, 1852-1987* (Sydney: Sydney University Monographs, no. 6, 1990), 74.

90   'Our University: VI: Science Research', *SMH*, 2 June 1905.

91   British Museum (Natural History), Mineralogical Department Archives, A.L. to Fletcher, 14 May 1907.

92   Senate Minutes, 6 May 1907; 10 June 1907. His motion failed.

93   *Ibid.*, 10 August 1910.

94   *Sydney University Calendar* (1916), 40-41; Clifford Turney, Ursula Bygott and Peter Chippindale, *Australia's First: A History of the University of Sydney* (Sydney: University of Sydney in association with Hale and Iremonger, 1991), vol. 1, 281.

95   Senate Minutes, Special Meeting, 13 May 1907.

96   Senate Minutes, 10 June 1907. Gurney, then in London, was also asked to serve on the committee, but declined owing to ill-health. *Ibid.*, 14 October 1907.

97   *Ibid.*, 9 December 1907.

98   Australian Museum Letterbook, 1907, Etheridge minute, 14 October 1907; f. 705-706, Sinclair to A.L., 8 October 1907. These included thirty-five volumes of the *American Journal of Science*, seven volumes of *The Geologist*, and lengthy runs of the *Journal and Proceedings of the Royal Society of NSW* and the *Comptes Rendus* of the Académie des Sciences.

99   Australian Museum Letterbook, 1907, f. 828, Etheridge to A.L., 25 November 1907.

100  Australian Museum Letterbook, A 23.79.1 A.L. to Etheridge, 16 December 1907.

101  Faculty of Science Minutes, 20 November 1907.

102  Senate Minutes, 14 October 1907.

103  *Ibid.*, 3 February 1908. Hufton took three months' leave when Liversidge left, and died on the job in April 1917. See C.E.F(awsitt), 'In Memoriam: Edward Hufton', *Hermes*, XXIII, ns. No.1 (May 1917), 40.

104  'Sydney University: The Science Side', *Daily Telegraph*, 10 May 1907.

105  *Ibid.*, 2 December 1907, 9 December 1907; University of Sydney Archives, G3/46/32, f. 955, Barff to A.L., 3 December 1907.

106  *SMH*, 9 December 1907.

107  'Our University: IV: The Science Schools', *SMH*, 19 May 1905.

108  By way of very approximate comparisons, in 1914, Leeds University had 1,065 students at all stages and 171 teachers at all levels (an admittedly inexact ratio of 6.2); Manchester, 1,655 students and 269 teachers (6.2); Birmingham, 874 students and 134 teachers (6.5); and Liverpool 814 students and 228 teachers (3.5). In 1919, Melbourne had 1,137 students and 174 staff, or a ratio of 6.5, while Sydney had 1,736 students and 178 staff (9.8). See Barff, *op.cit.*, 168 and Department of Repatriation and Demobilisation, A.I.F., *The Universities of Australia* (London: Ede and Townsend, 1919), 11. British figures are drawn from the records of the Universities Bureau of the British Empire.

109  Eric (later Lord) Ashby, 'Universities in Australia', *The Future of Education*, 5 (Melbourne: Australian Council for Educational Research, 1943), 19-20.

110  T.H. Easterfield, 'The Position of Chemical Research in Australasia', Section B, *Report of the AAAS*, 12 (Brisbane, 1909), 116.

111  For a synoptic view of the discipline, see Arthur J. Birch, 'Chemistry in Australia: 200 Years On', *Chemistry in Britain*, 24 (4), (1988), 359-362.

112  Senate Minutes, 18 May 1908.

113  'Our University: IV: The Science Schools', *SMH,* 19 May 1905.

114  'The Mineral Wealth of New South Wales', *Daily Telegraph*, 11 September 1906.

115  'The Mining Profession: Success of Sydney Graduates', *Daily Telegraph*, 4 September 1906; 'State Mineral Exhibits at the Royal Exchange: Our University Men: Noteworthy Achievements at Home and Abroad', *SMH*, 27 May 1911.

116  *Sydney Mail*, 25 December 1907.

117  *Ibid.*

118  Jules Verne, *The Tour of the World in Eighty Days* (Boston: James and Osgood, 1873), 44, 52.

# Chapter 13

1  Australian Museum Letterbook, A 23.79.1, A.L. to Etheridge, 4 June 1908; BM (Nat. Hist.), Mineralogy Department, A.L. to Lazarus Fletcher, 29 June 1908; Mitchell Library, John Le Gay Brereton Papers, MSS. 281/15X, T.H. Laby to Robert Dallan, 31 May 1908. For four weeks, he stayed at the imposing Canford Cliffs Hotel in Bournemouth – today, since a German bomb in 1941, only a fond memory. Derek Beamish *et al.*, *Poole and World War II* (Poole: Poole Historical Trust, 1980), 97-98.

2  Australian Museum Letterbook, A23.79.1, A.L. to Etheridge, 4 June 1908.

3  Liversidge was elected in 1901. For the University Club, see Ralph Nevill, *London Clubs* (London: Chatto and Windus, 1911), 239; Tom Girton, *The Abominable Clubman* (London: Hutchinson, 1924); and Charles Graves, *Leather Armchairs* (London: Cassell, 1963), 38. In 1972, the UUC premises were sold to the British School of Osteopathy.

4  BM (Nat. Hist.), Mineralogy Department, A.L. to L. Fletcher, 16 September 1908.

5  The presentation was made by Edgeworth David and J.H. Maiden, and Normand MacLaurin accepted for the University. See *Australasian Medical Gazette*, XXIX (20 August 1910), 462.

6  'This Week in London', reprinted in *SMH*, 31 October 1907.

7  'See Roy MacLeod, 'Science, Progressivism and "Efficient Imperialism"', in Roy MacLeod and Richard Jarrell (eds.), *Dominions Apart: Reflections on the Culture of Science and Technology in Canada and Australia, 1850-1945*, *Scientia Canadensis*, 17 (1,2), (1995), 8-25.

8  Faculty of Science Minutes, 25 June 1912.

9  *Ibid.*, 20 July 1909, 3 August 1909.

10  See Senate Report, 1910, 497; Senate Report, 1913, 516; Senate Report, 1918, 611; J. Pettigrew, 'Abercrombie Anstruther Lawson (1870-1927)', *Australian Dictionary of Biography*, vol. 10, 15-16.

11  Owing, it was said, to his father's illness. University of Sydney Archives, G.3/46/33 Letterbook, December 1907-June 1918, f. 139, R. Dallen to T.A. Coghlan, Agent-General for New South Wales, 3 June 1908.

12  University of Sydney Archives, G.3/46/33, f. 683, R. Dallen, Acting Registrar, to A.L., 7 April 1908; f. 247, Dallen to Schofield, 4 February 1908; f. 291, Dallen to H.C.L. Anderson, Department of Agriculture, 11 February 1908; f. 297, Dallen to Schofield, 11 February 1908.

13  *Ibid.*, f. 943, Dallen to Liversidge, 23 May 1908.

14  But not before Schofield had been asked to review the department and prepare proposals for future teaching. *Ibid.*, f. 299, Dallen to Schofield, 11 February 1908). Schofield remained an associate professor until his retirement in 1926. See *Journal and Proeedings of the Australian Chemical Institute*, I (8), (1934), 69. T.H. Laby, Liversidge's student, was also considered but, given that his first love was physics, was predictably not appointed. Senate Minutes, 12 October 1908.

15  K.J. Cable and Ursula Bygott, 'Charles Edward Fawsitt (1878-1960)', *Australian Dictionary of Biography*, vol. 8, 474-475.

16  Professorial Board Minutes, 30 October 1911.

17  *Ibid.*, 11 December 1912.

18  *Ibid.*, 15 April 1912.

19  Faculty of Science, By-laws, 1912, ch. 8.

20  H.E. Barff, 'The University of Sydney', in W.S. Dun (ed.), *Handbook for New South Wales*, British Association for the Advancement of Science (Sydney: Edward Lee, 1914), 165-169.

21  Faculty of Science Minutes, 28 October 1912.

22  Sydney University Archives, G.3/46/44, f. 114, Registrar to A.L., 22 October 1912. On this occasion, the London committee included Liversidge; Raphael Meldola of Finsbury College; W.H. Perkin Jr., of Manchester; Sir Edward Thorpe of Imperial College; and Col. (later Sir David) Prain, Director of the Royal Botanic Gardens at Kew.

23  See A.R. Todd, 'Sir Robert Robinson (1886-1975)', *Dictionary of National Biography, 1971-1980* (Oxford: Oxford University Press, 1986), 729-731; Sir Robert Robinson, *Memoirs of a Minor Prophet, 70 Years of Organic Chemistry* (Amsterdam: Elsevier, 1976), 79-80.

24  University of Sydney Archives, G.3/46, f. 421. H.E. Barff to A.L., 10 August 1915. In 1916, Robinson was followed by his runner-up John Reed, who was from W.J. Pope's department at Cambridge. In 1923, Reed left Sydney for St. Andrews, where he wrote a brilliant textbook and established a famous collection of alchemical texts. Reed was succeeded by James Kenner in 1923, and James Campbell Earl (1928-1948). Liversidge's chair in inorganic chemistry was held by Fawsitt until 1946, by R.J.W. Le Fèvre from 1946 to1970, by Hans Freeman from 1970 to 1995, and by Leonard Lindey from 1996 to 2005.

25  D.F. Branagan and H. Graham Holland (eds.). *Ever Reaping Something New: A Science Centenary* (Sydney: University of Sydney, Science Centenary Committee, 1985), 85.

26  Liversidge Papers, Box 19, f. 5, S. Sinclair to A.L., 10 February 1908; f. 6, A.L. to Dr. James Cox, 4 June 1908.

27  University of Sydney Archives, G.3/46/34, f. 909, Robert Dallen to Registrar, University of Cambridge, 8 December 1908; f. 910, Dallen to Liversidge, 8 December 1908. Gurney was asked, but declined, pleading ill-health. Alas, Liversidge was also ill and in the event did not attend. 'Darwin Centenary Celebration, 1909', *Chemical News,* 101 (18 February 1910), 82.

28  A.L., 'The Australasian Association for Advancement of Science', *Nature,* 82 (30 December 1909), 264-266; 'The Royal Society of New South Wales', *Nature,* 83 (23 June 1910), 502-503.

29  BM (Natural History), Mineralogy Department, A.L. to Spencer, 24 November 1911, concerning torbanite (or kerosene shale).

30  Royal Botanic Gardens, Kew (afterwards, Kew Archives), Australian and Tasmanian Letters, vol. 170, A.L. to David Prain, 1 December 1910.

31  Kew Archives, Thiselton-Dyer Letters, vol. 2, f. 165, A.L. to Thiselton-Dyer, 10 December 1910.

32  'Professor Liversidge, MA, FRS, through whose special efforts the Technological Museum was founded in 1880', Liversidge Papers, Box 19, R.J. Baker to A.L., 5 July 1911.

33  Liversidge Papers, Box 19, f. 5, Etheridge to A.L., 5 July 1910.

34  His collection of some 3,000 items was held in twenty-two cases. See Appendix IV. To his dismay, the Museum bought only his books and cabinets, for £38. Australian Museum Letterbook, 1909, f. 174, Sinclair to A.L., 18 March 1909. The Museum stored his minerals until 1911, when they were returned to him in London.

35  Liversidge Papers, Box 16, f. 2.

36  Australian Museum Letterbook, A 23.79.1, A.L. to Etheridge, 9 October 1910, 3 November 1910; Etheridge to A.L., 24 November 1910, 11 January 1911. No New South Wales presence was planned for the Empire Exhibition: 'The advent of the Labor Party', as Etheridge wrote Liversidge, 'has so complicated affairs, that one

cannot predict what may happen in any direction'. Australian Museum Letterbook, 1910, f. 799, Etheridge to A.L., 14 November 1910.

[37] Australian Museum Letterbook, 1914, f. A23.79.1, A.L. to Etheridge, 24 July 1914; BM (NH), Mineralogy Department, S.E. Chandler to A.L., 14 October 1915; W. Dunstan to A.L., 24 September 1917.

[38] *The Times*, 5 May 1908.

[39] Katherine Watson, 'Temporary Hotel Accommodation: The Early History of the Davy-Faraday Research Laboratory, 1894-1923', in Frank A.L. James (ed.), '*The Common Purposes of Life*': Science and Society at the Royal Institution of Great Britain (London: Ashgate, 2002), 207-208.

[40] Barbara Given, *The Davy-Faraday Research Laboratory, 1896-1923* (London: Royal Institution, 1977), 8, 13, 15. The Laboratory's superintendent, Alexander Scott, presided over a small army of six laboratory assistants, undreamt of at Sydney. However, four-fifths of Mond's capital went into the building, and only £1,700 remained for annual running expenses. Limited funds for individual researchers inclined the company towards supporting men who embodied the individualism of Cavendish, Joule and Darwin. Selection favoured men with money as well as ability.

[41] W.H. Brock, 'Exploring the Hyperarctic: James Dewar at the Royal Institution', in Frank James (ed.), *The Common Purposes of Life, op. cit.*, 169-190, esp. 188-189.

[42] 'Liquified Air: Demonstration by Professor Liversidge', *Daily Telegraph* (Sydney), 17 August 1899. The event is well remembered. See R.J.W. Le Fèvre, 'The Establishment of Chemistry in Australia,' in *A Century of Scientific Progress: The Centenary Volume of the Royal Society of New South Wales* (Sydney: Royal Society of NSW, 1968), 374.

[43] Gwendolyn Caroe, *The Royal Institution: An Informal History* (London: John Murray, 1985), 94.

[44] Between 1896 and 1923, 123 applied, seventy-two were approved and forty-eight were rejected. Of those who applied, five were women. Applicants came from thirteen different nationalities, and their average age was thirty-five. See Watson, *op. cit.* 209.

[45] In the Royal Institution's *Annual Report* for 1910, Liversidge was listed correctly as 'professor', but incorrectly as 'Arthur'. In 1910, he was joined by younger researchers: Dr Riko Majima, a Japanese, who spent one term on the oxidation of aniline; Dr Walter Wahl of Helsinki, on reactions at high pressures; and Dr Frédéric Schwers, of Belgium, on magnetic rotation.

[46] Royal Institution Archives, Minutes of the Davy-Faraday Laboratory, vol. 1, 247. For his life, see Ronald W. Clark, *Tizard* (London: Methuen, 1965).

[47] Joseph Petavel, an 1851 Exhibitioner from University College, moved to Manchester University in 1900 became Director of the Natural Physical Laboratory in 1919. Among his colleagues was John (later Sir John) Shields, another 1851 Exhibitioner, who later became a member of William Ramsay's team at UCL.

[48] Watson, *op. cit.* 216.

[49] Annual Reports were presented, although none for the period has survived.

[50] 'We are only too glad to be of service to one who has done so much to advance the interests of the Institution as you have'. Australian Museum Letterbook, 1915, f. 273, Etheridge to A.L., 3 May 1915.

[51] He found conifers were the most argentiferous. Kew Letters, f. 170, A.L. to Prain, 2 December 1910.

[52] Australian Museum Letterbook, A 23.79.1, A.L. to Etheridge, 24 July 1914; Letterbook, 1915, A.L. to Ro, 15 March 1915. Etheridge assured him that no silver-lined stills were used in preparing the spirit in which the specimens were preserved.

[53] Australian Museum Letterbook, 1916, A.L. to Etheridge, 18 April 1916.

[54] 'On the Internal Structure of some Gold Crystals' (2 October 1907), *JPRSNSW*, 41 (1907), 143-145.

[55] See F.W. Clarke, *The Data of Geochemistry* (Washington, DC: Government Printing Office, 1911).

[56] Australian Museum Letterbook, 1910, A.L. to Etheridge, 9 May 1910.

[57] See Roy MacLeod, 'Sir Alexander Pedler (1849-1918)', *Oxford Dictionary of National Biography* (Oxford: Oxford University Press, 2004), vol. 34, 390-391.

[58] *Proceedings of the Institute of Chemistry*, Pt.II (March 1910), 12.

[59] Perry, who became President of the Institution of Electrical Engineers, had earlier presided over the introduction of Western engineering technology to Japan. See John Perry, *England's Neglect of Science* (London: T. Fisher Unwin, 1900).

[60] For the Guild, see Roy MacLeod and Peter Collins (eds.), *The Parliament of Science: Essays in Honour of the British Association for the Advancement of Science, 1831-1981* (London: Science Reviews Ltd., 1981), especially Peter

Collins, 'The British Association as Public Apologist for Science, 1919-1946', 211-236; W.H.G. Armytage, *Sir Richard Gregory: His Life and Work* (London: Macmillan, 1957); A.J. Meadows, *Science and Controversy: A Biography of Sir Norman Lockyer* (London: Macmillan, 1972); and Roy MacLeod, 'Science and Imperial Efficiency: The British Science Guild, 1905-1936', *Public Understanding of Science*, 3 (1),(1994), 155-194.

[61] At the Guild's first Annual Dinner, held on 6 May 1910 in the rooms of the Royal Institution of Painters in Water Colours, in Piccadilly, Liversidge responded to Sir Boverton Redwood's toast to the 'British Science Guild in Greater Britain'. See *4th Annual Report, British Science Guild* (afterwards *ARBSG*), (April 1911), 21.

[62] *4th ARBSG* (1910), 11, 31-32.

[63] *5th ARBSG* (1911), 13.

[64] *Ibid*, 47. The memorialists included Sir J.J. Thomson, Master of Trinity College, Cambridge; Henry Armstrong of the Royal College of Science; William Pope, J. N. Langley and Arthur Shipley of Cambridge; W.J. Sollas of Oxford; and three dozen other scientists of Liversidge's acquaintance. His signature on the memorial preceeded Lockyer's.

[65] *4th ARBSG* (1910), 12, 23; *5th ARBSG* (1911), 15.

[66] For Rose, see D.S.L. Cardwell, *The Organisation of Science in England* (London: Heinemann, 1957, 1972), 169.

[67] *5th ARBSG* (1911), 55-57.

[68] Sir William Mather (1838-1920), head of a famous Manchester engineering firm, had been a Special Commissioner to the Royal Commission on Technical Education (the Samuelson Commission) in 1881. See *Who Was Who* (London: A&C Black 1991), vol. 2 (1916-1928).

[69] *7th ARBSG* (21 May 1913), 6, 32-43. After the war, the Education and Technical Committees were merged. Cf. *13th ARBSG* (1919), 7-8.

[70] *7th ARBSG* (1913), 32-43.

[71] Cf. A.E. Shipley, 'In Defence of the English', *The Times*, 18 October 1910, in Liversidge Papers, Box 18. See Liversidge's marginal comments on Bartle C. Frere, 'An Aristocracy of Character', *The Times*, 2 January 1911, *idem*.

[72] BM (NH), Mineralogy Department, A.L. to L. Fletcher, 29 June 1908. He was elected with David (later Sir David) Ferrier, the neurologist, and Sir Philip Watts, the naval architect.

[73] Details are recorded in the *RSC Signature Book* for 1905-10, held in the Royal Society's Archives. This was kept by Liversidge, and later given to the Archives by him. For an official account of the RSC in this period, see T.E. Allibone, *The Royal Society and its Dining Clubs* (Oxford: Pergamon Press, 1976), 281.

[74] Australian Museum Letterbook, A.23. 79.1, A.L. to Etheridge, 27 November 1908.

[75] Kew Archives, Thiselton-Dyer Letters, vol. 2, 10 December 1910.

[76] These were listed by the Royal Society and the Mitchell Library in the collections returned to Australia after Liversidge's death.

[77] Liversidge Papers, Box 19, f. 6, A.L. to Trustees, Australian Museum, 10 October 1910.

[78] R.H. Mathews, 'Notes on Some Published Statements with Regard to the Australian Aborigines', *Rep. of the AAAS*, 13 (Sydney, 1911), 449-453; 'Some Mourning Customs of the Australian Aborigines', *idem*., 13 (Sydney, 1911), 445-449; George Brown, 'On The Necessity for a Uniform System of Spelling Australian Proper Names', *idem*., 13 (Sydney, 1911), 436-441; Charles Daley, 'The Artistic Sense as Displayed in the Aborigines of Australia', *idem.,* 13 (Sydney, 1911), 427-436; C.H.S. Matthews, *A Parson in the Australian Bush* (London: Edward Arnold, 1908).

[79] Australian Museum Letterbook, 1910, A.L. to Etheridge, 9 May 1910.

[80] The house was sold to Arthur Algernon Johnston, and by him to H.S. Holt, a relative of Thomas Holt, the wool merchant and politician, for many years a member of the Royal Society of New South Wales. During the Great War, it served as a residence for wounded soldiers. All the house except the sandstone structure was demolished after the Second World War. The site is now occupied by Ascham School. Cf. M. Flower, *The Story of Ascham School (1886-1962)* (Sydney: G.W. Hall, 1962), 27-32. Land Titles Office, Sydney, Primary Application 14988, 22 August 1907; Certificate of title, vol. 1840, f. 24, 17 December 1907.

[81] 'Fieldhead' was the name of Joseph Priestley's birthplace in Yorkshire. The borrowing may not have been coincidental. I owe this reference to Dr. Graham Holland.

[82] This first appears in the Electoral Register of 1908, under the name of the Krupp family, owners and editors of the *Surrey Comet*.

83  Prompted by David Orme Masson and David Rivett, in 1912 the AAAS repeated to its 'sister' BAAS the invitation to visit Australia that had failed to materialize in 1888. The visit and its consequences are discussed in Roy MacLeod (ed.), *The Commonwealth of Science: ANZAAS and the Scientific Enterprise in Australasia, 1888-1988* (Melbourne: Oxford University Press, 1988).

84  Australian Museum Letterbook, A.23.79.1, A.L. to Etheridge, 24 July 1914.

# CHAPTER 14

1   Field Marshal Earl Haig, *Fourth Supplement to the London Gazette*, No. 4700 (8 April 1919), quoted in *Proceedings of the Institute of Chemistry* (April 1919), Part II, 46-47.

2   See Guy Hartcup, *The War of Inventions: Scientific Developments, 1914-18* (London: Brassey's, 1988).

3   The so-called 'Appeal to the Cultured Peoples of the World'. See Hans Wehberg. *Wider den Aufruf der 93!* (Charlottenburg: Deutsche-Verlagsgesellschaft für Politik und Geschichte, 1920), 1-39; John Heilbron, *The Dilemmas of an Upright Man: Max Planck as Spokesman for German Science* (Berkeley: University of California Press, 1968)

4   Liversidge Papers, Box 19, f. 18, Threlfall to A.L., 25 June 1916.

5   For background, see Roy MacLeod and Kay Andrews, 'The Social Relations of Science and Technology, 1914-1939', in C. Cipolla (ed.), *The Fontana Economic History of Europe vol. 5: The Twentieth Century*, Part I, (London: Collins, 1976), 301-335; Roy MacLeod, 'The Scientist Goes to War': *The Mobilisation and Direction of Allied Scientific Operations in Europe, 1914-1919* (in preparation).

6   The classic case was, of course, H.G.J. Moseley, physics lecturer at Manchester, who began important work on the x-ray spectra of chemical elements, and who was killed at Gallipoli. See John Heilbron, *H.G.J. Moseley: The Life and Letters of an English Physicist, 1887-1915* (Berkeley: California University Press, 1974).

7   See N.T.M. Wilsmore, 'Chemical Research and the State', *Report of ANZAAS*, 20 (Brisbane 1930), 546-569.

8   Cf. 'The War and British Chemical Industry', *Nature,* 95 (1 April 1915), 119-120; Raphael Meldola, 'Professional Chemists and the War', *Nature,* 95 (4 March 1915), 18-19.

9   Royal Commission for the Exhibition of 1851 Archives, Lancelot Harrison to Secretary, 23 June 1915.

10  E. Rutherford, "Henry Gwyn Jeffreys Moseley", *Nature*, 96 (9 September 1915), 33.

11  "The Waste of Brains: Young Scientists in the Fighting Lines", *The Times*, 24 December 1915; cf. [R.A. Gregory], 'Science in National Affairs', *Nature*, 96 (21 October 1915), 195-197.

12  'Annual General Meeting', *Proceedings of the Institute of Chemistry* (April 1919), Part II, 16.

13  Roy MacLeod and E.K. Andrews, 'The Origins of the DSIR: Reflections on Ideas and Men', *Public Administration,* 48 (1970), 23-48.

14  See Roy MacLeod and Kay Andrews, 'Scientific Advice for the War at Sea: The Board of Invention and Research', *Journal of Contemporary History,* VI (1971), 3-40.

15  R. MacLeod and E.K. Andrews, "The Origins of the DSIR: Reflections on Ideas and Men", *Public Administration*, 48 (1970), 23-48.

16  Cf. 'Science Committees', *ibid.*, 8.

17  Michael Sanderson, *The Universities and British Industry, 1850-1970* (London: Routledge and Kegan Paul, 1972).

18  See 'Science Committees and War Problems', *10th Annual Report of the BSG*, June 1916, 8-10; 'Government Committees', *11th Annual Report of the BSG*, (1917), 37-45.

19  *Annual Report of the University of Adelaide* (1916), 366, 390.

20  Cf. *History of the Ministry of Munitions* (London, 1919), vol. 2, pt. 6 (Australia), 9-10; cf. 'Calibration of Munitions Gauges', *Melbourne University Magazine, War Memorial Number* (July 1920), 157.

21  *The Argus*, 10 August 1914, 5.

22  When it was later discovered that Penck's luggage contained militarily useful contour maps of Australian capital cities, he was detained in Britain. However, he returned to Germany in 1915, where he published a vitriolic attack on the British: *Von England festgehalten: Meine Erlebnisse während des Krieges im Britischen Reich* (Stuttgart, 1915). Penck subsequently amplified his views on the war in *Was wir im Kriege gewonnen und was wir verloren haben* (Berlin, 1915). The story was retold from the German perspective in 'Die Rücke kehr der deutschen Teilnehmer von der Letzten Versammlung der British Association for the Advancement of

Science', *Sonderabdruck aus der Zeitschrift der Gesellschaft für Erdkunde* (Berlin, 1915), 62-64. Details of Penck's alleged espionage, related by the Melbourne *Age* and the *Daily Express* of London (18 February 1916), were later reported by the *Vossiche Zeitung* (1917). For the Australian side, see Ernest Scott, *The Official History of Australia During the War of 1914-18* (Sydney, 1936), vol. 11, *Australia During the War*, 34-35. Nearly seven thousand people, including women and children, were interned during the war; see Michael Dugan and Josef Szware, *There Goes the Neighbourhood: Australia's Migrant Experience* (Melbourne, 1984), 116.

23   Ernest Scott, *Australia During the War* (Sydney, 1937), 141-144.

24   Of the seventy-three who graduated from Sydney Medical School in 1915, fifty-one served in the forces. Of these, only three were killed, but many were wounded. In February 1917 there were 496 medical officers in the AIF, and about twenty per cent replacements were required annually. Cf. John Atherton Young *et. al., Centenary Book of the University of Sydney Faculty of Medicine* (Sydney, 1984), 189.

25   See W.A. Osborne, 'Remembrance of Things Past: How Biochemistry Came to Melbourne', *Meanjin*, 20 (2) (1961), 209-214.

26   See John Hilvert, *Blue Pencil Warriors: Censorship and Propaganda in World War II* (St. Lucia: University of Queensland Press, 1984).

27   *Report of the Senate of the University of Sydney*, 1916, 587.

28   Hansard, *Parliamentary Debates*, 5th series, LXXI (13 May 1915), col. 1905.

29   Cf. Rohan Rivett, *David Rivett: Fighter for Australian Science* (Melbourne, 1972), 59-62.

30   He was first made a censor in Brisbane, went to England as a temporary major before returning as general manager of the Australian Arsenal in 1918 – all the while retaining his chair, until 1919, when he moved to BHP and consulting practice. *An Account of the University of Queensland during its First Twenty Five Years* (St. Lucia, 1935), 37, 43.

31   Frederick Alexander, *Campus at Crawley* (Melbourne: F.W. Cheshire for the University of Western Australia, 1963), 92.

32   *Ibid.*, 94.

33   *Report of the Senate of the University of Sydney* (1915), 17.

34   D.A. Welsh, *The Great Opportunity* (Sydney: Sydney and Melbourne Publishing Co., 1915), 9.

35   See H.G.Wells, *Mr Britling Sees it Through* (London: Cassells, 1916).

36   See W.H. Brock, 'Exploring the Hyperarctic: James Dewar at the Royal Institution', in Frank James (ed.), *The Common Purposes of Life: Science and Society at the Royal Institution of Great Britain* (Aldershot: Ashgate 2002), 188.

37   W.J. Reader, *ICI: A History. The Forerunners, 1870-1926* (London: Oxford University Press, 1970-75), vol. 1, 269.

38   Roy MacLeod, 'The Chemists Go to War: The Mobilisation of Civilian Chemists and the British War Effort, 1914-1918', *Annals of Science*, 50 (1993), 455-448.

39   See Roy MacLeod, '"Full of Honour and Gain to Science": Munitions Production, Technical Intelligence and the Wartime Career of Sir Douglas Mawson, FRS', *Historical Records of Australian Science*, 7 (2), (1988), 189-203.

40   Roy MacLeod, 'The Industrial Invasion of Britain: Mobilising Australian Munitions Workers, 1916-1919', *Journal of the Australian War Memorial*, No. 27, October 1995, 37-46; and Roy MacLeod, '"The Other ANZACs": Australian Munitions Workers in the Great War', *Voices* (Canberra: National Library of Australia, 1995), 5 (4), (1995/6), 26-48.

41   National Library of Australia (Canberra), MS 646/3327/412. Edgeworth David to Sir Ronald Munro Ferguson, Governor-General, 21 November 1916. The contribution of Australian science was abundantly noted at the time, but subsequently forgotten. For the 'geological war', see Roy MacLeod, 'The Phantom Soldiers: Australian Tunnellers on the Western Front, 1916-18', *Journal of the Australian War Memorial*, No. 13 (1988), 31-43; and Roy MacLeod, '"*Kriegesgeologen* and Practical Men": Military Geology and Modern Memory 1914–1918', *British Journal of the History of Science*, 28 (4), (1995), 427-450.

42   Liversidge Papers, Box 10, *SMH*, 20 April 1918, MS notes.

43   *Report of the Senate of the University of Sydney* (1915), 584.

44   Dr Alexander Leeper, 'The Universities and the War', *The University Review* (Melbourne), No. 3 (September 1915), 22.

45   By this date, 300 Melbourne students had enlisted. G. Blainey, *A Centenary History of the University of*

*Melbourne* (Melbourne, 1957), 141.

46    Leeper, *op. cit.*

47    'The University's Record in the War', *Hermes*, 24 (1), (June 1918), 29.

48    *Ibid.*

49    John Jenkin, *William and Lawrence Bragg, Father and Son: The Most Extraordinary Collaboration in Science* (Oxford: Oxford University Press, 2008)

50    See Wilfrid Eggleston, *Scientists at War* (Toronto: Oxford University Press, 1950), and *National Research in Canada* (Toronto: Clarke Irwin, 1978).

51    For its history, see Roy MacLeod and E.K. Andrews, 'Scientific Advice in the War at Sea, 1915-1917', *Journal of Contemporary History*, VI (1971), 3-40.

52    H.T. Tizard, 'Sir Richard Threlfall, GBE, FRS (1861-1932)', *Journal of the Chemical Society*, Part 1 (1937), 194; Rod Home, 'First Physicist of Australia: Richard Threlfall at the University of Sydney, 1886-1898', *Historical Records of Australian Science,* 6 (3), (1986), 333-357. Threlfall later supervised Britain's belated efforts to produce mustard gas, which succeeded only as the war reached its concluding stages.

53    Sydney University Archives, M. 193, 'Biological Laboratory', A.L. to W.A. Haswell, 8 April 1917.

54    'Committees for 1917-18', *Proceedings of the Institute of Chemistry,* (April 1917), Part II, 6.

55    See Roy MacLeod, 'The "Arsenal in the Strand": Australian Chemists and the British Munitions Effort, 1916-1919', *Annals of Science*, 46 (1), (1989), 45-67.

56    George Currie and John Graham, *The Origins of the CSIRO: Science and the Commonwealth Government, 1901-1926* (Melbourne: CSIRO, 1966), 30-39.

57    Professor William Osborne, quoted in *ibid.,* 29.

58    *Ibid.,* 27.

59    *Ibid.,* 43.

60    Hughes, cited in *ibid.,* 35.

61    Lord Crewe, Arthur Henderson, and L.A. Selby-Bigge, 'Memorandum on the Suggestions made by the Governments of Victoria and New South Wales for making the Scheme for the Organisation and Development of Scientific and Industrial Research applicable to the whole Empire,' quoted in *ibid.*, 40.

62    *Ibid.,* 41.

63    Now, the Commonwealth Scientific and Industrial Research Organisation (CSIRO). For the CSIR/O's history, see C.B. Schedvin, *Shaping Science and Industry: A History of Australia's Council for Scientific and Industrial Research, 1926-1949* (Sydney: Allen and Unwin, 1987).

64    *10th Annual Report of the BSG* (17 May 1916), 50.

65    *Ibid.,* 51, 64.

66    *Ibid.,* 64.

67    *Ibid.,* 59.

68    *Ibid.,* 46.

69    See Frank Foden, *Philip Magnus: Victorian Educational Pioneer* (London: Vallentine Mitchell, 1970).

70    'Memorandum on the Encouragement of Teaching and Research in Science in British Universities', *Journal of the BSG*, No. 4 (November 1916), 12-16.

71    'Coordination of Technological Education and Research', *Journal of the BSG*, No. 6 (December 1917), 26; Resolutions concerning the Education Bill, 1917, *idem.*, 29; 'Memoranda issued by the Education Committee', *12th Annual Report of the Executive Committee of the BSG* (July 1918), 13-21.

72    See Roy MacLeod, 'Sight and Sound on the Western Front: Surveyors, Scientists and the 'Battlefield Laboratory, 1915-1918', *War and Society*, 18 (1), (May 2000), 23-46; see also Jenkin, *op. cit.*

73    See Roy MacLeod, 'Secrets among Friends: The Research Information Service and the "Special Relationship" in Allied Scientific Information and Intelligence, 1916-18', *Minerva*, 37 (3), (1999), 201-233.

74    T.E. Allibone, *The Royal Society and its Dining Clubs* (Oxford: Pergamon, 1976), 296.

75    *Ibid.,* 293.

76    *Ibid.,* 291.

77    C.R.M.F. Cruttwell, *A History of the Great War, 1914-1918* (Oxford: Clarendon Press, 1934, 2nd ed., 1982), 128.

78    Alfred Gollin, *No Longer an Island: Britain and the Wright Brothers, 1902-1909* (Stanford: Stanford University Press, 1984).

79  In fifty-three raids, involving about twenty-five airships, approximately 1,800 people were killed or wounded. See M.J.B. Davy, *Aeronautics: Lighter-than-Air Craft* (London: HMS), 1950), 67.

80  Cruttwell, *op. cit.*, 497.

81  Basil Collier, *The Airship: A History* (London: Hart Davis, 1974), 74, 83, 89. In 1916, fewer than 300 anti-aircraft guns were available for Britain's defence.

82  An incendiary bullet was invented by Commander F.A. Brock of the Intelligence Section of the Admiralty's Air Department, but the bullet chosen for use by the Munitions Inventions Department was invented by John Pomeroy, an Australian. Cf. Kenneth Poolman, *Zeppelins Over England* (London: White Lion Publishers, 1960), 138, 147.

83  Christopher Cole and E.F. Cheesman, *The Air Defence of Britain, 1914-1918* (London: Putnam, 1984), 168-170, 207.

84  The L15 was actually brought down in the sea off the mouth of the Thames, and its crew were taken to Chatham. The airship was taken in tow, but broke up and sank off Westgate. See Collier, *op.cit.*, 95.

85  DSIR 22/1. Advisory Committee for Aeronautics, *Minutes*, 12 May 1909. (Later the Aeronautics Research Committee.) Established in 1909, the ACA reported to the PM, the Services and the NPL. In 1918, it was absorbed by the Secretary of State for Air, and in 1940, into the Ministry of Aircraft Production. The original committee comprised Capt. R.H. Bacon of the Admiralty; Horace (later Sir Horace) Darwin, FRS; R.T. (later Sir Richard) Glazebrook, FRS, of the NPL; Sir A.G. Greenhill; Sir C.F. Hadden; F.W. Lanchester, the aerodynamical engineer; H.A. Mallock; Professor J.E. Petavel, FRS; and Dr. W. (later Sir William) Napier Shaw, Director of the Meteorological Office.

86  Liversidge Papers, Box 19, f. 18, Edward Liversidge to A.L., 28 April 1916.

87  Liversidge Papers, Box 19, Edward Liversidge to A.L., 4 May 1916.

88  Birmingham Public Library, Threlfall Papers, Airships Sub-Committee, *Correspondence and Reports relating to the Supply of Helium Gas for Airships* (London, 1914-15).

89  A. Kelly, 'Walter Rosenhain and Materials Research at Teddington', *Phil. Trans. Roy. Soc.*, A 282 (1976), 18.

90  The Sub-committee, under Rayleigh's son, Professor the Hon. R.J. Strutt, FRS, included G.S. Albright; Horace Darwin, FRS; Professor J.E. Petavel, FRS; and Richard Threlfall, FRS.

91  The BIR, under Lord Fisher, was advised by Sir J.J. Thomson, Sir Charles Parsons, and a panel of thirteen other scientists and engineers. The BIR had five sub-committees: Airships, Submarines and Mines, Naval Construction, Anti-aircraft, and Ordnance and Ammunition.

92  The Airships Sub-committee's membership overlapped with the Advisory Committee for Aeronautics: a not uncommon occurrence in the 'war of committees'.

93  Liversidge Papers, Box 19, f.18. Threlfall to A.L., 25 June 1916.

94  *Ibid.*

95  *Ibid.*

96  Cf. Michael Pattison, 'Scientists, Inventors and the Military in Britain, 1915-19: The Munitions Inventions Department', *Social Studies of Science*, 13 (4), (1983), 521-568; W. Heath, *Heath Robinson at War* (London: Duckworth, 1978).

97  PRO Adm. 116/1601B. Report on the Present Organisation of the BIR, August 1918, 15.

98  PRO Adm. 116/1431. Sub-Committee on Airships, 23 June 1916.

99  Liversidge Papers, Box 19, f. 18, A.L. to T.H. Hoste, BIR, 8 August 1916.

100  Poolman, *op. cit.*, 157-173.

101  See Collier, *op. cit.*, 109. The crew had set fire to their ship with flares, but she had lost so much hydrogen that she did not completely burn. This airship, 650 feet long and with 55,2000 cubic meters of gas, weighing 74,800 pounds and travelling at seventy miles an hour, became the model on which the British based the later R33 and R34, the first airships to fly the Atlantic in both directions. Intelligence obtained from the aircrew suggests that the Admiralty knew much about the construction of the airship before it was brought down. Whether Liversidge contributed to its analysis is unknown. See Capt Ernst A. Lehmann, *Zeppelin: The Story of Lighter-than-air Craft*, trans J. Dratler (London: Longmans, Green, 1937), 173.

102  An official account of the L33 is held in the Imperial War Museum. A report of it was published in H.W. Wilson and J.A. Hammerton, *The Great War* (London: Amalgamated Press, 1914-18), vol. 8, 220-222. Cf. Poolman, *op.cit.*, 182-187.

103  Liversidge Papers, Box 19, f. 18, Magnesium Metal Company to A.L., 26 October 1917.

[104] PRO Adm. 116/1431, Minutes, 15 February 1917.

[105] Ernest Scott, *The Official History of Australia in the War of 1914-1918,* vol. XI: *Australia During the War* (Sydney: Angus and Robertson, 1927), 241-242. For his work, see J.L Haughton, 'The Work of Walter Rosenhain', *Journal Inst. Metals,* LV (2), (1934), 17-32; *Obituary Notices of Fellows of the Royal Society,* 1 (1932-1935), 353-359.

[106] Published in 1914, Rosenhain's *Physical Metallurgy* defined the subject for a generation.

[107] Kelly, *op.cit.,* 23.

[108] Liversidge Papers, Box 19, f.18. A.L. to Rosenhain, 3 May 1917.

[109] Cf. *Eleventh Report of the Alloys Research Committee on Some Alloys of Aluminium (Light Alloys)* (London: Institution of Mechanical Engineers, 1921).

[110] Cf. *11th Annual Report of the BSG* (1917), 45.

[111] Kelly, *op.cit.,* 8.

[112] Liversidge Papers, Box 19, f. 18, Heckstall Smith to A.L., 22 June 1917.

[113] Kelly, *op.cit.,* 8.

[114] Liversidge Papers, Box 19, f. 18. Threlfall to A.L., 18 November 1916.

[115] *Ibid.*

[116] Sir Thomas Holland, 'Proposed Review of the Mineral Resources of the British Empire', *Nature,* 119 (30 April 1927), 640.

[117] *Ibid.*

[118] Michael T. Klare, 'A New Geography of Conflict', *The Australian Financial Review,* 27 April 2001, 2.

[119] Henry Louis, 'An Imperial Department of Mineral Production', *Nature,* 98 (5 October 1916), 91.

[120] Henry Lewis, 'The Iron Ore Resources of the British Empire', *Mining Magazine,* XXVII (3), (September 1922), 137.

[121] John E. Kendle, *The Colonial and Imperial Conferences, 1887-1911: A Study in Imperial Organisation* (London: Royal Commonwealth Society, 1967); Ian M. Drummond, *British Economic Policy and the Empire* (London: George, Allen and Unwin, 1972); and *Imperial Economic Policy, 1917-1939* (London: George, Allen and Unwin, 1974).

[122] American experience was praised, in particular the work of the Mineral Resources Division of the U.S. Geological Survey, whose reports were models of comparative data. H. Louis, 'National Interest in Mineral Resources', *Nature,* 97 (20 July 1916), 428.

[123] Liversidge Papers, Box 19, f.17. C. McDermid, Secretary, Institution of Mining and Metallurgy, to Secretary, Committee on Commercial and Industrial Policy, 22 January 1917.

[124] National Academy of Science (Washington, D.C.) NRC Papers, Confidential Report, Dr. A. Strahan to Maj. D.W. Johnson, U.S. Army, 24 April 1918.

[125] Louis, *op. cit.; Nature,* 98 (5 October 1916), 91.

[126] Liversidge Papers, Box 19, f.17. Memorandum to Sir William McCormick, 22 September 1916.

[127] 'Mineral Resources of the British Empire', *Nature,* 98 (4 January 1917), 361.

[128] Liversidge Papers, Box 19, f. 17, A.L. to Secretary, Board of Scientific Societies, 14 January 1917.

[129] *Final Report of the Royal Commission on the Natural Resources, Trade and Legislation of Certain Portions of His Majesty's Dominions,* 1917 [Cd. 8402]. 65-66.

[130] *Ibid.,* 70.

[131] 'Notes', *Nature,* 99 (15 March 1917), 50-51.

[132] 'British Minerals for the War', *The Times,* 2 April 1917, 8.

[133] 'Notes', *Nature,* 99 (5 April 1917), 109.

[134] J.W. Finch and J.W. Furness, *An Analysis of the Strategic Mineral Problem of the United States* (Washington, DC: U.S. Bureau of Mines, 1939), quoted in *Mining Magazine,* LX (1939), 324.

[135] J.E. Spurr (U.S. Shipping Board), 'War Minerals', *Economic Geology,* 13 (1918), 500-511.

[136] E. Bastin, 'War Time Mineral Activities in Washington', *Economic Geology,* 13 (1918), 524-537.

[137] *Final Report of the Committee on Commercial and Industrial Policy after the War,* 1918 [Cd. 9035]. xiii. 239, paras. 127-129.

[138] *Ibid.,* para. 230.

[139] In 1926, this was recast into the 'Empire Marketing Board' – an agency which, whatever it did to improve imperial trade until its demise in 1933, was to have little relevance to imperial minerals policy. See Drummond,

op.cit.; Stephen Constantine, *The Making of British Colonial Development Policy, 1914-1940* (London: Cass, 1984); and John M. MacKenzie, *Propaganda and Empire* (Manchester: Manchester University Press, 1984).

140   Liversidge Papers, Box 19, f. 17, G.C. Lloyd to A.L., 1 August 1917; G. Shaw Scott to A.L., 17 September 1917; L.T. O'Shea to A.L., 31 July 1917; Dunstan to A.L., 24 July 1917.

141   Liversidge Papers, Box 19, f. 17, W.W. Watts to A.L., 1 October 1917.

142   H. Louis, 'Imperial Mineral Resources Bureau', *Nature,* 100 (13 September 1917), 25-26.

143   *Report of the Controller of the Department for the Development of Mineral Resources in the United Kingdom,* 1919. [Cd. 9184]. See H. Louis, 'The Future of British Mineral Resources', *Nature,* 102 (9 June 1919), 366-367.

144   Sir Richard Redmayne, *Men, Mines and Memories* (London: Eyre and Spottiswoode, 1942).

145   See William Golant, *Images of Empire: The Early History of the Imperial Institute, 1887-1925* (Exeter: Exeter University Press, 1984); *The Imperial Institute: 1887-1956* (London: Imperial Institute, 1956).

146   Sydney University Archives, M193 'Biological Laboratory', A.L. to W.A. Haswell, 8 April 1917.

147   Roy MacLeod, '"Full of Honour and Gain to Science": Munitions Production, Technical Intelligence, and the Wartime Career of Sir Douglas Mawson, FRS', *Historical Records of Australian Science,* 7 (2), (1988), 189-203.

# CHAPTER 15

1   'British Scientific Products Exhibition', *12th Annual Report of the British Science Guild,* (*ARBSG*), (1918), 10-12.

2   'British Science in Industry', *The Times,* 1918; reprinted in *Science,* XLVIII (8 November 1918), 456-459.

3   *Ibid.*, 29-33.

4   Sydney University Archives, M. 193, Liversidge to Haswell, 26 August 1918.

5   Liversidge Papers, Box 10, *SMH,* 20 April 1918, MS notes.

6   J. E. Coates, 'The Haber Memorial Lecture', *Journal of the Chemical Society,* II (1939), 1661.

7   At first, he bubbled $SO^2$ through synthetic seawater 'spiked' with gold, and precipitated about thirty-five per cent of the mineral. Then he used lead acetate and ammonium sulphide; the sulphide acted as a collector for the gold, which was then separated. Lead formate and boric acid were added, and the gold determined by assay. Eventually, he also used sodium polysulphide, which precipitated gold-bearing particles of sulphur, which were then filtered off. See Joe B. Rosenbaum *et al.,* 'Gold in Sea Water- Fact or Fancy?' *The Mines Magazine,* 59 (1969), 14-17.

8   Morris Goran, *The Story of Fritz Haber* (Norman: University of Oklahoma, 1967), 94-95; see also Dietrich Stoltzenberg, *Fritz Haber: Chemist, Nobel Laureate, German, Jew* (Philadelphia: Chemical Heritage Press, 2nd ed., 2004), 241-249.

9   Coates, *op.cit.,* 1646-1648.

10   See, for example, P. Henderson, *Inorganic Chemistry* (London: Pergamon Press, 1982), 281. I am indebted to Dr. Jim Eckert for this reference.

11   F. Haber, 'Das Gold im Meerwasser', *Zeitschrift für Angewandte Chemie,* 40 (1927), 303-314.

12   Letter to the Editor, 'Gold in Sea Water', *Chem. News,* 65 (March 1892), 131.

13   Haber, *op.cit.,* 307.

14   J. Jaenicke, 'Haber's Forschungen über das Goldvorkommen im Meerwasser', *Die Naturwissenschaften,* 23 (1935), 57.

15   It is ironic that, where Liversidge failed, another Australian, Alan Walsh, would make amends, with the invention of the atomic-absorption spectrometer – an instrument said to be, in a scientific sense, worth more than its weight in gold. See Peter Hannaford, 'Alan Walsh, 1916-1998', *Historical Records of Australian Science,* 13 (2), (2000), 179-206.

16   William Caldwell, 'The Gold Content of Sea Water', *Journal of Chemical Education* (November 1938), cf. Rosenbaum, *op. cit.,* 14-17.

17   Coates, *op. cit.,* 1661.

18   T. Hodge Smith, 'Mineralogical Notes, No. 1', *Records of the Australian Museum,* XIV (2), (December 1923), 110.

19   See Sir Daniel Hall *et al., The Frustration of Research* (London: Allen and Unwin, 1935); Julian Huxley, *Scientific Research and Social Needs* (London: Watts, 1934).

20    *Daily Telegraph*, 7 August 1919.

21    *SMH*, 30 August 1920.

22    See E.R. Holme *et al., The Universities of Australia* (London: Ede and Townsend for the Department of Repatriation and Demobilisation, 1919), 11.

23    *SMH*, 26 February 1925, 8.

24    'The University', *SMH*, 1 January 1921.

25    G.V. Portus, 'The University in Australia', in Edward Bradby (ed.), *The University Outside Europe* (London: Oxford University Press, 1939), 178.

26    *Ibid.*, 177.

27    *Ibid.*, 179.

28    R.E. Priestley, 'Melbourne University: Its Finance and its Objectives', in *The University and the National Life: Three Addresses to Victorian Political Organizations* (Melbourne: Melbourne University Press, 1937), 18.

29    T. P. Anderson Stuart, 'A Review of University Life in Australasia, with its Conditions and Surroundings in 1891', *Proceedings of the Royal Colonial Institute,* XXIII (3), (1892), 93-141.

30    Priestley, 'A Free University?', in Priestley, *idem.*, 37.

31    See David R. Jones, 'A Century of Exoticism: Australian Universities, 1850-1950', *History of Education Review,* 14 (1), (1985),12-24.

32    S.H. Roberts, *The Challenge to the University* (Sydney: University of Sydney, 1947); Portus, *op. cit.* Table V, 183, citing the *Commonwealth Yearbook* No. 30 (London, 1937), 206.

33    In 1935, NSW voted £57,000 and Victoria £55,000 for their respective universities, or about sixpence per head of population. The same year, Birmingham University, with 17,000 students, had an income of £123,000 – or £70 per student vs. Sydney's £50 per student. In 1937, Melbourne's Vice-Chancellor complained that the Victorian Government was granting his university £7,500 a year less than it had received fourteen years earlier, despite the fact that it was enrolling sixty per cent more students. He pleaded for a doubling of the vote, so to make Melbourne at least a 'shilling university'.

34    Sydney made a similar plea, and was promised legislation that would double its grant. But this was not to come before the Second World War.

35    Bradby, *op. cit.,* 21, cited in Jones, *op.cit.*, 14.

36    Quoted by Roberts, *op.cit.,* 12.

37    See Eric Ashby, 'Universities in Australia', *The Future of Education*, No. 5 (Sydney: Australian Council for Educational Research, 1944); *Report of the Committee on Australian Universities, 1957* (Chairman: Sir Keith Murray), *Federal Parliamentary Papers, 1958* (12), vii, 823.

38    *SMH*, 23 June 1926.

39    See Barry W. Butcher, 'Science and the Imperial Vision: The Imperial Geophysical Experimental Survey, 1928-1930', *Historical Records of Australian Science,* 6 (1), (1984), 31-43.

40    The Royal School of Mines had, in the 1980s, an Australian as its Dean. For UCL, see Alan Maccoll, 'Australian Chemists at University College London, 1899-1988', *Ambix,* 36 (2), (1989), 82-90.

41    He was one of five organic chemists elected to the Royal Society at Sydney by 1962: Robert Robinson (1920), John Read (1935), Eustace Turner (1939), and Raymond Lèfevre (1962).

42    *SMH*, 25 February 1925, 16.

43    Liversidge Papers , Box 19, f.4, H. Forster Morley to Secretary, Royal Society of NSW, 6 December 1921; R.H. Cambage to Liversidge, 10 May 1922, 1 September 1922.

44    Liversidge Papers, Box 19, J.E.G. Harris to Liversidge, 11 June 1923; Liversidge to Cambage, 13 June 1923.

45    Cf. *Nature*, 111 (16 June 1923), 825; *idem.*, 112 (7 July 1923), 24.

46    John Jenkin, *William and Lawrence Bragg, Father and Son* (Oxford: Oxford University Press, 2008), 425.

47    Watson, Katherine, '"Temporary Hotel Accommodation?": The Early History of the Davy-Faraday Research Laboratory, 1894-1923', in Frank A.L. James (ed.), *'The Common Purposes of Life': Science and Society at the Royal Institution of Great Britain* (London: Ashgate, 2002), 222.

48    *SMH*, 17 November 1925. David Branagan, 'Putting Geology on the Map: Edgeworth David and the Geology of the Commonwealth of Australia', *Historical Records of the Australian Academy of Science,* 5 (2), (1981), 31-57.

49    Arthur Schuster and Arthur E. Shipley, *Britain's Heritage of Science* (London: Constable and Co., 1917).

50    *Christ's Magazine,* 36 (113), 1-8; (114), 76-79; John A. Benn, 'An Appreciation of Sir Arthur Shipley', *Discovery,*

VIII (1927), 366-367.

51  A.L., 'The Designation of Vitamines', *Nature,* 107 (10 March 1921), 45.

52  A.L., 'Vanishing Customs in the Fiji Islands', *Man,* 21 (81), (September 1921), 133-136.

53  The company went on to repair royal coach bodies through the Second World War, but its premises were badly damaged in the bombing of London. A descendant, Glover Webb, located near Southampton, went on to make armoured vehicles for the British Army for use in Northern Ireland. In 1994, this was taken over by GKN, and now forms part of BAE Land Systems., where (as of 2007) it makes armoured personnel carriers for export. See *Jane's Armour and Artillery* (London, 2007).

54  Charles E. F[awsitt], 'Archibald Liversidge (1897-1927)' ,*Union Recorder* (University of Sydney), I (26), (6 October 1927), 201.

55  Frontispiece, Liversidge's notes of Warington Smyth's *Lectures on Mining, 1869-70,* held in the Library of the Royal Society of New South Wales.

56  *The Times,* 12 December 1927.

57  *Australian Museum Magazine,* I (1922-23), 4.

58  See Appendix I: Tributes and Obituaries.

59  Sir Ernest Rutherford, 'Anniversary Address', *Proc. Roy. Soc.,* B. 102 (1927-28), 242-243.

60  Sir John Hammerton, 'Liversidge, Archibald (1847-1927)', *Concise Universal Biography: A Dictionary of the Famous Men and Women of All Countries and All Times* (London: Educational Book Co., 1935).

61  'Professor Liversidge', *Hermes,* 33 (3), (Michaelmas, 1927), 162.

62  Sir T. W. Edgeworth David, 'Prof. A. Liversidge, FRS', *Nature,* 120 (1927), 625-626; 'Archibald Liversidge, 1847-1927', *Proc. Roy. Soc.,* A 126 (1930), xii-xiv; 'Archibald Liversidge', *Journal of the Chemical Society,* (1931), Part I, 1039-1042.

## CONCLUSION

1  M. Foster, 'A Few More Words on Thomas H. Huxley', *Nature,* 52 (l August 1895), 318-320.

2  Sylvia Lawson, *The Archibald Paradox: A Strange Case of Authorship* (Ringwood: Penguin, 1983).

3  Harvey W. Becher, 'William Whewell's Odyssey: From Mathematics to Moral Philosophy', in Menachem Fisch and Simon Schaffer (eds.), *William Whewell: A Composite Portrait* (Oxford: Clarendon Press, 1991), 1.

4  Quoted in C.J. Pettigrew, 'Abercrombie Anstruther Lawson (1870-1927)', *Australian Dictionary of Biography,* vol. 10, 16.

5  Quoted in S. Murray-Smith, 'William Charles Kernot (1845-1909)', *Australian Dictionary of Biography,* vol. 5, 22.

6  Some of Liversidge's colleagues (including Badham and Pell, Thomson, Woolley, Warren, Anderson Stuart, and MacCallum) brought wives and families with them from England. Others, such as Smith, and many contemporaries in the Indian Army, the ICS, and the Colonial Service, married when 'home' in Britain on leave. T.W. Edgeworth David met his future bride on board ship. The Glaswegians Charles Fawsitt and Francis Anderson married after reaching Sydney, to women of similar background, and in the latter's case, of much younger years. A very few, like the Glaswegian H.S. Carslaw, might find a wealthy widow (but, alas, for only a year, leaving him a widower for the next thirty-seven years). After 1881, when women could matriculate at the University, some of the younger professors, like the historian George Arnold Wood, married former students. At the age of forty-two, Henry Barff, the Registrar, met and married Henry Russell's daughter Jane, among the earliest women graduates. Both women were twelve years younger than their husbands. Of the six pre-Challis professors, only three (Stephens, David and Gurney) married women of their own age (that is, in their late twenties and thirties). This pattern – late marriage, if at all, and then to younger women – was a feature of university life. W.A. Haswell is described, like Liversidge, as 'shy'; but he succeeded, at the age of forty, in marrying a former pupil. Liversidge had no such luck or interest or, by the 1890s, time.

7  Joseph Beete Jukes, *Letters and Extracts from the Addresses and Occasional Writings, edited by his Sister* (London: Chapman and Hall, 1871), 408, 410.

8  Quoted in John Tyndall, 'An Address to Students', University College London, *Fragments of Science* (London: Longmans, Green and Co., 6th ed., 1879), II, 91.

9  See his marginal notes on Smyth's *Lecture Notes on Mining, 1869-70,* held in the Library of the Royal Society

of NSW.

10   George Humphry, 'Introduction', *Journal of Anatomy and Physiology*, I (1867), i.

11   For a brief account of professional developments, see Roy MacLeod, 'The Social Framework of *Nature* in its First Fifty Years,' *Nature*, 224 (5218), (1 November 1969), 441-446; for the role of the scientist as 'public man', see Robert Kargon, *Science in Victorian Manchester: Enterprise and Expertise* (Baltimore: Johns Hopkins University Press, 1977), 32-45.

12   Leonie Foster, *High Hopes: The Men and Motives of the Australian Round Table* (Melbourne: Melbourne University Press, 1986), 241.

13   See R.M. Crawford, '*A Bit of a Rebel': The Life and Work of George Arnold Wood* (Sydney: Sydney University Press, 1975)**.**

14   Valerie Desmond, *The Awful Australian* (Melbourne: E.W. Cole, 1911), 38, 51, 83.

15   Beatrice Webb (1898) in A.C. Austin (ed.), *The Webbs' Australian Diary*, quoted in Kaye Harman (ed.), *Australia Brought to Book: Responses to Australia by Visiting Writers, 1836-1939* (Sydney: Boobook, 1985), 148-50.

16   'Science is, I believe, nothing but trained and organised common sense, differing from the latter only as a veteran may differ from a raw recruit: and its methods differ from those of common sense only so far as the guardsman's cut and thrust differ from the manner in which a savage wields his club.' T.H. Huxley, 'On the Educational Value of the Natural History Sciences' in his *Collected Essays, Volume 3: Science and Education* (London: Macmillan, 1893), 45.

17   D.P. Mellor,' Liversidge, Archibald (1846-1927)', *Australian Dictionary of Biography*, vol. 5, 94.

18   Bruno Latour, 'Give Me A Laboratory and I Will Raise the World', in Karin Knorr-Cetina and Michael Mulkay (eds.), *Science Observed: Perspectives on the Social Study of Science* (London: Sage, 1983), 141-170.

19   See Bob Beale, 'Going for Gold', *Good Weekend, Sydney Morning Herald,* 10 January 1998; 'The Mining Business: Towards a Viable Future', *Sydney Morning Herald,* 15 February 1999**.**

20   See Nicolaas A. Rupke, *Richard Owen, Victorian Naturalist* (New Haven: Yale University Press, 1994); Paul Wood (ed.), *Science and Dissent in England, 1688-1945* (Aldershot: Ashgate, 2004); James Endersby, 'A Garden Enclosed: Botanical Barter in Sydney, 1818-1839', *British Journal for the History of Science*, 33 (118), (2000), 313-334; Janet Browne, *Charles Darwin: Voyaging*, vol. 1 (London: Alfred Knopf and Jonathan Cape, 1995); *Charles Darwin: The Power of Place,* vol. 2 (London: Jonathan Cape and Alfred Knopf, 2002).

21   W.B. Clarke, 'Anniversary Address', *TRSNSW,* 1 (1867), 4

22   A.L., 'Captain Hutton, FRS' [obituary], *JPRSNSW,* 39 (1905), 139-141.

23   F.B. Smith, 'Academics In and Out of the *Australian Dictionary of Biography*', in F.B. Smith and Pamela Crichton (eds.), *Ideas for Histories of Universities in Australia* (Canberra: Australian National University, 1990), 10.

24   Michel Foucault, *The Archaeology of Knowledge* (London: Routledge, 1972, 1989); *Power/Knowledge: Selected Interviews and Other Writings, 1972-1977* ed. by Colin Gordon (New York: Pantheon Books, 1980).

25   George Basalla, 'Science, Scientists and Metaphor ' (Unpublished ms.). I am grateful to Professor Basalla for many insights into the history of Victorian science.

26   See David Hull, *Darwin and his Critics: The Reception of Darwin's Theory of Evolution* (Cambridge, Mass.: Harvard University Press, 1973).

27   Nicolaas A. Rupke, *The Great Chain of History: William Buckland and the English School of Geology, 1814-1849* (Oxford: Oxford University Press, 1983).

28   Coral Lansbury, *Arcady in Australia: The Evocation of Australia in Nineteenth-Century English Literature* (Melbourne: Melbourne University Press, 1970), 75.

29   Bernard Smith, *European Vision and the South Pacific, 1768-1850: A Study in the History of Art and Ideas* (Oxford: Clarendon Press, 1960; 2nd ed. New Haven: Yale University Press, 1985).

30   Esmond Wright, *Franklin of Philadelphia* (Cambridge, Mass.: Harvard University Press, 1986), 25, 358-359.

31   J.C. Beaglehole, *The Life of Captain James Cook* (London: A. and C. Black, 1974), 702.

32   W.H. Brock, W.H. 'Chemical Geology or Geological Chemistry?' in L.J. Jordanova and R. Porter (eds.), *Images of the Earth* (London: British Society for the History of Science, 1978), 147-170.

33   Roy Porter, reviewing Stephen J. Pyne, *Grove Karl Gilbert: A Great Engine of Research* (Austin: University of Texas Press, 1980), in *British Journal for the History of Science,* 15 (1982), 80-81.

34   Martin Rudwick, *The Great Devonian Controversy: The Shaping of Knowledge among Gentlemanly Specialists*

(Chicago: University of Chicago Press, 1985; Rudwick, *The Meaning of Fossils: Episodes in the History of Paleontology* (New York: Science History Publications, rev. ed., 1976); Charles C. Gillespie, *Genesis and Geology: A Study in the Relations of Scientific Thought: Natural Theology and Social Opinion in Great Britain, l790-1850* (Cambridge, Mass.: Harvard University Press, 1951).

35 See Stephen Drury, *Stepping Stones: The Making of Our Home World* (Oxford: Oxford University Press, 1999).

36 Reginald A. Daly, *The Changing World of the Ice Age* (New Haven: Yale University Press, 1934); Mott Greene, *Geology in the Nineteenth Century: Changing Views of a Changing World* (Ithaca: Cornell University Press, 1982); Mott Greene, 'History of Geology', *Osiris* (2nd series), l (1985), 97-116.

37 For its wider importance, see Warren E. Leary, 'Do Diamonds Rain on Uranus?', *International Herald Tribune,* 11 October 1999.

38 J.R. Partington, *History of Chemistry* (London: Macmillan, 1964), vol. 4; Alexander Findlay, *A Hundred Years of Chemistry* (New York: Macmillan, 1937); Alexander Findlay and William H. Mills, *British Chemists* (London: Chemical Society, 1947).

39 Gilbert T. Morgan, *Three Lectures Embodying a Survey of Modern Inorganic Chemistry* (London: Institute of Chemistry, 1936).

40 J.M. MacLaren, *Gold: Its Geological Occurrence and Geological Distribution* (London: The Mining Journal, 1908).

41 Whitman Cross *et al., Quantitative Classification of Igneous Rocks* (Chicago: University of Chicago Press, 1903); Henry Washington, *Manual of the Chemical Analysis of Rocks* (New York: John Wiley, 1904); and Joseph Iddings, *Rock Minerals: Their Chemical and Physical Characters and their Determination in Thin Sections* (New York: John Wiley, 1906).

42 See V.I. Vernadsky, *La Géochimie* (Paris: Librairie Félix Alcan, 1924), later published as *Ocherki geokhimiii (Sketches on Geochemistry)*, (Moscow, 1927-1934). See I.A. Fedoseyev, 'V.I.Vernadsky', *Dictionary of Scientific Biography*, XIII (New York: Scribners, 1976), 616-620; R.K. Balandin, *Vladimir Vernadsky* (Moscow: Mir, 1982, Chicago: Imported Publications, Inc.), and the biography by Kendall E. Bailes, *Science and Russian Culture in an Age of Revolutions: V.I.Vernadsky and his Scientific School, 1863-1945* (Bloomington: Indiana University Press, 1990).

43 Cf. V.M. Goldschmidt, 'The Principles of Distribution of Chemical Elements in Minerals and Rocks', *J.Chem.Soc.* (1937), 655; V.M. Goldschmidt, *Geochemistry*, edited by Alex Muir (Oxford: Clarendon Press, 1954 ed.).

44 Kalervo Rankama and T.G. Sahama, *Geochemistry* (Chicago: University of Chicago Press, 1950). 8-11. I owe this reference to Dr. R.W. Boyle, formerly of the Geological Survey of Canada.

45 Elizabeth Garber, 'Kelvin on Atoms and Molecules', in Raymond Flood, Mark McCartney and Andrew Whitaker (eds.), *Kelvin: Life, Labours and Legacies* (Oxford: Oxford University Press, 2008), 209.

46 Whitaker, 'Kelvin: The Legacy', in *ibid.*, 304.

47 See Rupert J. Best, *Discoveries by Chemists: The University of Adelaide* (Adelaide: University of Adelaide Foundation, 1986), 18.

48 See F.W. Clarke, *The Data of Geochemistry*, U.S. Geological Survey *Bulletin* No. 491 (Washington, D.C.: U.S. Government Printing Office, 1911), 508, 617.

49 Today, nearly one-third of the 2,600 mineral species known to science suffer from conflicting, sometimes incompatible, recording. Even with high resolution microscopy and computer-based techniques, it takes major efforts to resolve each species. The magnitude of the task, and the need to understand each mineral *in situ,* requires investment of an order of magnitude Liversidge's generation could not have realised.

50 See Edward Rennie's Presidential Address, 'The Chemical Exploitation, Past, Present and Future of Australian Plants,' *Report of the AAAS* (Adelaide, 1926).

51 Today, CSIRO uses bacteria to detect the presence of economic minerals in ore bodies, and to understand the role of biological agencies in the formation of mineral deposits. Liversidge's fascination with 'fungoid growths' in 1889 may have foreshadowed such modern techniques.

52 See Robert S. Jones, 'Gold in Igneous, Sedimentary and Metamorphic Rocks', *Geological Survey Circular* No. 610 (Washington, DC: US Department of the Interior, 1969), which cites Liversidge's work on gold in marine deposits.

53 See Charles R.M Butt and R.H. Hough, 'Archibald Liversidge and the Origin of Gold Nuggets, 100 Years

On,' in R.R. Pieron (ed.), *The History of Geology in the Second Half of the Nineteenth Century: The Story in Australia, and in Victoria, from Selwyn and McCoy to Gregory* (Melbourne: Earth Sciences History Group, Geological Society of Australia, Special Publication No. 1, 2007), 16-19

54    W.B. Clarke, 'Anniversary Address', *TRSNSW,* 1 (1867), 4.

55    William Macleay, 'Address to the Linnean Society of New South Wales', reported in *Sydney Morning Herald*, 2 February 1876.

56    Arthur Hutchinson, *Mineralogical Chemistry: Annual Reports on Progress for 1904* (London: Chemical Society, 1905).

57    Within his lifetime, many of Liversidge's descriptions were overtaken, sometimes by those who owed most to him – including Edward Pittman and E.C. Andrews. See Edward F. Pittman, *The Mineral Resources of New South Wales* (Sydney: Department of Mines/ Government Printer, 1901) and Ernest C. Andrews *et. al.., The Mineral Industry of New South Wales* (Sydney: Department of Mines/Government Printer, 1928). See also N.L. Markham and H. Basden (eds.), *The Mineral Deposits of New South Wales* (Sydney: Department of Mines, 1974).

58    David Oldroyd, *Thinking about the Earth: A History of Ideas in Geology* (London: Athlone, 1996), esp. 313-315.

59    See Richard Sennett's perceptive discussion of the concept, in *The Fall of Public Man* (New York: Vintage Books, 1978), esp. 197-218, 255-261.

# Bibliography

## THE LIVERSIDGE PAPERS

The Liversidge Papers, today held in the University Archives at the University of Sydney, are almost as well-travelled as their author. On returning to England in 1908, Liversidge took his books, journals, and papers with him. These included his student lecture notes, his medals and diplomas, and bound volumes of his own scientific papers. At his death in 1927, his papers and books were offered to Christ's College, Cambridge, with instructions that those not wanted were to be sent to the Royal Society of New South Wales. In 1935, some books, forty-three sets of journals, and several sets of offprints were donated to the Mitchell Library. In 1953, the correspondence and news clippings were deposited with the Chemistry School of the University of Sydney. These were transferred to the University Archives in 1956.

Taken as a whole, the Liversidge Papers represent the man as a scientist and public figure. There are few personal letters or family memorabilia. Liversidge was seldom given to introspection, or injudicious gossip, which might otherwise make a biographer's work lighter, if not also more entertaining. By the same token, there is little evidence of his educational or philosophical views. In reconstructing his life, it is therefore necessary to trace (and sometimes, to infer) his influence through published records and the correspondence of others.

Within the University, Liversidge's 'organising genius' can be seen in the minutes of Senate and its committees, the Board of Studies (later the Professorial, now the Academic Board), the Faculty of Science, and the Board (later, the Executive Committee of the Board) for Public Examinations. Outside the University, his presence is evident in the archives of the Australian Museum and the Museum of Applied Arts and Sciences, in the Council Minutes of the Royal Society of New South Wales, and in NSW parliamentary and departmental correspondence. His early life in Sydney is reflected in the Clarke Papers in the Mitchell Library, now made easily accessible by Ann Moyal, *The Web of Science: The Scientific Correspondence of W.B. Clarke* (Melbourne: Australian Scholarly Publishing, 2005). For general background, the biographer must look also to Sydney's daily and weekly press – beginning with the *Sydney Morning Herald*, the *Sydney Mail*, the *Evening News*, and the *Daily Telegraph,* and continuing with *Sydney Punch*, *The Australian Town and Country Journal* and *The Illustrated Sydney News*. The Liversidge Papers contain an extensive run of clippings from these periodicals, collected by Liversidge himself.

Like many scientists of his day – and since – Liversidge wished to be remembered by his scientific publications, which are listed below. The first section includes his technical articles, many of which were published long after delivery, and some of which were reprinted in summary form; for this reason, references are listed in order of first presentation (so far as is known), instead of strictly by date of publication.

Liversidge's fondness for collecting specimens of natural history dated from his early childhood in London. In Sydney, from December 1875 – when he first exhibited a specimen of manganese oxide lifted from the Pacific seabed by HMS *Challenger* – he made numerous presentations to the meetings and *conversazioni* of the Royal Society of New South Wales. Many of these were reported in the *Sydney Morning Herald*, and were recorded in the Society's *Journal and Proceedings*. These are listed in the second section, and may be read in conjunction with the ethnographical exhibits described in Appendix IV.

The third section lists Liversidge's reports to the colonial government, which were first published in the Annual Reports of the Department of Mines and in the *Votes and Proceedings* of the Legislative Assembly and of the Legislative Council of New South Wales. In the fourth category are Liversidge's reports to the Australasian Association for the Advancement of Science, and known obituaries written by him. The fifth category lists what have been found of his 'Letters to the Editor', chiefly to the *Sydney Morning Herald* and to *Nature*.

# I. Scientific Books and Articles

1. 'Detection of Magnesia in the Presence of Manganese (Notes and Queries)', *Chemical News*, 17 (24 January 1868), 49.
2. 'Dendritic Spots on Paper (Preliminary Note)', *Hardwicke's Science Gossip: An Illustrated Medium of Interchange and Gossip for Students and Lovers of Nature*, 5 (1 April 1869), 81.
3. 'Sulphur in Coal-Gas. (Notes and Queries)', *Chemical News*, 21 (29 April 1870), 204.
4. 'Nuclei and Supersaturated Saline Solutions', *Chemical News*, 22 (19 August 1870), 90-92; 22 (26 August 1870), 97.
5. 'Experiments upon Supersaturated Solutions of Sodic Sulphate', *Chemical News*, 22 (18 November 1870), 242-244; 22 (25 November 1870), 253-254; 22 (2 December 1870), 245; 22 (16 December 1870), 298.
6. 'Process for the Estimation of Fluorine', *Chemical News*, 24 (10 November 1871), 226.
7. 'Dendritic Spots on Paper', *J. Chem. Soc.*, 25 (16 May 1872), 646-648; *Chemical News*, 25 (14 June 1872), 284; correspondence in 26 (5 July), 11.
8. 'Note upon Kryptophanic Acid' (April 1872), *Journal of Anatomy and Physiology*, VI, (ii), (May 1872), 422-425.
9. 'On Supersaturated Saline Solutions' (20 June 1872), *Proc. Roy. Soc. of London*, XX (1872), 497-507; *Phil. Mag.*, XLV (January 1873), 67-76; *J. Chem. Soc.*, 26 (6 March 1873), 469-470.
10. 'On the Amylolytic Ferment of the Pancreas' (November 1872), *Journal of Anatomy and Physiology*, VIII (1), (November 1873), 23-29; reprinted in M. Foster (ed.), *Studies from the Physiological Laboratory* (Cambridge), I (1873), 45-51.
11. 'The Deniliquin or Barratta Meteorite' (18 December 1872), *TRSNSW*, 6 (1872), 97-103.
12. 'The Bingera Diamond Field' (1 October 1873), *TRSNSW*, 7 (1873), 91-103; reprinted in *Mines and Mineral Statistics of New South Wales* (Sydney: New South Wales Intercolonial and Philadelphia International Exhibition, 1875), 104-115.
13. 'Report on the Discovery of Diamonds at Bald Hill, near Hill End' (1 October 1873), *TRSNSW*, 7 (1873), 102-103; and in *Mines and Mineral Statistics of New South Wales* (Sydney: New South Wales Intercolonial and Philadelphia International Exhibition, 1875), 115-116.
14. 'Note on a New Mineral from New Caledonia' (7 May 1874), *J. Chem. Soc.*, 27 (July 1874), 613-615; *Chemical News*, 29 (15 May 1874), 212. Cf. correspondence in *Chemical News*, 36 (6 July 1877), 9.
15. 'Notes on the Bingera Diamond-field, with Notes on the Mudgee Diamond-field' (24 June 1874), *Quart. J. Geol. Soc. Lond.*, 31 (1875), 489-492.

16. 'Nickel Minerals from New Caledonia' (9 December 1874), *TRSNSW*, 8 (1874), 75-80.

17. 'Observations at Goulburn', Reports of Observations of the Transit of Venus (8-9 December 1874), made at Stations in New South Wales, communicated by H.C. Russell, *Memoirs of Royal Astronomical Society*, 47 (1882-83), 49-88 (at 74-76).

18. 'Iron Ore and Coal Deposits at Wallerawang, New South Wales' (9 December 1874), *TRSNSW*, 8 (1874), 81-91.

19. 'The Minerals of New South Wales' (9 December 1874), *TPRSNSW*, 9 (1875), 153-215; expanded into *The Minerals of New South Wales* (Sydney: Government Printer, 1876); a second edition appeared as *Description of the Minerals of New South Wales*, in *Mineral Products of New South Wales* (Sydney: Dept. of Mines, 1882); and a third edition as *The Minerals of New South Wales* (London: Trübner and Co., 1888).

20. 'Examples of pseudo-crystallization' (1 December 1875), *TPRSNSW*, 9 (1875), 151-152.

21. 'Notes on the Iron and Coal Deposits at Wallerawang, and on the Diamond Fields', in *Mines and Mineral Statistics of New South Wales* (Sydney: New South Wales Intercolonial and Philadelphia International Exhibition, 1875), 94-103.

22. 'Report upon the Sugar-cane Disease in the Mary River District, Queensland', *Queenslander*, no. I (6 May 1876), 21; no. II (13 May 1876), 20; no. III (20 May 1876), 20; no. IV (27 May 1876), 19; no. V (3 June 1876), 13; no. VI (10 June 1876), 10; series of reports reprinted in *The Sugar Cane*, 8 (1 August 1876), 429-435; 8 (1 September, 1876), 471-480; 8 (2 October 1876), 539-545; 8 (1 December 1876), 633-641; also reprinted in the *Moniteur de la Nouvelle-Caledonie, Journal Officiel* (October 1876) and reviewed in *Chemical News*, 34 (1876), 272.

23. 'On the Formation of Moss Gold and Silver' (6 September 1876), *JPRSNSW*, 10 (1876), 125-133, 134; *Chemical News*, 35 (16 February 1877), 68-71 *Jahrb. für Min.*, (1879), 622; abstracted in *J. Chem. Soc.*, 40, A (1881), 687.

24 'Note on a mineral from New South Wales, presumed to be Laumontite' (6 September 1876), *Mineralogical Magazine*, 1 (2), (November 1876), 54.

25 'Fossiliferous Siliceous Deposit from the Richmond River, NSW and the so-called Meerschaum from the Richmond River, NSW' (6 December 1876), *JPRSNSW*, 10 (1876), 237-240

26. 'On a Remarkable Example of Contorted Slate', (6 December 1876), *JPRSNSW*, 10 (1876), 241-242.

27. 'On the Occurrence of Chalk in the New Britain Group' (4 July 1877), *JPRSNSW*, 11 (1877), 85-91; *Geological Magazine*, 4 (December 1877), 529-534.

28. 'Notes on some of the New Zealand Minerals belonging to the Otago Museum, Dunedin' (Otago Institute, 23 October 1877), *Trans. NZI*, 10 (1877), 490-505.

29. 'Analyses of a Rock Specimen from New Zealand showing the Junction between Granite and Slate' (Otago Institute, 8 November 1877), *Trans. NZI*, 10 (1877), 505-506.

30. 'The International Congress of Geologists at Paris, 1878' (4 June 1879), *JPRSNSW*, 13 (1879), 35-42.

31. 'An Analysis of Moa Eggshell' (Philosophical Institute of Canterbury, 5 August 1880), *Trans. NZI*, 13 (1880), 225-227; *Geological Magazine*, 7 (December 1880), 546-548.

32. 'Water from a Hot Spring, New Britain' (1 September 1880), *JPRSNSW*, 14 (1880), 145; *Chemical News*, 42 (31 December 1880), 324; *J. Chem. Soc.*, 40, A (1881), 564, 1019-1020.

33. 'The Action of Sea-water upon Cast-iron' (1 September 1880), *JPRSNSW*, 14 (1880), 149-154; *Chemical News*, 43 (18 March 1881), 121-123.

34. 'Water from a Hot Spring, Fiji Islands' (1 September 1880), *JPRSNSW*, 14 (1880), 147-148; *Chemical News*, 42 (31 December 1880), 324-325; *J. Chem. Soc.*, 40 A (1881), 564, 1019-1020.

35. 'Notes upon some Minerals from New Caledonia' (1 September 1880), *JPRSNSW*, 14 (1880), 227-246.

36. 'On the Composition of some Coral Limestones, &c., from the South Sea Islands' (6 October 1880), *JPRSNSW*, 14 (1880), 159-162; abstr. in *J. Chem. Soc.*, 40 A (1881), 1011-1012.

37. 'The Alkaloid from Piturie' (3 November 1880), *JPRSNSW*, 14 (1880), 123-132; *Chemical News*, 43 (18 March 1881), 124-126, 138-140; *Moniteur Scientifique*, 23 (1881), 774-780.

38. 'On Some New South Wales Minerals' (3 November 1880), *JPRSNSW*, 14 (1880), 213-225; abstr. in *J. Chem. Soc.*, 40 A, (1881), 991-997.

39. 'On the Composition of some Wood enclosed in Basalt' (1 December 1880), *JPRSNSW*, 14 (1880), 155-157.

40. 'Upon the Composition of some New South Wales Coals' (8 December 1880), *JPRSNSW*, 14 (1880), 181-212; abstr. in *J. Chem. Soc.*, 40 A (1881), 980-983.

41. 'The Deniliquin or Barratta Meteorite (Second Notice)' (8 December 1880), *JPRSNSW*, 16 (1882), 31-33; *Jahrb. für Min.*, 2 (1885), Ref. 270-271; abstr. in *J. Chem. Soc.*, 50 A (1886), 134.

42. 'On the Bingera Meteorite, New South Wales (Preliminary Notice)' (8 December 1880), *JPRSNSW*, 16 (1882), 35-37; *Jahrb. für Min.*, 2 (1885), Ref. 271; abstr. in *J. Chem. Soc.*, 50 A (1886),133.

43. 'On the Chemical Composition of Certain Rocks, New South Wales, &c. (Preliminary Notice)' (8 December 1880), *JPRSNSW*, 16 (1882), 39-46; *Jahrb. für Min.*, 2 (1885), Ref. 60.

44. 'Rocks from New Britain and New Ireland (Preliminary Notice)' (8 December 1880), *JPRSNSW*, 16 (1882), 47-51; *Jahrb. für Min.*, 2 (1885), 63.

45. 'Analyses of Queensland Soils' (20 January 1881), *J. Chem. Soc.*, 39 (1881), 61-63; *Chemical News*, 43 (28 January 1881), 43.

46. 'Stilbite, from Kerguelen's Island' (7 February 1881), *Proc. Roy. Soc. Edinburgh*, 11 (1882), 117-119; abstr. in *J. Chem. Soc.*, 40 A (1881), 695.

47. 'On the Torbanite or "Kerosene Shale" of New South Wales', *American Journal of Science*, XXII, No. 127 (July 1881), 32.

48. 'A Peculiar Copper Ore from Coombing Copper Mine, Carcoar, New South Wales' (2 September 1881), *Mineralogical Magazine*, 5 (1882), 32-33.

49. *Tables for Qualititive Chemical Analysis Arranged for the Use of Students* (Sydney: J. Richards, Government Printer, 1881; second edition Angus & Robertson, 1903; London: Macmillan & Co., 1904). (Abstr. in *J. Soc. Chem. Indust.*, 23, No. 20, (31 October 1904), 1005); reviewed in 'Chemical Analysis for Beginners', *Nature*, 71 (3 November 1904), 4-5.

50. *The Minerals of New South Wales* (Sydney: Department of Mines, 1882).

51. 'On some New South Wales Minerals' (2 July 1884), *JPRSNSW*, 18 (1884), 43-48; *Jahrb. für Min.*, 2 (1886), Ref., 28-29; abstr. in *J. Chem. Soc.* 50 A (1886), 774.

52. 'Analysis of Slate in contact with Granite from Preservation Inlet, New Zealand' (with a note by Prof. Hutton), (Philosophical Institute of Canterbury, 27 November 1884), *Trans. NZI*, 17 (1884), 340-341.

53. 'President's Address', (5 May 1886), *JPRSNSW*, 20 (1886), 1-41 (also printed as a pamphlet).

54. 'Metallic Meteorite, Queensland (Preliminary Notice)', (1 December 1886), *JPRSNSW*, 20 (1886), 73.

55. 'Notes on Some Rocks and Minerals from New Guinea, etc.', (1 December 1886), *JPRSNSW*, 20 (1886), 227-230.

56. 'Notes on some New South Wales Silver and other Minerals', (1 December 1886), *JPRSNSW*, 20 (1886), 231-233, 285; *Jahrb. für Min.* (1889), Ref. 384.

57. 'On the Composition of some Pumice and Lava from the Pacific' (1 December 1886), *JPRSNSW*, 20 (1886), 235-238.

58. *The Minerals of New South Wales* (London: Trübner & Co., 1888).

59. 'The Proposed Chemical Laboratory, at the University of Sydney' (3 September 1888), *Rep. AAAS, 1888* (Sydney), 1 (1889), 168-182, Plates V - XVI; *The Building and Engineering Journal*, 5 (213), (6 October 1888), 282.

60. 'Notes on Some Australian Minerals' (3 September 1888), *Rep. AAAS, 1888* (Sydney), 1 (1889), 182.

61. 'Notes on the Thunda Meteorite' (3 September 1888), *Rep. AAAS, 1888* (Sydney), 1 (1889), 182.

62. 'Notes on some New South Wales Minerals (Note No. 5)' (5 December 1888), *JPRSNSW*, 22 (1888), 362-366.

63. 'Note upon the Hot Spring Waters, Fergusson Island, D'Entrecasteaux Group' (1890), Queensland (8 March 1889), *Ann. Rep. Brit. New Guinea* (1888-1889), 31-32; *Chemical News*, 62 (28 November 1890), 264-266. repeated in 'Note upon Samples of Hot Spring Waters, Fergusson Island, British New Guinea', procured by Sir William MacGregor, KCMG, in J.P. Thomson, *British New Guinea* (London: George Philip & Son, 1892), 215-217.

64. 'Note on the Foetal Membranes of *Mustelus Antarcticus*, by T.J. Parker; with an analysis of the Pseudamniotic Fluid, by A. Liversidge' (Otago Institute, 11 June 1889), *Trans. NZI.*, 22 (1889), 331-333.

65. 'Australian Meteorites' (January 1890) *Rep. AAAS. 1890* (Melbourne), *2* (1891), 387-388; *Chemical*

*News*, 62 (28 November 1890), 267; abstr. in *J. Chem. Soc.*, 60 A (1891), 279.

66. 'Notes on some Hot Spring Waters (New Guinea and Solomon Islands)' (January 1890), *Rep. AAAS. 1890* (Melbourne), 2 (1891), 388-394; *Chemical News*, 62 (28 November 1890), 264-266; abstr. in *J. Chem. Soc.*, 60 A (1891), 280.

67. 'On the Removal of Gold from Suspension and Solution by Fungoid Growths' (January 1890), *Rep. AAAS, 1890* (Melbourne), 2 (1891), 399-407; *Chemical News*, 62 (5 December 1890), 277-279; 290-291; abstr. in *J. Chem. Soc.*, 60 A (1891), 401; *American Journal of Science*, 43 (255), (March 1892), 245.

68. 'Chalk and Flints from the Solomon Islands' (January 1890), *Rep. AAAS, 1890* (Melbourne), 2 (1891), 417-420; *American Journal of Science*, 43 (254), (February 1892), 157-158.

69. 'Anniversary Address' (7 May 1890), *JPRSNSW*, 24 (1890), 1-38.

70. 'On Some New South Wales and other Minerals (Note No. 6)' (2 December 1891), *JPRSNSW*, 25 (1891), 234-241; 29 (1895), 316-325; *Chemical News*, 74 (4 September 1896), 113-116; abstr. in *J. Chem. Soc.*, 70 (2), (1896), A, 657-658.

72. 'On Iron Rust possessing Magnetic Properties' (January 1892) *Rep. AAAS, 1892* (Hobart), 4 (1893), 302-320; *Chemical News*, 66 (4 November 1892), 230-232, 239-243.

73. 'Note on some Bismuth Minerals, Molybdenite and Enhydros' (1892), *Aust. Mus. Records*, 2 (3), (August 1892), 33-36.

74. 'Notes on some Australasian and other Stone Implements' (5 December 1892), *JPRSNSW*, 28 (1894), 232-245.

75. 'The Specific Gravities of some Gem Stones' (28 September 1893), *Rep. AAAS, 1893* (Adelaide), 5 (1894), 404-407; *J. Amer. Chem. Soc.*, (23 August 1893), 16 (1894), 205-209.

76. 'Schedules for the Use of Students in Mineralogy' (28 September 1893), *Rep. AAAS, 1893* (Adelaide), 5 (1894), 408-409.

77. 'On the Origin of Moss Gold' (6 September 1893), *JPRSNSW*, 27 (1893), 287-298; *Rep. AAAS, 1893* (Adelaide), 5 (1894) 324; *Chemical News*, 69 (30 March 1894), 152-155; *Mineralogical Magazine*, 11 (50), (September 1895), 101.

78. 'On the Condition of Gold in Quartz and Calcite Veins' (6 September 1893), *JPRSNSW*, 27 (1893), 299-303; (September 1893), *Rep. AAAS, 1893* (Adelaide), 5 (1894), 324; *Chemical News*, 69 (6 April 1894), 162-163; abstr. in *J. Chem. Soc.*, 66 (2), (1894), A, 354; *Mineralogical Magazine*, 11 (50), (September 1895), 101.

79. 'On the Origin of Gold Nuggets' (6 September 1893), *JPRSNSW*, 27 (1893), 303-343; *Chemical News*, 69 (1 June 1894), 260-262; 69 (8 June), 267-268; 69 (15 June), 281-284; 69 (22 June), 296-298; 69 (29 June), 303-304; 70 (6 July 1894), 6-8; 70 (13 July), 21-22; *Mineralogical Magazine*, 11 (50), (September 1895), 101; (28 September 1893), *Rep. AAAS, 1893* (Adelaide), 5 (1894), 324. Abstr. in *J. Soc. Chem. Ind.*, 13 (1894), 398.

80. 'On the Crystallization of Gold in Hexagonal Forms' (6 September 1893), *JPRSNSW*, 27 (1893), 343-346; (28 September 1893), *Rep. AAAS, 1893* (Adelaide), 5 (1894), 324; *Chemical News*, 69 (13 April 1894), 172-173; *J. Chem. Soc.*, 66 (2), (1894), A, 353-354; *Mineralogical Magazine*, 11 (50), (September 1895), 101.

81. 'Gold Moire-Metallique' (6 September 1893), *JPRSNSW*, 27 (1893), 346-347; (28 September 1893), *Rep. AAAS, 1893* (Adelaide), 5 (1894), 325; *Chemical News*, 69 (4 May 1894), 210; *Mineralogical Magazine*, 11 (50), (September 1895), 102.

82. 'A Combination Laboratory Lamp, Retort, and Filter Stand' (6 September 1893), *JPRSNSW*, 27 (1893), 347-348; (28 September 1893), *Rep. AAAS, 1893* (Adelaide), 5 (1894), 325; *Chemical News*, 69 (11 May 1894), 219.

83. 'Nantokite from New South Wales' (10 April 1894), *Mineralogical Magazine*, 10 (48), (September 1894), 326-327; abstr. in *J. Chem. Soc.*, 70 (2), (1896), A, 31-32.

84. 'Boleite, Nantokite, Kerargyrite, and Cuprite from Broken Hill, N.S.Wales' (6 June 1894), *JPRSNSW*, 28 (1894), 94-98; abstr. in *J. Chem. Soc.*, 70 (2), (1896), A, 32; *Mineralogical Magazine*, 11 (51), (October 1896), 165.

85. 'Preliminary Note on the Occurrence of Gold in the Hawkesbury Rocks about Sydney' (3 October 1894), *JPRSNSW*, 28 (1894), 185-188.

86. 'Structure of Gold Nuggets' (5 September 1894), *JPRSNSW*, 28 (1894), 331; (3 October 1894),

*JPRSNSW,* 28 (1894), 333; *Chemical News,* 70 (1894), 199. Cf. 'On the Internal Structure of Gold Nuggets', (January 1895), *Report AAAS, 1895* (Brisbane), 6 (1896), 240.

87. 'On the Crystalline Structure of Gold and Platinum Nuggets and Gold Ingots' (3 October 1894), *JPRSNSW,* 31 (1897), 70-79; *J. Chem. Soc.,* 71 (1897), Pt. 2, 1125-1131; *Proc. Chem. Soc.,* 174 (4 February 1897), 22; *Chemical News,* 75 (19 March 1897), 139; Abst. in *J. Soc. Chem. Indust.,* 16 (3), (31 March 1897), 242; *Mineralogical Magazine,* 12 (55), (April 1899) 135.

88. 'Variation in the Amount of Free and Albuminoid Ammonia in Waters, on Keeping (Illustrated by curves)' (January 1895), *Rep. AAAS, 1895* (Brisbane), 6 (1896), 235-239; *Chemical News,* 71 (10 May 1895), 225-226; 71 (17 May) 237; 71 (24 May), 249.

89. 'On the Corrosion of Aluminium' (January 1895), *Rep. AAAS, 1895* (Brisbane), 6 (1896), 239-240; *Chemical News,* 71 (25 March 1895), 134; repeated in 77 (6 May 1898), 207.

90. 'Crystallised Carbon Dioxide' (January 1895), *Rep. AAAS, 1895* (Brisbane), 6 (1896) 240; *Chemical News,* 71 (29 March 1895), 152; repeated in 77 (13 May 1898), 216.

91. 'Contributions to the Bibliography of Gold' (January 1895), *Rep. AAAS, 1895* (Brisbane), 6 (1896), 240-256.

92. 'Abbreviated Names for Certain Crystal Forms' (January 1895), *Rep. AAAS, 1895* (Brisbane), 6 (1896), 320; *Chemical News,* 71 (22 March 1895), 140-141.

93. 'Models to Show the Axes of Crystals' (January 1895), *Rep. AAAS, 1895* (Brisbane), 6 (1896), 321.

94. 'Experiments on the Waterproofing of Bricks and Sandstones with Oils' (January 1895), *Rep. AAAS, 1895* (Brisbane), 6 (1896), 734-737; *Scientific American,* 72 (April 1895), 258.

95. 'Experiments upon the Porosity of Plasters and Cements' (January 1895), *Rep. AAAS, 1895* (Brisbane), 6 (1896), 737-740.

96. 'On the Amount of Gold and Silver in Sea-Water' (2 October 1895), *JPRSNSW,* 29 (1895), 335-349; *Chemical News,* 74 (18 September 1896), 146-148; 74 (25 September 1896), 160-161; 74 (2 October 1896), 166-168.

97. 'The Removal of Silver and Gold from Sea-Water by Muntz Metal Sheathing' (2 October 1895), *JPRSNSW,* 29 (1895), 350-366; Abst. *J. Soc. Chem. Indust.,* 15 (8), (1895), 598; *Chemical News,* 74 (9 October 1896), 182-184; 74 (16 October 1896), 191-194.

98. 'On Some New South Wales and Other Minerals (Note No. 7)' (6 November 1895), *JPRSNSW,* 29 (1895), 316-325; *Chemical News,* 74 (4 September 1896), 113-116.

99. 'On the Presence of Gold in Natural Saline Deposits and Marine Plants', (4 February 1897), *Proc. Chem. Soc.,* 13 (174), 22-23; *J. Chem. Soc.,* 71 (1897), 298-299; *Chemical News,* 75 (19 March 1897), 139; abst. in *J. Soc. Chem. Indust.,* 16 (3), (31 March 1897), 242.

100. 'President's Address', (6 January 1898), *Rep. AAAS, 1898* (Sydney), 7, (1899), 1-56.

101. 'Native Silver Accompanying Matte and Artificial Galena' (7 December 1898), *JPRSNSW,* 32 (1898), xli-xlii.

102. 'The Blue Pigment in Coral (*Heliopora coerulea*) and other Animal Organisms' (7 December 1898), *JPRSNSW,* 32 (1898), 256-268; *Chemical News,* 80 (21 July 1899), 29-31; 80 (28 July) 41-43; abstr. in *J. Chem. Soc.,* 78 (1), (1900), A, 70.

103. 'On the Crystalline Structure of Some Silver and Copper Nuggets' (3 August 1898), *JPRSNSW,* 34 (1900), 255-258; abstr. in *J. Chem. Soc.,* 80 (2), (1901), A, 662.

104. 'On the Crystalline Structure of Some Gold Nuggets from Victoria, New Zealand, and Klondyke' (1 November 1899), *JPRSNSW,* 34 (1900), 259-262; *J. Chem. Soc.,* 8 (2), (1901), A, 662.

105. 'Boogaldi Meteorite' (5 December 1900), *JPRSNSW,* 34 (1900), xlvi-xlvii; reported in *Nature,* 63 (11 April 1901), 579.

106. 'Presidential Address' (1 May 1901), *JPRSNSW,* 35 (1901), 1-29; reported in *Nature,* 64 (18 July 1901), 296.

107. 'The Metallurgical Laboratories of the School of Mines, in the University of Sydney' (January 1902), *Rep. AAAS, 1902* (Hobart), 9 (1903), 830-832, with plates.

108. 'Meteoric Dusts, New South Wales' (3 September 1902), *JPRSNSW,* 36 (1902), 241-285; reprinted in *Chemical News,* 86 (24 October 1902), 207; 88 (10 July 1903), 16-18; 88 (17 July 1903) ,32-34; 88 (24 July 1903), 41-45; 88 (31 July), 55-58; *Eng. Mining J.,* 74 (1902), 648-650; *Am. J. Sci,* 14 (4), (1902), 466-467; *J. Iron Steel Inst.* No. 2 (1903), 539; reprinted in *Nature,* 67 (13 November 1902), 47.

109. 'Gold in Meteorites' (3 September 1902), *JPRSNSW*, 36 (1902), xxiv; *Chemical News*, 86 (24 October 1902), 207-208.

110. 'On the Presence of Platinum and Iridium Metals in Meteorites' (3 September 1902), *JPRSNSW*, 36 (1902), xxix-xl; *Chemical News*, 87 (20 February 1903), 92; reported in *Nature*, 67 (26 February 1903), 408.

111. 'The Boogaldi, Barratta Nos. 2 and 3; Gilgoin Nos. 1 and 2, and Eli Elwah or Hay Meteorites, New South Wales' (5 November 1902), *JPRSNSW*, 36 (1902), 341-359; *J. Chem. Soc.*, 84 (1903), A (2), 658. *J. Iron Steel Inst.* No. 2, (1903), 539; *Am. J. Sci.*, 16 (4), (1903), 336; reported in *Nature*, 67 (29 January 1903), 312; *Zeitschrift Kristallographie,* 41 (1906), 407.

112. 'On the Narraburra Meteorite' (2 September 1903), *JPRSNSW*, 37 (1903), 234-242; *Chemical News*, 88 (6 November 1903), 233; abstr. in *J. Chem. Soc.*, 86 (2), (1904), A, 671.

113. *Tables for Qualitative Chemical Analysis* (London: Macmillan & Co., 1904).

114. 'International Catalogue of Scientific Literature' (3 May 1905), *JPRSNSW*, 39 (1905), xi-xv; *Chemical News*, 92 (8 September 1905), 114-115.

115. 'On so-called Gold-coated Teeth in Sheep' (7 June 1905), *JPRSNSW*, 39 (1905), 33-34; *Chemical News*, 92 (8 September 1905), 115-116; abst. in *J. Soc. Chem. Indust.*, 24 (19), (16 October 1905), 1039.

116. 'Gold Nuggets from New Guinea showing a Concentric Structure' (5 December 1906), *JPRSNSW*, 40 (1906), 161-162; (January 1907), *Rep. AAAS, 1907* (Adelaide), 11 (1908), 365; reported in *Nature*, 75 (14 March 1907), 480.

117. 'On the Internal Structure of some Gold Crystals' (2 October 1907), *JPRSNSW*, 41 (1907), 143-145; reported in *Nature*, 77 (16 January 1908), 263.

118. 'The Australasian Association for the Advancement of Science' (30 December 1909), *Nature*, 82 (1909-10), 264-266

119. 'The Royal Society of New South Wales' (23 June 1910), *Nature*, 83 (1910-11), 502-503.

120. 'Vanishing Customs in the Fiji Islands', *Man*, 21 (81), (September 1921), 133-136.

## II. Presentations to the Royal Society of New South Wales

1. 'On 'Black' and 'Ruby' Tin Ore' (8 November 1876), *JPRSNSW*, 10 (1876).

2. 'On the Composition of Lignite, Rewa River, Fiji Islands' (19 October 1877), *JPRSNSW*, 11 (1877) 261-262.

3. 'Silicious and other Deposits from Some of the Hot Springs in New Zealand' (16 November 1877), *JPRSNSW,* 11 (1877), 262-264.

4. 'On Gems from Berrima' (5 November 1884), *JPRSNSW*, 18 (1884), 139.

5. 'On the Decay of Eucalypts' (5 August 1885), *JPRSNSW*, 19 (1885), 95.

6. 'On Alleged Effects of Lightning during the late Volcanic Eruptions at Tarewera in New Zealand' (3 November 1886), *JPRSNSW*, 20 (1886), 294-295.

7. 'On Crystalline Structures of Ingots of Gold' (7 August l895), *JPRSNSW,* 29 (1895), 560.

8. 'Mineral Specimens' (2 December 1897), *JPRSNSW,* 31 (1897), xlviii.

9. 'An Account of Woad (*Isatis tinctoria*) and its Uses' (1 November 1905), *JPRSNSW*, 39 (1905), xxxiv-xxxvi.

## III. Government Reports

1. Sydney City and Suburban Sewage and Health Board, 'Report upon the Sydney City Suburban Water Supply' (6 July 1875), *J. Legis. Assembly NSW*, Session 1875-1876, vol. 4, 425-444; *J. Legis. Council NSW*, Session 1875-1876, vol. 26, Part II, 253-272.

2. Sydney City and Suburban Sewage and Health Board, 'Second Report upon the Sydney City and Suburban Water Supply' (17 August 1875), *J. Legis. Assembly NSW*, Session 1875-1876, vol. 5, 665-670; *J. Legis. Council NSW*, Session 1875-1876, 26, Part II, 273-278.

3. 'Dredgings from Farm Cove' (21 October 1875), *J. Legis. Assembly NSW*, Session 1875-1876, vol. 5, 386-387.

4. 'Analyses of Water' (23 October 1875), *J. Legis. Assembly NSW*, Session 1875-1876, vol. 4, 387-388.

5. Sydney City and Suburban Sewage and Health Board, 'Third Report upon the Quality of the Sydney City and Suburban Water Supply' (Sydney: Government Printer 1876), (reviewed in *Chemical News*, 35 (11 May 1876), 195).

6. Sydney City and Suburban Sewage and Health Board, Minutes of Evidence, 'Twelfth and Final Report of the Board Appointed on the 12th April, 1875, to inquire into and report as to the best means of disposing of the Sewage of the City of Sydney and its Suburbs, as well as of Protecting the Health of the Inhabitants thereof' (11 May 1877), *J. Legis. Assembly NSW*, Session 1876-1877, vol. 3, Mins. Evid. (9 October 1876), Qs. 3484-3499.

7. 'Report upon certain of the Coals, Iron Ores, Limestones, and Copper Ores of New South Wales', Part I, 127-147; 'Report upon certain of the Auriferous and Stanniferous Tailings, Auriferous and Argentiferous Copper Ores, and Other Minerals of New South Wales', Part II, 149-167, *Annual Report, NSW Department of Mines, 1875* (Sydney: Government Printer, 1876), 127-167.

8. 'Report upon Mineral and Other Specimens examined for the Mining Department during the year 1876', *Annual Report, NSW Department of Mines, 1876* (Sydney: Government Printer, 1877), 181-184.

9. 'Analyses of Auriferous Tailings from New South Wales', *Annual Report, NSW Department of Mines, 1877* (Sydney: Government Printer, 1878), 14-19.

10. 'Analyses of Samples of Coal from New South Wales', *Annual Report, NSW Department of Mines, 1877* (Sydney: Government Printer, 1878), 24-29.

11. 'Report on the Character and Value of Copper Ores of New South Wales', *Annual Report, NSW Department of Mines, 1877* (Sydney: Government Printer, 1878), 34-37.

12. 'Report upon the Theoretical Calorific Power of certain Samples of New South Wales Coal', *Annual Report, NSW Department of Mines, 1877* (Sydney: Government Printer, 1878), 209.

13. 'Report upon Mineral Specimens examined for the Mining Department during the year 1877', *Annual Report, NSW Department of Mines, 1877* (Sydney, 1878), 210-212.

14. 'Analyses of Samples of Quartz Tailings and other Waste Gold-bearing Products', *Annual Report, NSW Department of Mines, 1878* (Sydney, 1879), 12-13, 18-19.

15. 'Percentage Composition of Coals in the Northern, Western and Southern Districts', *Annual Report, NSW Department of Mines, 1879* (Sydney, 1880), 29-30.

16. 'Deodorization of Sewage Matter' (Correspondence with the Hon. John. Smith and C. Watt), (8 May 1880) *J.Legis. Assembly NSW* (22 June 1880), Session 1879-1880, vol. 5, 533-536.

17. 'Report upon Certain Museums for Technology, Science and Art, and also upon Scientific, Professional and Technical Instruction, and Systems of Evening Classes in Great Britain and on the Continent of Europe' (13 July 1880), *J. Legis. Assembly NSW*, Session 1879-1880, vol. 3, 787-1059; *J. Legis. Council NSW*, Session 1879-1880, vol. 2, 653-925.

18. 'Report of the Royal Commission to inquire into the Nature and Operations of and to classify Noxious and Offensive Trades in the City of Sydney and its Suburbs' (19 April 1883) *J.Legis. Assembly N.S.W.*, Session 1883, vol. 1, 93-217.

19. Wood Pavement Board, 'Report, Minutes of Proceedings and Appendix' (26 November 1884), *J.Legis. Assembly NSW*, Session 1884, vol.1, 489-503.

20. Wood Pavement Board, 'Preliminary Report - Wood Pavement Blocks', *Legis. Assembly NSW* (26 November 1884), Session 1884, vol. 1, 525-7.

21 'Recommendations respecting the Proposed Formation of Technical and Mining Schools in the Principal Mining Centres of New South Wales, for the Minister of Public Instruction ', *State Archives, Department of Public Instruction*, 10/14282 (23 June l886).

22. 'Report of the Parliamentary Standing Committee on Public Works, Minutes of Evidence, Report relating to the proposed Public Offices, Phillip, Bridge and Young Streets, Sydney', *J.Legis. Assembly. NSW* (3 August 1899) Third Session, 1899, vol.4, 937-954.

# IV. AAAS Reports and Obituaries

1. 'Report on the Ellsmore Tin Mine (2 July 1873)', *Australian Town and Country Journal*, 26 July 1873, 110-111, abstr. in *London Mining Journal,* 45 (1875), 1096.
2. 'Report of the Committee on a Mineral Census of Australasia' (January 1890), *Rep. AAAS, 1890 (*Melbourne), 2 (1891), 205-282.
3. 'Report of the Committee on the State and Progress of Chemical Science in Australasia, with Special Reference to Gold and Silver Appliances used in the Colonies and Elsewhere' (January 1890), *Rep. AAAS, 1890* (Melbourne), 2 (1891), 283-292.
4. 'Report of Committee consisting of Professor Archibald Liversidge, FRS, Mr. William Skey, FCS, and Mr George Grey, FCS, (Secretary) on the Composition and Properties of the Mineral Waters of Australasia' (January 1898), *Rep. AAAS, 1898* (Sydney), 7 (1899), 87-108; abstr. in *J. Chem. Soc.*, A (2), (1900), 288.
5. 'Captain Hutton, FRS' (obit.), *JPRSNSW, 39* (1905), 139-141.
6. 'Sir Lazarus Fletcher (1854-1921)' (obit.), *Proc. Roy. Soc. London*, Ser. A, 99 (1921), ix-xii.

# V. Letters

1. 'The London School of Mines', Letter to the Editor, *Sydney Morning Herald*, 11 December 1872, 3.
2. 'School of Mines', Letter II to the Editor, *Sydney Morning Herald*, 6 February 1873, 6.
3. 'School of Mines', Letter III to the Editor, *Sydney Morning Herald*, 27 February 1873, 6; Editorial, *Sydney Morning Herald*, 27 February 1873, 7.
4. 'Sydney Water', Letter to the Editor, *Sydney Morning Herald*, 30 January 1874, 5.
5. 'The British Association', Letter to the Editor, *Sydney Morning Herald*, 18 February 1884.
6. 'Australian Rubies', *Sydney Morning Herald*, 24 April 1888.
7. 'The Matriculation Examination', *Sydney Morning Herald*, 30 May 1888.
8. 'Darwin Centenary Celebration, 1909', *Chemical News*, 101 (1910), 82.
9. 'The Designation of Vitamines' (10 March l921) *Nature*, 107 (1921), 45.
10. 'The Movements of Molecules' (28 August l926) *Nature*, 118 (1926), 303.

# General Bibliography

The following sources are relevant to an understanding of Liversidge and the history of Australian science in the nineteenth century. *Primary Sources* include manuscripts, records and correspondence, and published works appearing before or during Liversidge's lifetime. Titles marked with an asterisk (*) were owned by him, and are now in the Old Library of Christ's College, Cambridge. *Secondary Sources* may be read usefully in conjunction with the bibliography of current work that appears twice yearly in *Historical Records of Australian Science*, published under the auspices of the Australian Academy of Science.

## I. Primary Sources

## Archives

## Australia and New Zealand

Australian Academy of Science, Adolph Basser Library, Anderson Stuart Papers, 1884 E.C. Andrews Papers, 1870-1948.

Alexander Turnbull Library, Wellington, James Hector Papers, 1865-1909; Sir Julius von Haast Papers, 1860-1886

Australian Museum Archives, Letterbooks, 1876-1880; Minute Books, 1879-1883; Major Correspondence Series, 1837-1949

Mitchell Library, Sydney, ANZAAS Archives, 1886-1952; John Le Gay Brereton Papers, 1851-1933; W.B. Clarke Papers, 1827-1951; Gerard Krefft Papers, 1856-1895; Records of the Linnaean Society of New South Wales, 1790-1870; Sir William Macarthur Papers, 1824-1882; H.N. MacLaurin Papers, 1886-1928; W.M. Manning Papers, 1829-1892; Henry Parkes Correspondence, 1833-1896; Edward Deas Thomson Papers, 1823-1883; Baldwin Spencer Papers, 1880-1929; Windeyer Papers, 1827-1928.

Museum of Applied Arts and Sciences, Inward Correspondence, 1890-1910; Trustees Correspondence, 1883-1886; Sydney Observatory Archives ('Transit of 1874 Papers')

National Museum of New Zealand Archives (James Hector Correspondence, 1865-1903)

New South Wales State Archives, Board of Technical Education, 1883-1889; Department of Mines, 1874-1891; Department of Public Instruction, 1880-1915; Sydney City and Suburban Sewerage and Health Board, 1875; Royal Mint (Sydney Branch), Copies of Letters Sent, 1854-1906.

Royal Society of New South Wales, Liversidge Collection; Council Minutes, 1872-1920.

Union Club of Sydney, Archives, House Committee Minutes, 1873-1878.

University of Sydney Archives, Registrar's Correspondence, 1855-1890; Chancellor's Addresses, 1852-1920; Faculty of Science Minutes, 1882-1920; Professorial Board Minutes, 1883-1975; Senate Minutes, 1851-1987; Liversidge Papers, 1874-1924; School of Chemistry Papers.

University of Sydney, Macleay Museum, Historic Photograph Collection.

University of Melbourne Archives, Thomas Laby Papers, 1908-1970.

# United Kingdom

British Museum (Natural History), Case and Policy Files, 1919-1972; Mineralogy Department and Mineralogy Library Archives, 1828-1957.

Cambridge University Archives, University Registry Guard Books, 1748-1891; Annual Reports, Cambridge Union Society, 1871, 1872; Reports, Cambridge Undergraduate Natural Sciences Club, 1863.

Christ's College Archives, Athletic Club Minutes, 1864-1898; Boat Club Minutes, 1867-1875; Scholarship Examinations, 1870.

Glasgow University Archives, Kelvin Papers, 1839-1969.

Imperial College Archives, Liversidge Papers; W.J. Sollas Papers; H. E. Armstrong Papers; London Union Society, 1924; Examination Books, 1866-1870.

Mineralogical Society (London) Archives, Correspondence, 1875-1876.

National Archives of the United Kingdom, Kew, Board of Invention and Research - Airship, Aeroplane and Seaplane Sub-Committees, 1915-1917; Board of Inventions and Research, 1917-1918; Department of Scientific and Industrial Research, Aeronautical Research Council, Minutes of Meetings, 1909-1977.

Royal Botanic Gardens Archives, Kew, NSW and Victorian Letters, 1865-1900.

Royal Commission for the Exhibition of 1851 (South Kensington) Archives, 1890-1918.

Royal Institution Archives, Minutes of the Davy-Faraday Laboratory.

Royal Society of London Archives, Council Minutes, 1870-1920; Candidates Books, 1870-1920; Miscellaneous Correspondence, 1882-1898.

St. John's College, Cambridge, Archives, Papers and Accounts relating to the College Chemistry Laboratory and Inter-collegiate Chemistry Lectures, 1853-1903.

University College London, Ramsay Papers, 1861-1961.

# United States

Smithsonian Institution Archives, Records of the US National Museum, 1850-1958.

# Published Works

Abercromby, Ralph. *Three Essays on Australian Weather* (Sydney: Frederick T.White, 1896).

Aguillon, Louis. 'L'École des Mines de Paris: Notice Historique', *Annales des Mines*, No. 3 (1889), 433-686.

*Allbutt, T. Clifford. *Notes on the Composition of Scientific Papers* (London: Macmillan, 1904).

Anon. 'The Function of an Australian University', *Hermes*, 33 (Trinity, 1927), 83-86.

Anon. 'The University and the Colleges', *Sydney University Magazine*, 1 (1), (January 1855), 98-120.

Anon. 'Exhibition of the Products of New South Wales', *Sydney University Magazine*, 2 (April 1855), 146-166.

Ashburner, R. 'The 75th Anniversary Appeal', *Hermes*, 33 (Trinity Term, 1927), 79-81.

Badham, Charles. 'University Studies', *Sydney University Review*, 1 (2), (April 1882), 161-170.

_____. *Speeches and Lectures delivered in Australia, by the late Charles Badham, DD, Professor of Classics in Sydney University* (Sydney: William Dymock, 1890).

Baker, R.T. and H.G. Smith, 'On an Undescribed Species of Leptospermum and its Essential Oil', *Journal and Proceedings of the Royal Society of New South Wales*, 39 (1905), 124-130.

Ball, Sir Robert. 'Visitors from the Sky; the Origin of Meteorites', *In the High Heavens* (London: Isbister and Company, 1893), 294-355.

Barrett, James W. *The Twin Ideals: An Educated Commonwealth* (London: H.K Lewis, 1918).

Barff, H.E. *A Short Historical Account of the University of Sydney in connection with the Jubilee Celebrations, 1852-1902* (Sydney: Angus and Robertson, 1902).

Bastion, H. Charlton, 'Spontaneous Generation: A Last Word', *Nineteenth Century*, 3 (1878), 497-508.

Barwick, G.F. 'International Exhibitions and their Civilising Influence', in J. Samuelson, *The Civilisation of Our Day* (London: Sampson, Low, Marston and Co., 1896).

Bather, F.A. 'How may Museums Best Retard the Advance of Science?', *Report of the Proceedings of the Seventh Annual Meeting of the Museums Association for 1896* (London: British Museum, 1897), 92-105; reprinted in *Science*, 5 (122), (30 April 1897), 677-683.

Bavin, J.W. 'The Society of Ignorance', *Hermes,* 30 (Michaelmas, 1924), 39-41.

_____. 'The Idea of a University', *Hermes,* 30 (Trinity, 1924), 59-60.

*Beale, Sir William Phipson. *An Amateur's Introduction to Crystallography (from Morphological Observations)* (London: Longmans, Green, 1915).

Bennett, George. *Gatherings of a Naturalist in Australia* (London: John Van Voorst, 1860).

*Bentham, George (revised by Sir J.D. Hooker). *Handbook of the British Flora: A Description of the Flowering Plants and Ferns Indigenous to, or Naturalised in, The British Isles. For the Use of Beginners and Amateurs* (London: Lovell Reeve and Co., 1908).

Bertie, Charles Henry. *Old Sydney* (Sydney: Angus and Robertson, 1911).

Bickerton, Alexander W. 'Old Times by an Old Boy', *Phoenix,* 23 (June 1911), 187-189.

Bischof, Gustav. *Elements of Chemical and Physical Geology* (London: Cavendish Society, 1854-59), 3 vols.

*Bolton, Henry. *A Select Bibliography of Chemistry, 1492-1892* (Washington, DC: Smithsonian Institution, 1893).

*Bower, E.O. *Plants and Man* (London: Macmillan, 1925).

Bragg, W.H. 'Bakerian Lecture: X-Rays and Crystal Structure', *Philosophical Transactions of the Royal Society of London*, Series A, CCXV (1915), 253-274.

Brereton, J. Le Gay. *Bibliographical Record of the University of Sydney, 1851-1913* (Sydney: University of Sydney, 1914).

Brewer, Mrs. and Edwin W. Streeter. *Gold: or, Legal Regulations for the Standard of Gold and Silver Wares in Different Countries of the World* (London: Chatto and Windus, 1877).

Bromby, Rev. J.E. 'Creation versus Development: A Lecture', in *Lectures delivered before the Early Closing Association, 1869-70* (Melbourne: Samuel Mullen, 1870).

Brough Smyth, R. *The Gold Fields and Mineral Districts of Victoria* (Melbourne: Government Printer, 1869).

* Brown, James Campbell. *A History of Chemistry from the Earliest Times till the Present Day* (London: J. and A. Churchill, 1913).

Brown, Walter Lee. *Manual of Assaying Gold, Silver, Copper and Lead Ores* (Chicago: E.H. Sargent, 1889).

Calverley, C.S. *Flyleaves* (Cambridge: Deighton, Bell, 1872).

*Calvert, John. *The Gold Rocks of Great Britain and Ireland: and a General Outline of the Gold Regions of the World, with a treatise on the Geology of Gold* (London: Chapman and Hall, 1853).

Cameron, A.M. 'Australian Meteorology', *Sydney University Magazine*, 1 (4), (1878), 228-235.

Carslaw, H.S. and John McLuckie. 'Professor A.A. Lawson', *Hermes,* 33 (Lent Term, 1927), 1-6.

Chapman, H.G. 'A History of Science in New South Wales', in R.H. Cambage (ed.), *Handbook for New South Wales* (Sydney: British Association for the Advancement of Science, 1914).

Clark, Donald. *Australian Mining and Metallurgy* (Melbourne: Critchley Parker, 1904).

Clarke, Edward Daniel. *The Life and Remains of the Rev. E.D. Clarke, Professor of Mineralogy at the University of Cambridge* (London: J.F. Dove, 1824; 2nd ed., G. Cowie, 1825).

*Clarke, Frank Wigglesworth. *The Constants of Nature: A Table of Specific Gravity for Solids and Liquids* (Washington, DC: Smithsonian Institution, 1888).

*Clarke, Frank Wigglesworth. *The Data of Geochemistry,* US Geological Survey Bulletin No. 491, (Washington, DC: Government Printing Office, 1911).

*Clarke, Rev. W.B. *Researches in the Southern Gold Fields of New South Wales* (Sydney: Reading and Wellbank, 1860).

_____. *Remarks on the Sedimentary Formations of New South Wales* (Sydney: Thomas Richards, 1871).

Coghlan, T.A. *The Wealth and Progress of New South Wales* (Sydney: Government Printer, 1886-87), 2 vols.

*Collins, J.H. *A Handbook to the Mineralogy of Cornwall and Devon* (London: William Collins, Sons and Co., 1876).

Combes, E. *Report on Technical Education to the Legislative Assembly of New South Wales* (Sydney: Government Printer, 1887).

*Coningsby, Robert James. *The Discovery of Gold in Australia* (London: William Milligan and Co., 1895).

Cooksey, T. 'The Nocoleche Meteorite, with Catalogue and Bibliography of Australian Meteorites", *Records of the Australian Museum*, 3 (3), (1897), 51-62.

Council of Technical and Further Education. *Spanners, Easels and Microchips. A History of Technical and Further Education in New South Wales, 1883-1983* (Sydney: NSW Government Printing Service, 1983).

Cox, Samuel Herbert and F. Ratte. *Mines and Minerals; A Guide for the Australian Miner* (Sydney: J. Woods and Co., 1885).

*Crook, Thomas. *Economic Mineralogy: A Practical Guide to the Study of Useful Minerals* (London: Longmans, 1921).

*Cross, Whitman, *et al. Quantitative Classification of Igneous Rocks, based on Chemical and Mineral Characters, with a Systematic Nomenclature* (Chicago: University of Chicago Press, 1903).

Cumming, Fred. *The Sydney Garden Palace: A Patriotic and Historical Poem* (Sydney: Stewart and Co., 1887).

Cumming, J.F. Gordon. *At Home in Fiji* (London: William Blackwood & Sons, 1901).

Curle, J.H. *The Gold Mines of the World* (London: Waterlow and Sons, 1899).

Curran, John Milne. *Geology of Sydney and the Blue Mountains* (Sydney: Angus and Robertson, 1899).

Daintree, Richard. *Notes on the Geology of the Colony of Queensland* (London: Geological Society, 1872).

*Dana, Edward Salisbury. *The System of Mineralogy of James Dwight Dana, 1837-68. Descriptive Mineralogy* (London: Kegan Paul, Trench, Trubner, 6th ed., 1894).

*_____ . *First Appendix to the Sixth Edition of Dana's System of Mineralogy Completing the Work to 1899* (New York: John Wiley and Sons, 1899).

*Dana, Edward Salisbury and William E. Ford. *Second Appendix to the Sixth Edition of Dana's System of Mineralogy* (New York: John Wiley and Sons, 1909).

*Darwin, Charles. 'Australia', in *Journal of Researches* (London: Henry Colburn, Wardlock, 1839), 408-426.

Darwin, Charles. *The Autobiography of Charles Darwin (1809-1882)*, ed. Nora Barlow (1958; New York: Norton, 1969).

David, T.W. Edgeworth. 'Geology and Mineralogy', *Australasian Home Reader,* 1 (2), (June 1892), 36-41.

_____. 'University Science Teaching', in *Record of the Jubilee Celebrations of the University of Sydney* (Sydney: William Brooks and Co., 1903), 93-121.

_____. 'The Aims and Ideals of Australasian Science', *Report of the Australasian Association for the Advancement of Science,* 10 (Dunedin, 1904), 1-43.

*Davison, Simpson. *The Discovery and Geognosy of Gold Deposits in Australia* (London: Longman, Green, Longman and Roberts, 1860).

*Dawkins, W. Boyd. *Cave Hunting, Researches on the Evidence of Caves Respecting the Early Inhabitants of Europe* (London: Macmillan, 1874).

Dawkins, W. Boyd. *Early Man in Britain and His Place in the Tertiary Period* (London: Macmillan, 1880).

Dayton, W.T. *Catalogue of the Scientific Serial Literature in the Following Libraries in Sydney, NSW: The Australian Museum, Free Public Library, Linnean Society of New South Wales, Observatory, Parliamentary Library, Royal Geographical Society of New South Wales, Royal Society of New South Wales, Technological Museum and the University of Sydney,* compiled under the direction of T.P. Anderson Stuart (Sydney: Government Printer, 1889).

Denison, Sir William. *Varieties of Vice-Regal Life* (London: Longmans, Green & Co., 1870).

*Desch, Cecil. *Metallography* (London: Longmans, Green, 1910).

De Morgan, Augustus. *Budget of Paradoxes* (London: Longman Green, 1872), 2 vols.; 2nd ed. by David Eugene Smith (New York: Books for Libraries Press, 1969).

De Vis, C.W. 'Presidential Address', *Proceedings of the Royal Society of Queensland,* 6 (1889), xviii-xxxi.

Dicken, Charles. *The Mineral Wealth of Queensland* (London: Unwin, 1884).

Doyle, Arthur Conan. *Memories and Adventures* (1924, London: Greenhill, 1988).

Eissler, M. *The Metallurgy of Gold: A Practical Treatise* (London: Crossy Lockwood & Son, 1896).

Epps, William. *Anderson Stuart, M.D. Physiologist, Teacher, Builder, Organiser, Citizen* (Sydney: Angus and Robertson, 1922).

Etheridge, Robert and Robert L. Jack. *Catalogue of Works, Papers, Reports and Maps on the Geology, Paleontology, Mineralogy, Mining and Metallurgy, etc, of the Australian Continent and Tasmania* (London: Edward Stanford, 1881).

*Faraday, Michael. *Chemical Manipulation, Being Instructions to Students in Chemistry on the Methods of Performing Experiments of Demonstration or of Research with Accuracy and Success* (London: John Murray, 2nd ed., 1880).

*Farrington, Oliver. *Meteorites: Their Structure, Composition and Terrestrial Relations* (Chicago: privately printed,

1915).

*Fitch, W.H. and W.G. Smith. *Illustrations of the British Flora* (London: Lovell Reeve and Co., 7th ed., 1908).

*Fletcher, Lazarus. *An Introduction to the Study of Meteorites with a List of the Meteorites represented in the Collection* (London: British Museum, 11th ed., 1914).

*Flight, Walter. *A Chapter in the History of Meteorites* (London: Dulau, 1887).

*Flower, Philip William. *A History of the Trade in Tin* (London: George Bell, 1880).

Ford, William E. *Third Appendix to the Sixth Edition of Dana's System of Mineralogy* (New York: John Wiley and Sons, 1915).

*Frankland, Edward. *Experimental Researches in Pure, Applied and Physical Chemistry* (London: John Van Voorst, 1877).

Franklyn, Henry M. *A Glance at Australia in 1880* (Melbourne: Victorian Review Publishing Co., 1881).

Geikie, Archibald. *Life of Sir Roderick I. Murchison* (London: John Murray, 2 vols., 1875).

_____. *Class-Book of Geology* (London: Macmillan, 1886).

_____. *Memoir of Sir Andrew Crombie Ramsay* (London: Macmillan, 1895).

Gellatly, F.M. 'Science and Industry: Establishment of Commonwealth Institute. How the University Can Assist', *Hermes*, 24 (1), (1918), 18-22.

Goode, George Brown. *The Principles of Museum Administration* (York: Coultas and Volans, 1895).

Gould, Nat. *Town and Bush: Stray Notes on Australia* (Ringwood, 1896; Penguin, 1974).

Haast, Julius von. 'On the Progress of Geology: An Opening Address delivered to the Students of Canterbury College on March 28th, 1883', *New Zealand Journal of Science*, 1 (9), (1883), 395-406.

Hamilton, A.G. 'Presidential Address', *Proceedings of the Linnean Society of N.S.W*, 42 (1917), 1-25.

*Hargreaves, Edward Hammond. *Australia and its Gold Fields* (London: H. Ingram, 1855).

*Haswell, W.A. and T. Jeffery Parker. *A Textbook of Zoology* (London: Macmillan, 1897; 4th ed. 1928, 7th ed. 1972).

Hector, James. 'Report on the Mineral Waters of New Zealand', *New Zealand Department of Lands and Surveys, New Zealand Thermal Springs Districts* (Wellington: Government Printer, 1882).

Hooper, G. 'Historical Sketch of Technical Education in NSW', *Australian Technical Journal,* 1 (1897), 1-6.

Hunt, Robert. *British Mining* (London: Crosby, Lockwood and Co., 1884).

*Hunt, Thomas Sterry. *Chemical and Geological Essays* (Boston: James Osgood, 1875).

Huxley, Thomas H. 'On the Educational Value of the Natural History Sciences (1854)', in Huxley, *Collected Essays*, vol. 3 (New York: Appleton, 1898; Greenwood, 1968), 38-65.

Huxley, Thomas H. 'On the Method of Zadig' (1880), 'Science and Hebrew Tradition', in Huxley, *Collected Essays*, vol. 4 (New York: Appleton, 1896; Greenwood, 1968), 1-23.

*Iddings, Joseph. *Gold and Silver: A Supplement to Strzelecki's Physical Description of New South Wales and Van Diemen's Land* (London: Longmans, 1856).

*_____. *Rock Minerals: Their Chemical and Physical Characters and their Determination in Thin Sections* (New York: John Wiley and Sons, 1906).

Jack, R.L. and R. Etheridge. *Geology and Palaeontology of Queensland and New Guinea* (Brisbane: Government Printer, 1892).

Jevons, William Stanley, *The Principles of Science: A Treatise on Logic and Scientific Method* (London: Macmillan, 1874; 2nd ed., 1877, 3rd ed., 1879; New York: Dover Edition, 1958).

Joubert, Jules. *Shavings and Scrapes in Many Parts* (Dunedin: J. Wilkie, 1890).

*Jukes, Joseph Beete *et al.* *Lectures on Gold: For the Instruction of Emigrants about to proceed to Australia* (London: David Brogue, 1852).

*Jukes, Joseph Beete (edited by his sister). *Letters and Extracts from the Addresses and Occasional Writings* (London: Chapman and Hall, 1871).

*King, C.W. *The Natural History of Precious Stones and of the Precious Metals* (London: Bell and Daldy,1870).

King, Georgina. *The Mineral Wealth of New South Wales and Other Lands and Countries* (Sydney: William Brooks & Co., 1911).

Knibbs, G.H. *Official Yearbook of the Commonwealth of Australia* (Melbourne: McCarron, Bird & Co., 1901).

Knox, G. *Vitality or Endowments? The Present Needs of the University of Sydney* (Sydney: John Woods and Co., 1880).

Kopp, Hermann. *On the Past and Present of Chemisty* (London: W. Clewes, 1868).

Krefft, Gerard. 'The Improvements effected in Modern Museums in Europe and Australia', *Transactions of the Royal Society of New South Wales*, 2 (1868), 15-25.

*Kunz, George Frederick. *Gems and Precious Stones of North America. A Popular Description* (New York: The Scientific Publishing Co., 1890).

Langley, J.N. and W. Lee Dickinson. 'Pituri and Nicotin', *Journal of Physiology, XI* (4 and 5), (1890), 265.

Lauterer, J. 'Progress of Science in Australia', *Proceedings of the Royal Society of Queensland,* 12 (1897), viii-xix.

Liebig, Justus von. 'Lord Bacon as Natural Philosopher', *Macmillan's Magazine*, 8 (1863), 237-249, 257-267.

Lindsay, W. Lauder. 'The Place and Power of Natural History in Colonisation with special reference to Otago (New Zealand)', *Edinburgh New Philosophical Journal*, 17 (n.s.), (1863), 125-146.

Lock, Alfred G. *Gold: Its Occurrence and Extraction* (London: E. and F.N. Spon, 1882).

MacCallum, M. 'The Federation of the Universities', *Hermes*, VIII (Jubilee Issue, 1902), 51-58.

_____. 'University Development', *The Australian Journal of Education*, 4 (12), (1907), 6-8.

*MacLaren, James Malcolm. *Gold: Its Geological Occurrence and Geographical Distribution* (London: The Mining Journal, 1908).

McCoy, Frederick. 'The Order and Plan of Creation: The Substance of Two Lectures delivered in connection with the Early Closing Association' in *Lectures Delivered Before the Early Closing Association, Melbourne, 1869-70* (Melbourne: Samuel Mullen, 1870), 24-32.

MacDonald, A.C. 'The Utility of, and Necessity for, a Geographical Society', *Proceedings of the Geographical Society of Australasia (N.S.W. and Victoria),* 1 (1883-1884), 133-136.

Maiden, J.H. 'Portraits of Scientific Men of New South Wales', *Proceedings of the Royal Society of New South Wales,* 46 (1912), 17-20.

_____. 'A Contribution to a History of the Royal Society of New South Wales (with Information in regard to other New South Wales Societies)', *Journal and Proceedings of the Royal Society of New South Wales,* 52 (1918), 215-361.

*Marr, H.E. and A.E. Shipley. *Handbook to the Natural History of Cambridgeshire* (Cambridge: Cambridge University Press, 1904).

Meyrick, Edward. 'Charles Darwin', *Sydney University Review*, 3 (July 1882), 244-253.

Merz, J. T. *A History of European Thought in the 19ᵗʰ Century*, 4 vols, (1896-1914; Edinburgh: Blackwood, 1923).

*Miers, Henry. *Mineralogy. An Introduction to the Scientific Study of Minerals* (London: Macmillan, 1902).

Moseley, H.N. *Notes by a Naturalist on the Challenger* (London: Macmillan, 1879).

Newman, A.K. 'A New Zealand Association of Science', *New Zealand Journal of Science*, 1 (1883), 145-150.

Owen, Richard. *On the Extent and Aims of a National Museum of Natural History* (London: Saunders, Oatley and Co., 1862).

Parke, James. *The Cyanide Process of Gold Extraction* (Auckland: Chamtaloup and Cooper, 1897).

Pearson, Karl. *The Grammar of Science* (London: Macmillan, 1892; 3ʳᵈ ed. 1911).

Penn, Granville. *Conversations on Geology Comprising a Familiar Explanation of the Huttonian and Wernerian Systems: The Mosaic Geology* (London: Samuel Maunder, 1828).

*Phipson, T.L. *Meteors, Aerolites and Falling Stars* (London: Lovell Reeve, 1867).

Porter, Whitworth. *History of the Corps of Royal Engineers* (London: Longmans Green & Co, 1889; 2nd ed. Chatham: Institution of Royal Engineers, 9 vols., 1951-58).

Prestwich, Joseph. *Geology Chemical, Physical and Stratigraphical* (Oxford: Clarendon Press, vol. 1, 1886, vol. 2, 1888)

Prestwich, Lady Grace. *Life and Letters of Sir Joseph Prestwich* (Edinburgh: Blackwood, 1899).

Prior, G.T. 'The Classification of Meteorites', *Mining Magazine, XIX* (90), (September 1920), 51-63.

Proctor, Richard A. *Wages and Wants of Science Workers* (London: Smith, Elder, 1876).

Power, Frederick Danvers. 'Australasian Views on the Origin of Gold Nuggets', *Australian Mining Standard*, 27 July 1895, 421-422; 3 August 1895, 434-435.

*Ramsay, A.C. *The Physical Geology and Geography of Great Britain* (London: Edward Stanford, 1878).

*Ranft, J.A.H. Theodore. *The Origin and Formation of Auriferous Rocks and Gold* (Sydney: Turner and Henderson, 1889).

Ratte, Felix A. *Descriptive Catalogue and General Collection of Minerals in the Australian Museum* (Sydney: Government Printer, 1885).

Read, John. 'The Study of Science Abroad', *Hermes, 32* (2), (Trinity Term, 1926), 85-87.

*Read, John. *A Textbook of Organic Chemistry: Historical, Structural and Economic* (London: G. Bell, 1926).

Reeks, Margaret. *Register of the Associates and Old Students of the Royal School of Mines and History of the Royal School of Mines* (London: Royal Society of Mines (Old Student's) Association, 1920).

Reid, Sir Thomas Wemyss. *Memoirs and Correspondence of Lyon Playfair, First Lord Playfair of St. Andrews* (London: Cassell and Co., 1899).

Richards, Henry C. 'Development of Petrology during the Present Century', (Presidential Address), *Proceedings of the Royal Society of Queensland, XXVI* (1914), 1-11.

Robertson, J.Steele. 'Natural Science in Australia', *Centennial Magazine*, 2 (1889-90), 523-527.

Roscoe, Henry E. *The Life and Experiences of Sir Henry Enfield Roscoe* (London: Macmillan, 1906).

Rose, T. Kirke. *The Metallurgy of Gold* (London: Charles Griffin, 5th ed., 1906).

Roth, Henry Ling. *A Report on the Sugar Industry in Queensland* (Brisbane: Gordon and Gotch, 1880).

Russell, H.C. (ed.). *Observations of the Transit of Venus, 9 December 1874 made at Stations in New South Wales* (Sydney: Government Observatory, 1892).

*Russell, H.C. *Climate of New South Wales* (Sydney: Government Printer, 1877).

*Rutley, Frank. *Mineralogy* (London: Thomas Mundy, 1876).

St. Julien, Charles and E.K. Silvester. *The Productions, Industry and Resources of New South Wales* (Sydney: J. Moore, 1853).

Seaver, Jonathan C.B.P. 'Origins and Mode of Occurrence of Gold-bearing Veins and of the Associated Minerals', *Journal and Proceedings of the Royal Society of New South Wales*, 21 (1887), 125-158.

*Selfe, Norman. 'Sydney and its Institutions, as They Are and Might Be, from an Engineer's Point of View', *Proceedings of the Engineering Association of New South Wales*, XV (1900), 19-44.

Shadwell, Arthur. *The London Water Supply* (London: Longmans, Green, 1899).

Smith, John. 'On the Results of the Chemical Examination of Waters for the Sydney Water Commission', *Transactions of the Royal Society of NSW*, 2 (1868), 146-156.

_____. *Wayfaring Notes: Sydney to Southhampton by way of Egypt and Palestine* (Sydney: Sherriff and Downing, 1865).

_____. 'On the Water Supply of Sydney', *Transactions of the Royal Society of NSW*, 2 (1868), 86-96.

_____. *Wayfaring Notes (Second Series): A Holiday Tour Round the World* (Aberdeen: A. Brown and Co., 1876).

Small, John. *A Hundred Wonders of the World in Nature and Art* (London: William P. Nimmo, 1878).

Smiles, Samuel. *Lives of the Engineers: Boulton and Watt, The Steam Engine* (London: John Murray, 1904).

Snyder, Carl. 'America's Inferior Position in the Scientific World', *North American Review*, DXLII (January 1902), 59-72.

Sorby, H.C. 'On the Microscopical Structure of Crystals Indicating the Origin of Minerals and Rocks', *Journal of the Geological Society*, 14 (1858), 453-500.

Spencer, Herbert. 'The Genesis of Science', *Essays: Scientific, Political and Speculative* (London: Williams and Norgate, 3 vols, 1891; 2nd ed., Osnabruck: Zeller, 1966).

*Spry, W.J.J. *The Cruise of HMS Challenger: Voyages over Many Seas, Scenes in Many Lands* (London: Sampson Low, 1877).

*Step, Edward. *Wayside and Woodland Blossoms* (London: Frederick Warne, 1906).

Stephens, William J. 'Notes and Queries about Artesian Prospects in New South Wales', *Sydney University Review*, 3 (July 1882), 207-224.

_____. 'Biological Science: A Necessary Factor in University Work', *Sydney University Review*, 4 (December 1882), 393-407.

*Stirling, Patrick James. *The Australian and Californian Gold Discoveries, and their Probable Consequences* (Edinburgh: Oliver and Boyd, 1853).

*Streeter, Edwin W. *Gold or Legal Regulations for the Standard of Gold and Silver Wares in Different Countries of the World* (London: Chatto and Windus, 1877).

*Strzelecki, P.E. de. *Gold and Silver: A Supplement to Strzelecki's Physical Description of New South Wales and Van Dieman's Land* (London: Longman, Brown, Green, and Longmans, 1856).

* Stuart, T. P. Anderson. 'A Review of University Life in Australia, with its Conditions and Surroundings in 1891', *Proceedings of the Royal Colonial Institute*, XXIII, Part III (12 January 1892), 93-141.

Symons, Brenton. *Genesis of Metallic Ores and the Rocks which Enclose Them* (London: Mining Journal, 1908).

Therry, Roger. *Reminiscences of Thirty Years' Residence in New South Wales and Victoria* (London: Samson, Low,

Son & Co, 1863; Sydney: Sydney University Press, 1974).

Thomson, Sir C. Wyville. *The Voyage of the Challenger*, Vol. 1 (London: Macmillan, 1877).

Thompson, G.M. 'Biographical Notes: Frederick Wollaston Hutton', *The New Zealand Journal of Science*, 2 (7), (1885), 301-306.

Thompson, J. Ashburton. 'A Record of the Sanitary State of New South Wales', *Transactions of the Intercolonial Medical Congress of Australasia*, 3 (1889), 434-456.

Thompson, Silvanus. *The Life of William Thomson, Baron Kelvin of Largs* (London: Macmillan, 1910), 2 vols.

Threlfall, Richard. 'The New Physical Laboratory at the University of Sydney,' *Building and Engineering Journal*, 2 (1889), 1-3.

Twopenny, Richard E.N. *Town Life in Australia* (London: Elliot Stock, 1883).

Vonwiller, O.U. 'The late Professor Pollock', *Hermes*, 28 (August 1922), 77-80.

Walkolm, A.B. *Linnean Society of New South Wales: Historical Notes on its First Fifty Years* (Sydney: Linnean Society of New South Wales, 1925).

Washington, Henry S. *Manual of the Chemical Analysis of Rocks* (New York: John Wiley and Sons, 1904).

Watson, J. Forbes. *The Imperial Museum for India and the Colonies* (London: W.H. Allen, 1876).

Watts, Isaac. *The Proposed Imperial Museum for India and the Colonies* (London: W.H. Allen, 1876).

Welsh, David A. 'The Great Opportunity', *Hermes*, XXI, No 3 (November 1915), 103-112; republished by the Sydney and Melbourne Publishing Co, 1915.

Wells, H. G. *A Modern Utopia* (London: Chapman and Hall, 1905).

_____. *Mr Britling Sees it Through* (London: Collins, 1916).

Wrixon, H.J. 'The Condition and Prospects of Australia as compared with Older Lands', in *Lectures delivered before the Early Closing Association, Melbourne, 1869-70* (Melbourne: Samuel Mullen, 1870), 1-23.

# II. Secondary Sources

# Theses and Unpublished MS

Barrett, Desmond. 'A Cuckoo in the Nest? Chemical Research in Museums, with special reference to the origin, development, and demise of the Phytochemical Research Programme at the Museum of Applied Arts and Sciences, Sydney, Australia, 1878-1979' (PhD dissertation, University of New South Wales, 1999).

Burnett, R.I. 'The Life and Work of James Hector with Special Reference to the Hector Collection' (MA Hons. thesis, University of Dunedin, 1936).

Campbell, D. 'Culture and the Colonial City: A Study in Ideas, Attitudes and Institutions: Sydney, 1870-1890' (PhD dissertation, University of New South Wales, 1982).

Ewing, L.J. 'The Work in Otago of Sir James Hector, Provincial Geologist, 1862-1865' (BA Hons. thesis, University of Otago, 1929).

Gerathy, Greta. 'The Role of the City City Council in the Development of the Metropolitan Area, 1842-1912' (MA thesis, University of Sydney, 1967).

Gilbert, L.A. 'Botanical Investigation of Eastern Seaboard Australia, 1788-1810' (BA Hons. thesis, University of New England, 1962).

Glover, Richard. 'Scientific Racism and the Australian Aboriginal, 1865-1915' (BA Hons. thesis, University of Sydney, 1982).

Hoare, Michael. 'Science and Scientific Association in Eastern Australia, 1820-1890' (Ph.D. dissertation, Australian National University, 1974).

Johnston, R.I. 'History of the Sydney Mechanics School of Arts' (MA thesis, Australian National University, 1968).

Ling, Edward J. 'William Denison: Reformer of Colonial Science' (BA Hons. thesis, University of New South Wales, 1973).

Liston, Carol. 'New South Wales under Governor Brisbane, 1821-1825,' (PhD dissertation, University of Sydney, 1981).

McKelleher, Brian G. 'The Contribution of Francis Anderson to Education in New South Wales' (MEd thesis,

University of Sydney, 1967).

Martin, Stephen. 'The Reverend William Branwhite Clarke, Colonial Scientist' (MScSoc thesis, University of New South Wales, 1981).

Melleuish, Gregory. 'The Sydney Intellectual Milieu, c.1850-1856' (MA thesis, University of Sydney, 1980).

Moran, Jean. 'Scientists in the Political and Public Arena: A Social-Intellectual History of the Australian Association of Scientific Workers, 1939-1949' (MPhil thesis, Griffith University, 1983).

Murray-Smith, Stephen. 'A History of Technical Education in Australia, with special reference to the period before 1914' (PhD dissertation, University of Melbourne, 1966).

O'Connor, D.J. 'Royal Australian Chemical Institute, 1917-67' (MEd thesis, University of Sydney, 1970).

Oppliger, Kurt. 'Science in New South Wales in the 1820's' (MA thesis, University of Sydney, 1983).

Orr, Kevin H. 'The Sydney International Exhibition, 1879' (BA thesis (Economic History), Macquarie University, 1972).

Radford, Wilma. 'Charles Badham and his Work for Education in New South Wales' (MEd thesis, University of Sydney, 1969).

Riley, Jean. 'The Mechanics' School of Arts Movement in New South Wales' (Unpublished essay, University of Sydney, 1983).

Smith, Miriam E. 'The History of New Zealand Exhibitions, with particular reference to the New Zealand and South Seas International Exhibitions, Dunedin, 1925-1926' (MA thesis, University of Auckland, 1974).

Taylor, R.S. 'Changes in the Curricula of the University of Sydney' (MA thesis, University of Sydney, 1970).

Willis, J.L. 'From Palace to Power House: The First One Hundred Years of the Sydney Museum of Applied Arts and Sciences' (unpublished MS, Powerhouse Museum Archives, 1982).

Winter, Gillian. '"For . . . the Advancement of Science": The Royal Society of Tasmania, 1843-1885' (BA thesis, University of Tasmania, 1972).

Zacharias, Kristen L. 'Construction of a Primate Order: Taxonomy and Comparative Anatomy in Establishing the Human in Nature, 1733-1916' (PhD dissertation, Johns Hopkins University, 1980).

# Books and Articles

Abbie, A.A. 'The History of Biology in Australia', *The Australian Journal of Science,* 17 (1), (1954), 1-9.

Adamson, John W. *English Education, 1789-1902* (Cambridge: Cambridge University Press, 1930), 387-414.

Aird, W.V. *The Water Supply, Sewerage and Drainage of Sydney* (Sydney: Halstead Press, 1961).

Alexander, A.E. 'Some Reflections on the Teaching of Chemistry in Australian Universities', *Vestes,* 3 (4), (1960), 36-40.

Allen, David. *The Naturalist in Britain: A Social History* (London: Allen Lane, 1976; 2nd ed., Princeton: Princeton University Press, 1994).

Allen, J. 'The Technology of Colonial Expansion: A Nineteenth Century Military Outpost on the North Coast of Australia', *Industrial Archaeology,* 4 (2), (1967), 111-137.

Allwood, John. *The Great Exhibitions* (London: Studio Vista, 1977).

Ames, Winslow. *Prince Albert and Victorian Taste* (London: Chapman and Hall, 1968).

Anderson, C. 'Foundations and Early History of the Australian Museum', *Journal of the Royal Australian Historical Society,* 20 (3), (1934), 196-200.

Andrade, E.N. da C. 'William Henry Bragg', *Obituary Notices of Fellows of the Royal Society,* 4 (1942-1944), 277-300.

Andrews, E.C. 'The Heroic Period of Geological Work in Australia', *Proceedings of the Royal Society of New South Wales,* LXXVI (1942), 96-128.

Archer, Mary and Christopher D. Haley (eds.). *The 1702 Chair of Chemistry at Cambridge: Transformation and Change* (Cambridge: Cambridge University Press, 2005).

Ashby, Eric. 'Universities in Australia' in *The Future of Education,* No. 5 (Sydney: Australian Council for Educational Research, 1944).

Auerbach, Jeffrey A. *The Great Exhibition of 1851: A Nation on Display* (New Haven: Yale University Press, 1999).

Australian Academy of Technological Sciences and Engineering, *Technology in Australia, 1788-1988: A*

*Condensed History of Australian Technological Innovation and Adaptation during the First Two Hundred Years* (Melbourne: Australian Academy of Technological Sciences and Engineering, 1988); J. E. Kolm, 'The Chemical Industry — Australian Contributions to Chemical Technology', 631-732; G. B. O'Malley, 'The Mineral Industries', 733-777.

Austin, A.G. *Australian Education, 1788-1900: Church, State and Public Education in Colonial Australia* (Melbourne: Sir Isaac Pitman and Sons, 1961).

Ayres, Philip. *Mawson: A Life* (Melbourne: Miegunyah Press, 1999).

Bacchi, Carol. 'The Nature-Nurture Debate in Australia, 1900-1914,' *Historical Studies*, 19 (75), (1980), 199-212.

Badger, Sir Geoffrey. 'The Role of Government in Australian Science', *Australian Physicist*, 17 (10), (1980), 157-160.

Bailes, Kendall E. *Science and Russian Culture in an Age of Revolutions: V.I. Vernadsky and his Scientific School, 1863-1945* (Bloomington: Indiana University Press, 1990).

Bailey, E. B. *Geological Survey of Great Britain* (London: Thomas Murby, 1952).

Baker, A.T. 'Archibald Liversidge: A Giant of Australian Science', *Chemistry in Australia*, 64 (9), (October 1997), 1-19.

Baragwanath, W. 'The Geological Survey, 1852-1952', *Victorian Mining and Geology Journal*, 5 (1953), 4-12.

Barber, Lynn. *The Heyday of Natural History, 1820-1870* (New York: Doubleday and Company, 1980).

Barcan, Alan. *A Short History of Education in New South Wales* (Sydney: Martindale Press, 1965).

Barnard, Alan. *Visions and Profits: Studies in the Business Career of T.S. Mort* (Melbourne: Melbourne University Press, 1961).

Barton, Ruth. '"An Influential Set of Chaps": The X-Club and Royal Society Politics, 1864-1885', *British Journal for the History of Science,* 23 (1), (1990), 53-81.

Basalla, George. 'The Spread of Western Science', *Science*, 156 (3775), (5 May 1967), 611-622.

_____. 'Science and the City before the Nineteenth Century', in Everett Mendelsohn (ed.), *Transformation and Tradition in the Sciences: Essays in Honour of I. Bernard Cohen* (Cambridge: Cambridge University Press, 1984), 513-529.

Basalla, George. William Coleman and Robert H. Kargon (eds.), *Victorian Science: A Self-Portrait for the Presidential Addresses to the British Association for the Advancement of Science* (Garden City: Doubleday, 1970).

Bastings, L. 'History of the New Zealand Society, 1851-1868: A Wellington Scientific Centenary', *Transactions of the Royal Society of New Zealand,* 80, (3 and 4), (February 1953), 359-366.

Beaumont, Jeanette and W. Vere Hole. *Letters from Louisa: A Woman's View of the 1890s, based on the Letters of Louisa Macdonald, first Principal of the Women's College, University of Sydney* (Sydney: Allen & Unwin, 1996).

Becher, Harvey W. 'William Whewell's Odyssey: From Mathematics to Moral Philosophy', in Menachem Fisch and Simon Schaffer (eds.), *William Whewell: A Composite Portrait* (Oxford: Clarendon Press, 1991), 1-29.

Beder, Sharon. 'Early Environmentalists and the Battle against Sewers in Sydney', *Journal of the Royal Australian Historical Society,* 76 (1), (1990), 27-44.

_____. 'From Sewage Farms to Septic Tanks: Trials and Tribulations in Sydney', *Journal of the Royal Australian Historical Society,* 79 (1-2), (1993), 72-95.

Bentley, Jonathan. 'The Chemical Department of the Royal School of Mines: Its Origins and Development under A.W. Hofmann', *Ambix,* 17 (1970), 153-181.

Berdahl, R.O. *British Universities and the State* (Berkeley: University of California Press, 1959).

Berman, Morris. *Social Change and Scientific Organisation: The Royal Institution, 1799-1844* (London: Heinemann, 1978).

Berry, R.J. (ed.). 'Happy is the Man that Findeth Wisdom', in 'Charles Darwin: A Commemoration, 1882-1902', *Biological Journal of the Linnean Society,* 17 (1), (1982), 1-8.

Best, Rupert J. *Discoveries by Chemists: The University of Adelaide* (Adelaide: University of Adelaide Foundation, 1986).

Betz, F. 'Geological Communication', in C.C. Albritton (ed.), *The Fabric of Geology* (Reading, Mass.: Addison-Wesley, 1963), 193-217.

Birch, Arthur J. 'Chemistry in Australia: 200 Years On', *Chemistry in Britain,* 24 (4), (1988), 359-362.

Birmingham, Judy *et al. Australian Pioneer Technology: Sites and Relics* (Sydney: Heinemann, 1979).

Blackburn, Charles Bickerton. 'The Life and Work of Sir Thomas Anderson Stuart', *Bulletin of the Post Graduate Committee in Medicine University of Sydney,* 4 (4), (1948), 105-134.

Blainey, Geoffrey. *A Centenary History of the University of Melbourne* (Melbourne: Melbourne University Press, 1957).

_____. *The Tyranny of Distance: How Distance Shaped Australia's History* (Melbourne: Sun Books, 1977).

_____. 'Reverberations of Perpetual Change: Technology's Leads and Lags', *B.H.P. Journal*, 2 (1981), 8-15.

Block, Ed., Jr. 'T.H. Huxley's Rhetoric and the Popularisation of Victorian Scientific Ideas, 1854-1874', in Patrick Brantlinger (ed.), *Energy and Entropy: Science and Culture in Victorian Britain* (Bloomington: University of Indiana Press, 1989), 205-228.

Bolton, Geoffrey. *Richard Daintree: A Photographic Memoir* (Brisbane: Jacaranda Press, 1965).

_____. 'The Idea of a Colonial Gentry', *Historical Studies*, 13 (1968), 307-328.

Bolton, H.C. 'J. J. McNeill and the Development of Optical Research in Australia', *Historical Records of Australian Science,* 5 (4), (1983), 55-70.

Bonython, Elizabeth and Anthony Burton. *The Great Exhibitor: The Life and Work of Henry Cole* (London Victoria and Albert Museum, 2003).

Borchardt, D.H. (ed.). *Some Sources for the History of Australian Science*, Historical Bibliography Monograph No. 12 (Sydney: University of New South Wales, History Project Incorporated, 1982).

Boyle, R.W. 'The Geochemistry of Gold and its Deposits', *Geological Survey Bulletin,* No. 280 (Washington, DC: US Geological Survey, 1979).

Branagan, D.F. 'The *Challenger* Expedition and Australian Science', *Proceedings of the Royal Society of Edinburgh, B,* 73 (10), (1971/2), 85-95.

_____.'History of Science in Australia: The Gentleman Scientists', *SCAN* (March 1968), 24-27.

_____. 'History of Science in Australia - Part 5: Sir Joseph and the Botanists', *SCAN* (September 1968), 17-22.

_____. 'History of Science in Australia - Part 8: Scientific Clubs, Museums and Men of Letters', *SCAN* (December 1968), 24-29.

_____. 'Words, Actions, People: 150 Years of the Scientific Societies in Australia', *Journal and Proceedings of the Royal Society of New South Wales*, 104 (1972), 123-141.

_____. 'The Geological Society of Australasia, 1885-1907', *Journal of the Geological Society of Australia,* 23 (2), (1976), 169-182.

_____. 'Putting Geology on the Map: Edgeworth David and the Geology of the Commonwealth of Australia', *Journal of the Geological Society of Australia*, 24 (1977), 279-305.

_____. 'Putting Geology on the Map', *Historical Records of Australian Science*, 5 (2), (1981), 30-57.

_____. 'J.W. Gregory, Traveller in the Dead Heart', *Historical Records of Australian Science,* 5 (2), (1981), 30-57.

_____. 'A.R.C. Selwyn: Pioneer of the Australian Geological Profession', *Abstracts of the Geological Society of Australia*, 12 (7th Australian Geological Convention (Sydney, 1984), 76-78.

_____. 'The History of Geological Mapping in Australia', in D. Borchardt (ed.), *Some Sources for the History of Australian Science* (Sydney: History Project Incorporated, 1984).

_____. 'Alfred Selwyn - Nineteenth Century Trans-Atlantic Connections Via Australia', *Earth Sciences History,* 9 (2), (1990), 143-157.

_____. 'Richard Owen, Thomas Mitchell and Australian Science: A Commemorative Symposium', *Journal and Proceedings, Royal Society of New South Wales,* 125 (3-4), (1992), 93-94.

_____. 'Samuel Stutchbury and the Australian Museum', *Records of the Australian Museum Supplement*, 15 (1992), 99-110.

_____. 'Samuel Stutchbury: A Natural History Voyage to the Pacific, 1825-27 and Its Consequences', *Archives of Natural History*, 20 (1), (1993), 69-91.

_____. *Samuel Stutchbury: Science in a Sea of Commerce: The Journal of a South Seas Trading Venture, 1825-27* (Northbridge, NSW: Privately Printed, 1996).

_____. 'Samuel Stutchbury (1798-1859)', *Oxford Dictionary of National Biography* (Oxford: Oxford University Press, 2004).

_____. *T.W. Edgeworth David: A Life: Geologist, Adventurer, Soldier and 'Knight in the Old Brown Hat'* (Canberra: National Library of Australia, 2005).

Branagan, D.F. (ed.). *Rocks - Fossils - Profs: Geological Sciences in the University of Sydney, 1866-1973* (Sydney: Science Press, 1973).

Branagan, D.F. and H. Graham Holland (eds.). *Ever Reaping Something New: A Science Centenary* (Sydney: University of Sydney, Science Centenary Committee, 1985).

Branagan, D.F. and K.A.Townley. 'The Geological Sciences in Australia - A Brief Historical Review', *Earth Science Reviews*, 12 (1976), 323-346.

Branagan, D.F., and T. Vallance. 'The Geological Society of Australasia (1885-1905)', *Journal of Geological Society of Australia*, 14 (2), (1967), 349-351.

Branagan, D.F. Samuel Stutchbury (1798-1859), *Australian Dictionary of Biography* (Melbourne: Melbourne University Press, 1976), vol. 6, 216-217.

Bray, R. J. 'Australia and the Transit of Venus', *Proceedings of the Astronomical Society of Australia,* 4 (1), (1980), 114-120.

Briggs, Asa. *Victorian Cities* (Harmondsworth: Penguin Books, 1982).

Brock, W.H. 'Chemical Geology or Geological Chemistry?' in L.J. Jordanova and R. Porter (eds.), *Images of the Earth* (London: British Society for the History of Science, 1978), 147-170.

_____. 'Exploring the Hyperarctic: James Dewar at the Royal Institution', in Frank James (ed.), *The Common Purposes of Life: Science and Society at the Royal Institution of Great Britain* (Aldershot: Ashgate, 2002), 169-190.

_____. *Sir William Crookes (1832-1919) and the Commercialization of Science* (Burlington, VT: Ashgate, 2007).

Brock, W.H. (ed.). *H.E. Armstrong and the Teaching of Science, 1880-1930* (London: Cambridge University Press, 1973).

Brock, W.H. and Roy MacLeod. 'The Scientists' Declaration: Reflexions on Science and Belief in the Wake of *Essays and Reviews*, 1864-1865', *British Journal of the History of Science*, IX Part 1, (31), (1976), 39-66.

Brockway, Lucile H. *Science and Colonial Expansion. The Role of the British Royal Botanic Garden* (New York: Academic Press, 1979).

Brooke, Christopher. *A History of the University of Cambridge*, vol. 4 (Cambridge: Cambridge University Press, 1992).

Brown, D.K. *Century of Naval Construction: The History of the Royal Corps of Naval Constructors, 1883-1983* (London: Conway Maritime Press, 1983).

Browne, Janet. *Charles Darwin: Voyaging*, vol. 1 (London: Alfred Knopf and Jonathan Cape, 1995).

_____. *Charles Darwin: The Power of Place*, vol. 2 (London: Jonathan Cape and Alfred Knopf, 2002).

Brown, Harrison Scott (ed.). *A Bibliography of Meteorites* (Chicago: University of Chicago Press, 1953).

Brown, W.R. and Ida A. Brown. 'The Royal Society of New South Wales: Scientific Societies in Australia', *Proceedings of the Royal Australian Chemical Institute*, 28 (6), (1961), 100-110.

Bryan, W.H. 'Samuel Stutchbury and Some of Those who Followed Him', *Queensland Government Mining Journal,* LV (1954), 641-646.

Buchwald, V.F. *Handbook of Iron Meteorites: Their History, Distribution, Composition, and Structure* (Berkeley: University of California Press, 1975).

Buckley, Vincent. 'Intellectuals', in Peter Coleman (ed.), *Australian Civilization* (Melbourne: F.W. Cheshire, 1962), 89-104.

Burke, John G. *Origins of the Science of Crystals* (Berkeley: University of California Press, 1966), 1-9, 10-51, 107-146.

Burn, W.L. *The Age of Equipoise: A Study of the Mid-Victorian Generation* (London: George Allen and Unwin, 1964).

Burdon, R.M. *Scholar Errant: A Biography of Professor A.W. Bickerton* (Christchurch: The Pegasus Press, 1956).

Butcher, Barry W. '"Adding Stones to the Great Pile"? Charles Darwin's Use of Australian Resources, 1837-1882', *Historical Records of Australian Science,* 8 (1), (December 1989), 1-14.

Butlin, N.G. *Investment in Australian Economic Development, 1861-1900* (Cambridge: Cambridge University

Press, 1964).

Butt, Charles R.M and R.H. Hough. 'Archibald Liversidge and the Origin of Gold Nuggets, 100 Years On', in R.R. Pieron (ed.), *The History of Geology in the Second Half of the Nineteenth Century: The Story in Australia, and in Victoria, from Selwyn and McCoy to Gregory* (Melbourne: Earth Sciences History Group, Geological Society of Australia, Special Publication No. 1, 2007), 16-19.

Bygott, Ursula and Ken Cable. *Pioneer Women Graduates of the University of Sydney, 1881-1921*, Sydney University Monographs No. 1 (Sydney: University of Sydney, 1985).

Byrnes, F.R. 'Other Days, Other Ways', *Quadrant*, 25 (May 1981), 63-65.

Cable, K.J. 'The University of Sydney and its Affiliated Colleges, 1850-1880', *The Australian University*, 2 (3), (1964), 183-214.

Cain, John and John Hewitt. *Off Course: From Public Place to Marketplace at Melbourne University* (Melbourne: Scribe, 2004).

Cairns, D. *Scientific Institutions in New Zealand, 1949* (Christchurch: C.S.W., 1949).

Caldwell, William.'The Gold Content of Sea Water', *Journal of Chemical Education*, 15 (11), (1938), 507-510.

Callaghan, F.R. (ed.). *Science in New Zealand* (Wellington: A.H. and A.W. Reed, 1957).

Cambie, R.C. and B.R. Davis. *A Century of Chemistry at the University of Auckland* (Auckland: Percival Publishing Company, 1982).

Cameron, H.C. *Sir Joseph Banks, KB, FRS: The Autocrat of the Philosophers* (London: Batchworth, 1952).

Candy, P.C. and J. Laurent (eds.). *Pioneering Culture: Mechanics' Institutes and Schools of Arts in Australia* (Adelaide: Auslib Press, 1994).

Cannon, W.F. 'Scientists and Broad Churchmen: An Early Victorian Intellectual Network', *The Journal of British Studies,* 4 (1), (1964), 65-88.

Cardwell, Donald S.L. *The Organisation of Science in England: A Retrospect* (London: Heinemann, 1957, 1972).

_____. 'Science and World War I', *Proceedings of the Royal Society of London,* A, 342 (1975), 447-456.

Caroe, G.M. *William Henry Bragg, 1862-1942* (Cambridge: Cambridge University Press, 1978).

_____. *The Royal Institution: An Informal History* (London: John Murray, 1985).

Chalmers, R.O. *Australian Rocks, Minerals and Gemstones* (Sydney: Angus and Robertson, 1967).

_____. 'History of the Department of Mineralogy, Australian Museum, Part 1: 1827-1901', *Journal of the Mineralogical Society of New South Wales*, 1 (June 1979), 13-21.

Chesneau, Gabriel. *L'École des Mines* (Paris: Association Amicale des Anciens Élèves de l'École Nationale Supérieure des Mines, 1931).

Chisholm, A.R. *Men Were My Milestones* (Melbourne: Melbourne University Press, 1958).

Chippindale, Peter. 'A Decade of Reform – The University of Sydney, 1876-1886', *Sydney University Archives Record* (1998), 7-26.

Clark, Betty. 'Men, Minerals and Museums: A Century (and more) of Mineral History in New South Wales. Parts 1-3', *Australian Lapidary Museum,* 11 (November 1974), 4-15; (December 1974), 27-34; (January 1975), 25-30.

Clark, David. '"Worse than Physic": Sydney's Water Supply, 1788-1888', in Max Kelly (ed.), *Nineteenth Century Sydney. Essays in Urban History* (Sydney: Sydney University Press, 1978), 54-65.

Clarke, F.G. *The Land of Contrarieties. British Attitudes to the Australian Colonies, 1828-1855* (Melbourne: Melbourne University Press, 1977).

Coates, J.E. 'The Haber Memorial Lecture', *Journal of the Chemical Society*, II (1939), 1642-1672.

Cobb, Joan. *Sweet Road to Progress* (Sydney: Department of Technical and Further Education, 2000).

Cockburn, Stewart and David Ellyard. *Oliphant: The Life and Times of Sir Mark Oliphant* (Adelaide: Axiom Books, 1981).

Collins, David J. 'An Explosion of Nitroglycerin in Sydney, 1866: A Window on Chemists and Chemistry in Mid-Nineteenth Century Australia', *Historical Records of Australian Science*, 13 (2), (2000), 131-149.

Collis, E.H. *Lost Years: A Backward Glance at Australian Life and Manners* (Sydney: Angus and Robertson, 1948).

Connell, R.W. 'The Colonial Mentality in Social Science', *Search*, 15 (1984), 100-111.

Correns, C.W. 'The Discovery of the Chemical Elements, the History of Geochemistry, Definitions of

Geochemistry', in C.W. Correns, D.M. Shaw, K.K. Turekian, and J. Zemann (eds.), *Handbook of Geochemistry* (New York: Springer-Verlag, 1969), vol. 1, 1-11.

Craig, D.P. 'Physical Chemistry at the University of Sydney', *Australian Journal of Science*, 16 (1954), 138-139.

Craig, Sir John. *The Mint: A History of the London Mint from A.D. 287 to 1948* (Cambridge: Cambridge University Press, 1953).

Cramp, K.R. 'Who First Discovered Payable Gold in Australia?', *Journal of the Royal Australian Historical Society*, 33 (V), (1947), 264-296.

Crawford, R.H. *'A Bit of a Rebel': The Life and Work of George Arnold Wood* (Sydney: Sydney University Press, 1975).

Crook, Thomas. *History of the Theory of Ore Deposits* (London: Thomas Murby, 1933).

Crosby, Alfred. *Ecological Imperialism the Biological Expansion of Europe, 900-1900* (Cambridge: Cambridge University Press, 1993).

Crowley, Frank (ed.). *A New History of Australia* (Melbourne: Heinemann, 1984).

Currie, Sir George and John Graham. *The Origins of CSIRO: Science and the Commonwealth Government, 1901-1926* (Melbourne: Melbourne University Press, 1966).

_____. 'Growth of Scientific Research in Australia: The Council for Scientific and Industrial Research and the Empire Marketing Board', *Records of the Australian Academy of Science*, 1 (3), (1968), 25-35.

_____. 'C.A. Julius and Research for Secondary Industry (CSIR)', *Records of the Australian Academy of Science*, 2 (1), (1970), 10-28.

_____. 'CSIR, 1926-1939', *Public Administration*, 33 (1974), 230-252.

Dale, Peter Allan. *In Pursuit of a Scientific Culture: Science, Art and Society in the Victorian Age* (Madison: University of Wisconsin, 1989).

Dallen, Robert. A. 'The Chancellors of the University of Sydney since its Foundation', *Journal of the Royal Australian Historical Society*, 19 (IV), (1933), 209-238.

_____. *The University of Sydney: Its History and Progress, 1852-1938* (Sydney: Angus and Robertson, 1938).

Darragh, T. 'The First Geological Maps of the Continent of Australia', *Journal of the Geological Society of Australia*, 24 (1977), 279-305.

Darragh, Thomas A. 'The Geological Survey of Victoria under Alfred Selwyn, 1852-1868', *Historical Records of Australian Science*, 7 (1), (1987), 1-25.

David, M.E. *Professor David. The Life of Sir Edgeworth David* (London: Edward Arnold, 1937).

Davies, Gordon L. *Earth in Decay: A History of British Geomorphology, 1578-1878* (New York: Neale Watson, 1969).

Davison, Graeme. 'Festivals of Nationhood: The International Exhibitions', in S.L. Goldberg and F.B. Smith (eds.), *Australian Cultural History* (Cambridge: Cambridge University Press, 1988), 158-177.

Davison, Graeme and Kimberley Webber (eds.). *Yesterday's Tomorrows: The Powerhouse Museum and its Precursors, 1880-2005* (Sydney: Powerhouse Publishing in association with UNSW Press, 2005)

Day, Alan A. 'The Development of Geophysics in Australia', *Proceedings of the Royal Society, New South Wales*, 100 (1966), 33-60.

_____. 'History of Science in Australia - Part 3: Early Scientific Expeditions', *SCAN* (June-July, 1968), 26-32.

Dean, Katrina. 'Inscribing Settler Sciences: Earnest Rutherford, Thomas Laby and the Making of Careers in Science', *History of Science*, XLI (2), (2003), 217-240.

Denholm, David. *The Colonial Australians* (Harmondsworth: Penguin, 1980).

Denmead, A.K. 'The Geological Survey of Queensland is 100 Years Old', *Queensland Government Mining Journal*, 69 (1968), 145-148.

Denoon, Donald. 'Understanding Settler Societies', *Historical Studies*, 18 (1979), 511-527.

_____. *Settler Capitalism. The Dynamics of Dependent Development in the Southern Hemisphere* (Oxford: Clarendon Press, 1983).

Desmond, Adrian. *Archetypes and Ancestors. Palaeontology in Victorian London, 1850-1875* (London: Blond and Briggs, 1982).

Dick, I.D. 'The History of Scientific Endeavour in New Zealand', *New Zealand Science Review*, 9 (9), (1951),

139-143.

Dixon, J.K. and F.R. Callaghan (eds.). *Science in New Zealand* (Wellington: A.H. and A.W. Reed, 1957).

Docker, John. *Australian Cultural Elites: Intellectual Traditions in Sydney and Melbourne* (Sydney: Angus and Robertson, 1974).

_____. *The Nervous Nineties: Australian Cultural Life in the 1890s* (Oxford: Oxford University Press, 1991).

Dolby, Alex. 'Debates over the Theory of Solutions: A Study of Dissent in Physical Chemistry in the English-speaking World in the late Nineteenth and early Twentieth-centuries', *Historical Studies in the Physical Sciences,* 7 (1976), 297-404.

Donath, E.J. *William Farrer* (Oxford: Oxford University Press, 1962).

Donnelly, J.F. 'Representations of Applied Science: Academics and Chemical Industry in late Nineteenth-Century England', *Social Studies of Science,* 16 (2), (1986), 195-234.

Drayton, Richard H. *Nature's Government: Science, Imperial Britain and the Modern World* (New Haven: Yale University Press, 2000).

Drury, Stephen. *Stepping Stones: The Making of Our Home World* (Oxford: Oxford University Press, 1999).

Dugan, K.G. 'Darwin and *Diprotodon*: The Wellington Caves Fossils and the Law of Succession', *Proceedings of the Linnean Society of New South Wales*, 104 (4), (1980), 265-272.

Duncan, R. 'Late Nineteenth Century Immigration to New South Wales from the United Kingdom', *Australian Economic History Review*, XIV (March 1974), 58-74.

Dunlap, Thomas R. 'Australian Nature, European Culture: Anglo Settlers in Australia', *Environmental History Review,* 17 (1), (1993), 25-48.

Dunstan, David. *Governing the Metropolis: Politics, Technology and Social Change in a Victorian City: Melbourne, 1850-1891* (Melbourne: Melbourne University Press, 1984).

Eckert, Jim. *The Early Years: For Chemistry Alumni* (Sydney: University of Sydney School of Chemistry, 2006).

Eddy, J.J. *Britain and the Australian Colonies, 1818-1831. The Techniques of Government* (Oxford: Clarendon Press, 1969).

Eggleston, Wilfrid. *Scientists at War* (Toronto: Oxford University Press, 1950).

_____. *National Research in Canada* (Toronto: Clarke Irwin, 1978).

Elkin, A.P. 'The Australian National Research Council', *Australian Journal of Science*, 16 (6), (1954), 203-211.

_____. 'Centenary Oration: Part 1; The Challenge to Science, 1866', *Journal and Proceedings of the Royal Society of New South Wales*, 100 (1966), 105-118.

Elkin, A.P. (ed.). *A Goodly Heritage* (Sydney: Australian and New Zealand Association for the Advancement of Science, 1962).

_____. (ed.). 'ANZAAS: A History', *Australian Journal of Science,* 25 (1962), 2-4.

Elkin, A.P. *et al. A Century of Scientific Progress: The Centenary Volume of the Royal Society of New South Wales* (Sydney: The Royal Society of New South Wales, 1968).

Emeléus, H.E. 'Early History of the Royal School of Mines and the Royal College of Science, 1837-1857', *The Record,* ser. III., no. 5 (1937), 59-62.

_____. 'Early Development of the Royal College of Science at South Kensington', *The Record,* ser. III, no. 6 (1937), 80-82.

Endersby, James. 'A Garden Enclosed: Botanical Barter in Sydney, 1818-1839', *British Journal for the History of Science,* 33 (118), (2000), 313-334.

Epps, Wiliam. *Anderson Stuart, MD: Physiologist, Teacher, Builder, Organizer, Citizen* (Sydney: Angus and Robertson, 1922).

Etherington, Norman. *Theories of Imperialism: War, Conquest and Capital* (London: Croom Helm, 1984).

Eve, A.S. and C.H. Creasey. *Life and Work of John Tyndall* (London: Macmillan, 1945).

Eyre, J. Vargas. *Henry Edward Armstrong, 1848-1937, the Doyen of British Chemists and Pioneer of Technical Education* (London: Butterworths Scientific Publications, 1958).

Farrer, K.T.H. *A Settlement Amply Supplied: Food Technology in Nineteenth Century Australia* (Melbourne: Melbourne University Press, 1980).

Fay, C.R. *Palace of Industry* (Cambridge: Cambridge University Press, 1951).

Fay, F.R. 'History of Medical Congresses in Tasmania', *Medical Journal of Australia,* 1 (9), (3 May 1980),

414-416, 437.

Feehan, H. Victor. 'Johann August Kruse, 1822-1895', *Victorian Historical Journal*, 52 (November 1981), 248-253.

Feehan, H.V. and G.N.Vaughan. '100 Years of "Modern" Pharmacy in Australia, 1881-1891', *Pharmacy International,* 2 (July 1981), 149-151.

Fenner, Frank. 'Scientific Societies in Australia: The Australian Academy of Science', *Proceedings of the Royal Australian Chemical Institute*, 27 (1960), 289-294.

Fenner, Frank and A.L.G. Rees (eds.). *The First Twenty Five Years: The Australian Academy of Science, 1954-1979* (Canberra: Australian Academy of Science, 1980).

Ferguson, Eugene S. 'Technical Museums and International Exhibitions', *Technology and Culture*, 6 (1), (1963), 30-46.

Findlay, Alexander. *A Hundred Years of Chemistry* (New York: Macmillan, 1937).

Findlay, Alexander and W. Hobson Mills. *British Chemists* (London: Chemical Society of London, 1947).

Fischer, G.L. *The University of Sydney, 1850-1975. Some History in Pictures to mark the 125th Year of its Incorporation* (Sydney: University of Sydney, 1975).

Fisher, Shirley. 'The Pastoral Interest and Sydney's Public Health', *Historical Studies,* 20 (78), (1982), 73-89.

Fitzgerald, Shirley. *Rising Damp: Sydney, 1870-90* (Melbourne: Oxford University Press, 1987).

Fleming, C.A. 'The Royal Society of New Zealand - A Century of Scientific Endeavour', *Transactions of the Royal Society of New Zealand*, 2 (6), (1968), 99-114.

_____. 'J.A. Thompson's Proposals for Reform of the New Zealand Institute in 1917 - A Chapter in the History of the Royal Society of New Zealand', *Transactions of the Royal Society of New Zealand,* 2 (8), (1969), 129-133.

_____. 'The Contribution of New Zealand Geoscientists to the Development of Scientific Institutions', *Journal of the History of the Earth Sciences Society*, 5 (1986), 3-11.

_____. *Science, Settlers and Scholars: Centennial History of the Royal Society of New Zealand* (Wellington: Royal Society of New Zealand, 1986).

Fleming, Donald. 'Science in Australia, Canada and the United States: Some Comparative Remarks', *Proceedings of the 10th International Congress of the History of Science,* 1 (1962), 179-196.

Fletcher, Brian H. 'Biography and the History of Colonial New South Wales', *Teaching History* (July 1981), 4-22.

_____. 'The Agricultural Society of New South Wales and its Shows in Colonial Sydney', *Journal and Proceedings, Royal Society of New South Wales*, 118 (1985), 195-208.

Fletcher, J.J. 'The Society's Heritage from the Macleays', *Proceedings of the Linnean Society of New South Wales,* XIV (1), (1920), 567-636; LIV (2), (1929), 185-272.

Flower, Cedric. *The Antipodes Observed: Prints and Print Makers of Australia, 1788-1850* (Melbourne: Macmillan, 1975).

Foster, Leonie. *High Hopes: The Men and Motives of the Australian Round Table* (Melbourne: Melbourne University Press, 1986).

Foster, Stephen G. *Colonial Improver: Edward Deas Thomson, 1800-1879* (Melbourne: Melbourne University Press, 1978).

Fowler, R.T. 'History of Chemical Engineering Education in Australia', *Chemical Engineering in Australia*, 5 (4), (1981), 40-48.

Fox, Dixon R. 'Civilization in Transit', *American Historical Review, 32* (1926-7), 753-768.

French, Yvonne. *The Great Exhibition, 1851* (London: Harwill Press, 1950).

Froggart, Walter W. 'The Curators and Botanist of the Botanic Gardens, Sydney', *Journal of the Royal Australian Historical Society*, 18 (1932), 101-133.

Gani, Joseph M. *et al. The Condition of Science in Australian Universities: A Statistical Survey, 1939-1960* (Canberra: Australian National University Press, 1962).

Gardner, W.J. *Colonial Cap and Gown: Studies in the Mid-Victorian Universities of Australasia* (Christchurch: University of Canterbury Press, 1979).

Garland, Martha McMackin. *Cambridge Before Darwin: The Ideal of a Liberal Education, 1800-1860* (Cambridge: Cambridge University Press, 1980).

Gay, Hannah. *The History of Imperial College London, 1907-2007: Higher Education and Research in Science, Technology and Medicine* (London: Imperial College Press, 2007).

Gibbs, C.R. Vernon. *British Passenger Liners of the Five Oceans: A Record of the British Passenger Liners and their Lines from 1838 to the Present Day* (London: Putnam, 1963).

Gibbs, W.J. *The Origins of Australian Meteorology* (Canberra: Australian Government Printing Service, 1975).

Geison, Gerald L. *Michael Foster and the Cambridge School of Physiology: The Scientific Enterprise in Late Victorian Society* (Princeton: Princeton University Press, 1978).

Gerathy, Greta. 'Sydney Municipality in the 1880s', *Journal of the Royal Australian Historical Society*, 58 (1), (1972), 23-54.

Gibbs, W.J. *The Origins of Australian Meteorology* (Canberra: Government Printing Service, 1975).

Gibbs-Smith, C.H. *The Great Exhibition of 1851* (London: H.M.S.O., 1981).

Gibson, Ross. *Diminishing Paradise: Changing Literary Perceptions of Australia* (Sydney: Angus and Robertson, 1984).

Gilbert, Lionel A. 'The Bush and the Search for a Staple in New South Wales, 1788-1810', *Records of the Australian Academy of Science,* 1 (1), (1966), 6-17.

_____. 'Plants, Politics and Personalities in Nineteenth Century New South Wales', *Journal of the Royal Australian Historical Society*, 56 (1), (March 1970), 15-35.

_____. 'Plants, Politics and Personalities in Colonial New South Wales', in D.J. and S.G.M. Carr (eds.), *People and Plants in Australia* (Sydney: Academic Press, 1981), 220-258.

_____. 'Plants and Parsons in Nineteenth Century New South Wales', *Historical Records of Australia*, 5 (3), (1982), 17-32.

_____. *The Little Giant: The Life and Work of Joseph Henry Maiden, 1859-1952* (Sydney: Kardoorair Press in association with the Royal Botanic Gardens, 2001).

Goddard, Roy H. *The Union Club, 1857-1957* (Sydney: Halstead Press, 1957).

Golder, Hilary. *Politics, Patronage and Public Works: The Administration of New South Wales, Vol 1: 1842-1900* (Sydney: University of New South Wales Press, 2005), 213.

Goldschmidt, V.M. *Geochemistry,* edited by Alex Muir (Oxford: Clarendon Press, 1954 ed.).

Goodman, Jordan. *The Rattlesnake: A Voyage of Discovery to the Coral Sea* (London: Faber and Faber, 2005).

Goodwin, Crafurd D.W. *The Image of Australia: British Perceptions of the Australian Economy from the Eighteenth to the Twentieth Century* (Durham: Duke University Press, 1974).

Grainger, Elena. *The Remarkable Reverend Clarke* (Melbourne: Oxford University Press, 1982).

Grant, Robert. *Representations of British Emigration, Colonisation and Settlement: Imagining Empire, 1800-1860* (London: Palgrave, 2005).

Greene, Mott T. *Geology in the Nineteenth Century: Changing Views of a Changing World* (Ithaca: Cornell University Press, 1982).

_____. 'History of Geology', *Osiris* (2nd series), 1 (1985), 97-116.

Greenhalgh, Paul. *Ephemeral Vistas: A History of the Expositions Universelles, Great Exhibitions and World's Fairs, 1851-1939* (Manchester: Manchester University Press, 1988).

Gregory, Cedric. *A Concise History of Mining* (London: Pergamon Press, 1982).

Griffiths, Tom. *Hunters and Collectors: The Antiquarian Imagination in Australia* (Cambridge: Cambridge University Press, 1996).

Griffiths, Tom and Libby Robin (eds.). *Empire and Ecology: Environmental History of Settler Societies* (Melbourne: Melbourne University Press, 1997).

Griggs, Peter. 'Rust Disease Outbreaks and their Impact on the Queensland Sugar Industry, 1870-1880', *Agricultural History*, 69 (3), (1995), 413-437.

_____. 'Australian Scientists, Sugar Cane Growers and the Search for New Gumnosis-resistant and Sucrose-rich Varieties of Sugar Cane, 1890-1920', *Historical Records of Australian Science*, 14 (3), (2003), 291-311.

_____. 'Defeating Cane Diseases: Plant Pathologists and the Development of Disease Control Strategies in the Australian Sugar Industry, 1920-1950', *Historical Records of Australian Science*, 18 (1) (2007), 43-73.

Groom, Barry and Warren Wickman. *Sydney - The 1850s: Lost Collections* (Sydney: Historic Photographs Collection, University of Sydney, 1982).

Guntau, Martin. 'The Mining Academy of Freiberg: A Centre of Geoscientific Teaching and Research', *Journal of Mines, Metals and Fuels (Calcutta)*, 22 (1974), 223-227.

_____. 'The History of the Origins of the Prussian Geological Survey in Berlin (1873)', *History and*

*Technology*, 5 (11), (1988), 51-58.

Haber, Lutz. *The Chemical Industry, 1900-1930: International Growth and Technological Change* (Oxford: Oxford University Press, 1971).

Hall, A. Rupert. *Science for Industry: A Short History of the Imperial College of Science and Technology and its Antecedents* (London: Imperial College, 1982).

Hall, Brian K. 'The Paradoxical Platypus', *Bioscience*, 49 (3), (1999), 211-218.

Hall, Catherine. *Civilising Subjects: Metropole and Colony in the English Imagination, 1830-1867* (London: Polity, 2002).

Hall, Lincoln. *Douglas Mawson: The Life of an Explorer* (Sydney: Hew Holland, 2000).

Hamlin, Christopher. 'Edward Frankland's Early Career as London's Official Water Analyst, 1865-1876: The Context of Previous Sewage Contamination', *Bulletin of the History of Medicine*, 56 (1982), 56-76.

_____. 'Providence and Putrefaction: Victorian Sanitarians and the Natural Theology of Health and Disease', *Victorian Studies*, 28 (3), (1985), 381-411.

Hanley, W.S. 'Griffith Taylor's Antarctic Achievement: a Geographical Foundation', *Australian Geographical Studies*, 18 (1), (1980), 22-36.

Hannaford, Peter. 'Alan Walsh, 1916-1998', *Historical Records of Australian Science*, 13 (2), (2000), 179-206.

Hannan, Frank (ed.). *Gold, Mineral and Gemstone Localities of New South Wales* (Kensington: Continental Atlas Agency, 1965).

Harman, Kaye (ed.). *Australia Brought to Book: Response to Australia by Visiting Writers, 1836-1939* (Sydney: Boobooks, 1985).

Harman, Peter. *Wranglers and Physicists: Studies on Cambridge Physics in the Nineteenth Century* (Manchester: Manchester University Press, 1985).

Hartog, Philip J. 'John Percy (1817-1889)', *Dictionary of National Biography*, XLIV, 425-426.

Hartcup, Guy. *The War of Invention: Scientific Developments, 1914-18* (London: Brassey's, 1988).

Haynes, Roslyn. *From Faust to Strangelove: Representations of the Scientist in Western Literature* (Baltimore: Johns Hopkins University Press, 1994).

Hays, J.N. 'Science in the City: The London Institution, 1819-40', *British Journal for the History of Science*, 7 (26), (1974), 146-162.

Head, Brian and James Walter (eds.). *Intellectual Movements and Australian Society* (Sydney: Oxford University Press, 1988).

Headrick, Daniel R. *The Tools of Empire: Technology and European Imperialism in the Nineteenth Century* (New York: Oxford University Press, 1981).

_____. *Tentacles of Progress: Technology Transfer in the Age of Imperialism* (New York: Oxford University Press, 1988).

Healy, J. 'The Hot Springs and Geothermal Resources of Fiji', *New Zealand Department of Scientific and Industrial Research,* Bulletin No. 136 (Wellington: Government Printer, 1960).

Heide, Fritz. *Meteorites* (Chicago: University of Chicago Press, 1964).

Henry, F.J.J. *The Water Supply and Sewerage of Sydney* (Sydney: Metropolitan Water Sewerage and Drainage Board, 1939).

Herbert, D.A. 'A Story of Queensland's Scientific Achievement, 1859-1959', *Proceedings of the Royal Society of Queensland*, LXXI (1), (1959), 1-15.

Hevesy, George von. 'Significance of Meteorites', in von Hevesy, *Chemical Analysis by X-Rays and its Applications* (New York: McGraw Hill, 1932), 280-291.

Heyck, T. W. *The Transformation of Intellectual Life in Victorian England* (London: Croom Helm, 1982).

Hicks, Neville. *'This Sin and Scandal'. Australia's Population Debate, 1891-1911* (Canberra: Australian National University Press, 1978).

Hill, Dorothy. *The First Fifty Years of the Department of Geology at the University of Queensland* (Brisbane: Department of Geology, University of Queensland, Paper No. 10, 1981).

Hill, J.P. 'J.T. Wilson: A Biographical Sketch of His Career', *Journal of Anatomy*, 76 (1941-42), 3-8.

Hind, Robert J. 'The Internal Colonial Concept', *Comparative Studies in Society and History*, 26 (1984), 543-568.

Hoare, Michael E. 'Learned Societies in Australia: The Foundation Years in Victoria, 1850-1860', *Records of the Australian Academy of Science,* 1 (2), (1967), 7-29.

_____. 'Doctor John Henderson and the Van Diemen's Land Scientific Society', *Records of the Australian*

*Academy of Science*, 1 (3), (1968), 7-24.

_____. 'The Challenge of Science Accepted in New South Wales', *Records of the Australian Academy of Science*, 1 (4), (1969), 32-37.

_____. '"All Things are Queer and Opposite": Scientific Societies in Tasmania in the 1840s', *Isis*, 60 (2), (1969), 198-209.

_____. 'Cook the Discoverer: An Essay by George Forster, 1787', *Records of the Australian Academy of Science,* 1 (4), (1969), 7-16.

_____. 'Some Primary Sources for the History of Scientific Societies in Australia in the Nineteenth Century', *Records of the Australian Academy of Science,* 1 (4), (1969), 71-76.

_____. '"The Half-Mad Bureaucrat" Robert Brough Smyth (1830-1889)', *Records of the Australian Academy of Science,* 2 (4), (1974), 25-40.

_____. 'The History of Australian Science: Prospect and Retrospect', *Australasian Association for the History and Philosophy of Science Newsletter*, No. 5 (1974), 21-36.

_____. 'The Intercolonial Science Movement in Australasia, 1870-1890', *Records of the Australian Academy of Science*, 3 (2), (1975), 7-28.

_____. 'Light in our Past: Australian Science in Retrospect', *Search,* 6 (7), (1975), 285-290.

_____. 'Some Recent Writings in the History of Australian Science: A Critical Review', *Australasian Association for the History and Philosophy of Science Newsletter,* No. 7 (1975-76), 1-22.

_____. 'The Relationship between Government and Science in Australia and New Zealand', *Journal of the Royal Society New Zealand*, 6 (3), (1976), 381-394.

_____. *Reform in New Zealand Science, 1880-1926* (Melbourne: Hawthorn Press, 1977).

_____. *Beyond the 'Filial Piety': Science History in New Zealand: A Critical Review of the Art* (Melbourne: The Hawthorn Press, 1977).

_____. 'Botany and Society in Eastern Australia', in D.J. and S.G.M. Carr (eds.), *People and Plants in Australia* (Sydney: Academic Press, 1981), 183-219.

Hoare, M.E. and L.G. Bell (eds.). 'In Search of New Zealand's Scientific Heritage', *The Royal Society of New Zealand,* Bulletin No. 21 (Wellington: Royal Society of New Zealand, 1984).

Hodge-Smith, T. *Australian Meteorites,* Memoir VII (Sydney: The Australian Museum, 1939).

Hoffenberg, Peter. *An Empire on Display: English, Indian and Australian Exhibitions from the Crystal Palace to the Great War* (Berkeley: University of California Press, 2001).

Hogarth, J.W. 'Time has Seen Many Changes', *Science Year Book* (Sydney, 1951), 33-35.

Holland, Julian. 'Diminishing Circles: W.S. Macleay in Sydney, 1839-1865', *Historical Records of Australian Science,* 11 (2), (1996), 119-147.

Home, R.W. 'Origins of the Australian Physics Community', *Historical Studies,* 20 (80), (1981-82), 383-400.

_____. 'Between Classroom and Industrial Laboratory: The Emergence of Physics as a Profession in Australia', *Australian Physicist*, 20 (7), (1983), 163-167.

_____. 'The Problem of Intellectual Isolation in Scientific Life: W.H. Bragg and the Australian Scientific Community, 1886-1909', *Historical Records of Australian Science*, 6 (1), (1984), 19-30.

_____. 'First Physicist of Australia: Richard Threlfall at the University of Sydney, 1886-1898', *Historical Records of Australian Science*, 6 (3), (1986), 333-357.

_____. 'Australian Science and its Public', *Australian Cultural History*, No. 7 (1988), 86-103.

_____. 'Learning From Buildings: Laboratory Design and the Nature of Physics', in Renato G. Mazzolini (ed.), *Non-Verbal Communication in Science prior to 1900* (Firenze: Leo S. Olschki, 1993), 587-608.

_____. 'The Royal Society and the Empire: the Colonial and Commonwealth Fellowship, Part 2: After 1947', *Notes and Records of the Royal Society of London*, 5 (1), (2003), 47-84.

_____ (ed.). *Australian Science in the Making* (Sydney: Cambridge University Press, 1988).

Home, R. W. and Sally Gregory Kohlstedt (eds.). *National Scientific Identity: Australia Between Britain and America* (Dordrecht: Kluwer Academic Publishers, 1991).

Hooker, Claire. *Irresistible Forces: Australian Women in Science* (Melbourne: Melbourne University Press, 2004).

Hooykaas, Reijer. 'René-Just Haüy (1743-1822)', *Dictionary of Scientific Biography* (New York: Charles Scribner's Sons, 1970), VI, 178-182.

Howarth, O.J.R. *The British Association for the Advancement of Science: A Retrospect, 1831-1931* (London: The Association, 1931).

Hull, David. *Science as a Process: An Evolutionary Account of the Social and Conceptual Development of Science* (Chicago: University of Chicago Press, 1988).

Hunt, Bruce J. 'Doing Science in a Global Empire: Cable Telegraphy and Electrical Physics in Victorian Britain', in Bernard Lightman (ed.), *Victorian Science in Context* (Chicago: University of Chicago Press, 1997).

Hutchinson, Arthur. *Mineralogical Chemistry: Annual Reports on Progress for 1904* (London: Chemical Society, 1905).

Inkster, Ian. 'Scientific Enterprise and the Colonial "Model": Observations on Australian Experience in Historical Context (Discussion Paper)', *Social Studies of Science*, 15 (4), (1985), 677-704.

Ingpen, R. *Australian Inventions and Innovations* (Adelaide: Rigby, 1982).

James, Frank A.J.L. 'The Practical Problems of New Experimental Science: Spectro-chemistry and the Search for Hitherto Unknown Chemical Elements in Britain, 1860-69', *British Journal for the History of Science,* 21 (Pt 2), (69), (1988), 181-194.

James, Frank (ed.). *Chemistry and Theology in Mid-Victorian London: The Diary of Herbert McLeod, 1860-1870* (London: Mansell, 1987).

_____.(ed.). *The Common Purposes of Life: Science and Society at the Royal Institution of Great Britain* (Aldershot: Ashgate 2002).

Jardine, Nicholas, Anne Secord, and Emma Spary (eds.). *Cultures of Natural History* (Cambridge: Cambridge University Press, 1996).

Jeans, D.N. *An Historical Geography of New South Wales to 1901* (Sydney: Reed Education, 1972).

Jenkin, John. 'The Appointment of W.H. Bragg, FRS, to the University of Adelaide', *Notes and Records of the Royal Society of London*, 40 (1), (1985), 75-99.

_____. 'W.H. Bragg and the Public Image of Science in Australia', *Search*, 18 (1), (1987), 34-37.

_____. 'British Influence on Australian Physics, 1788-1988', *Berichte zur Wissenschaftsgeschichte*, 13 (1990), 93-100.

_____. 'William Henry Bragg in Adelaide: Beginning Research at a Colonial Locality', *Isis,* 95 (1), (2004), 58-90.

_____. *William and Lawrence Bragg, Father and Son: The Most Extraordinary Collaboration in Science* (Oxford: Oxford University Press, 2008).

Jenkin, John and R. W. Home. 'Horace Lamb and Early Physics Teaching in Australia', *Historical Records of Australian Science*, 10 (4), (1995), 349-380.

Jervis, James. 'Rev. W.B. Clarke, MA, FRS, FGS, FRGS: The Father of Australian Geology', *Journal of the Royal Australian Historical Society,* 30 (VI), (1944), 345-459.

Johns, R.K. (ed.). *History and Role of Government Geological Surveys in Australia* (Adelaide: Government Printer, 1976).

Jones, Robert S. 'Gold in Igneous, Sedimentary and Metamorphic Rocks', *Geological Survey Circular* No. 610 (Washington, DC: US Department of the Interior, 1969).

Jordens, Ann-Mari. *The Stenhouse Circle: Literary Life in mid-Nineteenth Century Sydney* (Melbourne: Melbourne University Press, 1979).

Joy, William. *The Other Side of the Hill: 200 Years of Australian Exploration* (Sydney: Doubleday, 1984).

Kargon, Robert H. *Science in Victorian Manchester: Enterprise and Expertise* (Manchester: Manchester University Press, 1977).

Kelly, Farley (ed.). *On the Edge of Discovery: Australian Women in Science* (Melbourne: Text Publishing, 1993).

Kelly, Max. 'Picturesque and Pestilential: The Sydney Slum Observed, 1860-1900', in Max Kelly (ed.), *Nineteenth Century Sydney: Studies in Urban and Social History* (Sydney: Sydney University Press, 1978), 66-80.

Kelly, Max and Ruth Crocker. *Sydney Takes Shape: A Collection of Contemporary Maps from Foundation to Federation* (Sydney: Sydney University Press, 1977).

Kent, Christopher. *Brains and Numbers: Elitism, Comtism, and Democracy in mid-Victorian England* (Toronto: University of Toronto Press, 1978).

Kerr, Joan. 'The Architecture of Scientific Sydney', *Journal and Proceedings of the Royal Society of New South*

*Wales*, 118 (3-4), (1985), 181-193.

Kingsmill, A.G. *Witness to History: A Short History of the Colonial Secretary's Department* (Sydney: Alpha Books, 1972).

Kohlstedt, Sally Gregory. *The Formation of the American Scientific Community: the American Association for the Advancement of Science, 1848-1860* (Urbana: University of Illinois Press, 1976).

_____. 'Servants and Professionals: The American Association for the Advancement of Science, 1848-1860', in Alexander Oleson and Sanborn C. Brown (eds**.**), *The Pursuit of Knowledge in the Early American Republic* (Baltimore: Johns Hopkins University Press, 1976), 229-325.

_____ (ed.). *History of Women in the Sciences: Readings from Isis* (Chicago: University of Chicago Press, 1999).

Kynaston, Edward. *A Man on the Edge: A Life of Baron Sir Ferdinand von Mueller* (London: Ringwood, 1981).

La Nauze, J.A. 'Jevons in Sydney', *Economic Record*, 17 (1941), 31-45, reprinted in John Cunningham Wood (ed.), *William Stanley Jevons: Critical Assessments* (London: Routledge, 1988), 109-125.

_____. '"Other Like Services": Physics and the Australian Constitution', *Records of the Australian Academy of Science,* 1 (3), (1968), 36-44.

Laudan, Rachel. *From Mineralogy to Geology: The Foundations of a Science* (Chicago: University of Chicago Press, 1987).

Lane, E.A. and A.M. Richards. 'The Discovery, Exploration and Scientific Investigation of the Wellington Caves, New South Wales', *Hecate, 2* (1), (1963), 1-53.

Laurent, John. 'Bourgeois Expectations and Working Class Realities: Science and Politics in Sydney's Schools of Arts', *Journal of the Royal Australian Historical Society,* 75 (1), (June 1989), 33-50.

Lawson, Sylvia. *The Archibald Paradox* (Ringwood: Allen Lane, 1983).

Layton, David. *Science for the People: The Origins of the School Science Curriculum in England* (London: Allen and Unwin, 1973).

Le Fèvre, R.J.W. 'Chemistry at the University of Sydney', *Chemistry and Industry,* (1951), 415-416.

_____. 'The Development of Chemistry in Australia', *Chemistry and Industry,* (1953), 736-738.

_____. 'Chemistry at the University of Sydney', *Chemistry and Industry,* (1957), 551-553.

_____. 'New Chemical Laboratories in the University of Sydney', *Nature,* 187 (1960), 833-834.

_____. 'The Establishment of Chemistry within Australia-Contributions from New South Wales', in A.P. Elkin (ed.), *Century of Scientific Progress* (Sydney: Royal Society of New South Wales, 1968), 332-378.

Lenoir, Timothy. *Strategy of Life: Teleology and Mechanics in Nineteenth Century German Biology* (Dordrecht: Reidel, 1982).

Levere, Trevor H. 'Geology and Chemistry, the Inward Powers of Matter', in Trevor Levere, *Poetry Realized in Nature* (Cambridge: Cambridge University Press, 1981), 159-200.

Levine, Philippa. *The Amateur and the Professional: Antiquarians, Historians and Archaeologists in Victorian England, 1838-1886* (Cambridge: Cambridge University Press, 1986).

Leviton, Alan E. and Michele L. Aldrich. 'Impact of Travels on Scientific Knowledge: William Thomas Blandford, Henry Francis Blanford and the Geological Survey of India, 1851-1889', in Michael T. Ghiselin and Alan E. Leviton (eds.), *Impact of Travels on Scientific Knowledge* (San Francisco: California Academy of Science, 2005), 117-137.

Lewis, David. *On the Plurality of Worlds* (Oxford: Blackwell, 1986).

Lewis, M.J. 'The Royal Society of Australia: An Attempt to Establish a National Academy of Science', *Records of the Australian Academy of Science,* 4 (1), (1978), 51-62.

_____. 'The Idea of a National University: The Origins and Establishment of the Australian National University', *Journal of the Australian and New Zealand History of Education Society,* 8 (1), (1979), 40-55.

Lightman, Bernard. '"The Voices of Nature": Popularizing Victorian Science', in Bernard Lightman (ed.), *Victorian Science in Context* (Chicago: University of Chicago Press, 1997), 187-211.

Lines, William J. *Taming the Great South Land: A History of the Conquest of Nature in Australia* (Sydney: Allen & Unwin, 1991).

Livingstone, Stanley. 'History of Chemistry in Australia, 1788-1988', *Journal of the Indian Chemical Society,* 66 (August-October 1989), 735-742.

Lonsdale, K. and E.N. da C. Andrade. 'WH Bragg', *Obituary Notices of Fellows of the Royal Society of London*, 4 (12), (November 1943), 277-300.

Lougheed, Alan. 'The Discovery, Development and Diffusion of New Technology: The Cyanide Process for the Extraction of Gold, 1887-1914', *Prometheus*, 7 (1), (1989), 61-74.

Lucas, R.R. *History of Geological Mapping: New South Wales Department of Mines Centenary, 1874-1974* (Sydney: N.S.W. Department of Mines, 1974).

Luckhurst, Kenneth W. *The Story of Exhibitions* (London: Studio Publications, 1951).

Lumsdaine, Keith. 'Sydney College and Sydney Grammar School', *Journal of the Royal Australian Historical Society*, 31 (1945), 4-41.

Lynch, P.P. 'Teaching of Chemistry, or "From Here to There and Back Again"', *Chemistry in Australia*, 41 (4), (1981), 141-144.

Lyte, Charles. *Sir Joseph Banks: 18th Century Explorer, Botanist and Entrepreneur* (Newton Abbot: David and Charles, 1980).

Maber, J.M. *North Star to Southern Cross* (New York: Augustus M. Keley, 1971).

McCall, G.J. (ed.). *Anthropology in Australia: Essays to Honour Fifty Years of Mankind* (Sydney: Royal Anthropology Society, 1982).

Macarthy, P.G. 'Wages in Australia, 1891-1914', *Australian Economic History Review*, X (1), (1970), 56-76.

McCartney, P.J. *Henry De la Beche: Observations on an Observer* (Cardiff: Friends of the National Museum of Wales, 1977).

McCarty, J.W. and C.B. Schedvin (eds.). *Australian Capital Cities: Historical Essays* (Sydney: Sydney University Press, 1978).

MacColl, Allan. 'Australian Chemists at University College London, 1899-1988', *Ambix*, 36 (2), (1989), 82-90.

McDonald, D.I. 'The Diffusion of Scientific and Other Useful Knowledge', *Journal of the Royal Australian Historical Society*, 54 (2), (1968), 176-193.

Macintyre, Stuart and Richard Joseph W. Selleck. *A Short History of the University of Melbourne* (Melbourne: Melbourne University Press, 2003).

Mackay, David. *In the Wake of Cook: Exploration, Science and Empire, 1780-1801* (Wellington: Victoria University Press, 1985).

McKenzie, Bertha. *Stained Glass and Stone: The Gothic Buildings of the University of Sydney* (Sydney: University of Sydney, 1989).

MacLeod, Roy. 'The X-Club: A Scientific Network in Late-Victorian England', *Notes and Records of the Royal Society*, 24 (2), (1970), 305-322.

_____. 'The Support of Victorian Science: The Endowment of Science Movement in Great Britain, 1868-1900', *Minerva*, IX (1971), 197-230.

_____. 'The Resources of Science in Victorian England', in Peter Mathias (ed.), *Science and Society* (Cambridge: Cambridge University Press, 1972), 111-166.

_____. 'John Tyndall', *Dictionary of Scientific Biography* (New York: Charles Scribner's Sons, 1976), XIII, 521-524.

_____. 'Education: Scientific and Technical', in Gillian Sutherland (ed.), *Government and Society in Nineteenth-century Britain: Commentaries on Education, British Parliamentary Papers* (Dublin: Irish Academy Press, 1977), 196-225.

_____. 'On Visiting the Moving Metropolis: Reflections on the Architecture of Imperial Science', *Historical Records of Australian Science*, 5 (3), (1982), 1-6; reprinted in Nathan Reingold and Marc Rothenberg (eds.), *Scientific Colonialism: A Cross-Cultural Comparison* (Washington, DC: Smithsonian Institution Press, 1987), 217-249; subsequently reprinted as 'De Visita a la "Moving Metropolis": Reflexiones sobre la Arquitectura de la Ciencia Imperial' in Antonio Lafuente and Juan-José Saldana (eds.), *Historia de las Ciencias: Nuevas Tendencias* (Madrid: Consejo Superior de Investigaciones Cientificas, 1987), 217-240; and in William K. Storey (ed.), *Scientific Aspects of European Expansion* (London: Variorum, 1996), 23-55.

_____. '"Full of Honour and Gain to Science": Munitions Production, Technical Intelligence and the Wartime Career of Sir Douglas Mawson, FRS', *Historical Records of Australian Science*, 7 (2), (1988), 189-202.

_____. 'Gold from the Sea: Archibald Liversidge, FRS, and the "Chemical Prospectors", 1870-1928',

*Ambix*, 35 (2), (1988), 53-64.

_____. 'The "Practical Man": Myth and Metaphor in Anglo-Australian Science', *Australian Cultural History*, No. 8 (1989), 24-49.

_____. 'The "Arsenal in the Strand": Australian Chemists and the British Munitions Effort, 1916-1919', *Annals of Science*, 46 (1), (1989), 45-67.

_____. 'Science for Imperial Efficiency: Reflections on the British Science Guild, 1905-1936', *Public Understanding of Science*, 3 (1), (1994), 155-194.

_____. '"*Kriegesgeologen* and Practical Men": Military Geology and Modern Memory, 1914–1918', *British Journal of the History of Science*, 28 (4), (1995), 427-450.

_____. 'Colonial Engineers and the "Cult of Practicality": Themes and Dimensions in the History of Australian Engineering', *History and Technology*, 12 (2), (1995), 69-84.

_____. 'Instructed Men and Mining Engineers: The Associates of the Royal School of Mines and British Imperial Science, 1851-1920', *Minerva*, XXXII (4), (1995), 422-439.

_____. 'Science, Progressivism and "Efficient Imperialism"', in Roy MacLeod and Richard Jarrell (eds.), *Dominions Apart: Reflections on the Culture of Science and Technology in Canada and Australia, 1850-1945, Scientia Canadensis,* 17 (1, 2), (1995), 8-25.

_____. '"The Industrial Invasion of Britain": Mobilising Australian Munitions Workers, 1916-1919', *Journal of the Australian War Memorial,* No. 27 (October 1995), 37-46.

_____. '"The Other ANZACs": Australian Munitions Workers in the Great War', *Voices* (Canberra: National Library of Australia, 1995), 5 (4), (1995/6), 26-48.

_____. 'Science and Colonialism', in Peter Bowler (ed.), *Science and Society in Ireland: Papers on the Social Context of Science and Technology in Ireland, 1800-1950* (Belfast: Institute of Irish Studies, Queen's University of Belfast, 1997), 1-17.

_____. 'Reading the Discourse of Colonial Science', in Patrick Petitjean (ed.), *Les Sciences hors d'Occident au XXème Siècle, vol. 2, Les Sciences Coloniales: Figures et Institutions* (Paris: ORSTOM Editions, 1996), 87-98.

_____. 'Secrets among Friends: The Research Information Service and the "Special Relationship" in Allied Scientific Information and Intelligence, 1916-18', *Minerva*, 37 (3), (1999), 201-233.

_____. 'Sight and Sound on the Western Front: Surveyors, Scientists and the "Battlefield Laboratory", 1915-1918', *War and Society*, 18 (1), (May 2000), 23-46.

_____. 'Of Mines and Mining Education: The School of Mines at the University of Sydney, 1884-1894', *Earth Sciences History*, 19 (2), (2000) 192-215.

_____. 'Museums and the Cultivation of Knowledge in the Pacific', *Pacific Science*, 55 (4), (2001), 325-326.

_____. 'Sir Thomas Holland (1868-1947)', *Oxford Dictionary of National Biography* (Oxford: Oxford University Press, 2004), vol. 27, 700-703.

_____. 'Archibald Liversidge (1846-1927)', *Oxford Dictionary of National Biography* (Oxford: Oxford University Press, 2004), vol. 34, 52-53.

_____. 'Sir Alexander Pedler (1849-1918)', *Oxford Dictionary of National Biography* (Oxford: Oxford University Press, 2004), vol. 34, 390-391.

_____. 'Founding: South Kensington to Sydney', in Graeme Davison and Kimberley Webber (eds.), *Yesterday's Tomorrows: The Powerhouse Museum and its Precursors, 1880-2005* (Sydney: Powerhouse Publishing in association with UNSW Press, 2005), 42-54.

_____. 'Colonial Science under the Southern Cross: Archibald Liversidge, FRS, and the Shaping of Anglo-Australian Science', in Benedikt Stuchtey (ed.), *Science and Empire* (Oxford: Oxford University Press, 2005), 175-213.

MacLeod, Roy M. and E.K. Andrews. 'Scientific Careers of 1851 Exhibition Scholars', *Nature*, 218 (1968), 1011-1016.

_____. 'Scientific Advice in the War at Sea, 1915-1917', *Journal of Contemporary History,* VI (2), (1971), 3-40.

MacLeod, Roy M. and Milton Lewis. '"A Workingman's Paradise": Reflections on Urban Morality in Colonial Australia, 1860-1900', *Medical History*, 31 (4), (1987), 387-401.

MacLeod, Roy and Russell Moseley. 'Breadth, Depth and Excellence: Sources and Problems in the History of University Science Education in England, 1850-1914', *Studies in Science Education*, 5 (1978),

85-106.

_____. 'Fathers and Daughters: Reflections on Women, Science and Victorian Cambridge', *History of Education,* 8 (4), (1979), 321-333.

_____. 'The "Naturals" and Victorian Cambridge: Reflections on the Anatomy of an Elite, 1851-1914', *Oxford Review of Education,* 6 (2), (1980), 177-195.

MacLeod, Roy, James Friday and Philippa Shepherd. *John Tyndall, Natural Philosopher, 1820-1893: A Catalogue of Correspondence, Journals and Collected Papers* (London: Mansell, 1974).

MacLeod, Roy and Kimberley Webber. 'Empowering: Applied Research and the Commercial Museum', in Graeme Davison and Kimberley Webber (eds.), *Yesterday's Tomorrows: The Powerhouse Museum and its Precursors, 1880-2005* (Sydney: Powerhouse Publishing in association with UNSW Press, 2005), 96-109.

MacLeod, Roy (ed.). *University and Community in Nineteenth Century Sydney: Professor John Smith (1821-85),* (Sydney: University of Sydney Monographs No. 3, 1988).

_____ (ed.). *The Commonwealth of Science: ANZAAS and the Scientific Enterprise in Australasia, 1888-1988* (Melbourne: Oxford University Press, 1988).

_____ (ed.). *Nature and Empire: Science and the Colonial Experience, Osiris,* vol. 15 (Chicago: University of Chicago Press, 2001).

MacLeod, Roy and Peter Collins (eds.). *The Parliament of Science: The British Association for the Advancement of Science, 1831-1981* (London: Science Reviews, 1981).

MacLeod, Roy and Milton Lewis (eds.). *Disease, Medicine and Empire: Perspectives on Western Medicine and the Experience of European Expansion* (London: Routledge, 1988).

MacLeod, Roy and P.F. Rehbock (eds.), *'Nature in its Furthest Extent': Western Science in the Pacific* (Honolulu: University of Hawaii Press, 1988).

_____ (eds.) *Darwin's Laboratory: Evolutionary Theory and Natural History in the Pacific* (Honolulu: University of Hawaii Press, 1994).

Macmillan, D.S. 'The University of Sydney - The Pattern and the Public Reaction 1850-1870', *The Australian University,* 1 (1), (1963), 27-59.

McMinn, W.G. *Allan Cunningham, Botanist and Explorer* (Melbourne: Melbourne University Press, 1970).

McQueen, Humphrey. 'Australian Cultural Elites', *Arena,* 36 (1974), 40-49.

_____. *A New Britannia: An Argument Concerning the Social Origins of Australian Radicalism and Nationalism* (Harmondsworth: Penguin, 1978).

Mandelson, L.A. 'Norman Selfe and the Beginnings of Technical Education', in C. Turney (ed.), *Pioneers of Australian Education in the Australian Colonies, 1850-1900* (Sydney: Sydney University Press, 1972), vol. 2, 105-148.

Maor, Eli. *Venus in Transit* (Princeton: Princeton University Press, 2004).

Mashman, P. *A Resume of the History and Development of the Geological Survey of New South Wales* (Sydney: Geological Survey, N.S.W., 1974).

Mayne, A.J.C. *Fever, Squalor and Vice: Sanitation and Social Policy in Victorian Sydney* (Brisbane: University of Queensland Press, 1982).

Mellor, D.P. *The Role of Science and Industry: Australia in the War of 1939-1945,* Series 4, Civil, vol. 5 (Canberra: Australian War Memorial, 1959).

Melleuish, Gregory. 'Beneficent Providence and the Quest for Harmony: The Cultural Setting for Colonial Science in Sydney, 1850-1890', *Journal and Proceedings of the Royal Society of New South Wales,* 118 (1985), 167-180.

_____. *Cultural Liberalism in Australia: A Study in Intellectual and Cultural History* (Cambridge: Cambridge University Press, 1995).

Miller, David Philip. 'Hybrid or Mutant? The Emergence of the Chemical Engineer in Australia', *Historical Records of Australian Science,* 9 (4), (1993), 317-333.

Molella, Arthur P. *et al.* (eds.). *A Scientist in American Life: Essays and Lectures of Joseph Henry* (Washington, DC: Smithsonian Institution Press, 1980).

Moody, T. W. and J. C. Beckett. *Queens, Belfast, 1845-1949: The History of a University* (London: Faber & Faber for the Queen's University of Belfast, 1959).

Moran, Herbert M. *Viewless Winds: Being the Recollections and Digressions of an Australian Surgeon* (London: Peter Davies, 1939).

Morrell, Jack and Arnold Thackray. *Gentlemen of Science: Early Years of the British Association for the Advancement of Science* (Oxford: Oxford University Press, 1981).

Morris, D. 'Henry Parkes-Publicist and Legislator', in C. Turney (ed.), *Pioneers of Australian Education: A Study of the Development of Education in New South Wales in the Nineteenth Century* (Sydney: Sydney University Press 1969), 155-192.

Morrison, F.R. 'The Museum of Applied Arts and Sciences, Sydney', *Australian Journal of Science,* 24 (1962), 470-479.

Morison, Patricia. *J.T. Wilson and the Fraternity of Duckmaloi* (Amsterdam: Rodopi, 1997).

Morton, C. 'First 90 Years - The Development of Mechanical Cane Harvesting in Australia. Part 6', *Australian Canegrower,* 1 (10), (1980), 47-51.

Moyal, Ann. 'Collectors and Illustrators: Women Botanists in the Nineteenth Century', in D.J. and S.A.M. Carr (eds.), *People and Plants in Australia* (Sydney: Academic Press, 1981), 333-356.

_____. *Clear Across Australia: A History of Telecommunications* (Melbourne: Nelson, 1984).

_____. *A Bright and Savage Land: Science in Colonial Australia* (Sydney: Cassell Australia, 1986).

_____. *Platypus* (Sydney: Allen and Unwin, 2001).

Moyal, Ann Mozley. 'Science and the Press in Australia', *Search,* 4 (1973), 133-138.

_____. 'Sir Richard Owen and his Influence on Australian Zoological and Palaeontological Science', *Records of the Australian Academy of Science*, 3 (2), (1975), 41-56.

_____. 'The Australian Academy of Science: The Anatomy of a Scientific Elite. Part I (History and Sociology)', *Search*, 11 (1980), 231-238; Part II (Relations with Government), *Search*, 11 (1980), 231-238.

_____. 'Medical Research in Australia: A Historical Perspective', *Search*, 12 (9), (1981), 302-309.

_____. 'Scientific and Technological Change, 1939-1988', in *Australia, 1939-1988: Bicentennial History Bulletin*, 3 (1981), 58-67.

_____ (ed.), *Scientists in Nineteenth-Century Australia: A Documentary History* (Sydney: Cassell Australia, 1976).

Moyal, Ann. 'Invisible Participants: Women in Science in Australia, 1830-1950', *Prometheus,* 11 (2), (1993), 175-187.

Moyal, Ann (ed.). *The Web of Science: The Scientific Correspondence of the Rev. W.B. Clarke, Australia's Pioneer Geologist* (Kew: Australian Scholarly Publishing, 2003).

Mozley, Ann. 'The History of Australian Science', *Historical Studies of Australia and New Zealand,* 11 (42),(1964), 359-360.

_____. 'Checklist of Publications on the History of Australian Science', *Australian Journal of Science,* 25 (1962), 206-214; 27 (1964), 8-15.

_____. 'The Foundations of the Geological Survey of New South Wales', *Journal and Proceedings of the Royal Society of New South Wales*, 98 (1965), 91-100.

_____. 'Richard Daintree, First Government Geologist of Northern Queensland', *Queensland Heritage,* 1 (2), (1965), 11-16.

_____. *A Guide to the Manuscript Records of Australian Science* (Canberra: Australian Academy of Science with the Australian National University Press, 1966).

_____. 'Evolution and the Climate of Opinion in Australia, 1840-1876', *Victorian Studies, 10* (1967), 411-430.

_____. 'ANZAAS and the Public Communication of Science', *Search*, 11 (1974), 589-593.

Multhauf, Robert P. *History of Chemical Technology: An Annotated Bibliography* (New York: Garland, 1983).

Mulvaney, D.J. 'Anthropology in Victoria 100 Years Ago', *Proceedings of the Royal Society of Victoria*, 73 (1961), 47-50.

Mulvaney, D.J. and J.H. Calaby. *'So Much That Is New': Baldwin Spencer, 1860-1929. A Biography* (Melbourne: Melbourne University Press, 1985).

Murray-Smith, Stephen. 'Technical Education in Australia: A Historical Sketch', in E.L. Wheelwright (ed.), *Higher Education in Australia* (Melbourne: F.W. Cheshire, 1965), 170-191.

_____. 'Sir James Barrett (1862-1945)', *Australian Dictionary of Biography* (Melbourne: Melbourne University Press, 1979), vol. 7, 186-189.

Nadel, George. *Australia's Colonial Culture: Ideas, Men and Institutions in Mid-Nineteenth Century Eastern Australia* (Melbourne: Cheshire, 1957).

Neill, Norm. *Technically and Further: Sydney Technical College, 1891-1991* (Sydney: Hale and Iremonger, 1991).

Oldroyd, David. *Thinking about the Earth: A History of Ideas in Geology* (London: Athlone, 1996).

O'Connell, R.W. 'The Colonial Mentality in Social Science', *Search,* 15 (3-4), (April-May 1984), 110-111.

O'Neil, Bernard. *In Search of Mineral Wealth: The South Australian Geological Survey and Department of Mines to 1944* (Adelaide: Department of Mines and Energy, 1982).

Orchiston, Wayne. 'Illuminating Incidents in Antipodean Astronomy: Phillip Parker King and the Founding of Sydney Observatory', *Vistas in Astronomy,* 32 (1988), 285-301.

Organ, Michael. 'W.B. Clarke as Scientific Journalist', *Historical Records of Australian Science,* 9 (1), (June 1992), 1-16.

Organisation for Economic Cooperation and Development (OECD). *Reviews of National Science and Technology Policy: Australia* (Paris: OECD, 1986), 13.

Palmer, Vance. *The Legend of the Nineties* (Melbourne: Melbourne University Press, 1954).

Paneth, F.A. *The Origin of Meteorites* (Oxford: Oxford University Press, 1940).

Pang, Alex Soojung-Kim. 'The Social Event of the Season: Solar Eclipse Expeditions and Victorian Culture', *Isis,* 84 (2), (1993), 252-277.

_____. *Empire and the Sun: Victorian Solar Eclipse Expeditions* (Stanford: Stanford University Press, 2002).

Pasachoff, Jay M., Glenn Schneider, and Leon Golub. 'The Black-Drop Effect Explained,' in *Transits of Venus: New Views of the Solar System and Galaxy, Proceedings of the International Astronomical Union,* Colloquium No. 196 (2004), 1-12.

Parsons, T.G. 'Government Contracts and Colonial Manufacture: The Example of the Victorian Railways in the 1870s', *Australian Journal of Politics and History,* 26 (2), (1980), 242-253.

Partington, J. R. *History of Chemistry* (London: Macmillan, 1964).

Pascoe, Rob. *The Manufacture of Australian History* (Melbourne: Oxford University Press, 1979).

Penny, B.R. 'Australia's Reaction to the Boer War: A Study in Colonial Imperialism', *Journal of British Studies,* 7 (1), (1967), 97-130.

Pescott, R.T.M. *Collections of a Century. The History of the First Hundred Years of the National Museum of Victoria* (Melbourne: National Museum of Victoria, 1954).

_____. 'Royal Society of Victoria from then, 1854 to now, 1959', *Proceedings of the Royal Society of Victoria,* 73 (1961), 1-40.

Pettigrew, C.J. 'Abercrombie Anstruther Lawson (1870-1927)', *Australian Dictionary of Biography* (Melbourne: Melbourne University Press, 1986), vol. 10, 15-16.

Pevsner, Nikolaus. *High Victorian Design: A Study of the Great Exhibition of 1851* (London: Architectural Press, 1951).

Phillips, Patricia. *The Scientific Lady: A Social History of Women's Scientific Interests, 1520-1918* (New York: Collins and Brown, 1993).

Physick, John. *The Victoria and Albert Museum: The History of a Building* (London: Victoria and Albert Museum, 1982).

Plomley, N.J.B. *Several Generations* (Sydney: Wentworth Books, 1971).

Poiner, Gretchen and R. Burke. *No Primrose Path: Women as Staff at the University of Sydney* (printed at the University of Sydney, 1988).

Port, L.W. *Australian Inventors* (Sydney: Cassell, 1978).

Porter, Roy. 'Gentlemen and Geology: The Emergence of a Scientific Career, 1660-1920', *The Historical Journal,* 21 (4), (1978), 809-36.

_____. *The Making of Geology: Earth Science in Britain, 1660-1815* (Cambridge: Cambridge University Press, 1977)

_____. 'The Terraqueous Globe', in George Rousseau and Roy Porter (eds.), *The Ferment of Knowledge* (Cambridge: Cambridge University Press, 1980), 285-326.

_____. *History of the Earth Sciences: An Annotated Bibliography* (New York: Garland, 1981).

Potts, Eli Daniel. *Young America's Australian Gold: Americans and the Gold Rush of the 1850's* (St. Lucia: University of Queensland Press, 1974).

Powell, Alan. 'Explorer-surveyors of the Australian North Coast - Part 2: John Lort Stokes and the Crew of the *Beagle', Journal of the Royal Australian Historical Society,* 66 (1981), 225-236.

Praagh, G.Van (ed.). *H.E. Armstrong and Science Education* (London: John Murray, 1973).

Priestley, R.E. 'Sir Edgeworth David', *The Australian Quarterly*, X (2), (1938), 34-39.

Prince, J.H. *The First One Hundred Years of the Royal Zoological Society of N.S.W. 1879 to 1979* (Sydney: Surrey Beatty and Sons, 1979).

Radford, Joan. *The Chemistry Department of the University of Melbourne: Its Contribution to Australian Science, 1854-1959* (Melbourne: The Hawthorn Press, 1978).

_____. 'Chemistry in Nineteenth-Century New Zealand', *Chemistry in New Zealand*, 49 (1984), 35-37, 60-62, 125-129.

Rasmussen, Carolyn (ed.). *A Museum for the People: A History of Museum Victoria and its Predecessors, 1854-2000* (Melbourne: Scribe Publications, 2001).

Rankama, Kalervo and T.G. Sahama. *Geochemistry* (Chicago: University of Chicago Press, 1950).

Read, H.H. *Rutley's Elements of Mineralogy* (London: Thomas Murby, 1947).

Reingold, Nathan and Marc Rothernburg (eds.). *Scientific Colonialism: A Cross-Cultural Comparison* (Washington, DC: Smithsonian Institution Press, 1987).

Reynolds, John. *Men and Mines: A History of Australian Mining* (Melbourne: Sun Books, 1974).

Ritchie, J.M. 'The Development and Significance of Technical Education in New South Wales, 1880-1900', *Journal of the Royal Australian Historical Society*, 55 (1969), 245-261.

Ritvo, Harriet. *The Platypus and the Mermaid and other Figments of the Classifying Imagination* (Cambridge, Mass: Harvard University Press, 1997).

_____. 'Zoological Nomenclature and the Empire of Victorian Science', in Bernard Lightman (ed.), *Victorian Science in Context* (Chicago: University of Chicago Press, 1997), 334-353.

Rivett, Rohan. *David Rivett: Fighter for Australian Science* (Melbourne: Rohan Rivett, 1972).

Roberts, Gerrylynn K. 'The Establishment of the Royal College of Chemistry: An Investigation of the Social Context of Early-Victorian Chemistry', *Historical Studies in the Physical Sciences,* 7 (1976), 437-485.

Robertson, Peter. 'Coming of Age: The British Association for the Advancement of Science in Australia, 1914', *Australian Physicist*, 17 (1980), 23-27.

Robin, Libby. 'The Platypus Frontier: Eggs, Aborigines and Empire in 19th Century Queensland', in Deborah Bird Rose and Richard Davis (eds.), *Dislocating the Frontier: Essaying the Mystique of the Outback* (Canberra: ANU ePress, 2005), 99-120.

Robinson, Sir Robert. *Memoirs of a Minor Prophet: Seventy Years of Organic Chemistry* (Amsterdam: Elsevier, 1976).

Rocke, Alan J. *Chemical Atomism in the Nineteenth Century from Dalton to Cannizzaro* (Columbus: Ohio State University Press, 1984).

Roe, Jill (ed.). *Twentieth Century Sydney: Studies in Urban and Social History* (Sydney: Hale and Iremonger, 1980).

Roe, Michael. *Quest for Authority in Eastern Australia, 1835-1851* (Melbourne: Melbourne University Press, 1965).

_____. *Nine Australian Progressives: Vitalism in Bourgeois Social Thought, 1890-1960* (Brisbane: University of Queensland Press, 1984).

Rogers, Naomi. '"Any Decent Woman": A Study of Attitudes of the Medical Profession in 1873 towards Women as Colleague', *Women and Labour: Conference Papers*, 2 (1980), 385-398.

Rood, E.H. 'Henry Edward Armstrong (1848-1937)', *Oxford Dictionary of National Biography*, vol. 2, 422-424.

Ross, A. T. 'The Politics of Secondary Industry Research in Australia, 1926-1939', *Historical Records of Australian Science,* 7 (4), (June 1989), 373-392.

Rothblatt, Sheldon. *The Revolution of the Dons: Cambridge and Society in Victorian England* (London: Faber, 1968)

_____. *Tradition and Change in English Liberal Education: An Essay in History and Culture* (London: Faber and Faber, 1976).

_____. *The European and American University since 1800: Historical and Sociological Essays* (New York: Cambridge University Press, 1993)

_____. *Education's Abiding Moral Dilemmas: Merit and Worth in the Cross-Atlantic Democracies, 1800-2006* (Didcot: Symposium Books, 2007).

Rothenberg, Marc and Peter Hoffenberg. 'Australia at the 1876 Exhibition in Philadelphia', *Historical Records of Australian Science,* 8 (2), (June 1990), 55-62.

Rudwick, Martin J. S. *The Meaning of Fossils: Episodes in the History of Palaeontology* (New York: Science History Publications, 1972).

_____. *The Great Devonian Controversy: The Shaping of Scientific Knowledge among Gentlemanly Specialists* (Chicago: University of Chicago Press, 1985).

Rupke, Nicolaas A. *Richard Owen, Victorian Naturalist* (New Haven: Yale University Press, 1994).

Russell, A. *William James Farrer: A Biography* (Melbourne: F.W. Cheshire, 1949).

Russell, Colin A. *Science and Social Change, 1700-1900* (London: Macmillan, 1983).

_____. *Lancastrian Chemist: The Early Years of Sir Edward Frankland* (Milton Keynes: Open University, 1986).

_____. *Edward Frankland: Chemistry, Controversy and Conspiracy in Victorian England* (Cambridge: Cambridge University Press, 1996).

Russell, Colin A., Noel G. Coley and Gerrylynn K. Roberts. *Chemists by Profession: The Origins and Rise of the Royal Institute of Chemistry* (Milton Keynes: Open University Press, 1977).

Russell, Penelope. *A Wish of Distinction: Colonial Gentility and Femininity* (Melbourne: Melbourne University Press, 1994).

Rydell, Robert W. *All the World's a Fair: Visions of Empire at American International Expositions, 1876-1916* (Chicago: University of Chicago Press, 1984).

_____. *World of Fairs: The Century-of-Progress Expositions* (Chicago: University of Chicago Press, 1993).

Sanderson, Michael. *The Universities and British Industry, 1850-1970* (London: Routledge and Kegan Paul, 1972).

Saunders, Shirley. 'Sir Thomas Brisbane's Legacy to Colonial Science: Colonial Astronomy at the Parramatta Observatory, 1822-1848', *Historical Records of Australian Science,* 15 (2), (2004), 177-209.

Schedvin, C.B. 'The Culture of CSIRO', *Australian Cultural History,* 2 (1982/3), 76-89.

_____. 'Environment, Economy and Australian Biology, 1890-1939', *Historical Studies,* 21 (82), (1984), 11-28.

_____. *Shaping Science and Industry: A History of Australia's Council for Scientific and Industrial Research, 1926-49* (Sydney: Allen and Unwin, 1987).

Scherz, Gustav. 'Niels Stensen (1638-1686)', *Dictionary of Scientific Biography* (New York: Charles Scribner's Sons, 1976), XIII, 30-35.

Schrock, Robert Rakes. *Geology at M.I.T. 1865-1965: A History of the First Hundred Years of Geology at Massachusetts Institute of Technology, Departmental Operations and Products* (Cambridge, Mass: MIT Press, 1982).

Scott, Ernest. 'The History of Australian Science', *Australian Journal of Science,* 1 (4), (1939), 105-116.

Searby, Peter. *A History of the University of Cambridge,* vol. 3: 1750-1870 (Cambridge: Cambridge University Press, 1997).

Searle, G.R. *The Quest for National Efficiency: A Study in British Politics and Political Thought, 1899-1914* (Oxford: Basil Blackwell, 1971).

Secord, James A. 'King of Siluria: Roderick Murchison and the Imperial Theme in Nineteenth-Century British Geology', *Victorian Studies,* 25 (4), (1982), 413-442.

_____. 'The Geological Survey of Great Britain as a Research School, 1839-1855', *History of Science,* XXIV (1986), 224-275.

_____. *Controversy in Victorian Geology: The Cambrian-Silurian Dispute* (Princeton: Princeton University Press, 1986).

Seddon, George. 'Eurocentrism and Australian Science: Some Examples', *Search,* 12 (1981/82), 446-450.

Selleck, Richard J.W. *The Shop: The University of Melbourne, 1850-1939* (Melbourne: Melbourne University Press, 2003).

Serle, Geoffrey. *From Deserts the Prophets Come: The Creative Spirit in Australia, 1788-1972* (Melbourne: Heinemann, 2nd ed., 1974).

Shann, Edward. *An Economic History of Australia* (Cambridge: Cambridge University Press, 1930).

Sharpe, A. *Colonial New South Wales, 1853-1894* (Sydney: Harper and Row, 1979).

Shattock, Michael and Robert O. Berdahl. 'The British University Grants Committee 1919–83: Changing

Relationships with Government and the Universities', *Higher Education*, 13 (5), (1984), 471-499.

Sherington, Geoffrey. *Australia's Immigrants, 1788-1978* (Sydney: George Allen and Unwin, 1980).

Sheets-Pyenson, Susan. *Cathedrals of Science: The Development of Colonial Natural History Museums during the Late-Nineteenth Century* (Kingston: McGill-Queen's University Press, 1988).

Simpson, G.L. 'Reverend Dr John Woolley and Higher Education', in C. Turney (ed.), *Pioneers of Australian Education: A Study of the Development of Education in New South Wales in the Nineteenth Century* (Sydney: Sydney University Press, 1969).

Skeats, E.W. 'Some Founders of Australian Geology'. *David Lecture,* No. 1 (Sydney: Australian National Research Council, 1933).

Smith, F.B. 'Academics In and Out of the *Australian Dictionary of Biography*', in F.B. Smith and Pamela Crichton (eds.), *Ideas for Histories of Universities in Australia* (Canberra: Australian National University, 1990), 1-14.

Smith, Jonathan. *Fact and Feeling: Baconian Science and the Nineteenth-century Literary Imagination* (Madison: University of Wisconsin Press, 1994).

Souter, Gavin. *Lion and Kangaroo: Australia, 1901-1919: The Rise of a Nation* (Sydney: Fontana Collins, 1976).

Spicer, B.M. 'Physics and the University of Melbourne', *Australian Physicist*, 17 (7), (August 1980), 113-116.

Stafford, Barbara Maria. *Voyage into Substance. Art, Science, Nature, and the Illustrated Travel Account, 1760-1840* (Cambridge, Mass.: MIT Press, 1984).

Stafford, Robert A. 'Geological Surveys, Mineral Discoveries, and British Expansion, 1835-71', *Journal of Imperial and Commonwealth History*, XII (3), (1984), 5-32.

_____. 'Roderick Murchison and the Structure of Africa: A Geological Prediction and its Consequences for British Expansion', *Annals of Science*, 45 (1), (1988), 1-40.

_____. 'Preventing the "Curse of California": Advice for English Emigrants to the Australian Goldfields', *Historical Records of Australian Science,* 7 (3), (1989), 215-231.

_____. *Scientist of Empire: Sir Roderick Murchison, Scientific Exploration and Victorian Imperialism* (Cambridge: Cambridge University Press, 1989).

_____. 'A Far Frontier: British Geological Research in Australia during the Nineteenth Century', in R.W. Home and Sally Gregory Kohlstedt (eds.), *International Science and National Scientific Identity: Australia Between Britain and America* (Dordrecht: Kluwer, 1991), 75-96.

Stalker, R.J. *Comments on a Colonial Technology* (St. Lucia: University of Queensland Press, 1979).

Stanbury, Peter and Julian Holland. *Mr Macleay's Celebrated Cabinet* (Sydney: Macleay Museum, University of Sydney, 1988).

Stearn, William T. *The Natural History Museum at South Kensington* (London: Heinemann, 1981).

Stenhouse, John. 'Darwin's Captain: F.W. Hutton and the Nineteenth Century Darwinian Debates', *Journal of the History of Biology*, 23 (3), (1990), 411-442.

Stenning, Eve. 'John Hutchison: Australia's First Scientist', *Journal of the Royal Australian Historical Society,* 79 (1-2), (June 1993), 10-19.

Stephenson, Gabriela M. M. 'Australian Scientific Journalism in the Nineteenth Century: Then and Now', *Riverina Library Review,* 5 (4), (November 1988), 315-327.

Stephenson, Percy Reginald. *The History and Description of Sydney Harbour* (Adelaide: Rigby, 1966).

Stewart, Ken. 'The Colonial Literati in Sydney and Melbourne', in Susan Dermody, John Docker, and Drusilla Modjeska (eds.), *Nellie Melba, Ginger Meggs and Friends: Essays in Australian Cultural History* (Malmsbury: Kibble Books, 1982), 176-191.

Stock, John T. and Mary Virginia Orna (eds.). *The History and Preservation of Chemical Instrumentation* (Dordrecht: Reidel, 1986).

Stoltzenberg, Dietrich. *Fritz Haber: Chemist, Nobel Laureate, German, Jew* (Philadelphia: Chemical Heritage Pres, 2004).

Strahan, Ronald. *Rare and Curious Specimens: An Illustrated History of the Australian Museum, 1827-1979* (Sydney: Australian Museum, 1979).

Strick J. 'Darwinism and the Origin of Life: The Role of H.C. Bastian in the British Spontaneous Generation Debates, 1868-1873', *Journal of the History of Biology*, 32 (1), (1999), 51-92.

Summers, H.S. 'The Teachers of Geology in Australian Universities', *Journal and Proceedings of the Royal Society of New South Wales*, 81 (1947), 122-146.

Sumner, Rae. *A Woman in the Wilderness: The Story of Amalie Dietrich in Australia* (Sydney: University of New South Wales Press, 1993).

Taksa, Lucy. 'Instructing: The Museum and Technical Education', in Graeme Davison and Kimberley Webber (eds.), *Yesterday's Tomorrows: The Powerhouse Museum and its Precursors, 1880-2005* (Sydney: Powerhouse Publishing in association with UNSW Press, 2005), 82-95.

Taylor, F. Sherwood. *The Science Museum: The First Hundred Years* (London: HMSO, 1957).

Taylor, Peter. *An End to Silence: The Building of the Overland Telegraph Line from Adelaide to Darwin* (Sydney: Methuen, 1980).

Thépot, André. *Les Ingénieurs des Mines du XIXème Siècle: Histoire d'un Corps Technique d'Etat, tome 1: 1810-1914* (Paris: Editions ESKA, 1998).

Thornton, A.P. *Doctrines of Imperialism* (New York: John Wiley and Sons, 1965).

Tinkler, K. J. (ed.). *History of Geomorphology: From Hutton to Hack, The Binghamton Symposia in Geomorphology*, International Series, No. 19, 1989).

Todd, A.R. 'Sir Robert Robinson (1886-1975)', *Dictionary of National Biography, 1971-1980* (Oxford: Oxford University Press, 1986), 729-731.

Todd, Jan. 'Science at the Periphery: An Interpretation of Australian Scientific and Technological Dependency and Development Prior to 1914', *Annals of Science*, 50 (1), (1993), 33-58.

_____. *Colonial Technology: Science and the Transfer of Innovation to Australia* (Melbourne: Cambridge University Press, 1995).

Tomlin, S.C. 'William Henry Bragg, 1962-1942', *Australian Physicist*, 13 (1976), 76-99.

Turner, F.M. 'Public Science in Britain, 1880-1919', *Isis*, 71 (4), (1980), 589-608.

_____. 'Practicing Science: An Introduction', in Bernard Lightman (ed.), *Victorian Science in Context* (Chicago: University of Chicago Press, 1997), 283-289.

Turney, Clifford, Ursula Bygott and Peter Chippindale. *Australia's First; A History of the University of Sydney* (Sydney: University of Sydney in association with Hale and Iremonger, 1991), 2 vols.

Turtle, Alison M. 'Anthropometry in Britain and Australia: Technology, Ideology and Imperial Connection', *Storia della Psicologia*, 2 (2), (1990), 118-143.

Vallance, T.G. 'History of Science in Australia - Part 7: The Beginning of Geological System', *SCAN* (1968), 28-34.

_____. 'Origins of Australian Geology', *Proceedings of the Linnean Society of New South Wales*, 100 (1975), 13-43.

_____. 'The Fuss **a**bout Coal: Troubled Relations **b**etween Paleobotany and Geology', in D.J. and S.G.M. Carr (eds.), *Plants and Man in Australia* (Sydney: Academic Press, 1981), 136-178.

Vallance, T.G. and D.F. Branagan. 'New South Wales Geology - its Origin and Growth', in A.P. Elkin *et al.*, *Royal Society of NSW, A Century of Scientific Progress* (Sydney: Royal Society of NSW, 1968), 265-279.

_____. 'Origins of Australian Geology', *Proceedings of the Linnean Society of New South Wales*, 100 (1), (1975), 13-43.

_____. 'The Start of Government Science in Australia: A.W.H. Humphrey, His Majesty's Mineralogist in New South Wales, 1803-1812', *Proceedings of the Linnean Society of New South Wales*, 105 (1981), 107-146.

Vincent, E.A. 'W.J. Sollas (1849-1936)', *Oxford Dictionary of National Biography*, vol. 51, 534.

Verkade, Pieter Eduard. *History of the Nomenclature of Organic Chemistry by H.J.T. Bos* (Dordrecht: Reidel, 1985).

Vonwiller, O.U. 'Presidential Address. A Generation of Electron Theory', *Journal and Proceedings of the Royal Society of New South Wales*, 65 (6 May 1931), 14-36.

_____. 'The Social Relations of Science', *Australian Journal of Science*, 1 (1938), 30-32.

Wade, John (ed.). *The Sydney International Exhibition, 1879* (Sydney: Museum of Applied Arts and Sciences, 1979).

Wadham, Sir Samuel MacMahon. 'Science', in A.L. McLeod (ed.), *The Patterns of Australian Culture* (Melbourne: Oxford University Press, 1963), 70-94.

Walker, Mary Howitt. *Come Wind, Come Weather: A Biography of Alfred Howitt* (Melbourne: Melbourne University Press, 1971).

Walkom, A.B. *The Linnean Society of New South Wales: Historical Notes of its First Fifty Years* (Sydney: Australian

Medical Publishing Company, 1925).

Walshaw, Rodney. 'British Government Geologists Overseas - A Brief History', *Geoscientist,* 4 (2), (1994), 10-12.

Ward, John M. 'Historiography', in A.L. McLeod (ed.), *The Pattern of Australian Culture* (Ithaca: Cornell University Press, 1963), 195-251.

_____. *Empire in the Antipodes: The British in Australasia, 1840-1860* (London: Edward Arnold, 1971).

Ward, Russell. *The Australian Legend* (Melbourne: Oxford University Press, 1958).

_____. 'The Australian Legend Re-visited', *Historical Studies*, 18 (71), (1978), 171-190.

Warren, Perry. *The Science Museum of Victoria: A History of its First Hundred Years* (Melbourne: Science Museum of Victoria, 1972).

Warwick, Andrew. *Masters of Theory: Cambridge and the Rise of Mathematical Physics* (Chicago: University of Chicago Press, 2003).

Watson, Katherine. '"Temporary Hotel Accommodation"? The Early History of the Davy-Faraday Research Laboratory, 1894-1923', in Frank A.L. James (ed.), '*The Common Purposes of Life': Science and Society at the Royal Institution of Great Britain* (London: Ashgate, 2002), 191-223.

Watts, W.W. 'Progress of the Geological Survey', *Discovery*, XIII (1932), 152-156.

Webster, E.M. *Whirlwinds in the Plain: Ludwig Leichhardt, Friends, Foes and History* (Melbourne: Melbourne University Press, 1980).

Weickhardt, L.W. *Masson of Melbourne: The Life and Times of David Orme Masson, Professor of Chemistry, University of Melbourne, 1886-1923* (Melbourne: Royal Australian Chemical Institute, 1989).

Welch, Geoffrey B. *Geoffrey Baldwin Welch: His Life and Family: An Autobiography and Family History* (Sydney: Privately published, 1996).

Westbury, Ian. 'The Sydney and Melbourne Arts Courses, 1852-1861', in E.L. French (ed.), *Melbourne Studies in Education* (Parkville: Melbourne University Press, 1964), 256-284.

White, Richard. *Inventing Australia: Images and Identity, 1688-1980* (Sydney: Allen and Unwin, 1981).

Whitley, Gilbert P. *Early History of Australian Zoology* (Sydney: Royal Zoological Society of New South Wales, 1970).

Whitehead, A.N. *Science in the Modern World* (Harmondsworth: Penguin, 1926; Cambridge: Cambridge University Press, 1932, 1946).

Whitrow, Gerald J. *Centenary of the Huxley Building: Personalities and Events in the History of Imperial College* (London: Imperial College of Science and Technology, 1972).

Wiener, Martin. *English Culture and the Decline of the Industrial Spirit, 1850-1980* (London: Macmillan, 1981).

Wilcockson, W.H. 'The Geological Work of Henry Clifton Sorby', *Proceedings of the Yorkshire Geological Society,* 27 (1947), 1-22.

Williams, Bruce R. and David R.V. Wood. *Academic Status and Leadership in the University of Sydney, 1852-1987* (Sydney: Sydney University Monographs, No. 6, 1990).

Willis, J.H. 'Botanical Science in Victoria 100 Years Ago', *Proceedings of the Royal Society of Victoria*, 73 (February 1961), 41-46.

Willis, Margaret. *By Their Fruits: A Life of Ferdinand von Mueller, Botanist and Explorer* (Sydney: Angus and Robertson, 1949).

Windshuttle, Elizabeth. *Taste and Science: The Women of the Macleay Family, 1790-1850* (Sydney: Historic Houses Trust of New South Wales, 1988).

Winkler, Clemens. 'A Chemist at the Freiberg Mining Academy during the Second Phase of Industrialization in Germany', *History and Technology*, 7 (1), (1990), 51-61.

Wise, Tigger. *The Self-Made Anthropologist. A Life of A.P. Elkin* (Sydney: George Allen and Unwin, 1985).

Wood, Harley. 'The History of Science in Australia - Part 6. The Early Astronomers', *SCAN* (October 1968), 18-23.

Wood, Paul (ed.). *Science and Dissent in England, 1688-1945* (Aldershot: Ashgate, 2004).

Worster, Donald. *Nature's Economy: A History of Ecological Ideas* (Cambridge: Cambridge University Press, 1985).

Wrigley, C.W. 'William James Farrer (1845-1906)', *Australian Dictionary of Biography* (Melbourne: Melbourne University Press, 1981), vol. 8, 471-473.

Yeo, Richard. 'An Idol of the Market-Place: Baconianism in Nineteenth-Century Britain', *History of Science*, 23 (3), (1985), 251-298.

Young, John, Ann J. Sefton and Nina Webb (eds.). *Centenary Book of the University of Sydney Faculty of Medicine* (Sydney: Sydney University Press, 1984).

Young, Linda. 'The Sydney International Exhibition, 1879: Australia's First Expo', *Heritage Australia*, 7 (2),(1988), 26.

# Picture Credits

**Opposite page 1** Bust, unknown photographer, Historic Photograph Collection, by kind permission of the Macleay Museum, University of Sydney. Original of the bust has since been lost; **page 16** Newington Turnpike, Guildhall Library Print Room, Views in Surrey, p 9, Catalogue No p5398898, by courtesy of the Guildhall Library, City of London Corporation; **page 19** Elephant and Castle Inn, Guildhall Library Print Room, Main Print Collection, Pr.S2/NEW/but, Catalogue No p5398504 by courtesy of the Guildhall Library, City of London Corporation; **page 21** Family home, photograph by the author; **page 23** Lantern slide, reproduced with the permission of the University of Westminster Archive Services; **page 24** *Mirror of Literature*, reproduced with the permission of the University of Westminster Archive Services; **page 35** Interior of the Museum, negatives MN 26186, reproduced by kind permission of the Natural History Museum, London; **page 39** The Royal College of Chemistry, reproduced by kind permission of Archives, Imperial College, London; **page 40** Liversidge's teachers, montage by the author; **page 53** Liversidge's Certificate, Liversidge Papers, courtesy of the University of Sydney Archives; **page 56** King's Parade, Louise Rayner, ca 1887, from a postcard printed by Beric Tempest & Co Ltd; **page 61** Christ's College, reproduced by kind permission of the Cambridgeshire Collection, Cambridge Central Library; **page 62-63** Second Court, reproduced by kind permission of the Cambridgeshire Collection, Cambridge Central Library; **page 64** Slaughter House Lane, reproduced by kind permission of the Cambridgeshire Collection, Cambridge Central Library; **page 65** An undergraduate's room, reproduced by kind permission of the Cambridgeshire Collection, Cambridge Central Library; **page 67** The Geological Museum, Cambridge, reproduced by kind permission of the Syndics of Cambridge University Library; **page 72** George Downing Liveing (1827-1924), reproduced by kind permission of the University of Cambridge; **page 77** Site and Buildings, University Library Archives, reproduced by kind permission of the Syndics of Cambridge University Library; **page 84** The *Northumberland*, private collection; **page 87** Main Building, Liversidge Papers, courtesy of the University of Sydney Archives; **page 90** The Professors, courtesy of the University of Sydney Archives; **page 94** The Main Building and Great Hall, courtesy of the University of Sydney Archives; **page 102-103** Sydney Harbour, courtesy of the University of Sydney Archives; **page 105** Australian Museum, courtesy of the Australian Museum, reproduced from Ronald Strahan, *Rare and Curious Specimens: An Illustrated History of the Australian Museum, 1827-1979* (Sydney: Australian Museum, 1979) 31; **page 107** The Union Club, Small Picture File Collection, Mitchell Library, State Library of New South Wales; **page 110** The Royal Mint, SPF/320, Small Picture File Collection,

Mitchell Library, State Library of New South Wales; **page 112** School of Arts, by kind permission of the Sydney Mechanics School of Arts Archive; **page 117** Chemistry Laboratory, courtesy of the University of Sydney Archives; **page 124** Glass of Water, *Sydney Punch*, 13 February 1874; **page 152** Sydney Observatory, Historic Photograph Collection HP82.64.197, by kind permission of the Macleay Museum, University of Sydney; **page 153** Cameo Portraits, from the frontispiece of HC Russell's Observations of the Transit of Venus, 9 December 1874 published 1892, supplied by the Sydney Observatory, courtesy of the Powerhouse Museum, Sydney; **page 154** Seal of the Royal Society, courtesy of the Royal Society of New South Wales; **page 158** Inner circle, montage by the author; **page 170** The famous Pink Terraces, Liversidge Papers, courtesy of the University of Sydney Archives; **page 171** Tua Pua, Liversidge Papers, courtesy of the University of Sydney Archives; **page 175** Exposition, *Handbook of the Exposition Universelle of 1878* (Paris, 1878); **page 188** The Australian Museum, ML REF: GPO 1 frame number 05346, Government Printing Office collection, State Library of New South Wales; **page 191** Technological Museum, 00221187, courtesy of the Powerhouse Museum, Sydney; **page 192** Ladies reading room, *Illustrated Guide to Sydney and its Suburbs* (London: Gibbs, Shallard & Co, 1882); **page 195** The Intercolonial Exhibition, Historic Photograph Collection, HP82.64.13, reproduced by kind permission of the Macleay Museum, University of Sydney; **page 196-197** Garden Palace, Historic Photograph Collection, HP82.64.1384, reproduced by kind permission of the Macleay Museum, University of Sydney; **page 199** Interior of the Garden Palace, private collection; **page 204** Interior of the Museum in the Domain, 00220733, courtesy of the Powerhouse Museum, Sydney; **page 205** Joseph Maiden, courtesy of the Powerhouse Museum, Sydney; **page 215** Sydney's first Technical College, Department Technical Education; **page 220-221** The Technology Museum and the Technical College, courtesy of the Powerhouse Museum, Sydney; **page 226** Commemoration, *The Graphic*, 10 January 1880; **page 230** Elizabeth House, from A.H. Corbett, *The History of the Institution of Engineers, Australia* (Canberra, 1970); **page 231** Great Hall, *Illustrated Sydney News*, 1 November 1879; **page 237** University staff and third year students, courtesy of the University of Sydney Archives; **page 241** 'The Professor', by courtesy of the late Dr. Joan Kerr; **page 245** Chemistry Lecture, Historic Photograph Collection HP84.53.39 reproduced by kind permission of the Macleay Museum, University of Sydney; **page 264** The Main Corridor, Liversidge Papers, courtesy of the University of Sydney Archives; **page 267** Liversidge in 1885, Historic Photograph Collection, HP84.53.39, by kind permission of the Macleay Museum, University of Sydney; **page 269** Officers of the AAAS, Liversidge Papers, courtesy of the University of Sydney Archives; **page 273** Plan, *Report of the Australasian Association for the Advancement of Science*, 1 (Sydney, 1888), Plate V) **page 273** Main Chemical Laboratory, Liversidge Papers, courtesy of the University of Sydney Archives; **page 274** Office, Liversidge Papers, courtesy of the University of Sydney Archives; **page 274** Private Laboratory, Liversidge Papers, courtesy of the University of Sydney Archives; **page 279** Liversidge's Lecture, Liversidge Papers, courtesy of the University of Sydney Archives; **page 279** Preparation Room, Liversidge Papers, courtesy of the University of Sydney Archives; **page 280** Medical Students, courtesy of the University of Sydney Archives; **page 281**

Curriculum, *University of Sydney Calendar* (1898-99) **page 283** The 'Octagon', Liverside Papers, courtesy of the University of Sydney Archives; **page 286** Liversidge and his Colleagues, Liversidge Papers, courtesy of the University of Sydney Archives; **page 290-291** Threlfall's Physics Building, courtesy of the University of Sydney Archives; **page 294** Chemistry Laboratory, courtesy of the University of Sydney Archives; **page 296** Gold crystal, General Mineral Collection, 1921-2-7, Department of Geology, photograph reproduced from a postcard printed for Her Majesty's Stationery Office, Crown Copyright 53-2503, National Museum of Scotland, Edinburgh; **page 299** Map, private collection; **page 307** Mineral Collection, *The Australian Field,* VII (30), 27; **page 308** Plan, *Report of the Australasian Association for the Advancement of Science*, 4 (Hobart, 1902), Plate I **page 313** Cyanide vats, courtesy of the University of Sydney Archives; **page 318** The Huftons, Hufton Family Album, Historic Photograph Collection HP85.10.46, reproduced with the kind permission of the Macleay Museum, University of Sydney; **page 323** Students at work, Liversidge Papers, courtesy of the University of Sydney Archives; **page 325** *The Minerals of NSW* (London: Trübner, 3rd edition, 1888), from the original, Rare Book and Special Collections Library, University of Sydney; **page 327** The Assay Rooms, School of Mines, Liversidge Papers, courtesy of the University of Sydney Archives; **page 331** India by elephant, Liversidge Collection, by kind permission of the Master and Fellows, Christ's College, Cambridge; **page 336** Council Members of the Australasian Association for the Advancement of Science, Sydney, 1898, 38 photographs by J.H. Newman, ML524 by kind permission of the Mitchell Library, State Library of New South Wales; **page 343** President's Address, *Sydney Mail,*15 January 1898 **page 345** Picnic, *Tasmanian Mail*, 18 January 1902, 21; **page 347** Council Members and Officers, Hobart, *Tasmanian Mail*, 18 January 1902, 22; **page 353** Liversidge and Professor J.W. Gregory, Liversidge Papers, courtesy of the University of Sydney Archives; **page 361** The Geological Museum, Geological Museum negatives, reproduced by kind permission of the Natural History Museum, London; **page 364** Sir George Le Hunte, Liversidge Papers, courtesy of the University of Sydney Archives; **page 367** Sydney University, *University Calendar*, 1903; **page 371** Chemistry Class 1909, Historic Photograph Collection HP85.10.57, reproduced with the kind permission of the Macleay Museum, University of Sydney; **page 375** *Leptospermum Liversidgei*, from H.G. Smith and R.T. Baker, 'On an Undescribed Species of Leptospermum and its Essential Oil', *JPRSNSW*, 39 (1905), 124-130; **page 379** Liversidge and Dr. A.W. Howitt, Liversidge Papers, courtesy of the University of Sydney Archives; **page 383** Study, Historic Photograph Collection, HP84.53.6, reproduced with the kind permission of the Macleay Museum, University of Sydney; **page 383** Japanese urns, Historic Photograph Collection, HP85.10.16, reproduced with the kind permission of the Macleay Museum, University of Sydney; **page 385** The 'Octagon', Historic Photograph Collection, HP84.53.7, reproduced with the kind permission of the Macleay Museum, University of Sydney; **page 386** Geological Museum, Jermyn Street, London, 1907-20, Geological Museum negatives MN2758, reproduced by kind permission of the Natural History Museum, London; **page 390** The Royal Institution of Great Britain, courtesy of the Royal Institution with the kind assistance of Professor Frank James; **page 395** The Seating Plan, courtesy of the Archives of the Royal Society of London; **page 398** The skeleton of Zepplin L33, Imperial War Museum, Q 70185, reproduced with the kind permission of the Trustees of the Imperial War Museum, London; **page 416,** Archibald Liversidge, NPG x44536, courtesy of the National Portrait Gallery, London, London; **page 418** Fieldhead, photograph by the author; **page 420** Coach-building works, by courtesy of Glover and Liversidge, Ltd; **page 422** Advertisement by courtesy of Glover and Liversidge Ltd; **page 425** William and Sophia Liversdge, Liversidge Papers, courtesy of the University of Sydney Archives; **page 427** Liversidge's grave, photograph by the author; **page 441** Technological Museum key, courtesy of the Powerhouse Museum, Sydney.

# Index

## A

Abney, William, 356

Aboriginal culture, interest in, 433

academic competition: at Cambridge University, 100; at Royal School of Mines, 52–54

Acclimatisation Society of Mauritius, 128

Adams, Francis, 297

Adams, Henry, 197, 337

Addison, Christopher, 414, 415

Adelaide: AAAS Congress, 318, 324, 364, 379, 380; Intercolonial Medical Congress, 266

Adelaide International Exhibition, 266

Adelaide University, 298

Admiralty, wartime service for, 407–410

Advisory Committee for Aeronautics (ACA), 408

Advisory Council for Scientific and Industrial Research, 406

Advisory Council of Science and Industry, 405

Agassiz, Louis, 142

agricultural chemistry, 438

Agricultural Society of NSW, 108, 164, 184, 198

airships, World War I, 398, 404, 407–410

Airy, George, 58, 106, 151, 152

Albert, Prince, 34

Allen and Liversidge, 20

Allied Colonial Universities Conference, 357–358

alloys, research on, 409–410

Alloys Research Committee (ARC), 410

Anderson, Francis, 349, 365

Anderson, H.C.L., 344

Anderson, Henry, 350

Anderson Stuart, Thomas: AAAS Council, 270; appointments made by, 278, 366; educational reform efforts, 287, 349, 350; lecture on Australian universities, 315; Medical School building, 242, 276; at Octagon, 284; portrait of, 286; as professor of physiology, 236, 237; at Royal Society Club, 396; on selection committee, 382; travel to Australia, 263; on University of Sydney, 424; wartime service, 402

Andrews, Ernest C., 7, 372

Anglo-American Exhibition, 389, 397

Angus and Robertson, 376

Antarctic exploration, 302, 303, 312, 353, 387, 396

applied science. See practical science

apprenticeship schemes, 194, 214

architecture, at University of Sydney, 239, 300

*Argus* (Melbourne), 311, 346, 401

Armstrong, Henry E., 271, 280, 333, 369, 396, 407; on catalogue of scientific literature, 330; on 1851 Exhibition awards, 314; Liversidge's correspondence with, 208; at London Institution, 54; at Royal School of Mines, 36, 41, 45, 46, 49, 51; at Royal Society Club, 396; on social education, 55

# B

Bachelor of Science degree. *See* BSc degree

Backhouse, Alfred, 95, 317, 319

Badham, Charles, 114–116, 155, 219, 432; death of, 239; opinion of Liversidge, 234; opposition to science education, 121–123, 166–167, 212, 234, 238; portrait of, 90, 237; Sydney International Exhibition and, 199; University reform efforts, 172, 233

Baker, Richard, 211, 330

Baker, Richard T., 375, 380, 389

balance, 44, 50, 101, 130

Balfern: Caroline Sophia (sister), 18, 20, 21, 260, 389, 425, 428; John Edward, 21, 389; Percy (great-nephew), 20

Balfour, Arthur J., 334, 356–357, 359–360, 400

Balfour, Frank Maitland, 78, 81, 82, 100, 183, 254

Balfour Students, 254, 261

Ballarat School of Mines, 119, 207, 212, 306

Bancroft, Joseph, 128

Banks, Joseph, 4

Barff, Frederick, 204, 304

Barff, H.E., 275, 317, 332

Barnet, James, 198–199, 240

Barraclough, Henry, 314–315, 403, 419

Barratta meteorite, 352

Barrett, James, 360, 365

Barrow, Dora, 284

Bartley, Nehemiah, 282

Barton, Edmund, 91, 234, 285

Basalla, George, 5, 435

Bastian, Henry Charlton, 46

Bateson, W.H., 73

Bayly, Francis, 38

Beecher, Harvey, 430

Beilby, George, 394, 408

Bell, Edward, 125

Bellot, Hale, 78

Bendigo School of Mines, 119, 212, 306

Bennett, George, 8, 142, 146, 148, 266, 270

Bentham, George, 229

Bergakademie (Freiberg), 33, 36, 118, 130

Bernal, J.D., 392, 426, 438

Bernard, Claude, 80

Berthelot, Marcellin, 50, 74, 333

Bessemer process, 52

Bickerton, Alexander W., 38, 44, 51, 54, 169–170

biology: at Cambridge University, 75–80, 183; at University of Sydney, 254–255, 304, 338

Birkbeck, George, 25

Birkbeck College, 25, 27

Blacket, Edmund, 85, 88, 95

Blainey, Geoffrey, 431

Blanford, W.T., 361

Blomfield, C.J., 26

Bloxam, Charles, 74, 117

Board of Invention and Research (BIR), 400, 404; Airships Subcommittee, 408–410

Board of Studies (University of Sydney), 238–239, 291–292, 321, 366

Board of Technical Education (NSW), 218–224, 265, 304

Bonney, T.G., 49, 50, 64, 67

Boorstin, Daniel, 2

Borough of Hackney Microscopical and Natural History Society, 22

botanical chemistry, 109, 380

Botanic Gardens, 109, 201, 208, 319

botany: at Cambridge University, 58, 59, 70; at University of Sydney, 277

Botany Swamps, 126, 129

Boulton and Watt beam engine, 209, 211, 263

Boys, C.V., 396

Bragg, W.H., 245, 270, 332, 347, 388, 390, 404, 419, 426, 438

Bragg, W.L., 48, 407

Branagan, David, 4

Brande, W.T., 25, 29

# C

Society (CUNSS), 80
Cambridge University Rifle Volunteers, 69, 403
Canada, British Association Congress in, 252
Cannon, Michael, 108
Cannon, Walter B., 407
Canterbury University, 169
capitalism, frontier, 1–2, 7, 214
career (Liversidge): beginning of, 50; choices in, 52–55, 99–103, 262; final years, 370–372; middle years, 297–363; retirement, 378–385; in London, 386–397; research during, 389–392; social base during, 394–397, 406–407; success in, 173, 261–262
Carlton Club, 427, 428
Carpenter, W.B., 63, 70, 76
Carr, G.S., 428
Carruthers, Joseph A., 210, 216, 224, 304, 306, 381
Carslaw, H.S., 284, 341, 355, 423
Catalogue of Scientific Papers (Royal Society), 330
catastrophism, 143
Cavendish Laboratory, 59, 119, 245, 260, 272
Caves and Rivers expedition, 253
Cayley, George, 25
Centennial celebrations (NSW), 222
Centennial Magazine, 271
Central Foundation Boys' School, 20
Ceratodus, 183, 254, 255, 262
chalk, 172
Challenger expedition, 70, 145, 157, 159, 162, 252, 392
Challis, John Henry, 232; bequest, 232–233, 243, 272, 276, 277, 278, 312, 339, 351; appointments, 302; Chemistry Building, 272; Lectureship in Mining, 304; Professor of Biology, 302
Chamberlain, Joseph, 359

chemical doctorate, first at University of Sydney, 354
Chemical Laboratory, Cambridge University, 64
Chemical News, 45, 46, 50, 71, 271, 328–329, 353, 362
chemical nomenclature, paper on, 427
Chemical Society of London: first attendance at, 31; membership of, 161; papers presented to, 75, 139, 203, 329, 334, 429; as Vice-President, 393, 405
chemical warfare, 400
chemistry: agricultural, 438; botanical, 109, 380; at Cambridge University, 58, 59, 65, 70–75; at City of London College, 30; early laboratory work in, 42–47, 71, 75; history of, 436–437; imperial, 392–395; at Melbourne University, 301; mineral, 47–52; physiological, 79–80, 82, 301; professionalism of, 42–43, 372–376, 432; specialisation in, 43; at University of Sydney: appointments, 85, 96, 302–306, 372; compulsory requirement for, 119, 167, 236; curriculum, 281; funding for, 242–243, 247; lectures, 240, 245, 371, 424; research policy, 338–341; wartime applications, 402
Chemistry (Fownes), 30
Chemistry (Roscoe), 30
Chemistry, Laboratory Teaching, and Metals (Bloxam), 74
Chemistry Building (University of Sydney): building plan, 308; completion of, 278–280; increased funding for, 302, 382; proposals for, 116–123, 242–248, 271–278
Chemistry in its Relation to Agriculture and Physiology (Liebig), 34
Chevert expedition, 156, 251
Chimie Organique (Berthelot), 74

228, 270
Crick, Francis, 79
Crookes, William, 50, 134, 390, 395, 403
Crosby, Alfred, 1
Crossley, A.W., 393
crystallography, 47–48, 71, 130; of gold, 324–326, 340; research in, 422; at University of Sydney, 279
crystal spectrometer, 48
cultural agency, science as, 6, 9–10, 12, 106–109, 232, 358–359
Cumming, Constance Gordon, 169
Curran, J.A., 321
curriculum: Natural Sciences Tripos, 67, 70, 74, 81, 261; Royal School of Mines, 34, 41–42, 51; Sydney School of Mines, 310; University of Sydney (*See* University of Sydney curriculum)
Cuvier, Georges, 143, 255

# D

*Daily News* (London), 387
*Daily Telegraph,* 7, 239, 258, 365, 381, 384
Daintree, Richard, 54, 98, 134, 324–325, 328
Dana, James Dwight, 47, 48, 129, 131, 169
Danby, Thomas W., 54, 74
Darling Point (Sydney), 282, 381
Darwin, Charles, 8, 64, 99, 254; evolutionary theory, 52, 143; theory of coral atoll formation, 38, 334
Darwin, Francis, 184
Darwin, Horace, 408
Davidson, Simpson, 134
Davison, Graeme, 108, 199
Davy, S. Humphry, 23
Davy Faraday Laboratory, 389–392, 404, 420, 423

Dawkins, William Boyd, 253, 395
Deakin, Alfred, 381
Dean of Science (University of Sydney), 348; curriculum reform efforts, 285, 288, 317, 339; Edgeworth David acting as, 372; Liversidge elected as, 219, 236, 239, 316; Russell acting as, 330
Deas Thomson, Edward: Australian Club, 105; Australian Philosophical Society, 92, 95; Liversidge's association with, 156; mineralogical collection, 141; retirement of, 180; support for science education, 95–97, 121–122; university reform sought by, 114, 166–168, 172
de Beaumont, Elie, 133
De la Beche, Henry Thomas, 33, 94, 98
demonstratorships, junior, 354
dendritic spots, research on, 50, 75, 157
Denholm, David, 9
Deniliquin meteorite, 112, 138
Denison, William, 45, 93, 108, 202
D'Entrecasteaux Islands, 301
deoderizing agent, analysis of, 129
Department of Science and Art (UK), 28, 34, 38, 194, 212
Department of Scientific and Industrial Research (DSIR), 400, 404, 405, 415, 416
Descartes, René, 7
Des Cloizeaux, 131
*Descriptive Mineralogy* (Dana), 131
Desmond, Valerie, 433
Devonshire Commission, 60; Australian Museum and, 146; evidence to, 43, 48, 54, 65, 68, 76; Liversidge Report and, 193, 194, 195; University of Sydney reform and, 120, 167, 290
Dewar, James, 74, 260, 332, 337, 390, 403, 426
Dew-Smith, Arthur G., 78, 82, 183, 184, 334

diamonds, 138–139, 147

Dibbs, George, 208–209, 243, 319

*Dictionary of Chemistry* (Watt), 328

*Dictionary of National Biography,* 427

digestive enzymes, research on, 80, 82

Dilke, Charles, 11

*Diprotodon,* 144, 146, 253

discipline, student, 321, 370–372

Dixon, H.B., 382

Dixon, W.A., 219, 221

Docker, Joseph, 162–163, 168

Donnelly, John F., 28, 43

Douglass, Henry, 92

Dowling, Edward, 118, 213, 214, 217, 218

Drayton, Richard, 3

Duff, Robert, 211, 321

Dumas, Jean-Baptiste, 180

Dunstan, Wyndham R., 265, 412, 415

# E

Eastaugh, Frederick A., 354, 402

echidna, 255

*Echo,* 239

École des Mines (Paris), 33, 36, 118, 300

economic depression, 297, 308, 310–321

economy, colonial science and, 1–2, 11–13, 92

Edgeworth David, T.W., 286; as Acting Dean, 355, 369, 372; elected as FRS, 346; expeditions, 334, 344, 387, 396; as geology chair, 302–306; on importance of science, 423; Liversidge's succession handled by, 382; obituary written by, 387, 428; publications, 426; research culture encouraged by, 351–352; scientific apparatus vote, 380; on taxonomy research, 338; wartime service, 401, 403, 419

Edinburgh University, 89

education, science. *See* science education

education (Liversidge): as basis for Australian career, 101–103; Cambridge University, 56–83, 261; Central Foundation Boys' School, 20; City of London College, 25–31; Royal College of Chemistry, 32–55; Royal Polytechnic Institution, 22–26; Royal School of Mines, 32–55

Education Act of 1918 (UK), 394

Education Department (UK), 28–29

Electoral Act of 1858, 113

*Elementary Biology* (Klein and Foster), 82

*Elements* (Miller), 74

*Elements of Geology* (Lyell), 49

Elizabeth Street (Royal Society of NSW), 230

Elkin, A.P., 165

Ellery, Robert, 150, 207, 248, 250, 271

Elsmore Tin Mine, 138

Eltoft, Thomas, 28, 30

embryology, 254–255

Emerson, Ralph Waldo, 12, 440

emigration, 99–103

*Empire,* 107, 108, 111, 122, 123

Empire Exhibition, 389

empire federalism, 256–257, 265

Empire (Commonwealth) Universities Bureau, 358

employment: Cambridge University, Acting Demonstrator in Chemistry, 75; Frankland's laboratory, 46; Royal School of Naval Architecture, 46; University of Sydney, 81–82, 97; Dean of Science, 219, 236, 239, 316–317; Demonstratorship of Chemistry, 81–82, 97; Hovell Lecturer in Physical Geography and Geology, 168; Professor of Chemistry, 305; Professor of Chemistry and Mineralogy, 219, 236; Professor of Geology and Mineralogy, 122; Readership in Geology and Mineralogy, 81–82, 97

# M

204–211, 265, 319

Mann, E.A., 374

Manning, William Montague, 105, 180–183; AAAS Congress, 270, 300, 342; Arts Curriculum Committee, 285; death of, 320; Royal Society of NSW, 230, 247; School of Mines, 206, 310, 317; Sydney International Exhibition, 200; technical education reform, 214, 216; Technical Education Report, 181–182, 184–186, 189–196, 201, 211, 217, 300; Technological Museum, 203; University reform efforts, 232, 238–239, 243, 246–247, 272

*Manual of Chemical Physiology* (Thudichum), 80

*Manual of Chemistry* (Fownes), 129

*Manual of Geology* (Phillips), 49

*Manual of Mineralogy* (Dana), 129

Marcet, Alexander, 328

Marsh, Charles, 301

Marshite, 310

Martin, Allan, 155

Martin, Charles, 366, 393

Martin, Henry Newell, 52, 100, 184, 233; at Cambridge University, 66, 69, 75, 78, 81, 82

Martin, James, 174

Massachusetts Institute of Technology, 190

Masson, David Orme, 301, 303, 347, 373, 402, 403, 405

Masters, George, 142, 147

Mathematical Tripos, 58, 100

mathematics: at Cambridge University, 67–69, 89; at University of Sydney, 85, 167, 285, 341

Mather, William, 394, 406

Maurice, Frederick Denison, 25, 213

Mauritius, 128

Mawson, Douglas, 353, 377, 403, 407, 417, 419

Maybury, A.C., 28

Mayor, Joseph, 58

McCormick, William, 406, 412

McCoy, Frederick, 95, 131, 250

McGlone, Dominic, 370

McLeod, Herbert, 44, 45

mechanical drawing, 42

Mechanics' Institution, 25, 27

medals, 550; Clarke, 228–229, 250, 259, 423; inorganic chemistry, 38; mineralogy, 38; Paris Exhibition, 179

Medical School (University of Sydney), 242, 276

Melba, Nellie, 384

Melbourne: AAAS Congress, 311, 323; Industrial and Technological Museum, 131; travel to, 203–204

Melbourne Centennial Exhibition, 268

Melbourne Intercolonial Exhibition (1866-67), 108, 132, 175, 184, 212

Melbourne International Exhibition (1880-81), 198, 201, 258

*Melbourne Punch,* 133, 150, 255

Melbourne Technological Museum, 207

Melbourne University, 91, 95, 131, 276, 301, 306, 384, 401, 424

Mellor, J.W., 382

Mendeleev, Dimitri, 50

Merewether, Francis, 85

Merrifield, C.W., 68

metallurgy: at Royal School of Mines, 41–42, 51; at University of Sydney, 109, 277, 339

*The Metallurgy of Gold* (Rose), 329

meteorites, research on, 112, 138, 301, 329, 346, 352, 356, 439

Meteorological Society of Australasia, 257

meteorology, intercolonial cooperation in, 249

metric system, 349

Metropolitan Intercolonial Exhibition, 108

Metropolitan School of Science, 34, 37

Mond, Ludwig, 372, 389, 403
monotremes, 255
Montefiore, E.L., 159, 227, 282
Montefiore, George, 282, 284
Montefiore, Jacob, 179
Moore, Charles, 106, 109, 157, 201, 208,
    228, 250, 319
Moorehead, Alan, 10
moral agency, 9–10, 12
Moran, Herbert, 370
*Moreton Bay Courier,* 128
Morgan, Gilbert T., 382, 387, 393, 437
Morin La Meslée, Edmond, 251
Morley, H. Forster, 333
Morris, Edward, 276
Morris, John, 49
Mort, Edward, 282
Mort, T.S., 93, 136, 282
Morton, Alexander, 253
Moseley, H.G.J., 400
Moseley, H.N., 157, 252
moss gold, 161, 164, 324
mould experiments, 323
Moulton, John Fletcher, 66, 403
Mount Morgan gold mine, 299, 322, 324
Moyal, Ann, 4
Muir, Robert, 128
Müller, Hugo, 392
Münster, Emil, 328
Muntz metal, 328–329
Murchison, Roderick: at City of London
    College, 28; correspondence,
    93, 101; Geological Survey of
    NSW, 95; at Royal Polytechnic
    Institution, 25; at Royal School of
    Mines, 34, 35, 36, 37, 47, 96, 106
Murray, W.G., 165
Murray Commission, 425
museum(s). *See also specific museum:*
    collecting for, 183; exchanges
    between, 183; governance models,
    191–192; Liversidge Report on,
    190–193, 201
Museum of Applied Arts and Sciences, 201

Museum of Natural History, 183
Museum of Practical Geology, 22, 32, 33,
    35, 41
Museums and Lecture Rooms Syndicate,
    59, 73
Muspratt, Edmund, 372

# N

Nadel, George, 4, 106
Narraburra meteorite, 356
national efficiency, 356, 359, 365, 384,
    393
nationalism, colonial, 12, 251
National Physical Laboratory (NPL),
    408–410
natural history: at Cambridge University,
    70; at Royal School of Mines,
    51–52; at University of Sydney,
    219, 238, 277, 302
Natural History Museum, 334, 388
natural history societies, 22–23
Natural Sciences Tripos (NST), 22, 54,
    58, 60; admission requirements,
    67, 89; and career success, 100;
    curriculum, 67, 70, 74, 81,
    261; popularity of, 261; social
    acceptability of, 67
natural theology, 71
*Nature:* AAAS history in, 388; AAAS
    Presidential Address in, 343; on
    Australian science, 271; on BAAS
    visit to Australia, 258, 260; on
    Cambridge University, 63, 81;
    editor of, 70, 357, 419; imperial
    higher education statement in,
    359; on imperial minerals policy,
    414, 416; on Krefft affair, 148;
    Liversidge's gold papers in, 326,
    392; on Liversidge's honorary
    doctorate, 333; Liversidge's reports
    to, 173, 335; on mining education,
    309; review of Liversidge's

publication, 362; Royal Society of NSW history in, 388

*Neoceratodus forsteri,* 142

Neptunists, 132

Newbery, J.C., 131

New Britain, 172

New Caledonia, 139, 149, 159

Newcastle Commission (UK), 28

New Guinea, 251, 326, 389

Newman, Cardinal, 114

New South Wales. *See also* Sydney: Centennial celebrations, 222; Department of Mines and Geological Survey, 139; gold mining in (*See* gold rush (NSW)); mineralogical map of, 299; mining schools in, 298 (*See also specific school*); Ministry of Public Instruction: Board of Technical Education, 218–224, 265, 304; Technical Education Branch, 210, 224; new minerals discovered in, 310, 434; Paris Exhibition, 174–180; public education in, 113, 233, 288–289, 293, 349–350, 365; Royal Society (*See* Royal Society of NSW)

*New South Wales Magazine,* 11

New South Wales Mines Act of 1874, 298

newspapers. *See also specific paper*: on AAAS, 258, 271, 342; in Australian universities, 365; Krefft affair coverage, 146, 148; obituaries in, 429, 548–549; scientific coverage in, 107–108, 111

Newton, Alfred, 58, 70, 71

New Zealand: AAAS Congress in, 346; Paris Exhibition, 177; trips to, 168–172, 303; university science in, 169

New Zealand Institute, 170, 203, 250, 267, 303

New Zealand University, 169, 289

Nicholson, Charles, 85, 87, 92, 97, 98,

105, 173, 212

Nixon, Francis, 212

*Notes on the Physical Geology and Mineralogy of Victoria* (Selwyn and Ulrich), 132

Noumeïte, 149

Nuisances Prevention Act, 128

Nunn, T. Percy, 406

# O

'Octagon' (home), 274, 282–284, 330, 355, 379, 381, 383–385, 397

Oliver, Alexander, 234

Onslow, Arthur, 142, 145, 146, 147, 151

'Ordinary Subjects' examination (Cambridge University), 69

ore formation, theories of, 50, 132–136, 323–324, 327–328, 438

ore refining, 322

organic chemistry: education in, 46, 74; at University of Sydney, 341, 372, 388

organisational skills, development of, 75, 80–83, 111, 433

Osborne, Theodore, 401

Osborne, William A., 401, 405

Otago University, 169, 170

overseas leave. *See* travel

Owen, Richard, 106, 229, 253, 255; Krefft affair, 143–149

Owens College (Manchester), 42, 43, 66, 86, 89

Oxford Street laboratory (London), 42–47

oxidation theory of purification, 45

# P

palaeontology, 52, 228, 302, 340

pancreatic juice, research on, 80, 82

Papua New Guinea, 251, 389

elements, 427, 429; *Transactions of Royal Society of NSW,* 362; Transit of Venus observation, 153
public education, in New South Wales, 113, 233, 288–289, 293, 349–350, 365
Public Instruction Act of 1880, 233
public science, 11–12, 60; Liversidge's support of, 120, 340, 394, 433–434; students involved in, 45; University of Sydney contributions to, 92
'Purple-of-Cassius' solution, 328
Putney Vale Cemetery (London), 21, 427
Pyenson, Lewis, 6
pyrites, gold deposited from, 134–135

# Q

qualitative analysis, early work in, 43–44, 71, 75
*Qualitative Analysis* (Eltoff), 30
*Qualitative Analysis* (Fresenius), 117
quantitative analysis, early work in, 43–44, 71, 75
*Quantitative Analysis* (Thorpe), 129
*Quantitative Chemical Analysis* (Fresenius), 44
quartz, 131, 136, 327, 328
Queensland: AAAS Congress, 322, 326; Mount Morgan gold mine, 299, 322, 324; Royal Society, 259; soil analysis in, 203; sugar cane blight, 128
*Queenslander,* 128
Queensland lungfish, 142

# R

radioactivity, research on, 353–354
radium, 353

Ramsay, Andrew, 34, 38, 40, 48–49, 97, 98, 106
Ramsay, Edward, 128, 148, 156, 207, 379
Ramsay, William, 332, 347, 357, 359, 369, 382, 396
Ratte, Felix, 301
Rayleigh, Lord, 246, 408
Reader, W.J., 403
reading lists, educational, 30, 49, 74, 117, 129
recession, economic, 297, 308, 310–321
reconstruction, 2
Redmayne, Richard, 417
Regent Street Polytechnic (London), 26
Reid, George, 218, 222–223, 320, 344; University reform efforts, 233, 239, 240, 242
Reingold, Nathan, 10
Remsen, Ira, 233
Rennie, Edward, 219, 270, 401
Renwick, Arthur, 172, 208–209, 214, 222; Arts Curriculum Committee, 285; University reform efforts, 233, 234, 237–238, 242, 246, 272, 339
Renwick, Henry, 38
research. *See also specific topic of research*: AAAS support for, 271; at Cambridge University, 79–80; encouragement of, 315–316, 344, 351–354, 354–355; during retirement, 389–392; at Royal School of Mines, 42; wartime, 400, 409–410
research culture, 233; lack of, 315–316, 338–341, 351–352, 365
*Researches on the Southern Gold Fields* (Clarke), 132
research fellowships: imperial, 393; University-financed, 276, 314, 388
Research Lectureships, endowed by Liversidge, 428, 558
Reynolds, James Emerson, 392
Richard, Valentine, 428
Rivett, A.C.D., 402

334; *Proceedings,* 50, 81; War
Committee, 399

Royal Society of London Fellowships,
235–236; Australians, 91, 332,
346–347, 369, 425; Cambridge
University staff/students, 78, 81,
100; Clarke elected, 162; Liversidge
elected, 207, 219, 235–236, 262;
Royal School of Mines staff/
students, 38, 41

Royal Society of NSW, 97, 111, 112;
buildings, 165, 227–232, 312;
chemical society proposal,
373–374; *conversazioni,* 157–159,
162, 165, 228–231, 349;
Council, 156–165, 158; decline
of, 251–252, 311–312, 349,
377–378; funding for, 229, 252;
gold papers delivered to, 135–136,
324, 327–328, 340, 377, 380;
governance, 228; Honorary
Members, 160; incorporation,
229; intercolonial cooperation,
248–258, 376; *Journal,* 164, 173,
177, 229, 230, 326; library, 228,
229, 252, 311, 340; Liversidge
as representative of, 330;
'Liversidge Phase' in history of,
165; Liversidge's collection left
to, 429; membership, 159, 161,
163, 228, 311, 349, 378; mining
papers presented to, 299–300;
President, 257, 301–302, 311, 349;
prizes, 228–230; renaissance of,
156–165; resignation as Honorary
Secretary, 265, 311, 340; role in
AAAS, 269–270; support for Paris
Exhibition, 177; *Transactions,* 159,
160, 164, 362; Transit of Venus
observation, 150, 151, 153; water
supply papers, 127

Royal Society of Queensland, 259
Royal Society of South Australia, 252
Royal Society of Victoria, 250, 252

rubidium, 50
Rudler, F.W., 98
Rudwick, Martin, 437
Rupke, Nicolaas, 435
Rusden, H.K., 252
Russell, Colin, 42
Russell, Henry Chamberlain, 91, 112;
AAAS President, 266, 269; Acting
Dean of Science, 330; Australian
Museum Trustee, 148; Board of
Technical Education, 218, 222,
304; death of, 380; Deniliquin
meteorite, 138; on intercolonial
cooperation, 248; murder attempt,
165; portrait of, 158; retirement
of, 355; Royal Society Council,
156–165, 227, 229–230; Senate
member, 168, 182; transit of Venus
observation, 149–153; University
Buildings Committee, 237–238;
University reform effort, 233–234,
272

Russell, Jane, 243
Russell, Peter Nicol, 321, 338, 339, 366
Russell Fund, 339
Rutherford, Ernest, 426, 429
Rutherford, William, 79

# S

sabbaticals, 173–187. *See also* travel
saline solutions: gold in, 329;
supersaturated, 81
Salvin, Anthony, 59
Samuel, Saul, 263, 330, 334
Samuelson, Bernhard, 37
Samuelson Committee, 60, 72, 261
Samuel Whitbread and Co., 263
Sanderson, J.R. Burdon, 183
Sandhurst School of Mines, 207
sandstone, gold in, 324
San Francisco Geological Museum, 259
*Saturday Review,* 12, 271

discipline, 321
Scott, William, 149, 151, 157
search committee, for Liversidge's
    successor, 382
Seaver, Jonathon, 323
seawater, removal of gold from, 327–329,
    420–421
Sedgwick, Adam, 49, 58, 70–71, 76, 93,
    261
*Sedimentary Formations of New South Wales*
    (Clarke), 176, 179
See, John, 349
Seeley, H.G., 28, 49
Seeley, J.R., 64
self-discovery, in Australian science, 7
Selfe, Norman, 183, 213–215, 218,
    223–225
self-instruction, 29, 44
Selwyn, Alfred, 98, 131, 134, 324
Senckenberg Institute of Frankfurt, 161
Senter, George, 392
Serle, Geoffrey, 96
Service, James, 256
settler capitalism, 1–2, 214
sewerage, 129
Shackleton Expedition, 387, 396
Shann, Edward, 402
Shepherd, John, 222
Shipley, Arthur, 334, 362, 395, 426–427,
    428
Skey, William, 324, 328
Slatterly, T.M., 308
Smeeth, William Frederick, 306, 319
Smiles, Samuel, 263
Smith, F.B., 434
Smith, G.F. Herbert, 428
Smith, Henry G., 282, 319, 374, 375, 380
Smith, John, 85, 88, 91, 95–96, 97, 98,
    99; Board of Technical Education,
    218; Chair of Experimental
    Physics, 219; contributions
    to public science, 92; Dean of
    Arts, 239; death of, 236, 244;
    on intercolonial cooperation,

248; Liversidge's relations with,
    167–168; Philosophical Society of
    NSW, 111; portrait of, 90, 158,
    237; public lectures, 155; Royal
    Society Council, 156–165, 227,
    228; support for science education,
    122; technological museum
    and, 203; travel with, 103, 110;
    university reform efforts, 114–116,
    167–168, 172, 239–240, 244–245;
    water supply crisis and, 128–129
Smith, Thomas, 282
Smith, William, 435
Smithells, Arthur, 405, 407
Smithsonian Institution (US), 161, 173,
    259, 330
Smyth, Warington Wilkinson, 34, 40, 41,
    47, 48, 51, 98
Snelus, George, 52, 54
social class, educational opportunity and,
    42, 66
social clubs, 431, 552. *See also specific club*;
    in London, 394–397, 406–407
social education, at Cambridge University,
    55, 70
social life, 432–433. *See also* friends;
    during London retirement,
    394–397, 406–407, 426–427; in
    Sydney, 105–106, 112, 156, 284,
    378–379
society memberships, 172–173. *See also*
    *specific society*
Society of Arts (UK), 29, 30, 37, 194
Society of Chemical Industry (SCI), 43,
    372–374
Society of Dyers and Colourists, 43
Society of Public Analysts, 43
Society of the Chemistry Industry of
    Victoria, 373
solar parallax, 149
Sollas, William Johnson: application to
    University of Sydney, 302–304; at
    Cambridge University, 58, 62, 69,
    82, 100; expeditions, 334; at Royal

(SMSA), 92, 109, 112, 117, 183, 192, 194, 212, 213

Sydney Metropolitan Colonial Exhibition, 184

Sydney Metropolitan Water Board, 129

Sydney Mint, 109, 110, 177, 299

*Sydney Morning Herald:* on AAAS, 342; on Arts curriculum reform, 288–289; on Chemistry chair, 372; on chemistry laboratories, 280, 313; editor of, 248; on international telegraph, 155; on Liversidge's European trip, 331, 365; Liversidge's letters to: on science education, 118–121, 213, 309; on Sydney Geological Congress, 256; on water supply, 125; Liversidge's reading of, 387; on Liversidge's retirement, 381; Manning's public relations articles, 312–313; on mining education, 118; obituary in, 429; on pedagogical methods, 369; on practical science, 338, 350; review of Liversidge's book, 236; on Royal Society of NSW, 159, 164, 229–232; science coverage, 107, 111; on scientific circle, 106, 232; on Sydney Exhibition, 108; on transit of Venus observation, 151; on University of Sydney, 113, 116, 349, 366–368, 378; wartime coverage, 403, 419

Sydney Museum, 201

Sydney Observatory, 109, 152; transit of Venus observation, 149–153

*Sydney Punch,* 115, 124, 137, 140, 144, 146

Sydney Technical College: buildings, 201, 210–211, 214, 215, 220, 224, 387; distinction between University and, 217–218, 221–222, 224–225; enrolment numbers, 218–219; funding for, 217, 223; governance, 218; laboratory, 219; mining

education at, 222; scholarships, 221–222; University matriculation waiver, 308–309

*Sydney Times,* 107

Sydney University. *See* University of Sydney

*System of Mineralogy* (Dana), 47, 48

# T

*Tables for Qualitative Chemical Analysis* (Liversidge), 362

Tait, P.G., 244

Tank Stream, 125

Tasmania, AAAS Congress, 309, 315, 345, 346, 347

Tate, Ralph, 342

Taunton Commission, 37, 60, 189

Taylor, Edgar, 414

Taylor, James, 339

Teale, J.J. Harris, 81

Teall, Jethro, 407

technical education: funding for, 223; reform of, 211–219

Technical Education Branch, Ministry of Public Instruction (NSW), 210, 224

Technical Education Committee (BSG), 393–394

Technical Education Report (1880), 181–182, 184–186, 189–196, 201, 211, 217, 300

Technological, Industrial and Sanitary Museum of New South Wales (TISM), 203

Technological and Industrial Museum: Boulton and Watt engine, 209, 211, 263; buildings, 201–211, 220, 224, 380; collecting for, 185–186, 208–209; curator, 204–206, 211, 265, 319; experimental work at, 380; funding for, 202, 208–209; governance, 191–193; laboratory,

206; Liversidge's resignation as Trustee of, 265; Liversidge's vision of, 190–191, 203; organisation of collections, 193; proposal for, 184–185, 189–193; visitor numbers, 210

Tenison-Woods, Julian, 195, 301

Tennant, Harold, 400

Tennyson, Hallam, 337

Tennyson, Lady, 340

theory-dodging tendency, 9

Thiselton-Dyer, William, 70, 184, 278, 305, 311–312, 314, 389

Thompson, J. Ashburton, 209

Thomson, Alexander Morrison, 30, 97, 104, 141

Thomson, J.J.: Cavendish Laboratory, 245, 260, 315, 354; discovery of electron, 332, 337; on Threlfall, 246, 315; wartime service, 404, 407, 408

Thomson, William. *See* Kelvin, Lord

Thomson, Wyville, 157, 162

Thornton, A.P., 3

Thorpe, T.E., 306

Threlfall, Richard, 338; arrival at University of Sydney, 245–246, 275; on career choice, 388; curriculum reform effort, 287, 293–294; elected FRS, 346; as pioneer in electrical physics, 369; portrait of, 286; resignation of, 332, 341, 381; wartime service, 404, 406, 408, 411, 419

Thudichum, John, 79–80

Thunda meteorite, 352

Thurston, Robert, 263

*Thylacoleo carnifex,* 143, 253

Tilden, William A., 278, 382

Tilley, Cecil E., 402

*The Times,* 419, 429

tin, 138, 147, 161

Tizard, Sir Henry, 392

Todd, Charles, 249

Tomlinson, Charles, 50, 52

topaz, 341

Topley, William, 322

torbanites, 354

Tornaghi, Angelo, 151

Townsend, W.G., 402

trace elements, research on, 420, 422, 427, 429, 439

Trade and Labour Council (NSW), 213

trade schools, 194, 214, 223

Trades Union Congress, 214

Transit of Venus, 149–153, 249

travel, 435; to China and Japan, 209, 259; emigration to Australia, 85, 99, 103; to Europe, 174–187, 259–263, 265, 330–335, 355–363; to gold mines, 110; to India, 177, 330; to Melbourne, 203–204; to New Zealand, 168–172, 303; return to England, 387; to United States, 209, 259

*Treatise on Crystallography* (Miller), 71

Trickett, William, 243

Trollope, Anthony, 11, 99, 113, 114, 118, 139, 156, 249, 282

Trotter, Coutts, 73, 76, 261

trusteeship model, of museum governance, 191–192

Turner, Edward, 49

Turner, J.W., 350

Turnham Green (London), 18, 396

tutors: Cambridge University, 69; City of London College, 30–31

Twain, Mark, 330

Twopenny, Richard, 5

Tyndall, John, 37, 40, 42, 44, 46, 47

# U

Ulrich, George, 131, 135

Union Club, 104–107, 125, 282, 378, 385

United States: travel to, 209, 259;

universities in, 368; wartime service, 407, 414–415

United University Club, 387, 389, 396

University Amending Act of 1854, 88

University Amendment Act of 1861, 89

University Bill of 1878, 233

University College London, 30, 49

University Endowment Amendment Act of 1902, 351

University Extension Act of 1884, 235, 239

University Grants Committee (UK), 358

University Increased Endowment Bill, 182

University of Adelaide, 298

University of Canterbury, 169

University of Glasgow, 89, 332–333

University of London, 29, 35, 89

University of Melbourne, 91, 95, 131, 276, 301, 306, 384, 401, 424

University of New Zealand, 169, 289

University of Otago, 169, 170

University of Sydney: Arts Curriculum Committee, 285–295; Arts degree (*See* Arts degree); Bachelor of Mining Engineering degree, 309; Board of Studies (Professorial Board), 238–239, 291–292, 321, 366; British accreditations, 374; BSc degree (*See* BSc degree); Building Committee, 339; buildings and rooms, 85, 87, 88, 94, 95, 104, 116, 273–274, 279–281, 310, 323, 348, 367, 384; Buildings Committee, 237–238, 240–242; Challis Lectureship in Mining, 304; Challis Professor of Biology, 302; Chemistry Building (*See* Chemistry Building); curriculum (*See* University of Sydney curriculum); Dean of Science (*See* Dean of Science); demonstratorships (*See* demonstratorships); Department of Geology, 309; Department of History, 4, 248, 281, 366; Department of Mines, 309; Department of Natural History, 219; educational philosophy, 87–89, 113, 288, 297, 365, 368–369; enrolment numbers, 91, 113, 120, 181, 233, 289, 293, 306, 320, 337, 348, 368, 423; entrance requirements, 288–289, 308–309, 366; examinations, 88, 91–92, 113, 122–123, 234, 287, 288; Faculty of Arts, 234, 338; Faculty of Engineering, 382, 384; Faculty of Law, 234, 338; Faculty of Medicine, 234, 338, 366; Faculty of Science (*See* Faculty of Science); Finance Committee, 312, 339; first science doctorates, 354; foundation of, 22, 85–86; funding of (*See* University of Sydney funding); governance, 368; Honours Schools, 167, 238, 351; Hovell Lecturer in Physical Geography and Geology, 168; Jubilee celebrations, 351, 353, 365; laboratories (*See* University of Sydney laboratories); Liversidge's legacy at, 440; Liversidge's resignation from, 378, 381; London selection committees, 387–388; management structure, 86–87; Medical School, 242, 276; mineralogical collection, 95, 131, 141, 147, 280; mining and metallurgy lectureships, 304, 306, 339; notable graduates, 91; Physics Building, 246–247, 290; post-war expansion of, 423–424; Professor Emeritus, 384; Professor of Chemistry, 305, 387–388; Professor of Chemistry and Mineralogy, 219, 236; Professor of Geology and Mineralogy, 122; Professor of Geology and Palaeontology,

# V

Vaughan-Lee, C.L., 408
Venn, John, 68
Venus, Transit of, 149–153, 249
Vernadsky, V.I., 437
Verne, Jules, 385
Victoria. *See also* Melbourne: chemical
    industry society, 373; mining
    schools, 298; Royal Society, 250,
    252; technical education in, 212
Victoria, Queen, death of, 349
Victorian Geological Survey, 54, 94, 98,
    131, 132, 139, 324
Victorian Science Board, 160
Vigani, G.F., 73
Vincent, Edgar, 413
Vines, Sydney, 261
vitamins, research on, 427
volcanism, 134, 328
von Haast, Julius, 160, 169–70, 177
von Hofmann, August, 34, 41, 42, 43, 74,
    110
von Mueller, Ferdinand, 172, 250, 252,
    270, 271, 301, 380
Vonwiller, Oscar, 377
Voss, H.H., 330

# W

*Waldheimia,* 262
Walker, Thomas, 227
Walkom, A.B., 428
Wallerawang (NSW), 149, 159
Walton, Thomas, 374
war: South Africa, 347–348, 356; World
    War I (*See* World War I)
Ward, John, 93
Ward, Russel, 106
War Office, Committee on Explosives,
    403
Warren, T. Herbert, 360
Warren, William H., 219, 286, 349, 426
Washington, Henry S., 437
Water Pollution Prevention Bill, 128

water supply: Great Britain, 45, 127;
    Sydney, 125–128
Watson, James, 79
Watt, Henry, 328
Watt, John Alexander, 320
Watts, William, 412
*Wayfaring Notes* (Smith), 98
wealth, 397, 428–429
Webb, Beatrice, 338, 356–357, 433
Webb, Glover, 20
Webb, Sidney, 25, 26, 356–357, 394, 406
Webster, John, 169
Wellington Caves expedition, 253
Welsh, D.A., 351, 402
Wentworth, William Charles, 86, 88, 113,
    121
Werner, Abraham, 132
Westgarth, William, 268
Weston, E.A., 402
wheat, 'Federation' variety of, 305
Whewell, William, 37, 58, 59, 430
White, Patrick, 7
White, Robert Hoddle, 106, 209
Whitfield, H.E., 402
Whitmell, C.T., 81
Whitworth, Joseph, 28
Wilkinson, Charles S.: death of, 324;
    discovery of Laumontite, 162;
    Geological Survey of NSW, 139,
    159, 206, 250; gold solutions
    research, 134; Royal Society of
    NSW, 228; Sydney Technical
    College, 218; Wellington Caves
    expedition, 253
Wilkinson's Mining and Geological
    Museum, 200–201, 206
will (Liversidge), 428–429
Williams, John, 10
Williamson, Alexander W., 30, 54
willyamite, 310
Wilmott, Eliza, 46
Wilsmore, N.T.M., 352, 402
Wilson, James T., 30, 271, 286, 347, 350,
    366, 369, 401, 428